D1614218

Neuropsychiatry and Behavioral Neuroscience

JEFFREY L. CUMMINGS, M.D.

The Augustus S. Rose Professor of Neurology
Professor of Psychiatry and Biobehavioral Sciences
Director, UCLA Alzheimer's Disease Center
UCLA School of Medicine

MICHAEL S. MEGA, M.D.

Laboratory of Neuroimaging
UCLA School of Medicine

Neuropsychiatry and Behavioral Neuroscience

OXFORD
UNIVERSITY PRESS
2003

OXFORD
UNIVERSITY PRESS

Oxford New York
Auckland Bangkok Buenos Aires Cape Town Chennai
Dar es Salaam Delhi Hong Kong Istanbul Karachi Kolkata
Kuala Lumpur Madrid Melbourne Mexico City Mumbai
Nairobi São Paulo Shanghai Taipei Tokyo Toronto

Published by Oxford University Press, Inc.
198 Madison Avenue, New York, New York, 10016
http://www.oup-usa.org

Library of Congress Cataloging-in-Publication Data
Cummings, Jeffrey L., 1948–
Neuropsychiatry and behavioral neuroscience/Jeffrey L. Cummings, Michael S. Mega.
p.;cm.
Includes bibliographical references and index.
ISBN 13 : 978-0-19-513858-0 (cloth)
ISBN 0-19-513858-9 (cloth)
1. Neuropsychiatry. 2. Neurosciences I. Mega, Michael S. II. Title.
[DNLM: 1. Brain Diseases—diagnosis. 2. Mental Disorders—diagnosis. 3. Interview, Psychological. 4. Mental Disorders—therapy. 5. Neuropsychological Tests. WM 141 C97ln 2003]
RC341.C847 2003
616.8—dc21
2002025341

The science of medicine is a rapidly changing field. As new research and clinical experience broaden our knowledge, changes in treatment and drug therapy occur. The author and publisher of this work have checked with sources believed to be reliable in their efforts to provide information that is accurate and complete, and in accordance with the standards accepted at the time of publication. However, in light of the possibility of human error or changes in the practice of medicine, neither the author, nor the publisher, nor any other party who has been involved in the preparation or publication of this work warrants that the information contained herein is in every respect accurate or complete. Readers are encouraged to confirm the information contained herein with other reliable sources, and are strongly advised to check the product information sheet provided by the pharmaceutical company for each drug they plan to administer.

3 4 5 6 7 8 9

Printed in Hong Kong
on acid-free paper

To our wives,
Inese and Susan

To our daughters,
Juliana and Leda

To our fellows, colleagues, and
international collaborators
whose interest, enthusiasm, and dedication
have inspired and invigorated us

Preface

Clinical Neuropsychiatry, by Jeffrey L. Cummings, was published in 1985 and represented an integration of behavioral neurology and biological psychiatry into a single volume devoted to explicating brain–behavior relationships. The volume was clinically oriented and intended for practitioners caring for patients with neuropsychiatric disorders. Since *Clinical Neuropsychiatry* was originally published, there has been a tremendous explosion of information pertinent to neuropsychiatry ranging from molecular biology of neuropsychiatric disorders on one end of the spectrum to neuro-ethology and the neurobiological basis of social interactions and culture at the other. Advances in neurochemistry, neuroanatomy, genetics, neuroimaging, and neuropharmacology have progressed at an unprecedented rate. New treatments have evolved for nearly all neuropsychiatric illnesses and neuropsychopharmacology has become a demanding discipline in its own right. A successor to the book was badly overdue. This volume represents an attempt to integrate the most salient evolving information into a single comprehensive presentation. The clinical emphasis of its predecessor has been maintained and enriched by the integration with evolving neuroscience information.

It might be argued that books are obsolete, since information is evolving rapidly and is updated more quickly through electronic resources, thus obviating the need for books such as this that are frozen in time. The dramatic increase in information, however, has made the need for an approach to delivering care to patients and of integrating the expanding information base into a systematic clinical framework even greater. Thus, the emphasis of this volume is on approaching the patient, understanding brain regions rendered dysfunctional by disease, and optimizing care based on knowledge of brain–behavior relationships.

We hope that those who read and study this volume find that their understanding of brain–behavior relationships is enhanced, their clinical assessment and management enriched, and the quality of life of their patients and their families improved.

Los Angeles, California

J.L.C.
M.S.M.

Acknowledgments

This volume represents a progress report in the evolution of neuropsychiatry and of the authors' understanding of neuropsychiatric disease and treatment. As such it is another step in the long march toward understanding central nervous system function and disease. Many have contributed to our passion for neuropsychiatry; chief among these is the late D. Frank Benson, M.D. Dr. Benson's enthusiasm for behavioral neurology, neuropsychiatry, and for teaching had a profound and lasting influence on the authors and his vision of neuropsychiatry permeates the pages of this book.

Others to whom we owe a debt of gratitude for their support or instruction include Martin Albert, Michael Alexander, Simeon Locke, Paul Yakovlev, Robert Collins, Arthur Toga, and John Mazziotta. Our colleagues in the Behavioral Neuroscience program at UCLA also have contributed importantly to our ability to practice neuropsychiatry and develop this volume; among these are Donna Masterman, Tiffany Chow, Mario Mendez, Bruce Miller, David Sultzer, Seth Weingarten, Ron Saul, and Bud Ullman. National and international colleagues too numerous to list have encouraged us through their enthusiasm and by sharing their knowledge. The authors have benefited greatly from the stimulating interaction provided by the many fellows of the UCLA Dementia and Behavioral Neuroscience Research Fellowship as well as the international trainees who have been members of our training program. Financial support from the National Institute on Aging for our Alzheimer's Disease Center, from the State of California for our Alzheimer's Disease Research Center of California, and from the Willard K. and Patricia S. Shaw Memorial Fund has contributed importantly to our endeavors. The tremendous financial and emotional support provided by the late Katherine Kagan of the Sidell-Kagan Foundation was instrumental in allowing us to advance our research programs.

Finally, without the love and support of our families, none of the activities encompassed within this book would have been possible. Inese and Juliana (wife and daughter of J.L.C.) and Susan and Leda (wife and daughter of M.S.M.) have been unfailing in the support and sacrifice required for the time devoted to research, teaching, and writing on which this volume is based.

Contents

Neuropsychiatry and Behavioral Neuroscience

Chapter 1

Introduction

All human experience, emotion, motivation, behavior, and activity are products of brain function. This basic premise underlies contemporary approaches to understanding human behavior and the effects of brain dysfunction in the clinical discipline of neuropsychiatry. This approach does not deny the important influence of interpersonal relationships, social and cultural influences, and the modulating influence of the environment on human emotion and behavior; the brain-based approach acknowledges that all of these environmental influences are mediated through central nervous system (CNS) structures and function. For every deviant environmental event there will be a corresponding change in CNS function, and when CNS function is altered there will be corresponding changes in the behavior or experience of the individual.

Neuropsychiatry is the clinical discipline devoted to understanding the neurobiological basis, optimal assessment, natural history, and most efficacious treatment of disorders of the nervous system with behavioral manifestations.[1] Neuropsychiatry embraces the rich interplay between the environment and the nervous system both during the development of the individual and throughout adulthood and old age. Neuropsychiatrists seek to understand the disorders of the CNS responsible for abnormal behavior.

This volume presents a contemporary view of neuropsychiatry and the advances in neuroscience applicable to understanding and interpreting human behavior. This introductory chapter summarizes themes and perspectives that provide the philosophical and clinical framework for the book.

ADVANCES IN NEUROSCIENCE

The last few decades have seen an incredible advance in neuroscience applicable to neuropsychiatry. Studies in genetics and molecular biology have revealed mutations that cause major neuropsychiatric disturbances including familial Alzheimer's disease, familial Parkinson's disease, Huntington's disease, Wilson's disease, and many developmental disorders. Risk genes for some conditions such as Alzheimer's disease have also been identified. These do not by themselves cause disease, but they increase the likelihood that individuals will express the disorder in the course of their lifetime. Genetic testing, available for many conditions, allows specific diagnoses or risk assessments and raises important ethical issues for neuropsychiatry. Identification of a mutation in an asymptomatic individual, for example, reveals critical aspects of his or her ultimate fate, knowledge not to be taken lightly.

Advances in developmental neurobiology have informed our understanding of congenital malformations of the CNS, many with severe associated behavioral disturbances. Progress also has been made in understanding hyperactivity-attention deficit disorder, autism and pervasive developmental disorders, and childhood epilepsies and movement disorders. In addition, there has been an evolving integration of developmental and maturational perspectives to allow a life span approach to human neurological disease and neuropsychiatric conditions. Even late-onset disorders such as Alzheimer's disease interact with educational level and native intellectual abilities to determine the time of onset and duration of the adult disorder.[2] Comprehensive understanding of any neuropsychiatric illness requires a careful developmental history and integration of life span information.

Advances in neuroimaging have been important in the growth of neuropsychiatry. Structural imaging techniques such as magnetic resonance imaging (MRI) have revealed, for instance, that white matter abnormalities are present in many patients with late-onset depressions.[3] The occurrence of white matter disturbances and basal ganglia lesions diminishes these patients' responsiveness to pharmacotherapy and increases the likelihood of confusion following electroconvulsive therapy.[4–6] Thus, imaging findings have treatment implications and imaging results are increasingly incorporated in neuropsychiatric assessment and treatment planning.

Functional imaging such as positron emission tomography (PET) and single photon emission computed tomography (SPECT) provide critical information about brain function in neuropsychiatric illness. Patients with Alzheimer's disease, for example, have reduced glucose metabolism in the parietal lobes; when psychosis and agitation are present, frontal and anterior temporal hypometabolism is also evident.[7,8] Positron emission tomography has shown disturbances in frontal lobe activation in patients with schizophrenia,[9] as well as orbito-frontal hypermetabolism in patients with obsessive-compulsive disorder.[10] Functional MRI (fMRI) provides a "stress test" for the CNS, revealing abnormal patterns of activation in patients with brain disease. This approach may eventually prove sensitive to the earliest changes in incipient neurological conditions. Magnetic resonance spectroscopy provides information about the chemical composition of brain structures, reveals abnormalities in individual diseases, and may provide a window on CNS concentrations of some therapeutic agents.[11] Magnetic resonance angiography (MRA) allows noninvasive study of the brain vasculature and MR perfusion studies have shown remarkable sensitivity to the occurrence of recent ischemic brain injury. Together these technologies provide a diverse armamentarium of techniques for diagnosing CNS disease and understanding their pathophysiology.

Progress in neuropsychology also informs contemporary neuropsychiatry. There have been substantial advances, for example, in recognizing and characterizing memory subroutines including registration, consolidation, and retrieval.[12] Focal brain lesions and degenerative disorders differentially affect these processes, depending on the brain structures involved. Similarly, the "frontal lobe syndrome" has been divided into medial frontal, orbitofrontal, and dorsolateral prefrontal types, and neuropsychological mechanisms mediated by the dorsolateral prefrontal cortex including planning, sequencing, implementing, executing, and monitoring of behaviors have been identified.[13] Attentional mechanisms, visuospatial processes, and language have been studied and the results integrated into the practice of neuropsychiatry.

Many diseases are much better understood as a result of the application of basic science. Idiopathic neuropsychiatric illnesses such as schizophrenia have been the subject of intensive scientific scrutiny. Regional changes in brain structure have been identified and genetic and environmental contributors to the syndrome discovered. The pathophysiology of Alzheimer's disease has been revealed in substantial detail including processing of the amyloid precursor protein to free amyloid β protein leading to neurotoxicity and the formation of neuritic plaques.[14] Abnormal accumulation of α-synuclein is recognized as characteristic of Parkinson's disease and related conditions.[15] This improved understanding of basic disease processes facilitates identification and interpretation of the clinical syndromes and provides a basis for the development of therapeutic agents.

Marked progress has been made in pharmacotherapy. Neurology and neuropsychiatry have changed from clinical disciplines with few available treatments to clinical arenas with major neurotherapeutic options. Epilepsy, migraine, multiple sclerosis, Parkinson's disease, Alzheimer's disease, amyotrophic lateral sclerosis, psychosis, depression, obsessive-compulsive disorder, anxiety, sleep disorders, substance use disorders, and eating disorders have all been the subjects of development of new pharmacotherapeutic agents capable of ameliorating disease-related symptoms and restoring more normal function.

The advances in all of these areas of neuroscience provide the basis for the update of neuropsychiatry developed in this volume.

FOUNDATIONS OF NEUROPSYCHIATRY

Neuropsychiatry includes both the psychiatric manifestations of neurologic illness and neurobiology of idiopathic psychiatric disorders. Two disciplines have been instrumental in the development of neuropsychiatry—biological psychiatry and behavioral neurology. *Biological psychiatry* received its primary impetus from the success of biological treatment of psychiatric disorders. Drugs that increase the levels of monoamines and serotonin relieve depressive symptoms and drugs that block dopamine receptors reduce psychosis. These observations imply (but do not prove) that transmitter disturbances are involved in the mediation of these behavioral disorders. In biological psychiatry, the chemistry of behavior is emphasized, with increasingly precise characterization of transmitter receptors and signal transduction mechanisms.[16] Biological psychiatrist, however, have not emphasized a neuroanatomy of behavior or the relationship of focal lesions of the CNS to behavioral disorders.

Behavioral neurology is the other cornerstone of neuropsychiatry. This discipline was revived in the 1960s by Norman Geschwind with his description of the disconnection syndromes.[17] Aphasias and amnesias resulting from focal brain injuries were characterized and hemispheric specialization was investigated in patients with injuries to the corpus callosum. Drawing heavily on techniques derived from neuropsychology, behavioral neurology provided detailed descriptions of language disorders, memory disturbances, visuospatial abnormalities, agnosias, and dementias associated with focal brain damage or degenerative CNS disease. In behavioral neurology, a probing mental status examination is used to aid in neuroanatomical interpretation of deficit syndromes. Behavioral neurologists have investigated the deficit disorders of aphasia, amnesia, agnosia, alexia, agraphia, and amusia but do not focus on the positive symptoms of neuropsychiatric disorders such as depression, mania, personality alterations, or obsessive-compulsive disorder associated with brain dysfunction.

Thus neither biological psychiatry nor behavioral neurology provides a comprehensive view of brain–behavior relationships. Neuropsychiatry draws on both disciplines in addition to recent advances in neuroscience to provide a comprehensive understanding of the relationship of brain and behavior.

TERMINOLOGY

The words *organic* and *functional* are eschewed in this volume as they are misleading in their assumptions. Many disorders called "organic," such as epilepsy, produce significant functional alterations with few or no structural abnormalities, and many "functional" illnesses, such as psychosis and depression, are the products of neurologic disorders. Although imperfect, the term *idiopathic* will be used to describe psychiatric disorders whose etiologies and pathophysiology have yet to be revealed, and *neurologic* or *toxic-metabolic* will be used when specific types of brain disorders have been identified that account for behavioral changes. These terms escape some of the objectionable assumptions associated with the traditional terminology.

CLINICAL APPROACH

The focus of this volume is on clinical utility and the relationship of clinical observations to the evolving neuroscience. Assessment of signs and symptoms, differential diagnosis, application of technology to explore diagnostic hypotheses, and pharmacotherapy are emphasized. The mental-status examination is borrowed largely from behavioral neurology and is used to help characterize a patient's attention, verbal output, memory, constructional skills, and executive abilities. This approach is augmented by interview techniques taken from psychiatry that emphasize anamnesis and help disclose subjective phenomena such as delusions, hallucinations, and intrusive thoughts. Mental status examination is complemented by elementary neurological and general physical examinations. Occasionally objections are raised to the probing mental status examination as being offensive to patients and insufficiently sensitive to their feelings of failure. A detailed mental status examination can be successfully achieved, however, by an expert clinician without the loss of respect for the patient's sense of vulnerability and exposure. Most errors in neuropsychiatric diagnosis are of omission, not commission.

Dynamically oriented psychiatrists and psychotherapists may object to the absence of dynamic considerations in the neuropsychiatric approach proffered here. Some have charged that neuropsychiatry attempts to turn a brainless psychiatry into a mindless neurology. The past excesses of classical analytic psychiatry are now apparent, however, and balance is being restored with respect to the relative spheres of psychological, environmental, genetic, and structural factors in behavior. In neuropsychiatry the emphasis is an identification of diseases and on alliance building with patients and families to maximize the success of therapeutic interventions. Dynamic psychological issues emerge during the process of this alliance, and must be considered in both the interpretation of the patient's symptoms and the response to therapy.

UNIQUENESS OF NEUROPSYCHIATRY

The care of patients with neuropsychiatric illness differs from that provided to patients with medical illnesses. Brain disorders, unlike their medical counterparts, are manifest by alterations in the behavior and experience of the victim; in many ways they are disorders *of* the person rather than disorders that happen *to* the individual. Patients illness may *have* pneumonia or congestive heart failure, but those with neuropsychiatric illness *are* demented, psychotic, or depressed.

The difference in the way neuropsychiatric disturbances affect patients necessitates a change in the way the clinician responds. Many neuropsychiatric illnesses, although treatable, may be only partially reversible. After head trauma or stroke or following the onset of a schizophrenic illness, a patient is unlikely to be completely restored to the same "person" as he or she was in the premorbid state. Thus, the patient has become a "new" person. The clinician is obligated to respect this change and is responsible for helping the patient and family formulate goals appropriate to the new situation. The patient's goals must be reconciled with the limits imposed by altered brain function while not succumbing to the temptation to allow a disability to unnecessarily circumscribe the patient's life opportunities. Knowledge of the course and impact of neuropsychiatric illness will aid the clinician in brokering appropriate expectations. Guidance and support provided by the clinician will often be as important as the medications dispensed.

In addition to the role of the neuropsychiatrist in treating patients and advising patients and their families, the practice of neuropsychiatry affords an exciting opportunity to learn from the patient. The victim of a neuropsychiatric illness is traversing an uncharted landscape and each pilgrim–patient is a source of information that can be utilized to help guide others with CNS disease. The patient's observations, descriptions, motoric changes, and reported experiences are invaluable information that contributes to the science of neuropsychiatry.

NEUROPSYCHIATRY AND PHILOSOPHY

Determinism

Contemporary neuroscience has established a fundamental correlation between brain function and mental activity; the data support the basic monistic premise that human intellectual and emotional life is dependent on neuronal operations. This monistic perspective is associated with a philosophy of materialism. Two objections have frequently been raised against the monist position: (*1*) that it commits one to a determinism that disallows any role for free will, and (*2*) it undermines respect for human beings by reducing them to machines or automatons. Neither objection is necessarily true. Free will in human behavior is not the ability to have random activity; it is the ability to direct one's behavior according to one's preferences, and brain function provides a neurophysiological basis for preference-motivated behavior. Volitional action is the final product of a hierarchy of competing alternative activities determined by one's past history and present contingencies valued according to evolving schemata and chosen for execution. "Choice" is the subjective counterpart of this dynamic process. Behavioral choices are not determined by CNS structure but are consistently reevaluated; decisions occur at the intersection of a constantly changing environment and an evolving "self." The myriad of competing influences that determine choice prohibit complete modeling of human action and make fulfillment of a strict determinist position as proven by the predictability of behavior impossible.

Although many aspects of CNS structure and, consequently, CNS function and behavior are genetically influenced, it is obvious that behavior is modified by development and experience and that there is a constant commerce between the CNS and the environment. Genetics is not fate except in the cases of fatal mutations. Behavior is a summary product of genetic, historical, experiential, and environmental influences, and CNS structure provides a physical basis for integrating and mediating these multiple behavioral determinates. Once this potential for environmental influence and preference-motivated behavior is accommodated within the monist proposition, the free will objection to monism loses force.

Likewise, the ability to integrate ongoing experience and environmental interaction with monism deflates the objection that monism inevitably leads to treating human beings like machines. Indeed, monism can provide the basis for increasing respect for human individuality by emphasizing that each individual is the product of a unique blend of genetic, experiential, and environmental influences all mediated through a private CNS structure. A genetically determined CNS modified and enriched by ongoing environmental interactions unique to each individual provides the basis for human individuality. Reductionism to molecular neuropsychiatry is not the necessary outcome of the monist position; an integrative neuropsychiatry is required to understand the complex interplay of genetic and environmental influences on human behavior.

Neuropsychiatry and Epistemology

Neuropsychiatry provides principles of brain–behavior relationships (presented in Chapter 5) and a means of understanding human knowledge. The CNS is the organ of knowledge generation, accumulation, and dissemination. The dissemination of knowledge over time defines human culture. Knowledge may to be divided into two types: public and private. All knowledge begins as private and becomes public through speech and writing. The generation of public knowledge is represented as history. The record of private knowledge is biography. Public knowledge, when committed to memory, is semantic, or factual memory, whereas private knowledge comprises episodic memory (Chapter 7). Public knowledge includes science, mathematics, history, philosophy, politics, sociology, psychology, law, business, and language. Public knowledge also informs shared social perceptions as part of a culture. The historical continuity of personal knowledge comprises the "self." As long as it is personal, it remains subjective and instantaneous; it becomes continuous history only when written or verbalized and made public. Ongoing private knowledge is contained in working memory (Chapter 9). Personal knowledge is heavily infused with emotion, including motivation, intention, perception of threat or support, and subjective emotional states such as euphoria, depression, anxiety, and irritability. Public knowledge is without emotion until it is re-experienced privately. Private knowledge and awareness of one's own consciousness is the basis for empathy and the assumption that others are experiencing a similar conscious state and subject to similar emotions. This self-awareness and the related possibility of empathy are heavily dependent on frontal lobe function (Chapter 9), are compromised in frontal degenerative disorders (Chapter 10) and other frontal disturbances, and reflect the unique expansion of the frontal cortex in humans.

Study of brain–behavior relationships and neuropsychiatric syndromes provides insight into abstract concepts such as self and culture, allows understanding of the neurobiological basis of human behavior both individually and in the context of an evolving culture, and allows construction of a neuroepistemology.

CHALLENGES OF NEUROPSYCHIATRY

Many influences are propelling neuropsychiatry to the forefront of neurology, psychiatry, and neuroscience.[18] The graying of the population and the increase in the number of individuals with age-related brain disease increases the prevalence of neuropsychiatric illness. Alzheimer's disease, Parkinson's disease, and stroke are all major age-related disorders with profound neuropsychiatric consequences. Developmental disorders are also increasing in society. As more premature infants are sustained, the number of individuals who are developmentally impaired increases. Subnormal intelligence, behavior and conduct disorders, and seizures are among the adverse long-term neuropsychiatric outcomes observed among individuals who survive premature birth. Malnutrition is a worldwide phenomenon associated with lowered I.Q. and deficits in language, personal, and social skills. As the human immunodeficiency virus (HIV) global epidemic continues, the prevalence of HIV is increasing in many countries; HIV encephalopathy is associated with mood disorders, psychosis, hallucinations, executive dysfunction, agitation, and apathy. Substance abuse is second only to anxiety as the most common psychiatric disorder in the United States. Substance abuse causes both acute and chronic changes in brain function and produces a wide variety of neuropsychiatric syndromes including acute intoxication, dependence, delirium, dementia, psychosis, mood disorders, anxiety, and personality changes. Excessive ingestion of alcohol during pregnancy may lead to the fetal alcohol syndrome, characterized by mild to severe mental retardation, irritability, hyperactivity, and distractibility. Head trauma is another major cause of neuropsychiatric disability. Management of the behavioral consequences of head injuries associated with traffic accidents, wars, domestic violence, and urban aggression requires substantial neuropsychiatric expertise. Criminal behavior is a major social concern and is more frequent among neuropsychiatrically ill individuals than among those without CNS disease. The proliferation of neurotoxic compounds worldwide exposes developing and mature individuals to a variety of pollutants and contaminants capable of altering CNS function and producing neuropsychiatric symptoms. Institutional populations, including individuals in nursing homes, prisons, chronic mental hospitals, and residences for the mentally retarded, all contain substantial numbers of individuals with neuropsychiatric disorders and require the attention of those with neuropsychiatric skills. Thus, there are many situations where optimal care and management require neuropsychiatric expertise, and the number of victims of neuropsychiatric illness is growing.

CONCLUSIONS

Neuropsychiatry is a clinical discipline applicable to individual patients with specific neuropsychiatric signs and symptoms; it is relevant to the most pressing problems of contemporary human existence. The brain is the organ of all mental life, and the increasing revelation of brain processes underlying behavior provides insight into mental illness, neuropsychiatric disorders, and normal cognition and emotion. A dehumanizing, reductionistic, and mechanical view of humankind is not the obligatory impact of neuroscience on societies; rather, neuroscience can enhance our appreciation of each individual by revealing the neurobiologic underpinnings of the complex interplay of structural, chemical, genetic, environmental, developmental, social, and cultural influences on behavior. Neuropsychiatry enhances intellectual commerce among related disciplines (e.g., psychiatry, neurology, and neuroscience), facilitates the translation of scientific advances into clinical practice, and serves patient care.

REFERENCES

1. Cummings JL, Hegarty A. Neurology, psychiatry, and neuropsychiatry. Neurology 1994;44:209–213.
2. Cummings JL, Booss J, et al. Dementia Identification and Assessment: Guidelines for Primary Care Practitioners. Washington, DC: U.S. Department of Veterans Affairs and the University Health System Consortium, 1997.
3. Coffey CE, Figiel GS, et al. Subcortical hyperintensity on magnetic resonance imaging: a comparison of normal and depressed elderly subjects. Am J Psychiatry 1990;147:187–189.
4. Figiel GS, Krishnan KRR, et al. Radiologic correlates of antidepressant-induced delirium: the possible significance of basal-ganglia lesions. J Neuropsychiatry Clin Neurosci 1989;1:188–190.
5. Figiel GS, Coffey CE, et al. Brain magnetic resonance imaging findings in ECT-induced delirium. J Neuropsychiatry Clin Neurosci 1990;2:53–58.
6. Hickie I, Scott E, et al. Subcortical hyperintensities on magnetic resonance imaging: clinical correlates and prognostic significance in patients with severe depression. Biol Psychiatry 1995;37:151–160.
7. Hirono N, Mega MS, et al. Left frontotemporal hypoperfusion is associated with aggression in patients with dementia. Arch Neurol 2000;57:861–866.
8. Mega MS, Lee L, et al. Cerebral correlates of psychotic symptoms in Alzheimer's disease. J Neurol Neurosurg Psychiatry 2000;69:167–171.
9. Weinberger DR, Berman KF, Illowsky BP. Physiological dysfunction of dorsolateral prefrontal cortex in schizophrenia. Arch Gen Psychiatry 1988;45:609–615.
10. Baxter LR, Phelps ME, et al. Local cerebral glucose metabolic rates in obsessive-compulsive disorder: A comparison with rates in unipolar depression and normal controls. Arch Gen Psychiatry 1987;44:211–218.
11. Henry ME, Frederick BD, et al. Magnetic resonance spectroscopy in psychiatric illness. In: Dougherty DD, Rauch SL, eds. Psychiatric Neuroimaging Research, Contemporary Strategies. Washington, DC: American Psychiatric Publishing, 2001:291–333.
12. Squire LR, Zola-Morgan S. The medial temporal lobe memory system. Science 1991;253:1380–1386.
13. Boone KB. Neuropsychological assessment of executive functions—impact of age, education, gender, intellectual level, and vascular status on executive test scores. In: Miller BL, Cummings JL, eds. The Human Frontal Lobes, Functions and Disorders. New York: The Guilford Press, 1999:247–260.
14. Beyreuther K, Masters CL. Alzheimer's disease: physiological and pathogenetic role of the amyloid precursor protein (APP), its AB-amyloid domain and free AB-amyloid peptide. In: Beyreuther K, Christen Y, Masters CL, eds. Neurodegenerative Disorders: Loss of Function Through Gain of Function. New York: Springer-Verlag, 2001:97–117.
15. Lansbury PTJ. The role of α-synuclein in Parkinson's disease: a biophysical analogy to AB and Alzheimer's disease. In: Lee VM-Y, Trojanowski JQ, Buee L, Christen Y, eds. Fatal Attractions: Protein Aggregates in Neurodegenerative Disorders. New York: Springer-Verlag, 2000:1–9.
16. Trimble MR. Biological Psychiatry. Chichester: John Wiley & Sons, 1996.
17. Geschwind N. Disconnection syndromes in animals and man. Brain 1965;88:237–294.
18. Cummings JL. Neuropsychiatry and society. J Neuropsychiatry Sci 1996;8:104–109.

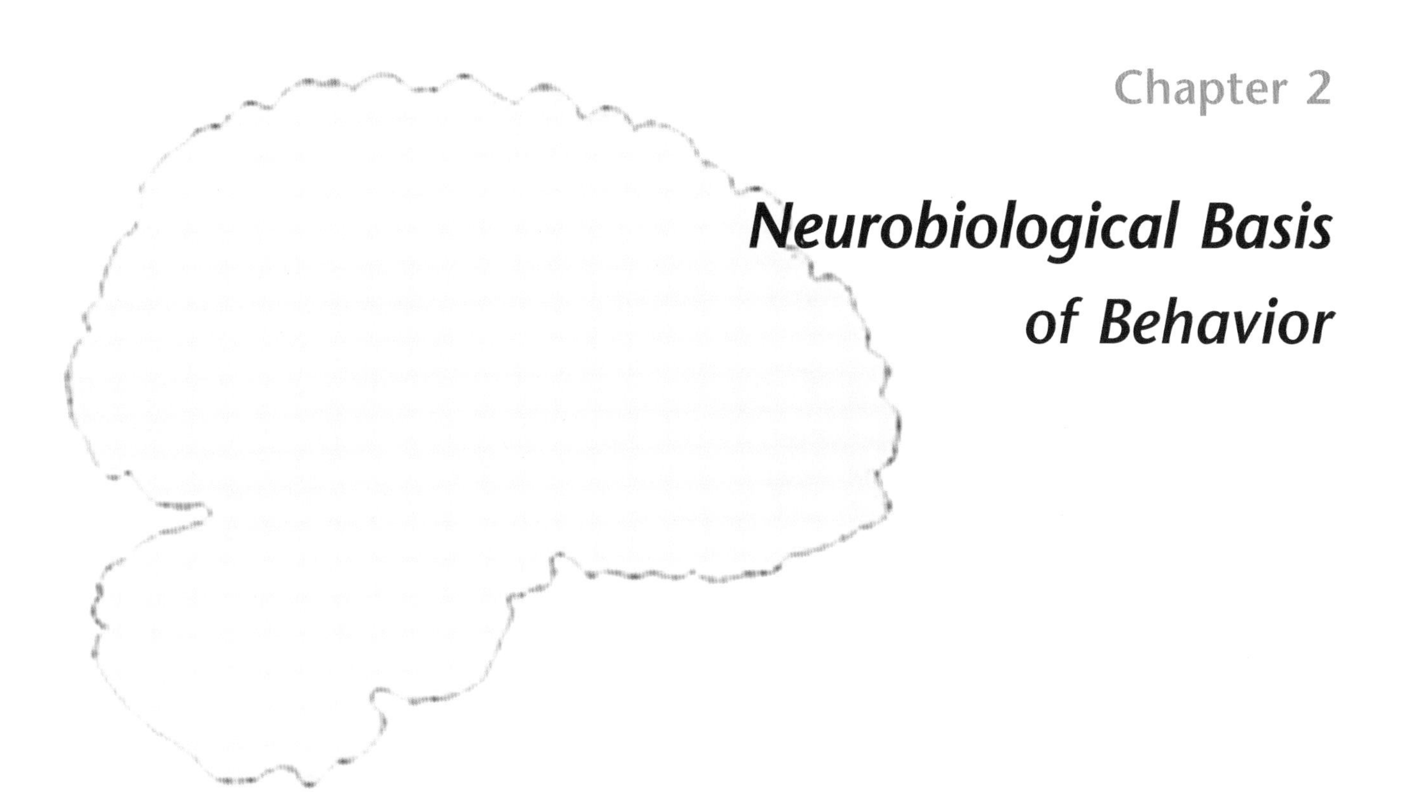

Chapter 2

Neurobiological Basis of Behavior

PHYLOGENETIC DEVELOPMENT

The evolution of the brain has produced increasingly refined systems for interacting with the environment. From reptilian, through lower mammalian, to human stages of evolution, successive elaborations of neuronal tissue have built upon and often elaborated systems of preceding phylogenetic epochs. Understanding the organization of these phylogenetic stages will aid our insight into the neurobiological basis of behavior. Yakovlev was the first to best describe three levels of central nervous system function[1] as a heuristic concept of brain–behavior relationships. A primitive inner core devoted to arousal and autonomic function is surrounded by a middle layer that includes the limbic system and basal ganglia; this in turn is encapsulated by the most recent phylogenetic layering of the neocortex and pyramidal system (Fig. 2.1). Each layer subserves different functions (Table 2.1). The inner layer contains the reticular core, a mesh of unmyelinated neurons controlling consciousness, cardiovascular, and respiratory function. The middle layer has partially myelinated organized cell groups, including the basal ganglia and limbic system; its functions concern arousal, communal activities, personality, and emotion. The outer neocortical layer of well-myelinated neurons enables fine motor control, detailed sensory processing, praxis, gnosis, and abstract cognition. Such abstract skills are contrasted to the more "emotionally charged" processing mediated by the middle layer or limbic system. With early mammalian development, rearing of the young was more interactive than in reptilian species. Thus, primal vocalization behavior emerged, such as the separation cry, while communal bonding and territorial behavior also mirrored the development of the middle limbic layer. During mammalian evolution, with the progressive expansion of the cortical mantle, a developmental progression from three-layered *allocortex* to six-layered *isocortex* occurred.

Comparisons of phylogenetic development across mammalian species have revealed two waves of increasing complexity from allocortex to isocortex first clearly described by Sanides.[2] These two waves originate from two primordial regions within the limbic ring (Fig. 2.2). The orbitofrontal region of the olfactory paleocortex spreads ventrolaterally up through the insula, temporal pole, and anterior parahippocampal area. The olfactory orbitofrontal spread is closely associated with the amygdala. The integration of appetitive drives with

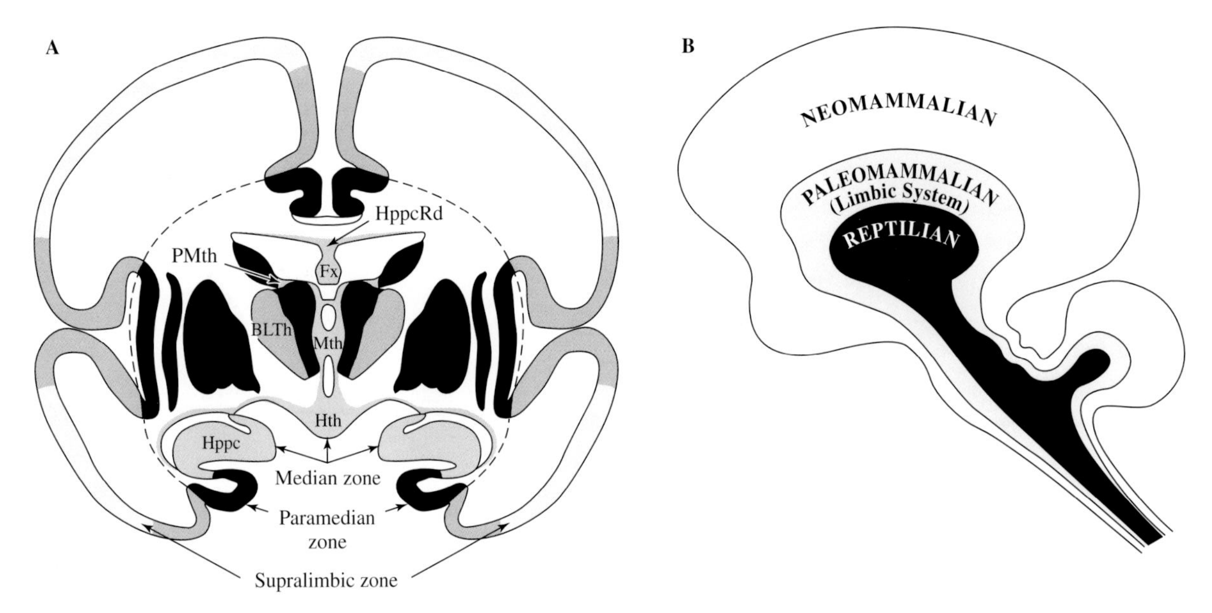

FIGURE 2.1 (*A*) Yakovlev's model of the three phylogenetic zones of brain development, as reflected by myelogentic stages.[151] BLTh, basolateral thalamus; Fx, fornix; Hppc, hippocampus; HppcRd, hippocampal radiations; Hth, hypothalamus; Mth, medial thalamus; PMTh, paramedial thalamus. (*B*) MacLean's rendition of this evolutionary layering produced a triune brain.[152]

aversion or attraction to stimuli dominates paleocortical function. A second center of cortical development grew out of the archicortex of the hippocampus and spread posteriorally through the entorhinal, posterior parahippocampal regions, and through the cingulate. The archicortex is largely concerned with the integration of information from different sensory modalities—the first step away from thalamic control, as seen in reptiles, toward cortical dominance. The hippocampal-centered spread emphasizes pyramidal cells and gives rise to the medial supplementary sensory and motor areas. In addition to the hippocampal archicortex this second limbic arm includes the cingulate gyrus, which is composed of four functional centers:[3] the visceral, cognitive, skeletomotor, and sensory processing regions (Fig. 2.3). The visceral control region, below the corpus

TABLE 2.1. *Yakovlev's Three-layered Model of Brain Organization, Structure, and Function*[1]

	Inner Layer	*Intermediate Layer*	*Outer Layer*
Neurons	Short, unmyelinated	Long, partially myelinated	Long, well myelinated
Organization	Diffuse	Ganglia, allocortex	Isocortex
Evolution	Invertebrate to reptile	Reptile to early mammal	Mammal to primate
Structure	Reticular core	Basal ganglia	Primary sensory cortices
	Cranial nerves	Limbic thalamus	Primary motor cortex
	Periaqueductal gray	Olfactory paleocortex	Corpus callosum
	Hypothalamus	Hippocampal archicortex	Association cortex
Function	Consciousness	Motor synergistic	Motor precision
	Metabolism	Arousal, motivation	Ppraxis
	Respiration	Mood, affect	Language
	Circulation	Personality	Gnosis

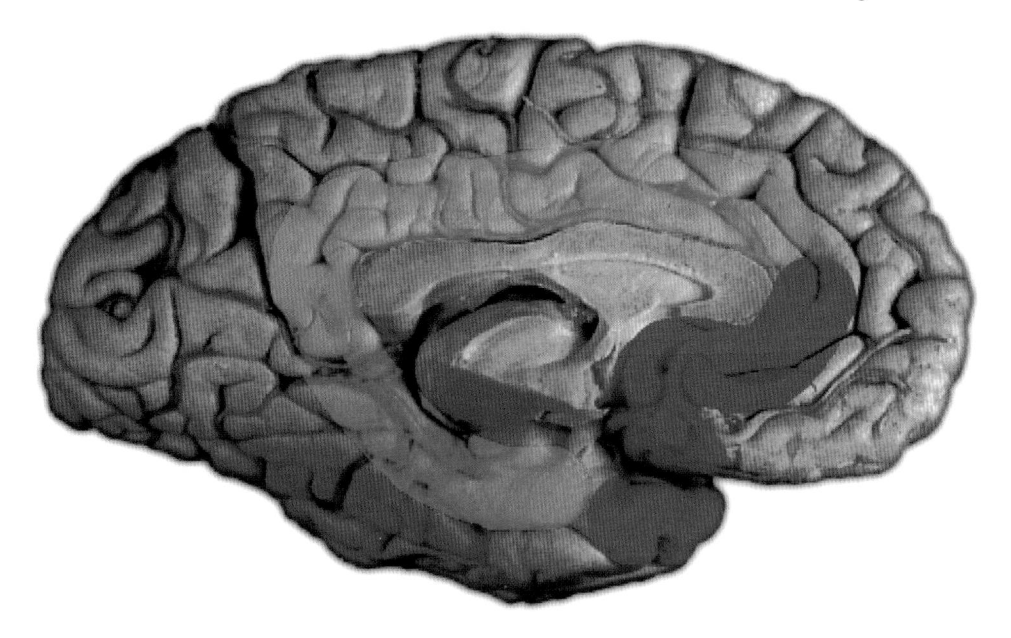

FIGURE 2.2 The paralimbic trends of evolutionary cortical development. The orbitofrontal-centered belt (red) extends into the subcallosal cingulate, temporal polar region, and the anterior insula (not shown). The hippocampal-centered trend (blue) extends its wave of cortical development dorsally through the posterior and anterior cingulate. Adapted from Mega et al. (1997).[50]

callosum, overlaps with the medial orbitofrontal limbic division. By virtue of the two limbic divisions' connections and the parallel development of other brain regions linked to the two limbic centers, the behavioral evolution of mammals mirrored the progressive trend toward cytoarchitectural complexity emanating from both the orbitofrontal and hippocampal centers.[4] Mesulam has described these two allocortical systems as paralimbic belts.[5] Table 2.2 provides an overview of the orbitofrontal and hippocampal paralimbic divisions.

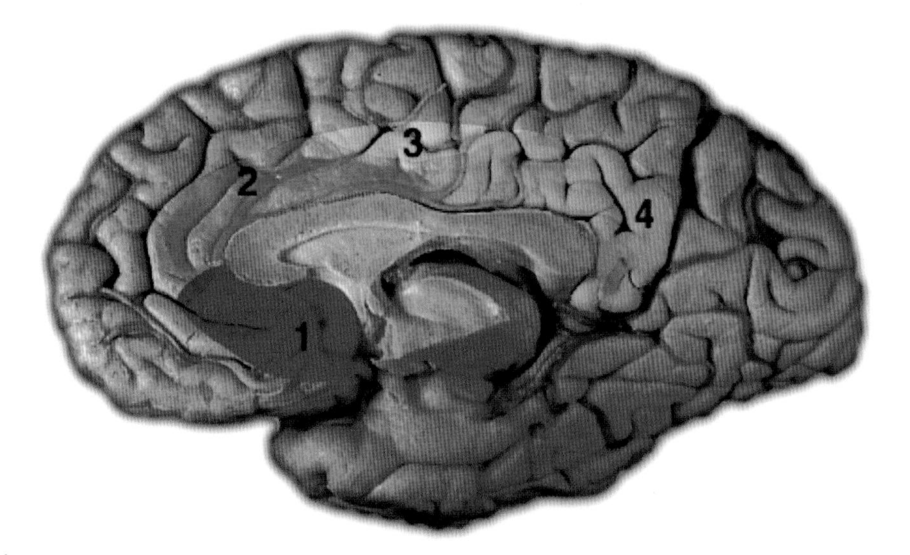

FIGURE 2.3 The four functional divisions of the cingulate: (*1*) the visceral effector region; (*2*) the cognitive effector region; (*3*) the skeletomotor effector region; and (*4*) the sensory-processing region. Adapted from Mega and Cummings (1996).[3]

TABLE 2.2. *The Two Paralimbic Divisions*

	Orbitofrontal Division	*Hippocampal Division*
Evolutionary trend	Paleocortical	Archicortical
Cell type	Granule cell	Pyramidal cell
Structures	Amygdala	Hippocampus
	Anterior parahippocampal	Posterior parahippocampal
	Insula	Retrosplenium
	Temporal pole	Posterior cingulate
	Infracallosal cingulate	Supracallosal cingulate
Function	Implicit processing	Explicit processing
	Visceral integration	Memory encoding
	Visual feature analysis	Visual spatial analysis
	Appetitive drives	Skeletomotor effector
	Social awareness	Attentional systems
	Mood	Motivation

ANATOMY AND NEUROTRANSMITTERS

The two paralimbic belts unite cortical and subcortical areas sharing phylogenetic and cytoarchitectural features common to the amygdala–orbitofrontal and hippocampal–cingulate limbic divisions. Because our growing refinement of the brain's connectional anatomy is derived from nonhuman primate tracer studies, all cortical anatomy described below is extrapolated from the Walker areas in nonhuman primates to their homologous Brodmann[6] areas on a human brain image. In cases where homology is not present, we interpret the animal-connectional data in reference to human clinical lesion data to extrapolate the cortical locations.

Both divisions of the limbic system work in concert. Processing in the amygdala–orbitofrontal division concerns the internal relevance that sensory stimuli have for the organism, thus facilitating intentional selection, habituation, or episodic encoding of these stimuli by the hippocampal–cingulate division. An understanding of the major reciprocal connections to a cortical region informs us about the functional system containing that region. There are also nonreciprocal connections, or open efferent (outgoing) and afferent (incoming) projections, associated with any given region. We limit our attention here to the reciprocally connected areas that segregate into general functional systems.

Orbitofrontal Paralimbic Division

The medial orbitofrontal cortex (Figure 2.4) and other regions reciprocally connected to it, when stimulated, effect visceral function,[7] probably through these regions' shared amygdalar connections. The rostral (agranular) insula, ventromedial temporal pole area 38, and medial subcallosal cingulate areas 25, 24, and 32 are also reciprocally connected to the medial division of the orbitofrontal cortex. The visceral control center of the subcallosal cingulate combines motivational input to the gustatory, olfactory, and alimentary information originating from anterior insula input converging on the medial orbitofrontal cortex. The anterior entorhinal area 36, also reciprocally connected to the medial orbitofrontal division, is a paleocortical extension with hippocampal connections allowing direct memory function to influence medial orbitofrontal processing. No visual information has direct access to the medial division of the orbitofrontal cortex that serves as an integrator of visceral drives while modulating the organism's internal milieu. The major cortical regions reciprocally connected with the medial division of the orbitofrontal paralimbic center are shown in Table 2.3 and Figure 2.5.

The lateral portion of the orbitofrontal paralimbic division is phylogenetically more developed than the medial portion. The lateral portion also has reciprocal

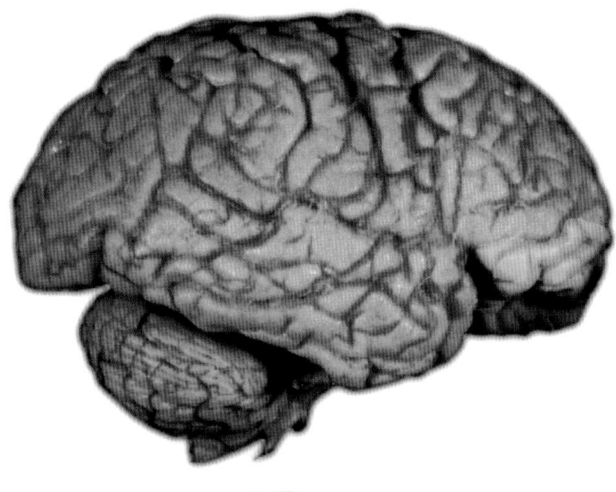

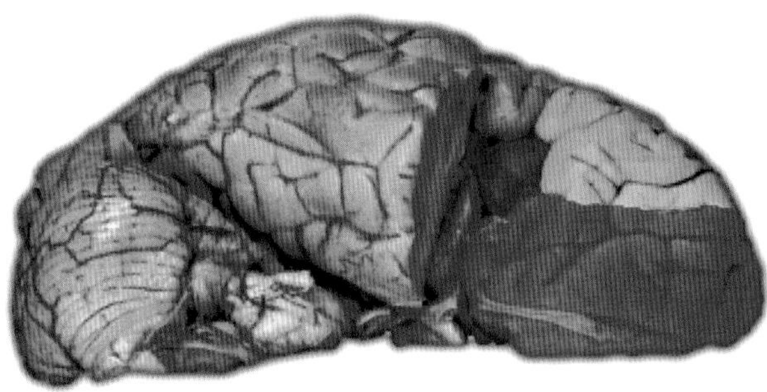

FIGURE 2.4 The two divisions of the orbitofrontal cortex. The medial division is in red and includes the gyrus rectus and medial orbital gyrus of area 11 in human. The lateral division is in green and includes the lateral orbital gyrus of area 11 and the medial inferior frontal gyrus of areas 10 and 47 in humans.

connections with the amygdala (more dorsal and caudal than the medial orbitofrontal cortex). The dorsal portion of the basal amygdala is the source of projections to the ventral visual processing system in the inferior temporal cortex. Reciprocal connections also occur with the supracallosal cingulate areas 24 and 32, a region that assists in the dorsolateral attentional system and effects cognitive engagement. The auditory association cortex of dorsolateral temporal pole area 38 is also reciprocally connected to the lateral orbitofrontal cortex. The lateral orbital cortex is a gateway for highly processed sensory information into the orbitofrontal paralimbic center. Reciprocal connections with the inferior temporal cortex area 20, the last processing step for the ventral visual system devoted to object feature analysis, and the supplementary eye fields in the dorsal portion of area 6 highlight the control over sensory processing occurring in the lateral orbitofrontal cortex. The major cortical regions reciprocally connected with the lateral portion of the orbitofrontal paralimbic division are shown in Table 2.3 and Figure 2.5.

Hippocampal Paralimbic Division

Papez[8] proposed that multimodal sensory information processed in the hippocampus projects to the mammillary bodies via the fornix and enters the anterior nucleus of the thalamus via the mammillothalamic tract. From there it projects to the cingulate gyrus, through the retrosplenial cortex, and then back to the hippocampus. Papez's circuit enables the conscious encoding of experience. Hippocampal archicortical development extends into the posterior cingulate and then forward into the anterior cingulate. The major reciprocal connections of these two cingulate regions provide two portals through which limbic influence can affect other regions; these diverse areas are organized into functional networks.

The posterior cingulate (Brodmann areas 23, and 29/30) is a nexus for sensory and mnemonic networks within the hippocampal paralimbic belt. Functional imaging data during episodic memory encoding tasks implicate the posterior cingulate in the consolidation of declarative memory,[9–12] and associative learning during classical conditioning.[13] The connections of the

TABLE 2.3. *Major Reciprocal Connections for Lateral and Medial Portions of Orbitofrontal Paralimbic Division*

Medial Orbitofrontal Portion	*Lateral Orbitofrontal Portion*
Medial basal amygdala[138–142]	Dorsal and caudal basal amygdala[138–142]
Infracallosal areas 25, 24, and 32[140,143,144]	Supracallosal areas 24 and 32[140,143,144]
Ventromedial temporal area 38[140,142,145]	Dorsolateral temporal area 38[140,142,145]
Rostral (agranular) insula[140,141,146]	Inferior temporal cortex area 20[142]
Anterior entorhinal area 36[142,147]	Supplementary eye field in dorsal area 6[142]

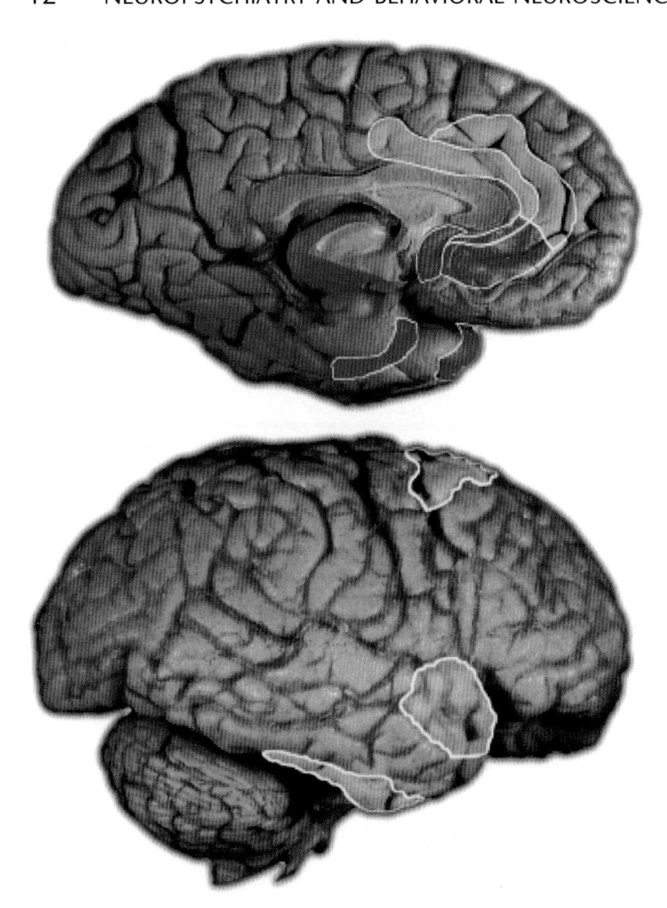

FIGURE 2.5 Major cortical reciprocal areas connected to the medial portion (shown in red), and lateral portion (shown in green) of the orbitofrontal limbic division.

posterior cingulate focus on the dorsolateral cortex. The posterior cingulate has major reciprocal connections with the posterior parahippocampal and perirhinal areas 36 and 35, as well as the presubiculum. These connections modulate the multimodal efferents entering the entorhinal layer III cells that give rise to the perforant pathway into the hippocampus. The dorsal visual system of the inferior parietal lobe 7a, dedicated to spatial processing,[14] and the frontal eye fields in area 8 also have bidirectional connections with the posterior cingulate. Reciprocal connections with lateral prefrontal area 46 allow an interaction between executive and sensory/mnemonic processing that may mediate perceptual working memory tasks. The major reciprocal connections of the posterior cingulate are shown in Table 2.4 and Figure 2.6a.

The orbitofrontal and hippocampal paralimbic belts intersect in the infracallosal cingulate region of Brodmann area 24. Cytoarchitectural development progresses from anterior to posterior and from inferomedial to dorsal across area 24.[15] The major pathway for information flow is the cingulum bundle. The cingulum contains the efferents and afferents of the cingulate to the hippocampus, basal forebrain, amygdala, and all cortical areas, as well as fibers of passage between the hippocampus and prefrontal cortex, and from the median raphé to the dorsal hippocampus.[16] Robust connections with the sensory processing region of the posterior cingulate overlap with the reciprocal amygdalar connections present in the anterior cingulate. Thus, area 24 is a nexus in the distributed networks subserving internal motivating drives and externally directed attentional mechanisms.[17] Papez's initial conception of the cingulate as the "seat of dynamic vigilance by which emotional experiences are endowed with an emotional consciousness"[8] is supported by this anatomic organization. The integration of both divisions of limbic processing in the anterior cingulate influences the dorsolateral executive cortex as Papez predicted: "the sensory excitations which reach the lateral cortex through the internal capsule receive their emotional coloring from the concurrent processes of hypothalamic origin which irradiate them from the gyrus cinguli."[8]

TABLE 2.4. *Major Reciprocal Connections for Posterior and Anterior Cingulate Cortex*

Posterior Cingulate	*Anterior Cingulate*
Posterior parahippocampal area 35/36[140,144, 48]	Anterior parahippocampal area 35/36[140,144,148]
Presubiculum[140,144,148]	Basal amygdala[138–141]
Prefrontal area 46[140,149]	Prefrontal areas 8, 9, 10, and 46[140,149]
Caudal parietal area 7[17,140,150]	Caudal orbitofrontal cortex area 47[140,143,144]
Frontal eye field area 8[140,149]	Inferior temporal pole area 38[140,145]
	Rostral insula[140,141,146]

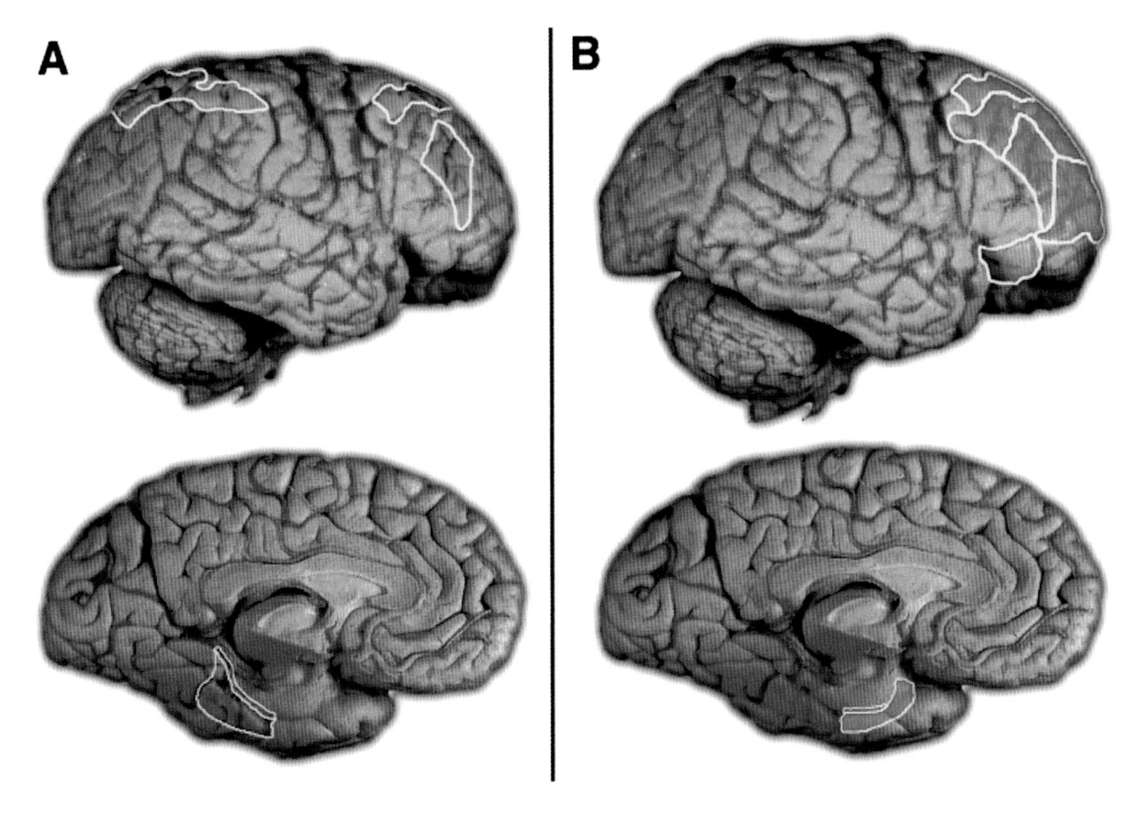

FIGURE 2.6 Major cortical reciprocal areas connected to the posterior (*A*; green) and anterior (*B*; blue) cingulate cortex.

Major reciprocal connections with the cognitive effector region (supracallosal areas 24 and 32) are with the basal amygdala (magno- and parvocellular divisions). The amygdala provides internal affective input to areas 24 and 32. Prefrontal areas 8, 9, 10, and 46 also have reciprocal connections. The anterior cingulate is more developed in its cytoarchitecture and has well-developed connections with phylogenetically more recent neocortex of dorsolateral prefrontal areas 8, 9, 10, and 46 devoted to executive function. Areas reciprocally connected to the anterior cingulate share a general similarity with the orbitofrontal center and include the caudal (dysgranular) orbitofrontal area 12 in monkeys (equivalent to area 47 in humans), which provides processed sensory information concerning object feature analysis; the anterior inferior temporal pole area 38, an auditory association area of the superior temporal gyrus; and the rostral (agranular) insula, a transitional paralimbic region integrating visceral alimentary input with olfactory and gustatory afferents,[18] while reciprocal connections with the anterior parahippocampal areas 35 and 36 allow attentional influence over multimodal sensory afferents entering the hippocampus. The major reciprocal connections of the anterior cingulate are shown in Table 2.4 and Figure 2.6b.

Subcortical Limbic Structures

Subcortical limbic structures have major connections with the amygdala, orbitofrontal, and anterior cingulate regions. Frontal–subcortical loops[19] interact with cortical regions and function in effector mechanisms, enabling the organism to act on the environment. Five main frontal–subcortical circuits have been initially described;[20] they define frontal–subcortical regions as functional modules. The two frontal–subcortical circuits with major limbic members are the anterior cingulate subcortical circuit, required for motivated behavior, and the orbitofrontal circuit, which allows the integration of emotional information into contextually appropriate behavioral responses. The amygdala links all subcortical limbic structures in a common system providing visceral input that potentiates emotional associations. In addition to the amygdala, the subcortical limbic structures include the ventral striatum (ventromedial caudate, ventral putamen, nucleus accumbens, and olfactory tubercle), ventral pallidum, and the ventral anterior, mediodorsal, and midline thalamic nuclei.

- **Anterior Cingulate Subcortical Circuit** The ventral striatum receives input from Brodmann area 24[21] and provides input to the rostromedial globus pallidus interna and ventral pallidum (the region of the globus pallidus inferior to the anterior commissure) as well as the rostrodorsal substantia nigra.[22] The ventral pallidum projects to the entire mediolateral range of the substantia nigra pars compacta, the medial subthalamic nucleus and its extension into the lateral hypothalamus,[23] and the midline nuclei of the thalamus and the magnocellular mediodorsal thalamus.[23] The anterior cingulate circuit closes with major projections from the midline thalamic nuclei[24] and relatively less from the dorsal section of the magnocellular mediodorsal thalamus[25,26] back to the anterior cingulate cortex. The midline thalamic nuclei provide reticular activating system relays to the medial frontal cortical wall.

- **Orbitofrontal Subcortical Circuit** The ventromedial caudate receives input from the medial and lateral orbitofrontal cortex.[21] This portion of the caudate projects directly to the most medial portion of the mediodorsal globus pallidus interna and to the rostromedial substantia nigra pars reticulata.[27] Neurons are sent from the globus pallidus and substantia nigra to the medial section of the magnocellular division of the ventral anterior thalamus as well as an inferomedial sector of the magnocellular division of the mediodorsal thalamus.[21,28] These thalamic regions close this frontal–subcortical limbic loop with projections back to the medial and lateral orbitofrontal cortex.[28] Figure 2.7 shows the subcortical limbic structures involved in both the anterior cingulate and orbitofrontal subcortical limbic circuits. Table 2.5 shows the subcortical limbic structures and their related cortical limbic regions.

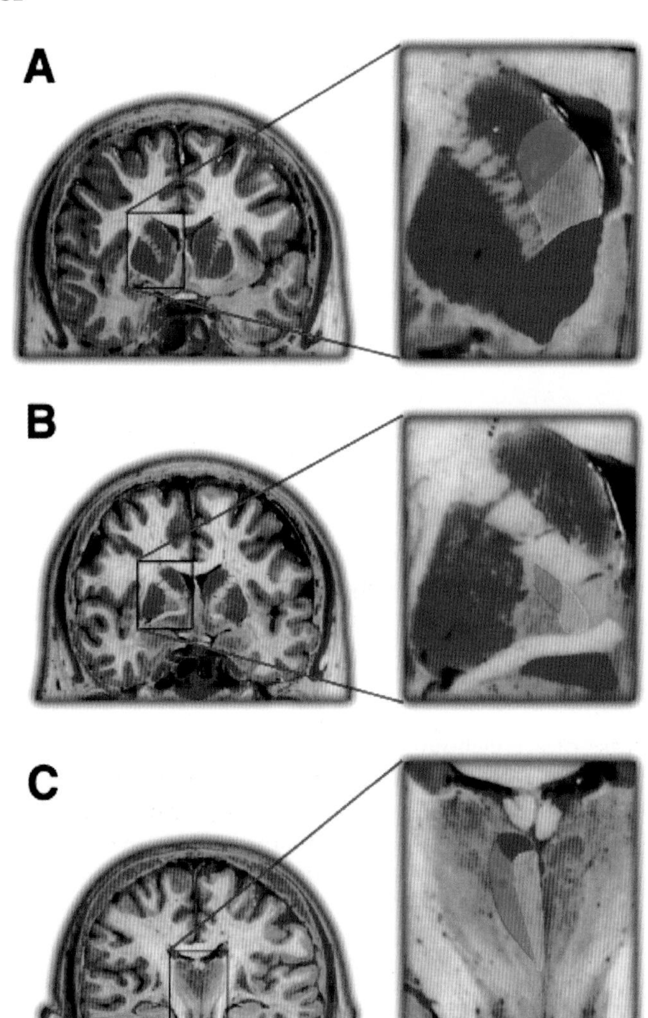

FIGURE 2.7 Subcortical limbic structures involved in both the anterior cingulate and orbitofrontal–subcortical limbic circuits at the level of the caudate and putamen (*A*), globus pallidus (*B*), and thalamus (*C*). Red, anterior cingulate members; green, orbitofrontal members; blue, dorsolateral frontal members.

Limbic Chemoarchitecture

Neurotransmitters central to limbic processing arise from the inner core of Yakovlev[1] and are modulatory in action. Serotonin, acetylcholine, dopamine, and norepinephrine are the primordial modulating transmitters, in contrast to the fast signaling transmitters concentrated in the outer layer (e.g., aspartate, glutamate, and γ-aminobutyric acid). The limbic distribution of the neuromodulators is currently being mapped and their receptor subtypes are targeted with "designer" pharmaceuticals.

- **Serotonergic System** Serotonin (5-hydroxytryptamine [5-HT]) is supplied by projections from the raphé nuclei. Synapses are well defined in the "basket system" innervating the limbic belt from M fibers of the median raphé (area B_8), while diffusely projecting D fibers to frontal cortex have indistinct synapses on the "varicose system" arising from the rostrolateral portion of area B_7 in the dorsal raphé. Of the many serotonin receptor subtypes, the 5-HT_{1A} and 5-HT_3 subtypes predominate in the limbic belt, while the 5-HT_2 subtype predominates in neocortex. The 5-HT_{1A} and 5-HT_2 receptors are G protein coupled to control cyclic adenosine monophosphate (cAMP) and open K^+

TABLE 2.5. *The Subcortical Limbic Structures and the Related Cortical Limbic Regions*

	Amygdala	*Striatum*	*Pallidum*	*Thalamus*
Orbitofrontal cortex	Basal nucleus (mc) Extended amygdala	Ventromedial caudate	Dorsomedial GP_i Rostromedial SN_r	Anterior (mc) Medial portion of mediodorsal (mc)
Hippocampus and posterior cingulate	Accessory basal nucleus (mc) Periamygdaloid cortex	Dorsal caudate	Dorsolateral GP_i	Mediodorsal anterior medial pulvinar
Anterior cingulate	Basal nucleus (pc, mc)	Ventral striatum	Ventral pallidum	Midline nuclei Dorsal portion of mediodorsal (mc)

GP_i, globus pallidus interna; mc, magnocellular; pc, parvocellular; SN_r, substantia nigra reticulata.

channels (5-HT_{1A} agonist: buspirone), or stimulate phosphoinositide hydrolysis via phospholipase C, a probable antagonistic focus for lithium,[29] and close K^+ channels (5-HT_2 agonist: LSD; antagonist: clozapine). The 5-HT_3 receptor (antagonist: ondansetron) is a ligand-gated cation channel that also modulates the release of acetylcholine and dopamine.

Serotonergic systems are involved in mood disorders and obsessional syndromes. Patients dying from violent suicide have increased serotonin receptors in the frontal cortex[30] while depressed patients at risk for suicide have decreased serotonergic metabolites in their cerebral spinal fluid.[31] Parkinson's patients with depression and frontal cognitive impairment, when successfully treated with selective serotonin reuptake inhibitors (SRIs) have an increase in metabolism in the basal medial frontal region on [^{18}F]-fluorodeoxyglucose positron emission tomography (FDG-PET).[32] The anatomy of depression implicates the medial orbitofrontal cortex in general and the serotonin system in particular.

• Cholinergic System Acetylcholine has two sources supplying the forebrain: one arising in the brainstem, the pedunculopontine and laterodorsal tegmental system, and the other from the basal forebrain, which includes the septum (Ch1), the vertical (Ch2) and horizontal (Ch3) diagonal band of Broca, and the basal nucleus of Meynert (Ch4).[33] The brainstem cholinergic system supplies the basal ganglia and most of the thalamus, while cholinergic fibers from the basal forebrain innervate portions of the ventroanterior, mediodorsal, and midline thalamus as well as the limbic structures and neocortex. Muscarinic (M_{1-5}) and nicotinic ($N\alpha_{2-8}$, β_{2-4}) receptors occur throughout the brain with the M_1 receptor predominating in the hippocampus. The muscarinic receptors are G protein coupled while nicotinic receptors are ligand-gated ion channels. Acetylcholine facilitates thalamic activation of the cortex and assists in the septal hippocampal pathway supporting mnemonic function. The basal nucleus of Meynert has an organized cortical projection pattern; by undergoing degeneration early in Alzheimer's disease, its cortical targets, even if spared by plaques or tangles, may become deactivated. Ablation of the posterior basal nucleus of Meynert may result in hypometabolism on FDG-PET in the parietal cortex due to the disconnection of cholinergic activation.[34] Behavioral improvement in patients with Alzheimer's disease, particularly a resolution of apathy[35–37] associated with anterior cingulate hypoperfusion,[38,39] occurs with cholinesterase inhibitor therapy. General cognitive improvement may also occur with drugs promoting cholinergic function.[40–43]

• Dopaminergic System Dopamine is manufactured in and projected from the substantia nigra pars compacta. The ventral tegmental area and medial portion of the ventral tier of the substantia nigra pars compacta provide dopaminergic innervation to the cortical and subcortical limbic forebrain.[44–46] The severity of dementia in Parkinson's disease is related to decreased cortical dopamine[47,48] and ventral tegmental area cell loss.[49] There are two dopamine receptor families. The D_1 family (D_1 and D_5 antagonist: bromocriptine) are G protein coupled and stimulate adenylyl cyclase; The D_2 family (D_{2-4} antagonist: butyrophenone neuroleptics) are probably all G protein coupled as well and inhibit adenylyl cyclase. The D_3 subtype is distributed to limbic regions similar to those for the D_4 receptor, which may have a greater concentration in the amygdala and frontal cortex. The D_5 receptor is found in the hippocampus and hypothalamus, while the D_2 subtype is found in sensorimotor striatal regions. This regional difference in dopamine receptor subtypes has instigated the development of dopamine receptor antagonists specific to limbic regions, such as the D_4 antagonist clozapine, with fewer motor side effects for the treatment of psychosis. Dopaminergic innervation of the anterior cingulate may be disrupted with subcortical lesions of the medial forebrain bundle or the ventral pallidum, resulting in profound apathy or even akinetic mutism, which may respond to agents such as bromocriptine. Motivation and attention rely on a distributed network involving the cingulate, dorsolateral prefrontal, and inferior parietal lobule; a balance between dopamine and norepinephrine may subserve the normal functioning of this distributed attentional network.

Norepinephrine System Norepinephrine is projected from a dorsal pathway that arises from the locus coeruleus to the entire cortex and hippocampus as well as to the spinal cord and cerebellum; a ventral pathway arises below the locus coeruleus to innervate the brainstem and hypothalamus. All the norepinephrine receptors are linked to G proteins. The $\alpha_{1A\text{–}C}$ receptors activate phospholipase C, the $\alpha_{2A\text{–}C}$ receptors inhibit adenylyl cyclase, while the $\beta_{1\text{–}3}$ receptors stimulate adenylyl cyclase. In the brain, β_2 receptors predominate in the cerebellum, while the β_1 receptor is found cortically. A balance between the levels of norepinephrine and dopamine may set the signal-to-noise ratio of the attentional system, enabling the selective filtering of sensory stimuli. Attention deficit disorders, either congenital or acquired by frontal injury, often respond to norepinephrine and dopaminergic pharmacological manipulation with bromocriptine or ritalin.

Clinical Syndromes

A central tenant in the biological basis of behavior is that behavioral disorders should have an anatomy that cuts across disease; a common neuronal network should be affected, regardless of the etiology, when similar signs and symptoms emerge. Neuropsychiatric disorders may be interpreted within a brain-based framework of limbic dysfunction divided into three general groups: decreased, increased, and distorted limbic syndromes.[50] Key imaging findings from patients suffering from these three limbic syndromes aid our understanding of functional brain defects and their resultant neuropsychiatric expressions. Table 2.6 shows the clinical manifestations and regional localization of hypo-, hyper-, and dysfunctional limbic syndromes.

Hypolimbic Syndromes Hypofunctioning of the medial division of the orbitofrontal cortex is associated with depressive symptoms in patients with Parkinson's disease, Huntington's disease, and primary affective disorders. The medial orbitofrontal cortex and the medial infracallosal region share similar efferent and afferent connections with visceromotor centers and the phylogenetically older magnocellular basolateral amygdala.[3] This amygdalar region may subserve the synthesis of internal mood and visceral functions.

Apathy is a common hypolimbic syndrome. Structural or functional lesions of the anterior cingulate above the corpus callosum, or a disconnection of its frontosubcortical circuit,[51] causes a loss of motivation, producing such profound apathy that patients may become akinetic and mute. A loss of motivation may man-

TABLE 2.6. *The Clinical Manifestations and Regional Localization of Hypo-, Hyper-, and Dysfunctional Limbic Syndromes*

HYPOLIMBIC SYNDROMES		HYPERLIMBIC SYNDROMES		DYSFUNCTIONAL LIMBIC SYNDROMES	
Syndrome	*Region*	*Syndrome*	*Region*	*Syndrome*	*Region*
Depression	Medial orbitofrontal circuit	Mania	Medial right diencephalon	Psychosis	Limbic system and dorsolateral frontal cortex
Apathy	Anterior cingulate circuit	Obsessions/compulsions	Orbitofrontal circuit	Social disdecorum	Lateral orbitofrontal circuit
Amnesia	Archicortical structures	Limbic epilepsy	Paleocortical structures	Anxiety/panic	Medial orbitofrontal cortex
Klüver-Bucy	Amygdala/temporal pole	Rage	Hypothalamus/amygdala	Utilization behavior	Lateral orbitofrontal cortex

ifest in the cognitive or motor domains. Anterior cingulate hypofunction produces transcortical motor aphasia[52] or a "loss of the will"[53] to engage in previously enjoyable pastimes;[54] when functional or structural lesions extend posteriorly into the skeletomotor effector region, a loss of spontaneous motor output will also occur.

The hippocampal–cingulate arm of the limbic system unites the dorsolateral spatial attentional network with Papez's circuit, which encodes relevant objects in the environment. *Amnesia*, the inability to encode new information, occurs when functional or structural lesions affect the hippocampus or other members of Papez's circuit. Encoding defects are present in patients with lesions of the fornix, mammillary bodies, anterior or dorsomedial thalamus, posterior cingulate, and entorhinal cortex—all members of Papez's classical medial circuit. The hypofunctioning of this more phylogenetically recent, archicortical limbic division results in declarative memory abnormalities; episodes of experience can no longer be encoded. This explicit process is contrasted to the implicit stimulus–response associations retained by the amygdala–orbitofrontal limbic arm.

When the amygdala, and probably the adjacent temporal cortex, are damaged the Klüver-Bucy syndrome[55–57] may ensue. Patients usually have only a portion of the symptoms associated with the classic lesion seen in nonhuman primates: placidity, hyperorality, hypersexuality, hypermetamorphosis (compulsive environmental exploration), hyperorality, and visual agnosia. Loss of premorbid aggressive tendencies is the common postsurgical sequel of bilateral amygdalectomy in humans.[58] The loss of implicit visceral, or affective, associations with sensory stimuli is the result of human[59] lesions of the amygdala.

• Hyperlimbic Syndromes Mania frequently results from right medial diencephalic lesions[60] that may disrupt hypothalamic circuits or disturb modulating transmitters traversing the adjacent medial forebrain bundle. Mania has also been observed in patients with orbitofrontal lesions, caudate dysfunction in basal ganglia disorders, and lesions of the thalamus.[61–63] Given the strong hypothalamic–amygdala–orbitofrontal connections and the increase in appetitive drives that often accompany mania, it may reflect hyperfunctioning of this paleocortical limbic arm.

Obsessive-compulsive disorder is characterized by involuntary recurrent ego-dystonic thoughts, images, or impulses (obsessions) and the performance of repetitive, stereotyped, involuntary acts or rituals (compulsions).[64] Obsessive-compulsive symptomatology frequently occurs in basal ganglia disorders including parkinsonism, encephalitis lethargica, manganese intoxication, and Gilles de la Tourette syndrome. Surgical resection of the subcallosal cingulate and the medial division of the orbitofrontal cortex provides the best treatment for medically refractory obsessive-compulsive patients.[65] Perhaps the most significant advances in understanding the neuroanatomy of neuropsychiatry have come from functional and structural imaging of both acquired and idiopathic obsessive-compulsive disorder. As stated above, a central tenant in the biological basis of behavior is that behavioral abnormalities should have an anatomy that cuts across disease; a common neuronal network should be affected, regardless of the etiology, when similar signs and symptoms emerge. Obsessive-compulsive disorder is the best example of this tenant.

The search for the structural signature of idiopathic obsessive-compulsive disorder began with computed tomography (CT) evidence of enlarged ventricles[66] and caudate abnormalities.[67] Subsequent studies showed conflicting results,[68–71] yet an abnormality in frontal and striatal structures, particularly the caudate nucleus, has emerged.[72–79] Idiopathic obsessive-compulsive disorder begins in late adolescence or early adulthood, exhibits a fluctuating course with exacerbations and remissions, and has a tendency to improve gradually over time.[80–83] There is an increased incidence of neuropsychological abnormalities, enlarged ventricles, and abnormal birth histories in patients with obsessive-compulsive disorder, suggesting a role for developmental insult in its pathogenesis.[66,84,85] Evidence for developmental abnormalities in idiopathic obsessive-compulsive disorder is also supported by the observation of caudate volume loss in drug-naive children[76] and an enlargement of the thalamus, which normalizes with paroxetine treatment[86] but not behavioral therapy in children.[87] The orbital frontal–striatal circuit is the likely focus of a developmentally aberrant positive feedback loop in idiopathic obsessive-compulsive disorder.[88,89]

Functional imaging data overwhelmingly support abnormal activity in the orbital frontal–subcortical circuit in obsessive-compulsive disorder in both resting state[90–96] and activation studies using provocative stimuli;[97–99] a reorganization of frontal systems probably occurs that also affects cognitive function and associated activation patterns.[93,100–102]

The most compelling evidence for determining the neuroanatomic basis of any complex behavior is to demonstrate the functional improvement of an aberrant anatomic network that corresponds to the clinical improvement resulting from successful treatment. This has been accomplished for obsessive-compulsive disorder. Although initial findings suggested a correlation between an increase in caudate metabolism with successful response to pharmacologic treatment,[90,103] subsequent studies have confirmed a decrease, or normalization, of the aberrant hyperactivity in the orbital frontal–subcortical circuit with both pharmacologically[104–111] and behaviorally[106,110–112] mediated symptom reduction. Responders to pharmacologic treatment may be different from nonresponders in that increased baseline orbitofrontal activity in some patients may be a compensatory mechanism; in those patients, administration of a serotonin autoreceptor agonist (e.g., sumatriptan) results in decreased frontal activity and exacerbation of symptoms. These patients are then less likely to respond to treatment with a serotonin reuptake inhibitor.[113] Thus, functional neuroimaging can serve as a useful tool in uncovering unique neurobiological mechanisms of disease in neuropsychiatry.

The interictal syndrome of Gastaut-Geschwind[114,115] comprises personality changes that include hypergraphia, hyposexuality, conversationality, obsessionality, and hyperreligiosity or an increased sense of individual destiny. Patients are viscous or "sticky," being unaware of social cues for discourse termination. Three subgroups have been described:[116] one with primarily altered physiologic drives, another with philosophical preoccupation, and a third with predominately altered interpersonal skills. These features may represent hyperfunctioning orbitofrontal centers, with such functioning of the medial division resulting in altered physiologic drives; the lateral division, in altered interpersonal skills; and the amygdala, in a sense of awe.

Rage attacks result from ventromedial hypothalamic lesions in humans.[117] Episodic explosive behavior may be related to hypothalamic or visceral-amygdalar abnormalities, since the best "calming" effect may be achieved from lesions involving the lateral part of the corticomedial nuclei.[118] Violent outburst responds to amygdalectomy[119] or posterior hypothalamotomy,[120] capitalizing on the placidity common to the Klüver-Bucy syndrome.

• Dysfunctional Limbic Syndromes The occurrence of psychotic symptoms in patients may involve dysfunction in the implicit integration of affect, drives, and object associations, supported by the orbitofrontal limbic arm, with explicit sensory processing, encoding, and attentional/executive systems, all supported by the hippocampal limbic arm and dorsolateral prefrontal cortex. Functional imaging studies of patients with pyschosis have implicated dysfunction in temporal limbic regions as well as in the dorsolateral frontal cortex bilaterally.[121–125]

Utilization and imitation behavior frequently result from lateral orbitofrontal pathology,[126] which is often a consequence of traumatic brain injury. The automatic imitation of the examiner's gestures and the dependence patients have upon objects in their environment result from dysfunction in appropriately integrating internal urges with ventral visual object processing, integration that occurs in the lateral division of the orbitofrontal paralimbic arm. The lateral orbital cortex also mediates empathic, civil, and socially appropriate behavior. Thus, patients with dysfunction in this region will also find it difficult to observe the rules of decorum during social interaction. The social inappropriateness they demonstrate is often sexual in nature, or other appetitive urges may be difficult to control, all behaviors implicating dysfunction at the level of externally directed lateral orbitofrontal integration.

Anxiety and panic disorders result from the dysfunction of integrating the visceral–amygdalar functions with the internal state of the organism; an integration occurring in the medial division of the orbitofrontal paralimbic center. The strong visceral components accompanying these disorders and the dominant role the amygdala plays in building stimulus–response associations implicate the paleocortical limbic system. Functional imaging studies reveal increased activity in the amygdala and temporal pole in patients with anxiety disorders. The autonomic aspects of anxiety are manifold and involve the cardiovascular system (tachycardia, palpitation, hypertension, syncope), respiratory system (dyspnea, hyperventilation), gastrointestinal system (anorexia, diarrhea), and genitourinary system (urinary frequency, urgency, hesitancy), as well as increased perspiration and pupillary dilatation. Gray[127] has constructed a neurophysiological model of anxiety, proposing that dysfunction in the limbic system overstimulates the behavioral system responsible for surveillance, vigilance, and stimulus evaluation. The dysfunction results secondarily in the perceptual, cognitive, motoric, and autonomic phenomena comprising the anxious state. The first resting functional imaging study of panic disorder patients[128] suggested a defect of greater left than right parahippocampal cerebral blood flow (CBF) in patients who were lactate inducible, compared to controls or noninducible patients. In normal controls, during anxiety provocation, activations of the insula,[129–131] anterior cingulate,[129,130,132,133] orbitofrontal,[130–132,134] and anterior temporal cortices occur.[129–132] During lactate induction of anxiety in patients with panic disorder, similar regions show increased CBF, compared to baseline,[135] although the temporal polar increase may be partially due to extracerebral increased flow secondary to jaw clenching.[129] Following successful treatment in patients with severe anxiety disorder with capsulotomy, which disrupts limbic–frontal connections, marked metabolic reductions in orbitofrontal and subcallosal cingulate cortex occur.[136] Resting metabolism in the left anterior cingulate and right parahippocampal regions are also significantly associated with the degree of anxiety in depressed patients, suggesting that the anatomic behavioral relationship cuts across disease and normal states.[137] This finding supports the central tenant that behavioral disorders should share a common neuroanatomy, regardless of the etiology, when similar signs and symptoms emerge.

SUMMARY

Modern neuropsychiatry is now drawing from diverse disciplines grounded in anatomy, physiology, and clinical neuroscience. In this chapter we provided a model for interpreting neuropsychiatric disorders as fundamentally limbic disorders that combines phylogenetic, anatomic, functional, and clinical data to interpret these diverse diseases of human behavior. An understanding of the development and organization of the limbic system will aid the reader's interpretation of the myriad disorders of human behavior described throughout this volume. With this neurobiological basis, the reunification of psychiatry and neurology will ultimately improve our treatment of the patients who suffer from these disorders.

REFERENCES

1. Yakovlev PI. Motility, behavior, and the brain. J Nerv Ment Dis 1948;107:313–335.
2. Sanides F. Comparative architectonics of the neocortex of mammals and their evolutionary interpretation. Ann NY Acad Sci 1969;167:404–423.
3. Mega MS, Cummings JL. The cingulate and cingulate syndromes. In: Trimble MR, Cummings JL, eds. Contemporary Behavioral Neurology. Boston: Butterworth-Heinemann, 1997: 189–214.
4. Pandya DN, Yeterian EH. Architecture and connections of cortical association areas. In: Peters A, Jones EG, eds. Cerebral Cortex. New York: Plenum Press, 1985:3–55.
5. Mesulam M-M. Patterns in behavioral neuroanatomy: association areas, the limbic system, and hemispheric specialization. In: Mesulam M-M, ed. Behavioral Neurology. Philadelphia: F.A. Davis Company, 1985:1–70.
6. Brodmann K. Vergleichende Lokalisationslehre der Grosshirnrinde in ihren Prinzipien dargestellt auf Grund des Zellenbaues. Leipzig: Barth, 1909.
7. Kaada BR, Pribram KH, Epstein JA. Respiratory and vascular responses in monkeys from temporal pole, insula, orbital surface and cingulate gyrus. J Neurophysiol 1949;12:347–356.
8. Papez JW. A proposed mechanism of emotion. Arch Neurol Psychiatry 1937;38:725–733.
9. Grasby PM, Frith CD, et al. Functional mapping of brain areas implicated in auditory–verbal memory function. Brain 1993; 116:1–20.
10. Grasby PM, Firth CD, et al. Activation of the human hippocampal formation during auditory–verbal long-term memory function. Neurosci Lett 1993;163:185–188.
11. Fletcher PC, Firth CD, et al. Brain systems for encoding and retrieval of auditory–verbal memory. An in vivo study in humans. Brain 1995;118:401–416.
12. Shallice T, Fletcher P, et al. Brain regions associated with acquisition and retrieval of verbal episodic memory. Nature 1994;368:633–635.
13. Molchan SE, Sunderland T, et al. A functional anatomical study of associative learning in humans. Proc Natl Acad Sci USA 1994;91:8122–8126.
14. Posner MI, Walker JA, et al. How do the parietal lobes direct covert attention? Neuropsychologia 1987;25:135–145.
15. Vogt BA, Nimchinsky EA, et al. Human cingulate cortex: surface features, flat maps, and cytoarchitecture. J Comp Neurol 1995;359:490–506.
16. Vogt BA. Structural organization of cingulate cortex: areas, neurons, and somatodendritic transmitter receptors. In: Vogt BA, Gabriel M, eds. Neurobiology of Cingulate Cortex and Limbic Thalamus: A Comprehensive Handbook. Boston: Birkhäuser, 1993:19–70.
17. Morecraft RJ, Geula C, Mesulam M-M. Architecture of connectivity within a cingulfronto-parietal neurocognitive network. Arch Neurol 1993;50:279–284.
18. Mesulam M-M, Mufson EJ. The insula of Reil in man and monkey: architectonics, connectivity, and function. In: Jones EG, Peters AA, eds. Cerebral Cortex. New York: Plenum Press, 1985:179–226.
19. Mega MS, Cummings JL. Frontal subcortical circuits and neuropsychiatric disorders. J Neuropsychiatry Clin Neurosci 1994; 6:358–370.
20. Alexander GE, DeLong MR, Strick PL. Parallel organization of functionally segregated circuits linking basal ganglia and cortex. Annu Rev Neurosci 1986;9:357–381.
21. Selemon LD, Goldman-Rakic PS. Longitudinal topography and interdigitation of corticostriatal projections in the rhesus monkey. J Neurosci 1985;5:776–794.
22. Haber SN, Lynd E, et al. Topographic organization of the ventral striatal efferent projections in the rhesus monkey: an anterograde tracing study. J Comp Neurol 1990;293:282–298.
23. Haber SN, Lynd-Balta E, Mitchell SJ. The organization of the descending ventral pallidal projections in the monkey. J Comp Neurol 1993;329:111–128.
24. Vogt BA, Pandya DN. Cingulate cortex of the rhesus monkey: I. Cytoarchitecture and thalamic afferents. J Comp Neurol 1987;262:256–270.
25. Giguere M, Goldman-Rakic PS. Mediodorsal nucleus: areal, laminar, and tangential distribution of afferents and efferents in the frontal lobe of rhesus monkey. J Comp Neurol 1988;277:195–213.
26. Goldman-Rakic PS, Porrino LJ. The primate mediodorsal (MD) nucleus and its projection to the frontal lobe. J Comp Neurol 1985;242:535–560.
27. Johnson TN, Rosvold HE. Topographic projections on the globus pallidus and substantia nigra of selectively placed lesions

in the precommissural caudate nucleus and putamen in the monkey. Exp Neurol 1971;33:584–596.
28. Ilinsky IA, Jouandet ML, Goldman-Rakic PS. Organization of the nigrothalamocortical system in the rhesus monkey. J Comp Neurol 1985;236:315–330.
29. Snyder SH. Second messengers and affective illness. Focus on the phosphoinositide cycle. Pharmacopsychiatry 1992;25:25–28.
30. Mann JJ, Stanley M, et al. Increased serotonin$_2$ and β-adrenergic receptor binding in the frontal cortices of suicide victims. Arch Gen Psychiatry 1986;43:954–959.
31. Brown GL, Linnoila MI. CSF serotonin metabolite(5-HIAA) studies in depression, impulsivity, and violence. J Clin Psychiatry 1990;51:31–41.
32. Mayberg H, Mahurin RK, et al. Parkinson's depression: discrimination of mood-sensitive and mood-insensitive cognitive deficits using fluoxetine and FDG PET. Neurology 1995;45 (Suppl 4):A166.
33. Mesulam M-M, Mufson EJ, et al. Cholinergic innervation of cortex by the basal forebrain: cytochemistry and cortical connections of the septal area, diagonal band nuclei, nucleus basalis (substantia innominata), and hypothalamus in the rhesus monkey. J Comp Neurol 1983;214:170–197.
34. Friedland RP, Brun A, Budinger TF. Pathological and positron emission tomographic correlations in Alzheimer's disease. Lancet 1985;1:228–230.
35. Kaufer DI, Cummings JL, Christine D. Effect of tacrine on behavioral symptoms in Alzheimer's disease: an open-label study. J Geriatr Psychiatry Neurol 1996;9:1–6.
36. Cummings JL, Cyrus PA, et al. Metrifonate treatment of the cognitive deficits of Alzheimer's disease. Metrifonate Study Group. Neurology 1998;50:1214–1221.
37. Mega MS, Masterman DM, et al. The spectrum of behavioral responses with cholinesterase inhibitor therapy in Alzheimer's disease. Arch Neurol 1999;56:1388–1393.
38. Craig HA, Cummings JL, et al. Cerebral blood flow correlates of apathy in Alzheimer's disease. Arch Neurol 1996;53:1116–1120.
39. Migneco O, Benoit M, et al. Perfusion brain SPECT and statistical parametric mapping analysis indicate that apathy is a cingulate syndrome: a study in Alzheimer's disease and nondemented patients. Neuroimage 2001;13:896–902.
40. Davis KL, Thal LJ, et al. A double-blind, placebo-controlled multicenter study of tacrine for Alzheimer's disease. N Engl J Med 1992;327:1253–1259.
41. Rogers SL, Farlow MR, et al. A 24-week, double-blind, placebo-controlled trial of Aricept® in patients with Alzheimer's disease. Neurology 1998;50:136–145.
42. Corey-Bloom J, Anand R, et al. A randomized trial evaluating the efficacy and safety of ENA 713 (rivastigmine tartrate), a new acetylcholinesterase inhibitor, in patients with mild to moderately severe Alzheimer's disease. Int J Geriatr Psychopharmacol 1998;1:55–65.
43. Tariot PN, Solomon PR, et al. A 5-month, randomized, placebo-controlled trial of galantamine in AD. Neurology 2000;54: 2269–2276.
44. Thierry AM, Tassin JP, et al. Studies on mesocortical dopamine systems. Adv Biochem Psychopharmacol 1978;19:205–216.
45. Moore RY, Bloom FE. Central catecholamine neuron systems: anatomy and physiology of the dopamine system. Annu Rev Neurosci 1978;1:129–169.
46. Oades RD, Halliday GM. Ventral tegmental (A10) system: neurobiology. 1. Anatomy and connectivity. Brain Res Rev 1987; 12:117–165.
47. Scatton B, Javoy-Agid F, et al. Reduction of cortical dopamine, noradrenaline, serotonin and other metabolites in Parkinson's disease. Brain Res 1983;275:321–328.
48. Agid Y, Rugerg M, et al. Biochemical substrates of mental disturbances in Parkinson's disease. In: Hassler RG, Christ JF, eds. Advances in Neurology. New York: Raven Press, 1984:211–218.
49. Rinne JO, Rummukainen J, et al. Dementia in Parkinson's disease is related to neuronal loss in the medial substantia nigra. Ann Neurol 1989;26:47–50.
50. Mega MS, Cummings JL, et al. The limbic system: an anatomic, phylogenetic, and clinical perspective. J Neuropsychiatry Clin Neurosci 1997;9:315–330.
51. Mega MS, Cohenour RC. Akinetic mutism: a disconnection of frontal–subcortical circuits. Neurol Neuropsychol Behav Neurol 1997;10:254–259.
52. Alexander MP, Benson DF, Stuss DT. Frontal lobes and language. Brain Lang 1989;37:656–691.
53. Damasio AR, Van Hoesen GW. Focal lesions of the limbic frontal lobe. In: Heilman KM, Satz P, eds. Neuropsychology of Human Emotion. New York: Guilford Press, 1983:85–110.
54. Devinsky O, Morrell MJ, Vogt BA. Contributions of anterior cingulate cortex to behavior. Brain 1995;118:279–306.
55. Klüver H, Bucy PC. "Psychic blindness" and other symptoms following bilateral temporal lobectomy in rhesus monkeys. Am J Physiol 1937;119:352–353.
56. Klüver H, Bucy PC. A analysis of certain effects of bilateral temporal lobectomy in the rhesus monkey, with special reference to "psychic blindness". J Psychol 1938;5:33–54.
57. Klüver H, Bucy PC. Preliminary analysis of functions of the temporal lobes in monkeys. Arch Neurol Psychiatry 1939;42: 979–1000.
58. Aggleton JP. The functional effects of amygdala lesions in humans: a comparison with findings from monkeys. In: Aggleton JP, ed. The Amygdala. New York: Wiley-Liss, 1992:485–503.
59. Bechara A, Tranel D, et al. Double dissociation of conditioning and declarative knowledge relative to the amygdala and hippocampus in humans. Science 1995;269:1115–1118.
60. Cummings JL, Mendez MF. Secondary mania with focal cerebrovascular lesions. Am J Psychiatry 1984;141:1084–1087.
61. Bogousslavsky J, Ferrazzini M, et al. Manic delirium and frontal-like syndrome with paramedian infarction of the right thalamus. J Neurol Neurosurg Psychiatry 1988;51:116–119.
62. Jorge RE, Robinson RG, et al. Secondary mania following traumatic brain injury. Am J Psychiatry 1993;150:916–921.
63. Starkstein SE, Pearlson GD, et al. Mania after brain injury. A controlled study of causative factors. Arch Neurol 1987;44: 1069–1073.
64. American Psychiatric Association. Diagnostic and Statistical Manual of Mental Disorders, 4th ed. Washington, DC: American Psychiatric Press, 1994.
65. Hay P, Sachdev P, et al. Treatment of obsessive-compulsive disorder by psychosurgery. Acta Psychiatr Scand 1993;87:197–207.
66. Behar D, Rapaport JL, et al. Computerized tomography and neuropsychological test measures in adolescents with obsessive-compulsive disorder. Am J Psychiatry 1984;141:363–369.
67. Luxenberg JS, Swedo SE, et al. Neuroanatomical abnormalities in obsessive-compulsive disorder detected with quantitative X-ray computed tomography. Am J Psychiatry 1988;145:1089–1093.
68. Kellner CH, Jolley RR, et al. Brain MRI in obsessive-compulsive disorder. Psychiatry Res 1991;36:45–49.

69. Stein DJ, Hollander E, et al. Computed tomography and neurological soft signs in obsessive-compulsive disorder. Psychiatry Res 1993;50:143–150.
70. Aylward EH, Harris GJ, et al. Normal caudate nucleus in obsessive-compulsive disorder assessed by quantitative neuroimaging. Arch Gen Psychiatry 1996;53:577–584.
71. Stein DJ, Coetzer R, et al. Magnetic resonance brain imaging in women with obsessive-compulsive disorder and trichotillomania. Psychiatry Res 1997;74:177–182.
72. Scarone S, Colombo C, et al. Increased right caudate nucleus size in obsessive-compulsive disorder: detection with magnetic resonance imaging. Psychiatry Res 1992;45:115–121.
73. Calabrese G, Colombo C, et al. Caudate nucleus abnormalities in obsessive-compulsive disorder: measurements of MRI signal intensity. Psychiatry Res 1993;50:89–92.
74. Robinson D, Wu H, et al. Reduced caudate nucleus volume in obsessive-compulsive disorder. Arch Gen Psychiatry 1995;52: 393–398.
75. Jenike MA, Breiter HC, et al. Cerebral structural abnormalities in obsessive-compulsive disorder. A quantitative morphometric magnetic resonance imaging study. Arch Gen Psychiatry 1996; 53:625–632.
76. Rosenberg DR, Keshavan MS, et al. Frontostriatal measurement in treatment-naive children with obsessive-compulsive disorder. Arch Gen Psychiatry 1997;54:824–830.
77. Bartha R, Stein MB, et al. A short echo 1H spectroscopy and volumetric MRI study of the corpus striatum in patients with obsessive-compulsive disorder and comparison subjects. Am J Psychiatry 1998;155:1584–1591.
78. Szeszko PR, Robinson D, et al. Orbital frontal and amygdala volume reductions in obsessive-compulsive disorder. Arch Gen Psychiatry 1999;56:913–919.
79. Giedd JN, Rapoport JL, et al. MRI assessment of children with obsessive-compulsive disorder or tics associated with streptococcal infection. Am J Psychiatry 2000;157:281–283.
80. Grimshaw L. The outcome of obsessional disorder. A follow-up study of 100 cases. Br J Psychiatry 1965;111:1051–1056.
81. Jenike MA. Obsessive compulsive disorder. Compr Psychiatry 1983;24:99–115.
82. Kringlen E. Obsessional neurotics. A long-term follow-up. Br J Psychiatry 1965;11:709–722.
83. Pollitt J. Natural history of obsessional states. BMJ 1957;1: 194–198.
84. Capstick N, Seldrup J. Obsessional states. A study in the relationship between abnormalities at the time of birth and the subsequent development of obsessional symptoms. Acta Psychiatr Scand 1977;56:427–431.
85. Insel TR, Donnelty EF, et al. Neurological and neuropsychological studies of patients with obsessive-compulsive disorder. Biol Psychiatry 1983;18:741–751.
86. Gilbert AR, Moore GJ, et al. Decrease in thalamic volumes of pediatric patients with obsessive-compulsive disorder who are taking paroxetine. Arch Gen Psychiatry 2000;57:449–456.
87. Rosenberg DR, Benazon NR, et al. Thalamic volume in pediatric obsessive-compulsive disorder patients before and after cognitive behavioral therapy. Biol Psychiatry 2000;48:294–300.
88. Modell JG, Mountz JM, et al. Neurophysiologic dysfunction in basal ganglia/limbic striatal and thalamocortical circuits as a pathogenetic mechanism of obsessive-compulsive disorder. J Neuropsychiatry Clin Neurosci 1989;1:27–36.
89. Rosenberg DR, Keshavan MS. A.E. Bennett Research Award. Toward a neurodevelopmental model of of obsessive-compulsive disorder. Biol Psychiatry 1998;43:623–640.
90. Baxter LR Jr, Phelps ME, et al. Local cerebral glucose metabolic rates in obsessive-compulsive disorder. A comparison with rates in unipolar depression and in normal controls. Arch Gen Psychiatry 1987;44:211–218.
91. Baxter LR Jr, Schwartz JM, et al. Cerebral glucose metabolic rates in nondepressed patients with obsessive-compulsive disorder. Am J Psychiatry 1988;145:1560–1563.
92. Swedo SE, Schapiro MB, et al. Cerebral glucose metabolism in childhood-onset obsessive-compulsive disorder. Arch Gen Psychiatry 1989;46:518–523.
93. Lucey JV, Burness CE, et al. Wisconsin Card Sorting Task (WCST) errors and cerebral blood flow in obsessive-compulsive disorder (OCD). Br J Med Psychol 1997;70:403–411.
94. Crespo-Facorro B, Cabranes JA, et al. Regional cerebral blood flow in obsessive-compulsive patients with and without a chronic tic disorder. A SPECT study. Eur Arch Psychiatry Clin Neurosci 1999;249:156–161.
95. Busatto GF, Buchpiguel CA, et al. Regional cerebral blood flow abnormalities in early-onset obsessive-compulsive disorder: an exploratory SPECT study. J Am Acad Child Adolesc Psychiatry 2001;40:347–354.
96. Busatto GF, Zamignani DR, et al. A voxel-based investigation of regional cerebral blood flow abnormalities in obsessive-compulsive disorder using single photon emission computed tomography (SPECT). Psychiatry Res 2000;99:15–27.
97. Rauch SL, Jenike MA, et al. Regional cerebral blood flow measured during symptom provocation in obsessive-compulsive disorder using oxygen ^{15}-labeled carbon dioxide and positron emission tomography. Arch Gen Psychiatry 1994;51:62–70.
98. Cottraux J, Gerard D, et al. A controlled positron emission tomography study of obsessive and neutral auditory stimulation in obsessive-compulsive disorder with checking rituals. Psychiatry Res 1996;60:101–112.
99. Breiter HC, Rauch SL, et al. Functional magnetic resonance imaging of symptom provocation in obsessive-compulsive disorder. Arch Gen Psychiatry 1996;53:595–606.
100. Martinot JL, Allilaire JF, et al. Obsessive-compulsive disorder: a clinical, neuropsychological and positron emission tomography study. Acta Psychiatr Scand 1990;82:233–242.
101. Rauch SL, Savage CR, et al. Probing striatal function in obsessive-compulsive disorder: a PET study of implicit sequence learning. J Neuropsychiatry Clin Neurosci 1997;9:568–573.
102. Pujol J, Torres L, et al. Functional magnetic resonance imaging study of frontal lobe activation during word generation in obsessive-compulsive disorder. Biol Psychiatry 1999;45:891–897.
103. Baxter LR Jr, Thompson JM, et al. Trazodone treatment response in obsessive-compulsive disorder—correlated with shifts in glucose metabolism in the caudate nuclei. Psychopathology 1987;20:114–122.
104. Benkelfat C, Nordahl TE, et al. Local cerebral glucose metabolic rates in obsessive-compulsive disorder. Patients treated with clomipramine. Arch Gen Psychiatry 1990;47:840–848.
105. Swedo SE, Pietrini P, et al. Cerebral glucose metabolism in childhood-onset obsessive-compulsive disorder. Revisualization during pharmacotherapy. Arch Gen Psychiatry 1992;49:690–694.
106. Baxter LR Jr, Schwartz JM, et al. Caudate glucose metabolic rate changes with both drug and behavior therapy for obsessive-compulsive disorder. Arch Gen Psychiatry 1992;49:681–689.
107. Azari NP, Pietrini P, et al. Individual differences in cerebral metabolic patterns during pharmacotherapy in obsessive-

compulsive disorder: a multiple regression/discriminant analysis of positron emission tomographic data. Biol Psychiatry 1993;34:798–809.
108. Rubin RT, Ananth J, et al. Regional ^{133}xenon cerebral blood flow and cerebral 99mTc-HMPAO uptake in patients with obsessive-compulsive disorder before and during treatment. Biol Psychiatry 1995;38:429–37.
109. Molina V, Montz R, et al. Drug therapy and cerebral perfusion in obsessive-compulsive disorder. J Nucl Med 1995;36: 2234–2238.
110. Brody AL, Saxena S, et al. FDG-PET predictors of response to behavioral therapy and pharmacotherapy in obsessive compulsive disorder. Psychiatry Res 1998;84:1–6.
111. Saxena S, Brody AL, et al. Localized orbitofrontal and subcortical metabolic changes and predictors of response to paroxetine treatment in obsessive-compulsive disorder. Neuropsychopharmacology 1999;21:683–693.
112. Schwartz JM, Stoessel PW, et al. Systematic changes in cerebral glucose metabolic rate after successful behavior modification treatment of obsessive-compulsive disorder. Arch Gen Psychiatry 1996;53:109–113.
113. Stein DJ, Van Heerden B, et al. Single photon emission computed tomography of the brain with Tc-99m HMPAO during sumatriptan challenge in obsessive-compulsive disorder: investigating the functional role of the serotonin auto-receptor. Prog Neuropsychopharmacol Biol Psychiatry 1999;23:1079–1099.
114. Gastaut H. Étude electroclinique des episodes psychotiques survenant en dehors des crises clin iques chez les épileptiques [Electroclinical study of interictal psychotic episodes in epileptics]. Rev Neurol 1956;94:587–594.
115. Waxman SG, Geschwind N. Hypergraphia in temporal lobe epilepsy. Neurology 1974;24:629–636.
116. Nielsen H, Christensen O. Personality correlates of sphenoidal EEG foci in temporal lobe epilepsy. Acta Neurol Scand 1981;64:289–300.
117. Reeves AG, Plum F. Hyperphagia, rage, and dementia accompanying a ventromedial hypothalamic neoplasm. Arch Neurol 1969;20:616–624.
118. Narabayashi H, Shima F. Which is the better amygdala target, the medial or lateral nucleus for behavioral problems and paroxysms in epileptics. In: Laitinen LV, Livingston KE, eds. Surgical Approaches in Psychiatry. Baltimore: University Park Press, 1973:129–134.
119. Kiloh LG, Gye RS, et al. Stereotactic amygdaloidotomy for aggressive behavior. J Neurol Neurosurg Psychiatry 1974;37: 437–444.
120. Schvarcz JR, Driollet R, et al. Stereotactic hypothalamotomy for behavior disorders. J Neurol Neurosurg Psychiatry 1972;35: 356–359.
121. Wik G, Wiesel FA. Regional brain glucose metabolism: correlations to biochemical measures and anxiety in patients with schizophrenia. Psychiatry Res 1991;40:101–14.
122. Woods SW. Regional cerebral blood flow imaging with SPECT in psychiatric disease: focus on schizophrenia, anxiety disorders, and substance abuse. J Clin Psychiatry 1992;53 Suppl: 20–5.
123. Raine A, Sheard C, et al. Prefrontal structural and functional deficits associated with individual differences in schizotypal personality. Schizophr Res 1992;7:237–247.
124. Dickey CC, McCarley RW, et al. Schizotypal personality disorder and MRI abnormalities of temporal lobe gray matter. Biol Psychiatry 1999;45:1393–402.
125. Downhill JE Jr, Buchsbaum MS, et al. Temporal lobe volume determined by magnetic resonance imaging in schizotypal personality disorder and schizophrenia. Schizophr Res 2001;48: 187–199.
126. Lhermitte F, Pillon B, Serdaru M. Human autonomy and the frontal lobes, part I: imitation and utilization behavior: a neuropsychological study of 75 patients. Ann Neurol 1986;19: 326–334.
127. Gray JB. The Neuropsychology of Anxiety. New York: Oxford University Press, 1982.
128. Reiman EM, Raichle ME, et al. A focal brain abnormality in panic disorder, a severe form of anxiety. Nature 1984;310:683–685.
129. Benkelfat C, Bradwejn J, et al. Functional neuroanatomy of CCK4-induced anxiety in normal healthy volunteers. Am J Psychiatry 1995;152:1180–1184.
130. Chua P, Krams M, et al. A functional anatomy of anticipatory anxiety. Neuroimage 1999;9:563–571.
131. Liotti M, Mayberg HS, et al. Differential limbic–cortical correlates of sadness and anxiety in healthy subjects: implications for affective disorders. Biol Psychiatry 2000;48:30–42.
132. Kimbrell TA, George MS, et al. Regional brain activity during transient self-induced anxiety and anger in healthy adults. Biol Psychiatry 1999;46:454–465.
133. Simpson JR Jr, Drevets WC, et al. Emotion-induced changes in human medial prefrontal cortex: II. During anticipatory anxiety. Proc Natl Acad Sci USA 2001;98:688–693.
134. Fredrikson M, Fischer H, Wik G. Cerebral blood flow during anxiety provocation. J Clin Psychiatry 1997;58:16–21.
135. Reiman EM, Raichle ME, et al. Neuroanatomical correlates of a lactate-induced anxiety attack. Arch Gen Psychiatry 1989;46: 493–500.
136. Mindus P, Ericson K, et al. Regional cerebral glucose metabolism in anxiety disorders studied with positron emission tomography before and after psychosurgical intervention. A preliminary report. Acta Radiol Suppl 1986;369:444–448.
137. Osuch EA, Ketter TA, et al. Regional cerebral metabolism associated with anxiety symptoms in affective disorder patients. Biol Psychiatry 2000;48:1020–1023.
138. Amaral DG, Price JL, et al. Anatomical organization of the primate amygdaloid complex. In: Aggleton JP, ed. The Amygdala. New York: Wiley-Liss, 1992:1–66.
139. Amaral DG, Price JL. Amygdalo-cortical projections in the monkey (*Macaca fascicularis*). J Comp Neurol 1984;230:465–496.
140. Vogt BA, Pandya DN. Cingulate cortex of the rhesus monkey: II. Cortical afferents. J Comp Neurol 1987;262:271–289.
141. Müller-Preuss P, Jürgens U. Projections from the "cingular" vocalization area in the squirrel monkey. Brain Res 1976;103:29–43.
142. Carmichael ST, Price JL. Limbic connections of the orbital and medial prefrontal cortex in Macaque monkeys. J Comp Neurol 1995;363:615–641.
143. Morecraft RJ, Geula C, Mesulam M-M. Cytoarchitecture and neural afferents of orbitofrontal cortex in the brain of the monkey. J Comp Neurol 1992;323:341–358.
144. Pandya DN, Van Hoesen GW, Mesulam M-M. Efferent connections of the cingulate gyrus in the rhesus monkey. Exp Brain Res 1981;42:319–330.
145. Moran MA, Mufson EJ, Mesulam M-M. Neural inputs to the temporopolar cortex of the rhesus monkey. J Comp Neurol 1987;256:88–103.
146. Mufson EJ, Mesulam M-M. Insula of the old world monkey.

II. Afferent cortical input and comments on the claustrum. J Comp Neurol 1982;212:23–37.

147. Insausti R, Amaral DG, Cowan WM. The entorhinal cortex of the monkey: II. Cortical afferents. J Comp Neurol 1987;264: 356–395.
148. Baleydier C, Mauguière F. The duality of the cingulate gyrus in monkey. Neuroanatomical study and functional hypothesis. Brain 1980;103:525–554.
149. Morecraft RJ, Van Hoesen GW. A comparison of frontal lobe afferents to the primary, supplementary and cingulate cortices in the rhesus monkey. Soc Neurosci Abstr 1991;17:1019.
150. Cavada C, Glodman-Rakic PS. Posterior parietal cortex in rhesus monkey: I. Parcellation of areas based on distinctive limbic and sensory corticocortical connections. J Comp Neurol 1989; 287:393–421.
151. Yakovlev PI, Lecours A-R. The myelogenetic cycles of regional maturation of the brain. In: Minkowski A, ed. Regional Development of the Brain in Early Life. Edinburgh: Blackwell Scientific Publications, 1967:3–70.
152. MacLean PD. The triune brain, emotion and scientific bias. In: Schmitt FO, ed. The Neurosciences; Second Study Program. New York: Rockfeller University Press, 1970:336–349.

Neuropsychiatric Assessment

The neuropsychiatric interview and mental status examination are the principal sources of information on which neuropsychiatric and neurobehavioral diagnoses are based. A revealing clinical assessment requires experience, understanding of brain function, and knowledge of examination techniques. When well done, the examination is exciting for the clinician, educational for the patient, and enlightening for any fortunate observer. The purposes of the evaluation include establishing a relationship with the patient and garnering information to make a tentative diagnosis and guide further testing and treatment. The clinical history, observed behavior, and errors made on tests provide information critical to localizing brain dysfunction and establishing an etiologic diagnosis.[1–3] The examination begins as soon as the clinician encounters the patient and continues as long as the two are together. Often, the patient's spontaneous behavior gives as much insight into a neuropsychiatric disorder as the more formal aspects of mental state testing. Appearance and behavior yield information that may not be accessible through discussion and testing. Observations concerning the patient's behavior as well as the content and form of the spontaneous conversation will generate hypotheses about the patient's mental functioning that can be probed and extended with mental status testing. Any separation between the interview and mental status examination is artificial, therefore, as the latter is simply a more explicit and standardized method of exploring specific aspects of intellectual function.

For convenience, this chapter begins by presenting those observations usually made during the initial portion of the interview, such as the appearance and behavior of the patient, the patient's affect, and the form and content of the patient's spontaneous conversation. The mental status examination is then presented with a discussion of each specific area to be explored, including language, memory, visuospatial skills, calculation, abstraction, praxis, and executive functions. Finally, the aspects of the neurological examination of greatest importance to neuropsychiatric and neurobehavioral diagnosis are presented. The role of neuroimaging and neuropsychological testing is briefly prevented. These are expanded in each chapter describing specific disorders. Details of localization and additional information regarding the recognition and characterization of each behavioral disorder observed are provided in chapters addressing specific neurobehavioral and neuropsychiatric syndromes.

TABLE 3.1. *Components of the Neuropsychiatric Interview and Mental Status Examination*

Interview	*Mental Status Examination*
Appearance	Attention and concentration
Motoric behavior	Language
Mood and affect	Memory
Verbal output	Constructions
Thought	Calculation skills
Perception	Abstraction
	Insight and judgment
	Praxis
	Frontal system tasks
	Miscellaneous tests
	Right–left orientation
	Finger identification

NEUROPSYCHIATRIC INTERVIEW

Table 3.1 lists the major components of the neuropsychiatric interview and mental status examination. Each of these elements is described and potential abnormalities are discussed below.

Appearance and Behavior

An assessment of the patient's general appearance is the first observation made in the neuropsychiatric examination. Patients reveal an enormous amount by the way in which they are dressed, their emotional display and attitude toward the examination, and their motor activity. Dishevelment reflecting a lack of self-care is most striking in frontal lobe syndromes and schizophrenic illnesses. Unilateral dressing disturbances occur in conjunction with hemispatial neglect. Specific dressing disturbances in which patients are unable to correctly orient themselves with regard to their garments and thus fail to dress appropriately occur with right parietal lesions. Dressing with multiple layers of clothing may be seen in the dementias, acute confusional states, and, occasionally, in schizophrenia. Unusual combinations of styles and colors of dress occur in schizophrenic and manic syndromes.[4]

The attitude of the patient should also be noted.[3] Patients may be cooperative and engaged, aloof, uncooperative, hostile and belligerent, vacuous, or guarded. These behaviors have important diagnostic implications and also inform the clinician's judgment about the quality of the information obtained. Psychotic patients tend to be guarded; some dementia patients have an empty quality; manic and disinhibited patients tend to be uncooperative, irritable, or hostile. Many patients with dementia are uncertain of their responses and turn to their spouse or other family member to provide answers to questions. Personality features are also often revealed by the patient's attitude toward the examination and the clinician.

Disturbances of motor function are among the most revealing of all aspects of the neuropsychiatric examination. No interview is without its behavioral components, and the observed motor characteristics should be part of all diagnostic formulations. Characteristic abnormalities of gait, posture, and spontaneous movement occur in most neuropsychiatric disorders (Table 3.2). Retarded depression is characterized by psychomotor slowing, long latency to replies, paucity of verbal output, hypophonia, and bowed posture. Agitated depressions produce abnormal pacing, hand wringing, and an akathisia-like inability to sit calmly. The motor manifestations of mania include psychomotor hyperactivity, pressured speech, and rapid talking (tachyphemia) (Chapter 14). Catatonic behav-

TABLE 3.2. *Motor Disturbances Characteristic of Neuropsychiatric Syndromes*

Hypokinesias
Bradykinesia (psychomotor retardation)
Paresis
Catatonia (waxy flexibility, passivity, negativism, and sustained posturing)
Dystonia
Hyperkinesias
Akathisia
Tremor
Tardive dyskinesia
Ballismus
Chorea
Athetosis
Tics
Myoclonus
Mannerisms and stereotypy
Agitation
Psychomotor hyperactivity
Compulsive acts and rituals

ior (stereotypy, mannerisms, waxy flexibility, passivity, negativism, sustained posturing) may occur in affective disorders, schizophrenic syndromes, and neurological and toxic-metabolic disturbances (Chapter 20). Anxiety is reflected by a rigid posture, widened palpebral fissures, dilated pupils, and action tremor.[4] Obsessions and compulsions are manifested by compulsory stereotyped acts, checking, cleaning, and rituals. Drugs used in the treatment of psychiatric disorders commonly produce motor system abnormalities such as tremor, dystonia, parkinsonism, dyskinesias, and akathisia (Chapter 20). The extrapyramidal diseases manifest rigidity, tremors, ballismus, athetosis, chorea, dystonia, tics, myoclonus, or bradykinesia (Chapter 18, 19).

Mood and Affect

Mood and affect, like motor activity, permeate the entire neuropsychiatric interview and are assessed in an ongoing manner throughout the examination. *Mood* refers to emotion as experienced and reported by the patient, whereas *affect* refers to the emotion manifested by the patient in speech, facial expression, and behavioral demeanor. The two aspects of emotion are usually congruent but may become dissociated in pathologic states such as pseudobulbar palsy, when the patient may laugh in spite of feeling depressed or cry even when in a good mood (Chapter 14). It is therefore necessary to verify observations of affect by inquiring specifically about the patient's mood. Euphoria, elevated mood, and grandiose expansiveness are common features of mania; sadness, reduced range and responsiveness of affect, and blunted emotional expression are noted in depressed patients.[3,4] Euphoria with silly facetiousness occurs with certain frontal lobe disturbances (Chapter 9). Eutonia, a feeling of physical well-being, is particularly common among patients with multiple sclerosis.[5] Uncontrollable anger and rage are manifestations of some dyscontrol syndromes (Chapter 24). Apathy and emotional blunting characterize the mood state of many patients with frontal lobe disturbances, schizophrenic syndromes, and extrapyramidal disorders. Patients with epileptogenic lesions of the limbic system may experience a heightening and intensification of their emotional states (Chapter 21).

Verbal Output

The patient's verbal output is one of the principal means of assessing mood, thought, and cognitive abilities. The clinician must attend to motor aspects (dysarthria, mutism), linguistic features (fluent, nonfluent), and content (delusions, perseveration, obsessions) of the verbalizations. Table 3.3 lists the principal abnormalities of verbal output encountered during the neuropsychiatric interview, and these are presented in more detail in Chapter 6. *Mutism* occurs in a wide variety of clinical circumstances, including catatonic states, conversion reactions, pseudobulbar syndromes, early in the course of some aphasic syndromes, and in advanced stages of many neurological diseases.[6,7] Speech disturbances include abnormally rapid speech occurring in mania and in many fluent aphasias; slow speech characteristic of depression and nonfluent aphasias; dysarthria secondary to mechanical disruption of articulation; and abnormalities of loudness, particularly hypophonia, in depression and many extrapyramidal syndromes. *Aprosodia* (also termed *dysprosody*) refers to the loss of melody, rhythm, and emotional inflection that accompanies nonfluent aphasia, extrapyramidal disturbances, and anterior right hemispheric lesions.[8,9]

Abnormalities of verbal fluency occur in the aphasias. Nonfluent aphasias are characterized by a halting, sparse output with dysarthria, whereas fluent aphasias have a normal or increased verbal output with promi-

TABLE 3.3. *Disorders of Verbal Output*

Disorder
Mutism
Speech disorders
Dysarthria
Hypophonia
Slow speech (bradyphemia)
Rapid speech (tachyphemia; press of speech)
Aprosodia
Aphasic syndromes
Nonfluent
Fluent
Reiterative disturbances
Stuttering
Echolalia
Palilalia
Verbigeration
Miscellaneous disorders
Word salad
Coprolalia

nent paraphasia. Nonfluent aphasias correlate with lesions of the anterior left hemisphere and fluent aphasias, with lesions located posteriorly in the left hemisphere (Chapter 6).[10,11]

Reiterative speech disturbances occur in a variety of clinical settings. *Stuttering*, the repetition of single syllables, is seen as a congenital abnormality, in extrapyramidal syndromes and advanced dementia patients, in the recovery phases of aphasia, and with bilateral cerebral insults. *Palilalia*, the repetition of the patient's own output, and *echolalia*, the repetition of the examiner's output, occurs in some aphasic syndromes, dementia, and Gilles de la Tourette's syndrome (Chapter 19). *Verbigeration* refers to the constant repetition of a word or phrase sometimes noted in schizophrenic disorders.[1,2]

Word salad is a rare disorder occurring in schizophrenia when the derailment of thought and loosening of associations become so profound that the individual words in a sentence bear little relationship to each other.[4] *Coprolalia*, the involuntary utterance of curse words, occurs primarily in Gilles de la Tourette syndrome, where it is usually accompanied by other involuntary vocalizations such as grunting, snorting, and barking (Chapter 19). Coprolalia is occasionally reported in other clinical disorders such as choreic syndromes, Lesch-Nyhan syndrome, and schizophrenia.

Thought Characteristics

No absolute distinctions can be drawn between disorders of verbal output and thought disorders, since the latter must necessarily be inferred from abnormalities of what the patient says. Nevertheless, there are a number of disorders that appear to reflect disturbances in the form or content of thought and are independent of speech or language disorders. Abnormalities in the form of thought are presented first, followed by a discussion of disturbances of thought content.

Disorders of the Form of Thought

Table 3.4 lists disorders in the form and content of thought. *Disturbed thought form* refers to abnormal relationships between ideas in the flow of conversation (loose associations, flight of ideas, perserveration, abnormally slow or fast thinking). Autistic thinking (personally idiosyncratic thought unrelated to reality) and loosening of associations are classical findings in schizophrenia and schizophrenia-like disorders. However, psychotic conversation may also reveal tangentiality

TABLE 3.4. *Disturbances of Thought*

Alterations in Form of Thought	*Alterations in Thought Content*
Autistic thinking	Delusions
Loosening of associations	Obsessions
Poverty of thought (small quantity or vague quality)	Phobias
Thought blocking	Hypochondria
Tangentiality	Confabulation
Circumstantiality	Approximate answers
Derailment	
Condensation	
Illogicality	
Neologisms	
Word salad	
Flight of ideas	
Clang association (association by rhyming)	
Assonance (association by similar sounds)	
Punning (association by double meaning)	
Word association (association by semantic meaning)	
Racing thoughts	
Incoherence	
Thought retardation	
Perseveration	

(digression without returning to the point of departure), thought blocking, vagueness and poverty of thinking, self-reference, derailment, illogicality, condensation of thoughts and sentences, and abnormal word and sentence construction with resulting neologisms or word salad.[4,12,13]

Circumstantiality (digression from the topic with eventual return to the intended point) is a frequent finding in some patients with epileptogenic lesions in the limbic system and associated personality alterations. Circumstantiality must be distinguished from circumlocution (talking around a word or defining without naming it because of word-finding difficulties).

Flight of ideas is characteristic of mania and is characterized by a rapid flow of ideas in which the direction is determined by specific word characteristics such as rhyming, assonance, punning, or sematic meaning.[4,13] Manic patients may have tachyphemia (rapid speech).

Perseveration is seen in many disorders that disrupt normal thought patterns, including dementia, frontal lobe syndromes, aphasia, and acute confusional states. Likewise, incoherence of thought occurs in schizophrenia and dementia and may be particularly striking in toxic-metabolic confusional states. Retardation of thought (bradyphrenia) occurs in depression and extrapyramidal syndromes.

• Disorders of Thought Content Delusions are the most flagrant abnormalities of thought content. They reflect the patient's loss of ability to correctly assess external reality. Ideas of reference, delusions of passivity, mind reading, thought broadcasting, grandiose beliefs, persecutory beliefs, and theme-specific delusions such as the misidentification syndromes occur in schizophrenic disorders, in neurological and toxic delusional conditions, and during some manic and depressive episodes. Delusions should be specifically sought in the course of the interview, but the patient may be guarded about revealing them. Delusions are usually either persecutory or grandiose in nature, although apparently benign delusions are occasionally reported. Patients should be questioned about fears of surveillance, threats against their lives, or special powers that they or others may possess. Concerns spontaneously voiced by the patient should be explored to determine whether they are based on verifiable observations and realistic possibilities or are the product of delusional fears and misinterpretations. Delusions are discussed in more detail in Chapter 12. Less severe disturbances of thought content include abnormal preoccupations or ruminations, obsessions, phobias, and hypochondriasis. *Confabulation* refers to fabrication of responses concerning situations that are unrecalled because of an impaired memory. The facts may be benign and trivial or fantastic productions generated without restraint by the patient (Chapter 7).[14–16] Confabulations, unlike delusions, lack stability and vary from day to day. They also lack the affective investment characteristic of many delusional beliefs.

The syndrome of approximate answers, or the Ganser syndrome, is an hysterical pseudodementia that usually occurs in patients with head trauma or toxic-metabolic encephalopathies but may also occur in schizophrenia. The pathognomonic sign of the syndrome is the approximate answer given in response to even trivially simple questions (e.g., "How many legs does a dog have?").[17]

Perception

Abnormalities of perception may be classified according to modality (visual, auditory, touch, olfactory, gustatory) or according to positive or negative abnormalities. Table 3.5 presents positive and negative visual perceptual disturbances. Positive visual phenomena include hallucinations and illusions (Chapter 13). The former may be either formed or unformed and occur without a corresponding external stimulus; the latter are distortions or misinterpretations of existing stimuli.[18,19] Negative visual phenomena include unilateral neglect in which one-half of the visual universe is ignored; blindness; central color blindness or achromatopsia; and agnosia, or the inability to recognize ob-

TABLE 3.5. *Abnormalities of Visual Perception*

Positive phenomena
Hallucinations
Illusions (metamorphopsia, macropsia, micropsia)
Palinopsia
Negative phenomena
Unilateral neglect
Blindness
Achromatopsia (central color blindness)
Agnosia
Visual object agnosia
Prosopagnosia (agnosia for familiar faces)
Environmental agnosia (agnosia for familiar places)
Simultanagnosia (inability to perceive two objects at once)
Color agnosia

jects, faces, or places, despite intact perceptual and naming functions (Chapter 9).[20–24] Agnosias may occur in auditory as well as visual modalities.[21]

Hallucinations may occur in all sensory modalities, including hearing, touch (formication hallucinations), smell, and taste, as well as vision. They may be formed (persons, objects) or unformed. Some specific types of auditory hallucination, such as hearing two voices discussing one or hearing one's own thoughts spoken aloud, occur primarily in schizophrenic conditions.

MENTAL STATUS EXAMINATION

The mental status examination augments and refines the observations made during the neuropsychiatric interview. The emphasis changes from an anamnesis concerning the patient's past and information derived from spontaneous conversation and behavior to a systematic testing of individual neuropsychological functions. Information allowing localization of lesions in the nervous system is based primarily on observing the errors made by the patient in response to questions on the mental status examination. Not all the tests presented in the following paragraphs need be done with each patient; clinical experience can guide the decision as to which aspects of the mental status require the most thorough exploration in any particular patient. Nevertheless, the major areas of neuropsychological function should be assessed at least briefly in most patients. In patients who appear to be cognitively intact, a timesaving strategy is to ask the patient to perform a difficult task derived from each of the five principal domains of neuropsychological function (attention, language, memory, visuospatial skills, and executive function); if the patient is able to perform these tasks, no further evaluation is required. If a failure occurs, that domain of function should be explored in greater detail.

Interpretation of the patient's responses to mental status questions requires that the examiner be cognizant of the patient's educational level, sociocultural background, and engagement with the testing. Patients who are uncooperative, inattentive, or anxious may fail tests that they would be able to perform under other circumstances; individuals with limited education will likely have smaller vocabularies, less skill in mathematics, and less familiarity with test-type questions; persons from a minority culture likewise may fail tasks because of unfamiliarity with the majority culture language, differing historical or social exposures, or distrust of the examiner's intentions. In these situations, the clinician should avoid overinterpretation of the test results and failures should be attributed to brain dysfunction only when there is a consistent pattern of behavior and supporting evidence of a neuropsychiatric disorder.

Attention and Concentration

Attention and concentration must be assessed in all patients since any disturbance in this sphere will result in failures throughout the mental status examination. For example, the drowsy or inattentive patient is likely to have difficulty with memory tests, calculations, and tests of language comprehension that might be misinterpreted as amnesia, acalculia, or aphasia if the attentional deficit is unrecognized. Three major types of attentional disturbance may be identified: (*1*) drowsiness or deficits of alertness, (*2*) deficits in concentration with distractibility and fluctuating attention, and (*3*) unilateral neglect or hemispatial inattention (Table 3.6). The neurophysiological and neuroanatomic aspects of attention are normally intergrated into a functional unit mediating arousal, concentration, and sensory awareness.[25] Deficits of alertness, if present, are evident during the interview and manifested by drowsiness and the need for repeated stimulation to keep the patient engaged with the examiner. Several degrees of reduced arousal are recognized, including clouding (mildly reduced wakefulness or awareness), obtundation (mildly to moderately reduced alertness with lessened interest in the environment, drowsiness while awake, and increased sleep), stupor (deep sleep or similar unresponsive state from which the patient can be aroused only with vigorous and repeated stimuli), and coma (unarousable unresponsiveness).[26] Reduced arousal reflects dysfunction of both cerebral hemispheres or of the reticular activating system on a structural or toxic-metabolic basis. The digit span test, in

TABLE 3.6. *Major Types of Attentional Deficit and Their Anatomic Correlates*

Type of Attentional Deficit	*Anatomic Basis*
Drowsiness	Reticular activating system, both hemispheres
Distractibility	Frontal lobe
Unilateral neglect	
Sensory (hemi-inattention)	Thalamus, parietal lobe
Motor (hemi-inintention)	Caudate nucleus, frontal lobe

which the patient is asked to repeat a list of numbers dictated by the examiner (normal, 7 ± 2 digits forward and 5 ± 1 digits in reverse), is a useful test of alertness and arousal.[1]

The ability to consistently sustain vigilance is best assessed by a continuous performance task such as the *A* test, in which the patient is asked to respond by lifting his or her hand whenever the letter *A* is heard in a list of letters read aloud by the examiner.[1] Errors of omission usually reflect distractibility or loss of set for the task; errors of commission are usually perseverative or due to impulsiveness or anxiety, with patients raising their hands for letters other than the letter *A*. Disturbances of sustained concentration and vigilance most often reflect frontal lobe dysfunction or toxic-metabolic encephalopathy. In addition to structural and toxic disorders, other neuropsychiatric syndromes, including dementia with Lewy bodies, depression, mania, anxiety, and schizophrenia, may produce prominent concentration and attentional disturbances (Chapter 11).

Unilateral disturbances of attention or hemispatial neglect may involve unilateral sensory attention (hemi-inattention) or unilateral motor activity (hemi-inintention) (Chapter 8). Behaviorally, patients may ignore all stimuli on the side contralateral to the lesion and will perceive or respond only to those stimuli in the hemispatial universe receiving their attention. They will perceive only one of two simultaneous auditory clicks (finger snaps) and will feel only one of two simultaneous somatosensory stimuli (touching each side of the body at the same time). Visually, they may read only half of written words (*northwest* as *north* or *west*, depending on which side is neglected), will draw only half of constructions they are asked to copy, and will ignore half of vertically written multiple-digit mathematical problems. A useful test of visual neglect is the line-crossing test in which the patient is given a piece of paper on which there are a number of short straight lines scattered across the page in a variety of orientations and is asked to cross each line in the middle.[27] All or a portion of the lines in the neglected field will be missed, and the middle of each line may be misjudged, with the line crossings systematically displaced away from the neglected field (Chapter 8). The occurrence of a hemianopa is independent of hemispatial neglect: patients with hemianopias may or may not have unilateral neglect, and neglect may occur with or without a hemianopia. Unilateral sensory neglect occurs primarily with parietal lobe lesions and is more profound and more persistent in patients with right parietal lesions than in those with left-sided insults.[23] When motor neglect is present (hemi-inintention), the patient may appear to be hemiparetic because of lack of use of an extremity, but normal motor and sensory function can be demonstrated when the patient's attention is specifically directed to the neglected limb.[23] Motor neglect occurs with frontal lobe and striatal lesions.

Mental Control

Mental control tasks are tests of complex attention that require the interaction of several neuropsychological functions including sustained attention, recent memory, and executive function. Mathematical skills are necessary to perform some mental control tests. Serial subtraction of 7's from 100, serial subtraction of 3's from 20, spelling *world* backwards, and saying the months of the year in reverse order are tests of mental control. Serial subtraction tasks require mathematical skills and may be impossible for individuals with limited education.

Language

Language disturbances, like attentional deficits, may profoundly influence the patient's ability to perform many aspects of the mental status examination. Memory testing, calculation, abstraction, and comprehension of instructions for all other testing depend on intact language capabilities, and the integrity of linguistic capabilities must be determined early in the course of assessing the patient's mental state. A systematic approach to language evaluation and interpretation is presented in more detail in Chapter 6; the principal areas to be tested are presented here and summarized in Table 3.7. Language disturbances (aphasia) must be distinguished from alterations in the mechanical aspects of sound production (dysarthria) and from thought disturbances (outlined above).

• Spontaneous Speech

A major change in the spontaneous speech of language-disordered patients is a disturbance in language fluency. *Nonfluent aphasias* are characterized by decreased verbal output, effortful speech, dysarthria, decreased phrase length, dysprosody (loss of speech rhythm), and agrammatism (omission of the small relational or "functor" words). *Fluent aphasias* have a normal or increased verbal output, normal articulation, normal phrase length, preserved prosody, empty speech, circumlocution, and paraphasia.[10,11] In nearly all right-handed individuals and a majority of left-handed subjects, nonfluent aphasia reflects structural changes in the left frontal lobe, whereas fluent aphasia is indicative of damage to the left posterior temporal, inferior parietal, or tem-

TABLE 3.7. *Principal Language Functions Tested or Observed in the Mental Status Examination*

Spontaneous speech
Comprehension
Repetition
Naming
Reading
Aloud
Comprehension
Writing
Word list generation
Speech prosody
Miscellaneous
Automatic speech
Completion phenomenon
Singing

poroparietooccipital junction region.[10,11] Speech prosody also may be disrupted by right-sided frontal lobe lesions and by subcortical dysfunction in extrapyramidal disturbances.[8,9,28]

● Comprehension Language comprehension is difficult to assess with precision. It is heavily dependent on attention, concentration, and cooperation as well as linguistic abilities. Comprehension should be tested in a graded fashion beginning with one-, two-, and then three-step pointing to room objects ("Point to the door, the window, and the chair.") Then a series of easy to difficult yes/no questions should be administered (easy: "Is your name 'Green'?"; difficult: "Do you put your shoes on before your socks?"). Finally, comprehension of more sophisticated linguistic structures can be assessed, such as sentences with passive constructions ("If a lion and a tiger are in a fight and the lion is killed by the tiger, which animal is dead?") and possessives ("Is my wife's brother a man or a woman?").[29] Disruption of language comprehension occurs with involvement of the left posterior temporal and temporoparietooccipital junction regions (Chapter 6). Left-sided anterior lesions generally spare most aspects of language comprehension, although patients may have difficulty following commands that depend on correct interpretation of sequential information ("Touch the pen with the pencil," in contrast to "With the pen touch the pencil").

● Repetition Repetition, like comprehension, should be tested with a graded series of phrases and sentences of increasing complexity. The patient is requested to repeat each sentence exactly as spoken by the examiner. From simple phrases such as "he is here," the examiner proceeds to longer and more difficult sentences, such as "The quick brown fox jumped over the lazy dog," and finally includes more complex and linguistically irregular phrases such as "no ifs, ands, or buts."[29] Again, concentration and attention deficits may interfere with all but the most simple tests of repetition, and interpretation of repetition failures must take this into account. Repetition span (the number of words that can be repeated) is typically two words longer than the patient's digit span; thus, patients with reduced digit span because of a toxic-metabolic encephalopathy will have a correspondingly limited ability to repeat long phrases. Hearing abnormalities also limit the patient's ability to cooperate with repetition testing. Patients with disturbances of repetition as a result of aphasia characteristically omit words, alter the word sequence, and have paraphasic intrusions when trying to reproduce the test sentence. Anatomically, failure of repetition occurs in aphasias with lesions situated adjacent to the left Sylvian fissure. Aphasics with preserved repetition have intact peri-sylvian structures.[10,11]

● Naming Naming disturbances are a sensitive indication of language impairment but lack specificity for the type of linguistic compromise. Anomia is most often one aspect of an aphasia syndrome but may occasionally be evident in toxic-confusional states, with increased intracranial pressure, and with other nonfocal disturbances.[30,31] Anomia may be manifested in spontaneous speech by word-finding pauses, emptiness, and circumlocution or may be identified by tests of confrontation naming. High- and low-frequency names should be tested (in general, object names, e.g., *wristwatch*, are high frequency, and object parts, e.g., *stem, crystal, band,* are lower frequency), as well as names in several linguistic categories (e.g., colors, body parts, room objects, actions). Naming errors may take the form of literal paraphasias (phonemic substitutions such as *greel* for *green*), verbal paraphasias (semantic substitutions such as *blue* for *green*), neologisms (completely new constructions), failure to make any response, or circumlocutions (descriptions of the object or its use without naming it).

● Reading Reading is a complex neurological function that must be learned and is subject to many cultural and educational influences. Literacy is still uncommon in much of the world, and the significance of an individual's reading difficulties can be judged only after considering the person's level of educational achievement. The ability to read aloud and reading comprehension must be tested separately, since some lesions may disrupt oral reading, leaving reading comprehension intact, whereas other disorders may impair reading comprehension but spare the ability to read aloud. Letter, word, and sentence reading should be tested. Failures may include an inability to read letters, an inability to read words, ignoring one-half of the word (hemialexia), or substitution of one word for another.[10,24] Alexias may occur with anterior or posterior left-sided lesions (Chapter 6).

● Writing Writing also is an acquired task heavily dependent on one's educational experience and occupational demands. *Agraphia*, an acquired disturbance of writing, may occur on an aphasic basis secondary to an interruption of linguistic function or on a nonaphasic basis produced by an impairment of the motor system and mechanical aspects of writing. All aphasics will make errors in their written as well as their oral productions, and the characteristics of the written language resemble those of the spoken output.[32] Peripheral, corticospinal, extrapyramidal, and cerebellar disturbances all disrupt the motoric aspects of writing and produce distinctive agraphic syndromes. The differential diagnosis of agraphia is presented in Chapter 6.

● Word List Generation Word list generation, sometimes referred to as *verbal fluency*, is a sensitive but nonspecific test of language dysfunction. In the two common versions of the test, the patient is asked to name as many animals as possible in 1 minute (normal, 18 ± 6)[29] or as many words beginning with the letter *F*, then *A*, and then *S*, with 1 minute allowed per letter (mean of 15 ± 5 words per letter or mean total of 45 for the test). All aphasics do poorly on the test, and, in addition, patients with prefrontal lesions, basal ganglia disorders, or psychomotor retardation fail to produce the expected number of words per minute (see Chapter 9, on frontal lobe syndromes).

● Prosody *Prosody* refers to the melodic, rhythmic, and inflectional elements of speech, and aprosodic output is typically monotonic, amelodic, and affectless. Two aspects of prosody should be assessed: spontaneous prosody and prosodic comprehension. Spontaneous prosody is judged simply by listening to verbal utterances occurring during the course of conversation. Prosodic comprehension is tested by having the patient, with eyes closed, listen to a neutral sentence executed by the examiner in four prosodic styles (surprised, happy, angry, sad). The patient is then asked to identify the emotional state of the speaker based on the way the sentence was inflected. Impaired spontaneous prosody is produced by right frontal lesions and extrapyramidal disturbances, whereas prosodic comprehension is most disturbed by right temporoparietal injuries.[8,9,28,33,34]

● Miscellaneous Language Tests In addition to these standard language tests, other types of linguistic probes may be used in specific circumstances. In profoundly aphasic patients, remnants of intact language function may be elicited by having the patient produce automatic speech (counting, reciting the alphabet, naming the days of the week, reciting the months of the year), attempting to complete overlearned sequences (filling in the last line of nursery rhymes or prayers), or singing. Rhythmic aspects of singing appear to be under left hemisphere control whereas tone and pitch are mediated by the right hemisphere.[35] Aphasic patients can often reproduce the tune of a familiar song (e.g., "Happy Birthday") even when they cannot sing the words. Severely aphasic patients retain knowledge of the iconic appearance of writing and can identify upside-down writing as abnormal.

Memory

Memory is often divided into three functions—immediate, recent, and remote—with immediate memory representing the ultra-short-term memory tested with digit span, recent memory representing the ability to learn new information and remote memory representing the recall of material learned in the past. Immediate memory is best considered an attentional capacity, since the information is not memorized or committed to memory for later recall. Working memory and mental control tasks utilize this aspect of memory. Attention is a necessary prerequisite for all aspects of memory, and the presence of intact attention must be demonstrated before conclusions about memory can be drawn.

Recent memory refers to the ability to learn and recall new information. Two types of test are commonly used to assess recent memory: orientation and the recall of recently presented verbal or nonverbal information. Orientation in time and space must be learned on a daily basis and inquiring whether one knows the correct year, month, day, date, and time of day as well as one's current location will reveal important information about learning and recent memory abilities. A more structured assessment of verbal learning can be performed by asking patients to learn three or four unrelated words and then asking them to recall the words after 3 minutes.[1] There are two types of recent memory disturbances: amnesias and retrieval deficit syndromes. In amnestic disorders, there is a failure to store new information, whereas in the retrieval deficit syndrome there is an impairment of timely recall of stored data. The two types of memory failure can be distinguished by the response of the patients to recall clues. Patients with amnesia cannot spontaneously recall recently learned information, are not helped by clues ("One of the words I asked you to remember was the name of a vegetable" as a clue for remembering the word *cabbage*), and cannot recognize recently learned words in a list that contains both targets (the recently learned words) and foils (similar words that were not part of the recent memory test). Patients with the retrieval deficit syndrome also have limited spontaneous recall but benefit from clues and are able to distinguish between targets and foils on tests of recognition memory.[36] Amnestic patients have disorders affecting the hippocampus in the medial temporal region, fornix, marnmillary bodies, mammillothalamic tract, or medial thalamus (Chapter 7). The retrieval deficit syndrome is associated with disturbances affecting the dorsolateral prefrontal cortex or caudate nucleus.

Testing nonverbal memory is more difficult than assessing verbal memory, but an assessment can be made by warning patients when they are copying constructions (discussed in the next section) that they will be asked to reproduce them later. After a 3-minute delay, the patient is instructed to redraw the figures from memory. If failures occur, recognition of forgotten figures can be tested by presenting a number of constructions and asking which ones had been shown previously. Unfortunately, many patients spontaneously develop verbal descriptions of the drawings, converting the test into an assessment of verbal memory. A more precise evaluation of nonverbal learning demands use of more specialized neuropsychological techniques and nonsense figures not easily verbally encoded. Nonverbal memory is mediated primarily by right hemisphere structures and verbal memory, by the left hemisphere.

Remote memory abilities are revealed during the neuropsychiatric interview by the patient's ability to recapitulate personal history. Such historical information, however, may be unverifiable by the clinician, and a review of public knowledge (past political leaders, dates of historical events, sports facts, etc.) may add another dimension to the testing. Information of this type is dependent on the educational background of the patient and intact language abilities.

Constructions

Assessment of visuoconstructive abilities is performed by asking the patient to copy drawings provided by the examiner (e.g., circle, cross, cube). The drawings should be a graded series of figures of increasing complexity and should include at least one three-dimensional representation. Relatively normal motor skills are an obvious prerequisite for performance of the task. Tests of constructional abilities are an excellent method of screening for acquired brain dysfunction. Most idiopathic psychiatric disorders spare constructional skills, whereas lesions of the frontal or parietooccipital regions of either hemisphere may disrupt visuoconstructive abilities.[21,37] Neglect of one side of the figures is most consistent with a posterior hemispheric lesion contralateral to the neglected hemispace, whereas a fragmented and disorganized approach to complex constructions occurs most often with frontal lobe lesions or in acute confusional states.[38–40]

Clock drawing is frequently used as a screening test of visuospatial abilities and can provide substantial information. The patient is first asked to draw the face of a clock including the numbers; if the patient draws a clock face too small to accommodate the numbers, there is preliminary evidence of a planning and foresight deficit. Once the circle of the clock face is drawn, the patient must place the numbers accurately and disturbances of planning or evidence of unilateral neglect may be apparent (patients may place all the letters on one side of the clock with either condition and additional testing will be required to distinguish the two). Next, the patient is asked to set the clock for 11:10. Patients with executive dysfunction and "stimulus boundedness" may put one hand on the 10 and one on the 11, thus setting the clock inaccurately (10:50). Clock drawing provides insight into both visuoconstructive and executive abilities and is compromised in patients with parietal or frontal lobe disorders.[41,42]

Calculation

Calculating abilities are tested by asking the patient to solve arithmetic problems (usually addition and multiplication) presented either orally or in written form. Attention must be intact, and previous competency in calculation must be assured before failures in calculating abilities can be interpreted. At least three types of acalculia are described: (*1*) patients with fluent aphasias may make semantic paraphasic errors when reading, writing, or saying numbers, making correct calculation impossible; (*2*) patients with right-sided parietal lesions may have a visuospatial acalculia, resulting from an inability to correctly align columns of written numbers; and (*3*) a primary anarithmetria may occur with left-sided posterior hemisphere lesions. This last type of acalculia may occur as an isolated disorder or as part of Gerstmann's syndrome and reflects an inability to perform the actual numerical manipulations.[24,43]

Abstraction

The ability to abstract provides a good index of general intellectual function and, like calculation, is dependent on the patient's level of educational achievement and cultural experience. Most English-speaking individuals will be able to abstract idioms such as "cold shoulder," "heavy heart," and "level-headed" regardless of their educational history, and impairment of this skill usually indicates a disturbance of abstracting abilities. Proverbs can be understood by most individuals with a high school education, and an inability to interpret proverbs by patients with more advanced educational achievement is evidence of compromised intellectual function. Both simple proverbs ("Don't cry over spilled milk") and complex proverbs ("People who live in glass houses shouldn't throw stones") should be tested.[7] In addition to testing the ability to abstract, proverb interpretation often elicits bizarre, paranoid, or idiosyncratic responses from patients with psychoses and macabre, pessimistic, hopeless interpretations from depressed patients. The ability to abstract depends on intact language function as well as higher-order comparative and extrapolation skills mediated by frontal–subcortical circuits.

Insight and Judgment

The traditional "insight" questions such as "What would you do if you found a stamped addressed envelope?" or "What would you do if you were the first to see a fire in a crowded theater?" are overly simplistic and provide little information in most circumstances. More insight can be achieved by inquiring about what the patient understands of his or her illness, intends to do after leaving the hospital, or perceives about current personal medical and psychosocial needs. Answers reveal the patient's judgment, foresight, motivation, and depth of insight.

Praxis

Ideomotor apraxia refers to the inability to perform volitional acts on command despite intact motor and sensory functions and preserved language comprehension.[44] To test praxis, the patient is asked to perform limb, whole-body, and oral-lingual movements. Limb commands include "Show me how you comb your hair" or "Show me how you brush your teeth"; whole-body commands include "Take a bow" and "Show me how you swing a golf club"; and oral-lingual commands include "Show me how you blow out a match," "Show me how you suck through a straw," and "Cough." The presence of ideomotor apraxia reflects focal dysfunction in the left hemisphere of right-handed individuals (Chapter 6).

Ideational praxis refers to the ability to synthesize a series of individual actions into a complex activity. It is tested by asking the patient to pantomime the steps of a complicated task such as putting a letter into an envelope, sealing and addressing the envelope, and adding a stamp. Ideational praxis is disrupted in patients with diffuse or multifocal brain lesions or dementia syndromes.

Executive Function

Executive functions are higher-order cognitive abilities mediated primarily by the dorsolateral prefrontal cortex and subcortical structures linked to this region (Chapter 9).[45] Dysexecutive syndromes are present in diverse neuropsychiatric conditions including frontotemporal dementias, focal frontal lobe disorders, basal ganglia diseases, schizophrenia, and some patients with depression.

Performance of many of the tests described above depends on intact executive function and frontal–subcortical circuit function. Attention and concentration depend on frontal–subcortical circuits, and distractibility is evident in many patients with executive dysfunction. Word list generation requires normal frontal circuits, and reduced generative capacity in a patient without aphasia impugns the integrity of frontal–subcortical circuitry. Retrieval of stored information is mediated by frontal–subcortical circuits, and a retrieval deficit syndrome is present in many patients with executive deficits. Complex constructions are approached in a segmented and fragmented manner by patients with frontal–subcortical dysfunction who lack the ability to develop an effective drawing strategy. As described above, clock drawing and time setting can provide insight into stimulus boundedness, a frequent manifestation of executive dysfunction. Abstraction, insight, planning, and judgment are also compromised in patients with frontal–subcortical circuit disorders and executive deficits.

Motor programming skills also depend on intact frontal–subcortical circuit function. Tests used to assess this ability include alternating programs, reciprocal programs, multiple loops, serial hand sequences, and rhythm tapping.[37,39,46,47] All these tests have in common a dependence on the patient's ability to program motor responses in specific repeating or alternating sequences. Failures include an inability to achieve the specific behavioral set required to perform the task, rapid extinction of the appropriate set with consequent deterioration of the expected sequence, or inability to change set with perseveration or intrusion of unwanted activities. The patient is asked to copy the alternating sequences or looped figures provided by the examiner and then to continue the same repeating pattern across the page. Reciprocal programs are performed by having the patient and examiner execute reciprocal maneuvers. For example, the patient is requested to tap once each time the examiner taps twice and twice whenever the examiner taps once. The examiner then proceeds to tap once or twice in random order and observes whether the patient is able to respond reciprocally. Similarly, rhythm tapping abilities are assessed by asking the patient to imitate a rhythmic sequence of taps performed by the clinician. Serial hand sequences are tested by requiring that the patient execute a series of repeating hand postures while announcing aloud the name of each posture (fist-slap-side) (Fig. 3.1).

Additional Tests and Miscellaneous Observations

● **Right–Left Orientation** The inability to distinguish right from left usually occurs as part of a symptom complex with dysgraphia acalculia, and finger agnosia (Gerstmann's syndrome), a reliable indication of a lesion in the dominant angular gyrus.[48,49] When right–left confusion occurs in other circumstances, its localizing value is less certain. Right–left orientation is tested by asking the patient to raise the right or left hand, touch the right ear (or other body parts) with the left or right hand, and then indicate the examiner's right or left hand as the examiner and the patient face each other. Deficits in attention or in language comprehension must be excluded before a failure can be attributed to a specific impairment of right–left orientation.

● **Finger Agnosia** Finger agnosia, like right–left disorientation, has localizing significance when it occurs as part of the Gerstmann syndrome.[48] In its simplest form, finger agnosia can be detected by asking the patient to point to the little finger, middle finger, ring finger, or index finger. More subtle forms can be de-

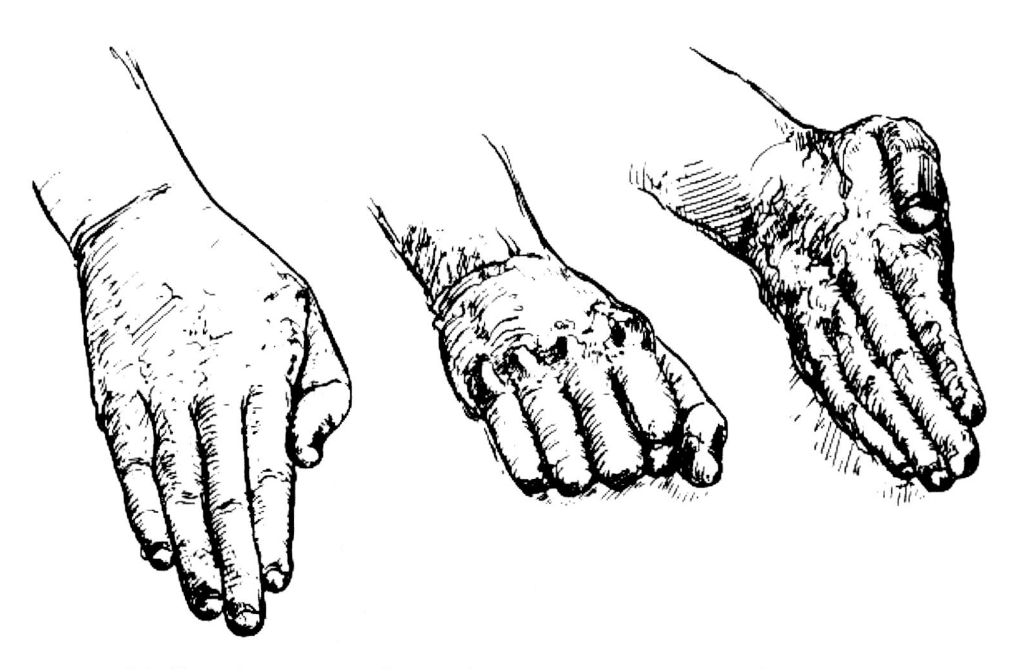

FIGURE 3.1 Alternating sequences often reveal perseveration in patients with frontal system defects.

tected by touching an individual finger on the patient's hand while it is held out of sight above head level and asking the patient to indicate the corresponding finger on the other hand. Alternatively, the examiner may touch any combination of two fingers on one of the patient's hands and ask the patient how many digits there are between those touched.[50]

• Miscellaneous Observations During the course of the interview and testing, many additional observations will be made that may require more detailed exploration or may, in themselves, have localizing significance. For example, inability to perceive two objects simultaneously occurs with bilateral parietal lesions. Any deviation from expected behavior may be useful in localization and diagnosis. Unusual responses should be followed up by supplementary observations. Tasks that appear too difficult for the patient should be simplified and repeated, while tests performed rapidly suggest that more taxing assessments are required. The neurobehavioral examination is a process of hypothesis formation and testing as the examiner attempts to understand the patient's difficulty; it resembles a complex choreography with each act of the patient eliciting a corresponding response from the examiner as conceptual and diagnostic clarification is sought.

NEUROLOGICAL EXAMINATION

The neurological examination can provide valuable information relevant to the diagnosis of neuropsychiatric and neurobehavioral examination.[51] Each patient presenting for neuropsychiatric assessment requires a comprehensive neurologic evaluation. The examination is guided by diagnostic hypotheses formulated during the course of the neuropsychiatric interview and mental status testing. The usual evaluation includes assessment of cranial nerves, motor system, sensory abilities, muscle stretch reflexes, and primitive reflexes.[52]

Cranial Nerve Examination

Abnormalities of the *first cranial nerve (olfaction)* are detected by asking the patient to identify a variety of distinctive odors (e.g., perfume, coffee, chocolate, soap, spices). The olfactory nerves, bulbs, and tracts are located on the inferior service of the frontal lobe, where they may be damaged by closed head trauma or compressed by subfrontal tumors. Olfactory abnormalities are not specific to frontal damage since smoking, rhinitis, and nasal trauma also adversely affect the ability to detect odors. Patients with Alzheimer's disease and Parkinson's disease have impaired odor detection and recognition.

Examination of the *second cranial nerve (optic)* is critical to the neuropsychiatric evaluation and includes visualization of the nerve head, testing of visual acuity, and assessing the visual fields. Examination of the optic nerve may reveal optic atrophy (a pale disc), papilledema reflecting increased intracranial pressure, or papillitis associated with optic nerve ischemia or inflammatory conditions such as multiple sclerosis. Ophthalmoscopy may also reveal macular degeneration, pigmentary retinopathy, retinal infarction, narrowing of retinal vessels, and retinal hemorrhages or exudates. Visual acuity of each eye is tested with a Snellen chart held 18 inches from the eyes. Patients should wear their glasses to minimize visual disturbances produced by abnormalities of the shape of the optic globe (astigmatism). Acuity should be correctable to 20/20 if optic nerve function is normal, and continued impairment after correction implies the presence of prechiasmatic optic nerve pathology. Visual field defects are identified by systematically bringing a small target in from the periphery toward the center of the patient's visual field and asking them to indicate when the target is first seen. Damage to an optic nerve produces monocular visual field defects of the ipsilateral eye. Postchiasmatic injuries typically produce a homonymous visual field defect. Occipital damage is associated with congruent field defects of similar size and shape in each field, whereas more anterior lesions (damage to the geniculocalcarine radiation in the temporal lobe) produce incongruent field defects of differing size in the two fields. Visual field defects are often associated with release hallucinations and other visual phenomena (Chapter 13).

Cranial nerves III, IV, and VI (oculomotor, trochlear, and abducens) mediate ocular motility and pupillary responses. The external rectus muscle turns the globe laterally and is innervated by the abducens nerve; the superior oblique moves the globe downward when the eye is adducted and is innervated by the trochlear nerve; all the remaining ocular muscles, the ciliary muscles mediating pupillary constriction, and some of the muscles of the eyelid are innervated by the oculomotor nerve. Pursuit movements and volitional eye movements should be tested. Pursuit is assessed by having the patient follow the examiner's finger through seven positions (adduction, adducted and up, abducted and up, abducted, abducted and down, adducted and down, convergence). Volitional eye movements are assessed by asking the patient to look right, left, up, and down without following a visual target.

Conjugate gaze is mediated by gaze-coordinating centers in the pons (lateral movements) and the midbrain (vertical movements). Abnormalities resulting from dysfunction of cranial nerves III, IV, and VI include ocular paresis, gaze-paretic nystagmus, and pupillary and lid disturbances. Brainstem lesions, extrapyramidal movement disorders, and lesions of the premotor regions of the frontal cortex may produce supranuclear gaze palsies manifested by impaired volitional eye movements. Gaze-paretic nystagmus results from inadequate ocular deviation; the quick or jerk phase of the nystagmus is in the direction of the weakened muscle. Changes in pupillary size occur in response to light stimulation and in conjunction with convergence (constriction of the pupil with near vision). Cranial nerve III is responsible for pupillary constriction through its parasympathetic branches; pupillary dilatation is a function of the sympathetic nervous system. Lid position is determined by input from the third nerve and sympathetic influences responsible for eye opening and from the seventh cranial nerve mediating lid closure. Third nerve palsy results in *ptosis*, a nonreactive pupil and an inability to adduct, elevate or depress the eye. Pupillary abnormalities occur with brainstem lesions, syphilis (Argyll-Robertson pupil in which the reaction to light is lost while the convergence mechanism is intact), diabetes, and nerve compression by aneurysms or other mass lesions.

The neuro-ophthalmologic examination may contribute essential information to neuropsychiatric diagnosis (see Table 3.8). Inspection of the cornea may reveal a Kayser-Fleischer ring in Wilson's disease. Retinal assessment may show macular degeneration (associated with blindness and visual hallucinations) or retinal degenerations (produced by phenothiazines, particularly thioridazine, and some hereditary brain disorders). Cataracts are associated with diabetes, hypoparathyroidism, and Wilson's disease and may occur with chlorpromazine and corticosteroid therapy.

The *fifth cranial nerve (trigeminal)* innervates the muscles of the jaw involved in mastication and is responsible for facial and corneal sensation. *Cranial nerve VII (facial)* supplies the facial musculature including the platysma muscles of the anterior neck, orbicularis

TABLE 3.8. *Neuro-ophthalmologic Features of Common Neuropsychiatric Disorders*

Conditions	*Neuro-ophthalmologic Manifestations*
Alzheimer's disease	Inadequate visual exploration; "visual grasp"; optic ataxia in posterior variant with Balint's syndrome
Frontotemporal dementia	Deficient anti-saccades
Dementia with Lewy bodies	Visual hallucinations
Parkinson's disease	Poor upgaze and convergence; diminished blinking; saccadic pursuit; visual hallucinations following dopaminergic therapy
Progressive supranuclear palsy	Supranuclear gaze palsy with deficient vertical gaze (down-gaze lost first); microsquare wave jerks; diminished blinking; deficient antisaccades; saccidic pursuit
Huntington's disease	Supranuclear gaze palsy; deficient anti-saccades
Corticobasal degeneration	Supranuclear gaze palsy
Wilson's disease	Kayser-Fleischer corneal rings; cataracts
Gilles de la Tourette syndrome	Blinking tics; sudden gaze tics; obsessional eye mutilation
Frontal lobe lesions	Ipsilateral gaze deviation in acute phase; contralateral gaze impersistence with chronic lesions
Geniculocalcarine lesions	Homonymous hemianopia; release hallucinations
Midbrain lesion	Peduncular hallucinations (benign, often Lilliputian hallucinations, occurring most commonly in the evenings
Basilar bifurcation occlusion	"Top of the basilar syndrome" with dream-like state and visual hallucinations
Multiple sclerosis	Optic neuritis, internuclear ophthalmoplegia, phosphenes
Delirium	Silent formed, visual hallucinations
Migraine	Scintillating scotomata, fortification spectra, formed hallucinations
Narcolepsy	Hypnagogic and hypnopompic hallucinations
Mass lesions	Papilledema; transient visual obscurations

oris and buccinator muscles around the mouth, orbicularis oculi of the eye, and the frontalis of the forehead. Its branches innervate the tear and salivary glands and stapedius muscle of the ear, and it provides taste fibers to the anterior two-thirds of the tongue. The upper facial muscles receive bilateral cortical input, whereas the lower facial muscles receive contralateral innervation only. Thus, upper motor neuron lesions produce only contralateral lower facial paresis, whereas peripheral nerve lesions cause ipsilateral paralysis of both upper and lower face. The facial nerve nucleus receives descending input from both pyramidal neurons mediating volitional movements and nonpyramidal limbic system connections mediating emotional responses. When one system is affected and the other is spared, the patient will exhibit either retained volitional grimacing but a partially paralyzed smile, or an intact emotional smile with hemifacial paresis of volitional grimacing.[53]

The *eighth cranial nerve* consists of an *auditory* branch and a *vestibular* branch. Auditory compromise causes deafness and is sometimes associated with auditory hallucinations. Vestibular end-organ or nerve lesions produce horizontal or combined horizontal-rotatory nystagmus and vertigo, whereas lesions disrupting vestibular connections within the central nervous system can produce nystagmus in any direction and usually are not associated with vertigo or nausea.

Cranial nerves IX and X (glossopharyngeal and vagus) control pharyngeal and laryngeal motor function, taste, and the gag reflex. The glossopharyngeal nerve conveys sensory information (including taste sensation from the posterior third of the tongue), while the vagus nerve is primarily responsible for the motor aspects of these structures. Dysfunction of the vagus nerve results in a hoarse voice, aphonia, or dysphagia. Supranuclear lesions disinhibit local nerve function and produce an exaggerated gag reflex. This is a common manifestation of pseudobulbar palsy (Chapter 14).

The *eleventh cranial nerve (spinal accessory)* innervates the upper half of the trapezius muscle and the sternocleidomastoid muscle; the *twelfth cranial nerve (hypoglossal)* innervates the tongue.

Motor System Examination

Motor system examination includes evaluation of muscle bulk, strength, tone, and coordination. Normal motor function depends on the successful integration of pyramidal, extrapyramidal, cerebellar, spinal, peripheral, myoneural junction, and muscular function. Assessment of the motor system is particularly important in neuropsychiatry: psychiatric disturbances are often accompanied by motor system manifestations (retardation, agitation, catatonia), movement disorders are often accompanied by neuropsychiatric abnormalities (Parkinson's disease, Huntington's disease), and many of the agents used to treat psychiatric diseases adversely affect the motor system (tremor, tardive dyskinesia). Extrapyramidal motor disorders are described in Chapter 18.

Muscle bulk is assessed by visual inspection, palpation and measurement. Muscle atrophy occurs with disuse, muscle diseases, and nerve or spinal diseases, and in generalized weight loss secondary to malnutrition or systemic illness.

Strength is evaluated by having the patient exert maximal power of specified muscle groups. Strength is graded as 0, no evidence of muscle contraction; 1, muscle contraction without movement of the limb; 2, limb movement after gravity eliminated; 3, limb movement against gravity; 4, limb movement against partial resistance; and 5, normal strength. Pyramidal tract lesions produce a "predilection" pattern of weakness affecting the extensor muscles of the upper limbs and the flexor muscles of the lower limbs with resulting flexion of the arm and extension of the leg; this is known as a *decorticate posture* and imitates the antigravity posture of a bipedal animal. Pyramidal lesions of the pons result in extensor posturing of all limbs, the antigravity posture of the quadrupedal animal.

Muscle tone is evaluated by passively manipulating the limbs or neck. Additional information can be garnered by asking the patient to perform motor activities with the contralateral limb while tone in the ipsilateral limb is determined (the patient is asked to draw a square in the air with one arm while the examiner is testing the tone of the other arm). Activation of tone by this maneuver (Froment's sign) is characteristic of parkinsonian-type extrapyramidal rigidity. Muscle tone is decreased in muscle and peripheral nerve disease, cerebellar syndromes, early in the course of many choreiform disorders, and acutely following an upper motor neuron lesion. Increased muscle tone (rigidity) occurs with pyramidal disorders (spasticity and clasp knife rigidity) and in extrapyramidal syndromes (plastic rigidity or cogwheel rigidity). *Gegenhalten* (also known as *paratonia*) refers to resistance to movement encountered in patients with advanced brain diseases. It is characterized by active resistance to all limb movements and may result from release of primitive protective mechanisms. *Waxy flexibility* is the tone change classically associated with catatonia; there is moderate resistance to passive movement and maintenance of the final position of the movement (like candle wax). Catatonia occurs in a variety of psychiatric, frontal lobe, extrapyramidal, and toxic-metabolic disorders (Chapter 20).

Coordination may be disrupted by many types of motor and sensory abnormalities but is dependent primarily on cerebellar function. Coordination is assessed by asking the patient to perform rapid alternating movements (alternating supination and pronation movements of the hand), fine finger movements (repeated apposition of the thumb and first finger), finger-to-nose movements (the patient alternates between touching his or her own nose and the examiner's finger), heel–knee–shin maneuvers (the patient touches the knee of one leg with the heel of the other and then gently slides the heel down the shin), and rebound check tests (the clinician suddenly releases as the patient is pushing his or her extended arms upward against the examiner; the clinician observes whether the patient can arrest the upward movement after the release). Dysdiadokokinesia (abnormal rapid alternating movements), ataxia, loss of rebound check, and intention tremor are characteristic of cerebellar disturbances.

Gait and posture are dependent on successful sensory motor integration. Gait analysis should include observation of step initiation, stride length, step height, base width, step symmetry, path deviation, trunk stability, speed, turning, and adventitious movements. Tandem toe-to-heel walking makes cerebellar, sensory, and vestibular disturbances of gait and balance more evident. Posture may be extended, flexed, or unstable. Postural stability is assessed by asking the patient to stand with legs comfortably apart and eyes open; the examiner then stands behind the patient and pulls backward on the patient's shoulders with a short, rapid jerk. Patients with extrapyramidal disorders fall backward or take a compensatory backward step to maintain balance. Sensory input to posture can be tested by asking the patient to stand with feet together with eyes closed. Patients with sensory abnormalities from peripheral neuropathy or dysfunction of the posterior columns of the spinal cord can stand with eyes open but not with eyes closed (Romberg's sign); patients with cerebellar ataxia cannot stand with feet together, regardless of whether the eyes are open or closed.

Soft signs are minor motor abnormalities that are normal early in the course of development but abnormal when they persist beyond childhood. They have been reported with increased frequency in a variety of neurosychiatric disorders. Table 3.9 lists the main soft signs elicited in the course of the neurological examination.[51,54]

TABLE 3.9. *Neurological Soft Signs*

Test	*Soft Sign*
Articulation	Mild dysarthria
Finger tapping	Slow, clumsy, irregular
Foot tapping	Slow, clumsy, irregular
Rapid alternating movements	Slow, clumsy, irregular
Eye closure	Cannot maintain on command
Tongue extrusion	Cannot maintain on command
Arms extended	Minor choreiform movements
Finger-to-nose	Jerky, clumsy
Heel-to-shin	Jerky, clumsy
Heel walking	Unsteady, posturing of upper limbs
Toe walking	Unsteady, posturing of upper limbs
Standing on one foot	Cannot balance
Hopping on one foot	Unsteady
Tandem walking	Unsteady
Tandem walking backwards	Unsteady
Face–hand test	Distal stimulus extinguished when distal and proximal stimuli delivered simultaneously
Two-point discrimination	Perceives two separate points only when more widely separated than normal
Graphesthesia	Errors
Romberg sign	Present (cannot stand with feet together and eyes closed)

Sensory Examination

Examination of sensory function includes testing of primary sensory modalities mediated at the thalamic level as well as secondary or cortical sensation. Primary sensory modalities include light touch, pain (tested by examining the patient's ability to distinguish a sharp from a dull point), temperature, and vibration. Cortical sensory modalities include joint position sense, two-point discrimination (ability to discriminate two closely spaced points of stimulation from a single stimulation), graphesthesia (ability to identify numbers "written" on the tips of the fingers), and stereognosis (ability to recognize objects by touch when placed in the hand). Double simultaneous stimulation is a sensitive means of detecting mild unilateral neglect. Patients with neglect fail to perceive stimuli on one side of the body when they are simultaneously stimulated on both sides.

Muscle Stretch Reflexes

Muscle stretch reflexes are monosynaptic spinal cord reflexes modulated by descending connections from the cortex. Muscle stretch reflexes are diminished in muscle, peripheral nerve, and nerve root disorders and increased in upper motor neuron syndromes. Reflexes are graded as 0, absent or unobtainable; 1, decreased; 2, normal; 3, brisk with spread of the contraction response to adjacent muscles; and 4, clonus.

Pathological and Primitive Reflexes

Babinski's sign is the most important pathological reflex of the neurological examination. It is elicited by stroking the lateral aspect of the plantar surface of the foot from back to front with a semisharp object. A normal response consists of plantar flexion of the great toe; a pathological response (Babinski's sign) is characterized by dorsiflexion of the great toe with or without fanning of the other toes. Babinski's sign is produced by upper motor neuron lesions and is a fragment of a triple flexion reflex. In its fully expressed form (observed in infants and in some patients with spinal cord injuries), there is a triple flexion synergy including flexion of hip, knee, and toe removing the foot from the noxious stimulus.

The *glabella tap reflex* (Myerson's sign) is an extrapyramidal sign that consists of continued blinking with repeated tapping of the glabellar region in the mid-forehead between the eyebrows. To be abnormal, the blinking must continue after four or more glabella taps.

The *grasp reflex* refers to involuntary gripping of objects. Three types of grasp response are observed: (*1*) the *simple grasp* is a gripping response observed when the patient's thenar eminence is stimulated; (*2*) the *traction grasp* is evident when the examiner withdraws his or her hand from the patient and the patient grips and exerts a counterpull; (*3*) the *groping response* or *magnetic grasp* is elicited by the sight of the examiner's hand or by a light touch of the patient's hand and consists of the patient moving his or her hand to grasp the examiner's hand. Grasp reflexes occur in patients with advanced brain disease and with lesions of the medial frontal lobes.

The *sucking reflex* consists of sucking movements of the lips, tongue, and jaw elicited by gentle stimulation of the patient's lips. Like the grasp reflex, the suck response is a primitive motor program that occurs normally in infants and reappears as an age-inappropriate sign in patients with frontal lobe and diffuse brain dysfunction. In some cases, a suck response can be elicited by the appearance of an object to be placed in the mouth or by approaching the patient's mouth with a reflex hammer handle.

The *palmomental reflex* consists of ipsilateral contraction of the mentalis muscle of the chin in response to stroking of the thenar eminence of the hand. This reflex can be seen in normal aged individuals and may be regarded as pathological only when it is unilateral or when it does not fatigue with repeated elicitation.

Selected Aspects of the Physical Examination

Selected aspects of the general physical examination are particularly important in neuropsychiatric assessment. Examination of the head is essential. Patients may have enlarged heads from compensated hydrocephalus or pathologically enlarged brains (adult males should have a head circumference <58.4 cm and adults females, <57.5 cm) or small heads from inadequate brain development. Palpation of the head may indicate a previous craniotomy not revealed or remembered by the patient. Cardiac and carotid auscultation as well as blood pressure measurement are necessary in most examinations to aid in assessing the risk of cerebrovascular disease. Dysmorphological features may lead to the recognition of congenital and chromosomal syndromes.

Neuropsychological Assessment, Laboratory Tests, and Neuroimaging

Neuropsychological testing, laboratory test, and neuroimaging complement a thorough clinical assessment. They allow the clinician to gather additional information relevant to diagnostic hypotheses formulated in the course of the exam. Neuropsychological testing will aid in the identification and characterization of intellectual deficits and provide quantitative data that can be followed during the course of the patient's illness. Projective testing may reveal information concerning the patient's mood and thought processes, and personality testing yields information regarding the patient's personality functioning.

Laboratory tests, including EEG, neuroendocrinologic testing, and tests for systemic illness and intoxications will be useful in specific circumstances. Recent developments in structural and functional brain imaging have provided another dimension of evaluation to neuropsychiatric disorders. Often structural brain imaging studies of patients with neuropsychiatric disorders are normal or only mildly abnormal with nonspecific changes even on high-resolution magnetic resonance imaging (MRI) studies. Functional imaging evaluation however can show gross abnormalities in the absence of significant structural lesions. The application of functional studies such as positron emission tomography (PET) or single-photon emission computed tomography (SPECT) is not yet routine in the workup of patients presenting with neuropsychiatric complaints because the diagnostic sensitivity and specificity of prospectively established regional brain abnormalities have not been established for many neuropsychiatric disorders. Perhaps the only significant advance in the

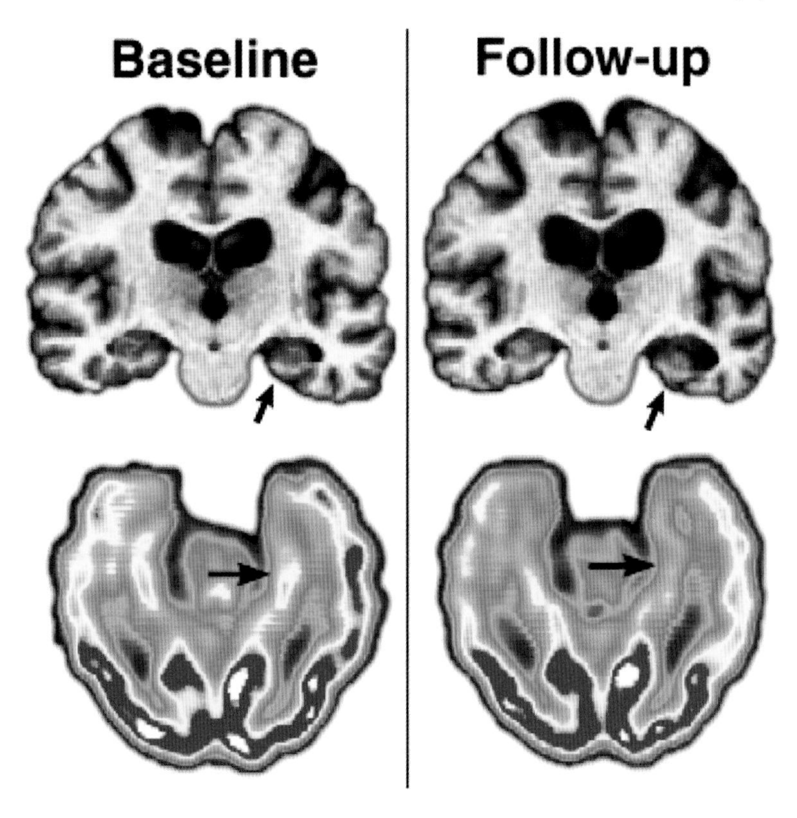

FIGURE 3.2 Coronal three-dimensional magnetic resonance images (MRI) and oblique axial slices through the long axis of the hippocampus of [^{18}F]-fluorodeoxyglucose positron emission tomography (FDG-PET) at baseline evaluation and at follow-up 18 months later showing progressive atrophy and metabolic decline in left hippocampus (arrows) for a patient presenting with mild memory impairment who later met clinical criteria for Alzheimer's disease.

use of functional imaging to diagnosis preclinical disease is within the dementia syndrome. The combination of medial temporal atrophy on structural imaging and functional defects in the parietal and temporal regions on PET or SPECT in patients with isolated memory complaints not sufficient to meet clinical criteria for dementia may predate the development of clinically diagnosed Alzheimer's disease by several years (Fig. 3.2).[56] Yet even in the evaluation of dementia few functional imaging studies have yet to prospectively test the diagnostic accuracy of specific regional deficits in individuals with mild complaints followed prospectively. Throughout the following chapters we will provide the current understanding of those regional defects on functional or structural imaging studies that best correlate to behavioral abnormalities but emphasis should always be placed on the clinical examination described above.

REFERENCES

1. Strub RL, Black FW. The Mental Status Examination in Neurology. Philadelphia: F.A. Davis, 1993.
2. Taylor MA. The Neuropsychiatric Guide to Modern Everyday Psychiatry. New York: The Free Press, 1993.
3. Trzepacz PT, Baker RW. The Psychiatric Mental Status Examination. New York: Oxford University Press, 1993.
4. Leff JP, Isaacs AD. Psychiatric Examination in Clinical Practice. London: Blackwell Scientific Publications, 1990.
5. Trimble MR, Grant I. Psychiatric aspects of multiple sclerosis. In: Benson DF, Blumer D, eds. Psychiatric Aspects of Neurologic Diseases. New York: Grune & Stratton, 1982:279–298.
6. Altshuler LL, Cummings JL, Mills MJ. Mutism: review, differential diagnosis, and report of 22 cases. Am J Psychiatry 1986; 143:1408–1414.
7. Cummings JL, Benson DF. Dementia: A Clinical Approach. Boston: Butterworth-Heinemann, 1992.
8. Cancelliere AEB, Kertesz A. Lesion localization in acquired deficits of emotional expression and comprehension. Brain Cogn 1990;13:133–147.
9. Ross ED. The aprosodias. Arch Neurol 1981;38:561–569.
10. Benson DF. Aphasia, Alexia, and Agraphia. New York: Churchill Livingston, 1979.
11. Damasio AR. Aphasia. New Engl J Med 1992;326:531–539.
12. Andreasen NC. Thought, language, and communication disorders, 1: Clinical assessment, definition of terms, and evaluation of their reliability. Arch Gen Psychiatry 1979;36:1315–1321.
13. Solovay MR, Shenton ME, Holzman PS. Comparative studies of thought disorders, I: Mania and schizophrenia. Arch Gen Psychiatry 1987;44:13–20.

14. Berlyne N. Confabulation. Br J Psychiatry 1972;120:31–39.
15. Shapiro BE, Alexander MP, et al. Mechanisms of confabulation. Neurology 1981;31:1070–1076.
16. Kopelman MD. Two types of confabulation. J Neurol Neurosurg Psychiatry 1987;50:1482–1487.
17. Whitlock FA. The Ganser syndrome. Br J Psychiatry 1967;113: 19–29.
18. Assad G, Shapiro B. Hallucinations: theoretical and clinical overview. Am J Psychiatry 1986;143:1088–1097.
19. Cummings JL, Miller BL. Visual hallucinations: clinical occurrence and use in differential diagnosis. West J Med 1987;146:46–51.
20. Cummings JL, Landis T, Benson DF. Environmental disorientation: clinical and radiologic findings. Neurology 1983b;33:103–104.
21. Bauer RM. Agnosia: visuoperceptual, visuospatial, and visuoconstructive disorders. In: Heilman KM, Valenstein E, eds. Clinical Neuropsychology, 3rd ed. New York: Oxford University Press, 1993:215–278.
22. Benton A, Tranel D. Visuoperceptual, visuospatial, and visuoconstructive disorders. In: Heilman KM, Valenstein E, eds. Clinical Neuropsychology, 3rd ed. New York: Oxford University Press, 1993:165–213.
23. Heilman KM, Watson RT, Valenstein E. Neglect and related disorders. In: Heilman KM, Valenstein E, eds. Clinical Neuropsychology, 3rd ed. New York: Oxford University Press, 1993:279–336.
24. McCarthy RA, Warrington EK. Cognitive Neuropsychology: A Clinical Introduction. New York: Academic Press, 1990.
25. Mesulam M-M. A cortical network for directed attention and unilateral neglect. Ann Neurol 1981;10:309–325.
26. Plum F, Posner JB. The Diagnosis of Stupor and Coma. Philadelphia: F. A. Davis, 1982.
27. Albert ML. A simple test of visual neglect. Neurology 1973;23: 658–664.
28. Ross ED, Mesulam M-M. Dominant language functions of the right hemisphere? Prosody and emotional gesturing. Arch Neurol 1979;36:144–148.
29. Goodglass H, Kaplan E. The Assessment of Aphasia and Related Disorders. Philadelphia: Lea and Febiger, 1985.
30. Benson DF. Neurologic correlates of anomia. In: Whitaker H, Whitaker HA, eds. Studies in Neurolinguistics. New York: Academic Press, 1979:293–328.
31. Cummings JL, Hebben NA, et al. Nonaphasic misnaming and other neurobehavioral features of an unusual toxic encephalopathy: case study. Cortex 1980;16:315–323.
32. Benson DF, Cummings JL. Agraphia. In: Frederick JAM (ed). Clinical Neuropsychology. Vol 45. Handbook of Clinical Neurology. New York: Elsevier, 1985, pp. 457–472.
33. Heilman KM, Bowers D, et al. Comprehension of affective and nonaffective prosody. Neurology 1984;34:917–921.
34. Tucker DM, Watson RT, Heilman KM. Discrimination and evocation of affectively intoned speech in patients with right parietal disease. Neurology 1977;27:947–950.
35. Borchgrevink HM. Prosody and musical rhythm are controlled by the speech hemisphere. In: Clynes M, ed. Music, Mind, and Brain. London: Plenum Press, 1982:151–157.
36. Cummings JL. Amnesia and memory disturbances in neurologic disorders. In: Oldham JM, Riba MB, Tasman A, eds. Review of Psychiatry. Washington, DC: American Psychiatric Press, 1993:725–745.
37. Holland AL, Fromm D, Greenhouse JB. Characteristics of recovery of drawing ability in left and right brain-damaged patients. Brain Cogn 1988;7:16–30.
38. Albert MS, Kaplan E. Organic implications of neuropsychological deficits in the elderly. In: Poon LW, Foard JL, Cermak LS, Arenberg D, Thompson LW, eds. New Directions in Memory and Aging. Hillsdale, NJ: Lawrence Erlbaum Associates, 1980: 403–432.
39. Carlesimo GA, Fadda L, Caltagirone C. Basic mechanisms of constructional apraxia in unilateral brain-damaged patients: role of visuo-perceptual and executive disorders. J Clin Exp Neuropsychol 1993;15:342–358.
40. Luria AR. Higher Cortical Functions in Man. New York: Basic Books, 1980.
41. Freedman M, Leach L, et al. Clock Drawing: A Neuropsychological Analysis. New York: Oxford University Press, 1994.
42. Watson YI, Arfken CL, Birge SJ. Clock completion: an objective screening test for dementia. J Am Geriatr Soc 1993;41:1235–1240.
43. Levin HS, Goldstein FC, Spiers PA. Alcalculia. In: Heilman KM, Valenstein E, eds. Clinical Neuropsychology, 3rd ed. New York: Oxford University Press, 1993:91–122.
44. Geschwind N. The apraxias: neural mechanisms of disorders of learned movement. Am Sci 1975;63:188–195.
45. Cummings JL. Frontal–subcortical circuits and human behavior. Arch Neurol 1993;50:873–880.
46. Malloy PF, Richardson ED. Assessment of frontal lobe functions. J Neuropsychiatry Clin Neurosci 1994;6:399–410.
47. Stuss DT, Benson DF. The Frontal Lobes. New York: Raven Press, 1986.
48. Gerstmann J. Syndrome of finger agnosia, disorientation for right and left, agraphia, and alcalculia. Arch Neurol Psychiatry 1940; 44:398–408.
49. Roeltgen DP, Sevush S, Heilman KM. Pure Gerstmann's syndrome from a focal lesion. Arch Neurol 1983;40:46–47.
50. Kinsbourne M, Warrington EK. A study of finger agnosia. Brain 1962;85:47–66.
51. Cummings JL. Neuropsychiatry: clinical assessment and approach to diagnosis. In: Kaplan HI, Sadock BJ, eds. Comprehensive Textbook of Psychiatry VI. Baltimore: Williams and Wilkins, 1995:167–187.
52. DeJong RN. The Neurologic Examination. New York: Harper and Row, 1979.
53. Besson G, Bogousslavsky J, et al. Acute pseudobulbar palsy and suprabulbar palsy. Arch Neurol 1991;48:501–507.
54. Stokman CJ, Shafer SQ, et al. Assessment of neurological 'soft signs' in adolescents: reliability studies. Dev Med Child Neurol 1986;28:428–439.
55. Hall RCW, Beresford TP, et al. The medical care of psychiatric patients. Hosp Commun Psychiatry 1982;33:25–34.
56. Mega MS, Thompson PM, et al. Neuroimaging in dementia. In: Mazziotta JC, Toga AW, Frackowiak R, eds. Brain Mapping: The Disorders. San Diego: Academic Press, 2000:217–293.

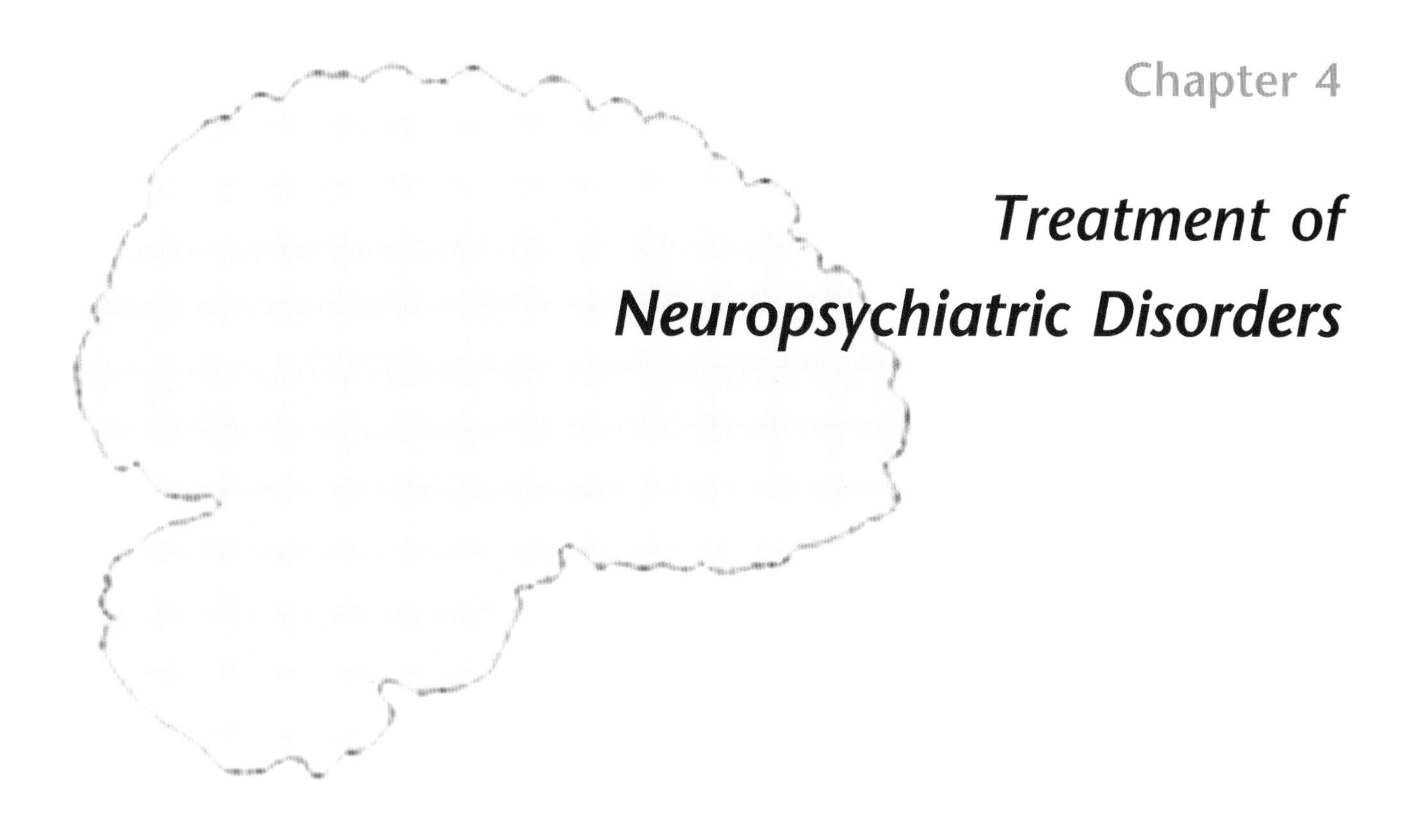

Chapter 4

Treatment of Neuropsychiatric Disorders

Neuropsychiatric disorders are generally underrecognized and undertreated, thus already compromised patients carry the added burden of disabling depression, psychosis, anxiety, or other symptoms. Many neuropsychiatric disorders are treatable with conventional psychotropic agents or with drug therapy specific for individual neurologic disorders. Although patients with neuropsychiatric disorders are more likely to develop side effects from drug therapy, this should not dissuade the practitioner from attempting therapy with due caution and a commitment to close surveillance. An alliance with the patient and the patient's family is critical to the success of any therapeutic program to achieve compliance and report both the effect sought and any adverse events. Each treatment episode must be considered both an intervention and a probe; each patient is likely to respond to a specific agent with a unique combination of responses and side effects, and these serve to inform the alert clinician's diagnostic and therapeutic impressions.

Treatment approaches in neuropsychiatry are too numerous and complex to be presented in detail in a single chapter; therapies unique to specific disorders are presented in the chapter in which that disorder is discussed (e.g., epilepsy, Parkinson's disease, Alzheimer's disease, etc.). Discussed here are the major classes of agents and their neuropsychiatric side effects, as well as treatment of those neuropsychiatric syndromes, such as depression, psychosis, mania, and anxiety, that occur in several conditions. In addition, some agents such as anticonvulsants are used in a variety of nonepileptic conditions and the range of applications of these agents is noted here. The unique position of cholinomimetic agents as examples of disease-specific psychotropic agents is also addressed. Treatment in neuropsychiatry is not confined to pharmacotherapy; electroconvulsive therapy, psychotherapy, neurosurgery, and rehabilitative therapies also have important roles in the management of some patients; the place of these therapies in a comprehensive management strategy is described briefly. Finally, some general principles of treatment planning and patient management are summarized. Specific sources regarding drug pharmacokinetics, interactions, teratogenicity, side effects, and doses should be consulted prior to prescribing any of the agents discussed in this chapter.

PRINCIPLES OF PHARMACOTHERAPY

Neurobiological Effects of Pharmacologic Agents

Pharmacotherapy in neuropsychiatry represents the attempt to use chemicals derived from external sources to influence the function of the nervous system and reduce aberrant behaviors (e.g., agitation), limit abnormal experiences (e.g., delusion, hallucinations), or enhance deficient activities (e.g., improve motivation to overcome apathy, enhance memory). Neurotransmitters and synaptic function are the principal targets of these interventions.[1] Administration of drugs that affect neurotransmitter synthesis, release, reuptake, or degradation as well as interaction with transmitter receptors with agonists or antagonists is the most common strategy for pharmacotherapy in neuropsychiatry (Table 4.1). Most drugs influence the modulating transmitters (dopamine, norepinephrine, serotonin, acetylcholine), whereas only a few affect the fast-acting excitatory and inhibitory amino acid transmitters (γ-aminobutyric acid [GABA] glutamate, aspartate). Excitatory and inhibitory amino acid transmitters generally have ligand-gated ion-dependent receptors, whereas modulatory transmitters use G protein–linked receptors that couple with second messengers such as cyclic AMP (cAMP) or cyclic GMP (cGMP).[2] Both types of receptors convert the drug–receptor interactions into a neural signal that produces the drug effect.

Neuropsychiatric symptoms can sometimes be reduced through disease-specific interventions in which the pharmacologic targets are epileptic discharges, infectious agents, inflammatory processes, neoplastic cells, or neurodegenerative mechanisms. More often, neuropsychiatric symptoms are ameliorated through the use of syndrome-specific rather than disease-specific psychotropic medications.

TABLE 4.1. *Effects on Neurotransmitters of Agents Commonly Used in Treatment of Neuropsychiatric Disorders*

Transmitter Effect	*Transmitter*	*Agents or Class of Agents*
Synthesis	Dopamine	Levodopa
	Serotonin	Tryptophan
	Acetylcholine	Choline, lecithin
Release (facilitation)	Dopamine	Amantadine, amphetamine
Release (inhibition)	Norepinephrine	Clonidine
	Glutamate	Lamotrigine
Reuptake inhibition	Serotonin	Fluoxetine, sertraline, citalopram, fluvoxamine, paroxetine, venlafaxine, nefazodone, tricyclic antidepressants
	Norepinephrine	Venlafaxine, nefazodone, maprotiline, nomifensine, tricyclic antidepressants, roboxetine
Degradation	Dopamine	Selegiline (MAO-B inhibitor) talcapone, entacapone (COMT inhibitors)
	Acetylcholine	Tacrine, donepezil, rivastigmine, galantamine
	Norepinephrine, serotonin	Phenelzine, pargyline (MAO-A,B inhibitors)
	Norepinephrine, serotonin	Moclobemide (MAO-A inhibitor)
	GABA	Vigabatrin (GABA transaminase inhibitor)
Depletion	Dopamine, norepinephrine	Reserpine, tetrabenazine
Receptor agonists	Dopamine	Bromocriptine, pergolide, ropinerole, pramipexole
	Acetylcholine	Milameline, xanomeline
	Adrenaline	Dextroamphetamine, clonidine
Receptor antagonists	Dopamine	Neuroleptic agents
	Adrenaline	Propranolol, yohimbine
Receptor facilitation	GABA	Benzodiazepines, barbiturates, alcohol

GABA, γ-aminobutyric acid; MAO, monoamine oxidase.

Drugs exert their principal effects through interactions with specific receptors. Receptors have differential distributions in the central nervous system (CNS), thus resulting in regional effects on a chemically defined brain anatomy. The behavioral specificity of drug effects is largely ascribable to the regional distribution of the receptors affected.

Pharmacokinetics and Pharmacodynamics

The *pharmacokinetic* profile of a drug relates to the relationship between drug dosage and its resulting plasma concentration and is influenced by the agent's absorption, distribution, metabolism, and elimination. After oral administration, a substantial amount of absorbed drug enters the portal circulation of the liver and is deactivated before entering the circulatory system. This is known as *first-pass elimination*. The amount of drug absorbed from the gastrointestinal tract and reaching the main cardiovascular system defines its *bioavailability*. After absorption, drug is distributed between a central compartment consisting of the circulation, brain, heart, lungs, and liver and a peripheral compartment comprised largely of lipophilic tissue reservoirs.[3] Binding to serum proteins influences the tissue distribution of the drug, and competitive binding is one of the major sources of drug–drug interactions. Drug metabolism occurs primarily through the liver; hydroxylation, desmethylation, oxidation, and deamination followed by conjugation and glucuronidation are the major forms of metabolic transformation. Oxidation occurs predominantly via the cytochrome P450 system with its many isoenzymes that are the substrates for a major set of drug–drug interactions.[4] Drug elimination is done principally through the kidney. The rate of elimination of a drug, known as its *clearance*,[5] usually depends on a combination of hepatic metabolism and renal excretion.

Serum levels of a drug are measures of the total serum concentration both bound and unbound; the amount of free drug in serum will be influenced by disorders affecting serum proteins. Serum levels provide pharmacokinetic but not pharmacodynamic (described below) information and are most useful when a *therapeutic window* relating serum concentration to therapeutic effect has been determined. Serum levels may also provide insight into compliance. The *steady state* of a drug is reached when the plasma concentration is constant as long as the dose and dosing regimen are unchanged.[3] Steady state is generally reached after five half-lives of a drug. A *half-life* is defined as the time required for the drug to obtain 50% of the steady-state level or to decay 50% from the steady-state concentration after drug discontinuation.[5]

Pharmacodynamics concern the relationship between plasma concentration and the agent's physiologic effects.[1] Pharmacodynamic effects are mediated by receptors that exhibit saturability (there are a finite number of receptors), specificity (receptors selectively interact with some compunds and not others), and reversibility (some interactions are irreversible, others are reversible).[2] Receptors have different affinities for compounds; high-affinity receptors interact at low doses and low-affinity receptors require higher doses to achieve an effect. A *dose–response curve* describes the relationship of the effect of a drug to its concentration. The *potency* of an agent is the dose or concentration required to produce an effect.[2]

Special Issues in Neuropsychiatry

There are many special issues that arise when prescribing for patients with neuropsychiatric disorders. One of the most problematic is the dearth of available information regarding treatment efficacy in patients with CNS disorders. There are few randomized clinical trials of the use of psychotropic agents in neuropsychiatric conditions. Changes in receptor number, sensitivity, and affinity that might impact efficacy can be anticipated, but little is known about pharmacodynamic relationships in these patients. The clinician must extrapolate from existing information derived from psychiatrically ill patients without brain injury and proceed cautiously.

In neuropsychiatry, drug–disease, drug–drug, and drug–individual interactions all have substantial importance. Some drugs, for example, have disease-related side effects, such as the sensitivity to anticholinergics exhibited by patients with cholinergic deficiency states (i.e., Alzheimer's disease) or the neuroleptic sensitivity shown by patients with parkinsonian syndromes. Patients with stroke, trauma, or other forms of focal cortical injury may have lowered seizure thresholds and an increased vulnerability to drug-induced seizures. Drug–drug interactions are important because in many cases both the underlying disease and the accompanying behavior will require psychopharmacologic treatment. This requires a complex neuropsychopharmacology with many potential drug interactions. Drug–individual interactions are also common in neuropsychiatry because neuropsychiatric disorders are most common in special populations such as children or the elderly.

Although polypharmacy is generally avoided to minimize the opportunity for drug–drug interactions, it is often necessary in neuropsychiatry. Many patients require both treatment for an underlying neurological disorder (i.e., epilepsy) and concomitant neuropsychiatric manifestations of the condition (i.e., interictal psychosis or depression). In Alzheimer's disease, drugs that improve cognitive function, such as anticholinesterase agents, do not slow the progress of the disease and those that slow progression, such as antioxidants (i.e., vitamin E, selegiline), do not improve existing symptoms. To optimize treatment, both types of agents must be used together, sometimes in conjunction with a psychotropic agent. Several neuropsychiatric conditions may coexist and require combined therapy; treatment of depression with psychosis, for example, usually requires using both an antidepressant and an antipsychotic agent. Thus, the dictum to avoid polypharmacy often cannot be adhered to in neuropsychiatry. This imposes an obligation on the clinician to be aware of potential drug–drug interactions when using multiple agents simultaneously.

Drug–individual interactions must be monitored in neuropsychiatry. Children, geriatric patients, pregnant women, and women who are nursing may require adjustment of treatment approaches when pharmacotherapy is considered. Dosage levels must be reduced in children and older patients. Children have proportionately less body fat and smaller volumes of drug distribution of fat-soluble drugs. Hepatic activity is higher is children than adults, receding to adult levels during puberty. This results in faster drug metabolism and lower serum levels in prepubertal children.[2]

Geriatric patients show more variability in pharmacokinetics and pharmacodynamics than younger individuals and require cautious dosing regimens (dosage should start low and go slow, but not be stopped until the desired effect occurs or side effects limit further dosage increases). With aging, body water and lean body mass are reduced and body fat increases, resulting in higher levels of water-soluble drugs and lower levels of fat-soluble drugs for any specific dose administered.[6] Hepatic and renal blood flow are reduced in the course of aging, often resulting in diminished drug metabolism and excretion. Some older individuals, however, metabolize and excrete drugs at rates comparable to those in younger patients and dosage regimens must be individualized.

Pregnant and lactating women comprise another critically important population in neuropsychiatry. Information on use of pharmacotherapy in this group is lacking for many drugs, but teratogenicity is increased, albeit to only low levels, with first-trimester exposure to low-potency phenothiazines, lithium, anticonvulsants, anticholinergic agents, and benzodiazepines. Antiepilpetic drugs cause neural tube defects such as spina bifida as well as cardiac abnormalities, cleft lip and palate, and microcephaly.[7] Polypharmacy is more freqently associated with congenital malformations than monotherapy. Organ dysgenesis has not been reported to be increased with in utero exposure to tricyclic antidepressants.[8,9]

Neuroleptics and antidepressants appear in breast milk at approximately the same levels as those in maternal serum; benzodiazepines, by contrast, have low milk–to–maternal serum ratios. Scheduling feeding times to coincide with lowest maternal serum levels will help minimize the exposure of the neonate to psychotropic medications.[9,10] Valproate therapy, particularly when administered to women under age 20, is associated with an increased frequency of polycystic ovaries, hyperandrogenism, and obesity.[11]

Side-Effect Monitoring

Adverse events are common in the course of pharmacotherapy and are a major source of morbidity for neuropsychiatric patients. In some cases, drugs produce only minor inconveniences such as constipation or dry mouth whereas in others, orthostatic hypotension in elderly individuals may lead to falls and hip fractures with substantial morbidity or even mortality. Neuroleptic agents have particularly profound effects on brain function, leading to acute dystonic reactions, parkinsonism, and tardive dyskinesia. These are discussed in detail in Chapter 20. The clinician should be familiar with the range of side effects associated with prescribed medications and should review these with the patient and caregiver when the drug is prescribed and on return visits. In addition, symptoms that emerge in concert with initiating a new drug or increasing the dose must be considered potential drug-induced side effects and dosage adjustments should be made accordingly. The principal side effects of the major classes of agents used in neuropsychiatry are described below.

Therapeutic Alliance

Drugs can be effective in relieving neuropsychiatric symptoms only if the patient takes them. The rate of compliance with medication regimens is low even among well-intentioned patients. To optimize proper medication use, the patient and the caregiver must enter a therapeutic alliance with the clinician. This alliance is complex and evolves as the patient and physician collaborate over time. It begins with education;

the physician must inform the patient and caregiver about the potential treatability of symptoms, the drugs available for use, the side effects, and the dose and regimen. In some cases, treatment will be declined, and this must be respected except when treatment is compelled by the law (e.g., when patients are dangerous to themselves or others without treatment). If treatment is initiated, the maximal effect of the drug is pursued and side effects must be monitored. If the treatment fails to accomplish its goals or intolerable side effects emerge, then treatment must be re-evaluated and a second agent chosen if the patient and caregiver desire to continue with attempts at treatment. The patient and physician will learn to trust each other and the therapeutic alliance should mature in the course of this process. More difficult decisions regarding driving, conservatorship, and other issues that arise in the course of caring for patients with neuropsychiatric conditions can be approached more successfully once a trusting relationship has been established.

CLASSES OF DRUGS USED IN NEUROPSYCHIATRY

Antidepressants

There are several classes of antidepressants available for use in neuropsychiatry. The classes, agents, usual doses, and common side effects of these agents are shown in Table 4.2. Antidepressant agents are used in a wide array of neuropsychiatric disorders, including depression, anxiety disorders, obsessive-compulsive disorder, syndromes with repetitive stereotyped behavior, attention deficit-hyperactivity disorder, self-injurious behavior, chronic pain, pseudobulbar palsy, insomnia, apathy, migraine, and agitation. Most of these agents exert their effect on serotonergic systems, but some agents have mixed adrenergic and serotonergic actions. Venlafaxin is a combined reuptake inhibitor affecting serotonin and norepinephrine; roboxetine is a specific noradrenergic reuptake blocker. Tricyclic antidepressants have anticholinergic side effects and may exacerbate confusion and memory loss in neuropsychiatric patients; agents with the least anticholinergic activity (Table 4.2) should be used in this population.

Lithium and Antimanic Agents

Lithium is the most widely used drug for the treatment of mania and hypomania. Lithium, however, has a low therapeutic index with side effects appearing soon after the therapeutic serum level of lithium is exceeded. In neuropsychiatric disorders, other antimanic agents such as valproate and carbamazepine are often preferable (Table 4.3). In addition to their utility in the treatment of mania, these agents can be effective in the therapy of self-injurious behavior, irritability, anxiety, agitation, and aggression. Polycystic ovaries, hyperandrogenism, and obesity may occur with valproate therapy, particularly when the drug is administered to women under age 20.[11]

Antipsychotic Agents

Antipsychotic agents are used for the treatment of delusional disorders as well as a wide array of other neuropsychiatric conditions. There are several classes of antipsychotic agents available; from a neuropsychiatric perspective they can be divided into those that have prominent D_2 dopamine receptor blocking effects and those that exert less marked D_2 blockade or D_2 blockade in conjunction with other receptor effects that provide antipsychotic efficacy with lower D_2 receptor occupancy. Agents producing marked effects on D_2 receptors (conventional antipsychotics, neuroleptics) at doses producing antipsychotic effects produce parkinsonism, tremor, dystonia, and akathisia in the early phases of treatment and frequently cause tardive dyskinesia with chronic treatment. Agents with relative sparing of the D_2 receptors (atypical or novel antipsychotics) are less likely to cause these effects; in addition, they may have more marked beneficial effects on negative symptoms frequently present in psychotic states, such as apathy, anhedonia, and asociality.

Conventional antipsychotics include phenothiazines (i.e., chlorpromazine, thioridazine, fluphenazine), butyrophenones (i.e, haloperidol), thioxanthenes (thiothixene), dihydroindolones (molindone), and dibenzoxazepines (i.e., loxapine) (Table 4.4).[12] Lower-potency phenothiazines such as thioridazine produce fewer extrapyramidal side effects and have more anticholinergic side effects, cause greater sedation, and are more likely to produce orthostatic hypotension than high-potency agents such as fluphenazine. Haloperidol has a side effect profile similar to that of the high-potency phenothiazines.

Antipsychotic agents that have relatively less D_2 blocking activity or less of a tendency to induce extrapyramidal effects include risperidone, clozapine, olanzapine, quetiapine, and ziprasidone (Table 4.4). Risperidone is a benzisoxazole agent whose antipsychotic properties depend on a combination of D_2 and 5-HT_2 blockade. A lower dose of drug is required for

TABLE 4.2. *Classes, Agents, Usual Doses, and Common Side Effects of Antidepressants*[2,69–71]

Class *Agent (Brand Name)*	*Usual Dosage Range (mg)*	*Elimination Half-life (hr)*	*Common Side Effects*
Selective serotonin reuptake inhibitors			
Fluoxetine (Prozac)	10–80	24–330	Insomnia, anxiety, amphetamine-like effects, tremor, orthostatic hypotension, gastrointestinal distress, sexual function disturbances
Sertraline (Zoloft)	50–200	24–30	Similar to above
Paroxetine (Paxil)	10–60	3–65	Similar to above
Citalopram (Celexa)			Similar to above
Fluvoxamine (Luvox)	50–300	17–22	Similar to above
Combined reuptake inhibitors			
Nefazodone (Serzone)	100–600	2–5	Dry mouth, constipation, drowsiness, headache, fatigue, orthostatic hypotension
Venlafaxin (Effexor)	75–225	3–7 (parent; 9–13 for metabolite)	Dry mouth, constipation, sweating, drowsiness, headache, orthostatic hypotension, sexual function disturbances
Tricyclic antidepressants			
Nortriptyline (Aventyl, Pamelor)	50–200	13–88	Dry mouth, constipation, drowsiness, orthostatic hypotension, tachycardia, weight gain, nausea, headache
Desipramine (Norpramine)	75–300	12–76	Similar to above
Protriptyline (Vivactil, Triptil)	20–60	54–124	Similar to above
Clomipramine (Anafranil)	75–300	17–37	Similar to above
Monoamine oxidase inhibitors			
Phenelzine (Nardil)	45–90	1.5–4	Dry mouth, constipation, drowsiness, insomnia, excitement, tremor, orthostatic hypotension, tachycardia, weight gain, sexual disturbances
Tranylcypromine (Parnate)	20–50	2.4	Dry mouth, drowsiness, insomnia, excitement, orthostatic hypotension, tachycardia
Isocarboxazid (Marplan)	30–60		Dry mouth, headache, tremor, orhtostatic hypotension, nausea
Moclobemide (Manerix)	300–600	4–9	Dry mouth, insomnia, headache, orhtostatic hypotension, nausea
Miscellaneous agents			
Amoxapine (Asendin)	100–600	8	
Trazodone (Desyrel)	150–600	4–9	Dry mouth, drowsiness, fatigue, orhtostatic hypotension
Maprotiline (Ludiomil)	100–225	27–58	Dry mouth, constipation, tremor, weight gain
Bupropion (Wellbutrin)	75–400	10–14	Dry mouth, constipation, insomnia, excitement, headache, tremor
Mirtazapine (Remeron)	15–45	20–40	Drowsiness, weight gain, dizziness, dry mouth, constipation, agranulocytosis (rare)

TABLE 4.3. *Agents, Usual Doses, and Common Side Effects of Drugs Used to Treat Mania*[2,69–71]

Agent (Brand Name)	*Usual Dosage Range (mg)*	*Elimination Half-life (hr)*	*Common Side Effects*
Lithium (Eskalith, Lithonate, Lithobid)	300–2400	8–35	Nausea, vomiting, diarrhea, weight gain, tremor, poyluria and polydipsia
Carbamazepine (Tegretol)	300–1600	12–17	Dizziness, ataxia, drowsiness, tremor, chorea, rare leukopenia
Valproic acid (Depakote)	750–6000	9–20	Weight gain, hyperammonemia, drowsiness, tremor, alopecia, rare hepatic toxicity
Clonazepam (Klonopin, Rivotril)	1–6	20–40	Ataxia, drowsiness, behavioral disinhibition, sexual dysfunction
Lamotrigine (Lamictal)	100–400* 300–500	15–24	Dizziness, ataxia, blurred vision, vomiting, rash (may be severe)
Gabapentin (Neurontin)	900–4800	5–7	Somnolence, fatigue, dyspepsia, dizziness, tremor, diplopia
Topiramate (Topamax)	200–600	18–30	Somnolence, dizziness, memory and speech difficulties

*100–400 when administered with valproic acid; 300–500 when administered without valproic acid.

TABLE 4.4. *Agents, Usual Doses, and Common Side Effects of Drugs Used in Management of Psychosis*[2,69–71]

Agent (Brand Name)	*Usual Dosage Range (mg)**	*Elimination Half-life (hr)*	*Common Side Effects*
Atypical antipsychotics			
Clozapine (Clozaril)	12.5–100 (900)	5–16	Drowsiness, orthostatic hypotension, tachycardia, anticholinergic effects, weight gain, sialorrhea
Risperidone (Risperdal)	0.5–6 (12)	20–24	Drowsiness, parkinsonism, akathisia, orthostatic hypotension, tachycardia, weight gain
Olanzapine (Zyprexa)	5–20 (200)		Sedation, orthostatic hypotension, weight gain
Quetiapine (Seroquel)	25–400 (800)	6	Somnolence, agitation, weight gain, transient liver enzyme
Ziprasidone (Geodon)	20–80 (160)	10	Somnolence, cardiac conduction disturbances
Neuroleptics			
Fluphenazine (Prolixin, Moditen)	1–20 (40)	13–58	Parkinsonism, akathisia, dystonia, tachycardia, galactorrhea
Molindone (Moban)	25–225 (300)	6.5	Drowsiness, parkinsonism, akathisia, dystonia, anticholinergic effects
Loxapine (Loxitane, Loxapac)	30–100 (250)	8–30	Drowsiness, parkinsonism, akathisia, dystonia, orthostatic hypotension, tachycardia
Thiothixene (Navane)	4–30 (60)	30–40	Tachycardia, hypotension, sedation, parkinsonism, restlessness, hematologic effects, rash, lactation, amenorrhea, dry mouth, blurred vision, impotence
Haloperidol (Haldol)	0.5–8 (40)	12–36	Parkinsonism, akathisia, dystonia
Pimozide (Orap)	2–20 (40)	29–55	Drowsiness, parkinsonism, akathisia

*Lower doses are commonly effective in patients with neurologic disorders. Dose in parentheses refers to upper limit of range in idiopathic psychoses.

antipsychotic efficacy and fewer side effects are produced.[13] Extrapyramidal signs and symptoms emerge with higher doses of risperidone. Clozapine, olanzapine, ziprasidone and quetiapine have variable D_2 effects and are antagonists of D_1, 5-HT_2, and α_2 receptors.[14–16] Clozapine and olanzapine have marked antimuscarinic receptor effects, although their clinical anticholinergic effects are modest.[14,17] Clozapine may induce agranulocytosis and requires weekly monitoring of the white blood count; it may also produce seizures in previously seizure-free individuals and precipitate them in patients with epilepsy.[18]

Antipsychotic agents with D_2 blocking effects are useful in suppressing chorea and tics. They are also beneficial in patients with obsessive-compulsive disorder with concomitant tics. Antipsychotic agents that spare the D_2 receptors are particularly beneficial in patients with parkinsonism and psychosis, such as those with levodopa-induced psychotic episodes.[19,20] Either type of antipsychotic agent may be useful in ameliorating agitation and psychosis in patients without coexisting extrapyramidal disorders. Antipsychotics are the principal agents used to treat agitation (with or without psychosis) in patients with dementia syndromes.

Anxiolytics

Benzodiazepine anxiolytics cause a reduction in anxiety through agonist effects on GABA (subtype A) receptors.[21] These agents typically enhance the ability of this inhibitory transmitter to increase the activity of the choloride channel. Metabolism of benzodiazepines is markedly affected by age; lorazepam and oxazepam are well metabolized by elderly individuals. Benzodiazepines are used in neuropsychiatry to reduce anxiety, control seizures, ameliorate tics and chorea, control myoclonus, relieve catatonia, reduce self-injurious behavior in mentally retarded individuals, and induce sedation and sleep. Benzodiazepines may be used to treat agitation but they may increase confusion in elderly individuals and usually should be avoided if continuous administration is required.

Benzodiazepines have marked effects on sleep. They decrease sleep-onset latency and increase total sleep time; they reduce stage 1 sleep and markedly reduce or abolish stage 4 sleep; stage 2 sleep is increased; rapid eye movement (REM) sleep is decreased in the beginning of the night.[22] Benzodiazepines are helpful in the treatment of restless legs syndrome and period limb movements during sleep.[23]

Buspirone, the other major anxiolytic used in neuropsychiatric practice, exerts antianxiety effects by blocking serotonin (5-HT_{1A}) receptors.[24] Buspirone may have less confusion-inducing actions and may be useful as an antiagitation agent. Table 4.5 summarizes information about the principal anxiolytics.

Selective serotonin reuptake inhibitors (SSRIs) have anxiolytic effects and are the agents of choice for chronic anxiety syndromes.

Stimulants

Stimulants are an important part of the armamentarium of the neuropsychiatrist and are useful in a variety of neuropsychiatric conditions. The four principle agents used are dextroamphetamine, methylphenidate, pemoline, and modafinil (Table 4.6).[25] The drugs are used to treat apathy, attentional disorders in adults, hyperactivity-attention deficit disorder in children, depression, chronic fatigue syndrome, fatigue in multiple sclerosis, and fatigue in human immunodeficiency virus encephalopathy. Adverse effects of these drugs include anxiety, tremor, appetite suppression and weight loss, hypertension, and occasional psychosis. Intravenous amphetamines may cause a cerebral angiitis with intracerebral hemorrhage. Modafonil has been used pri-

TABLE 4.5. *Agents,* Usual Dosage, and Common Side Effects of Anxiolytics*[2,69–71]

Agent (Brand Name)	*Usual Dosage Range (mg)*	*Elimination Half-life (hr)*	*Common Side Effects*
Clonazepam (Klonopin, Rivotril)	1–6	20–40	Ataxia, drowsiness, behavioral disinhibition, sexual dysfunction
Lorazepam (Ativan)	0.5–10	8–24	Drowsiness, behavioral disinhibition, sexual dysfunction
Oxazepam (Serax)	30–120	3–25	Drowsiness, behavioral disinhibition, sexual dysfunction
Buspirone (BuSpar)	15–60	1–11	Dizziness, headache, nervousness, nausea

*Selective serotonin reuptake inhibitors are the agents of choice for chronic anxiety disorders (Table 5.2).

TABLE 4.6. *Psychostimulants and Drugs Used to Treat Apathy in Neuropsychiatric Disorders*[2,69–71]

Agent (Brand Name)	*Usual Dosage Range (mg)*	*Elimination Half-life (hr)*	*Common Side Effects*
Psychostimulants			
Methylphenidate (Ritalin)	5–60	1–7	Nervousness, insomnia, irritability, headache, tics, gastrointestinal distress, anorexia
Dextroamphetamine (Addorall)	5–40	6–11	Similar to above
Magnesium pemoline (Cylert)	37.5–112.5	7–12	Similar to above
Modafinil (Provigil)	200–400	15	Insomnia, agitation, nervousness
Other agents with stimulant effects			
Fluoxetine (Prozac)	10–80	24–330	Insomnia, anxiety, amphetamine-like effects, tremor, orthostatic hypotension, gastrointestinal distress, sexual function disturbances
Protriptyline (Vivactil, Triptil)	20–60	54–124	Dry mouth, constipation, drowsiness, orthostatic hypotension, tachycardia, weight gain, nausea,headache
Bromocriptine (Parlodel)	5–40	2–8	Nausea, orthostatic hypotension, dyskinesias, nightmares, hallucinations, delusions
Pergolide (Permax)	0.25–5	2–4	Nausea, orthostatic hypotension, dyskinesias, nightmares, hallucinations, delusions
Tacrine (Cognex)*	80–160	2–6	Nausea, diarrhea, anorexia, liver toxicity
Donepezil (Aricept)*	5–10	75–90	Nausea, diarrhea, anorexia
Rivastigmine (Exelon)	6–12	8	Nausea, diarrhea, anorexia, weight loss
Galantamine (Reminyl)	16–24	6	Nausea, diarrhea, anorexia

*Reduces apathy in disorders with cholinergic deficiency such as Alzheimer's disease.

marily in narcolepsy and excessive daytime sleepiness; its use in other circumstances is being explored. Pemoline requires a longer period (often several weeks) to begin its effects, and this agent also has less of a tendency to produce addiction and abuse.

Dopaminergic Agents

Dopaminergic agents include levodopa, levodopa plus carbidopa (a peripheral decarboxylase inhibitor that reduces conversion of levodopa to dopamine outside of the brain), pergolide, bromocriptine, ropinirole, pramipexole, tolcapone, entacapone, and amantadine hydrochloride (Table 4.7). Levodopa is a dopamine precursor that is converted to dopamine by cells in the substantia nigra or ventral tegmental area. Dopamine is then transported via axonal transport to receptor sites in the striatum, amygdala, nucleus accumbens, medial frontal cortex, and anterior cingulate cortex.[26] Amantidine facilitates the release and inhibits the reuptake of dopamine from the presynaptic terminal, thus increasing the amount of intrasynaptic dopamine available. Pergolide, bromocriptine, ropinirole, and pramipexole are dopamine receptor agonists that directly stimulate the postsynaptic receptors in these target structures. Ropinirole and pramipexole have relatively selective effects on dopamine D_3 receptors. These agents may have antidepressive as well as antiparkinsonian activity.[27] Tolcapone and entacapone are catechol-O-methyltransferase inhibitors that reduce the metabolism of levodopa and extend its action.[28]

Dopaminergic agents improve motor function in Parkinson's disease and a beneficial response to these agents is often considered a criterion for the diagnosis of this disease (Chapter 18). Other parkinsonian syndromes such as progressive supranuclear palsy and vascular parkinsonism may have partial responses to dopaminergic therapy. In addition, akinetic mutism may respond to dopamine receptor agonists[29] and apathetic syndromes have also been reported to improve with dopaminergic therapy.[30] Restless legs syndrome[31] and pseudobulbar palsy[32] are two additional syndromes reported to be responsive to dopaminergic therapy. Amantadine has an important role in treating the

TABLE 4.7. *Agents Used in Treatment of Parkinson's Disease and Parkinsonism*[2,69–71]

Agent (Brand Name)	*Usual Dosage Range (mg)*	*Elimination Half-life (hr)*	*Common Side Effects*
Benztropine mesylate (Cogentin)	1–6	3–24	Dry mouth, blurred vision, constipation, urinary retention, rash, sexual dysfunction, memory impairment, confusion
Trihexyphenidyl (Artane)	1–15	3–4	Similar to above
Amantadine hydrochloride (Symmetrel)	100–400	15–24	Orthostatic hypotension, livedo reticularis, ankle edema, vivid dreams, hallucinations, delusions, insomnia urinary retention, dry mouth
Levodopa (Sinemet, Larodopa)	10/100 tid to 25/250 × 2 qid	2–4	Nausea, orthostatic hypotension, dyskinesias, nightmares, hallucinations, delusions
Selegiline (Eldepryl, Deprenyl)	5–10		Insomnia, dyskinesias, nightmares, hallucinations, delusions
Bromocriptine (Parlodel)	5–40	2–8	Nausea, orthostatic hypotension, dyskinesias, nightmares, hallucinations, delusions
Pergolide (Permax)	0.25–5	2–4	Nausea, orthostatic hypotension, dyskinesias, nightmares, hallucinations, delusions
Ropinerole (Requip)	0.75–3	6	Syncope, hypotension, hallucinations, sleep attacks
Pramipexole (Mirapex)	0.375–4.5	8–12	Syncope, hypotension, hallucinations, sleep attacks
Tolcapone (Tasmar)	300–600	2–3	Syncope, hallucinations, dyskinesia, hepatotoxicity
Entacapone (Comtan)	200–1600	1–2	Syncope, hallucinations, dyskinesia

fatigue syndrome of multiple sclerosis.[33] Neuropsychiatric side effects of dopaminergic agents include nightmares, hallucinations, delusions, hedonistic homeostatic dysregulation disorder, mania, and sexual behavior changes. Chorea may occur with chronic administration. Anticholinergic agents may ameliorate tremor in parkinsonism. These agents have many adverse side effects and should be avoided in elderly patients.

Cholinomimetic Agents

Cholinesterase inhibitors are available for the treatment of diseases with cholinergic deficiencies, primarily Alzheimer's disease. Four cholinesterase inhibitors—tacrine (Cognex), donepezil (Aricept), galantamine (Reminyl), and rivastigmine (Exelon) (Table 4.8)—are currently available and others are being developed. These agents have beneficial effects on cognition where they produce modest improvement of memory and other cognitive functions.[34–38] They also ameliorate some behavioral disturbances, including psychosis, agitation, apathy, disinhibition, and aberrant motor behavior (pacing, etc.).[39,40] They delay decline in activities of daily living. Alzheimer's disease is the most common disorder with a cholinergic deficiency, but other conditions also have deficits of choline acetyltransferase and cortical

TABLE 4.8. *Cholinomimetic Agents*

Agent (Brand Name)	*Usual Dosage Range (mg)*	*Elimination Half-life (hr)*	*Common Side Effects*
Tacrine (Cognex)	80–160	2–6	Nausea, diarrhea, anorexia, liver toxicity
Donepezil (Aricept)	5–10	75–90	Nausea, diarrhea, anorexia, vomiting
Rivastigmine (Exelon)	6–12	8	Nausea, diarrhea, anorexia, vomiting, weight loss
Galantamine (Reminyl)	16–24	6	Nausea, diarrhea, anorexia, vomiting

cholinergic abnormalities including Down's syndrome, Parkinson's disease with dementia, dementia with Lewy bodies, some cases of Creutzfeldt-Jakob disease, and Guamanian dementia–parkinsonism complex.[41] Cholinersterase inhibitors may be useful in these disorders and preliminary evidence suggests that some patients with multiple sclerosis, traumatic brain injury, and bipolar mood disorders benefit from treatment with these agents. Cholinergic receptor agonists and nerve growth factors with cholinoprotectic effects are being tested for clinical utility for the treatment of cholinergic deficiency states. Side effects of cholinesterase inhibitors include gastrointestinal cramps, diarrhea, vomiting, nausea, leg cramps, and rare cases of agitation.

Anticonvulsants

Anticonvulsants are used in neuropsychiatry for a wide variety of conditions. Their use in epilepsy is described in Chapter 21. Carbamazepine, valproate, gabapentin, topiramate, and lamotrigine are the agents most used in treating neuropsychiatric syndromes. They are used in patients with mood disorders, particularly rapid-cycling and atypical bipolar disorders, and conditions with mixed mood states. Patients with mental retardation and mood disorders may respond well to anticonvulsant agents. They have proven to be of benefit in patients with episodic dyscontrol and intermittent explosive disorder. Agitation in patients with dementia often responds to treatment with these agents. Self-injurious behavior occurring in a wide variety of clinical circumstances may also improve following treatment with these compounds. Table 4.9 presents the anticonvulsants most often used in neuropsychiatric syndromes.

Drugs Used to Treat Addictions

Progress has been made in identifying agents that reduce craving or interfere in chronic addiction syndromes through other means.[42,43] Table 4.10 summarizes the principle agents used to treat chronic addiction disorders (excluding the abstinence syndromes). The likelihood of alcohol ingestion can be reduced by the use of the dopamine β-hydroxylase inhibitor, disulfiram, and naltrexone has been shown to reduce the rate of relapse in recently abstinent alcoholics.[44] Acromposate reduces alcohol-related neuronal hyperexcitability and may have a therapeutic role. Nicotine in the form of chewing gum, nasal spray, or patch offers a substitute source of the compound for those wishing to relinquish smoking. Bupropion is an antidepressant with dopaminergic and noradrenergic effects that reduces relapse in patients trying to abstain from smoking.[45] Clonidine also helps prevent relapse to nicotine use. Opioid dependence may be treated with μ-receptor agonists (methadone, naltrexone, L-α-acetylcethadyl) or μ antagonists (naltrexone).[46] Cocaine dependence is treatment resistant but may respond to desipramine or amantadine.

These agents can be considered in neuropsychiatrically ill patients with substance use disorders and may be useful in behaviors possibly related to addictions such as self-injury syndromes. Selective serotonin reuptake inhibitors and atypical antipsychotics have also reduced self-injury in some patients.

ELECTROCONVULSIVE THERAPY

Electroconvulsive therapy (ECT) offers another means of improving the function of patients with neuropsychiatric disorders. This therapy has been shown to be efficacious for depression, mania, schizoaffective disorders, and some types of schizophrenia. It should be considered when the patient has a potentially responsive condition, has proven to be refractory to other therapies, is intolerant of alternative treatments, or requires a rapid response.[47] It should also be considered in treatment of refractory cases of Parkinson's disease, intractable seizures, and catatonia.[47,48]

The use of ECT in patients with epilepsy requires special consideration since the induced seizure may precipitate status epilepticus or, conversely, the presence of anticonvulsants may make it difficult to induce a therapeutic convulsion. Given the low rate of complications, it is currently recommended that patients with epilepsy continue on a stable dose of anticonvulsants through the course of ECT. Unless they have a history of status epilepticus or of recent seizures, they should not receive their antiepileptic drugs during the day on which the ECT is administered until after the treatment is complete. The dose of anticonvulsant should be the minimum necessary to control seizures during this period. The ECT dose should be titrated to the seizure threshold at the time of the first treatment and administered at slightly above threshold for the rest of the course.[47]

Raised intracranial pressure during ECT makes the presence of a brain tumor a relative contraindication, and because of the increased blood pressure occurring with seizures, caution is required when applying the technique to individuals with cerebrovascular disease. Skull defects may increase the risk of ECT-associated cognitive impairment and placement of electrodes over the area of bone loss should be avoided.

TABLE 4.9. *Anticonvulsants*[2,69–72]

Agent (Brand Name)	*Usual Dosage Range (mg)*	*Elimination Half-life (hr)*	*Common Side Effects*
Phenobarbital (Luminal)	60–240	55–140	Drowsiness, cognitive dulling, rash, hyperactivity, depression
Primidone (Mysoline)	750–2000	8–24, primidone 10–25, phenylethyl malonamide 55–140, phenobarbital	Drowsiness, cognitive dulling, rash, hyperactivity, depression, sexual dysfunction, hematopoetic effects
Phenytoin (Dilantin)	300–600	6–42	Nystagmus, ataxia, drowsiness, rash, gingival hyperplasia, hirsuitism, chorea, lupus-like syndrome, lymphadenopathy, peripheral neuropathy
Gabapentin (Neurontin)	900–4800	5–7	Somnolence, dizziness, ataxia, fatigue, nystagmus, tremor, diplopia
Lamotrigine (Lamictal)	300–500	12–62 (25 usual)	Diplopia, drowsiness, dizziness, ataxia, headache, nausea
Vigabatrin (Sabril)	1500–3000	5–7	Drowsiness, dizziness, headache, ataxia, irritability, depression, psychosis, weight gain
Carbamazepine (Tegretol)	400–2400	16–24	Dizziness, ataxia, drowsiness, tremor, chorea, rare leukopenia
Oxcarbazepine (Trileptal)	1200–3000	8–10	Dizziness, fatigue, headache, ataxia, hyponatremia, transient liver enzyme elevations, rash
Valproic acid (Depakote)	750–4000	9–20	Weight gain, hyperammonemia, drowsiness, tremor, alopecia, rare hepatic toxicity
Topiramate (Topamax)	200–600	18–30	Asthenia, dizziness, nausea, weight loss, somnolence, psychomotor slowing, decreased concentration, nervousness
Clonazepam (Klonopin, Rivotril)	2–20	20–40	Ataxia, drowsiness, behavioral disinhibition, sexual dysfunction
Ethosuximide (Zarontin)	500–1500	50–60	Anorexia, nausea, weight loss, gum leukopenia, eosinophilia, drowsiness, irritability, hyperactivity, ataxia, depression, rash
Levetiracetam (Keppra)	1000–3000	6–8	Somnolence, asthenia, dizziness, nervousness, headache
Tiagabine (Gabitril)	32–56	7–9	Dizziness, asthenia, tremor, irritability, decreased concentration
Zonisamide (Zonegran)	200–600	50–70	Somnolence, dizziness, anorexia, agitation, difficulty communicating, renal stones

Electroconvulsive therapy induces temporary amnesia and patients are often unable to recall part or most of the hospitalization during which the ECT was administered. There may be a slight decrement in cognitive function (up to three points on the Mini-Mental State Examination), which reverses fully when the course is complete.[49] Cognitive impairment associated with depression is relieved by ECT, and improvement in neuropsychological test performance commonly follows ECT treatment.[50] Memory impairment in patients with dementia may be temporarily exaggerated following ECT; those with preexisting cognitive compromise and those experiencing prolonged disorientation in the postictal period are most likely to manifest per-

TABLE 4.10. *Agents Used to Treat Chronic Addiction Syndromes*

Addiction	*Therapeutic Agent*	*Therapeutic Action*
Alcoholism	Disulfiram	Inhibits dopamine β-hydroxylase and causes an adversive syndrome when alcohol is ingested
	Naltrexone	Opioid antagonist
	Acromprosate	Reduces alcohol-induced neuronal hyperexcitability
Smoking	Nicotine therapy	Nicotine substitution
	Bupropion	Enhances mesolimbic dopamine levels and CNS noradrenergic levels
	Clonidine	α_2-Adrenergic agonist
Opioid dependence	Buprenorphine	Partial μ agonist
	Naltrexone	Opioid antagonist
	Methadone	μ Agonist
	L-α-acetylmethadyl	μ Agonist
Cocaine	Desipramine	Reduces norepinephrine reuptake
	Amantadine	Enhances dopamine release

sistent retrograde amnesia following ECT.[51] Careful studies with brain imaging reveal no changes in brain structure following ECT.[52,53]

PSYCHOTHERAPY

Every interaction with the patient is important to the patient. The clinician's demeanor, tone of voice, and actions are all of the greatest importance to the patient.[54] Every interaction can have therapeutic or counter therapeutic effects, allaying or increasing the concerns of the anxious patient, diminishing or exacerbating the fearfulness of the paranoid individual, enhancing or further decreasing the self-esteem of the depressed. The clinician's behavior can reassure or alarm the caregiver. The clinician must be cognizant of these effects, using them for their therapeutic impact and judging their effectiveness as part of the therapeutic alliance with the patient.

Neuropsychiatric patients must be assessed thoroughly before deciding on a psychotherapeutic approach. Patients with neurological disorders may be cognitively impaired, have anosognosia for or denial of deficits, or be unable to remember their therapeutic encounters. The impact of these conditions on the psychotherapeutic process must be anticipated before the patient in therapy. Since each neuropsychiatric patient has a unique combination of altered and retained abilities, each therapeutic plan must also be individualized. While accounting for the patient's deficits in a treatment plan, it is important to retain a commitment to maximizing the patient's autonomy and to avoid falling into the error of making decisions for the "organic" patient. Most patients have some aspects of cognition retained and personal choice should be honored within these bounds. A challenge to the therapist is to encourage patients to exercise maximum control of their life, while integrating the limitations that may be imposed by neurological illness into the discussions and patient and family expectations. The natural history of the neuropsychiatric disorder must also be considered in the treatment plan: degenerative diseases will require increasingly simple and environmentally oriented approaches, whereas the patient recovering from traumatic brain injury or stroke may be able to entertain progressively more complex interventions demanding personal insight and self-control.

Psychotherapy of the neuropsychiatric patient will likely involve an eclectic mixture of therapeutic approaches modified to fit individual circumstances. Classic insight-oriented psychotherapy is appropriate only for a few patients with minimal deficits and retained self-awareness. Among the interventions that may broadly be conceived as psychotherapy are specialized training, structuring of the environment (home or care facility) and supportive, behavioral, cognitive, and psychodynamic approaches.[55] Therapy should address three key areas—cognition, emotion, and behavior—in all neuropsychiatric patients.[55] Each of these dimensions may require a somewhat different psychotherapeutic technique, and the resulting treatment program will be an amalgamation of the best approaches to the three areas individualized for each specific patient. Educational approaches can help patients understand what has happened to them, while cognitive, supportive, and at least limited psychodynamic therapies may help the patient respond to the question, "why has this happened to me?"[56]

Psychotherapy may be integrated with traditional rehabilitation approaches including physical therapy, occupational therapy, and speech and language therapy. Most patients will require combinations of psychotherapy, rehabilitation, and the judicious use of medications to optimize treatment outcome.

Patients with neuropsychiatric disorders are often inordinately reliant on family members for their care or well-being. Many caregivers experience this dependency as a burden and a source of distress, anxiety, and depression. Assessment of the caregiver and referral for appropriate support or therapy is an essential part of the therapeutic process. In some cases, engagement of the patient and caregiver or family members in family therapy may be appropriate. Couples or families with interpersonal problems prior to the onset of the patient's neuropsychiatric disorder are more likely to have significant difficulty responding to the patient's needs than those with good premorbid function and interactions. Many disease-oriented lay groups provide educational information and seminars, family support groups, and services for patients. These groups are active advocates for neuropsychiatric patients.

NEUROSURGERY FOR NEUROPSYCHIATRIC DISORDERS

Neurosurgical treatments are not commonly used to relieve neuropsychiatric symptoms but should be considered for patients who have syndromes known to improve following surgery and who have proven to be unresponsive to other types of treatment. The patients must have failed several types of pharmacologic therapy as well as combined pharmacotherapy and psychotherapy before neurosurgery is considered as an alternative treatment.

All types of neurosurgery in which the brain is surgically manipulated may be accompanied by behavioral changes and should be regarded as psychosurgery. In some cases, however, behavioral changes are the principal reason for the surgical intervention. Conditions currently considered potential candidates for behavioral neurosurgery and the anatomic targets are listed in Table 4.11.

Pain and medically intractable obsessive-compulsive disorder and depression are the current principal con-

TABLE 4.11. *Conditions Reported to Have Responded to Behavioral Neurosurgical Procedures and Anatomic Target Affected*

Neuropsychiatric Condition	*Type of Surgery*	*Surgical Target*
Depression	Leucotomy	Projections from orbitomedial frontal lobe
	Vagus nerve stimulation	Vagus nerve nucleus
	Subcaudate tractotomy	Undercutting the orbital cortex
Obsessive-compulsive disorder	Anterior capsulotomy	Projections from thalamus to the frontal lobes
	Cingulumotomy	Cingulate fibers
Anorexia nervosa	Orbitomedial tractotomy	Projections from the orbitomedial frontal lobe
Anxiety	Cingulumotomy	Cingulate fibers or connections between the cingulum and contralateral basal ganglia
	Anterior capsulotomy	Projections from thalamus to the frontal lobes
Pain	Cingulotractotomy	Cingulate fiber projections
Addiction	Cingulotractotomy	Cingulate fiber projections
Violence/rage	Cingulumotomy	Cingulate fibers
	Obitomedial tractotomy	Projections from the orbitomedial frontal lobe
	Amygdalotomy	Amygdalae bilaterally
Sexual aggression/pedophilia	Posterior hypothalamotomy	Posteromedial hypothalamic nuclei

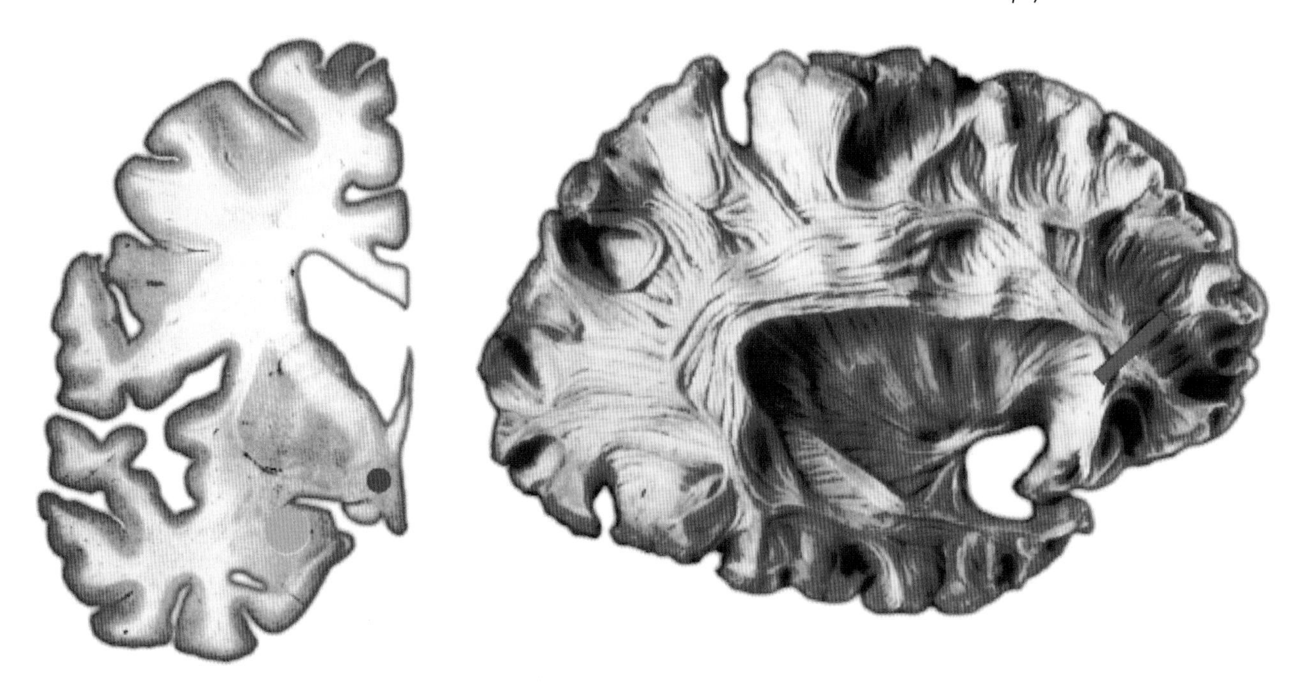

FIGURE 4.1 Locations of cingulumotomy for depression (blue, right), amygdalatomy for violent behavior (green, left), and hypothalamotomy for sexual aggression (red, left).

ditions in which psychosurgery is applied. Pain and depression may improve following cingulumotomy.[57,58] Vagus nerve stimulation also ameliorates depression in some patients with mood disorders. Obsessive-compulsive disorder is treated with anterior capsulotomy, cingulumotomy, or combined surgeries. Uncontrollable violence or rage may decrease following bilateral amygdala surgery, and sexual aggression has been treated with posterior hypothalamotomy.[57,58] While anorexia nervosa, addiction, and anxiety have been reported to benefit from psychosurgery, ablative techniques are rarely used for these conditions unless they become life threatening and all other approaches have been exhausted. Schizophrenic patients may evidence amelioration of any concomitant mood disorder but the psychotic process is typically unaffected by the available surgeries. Figure 4.1 shows the anatomic locations of lesions placed in psychosurgical procedures.

There are relatively few complications from contemporary psychosurgical interventions, although postoperative investigations have often been limited in terms of sophistication and duration of follow-up. Postoperative seizure frequency is approximately 2%. When neuropsychological deficits appear, they typically involve executive function and are evident on tests such as the Wisconsin Card Sort Test or the Picture Arrangement Test of the Wechsler Adult Intelligence Test. Lethargy or impulsiveness characteristic of frontal lobe dysfunction (Chapter 9) are the most common personality changes following psychosurgical procedures.[59]

Refinements of the neurosurgical approach to behaviorally relevant targets have decreased the operative morbidity and mortality. Stereotactic approaches allow more exact placement of lesions with limited collateral damage, and use of external radio frequency devices to generate intracranial ablative lesions has eliminated the need for neurosurgery for some type of treatments such as anterior capsulotomy.[57]

Neurosurgical procedures that are not overtly psychosurgical in intent may have behavioral consequences. Adrenal medullary transplantation for Parkinson's disease have commonly led to depression, delusions, hallucinations, diurnal rhythm disturbances, and confusion.[60] Pallidotomy for Parkinson's disease (Chapter 18) has limited cognitive and behavioral effects when limited to the ventral posterior region of the globus pallidus,[61] but misplacement may result in cognitive and behavioral abnormalities.[62] Unilateral pallidal stimulation is reported to produce mild decrements in verbal fluency and visuoconstructive abilities as well as a decrease in anxiety and depression.[63] Temporal lobectomy for treatment-refractory epilepsy may lead to diminished violence and aggression in those exhibiting these behaviors preoperatively.[57]

BRAIN STIMULATION AND NEUROAUGMENTATION STRATEGIES

Deep brain stimulation, transcranial magnetic stimulation, and vagus nerve stimulation are three new treatment strategies with neuropsychiatric applications. These approaches are relatively new and only a modest amount of information has accumulated regarding their efficacy and side effects. They represent a conceptual advance in treatment with stimulation of structures representing an alternative to physical intervention and structural ablation. In contrast to psychosurgical interventions, the functional changes are temporary, although the long-term functional adjustments of the nervous system to chronic implantation or stimulation are not known. Transcranial magnetic stimulation is the application of a localized pulse magnetic field to the surface of the head to cause a depolarization of neurons in the underlying cerebral cortex.[64] Nerve depolarization results in an action potential and if the stimulation occurrs near the motor cortex, there is a resulting topographically specific motor response. Application of transcranial magnetic stimulation to the dorsolateral prefrontal cortex has resulted in remission of depression.[65] Preliminary results suggest a beneficial effect in bipolar mood disorders.[66] Transcranial stimulation in conjunction with brain mapping provides an important tool for exploring brain circuitry relevant to understanding neuropsychiatric disorders, as well as providing a potential new therapeutic intervention.

Deep brain stimulation involves stereotactically placing electrodes in subcortical structures. Stimulation of the ventro-intermediate nucleus of the thalamus suppresses rest tremor, and stimulation of the subthalamic nucleus alleviates the akinesia and rigidity of parkinsonism.[67] In addition, the technique is being used to treat intention tremor, pain syndromes, intractable dystonia, and some aspects of epilepsy.[68] High-frequency stimulation has the same effect as ablation but is reversible and neuropsychiatric consequences have been observed with deep brain stimulation, including acute depression,[69] and some patients have executive deficits and behavioral changes consistent with frontal–subcortical circuit dysfunction.[70]

Vagus nerve stimulation is an approved therapy for treatment-resistant partial-onset seizures. Vagal nerve stimulation results in activation of brainstem nuclei and secondary stimulation of the cerebral cortex. Preliminary observations suggest that vagal nerve stimulation may also improve mood and represents an alternative therapeutic intervention for patients with treatment-resistant depression.[71,72]

TREATMENT PLANNING

A plethora of treatment interventions are available to aid the patient with a neuropsychiatric disorder. Pharmacotherapy in conjunction with supportive psychotherapy are the mainstay of neuropsychiatric intervention but many other treatments, including behavioral therapy, cognitive therapy, and psychosurgery, are available if needed. The ideal model of neuropsychiatric care involves multidisciplinary assessment and interdisciplinary treatment planning. Nurses, social workers, physical therapists, occupational therapists, speech therapists, and psychologists/neuropsychologists as well as physicians may all have relevant input to the evaluation and treatment of patients with neuropsychiatric disorders. Treatment plans should be logical algorithms that allow the clinician to work through several steps serially as the patient is treated with specified interventions and the response is assessed. Outcomes should be measurable and feasible and should be jointly accepted by patient, family, and clinician. Reduction of behavioral disturbances, increased independence, return to work, improved quality of life, and deferral or avoidance of placement in a long-term care facility are all goals to be considered in patients with neuropsychiatric disturbances.

REFERENCES

1. Johnston MV, Silverstein FS. Fundamentals of drug therapy in neurology. In: Johnston MV, Macdonald RL, Young AB, eds. Principles of Drug Therapy in Neurology. Philadelphia: F.A. Davis, 1992:1–49.
2. Paxton JW, Dragunow M. Attention deficit-hyperactivity disorder. In: Werry JS, Aman MG, eds. Practitioner's Guide to Psychoactive Drugs for Children and Adolescents. New York: Plenum Medical Book Company, 1993:23–74.
3. Greenblatt DJ. Principles of pharmacokinetics and pharmacodynamics. In: Schatzberg AF, Nemeroff CB, eds. The American Psychiatric Press Textbook of Psychopharmacology. Washington, DC: American Psychiatric Press, 1995:125–136.
4. Nemeroff CB, DeVane CL, Pollock BG. Newer antidepressants and the cytochrome P450 system. Am J Psychiatry 1996;153:311–320.
5. Leonard BE. Fundamentals of Psychopharmacology. New York: John Wiley and Sons, 1992.
6. Jenicke MA. Geriatric Psychiatry and Psychopharmacology: A Clinical Approach. Chicago: Mosby Year Book, 1989.
7. Dravet C, Julian C, et al. Epilepsy, antiepileptic drugs, and malformations in children of women with epilepsy: a French prospective cohort study. Neurology 1992;42:75–82.
8. Altshuler LL, Cohen L, et al. Pharmacologic management of psychiatric illness during pregnancy: dilemmas and guidelines. Am J Psychiatry 1996;153:592–606.
9. Stowe ZN, Nemeroff CB. Psychopharmacology during pregnancy and lactation. In: Schatzberg AF, Nemeroff CB, eds. The

American Psychiatric Press Textbook of Psychopharmacology. Washington, DC: American Psychiatric Press, 1995:823–837.
10. Hendrick V, Burt VK, Altshuler LL. Psychotropic guidelines for breast-feeding mothers. Am J Psychiatry 1996;153:1236–1237.
11. Isojarvi JI, Laatikainen TJ, et al. Obesity and endocrine disorders in women taking valproate for epilepsy. Ann Neurol 1996;39:579–584.
12. Marder SA, Van Putten T. Antipsychotic medications. In: Schatzberg AF, Nemeroff CB, eds. The American Psychiatric Press Textbook of Psychopharmacology. Washington, DC: American Psychiatric Press, 1995:247–261.
13. Owens DGC. Extrapyramidal side effects and tolerability of risperidone: a review. J Clin Psychiatry 1994;55:29–35.
14. Bymaster FP, Calligaro DO, et al. Radioreceptor binding profile of the atypical antipsychotic olanzapine. Neuropsychopharmacology 1996;14:87–96.
15. Goldstein JM, Arvantis LA. ICI 204,636 (Seroquel): a dibenzothiazepine atypical antipsychotic. CNS Drug Rev 1995;1:50–73.
16. Nordstrom A-L, Farde L, et al. D1, D2 and 5-HT2 receptor occupancy in relation to clozapine serum concentration: a PET study of schizophrenic patients. Am J Psychiatry 1995;152: 1444–1449.
17. Baldessarini RJ, Frankenburg FR. Clozapine. N Engl J Med 1991;324:746–754.
18. Pacia SV, Devinsky O. Clozapine-related seizures: experience with 5,629 patients. Neurology 1994;44:2247–2249.
19. Factor SA, Brown D, et al. Clozapine: a 2-year open trial in Parkinson's disease patients with psychosis. Neurology 1994;44: 544–546.
20. Wolters EC, Jansen ENH, et al. Olanzapine in the treatment of dopaminomimetic psychosis in patients with Parkinson's disease. Neurology 1996;47:1085–1087.
21. Bellenger JC. Benzodiazepines. In: Schatzberg AF, Nemeroff CB, eds. The American Psychiatric Press Textbook of Psychopharmacology. Washington, DC: American Psychiatric Press, 1995:215–280.
22. Gaillard J-M. Benzodiazepines and GABA-ergic transmission. In: Kryger MH, ed. Principles and Practice of Sleep Medicine, 2nd ed. Philadelphia: W.B. Saunders, 1994:349–354.
23. Montplaisir J, Godbout R, et al. Restless legs syndrome and periodic limb movements during sleep. In: Kryger MH, Roth T, Dement WC, eds. Principles and Practice of Sleep Medicine, 2nd ed. Philadelphia: W.B. Saunders, 1994:589–597.
24. Cole JO, Yonkers KA. Nonbenzodiazepine anxiolitics. In: Schatzberg AF, Nemeroff CB, eds. The American Psychiatric Press Textbook of Psychopharmacology. Washington, DC: American Psychiatric Press, 1995:231–244.
25. Taylor F, Russo J. Efficacy of modafinil compared to dextroamphetamine for the treatment of attention-deficit-hyperactivity disorder in adults. J Child Adolesc Psychopharmacol 2000;10:311–320.
26. Nieuwenhuys R. Chemoarchitecture of the Brain. New York: Springer-Verlag, 1985.
27. Piercey MF. Pharmacology of pramipexole, a dopamine D3-preferring agonist useful in treating Parkinson's disease. Clin Neuropharmacol 1998;21:141–151.
28. Kaakola S, Gordin A, Mannisto PT. General properties and clinical possibilities of new selective inhibitors of catechol-O-methyl transferase. Gen Pharmacol 1994;25:813–824.
29. Ross ED, Stewart RM. Akinetic mutism from hypothalamic damage: successful treatment with dopamine agonists. Neurology 1981;31:1435–1439.
30. Marin RS, Fogel BS, et al. Apathy: a treatable syndrome. J Neuropsychiatry Clin Neurosci 1995;7:23–30.
31. Tarsy D. Restless legs syndrome. In: Joseph AB, Young RR, eds. Movement Disorders in Neurology and Psychiatry. Boston: Blackwell Scientific Publications, 1992:397–400.
32. Udaka F, Yamao S, et al. Pathologic laughing and crying treated with levodopa. Arch Neurol 1984;41:1095–1096.
33. Krupp LB, Coyle PK, et al. Fatigue therapy in multiple sclerosis: results of a double-blind, randomized, parallel trial of amantadine, pemoline, and placebo. Neurology 1995;45:1956–1961.
34. Farlow M, Gracon SI, et al. A controlled trial of tacrine in Alzheimer's disease. JAMA 1992;268:2523–2529.
35. Knapp MJ, Knopman DS, et al. A 30-week randomized controlled trial of high-dose tacrine in patients with Alzheimer's disease. JAMA 1994;271:985–991.
36. Rogers SL, Friedhoff LT, Aricept Study Group. The efficacy and safety of donepezil in patients with Alzheimer's disease: results of a US multicentre, randomized, double-blind, placebo-controlled trial. Dementia 1996;7:293–303.
37. Corey-Bloom J, Anand R, Veach J, for the ENA 713 B352 Study Group. A randomized trial evaluating the efficacy and safety of ENA 713 (rivastigmine tartrate), a new acetylcholinesterase inhibitor, in patients with mild to moderately severe Alzheimer's disease. Int J Geriatr Psychopharmacol 1998;1:55–65.
38. Tariot PN, Solomon PR, et al. A 5-month, randomized, placebo-controlled trial of galantamine in AD. Neurology 2000;54:2269–2276.
39. Cummings JL. Cholinesterase inhibitors: a new class of psychoactive agents. Am J Psychiatry 2000;157:4–15.
40. Kaufer DI, Cummings JL, Christine D. Effect of tacrine on behavioral symptoms in Alzheimer's disease: an open label study. J Geriatr Psychiatry Neurol 1996;9:1–6.
41. Cummings JL, Benson DF. The role of the nucleus basalis of Meynert in dementia: review and reconsideration. Alzheimer Dis Assoc Disord 1987;1:128–145.
42. Cornish JW, McNicholas LF, O'Brien CP. Treatment of substance-related disorders. In: Schatzberg AF, Nemeroff CB, eds. Textbook of Psychopharmacology, 2nd ed. Washington, DC: American Psychiatric Press, 1998:851–867.
43. O'Brien CP. Principles of the pharmacotherapy of substance abuse disorders. In: Charney DS, Nestler EJ, Bunney BS, eds. Neurobiology of Mental Illness. New York: Oxford University Press, 1999:627–638.
44. Galant D. Alcohol. In: Galanter M, Kleber HD, eds. Textbook of Substance Abuse Treatment, 2nd ed. Washington, DC: American Psychiatric Press, 1999:151–164.
45. Ockene JK, Kristeller JL, Donnelly G. Tobacco. In: Galanter M, Kleber HD, eds. Textbook of Substance Abuse Treatment, 2nd ed. Washington, DC: American Psychiatric Press, 1999:215–238.
46. Kleber HD. Opioids: detoxification. In: Galanter M, Kleber HD, eds. Textbook of Substance Abuse Treatment. Washington, DC: American Psychiatric Press, 1999:251–269.
47. Krstal AD, Coffey CE. Neuropsychiatric considerations in the use of electroconvulsive therapy. J Neuropsychiatry Clin Neurosci 1997;9:283–292.
48. Fink M. Electroshock. New York: Oxford University Press, 1999.
49. Rubin EH, Kinscherg DA, et al. The nature and time course of cognitive side effects during electroconvulsive therapy in the elderly. J Geriatric Psychiatry Neurology 1993;6:78–83.
50. Stoudemire A, Hill CD, et al. Improvement in deperession-related cognitive dysfunction following ECT. Journal of Neuropsychiatry and Clinical Neurol 1995;7:31–34.

51. Sobin C, Sackeim HA, et al. Predictors of retrograde aamnesia following ECT. Am J Psychiatry 1995;152:995–1001.
52. Coffey CE, Weiner RD, et al. Brain anatomic effects of electroconvulsive therapy. Arc Gen Psychiatry 1991;48:1013–1021.
53. Devanand DP, Dwork AJ, et al. Does ECT alter brain structure? Amer J Psychiatry 1994;151:957–970.
54. Taylor MA. The Neuropsychiatric Guide to Modern Everyday Psychiatry. New York: The Free Press, 1993.
55. Miller L. Psycotherapy of the Brain-Injured Patient. New York: W.W. Norton and Company, 1993.
56. Moes E. Neuropsychiatric aspects of head injury. In: Ellison JM, Weinstein CS, Hodel-Malinofsky T, eds. The Psychotherapist's Guide to Neuropsychiatry. Washington, DC: American Psychiatric Press, 1994:217–254.
57. Kiloh LG, Smith JS, Johnson GF. Physical Treatments in Psychiatry. London: Blackwell Scientific Publications, 1988.
58. Laitinen LV. Psychosurgery today. In: Brihaye J, Calliauw L, Leow F, can den Bergh R, eds. Personality and Neurosurgery. New York: Springer-Verlag, 1988:158–162.
59. Valenstein ES. Review of the literature on postoperative function. In: Valenstein ES, ed. The Psychosurgery Debate. San Francisco: W.H. Freeman and Company, 1980:141–163.
60. Stebbins GT, Tanner CM. Behavioral effects of intrastriatal adrenal medullary surgery in Parkinson's disease. In: Huber S, Cummings JL, (eds). Parkinson's Disease: Neurobehavioral Aspects. New York: Oxford University Press, 1992:328–345.
61. Masterman D, DeSales A, et al. Motor, cognitive, and behavioral performance following unilateral ventroposterior pallidotomy for Parkinson's disease. Arch Neurol 1998;55:1201–1208.
62. Trepanier LL, Saint-Cyr JA, et al. Neuropsychological consequences of posteroventral pallidotomy for the treatment of Parkinson's disease. Neurology 1998;51:207–215.
63. Troster AI, Fields JA, et al. Unilateral pallidal stimulation for Parkinson's disease: neurobehavioral functioning before and 3 months after electrode implantation. Neurology 1997;49:1078–1083.
64. Bohning DE. Introduction and overview of TMS physics. In: George MS, Belmaker RH, eds. Trancranial Magnetic Stimulation in Neuropsychiatry. Washington, DC: American Psychiatric Press, 2000:13–44.
65. Lisanby SH, Sackeim HA. TMS in major depression. In: George MS, Belmaker RH, eds. Transcranial Magnetic Stimulation in Neuropsychiatry. Washington, DC: American Psychiatric Press, 2000:185–200.
66. Grisaru N, Yaroslavsky Y, Belmaker RH. Is TMS an antibipolar treatment? In: George MS, Belmaker RH, eds. Transcranial Magnetic Stimulation in Neuropsychiatry. Washington, DC: American Psychiatric Press, 2000:201–208.
67. Benazzouz A, Hallett M. Mechanism of action of deep brain stimulation. Neurology 2000;55:S13–S16.
68. Kopell BH, Rezai AR. The continuing evolution of psychiatric neurosurgery. CNS Spectrums 2000;5:20–31.
69. Bejjani BP, Damier P, et al. Transient acute depression induced by high-frequency deep-brain stimulation. N Engl J Med 1999;340:1476–1480.
70. Trepanier LL, Kumar R, et al. Neuropsychological outcome of GP_i pallidotomy and GP_i or STN deep brain stimulation in Parkinson's disease. Brain Cogn 2000;42:324–347.
71. George MS, Nahas Z, et al. Vagus nerve stimulation: a new form of therapeutic brain stimulation. CNS Spectrums 2000;5:43–52.
72. Sackeim HA, Keilp JG, et al. The effects of vagus nerve stimulation on cognitive performance in patients with treatment-resistant depression. Neuropsychiatry Neuropsychol Behav Neurol 2001;14:53–62.

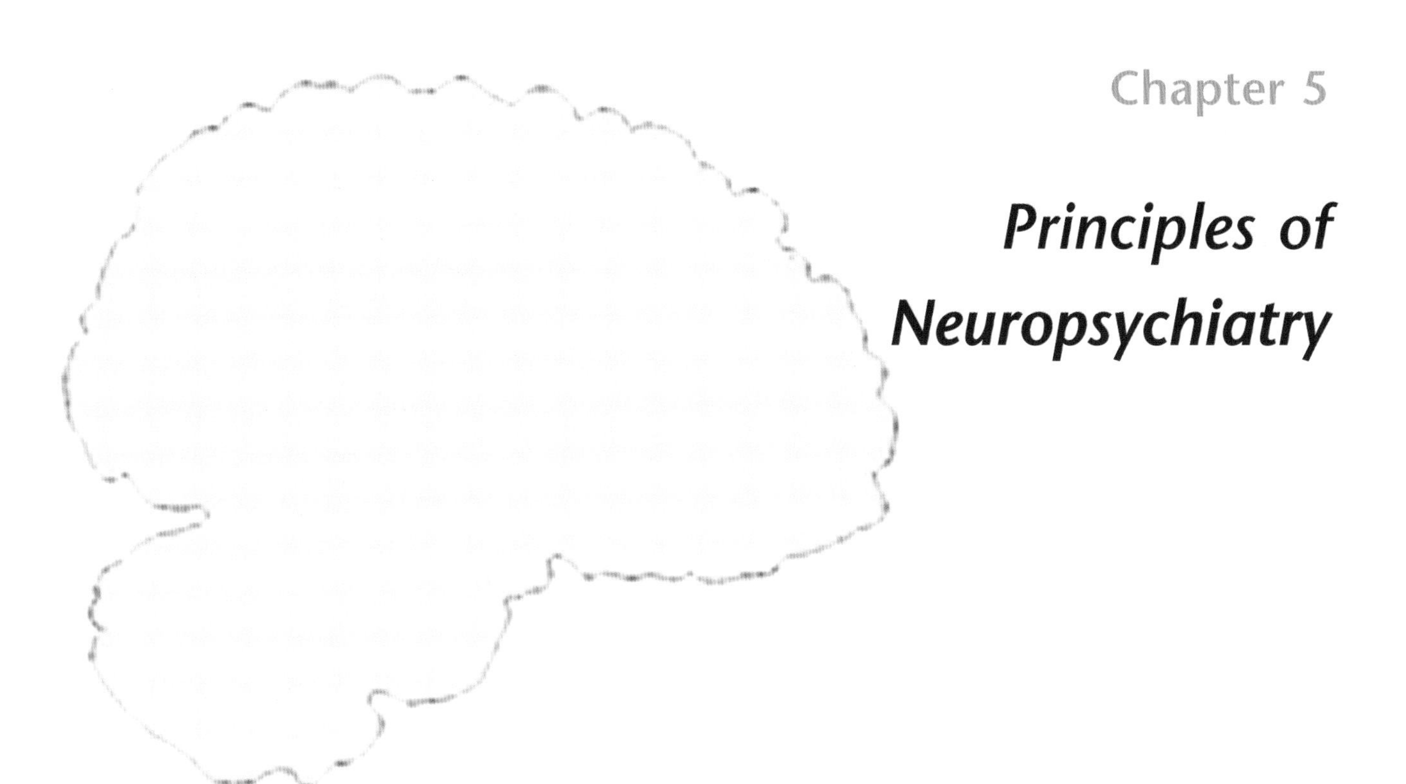

Chapter 5

Principles of Neuropsychiatry

Study of neuropsychiatric syndromes reveals predictable brain–behavior relationships across individuals and across different brain disorders. These regularities can be distilled into principles of neuropsychiatry useful in constructing models of brain–behavior relationships and guiding practitioners in differential diagnosis and neuropsychiatric therapeutics.[1] The principles of neuropsychiatry are sufficiently flexible to regularize neuropsychiatric concepts while allowing individual differences to be understood. The principles of neuropsychiatry will evolve as additional disorders are studied, neuroimaging technologies are applied to neuropsychiatric syndromes, advances in neuropsychopharmacology are integrated into neuropsychiatric care, and genetic and molecular biological studies reveal the molecular underpinnings of neuropsychiatric disorders.

PRINCIPLES

● Brain–Behavior Relationships Underlying Neuropsychiatric Syndromes Are Rule-Governed and Reproducible across Individuals This universality assumption implies that an individual's brain is representative of human brains in general and that deviations from normative behavior have corresponding neurobiological correlates. The verity of this observation reflects the shared evolutionary and genetic determinants of brain structure and function.[2]

● All Mental Processes Derive from Brain Processes Abnormal psychological states have correspondingly unusual or deviant brain states. This applies to behavioral abnormalities that are based on structural brain changes, genetic abnormalities, and environmentally induced alterations.

● Neuropsychiatric Symptoms Are Manifestations of Brain Dysfunction Neuropsychiatric symptoms reflect abnormalities of underlying brain function, whether produced by genetic, structural, or environmental influences. These symptoms are not variable situational reactions to the occurrence of neurologic disease. Thus, depression has only limited correlation with disability in disorders such as Parkinson's disease[3–7] and psychosis is largely independent of any co-occurring neuropsychological deficits.[8–12]

• Behavioral Features Unique to Individuals Are More Likely to Be Environmentally Influenced and Representative of Social and Cultural Factors; Aspects of Behavior That Are Invariant across Individuals Are More Likely to Be Neurologically Determined The individual's psychological environment interacts with brain function to influence the occurrence, nature, persistence and content of behavioral syndromes. Delusional disorders are common products of brain dysfunction and occur with similar frequencies in different world populations. However, the content of delusional disorders varies among cultures and reflects the social and cultural setting of the delusional individual.

• Brain Structure Is Genetically and Environmentally Influenced Genes determine the structure and connectivity of the brain. Therefore, they exert significant control over normal and abnormal behavior. Brain structure and function are influenced by environmental factors, thus genes are not the only determinant of behavior and genetic influences account for only a portion of the variance of human behavior.

While the structure and connectivity of the brain are determined by genetic factors, they may be modified by environmental circumstances. For example, an enriched environment produces increased synaptic connectivity between neurons.[12a] Thus, memory recall and associations among experiences will be increased by learning and involvement. These environmental influences contribute to individual differences associated with variable social, cultural, familial, and educational backgrounds.

• Neuropsychiatric Symptoms Are Related to Regional Brain Dysfunction and Are Not Disease-Specific Any of a variety of diseases that affect a specific circuit or brain regions will produce similar neuropsychiatric manifestations. Thus, disinhibition may be seen with orbitofrontal dysfunction in the context of vascular dementia, traumatic brain injury, degenerative brain disease, or cerebral infections; the symptom complex reflects the area of the brain involved rather than the specific disease present.[13] Likewise, apathy may occur with multiple sclerosis, human immunodeficiency virus (HIV), infection, vascular disease, or degenerative changes involving the anterior cingulate and related subcortical structures and is not disease-specific.[13]

• Brain Dysfunction May Produce Deficit Syndromes or Productive Syndromes Two types of behavioral changes have been observed in conjunction with brain disorders: (*1*) deficit syndromes characterized by the loss of cognitive function such as language, memory, praxis, and gnosis and resulting in aphasia, amnesia, apraxia, and agnosia;[14,15] and (*2*) productive syndromes in which patients exhibit behaviors not previously present including depression, mania, psychosis, anxiety, obsessive-compulsive behavior, personality alterations, and disturbances of sleep, appetite, or sexual behavior. Deficit syndromes traditionally have been the province of behavioral neurology whereas productive syndromes have been the subject of neuropsychiatry. This separation is artificial; deficit and productive syndromes frequently co-occur. The detection of a deficit disorder is an important clue to the presence of a brain disorder in patients with productive syndromes.

• Neuropsychiatric Disorders Typically Reflect Disruption of a System or Circuit Neuropsychiatric disorders are most common in abnormalities of the limbic system or the frontal–subcortical circuits.[13,16,17] Symptoms are similar regardless of where the lesion is located in the circuit. Thus, apathy may be associated with dysfunction of the anterior cingulate cortex, nucleus accumbens, globus pallidus, or medial thalamus, structures that are linked in a frontal–subcortical system.[13,16] Disinhibition may follow lesions of the orbital frontal cortex, ventral caudate nucleus, anterior globus pallidus, or medial dorsal thalamus.[13,16] Signature syndromes with obligatory anatomic relationships characteristic of neurologic and neurobehavioral syndromes are not common in neuropsychiatry. Signature syndromes reflect interruption of modules serially connected at the level of the cortex.

• Neuropsychiatric Disorders Reflect Abnormalities of Fundamental Functions Whereas Neurobehavioral and Deficit Disorders Reflect Abnormalities of Instrumental Functions Neuropsychiatric disorders arise from disruption of systems such as the limbic system and frontal–subcortical circuits mediating fundamental functions. Fundamental functions include emotion, motivation, mood, self-protection, and social affiliation. Deficit syndromes reflect lesions of brain modules mediating instrumental functions such as language, praxis, and gnosis. Neuropsychiatric disorders resulting from abnormalities of fundamental functions include mood disorders, apathy, paranoia, and disinhibition, whereas instrumental dysfunction produces aphasia, apraxia, and agnosia. Fundamental functions are mediated by phylogenetically more primitive brain systems such as the limbic system

TABLE 5.1. *Instrumental Disorders and Fundamental Disorders*

Instrumental Disorders	*Fundamental Disorders*
Abnormalities of language, praxis, gnosis	Abnormalities of mood, motivation, emotion, social attitude
Deficit syndromes: aphasia, apraxia, agnosia	Productive syndromes: psychosis, mania, depression, OCD, disinhibition
Signature syndromes indicating local dysfunction	Circuit syndromes (frontal–subcortical, circuit, limbic system) reflecting disruption
Serial organization of modules connected by tracts	Parallel organization into circuits
Channel function	State function
Disconnection syndromes common	Few disconnection syndromes
Marked lateralization of function	Limited lateralization of function
Cortical level disturbances	Subcortical or limbic system disturbances
Heteromodal neocortex	Paralimbic and limbic cortex; subcortical structures
Phylogenetically recent	Phylogenetically more primitive
Later ontogenetic function	Earlier ontogenetic function
Fast-acting phasic transmitters	Modulatory, tonic transmitters
Primary transmitters: GABA, glutamate	Primary transmitters: dopamine, serotonin, norepinephrine, acetylcholine
Deficits resistant to neurochemical therapy	Disorders responsive to neurochemical therapy

GABA, γ-aminobutyric acid; OCD, obsessive-compulsive disorder.

and the basal ganglia, whereas instrumental functions are mediated primarily by the more recently evolved heteromodal cortex (Table 5.1).[17] Fundamental functions can also be seen as "state" functions that have important biochemical substrates, whereas instrumental functions may be seen as "channel" functions more determined by their underlying anatomy (Table 5.1).[17]

• Any Psychiatric Condition May Occur as a Product of a Central Nervous System Disorder Depression, mania, psychosis, anxiety, obsessive-compulsive disorder (OCD), and sexual behavioral changes have occurred in the setting of neurological illness with symptoms indistinguishable from those of idiopathic psychiatric disorders.[18–22] This principle has the practical implication that neurologic disorders must be considered in a differential diagnosis of any patient presenting with a psychiatric syndrome.

• There Is a Convergence of Evidence Regarding Regional Brain Dysfunction from Studies of the Psychiatric Aspects of Neurologic Disease and Investigations of the Neurobiology of Psychiatric Disorders There is dysfunction of similar anatomic regions when disorders share behavioral manifestations. Thus, similar regional abnormalities have been detected in acquired and idiopathic depression, acquired and idiopathic mania, acquired and idiopathic psychosis, and acquired and idiopathic OCD.[18,23–30]

• The Occurrence and Type of Neuropsychiatric Disorders Are Contingent on Which Brain Regions Are Affected Different neuropsychiatric syndromes are associated with distinct patterns of anatomic involvement. Disturbances of the frontal lobes, temporal lobes, caudate nucleus, and globus pallidus are implicated in many neuropsychiatric syndromes but distinctive regions within these structures are affected in different neurological conditions. Table 5.2 provides a summary of the principal relationships between regional brain dysfunction and corresponding neuropsychiatric symptoms.

• The Laterality of Brain Lesions Influences the Associated Neuropsychiatric Syndrome Secondary mania has a nearly unique association with right hemisphere disorders.[19,21,31] Depression is usually accompanied by bilateral brain dysfunction; left-sided lesions may produce depression in the acute post-stroke period.[32,33] Psychosis with Schneiderian first-rank symptoms is typically associated with a left temporal lobe epileptic focus, whereas psychosis with misidentification symptoms is more often associated with right hemisphere dysfunction.

TABLE 5.2. *Regional Correlates of Neuropsychiatric Symptoms*

Neuropsychiatric Symptom	*Regional Dysfunction*
Mania	Right inferomedial cortex, caudate nucleus, thalamus, temporal–thalamic projection
Depression	Left anterior frontal cortex, left caudate (in the acute post-stroke period)
Psychosis with first-rank symptoms	Left temporal cortex
Psychosis with misidentification	Right hemisphere
Obsessive-compulsive disorder	Orbital or medial frontal cortex, caudate nucleus, globus pallidus
Apathy	Anterior cingulate gyrus, nucleus accumbens, globus pallidus, thalamus
Disinhibition	Orbitofrontal cortex, hypothalamus, septum
Paraphilia	Medial temporal cortex, hypothalamus, septum, rostral brainstem

• The Developmental Phase of the Individual Must Be Considered as Part of the Formula That Determines the Frequency and Nature of the Neuropsychiatric Syndromes Occurring in Conjunction with Brain Disease When Huntington's disease, idiopathic basal ganglia calcification, metachromatic leukodystrophy, or temporal lobe epilepsy begins in adolescence, the patient is more likely to develop psychosis than if these diseases begin later in life.[22,34–36] In post-encephalitic disorders following epidemic encephalitis, adults manifest parkinsonism and mood disturbances whereas children develop tics and conduct disorders.[37] Thus, the developmental state of the brain is one of the determinants of the type of behavioral disorder emerging with brain dysfunction.

• Gender Exerts Important Influences on Neuropsychiatric Symptoms The individual's gender exerts substantial influence on the type of neuropsychiatric symptoms exhibited following brain injury. Adolescent girls are more likely than adolescent boys to exhibit psychosis in association with epilepsy[22] and women are more likely than men to manifest depression following stroke.[33]

• Different Neurological Disorders Have Distinctive Profiles of Associated Neuropsychiatric Disturbances Alzheimer's disease is characterized by high frequencies of indifference, agitation, and dysphoria;[16] vascular dementia manifests depression and psychosis;[38] Parkinson's disease produces depression, apathy, and anxiety;[39] epilepsy patients have increased prevalence rates of depression and psychosis;[22,40] Huntington's disease manifests irritability, lability, and mood disorders;[41] Gilles de la Tourette's syndrome produces OCD and childhood hyperactivity-attention deficit disorder;[42,43] dementia with Lewy bodies has prominent visual hallucinations and delusions;[44] frontotemporal dementias manifest compulsions, disinhibition, and apathy;[45,46] patients with progressive supranuclear palsy manifest apathy and mild disinhibition;[47] corticobasal degeneration is associated with high rates of depression;[48] and multiple sclerosis is associated with prominent depression, anxiety, eutonia, and irritability.[49] These distinctive profiles reflect the differential topography of brain dysfunction associated with each disease (Table 5.3). The brain dysfunction may reflect structural or neurochemical abnormalities.

• Differences in Symptom Profiles Exist within Neuropsychiatric Disturbances Associated with Differing Neurological Disorders Suicide is common in depressive syndromes associated with epilepsy and Huntington's disease and rare in depressed patients with Parkinson's disease.[40,50–52] Psychotic depression is common in epilepsy and rare in Parkinson's disease.[40,52] Self-deprecatory thoughts are common in post-stroke depression and rare in the depression of Parkinson's disease.[53,54] Likewise, each neurologic disorder with depression produces a somewhat different profile of depressive symptoms.

Psychosis in Alzheimer's disease is typically associated with delusions of theft, infidelity, and misidentification, whereas schizophrenia has bizarre and religious

TABLE 5.3. *Neurological Disorders and Associated Behavioral Disorders*

Neurologic Disorder	*Associated Behavioral Disturbances*
Alzheimer's disease	Apathy, agitation, depression, irritability, anxiety, psychosis
Dementia with Lewy bodies	Hallucination, delusions, depression
Frontotemporal dementia	Disinhibition, apathy
Vascular dementia	Depression, apathy, psychosis
Traumatic brain injury	Depression, disinhibition, apathy
Huntington's disease	Depression, OCD, irritability, apathy
Parkinson's disease	Depression, anxiety, psychosis (drug-associated)
Progressive supranuclear palsy	Apathy, disinhibition
Corticobasal degeneration	Depression
Gilles de la Tourette syndrome	OCD, hyperactivity-attention deficit disorder
Multiple sclerosis	Depression, eutonia, irritability, anxiety
Epilepsy (partial complex)	Depression, psychosis
HIV encephalopathy	Apathy

HIV, human immunodeficiency virus; OCD, obsessive-compulsive disorder.

delusions and mania is associated with grandiose delusions. Investigation of these subtle differences in neuropsychiatric symptomatology may provide insight into the unique contribution of different brain structures to neuropsychiatric symptom complexes.

• Specific Regional Brain Changes Are Necessary but Not Sufficient to Produce Neuropsychiatric Symptoms Neuropsychiatric disorders exhibit anatomic contingency and are more likely to occur after lesions of some regions than with others, but they do not have obligate anatomic relationships with specific lesions consistently dictating the occurrence of unique neuropsychiatric symptoms. For example, depression is common in the acute period following a left anterior or left subcortical stroke, but not all patients with lesions in this area manifest depressive symptoms.[20,33] The observation that not all patients with specific lesions manifest the commonly associated behavioral syndromes indicates that additional factors contribute to determining which patients with the appropriate lesions will exhibit the neuropsychiatric symptoms.[1] These additional factors may be related to the characteristics of the lesion or of the host. *Lesion-related factors* include location, size, and laterality of the injury as well as involvement of neurotransmitter systems. *Host factors* that condition the occurrence of neuropsychiatric symptoms include cerebral atrophy, genetic factors, patient gender, and developmental phase of the brain at the time of the insult. In addition, family support and psychosocial milieu will condition the emergence of neuropsychiatric symptoms following a cerebral insult. Thus, nonlesional influences allow substantial individual variability in the expression of neuropsychiatric symptoms among patients with similar brain lesions or disorders.

• Neurologic Disorders Commonly Produce Multiple Simultaneous Neuropsychiatric Symptoms Diseases such as Alzheimer's disease, dementia with Lewy bodies, or frontotemporal dementia frequently have multiple simultaneous neuropsychiatric symptoms.[16] This complicates management and distinguishes neuropsychiatric from idiopathic psychiatric disorders in which more monosymptomatic states may prevail.

• Once Present, Neuropsychiatric Symptoms Tend to Be Persistent or Recurrent Although neuropsychiatric symptoms fluctuate more than cognitive deficits in neurologic disorders, their occurrence is not haphazard. For example, once neuropsychiatric symptoms emerge in Alzheimer's disease, they tend to persist or recur.[55]

• In Progressive Neurologic Disorders, Neuropsychiatric Symptoms Tend to Emerge as the Disease Progresses In Alzheimer's disease and other dementias, there are relatively few neuropsychiatric symptoms at disease onset and they become more common as the disease progresses.[16] Prominent behavioral changes are a common precipitant of nursing home admission as the diseases become severe. The emergence of neuropsychiatric symptoms reflects worsening brain disease and greater dysfunction of behaviorally relevant brain regions.

• Genetic and Environmental Factors Determine a Cerebral Reserve Which Affects the Likelihood That the Individual Will Manifest Neuropsychiatric Symptoms in Conjunction with Brain Dysfunction Individuals with greater native intellectual endowment and higher educational levels are less likely to exhibit Alzheimer's disease than patients with less robust intellectual function and lower educational levels.[56] These factors determine cerebral reserve and make it possible for an individual to sustain higher levels of a pathologic burden without manifesting cognitive dysfunction. Similar effects have been demonstrated in vascular dementia and HIV encephalopathy. Patients with cerebral atrophy are more likely to develop post-stroke depression than those without underlying atrophic changes.[33] Thus, cerebral reserve is one of the mediating host factors conditioning the development of neuropsychiatric symptoms in association with neurologic disorders.

• Neuropsychiatric Syndromes Often Have a Deferred Onset Psychosis in temporal lobe epilepsy may be delayed for several years after onset,[22] and a period of several years often separates the behavioral changes following epidemic encephalitis from the acute brain infection.[37] Depression becomes increasingly frequent for at least 2 years following stroke.[20,33] This indicates that the behavioral morbidity of neurologic disorders may not be evident immediately at the time of onset of the neurologic condition and the relationship of the behavioral disturbance to the brain disorder may be obscure if the brain injury was temporally remote. The deferred onset of neuropsychiatric symptoms may also lead to the underestimation of the frequency of the relationship between brain dysfunction and behavioral abnormalities. The neurobiological mechanisms underlying the deferred onset require further investigation but likely include dendritic remodeling and other changes in connectivity, changes in receptor density, and alterations in transmitter availability.

• Neurologic Disorders Produce Both Structural and Chemical Alterations and Neuropsychiatric Disturbances May Be Influenced by Either of These Post-stroke depression is initiated by the structural brain injury, but the subset of patients with the appropriate lesion most likely to become depressed are those with disturbed serotonergic function.[32] Patients with Alzheimer's disease have neuritic plaques and neurofibrillary tangles in the hippocampus and neocortex as well as a cortical cholinergic deficit. Memory, language, and visuospatial disorders may be determined primarily by the structural cortical pathology, whereas apathy and hallu-cinations may be more related to biochemical changes.[56,57] Cholinergic therapy in Alzheimer's disease has a greater influence on neuropsychiatric manifestations of the disorder than on the associated cognitive abnormalities. An interaction between structural and biochemical changes occurs in brain disorders.

• Disturbances in Transmitters or Transmitter Systems Have Specific Associated Neuropsychiatric Symptoms Abnormalities of dopamine have been implicated in psychosis and, to a lesser extent, in depression. Serotonergic abnormalities mediate abnormalities of mood (particularly depression) and OCD. Serotonin has also been implicated in impulsivity, aggression, and psychosis. Noradrenergic abnormalities have been associated with disturbances of mood and anxiety. Cholinergic disturbances have been associated with apathy and visual hallucinations.[56] Focal lesions can produce neurochemical deficits. Neurologic disorders producing transmitter deficits (e.g., Parkinson's disease) produce regional syndromes by depriving specific brain regions of transmitter input; and the specific regional distribution of neurotransmitter receptors will determine the regional effect of transmitter deficits and of pharmacotherapy. Thus, there is a dynamic integration of anatomic and biochemical aspects of brain function and dysfunction (Table 5.4).

TABLE 5.4. *Behavioral Abnormalities Associated with Transmitter Deficits*

Neurotransmitter	*Associated Behavioral Syndrome*
Serotonin	Depression, OCD, impulsivity/aggression, psychosis
Dopamine	Psychosis, depression
Norepinephrine	Depression, anxiety
Acetylcholine	Apathy, visual hallucinations

OCD, obsessive-compulsive disorder.

• Pharmacologic Interventions in Neuropsychiatric Syndromes May Be Disease-Specific, Transmitter-Specific, or Symptom-Specific *Disease-specific treatments* include anticonvulsants in the treatment of epilepsy, use of anti-inflammatory agents in collagen vascular and inflammation-related disorders and administration of antibiotics to patients with brain infections. Amelioration of the underlying brain disorder frequently results in a reduction of neuropsychiatric symptoms, although paradoxical effects may occur, as in the behavioral deterioration that may occur following "forced normalization" in epilepsy.[22]

Transmitter-specific pharmacologic strategies include the use of dopaminergic agents in Parkinson's disease and cholinergic agents in Alzheimer's disease to address the underlying neurotransmitter deficits. These transmitter-specific therapies may produce neuropsychiatric benefit (i.e., reduction of apathy and visual hallucinations in Alzheimer's disease) or neuropsychiatric side effects (i.e., hallucinations and delusions in patients with Parkinson's disease).

Symptom-specific pharmacotherapy involves the use of antidepressants, antipsychotics, and anxiolytics to treat depression, psychosis, and anxiety in conjunction with neurologic disorders. The success of similar pharmacotherapies in symptomatic and idiopathic behavioral disturbances supports the presence of shared mechanisms in these conditions.

• The Beneficial Effects of Psychotherapy Are Mediated through Changes in Brain Function The universality assumption that there is deviant brain function in association with all behavioral abnormalities implies that resolution of these disturbances through any means including pharmacotherapy, psychotherapy, and surgical therapy will have an effect on underlying brain function. This has been demonstrated in OCD, where behavioral therapy reduces orbitofrontal and caudate hypermetabolism.[58]

• Neuropsychiatric Disturbances Increase the Morbidity of Neurologic Disease Depression exacerbates disability associated with neurological illnesses, amplifies coexisting cognitive deficits, and contributes to diminished quality of life.[59–61]

• Neuropsychiatric Disorders Increase Caregiver Distress and May Precipitate Institutionalization of Patients with Neurologic Conditions Behavioral abnormalities in patients with neurological disorders have been related to the severity of caregiver distress.[62] Patients with behavioral disturbances are also more likely to be institutionalized and behavioral disorders are more common among nursing home residents than among community-dwelling patients with the same neurologic conditions.[63]

• Neuropsychiatric Disorders Involve Abnormalities of the Self Neuropsychiatric symptoms differ from those of other disorders in that they affect the individual's sense of self. Thus, one can *have* pneumonia or arthritis, but in a fundamental sense one *is* demented, psychotic, or depressed. Clinicians participating in the care of individuals with neuropsychiatric symptoms must be aware of this fundamental existential shift with its impact on the patients and the patients' family.

SUMMARY

These principles regularize considerations of brain–behavior relationships in neuropsychiatric disorders. They relate both to symptomatic and idiopathic neuropsychiatric conditions and standardize expected brain–behavior relationships while allowing for individual variability. They comprise an emerging epistemology of neuropsychiatry.

REFERENCES

1. Cummings JL. Principles of neuropsychiatry: towards a neuropsychiatric epistemology. Neurocase 1999;5:181–188.
2. David AS. Cognitive neuropsychiatry? Psychol Med 1993;23:1–5.
3. Stern RA, Bachman DL. Depressive symptoms following stroke. Am J Psychiatry 1991;148:351–356.
4. Robinson RG, Price TR. Post-stroke depressive disorders: a follow-up study of 103 patients. Stroke 1982;13:625–641.
5. Ehmann TS, Beninger RJ, et al. Depressive symptoms in Parkinson's disease: a comparison with disabled control subjects. J Geriatr Psychiatry Neurol 1990;2:3–9.
6. Gotham A-M, Brown RG, Marsden CD. Depression in Parkinson's disease: a quantitative and qualitative analysis. J Neurol Neurosurg Psychiatry 1986;49:381–389.
7. Menza MA, Mark MH. Parkinson's disease and depression: the relationship to disability and personality. J Neuropsychiatry Clin Neurosci 1994;6:165–169.
8. Bylsma FW, Folstein MF, et al. Delusions and patterns of cognitive impairment in Alzheimer's disease. Neuropsychiatry Neuropsychol Behav Neurol 1994;7:98–103.
9. Flynn FG, Cummings JL, Gornbein J. Delusions in dementia syndromes: investigation of behavioral and neuropsychological correlates. J Neuropsychiatry Clin Neurosci 1991;3:364–370.
10. Jeste DV, Wragg RE, et al. Cognitive deficits of patients with Alzheimer's disease with and without delusions. Am J Psychiatry 1992;149:184–189.
11. Kotrla KJ, Chacko RC, et al. Clinical variables associated with psychosis in Alzheimer's disease. Am J Psychiatry 1995;152:1377–1379.

12. Lopez OL, Becker JT, Brenner RP. Alzheimer's disease with delusions and hallucinations: neuropsychological and electroencephalophic correlates. Neurology 1991;41:906–912.
12a. Lund JS, Holbach SM, Chung WW. Postnatal development of thalamic recipient neurons in the monkey striate cortex. II. Influence of afferent driving on spine acquisition and dendritic growth of layer 4C spiny stellate neurons. J Comp Neurol 1991;309:129–140.
13. Cummings JL. Frontal–subcortical circuits and human behavior. Arch Neurol 1993;50:873–880.
14. Filley CM. Neurobehavioral Anatomy. Niwot, CO: University of Colorado Press, 1995.
15. Kirshner HS. Behavioral Neurology: A Practical Approach. New York: Churchill Livingstone, 1986.
16. Mega M, Cummings JL, et al. The spectrum of behavioral changes in Alzheimer's disease. Neurology 1996;46:130–135.
17. Mesulam M-M. Patterns of behavioral neuroanatomy: association areas, the limbic system, and hemispheric specialization. In: Mesulam M-M, ed. Principles of Behavioral Neurology. Philadelphia: F.A. Davis, 1985:1–70.
18. Cummings JL, Cunningham K. Obsessive-compulsive disorder in Huntington's disease. Biol Psychiatry 1992a;31:263–270.
19. Cummings JL, Mendez MF. Secondary mania with focal cerebrovascular lesions. Am J Psychiatry 1984;141:1084–1087.
20. Robinson RG, Travella JI. Neuropsychiatry of mood disorders. In: Fogel BS, Schiffer RB, Rao SM, eds. Neuropsychiatry. Philadelphia: Williams and Wilkins, 1996:287–305.
21. Starkstein SE, Pearlson GD, et al. Mania after brain injury. A controlled study of causative factors. Arch Neurol 1987;44: 1069–1073.
22. Trimble MR. The Psychoses of Epilepsy. New York: Raven Press, 1991.
23. Baxter LR, Phelps ME, et al. Local cerebral glucose metabolic rates in obsessive-compulsive disorder: a comparison with rates in unipolar depression and normal controls. Arch Gen Psychiatry 1987;44:211–218.
24. Baxter LR, Schwartz JM, et al. Reduction of prefrontal cortex glucose metabolism common to three types of depression. Arch Gen Psychiatry 1989;46:243–250.
25. Lesser IM, Mena I, et al. Reduction in cerebral blood flow in older depressed patients. Arch Gen Psychiatry 1994;51:677–686.
26. Machlin SR, Harris GJ, et al. Elevated medial–frontal cerebral blood flow in obsessive-compulsive patients: a SPECT study. Am J Psychiatry 1991;148:1240–1242.
27. Rauch SL, Jenike MA, et al. Regional cerebral blood flow measured during symptom provocation in obsessive-compulsive disorder using oxygen ^{15}-labeled carbon dioxide and positron emission tomography. Arch Gen Psychiatry 1994;51:62–70.
28. Roberts GW, Leigh PN, Weinberger DR. Neuropsychiatric Disorders. Chicago: Wolfe, 1993.
29. Schwartz JM, Baxter LR, et al. The differential diagnosis of depression: relevance of positron emission tomography studies of cerebral glucose metabolism to the bipolar–unipolar dichotomy. JAMA 1987;258:1368–1374.
30. Swedo SE, Pietrini P, et al. Cerebral glucose metabolism in childhood-onset obsessive-compulsive disorder. Arch Gen Psychiatry 1992;49:690–694.
31. Robinson RG, Boston JD, et al. Comparison of mania and depression after brain injury: causal factors. Am J Psychiatry 1988; 145:172–178.
32. Mayberg HS. Frontal lobe dysfunction in secondary depression. J Neuropsychiatry Clin Neurosci 1994;6:428–442.
33. Robinson RG. The Clinical Neuropsychiatry of Stroke. Cambridge, UK: Cambridge University Press, 1998.
34. Cummings JL, Gosenfeld LF, et al. Calcification of the basal ganglia: case report and review. Biol Psychiatry 1983;18:591–601.
35. Hyde TM, Ziegler JC, Weinberger DR. Psychiatric disturbances in metachromatic leukodystrophy. Arch Neurol 1992;49:401–406.
36. Morris M. Psychiatric aspects of Huntington's disease. In: Harper PS, ed. Huntington's Disease. Philadelphia: W.B. Saunders, 1991.
37. Schaette S, Cummings JL. Encephalitis lethargica: lessons for contemporary neuropsychiatry. J Neuropsychiatry Clin Neurosci 1995;7:125–134.
38. Sultzer DL, Levin HS, et al. A comparison of psychiatric symptoms in vascular dementia and Alzheimer's disease. Am J Psychiatry 1993;150:1806–1812.
39. Aarsland D, Larsen JP, et al. Range of neuropsychiatric disturbances in patients with Parkinson's disease. J Neurol Neurosurg Psychiatry 1999;67:492–496.
40. Mendez MF, Cummings JL, Benson DF. Depression in epilepsy. Significance and phenomenology. Arch Neurol 1986;43:766–770.
41. Folstein SE. Huntington's Disease: A Disorder of Families. Baltimore: Johns Hopkins University Press, 1989.
42. Frankel M, Cummings JL, et al. Obsessions and compulsions in Gilles de la Tourette's syndrome. Neurology 1986;36:378–382.
43. Kurlan RM. Tourette's syndrome in movement disorders. In: Watts RL, Koller WC, eds. Neurological Principles and Practice. New York: McGraw-Hill, 1997:569–575.
44. McKeith IG, Galasko D, et al. Consensus guidelines for the clinical and pathologic diagnosis of dementia with Lewy bodies (DLB): report of the Consortium on DLB International Workshop. Neurology 1996;47:1113–1124.
45. Ames D, Cummings JL, et al. Repetitive and compulsive behavior in frontal lobe degenerations. J Neuropsychiatry Clin Neurosci 1994;6:100–113.
46. Levy ML, Miller BL, et al. Alzheimer's disease and frontotemporal dementias: behavioral distinctions. Arch Neurol 1996a;53: 687–690.
47. Litvan I, Mega MS, et al. Neuropsychiatric aspects of progressive supranuclear palsy. Neurology 1996;47:1184–1189.
48. Litvan I. Neuropsychiatric features of corticobasal degeneration. J Neurol Neurosurg Psychiatry 1999;46:876–883.
49. Diaz-Olavarrieta C, Cummings JL, et al. Neuropyschiatric manifestations of multiple sclerosis. J Neuropsychiatry Clin Neurosci 1999;11:51–57.
50. Hawton K, Fagg J, Marsack P. Association between epilepsy and attempted suicide. J Neurol Neurosurg Psychiatry 1980;43:168–170.
51. Schoenfeld M, Myers RH, et al. Increased rate of suicide among patients with Huntington's disease. J Neurol Neurosurg Psychiatry 1984;47:1283–1287.
52. Cummings JL. Depression and Parkinson's disease: a review. Am J Psychiatry 1992;149:443–454.
53. Brown RG, MacCarthy B, et al. Depression and disability in Parkinson's disease: a follow-up study of 132 cases. Psychol Med 1988;18:49–55.
54. Lipsey JR, Spencer WC, et al. Phenomenological comparison of poststroke depression and functional depression. Am J Psychiatry 1986;143:527–529.
55. Levy M, Cummings JL, et al. Longitudinal assessment of symptoms of depression, agitation, and psychosis in 181 patients with Alzheimer's disease. Am J Psychiatry 1996;153:1438–1443.

56. Cummings JL, Kaufer D. Neuropsychiatric aspects of Alzheimer's disease: the cholinergic hypothesis revisited. Neurology 1996;47:876–883.
57. Kaufer D, Cummings JL, Christine D. Differential neuropsychiatric symptom responses to tacrine in Alzheimer's disease: relationship to dementia severity. J Neuropsychiatry Clin Neurosci 1998;10:55–63.
58. Baxter LR, Schwartz JM, et al. Caudate glucose metabolic rate changes with both drug and behavior therapy for obsessive-compulsive disorder. Arch Gen Psychiatry 1992;49:681–689.
59. Forsell Y, Winblad B. Major depression in a population of demented and nondemented older people: prevalence and correlates. J Am Geriatr Soc 1998;46:27–30.
60. Parikh RM, Robinson RG, et al. The impact of poststroke depression on recovery in activities of daily living over a 2-year follow-up. Arch Neurol 1990;47:785–789.
61. Perrine K, Hermann BP, et al. The relationship of neuropsychological functioning to quality of life in epilepsy. Arch Neurol 1995;52:997–1003.
62. Kaufer DI, Cummings JL, et al. Assessing the impact of neuropsychiatric symptoms in Alzheimer's disease: The Neuropsychiatric Inventory Caregiver Distress Scale. J Am Geriatr Soc 1998;46:210–215.
63. Steele C, Rovner B, et al. Psychiatric problems and nursing home placement of patients with Alzheimer's disease. Am J Psychiatry 1990;147:1049–1051.

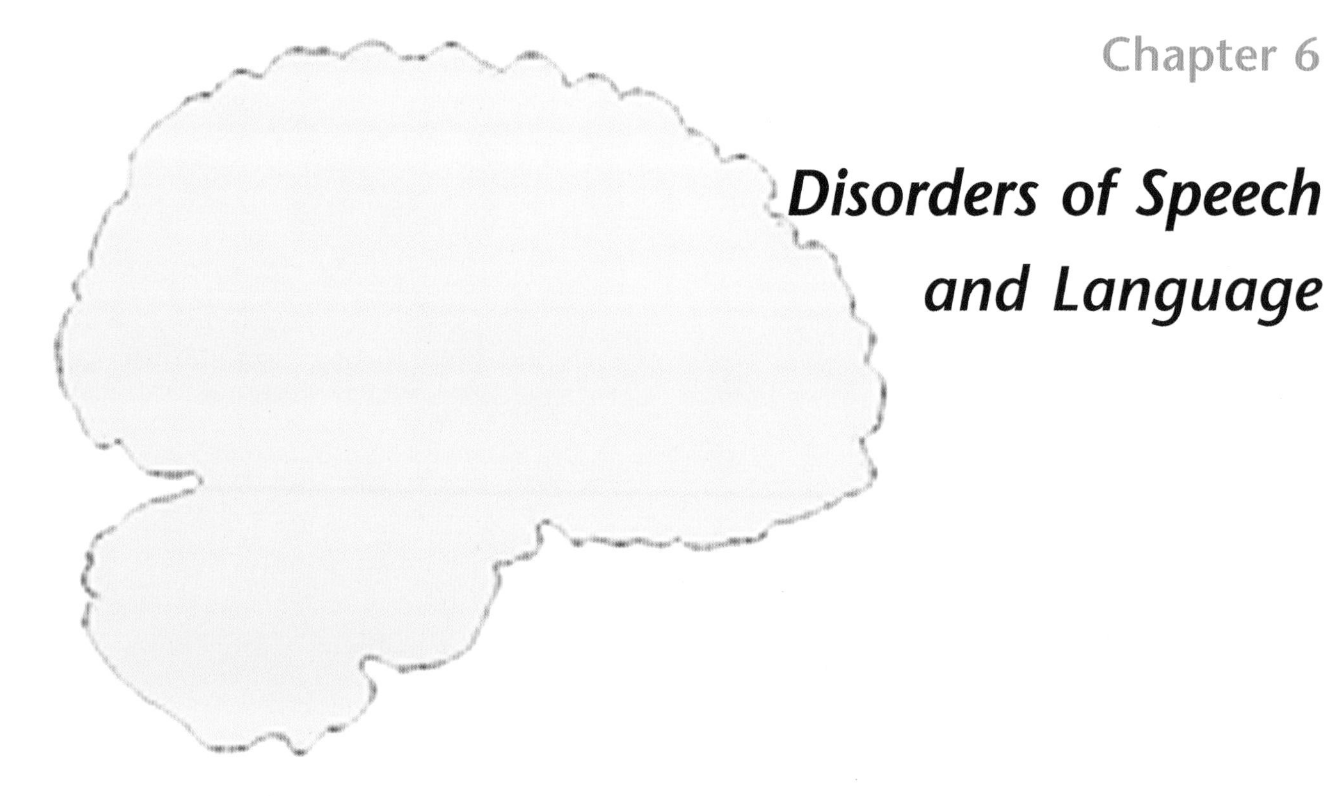

Chapter 6

Disorders of Speech and Language

Verbal interchange accounts for a major part of all neuropsychiatric interviews, and characteristics of the patient's verbal output are among the most diagnostically revealing clues available to the clinician. In this chapter, three principal disorders of verbal output are presented: mutism, aphasia and related disturbances, and psychotic speech. Dysarthria, aprosodia, stuttering, echolalia, and palilalia are also described. Coprolalia is discussed in Chapter 19 along with Gilles de La Tourette's syndrome.

MUTISM

The term *mutism* has been used to refer both to loss of propositional speech (but with the retained ability to grunt, cough, sing, etc.) and the complete obliteration of all sound-producing abilities in patients who are alert. Table 6.1 presents the differential diagnosis of mutism and lists the wide variety of disorders that may dramatically reduce verbal output. Examination of the function of the jaw, face, and tongue are critical in determining the cause of the mutism, and assessing the patient's attitude toward the loss of speech can also provide important insights.

Structural Disturbances Producing Mutism

The most common cause of loss of vocalization is the aphonia associated with laryngitis, in which there is local inflammation and throat pain as well as mutism. Laryngeal neoplasms and disruption of vocal cord function by myopathies, neuromuscular junction disorders, peripheral neuropathies, or lower motor neuron involvement in the brainstem (amyotrophic lateral sclerosis, polioencephalitis, cerebrovascular disease, neoplasm) are also capable of' producing complete mutism by interfering with laryngeal function. In many cases the mutism will be preceded by hoarseness, and laryngoscopy will reveal local inflammation, neoplastic involvement of the vocal cords, or paralysis of cord movement. Aphonic patients have no other neurologic abnormalities and have the expected frustrated reaction to soundlessness. They readily adopt alternate means of communication such as writing notes. When bulbar musculature or brainstem neurons are involved, dysphagia and tongue and facial weakness are often prominent.

Central nervous system (CNS) lesions rostral to the brainstem (pseudobulbar) can produce mutism when the corticobulbar tracts are involved bilaterally. In most cases, verbal output is preferentially affected, whereas

TABLE 6.1. *Differential Diagnosis of Mutism*

Peripheral disorders (aphonia)
Local laryngeal disturbances
Myopathies
Neuromuscular junction disorders (myasthenia)
Neuropathies
Brain disorders
Bulbar (brainstem) disturbances
Polioencephalitis
Amyotrophic lateral sclerosis
Cerebrovascular disease
Neoplasm
Pseudobulbar palsy
Cerebrovascular disease
Amyotrophic lateral sclerosis
Neoplasms
Multiple sclerosis
Traumatic brain injury
Advanced extrapyramidal disorders
Parkinson's disease
Progressive supranuclear palsy
Wilson's disease
Torsion dystonia
Dementias
Advanced Alzheimer's disease
Frontotemporal dementias
Transient mutism with aphasia
Broca's aphasia
Transcortical motor aphasia
Global aphasia
Aphasia with subcortical lesions
Childhood aphasia
Miscellaneous neurologic syndromes
Aphemia
Supplementary motor area lesions
Akinetic mutism
Cerebellar disorders
Psychiatric illnesses
Depression with catatonia
Mania with catatonia
Schizophrenia with catatonia
Conversion disorder
Malingering
Selective mutism
Autism
Drug-induced mutism
Cyclosporine (immunosuppressive agent)
FK506 (immunosuppressive agent)

nonverbal vocalization such as laughing and crying is spared or may even be pathologically exaggerated (Chapter 14).[1–5] Deficits commonly accompanying the mutism of the pseudobulbar syndrome include dysphagia, facial weakness, brisk jaw jerk and facial muscle stretch reflexes, impaired tongue movements, and an exaggerated gag reflex. The bilateral supranuclear lesions responsible for pseudobulbar palsy may be produced by cerebrovascular disease, neoplasms, multiple sclerosis, inflammatory or infectious disorders, or amyotrophic lateral sclerosis. When the lesions involve limbic as well as corticobulbar connections, the mutism may include both verbal and nonverbal emotional vocalizations.[6] Mutism is combined with limb paralysis in the locked-in syndrome.

Advanced neurological disease of practically any type can produce mutism but it is particularly common with the bradykinetic and dystonic extrapyramidal syndromes, in which progressive dysarthria and hypophonia eventually lead to mutism. Extrapyramidal syndromes capable of producing mutism include Parkinson's disease, progressive supranuclear palsy, Wilson's disease, and the dystonias. Occasionally, the facial and laryngeal involvement is out of proportion to limb and truncal involvement, resulting in mutism early in the course of the disease. In some cases the mutism may be overcome when the patient is extremely excited or angry,[6] and singing may be possible even when expository vocalization is not.

Cerebellar disorders are also associated with mutism, particularly in children. The syndrome typically occurs after removal of a cerebellar tumor and lasts from a few weeks to a few months. In some cases, the mutism does not begin immediately after surgery but is deferred for 1 or 2 days. The mute period is followed by the development of dysarthric speech and eventual return to normal speech. The syndrome is most common in children who had hydrocephalus at presentation and who have had a tumor adjacent to the fourth ventricle.[7–9] Postsurgical meningitis is also a risk factor for the disorder. Cerebellar mutism occurs almost exclusively in children under 10 years of age and has occurred after surgery for medulloblastomas, astrocytomas, and ependymomas.

Mutism precludes the recognition of aphasia since mutism is the absence of language production, whereas *aphasia* refers to the presence of an abnormal language output. Nevertheless, mutism may he present in the initial phases of patients with nonfluent aphasia (Broca's aphasia, global aphasia) and is a standard feature early in the course of aphasia associated with subcortical lesions (thalamus, basal ganglia) and of transcortical motor aphasia.[10] In these cases the mutism is transient, and aphasic agraphia is evident in the patient's writing. Childhood aphasia, unlike aphasic disturbances in adults, is commonly associated with an initial mute period regardless of the type of aphasia.[11,12]

Mutism may occur in the course of advanced dementia syndromes; it is more common with frontotemporal dementias than Alzheimer's disease (Chapter 10). Mutism has also been associated with closed head injury, particularly lesions of the basal ganglia or diffuse axonal injury.[13] *Aphemia* is an uncommon disorder that presents with an acute right hemiparesis and mutism, but with preserved ability to write. When speech is restored, it has a hoarse, breathy, dysarthric, and often hypophonic quality, but there is no aphasia.[14,15] The recovered speech of the aphemic patient often has dysprosodic qualities with changes in pitch and syllable stress that make the speech sound like that of a non-native speaker of the language. This disorder has been called the *foreign accent syndrome* and is uniquely associated with aphemia.[16–18] Broca's type aphasia may be present in the early phases of the disorder following recovery from the mutism but is not sustained. The lesion producing aphemia is usually an infarction limited to Broca's area in the left precentral frontal cortex or the white matter immediately subtending it.[19,20] The most common cause of the syndrome is embolic infarction associated with an embolus of cardiac origin.

Lesions of the supplementary motor area on the medial aspect of the left hemisphere commonly produce mutism in the acute stages. During recovery, the patient may manifest a transcortical motor type aphasia (discussed below) or may exhibit a paucity of speech output without aphasia.[21,22] The lesion associated with transcortical motor aphasia and early mutism has usually been occlusion of the left anterior cerebral artery with infarction of the medial frontal region including the supplementary motor area. Disruption of white matter connections underlying the supplementary motor area and connecting this region with Broca's area may be essential to the occurrence of the disorder (Fig. 6.1).[23] The syndrome has also been observed with left medial corticectomies, left medial subdural hematomas and neoplasms, and subcortical infarctions involving the left thalamus and putamen.[22,24–26]

Akinetic mutism is a state of nearly complete motionlessness combined with total mutism. Two varieties of akinetic mutism have been distinguished: a "vigilant coma" variety in which the patient is immobile yet seemingly alert, has full extraocular movements, and can occasionally be aroused to move or may even have brief agitated periods; the other is a somnolent variant in which the patient appears asleep most of the time and has oculomotor abnormalities. In the former, the lesion is usually situated at the base of the brain anteriorly in the region of the optic chiasm, medial forebrain bundle, anterior hypothalamus, or the anterior cingulate bilaterally (Fig. 6.2); in the later, the lesion is located more posteriorly in the anterior midbrain or mesodiencephalic junction.[27] The syndrome may be produced by vascular lesions, trauma, multiple sclerosis, or tumors in the region of the third ventricle.[28–32] Akinetic mutism has been treated successfully with dopamine receptor agonists (bromocriptine), which suggests that interruption of ascending dopamine projections may contribute to the marked reduction in spontaneous activity.[31,33]

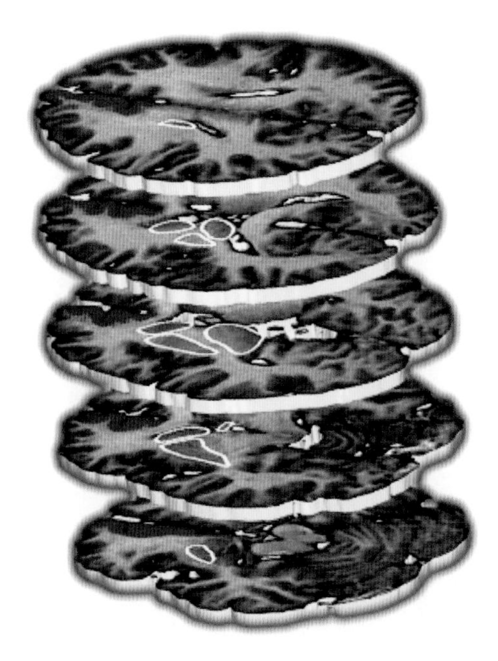

FIGURE 6.1 Disconnection of the white matter underlying the supplementary motor area (shown in red) and connecting this region with Broca's area is associated with transcortical motor aphasia.

Idiopathic Neuropsychiatric Disorders with Mutism

Mutism may be a manifestation of extreme retardation and fatigue in depression, and it is a common manifestation of catatonia occurring with mood disorders or schizophrenia[28] (Chapters 12, 14, and 20).

Mutism may occur as a conversion symptom and can take the form of either aphonia with whispered speech or complete mutism, sometimes with apparent inability to comprehend as well as to produce spoken language. In most cases the patient is able to read and write normally, and normal vocal cord function can be demonstrated by having the patient cough.[34] Normal

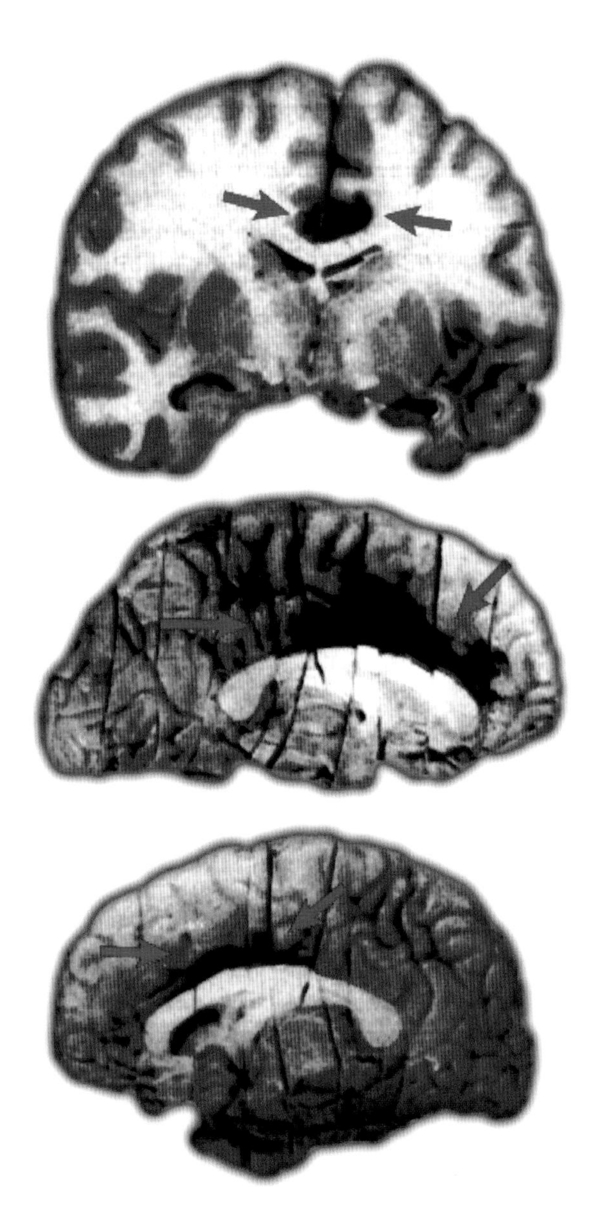

FIGURE 6.2 The lesion responsible for the "vigilant coma" variety of akinetic mutism is usually situated at the base of the brain anteriorly in the region of the optic chiasm, medial forebrain bundle, anterior hypothalamus, or the anterior cingulate bilaterally (shown by arrows).

coughing is impossible if the vocal cords are paretic, and patients with paretic cords produce a distorted, bovine sound. Hysterical mutism is more common in children than in adults and, like all conversion reactions, may be the harbinger of a major neurological or psychiatric illness (Chapter 22). Mutism is an occasional manifestation of malingering.

Selective mutism is a syndrome of preadolescent children who evidence language competence in some circumstances, such as among family members and friends, but do not speak in other social situations, particularly school. The syndrome is rare (7 out of 1000 children have transient selective mutism and in 0.7/1000 it persists for more than 6 months) and is more common among children with some other type of communication disorder (articulation defect) and recent immigrants learning a new language. Children with selective mutism often have associated behavioral disturbances such as negativism, defiance, shyness, poor peer relations, and social isolation. The prognosis for the syndrome is usually excellent.[35,36] Selective serotonin reuptake inhibitors (SSRIs) such as fluoxetine have hastened resolution of the syndrome.[37] Mutism may also occur as part of the childhood autism syndrome.

Mutism is occasionally observed in liver transplant patient recipients being treated with cyclosporine or FK506.[38]

DYSARTHRIA

Dysarthria refers to impairment in the motor aspects of speech. Dysarthric abnormalities include disturbances in speech rate (too slow, too fast, bursts), volume (hypophonia, megaphonia), tone, pitch, timing, and accuracy of articulation. Six basic types of dysarthria have been described (Table 6.2). Each dysarthria corresponds to a predominant motor disorder: flaccid, spastic, ataxic, hypokinetic (parkinsonian), hyperkinetic (choreic), and dystonic (spasmodic dysphonia).[39] In addition, mixed dysarthrias occur in disorders that have more than one type of motor dysfunction such as the mixed spastic–ataxia of multiple sclerosis or the mixed spastic–flaccidity of amyotrophic lateral sclerosis. Speech therapy may be of substantial benefit to many dysarthric patients.

APHASIA AND RELATED DISORDERS

Aphasia refers to an impairment of language produced by brain dysfunction.[40] It is an acquired syndrome that can be caused by a wide variety of cerebral insults, including cerebrovascular accidents, intracranial neoplasms, cerebral trauma, and degenerative conditions. Aphasia must be distinguished from mutism, disorders of speech volume and articulation (dysarthria), disturbances of speech melody and inflection (dysprosody), and thought disorders with abnormal verbal output. Several different patterns of aphasic output have been identified and correlated with lesions in specific anatomic areas. Individual aphasias have different neu-

TABLE 6.2. *Classification of Dysarthria*

Type of Dysarthria	*Principal Features*	*Typical Diseases*
Flaccid	Hypernasality, breathiness, imprecise consonants, monopitch	Myasthenia gravis
Spastic	Strained-strangled voice, slow rate, imprecise consonants, reduced stress, harsh voice, low pitch, monoloudness	Stroke with pseudobulbar palsy
Ataxic	Excess stress, irregular rate and rhythm, distorted vowels, imprecise consonants	Multiple sclerosis
Hypokinetic	Monopitch, monoloudness, reduced stress, variable rate, short rushes of speech, inappropriate silences, breathiness	Parkinson's disease
Hyperkinetic	Imprecise consonants, prolonged intervals, variable rate, monopitch, harsh voice, excess loudness variation, inappropriate silences, distorted vowels	Huntington's disease
Dystonic	Strained-strangled voice, monopitch, irregular articulation, monoloudness, harsh voice, distorted vowels	Spasmodic dysphonia
Mixed	Features of more than one type of dysarthria	Multiple sclerosis, ALS with upper and lower motor neuron dysfunction, Wilson's disease

ALS, amyotrophic lateral sclerosis.

ropsychiatric complications, prognosis, treatment, and etiology. The discussion that follows focuses on aphasia syndromes associated with relatively discreet CNS lesions associated with infarctions, neoplasms, or trauma; aphasia associated with dementia syndromes is presented in Chapter 10.

Types of Aphasia and Their Anatomic Correlates

Examination techniques for detecting and assessing language abnormalities were presented in Chapter 3. The observations concerning language function made during the neuropsychiatric interview and mental state assessment are utilized in this chapter to identify individual types of aphasia and to infer the location and etiology of the underlying lesion.

The first localization principle to be invoked with regard to aphasia concerns lateralization of language function. In right-handed individuals, aphasia will be correlated with a left hemispheric lesion 99% of the time. In left-handed patients, the situation is more complex and variable. Approximately 60% of left-handed individuals will have a dominance pattern similar to that of right-handers with language represented in the left hemisphere. A considerable number of left-handers, however, particularly those with a family history of left-handedness, appear to have cerebral ambilaterality with some degree of language competence present in both hemispheres.[15,41] This relative lack of lateralization among non–right-handers is manifested by their tendency to develop an aphasia regardless of which hemisphere is injured, a better prognosis for language recovery, and a poor correlation between pattern of aphasic deficits and lesion site.[15] The aphasia patterns presented in Figure 6.3 and Table 6.3 are those observed in right-handed individuals.

Figure 6.3 presents a systematic approach to aphasia diagnosis based on the cardinal observations of fluency of spontaneous speech, integrity of linguistic comprehension, and ability to repeat phrases and sentences. The first step in aphasia assessment involves determination of the fluency of verbal output.[15,42,43] *Nonfluent* output is characterized by a paucity of verbal output (usually 10–50 words per minute), whereas *fluent* aphasics have a normal or even exaggerated verbal output (up to 200 words or more per minute) with tachyphemia, logorrhea, press of speech, and an intrusive lack of regard for the usual rules of conversational interchange. Nonfluent aphasics have difficulty initiating and producing speech and tend to produce one-word replies or to utilize short phrases. Short grammatical connecting words are omitted; the output has an amelodic, dysrhythmic loss of prosody. The words produced are usually meaningful, and there is little paraphasia. Fluent aphasics have nearly the opposite pattern of verbal output, producing large amounts of well-articulated, prosodic phrases of normal length but

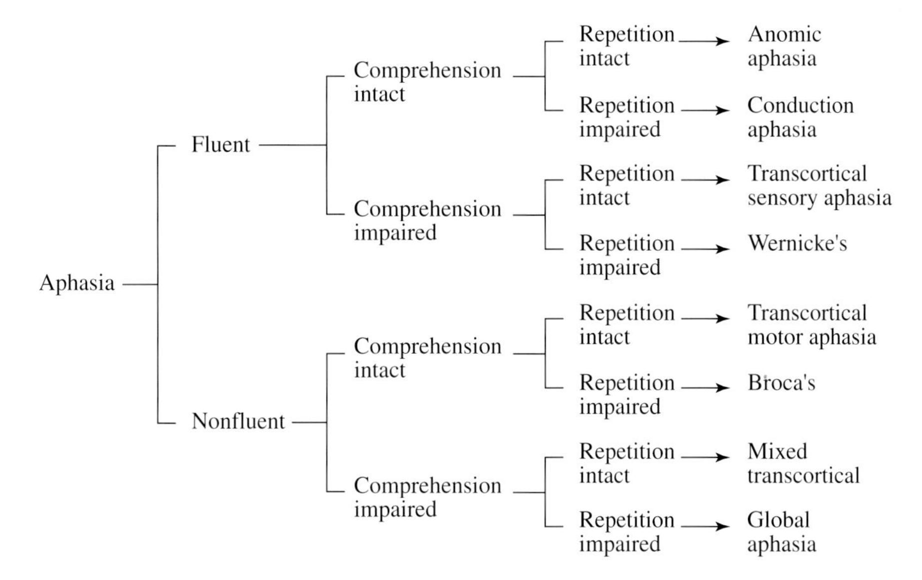

FIGURE 6.3 A systematic approach to aphasia diagnosis based on the cardinal observations of fluency of spontaneous speech, integrity of linguistic comprehension, and ability to repeat phrases and sentences is the best taxonomic aid for assessing language dysfunction.

with little information content. Grammatical words are not omitted, but errors in the use of grammar (paragrammatism) may occur. Paraphasias are prominent, and in the acute phases there is a tendency for the patient to be unaware of or to deny any language deficit. Not all of these distinguishing features are present in every case, and phrase length, agrammatism, and paraphasia are the most useful parameters for differentiation of fluent from nonfluent verbal output. In adults, fluent aphasias correspond to lesions in the posterior left hemisphere, whereas nonfluent aphasias are produced by lesions that include the left frontal lobe. In children, however, nonfluent aphasia may occur regardless of the location of the lesion.[44]

Assessment of language comprehension is the second step in evaluation of the aphasic patient. Comprehension deficits may be mild or severe, and a hierarchy of increasingly difficult tests of linguistic understanding must be utilized in aphasia testing (Chapter 3). In general, patients with focal lesions limited to the left frontal lobe have preserved comprehension (Broca's and transcortical motor aphasia), whereas patients with left posterior temporal or parietal involvement suffer some degree of comprehension impairment (Wernicke's, global, transcortical sensory, and isolation aphasias).

The third step in a systematic examination of the aphasic patient is to determine the patient's ability to correctly repeat words and sentences. For accurate assessment of this ability, the length of the presented material must not exceed the attention span of the subject. Failures in repetition take the form of paraphasic intrusions, alterations in the presented sequence of

TABLE 6.3. *Characteristics of Nonfluent and Fluent Aphasic Output*

Nonfluent Aphasia	*Fluent Aphasia*
Decreased output	Normal or increased verbal output
Effortful; difficulty with speech initiation	Little or no effort in initiating or producing speech
Dysarthria	Normal articulation
Decreased phrase length	Normal phrase length
Dysprosody	Melody and inflection preserved
Agrammatism	Low information content (empty speech)
High information content	Paraphasias present
Paraphasias are rare	Literal pharaphasias (phonemic substitution)
	Verbal paraphasia (semantic or word substitution)

TABLE 6.4. *Characteristics of the Aphasia Syndromes*

Aphasia Type	*Fluency*	*Comprehension*	*Repetition*	*Naming*	READING *Aloud*	READING *Comprehension*	*Writing*	*Special Features*
Broca's	Nonfluent	+[a]	−[b]	−	−	±	−	Comprehension of grammatically dependent constructions is impaired; apraxia is common
Transcortical motor	Nonfluent	+	+	−	+	±	−	Onset with mutism is typical
Global	Nonfluent	−	+	−	±	−	−	Automatic speech (e.g., counting) and singing may be preserved; right hemiparesis in nearly all cases
Mixed Transcortical (isolation aphasia)	Nonfluent	−	+	−	−	−	−	Echolalia
Wernicke's	Fluent	−	−	−	−	−	−	Semantic and neologistic paraphasia; no hemiparesis
Transcortical sensory	Fluent	−	+	−	±	−	−	Repeat without comprehending
Conduction	Fluent	+	−	−	±	+	±	Literal paraphasia is most common; limb apraxia is present
Subcortical (nonthalamic)	Fluent	+	+	±	−	+	±	Dysarthria; reduced generative language
Thalamic	Fluent	±	+	−	−	±	−	Onset with mutism; rapid fatigue of fluent output; dysarthria is common

words, omission of words, or a tendency to alter the content of the presented sentence. Aphasic syndromes with impaired repetition (Wernicke's, Broca's, and conduction aphasias) have lesions involving structures immediately adjacent to the sylvian fissure of the left hemisphere, whereas syndromes with intact repetition abilities (transcortical aphasias) are associated with lesions that spare the peri-sylvian areas.[45,46]

After the initial observations have been made regarding fluency, comprehension, and repetition, further refinements in characterization of the aphasic syndromes and lesion localization are made through use of tests of naming, reading, and writing. Table 6.4 summarizes the principal clinical features of each major type of aphasia.

- **Broca's Aphasia** *Broca's aphasia* is a nonfluent aphasic syndrome characterized by effortful, dysarthric, dysprosodic, and agrammatical verbal output. Although comprehension is largely intact in patients with Broca's aphasia, patients often have difficulty in mastering material that involves sequential manipulation based on specific grammatical relationships ("put the blue square on top of the red circle"). This deficit in grammatical comprehension is present for both oral and written material and corresponds to the agrammatism of the individual's spoken language.[47] Repetition, reading aloud, naming, and writing are also impaired. The lesion responsible for Broca's aphasia usually involves the inferior frontal gyrus and adjacent areas of the operculum and insula in the territory of the upper division of the middle cerebral artery.[15,46,48,49] The extent of the lesion determines the features of the aphasic syndrome. Damage to the frontal operculum produces difficulty with speech initiation; injury to the lower motor cortex results in dysarthria and dysprosody; damage that extends more posteriorly to involve the temporoparietal to opercular connections leads to phone-

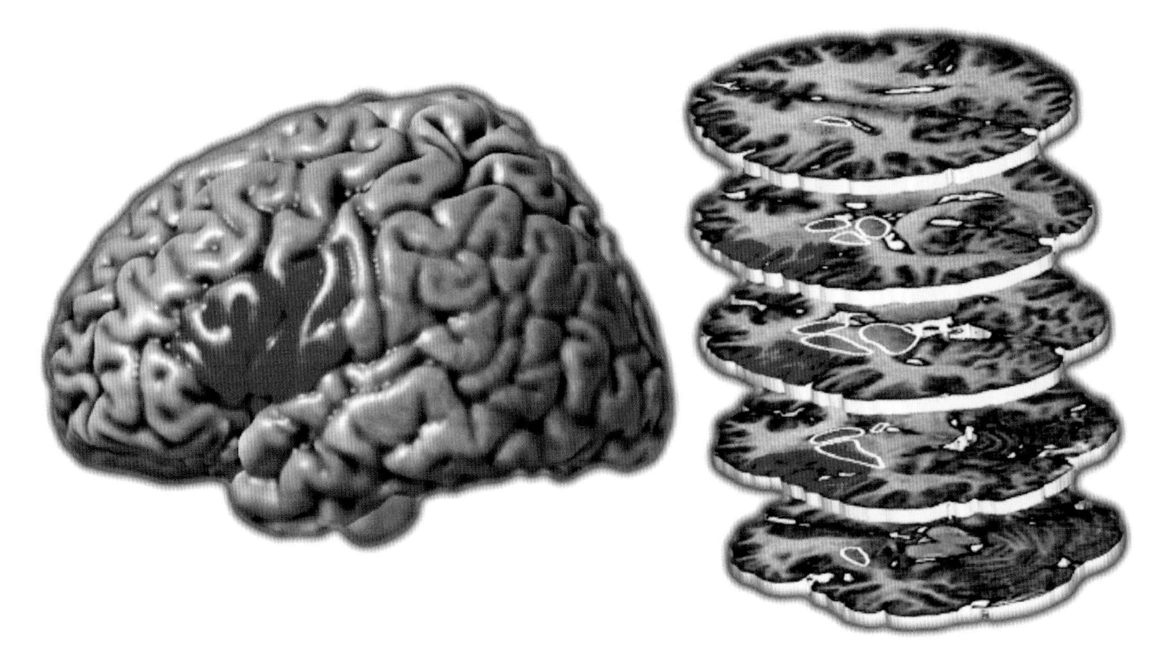

FIGURE 6.4 Classic Broca's aphasia is evident when the cortical lesion is accompanied by white matter adjacent to the ventricles and containing limbic–frontal periventricular pathways are included in the lesion (red).

mic paraphasias similar to those of conduction aphasia (described below). Classic Broca's aphasia combining all these features with agrammatical, shortened utterances is evident when the above regions plus the white matter adjacent to the ventricles and containing limbic–frontal periventricular pathways are included in the lesion (Fig. 6.4).[50] When the frontal lesion involves the premotor area and frontal operculum, a right-sided hemiparesis involving the face and the arm more than the leg usually accompanies the aphasia, and a sympathetic apraxia (discussed below) may affect buccolingual and left-sided limb function.[51–53]

Transcortical Motor Aphasia *Transcortical motor aphasia* is characterized by nonfluent verbal output, intact auditory comprehension, preserved repetition despite nonfluent spontaneous speech, intact ability to read aloud, variable reading comprehension, and poor naming and writing ability.[15,45] Echolalia may occur and there may be phonemic paraphasias in the patient's utterances.[48] Mutism is common in the early phases of the disorder. The syndrome resembles Broca's aphasia except that repetition is preserved and reading aloud is less impaired.

The usual lesion associated with transcortical motor aphasia involves infarction of the supplementary motor area and adjacent cingulate gyrus in the distribution of the anterior cerebral artery of the left frontal lobe,[21] but occasional cases with lesions of the frontal convexity sparing the Broca area or of the left putamen or thalamus have been reported.[15,22,25,45,46,54] The critical lesion may be disconnection of white matter tracts between frontal opercular language-related areas and supplementary motor area mediating speech initiation.[23] Lesions at either end of the connecting tracts may produce a similar clinical syndrome. In most cases, a right hemiparesis involving the leg more than the arm and face is present.

Global Aphasia Patients with global aphasia have impairments in virtually all aspects of language function, including spontaneous verbal output, comprehension, repitition, naming, reading aloud, reading comprehension, and writing.[15,45] Often the only spontaneous verbalizations will be stereotyped nonsense productions such as "za, za, za," although some patients may have a small repertoire of overlearned stock phrases ("hello," "I can't," etc.) that can be uttered fluently, and many global aphasics can curse with ease when angered. Automatic speech (counting, reciting the days of the week or months of the year), and humming of learned tunes ("Happy Birthday," "Jingle Bells") may be possible despite the severe defect in expressive propositional language. Poor comprehension of language distinguishes global aphasia from Broca's aphasia, and poor repetition distinguishes it from mixed transcortical aphasia (isolation aphasia). Even when comprehension is severely impaired, many global aphasics will be able to follow whole-body commands ("stand up," "sit down"), can distinguish foreign lan-

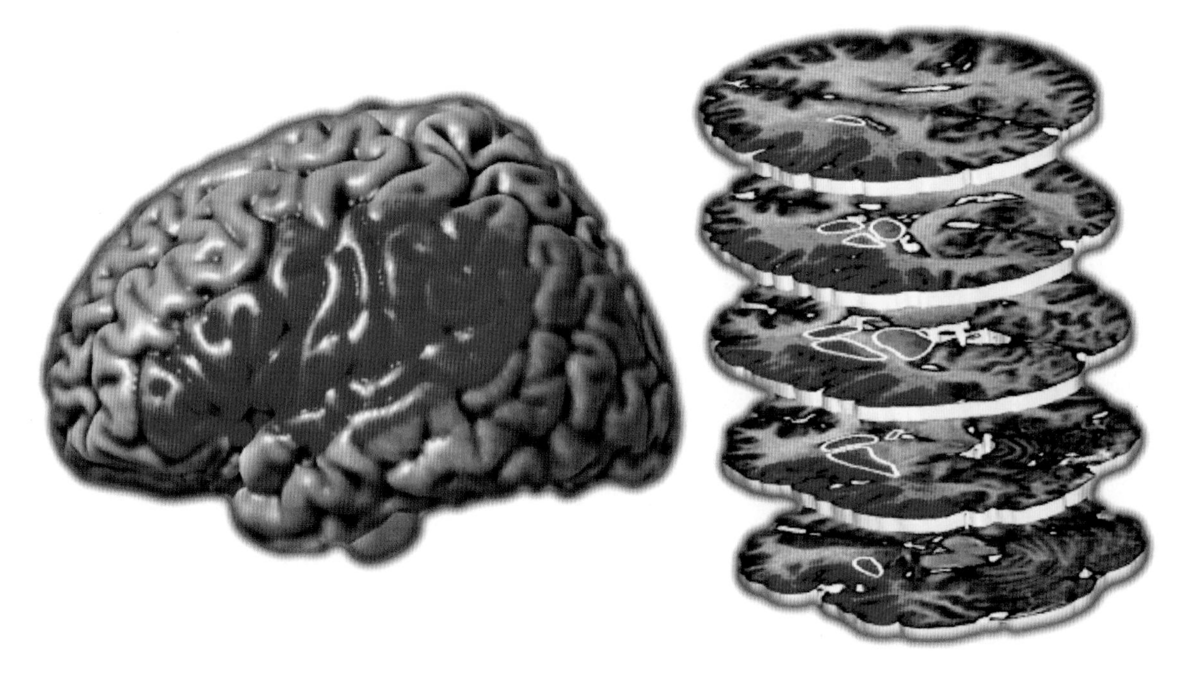

FIGURE 6.5 Global aphasia occurs when a large, left-sided lesion involves the entire territory of the middle cerebral artery (red).

guage and nonsense speech from their native tongues, can judge inflection sufficiently to differentiate questions and commands, can recognize spoken or written names of personally relevant persons and landmarks,[55] and will reject written language that is presented upside down.[46,56]

Pathologically, the usual lesion producing global aphasia is a large, left-sided infarction involving the entire territory of the middle cerebral artery (Fig. 6.5).[46] There is typically an accompanying hemiparesis, hemisensory defect, and homonymous hemianopsia. Rarely, multiple emboli to anterior and posterior language-mediating areas will produce global aphasia without major motor deficits.[57]

• **Mixed Transcortical (Isolation) Aphasia** *Mixed transcortical* or *isolation aphasia* is a rare aphasic syndrome in which the findings of transcortical motor aphasia and transcortical sensory aphasia are combined, leaving only a paradoxical preservation of the ability to repeat. In some cases, repeating whatever the examiner says (*echolalia*) is virtually the sole verbal output, whereas in others, nonfluent verbalization and even some naming may be intact.[58,59] Three types of lesions have been associated with mixed transcortical aphasia. In some patients there is damage in a sickle-shaped region involving the lateral aspects of the hemispheres but sparing the perisylvian cortex. A second type of lesion occurs with infarction in the territory of the anterior cerebral artery, again involving widespread cortical regions and sparing the peri-sylvian cortex.[59] A third type of lesion simultaneously affects posterior linguistic regions and the frontal lobes (Fig. 6.6) or frontal–subcortical circuits.[60,61] Involvement of the frontal lobes may produce environmental dependency and stimulus boundedness that contributes to the patient's paradoxical reduction of spontaneous speech (internally generated) with preserved repetition and echolalia (externally initiated).[60]

• **Wernicke's Aphasia** *Wernicke's aphasia* entails fluent, paraphasic output with poor comprehension, repetition, and naming. The patient's colorful, often nonsensical, logorrhea, frequently combined with an unawareness or denial of any deficit, creates one of the most striking syndromes in clinical neurology.[15,45] The patient exhibits press of speech with accelerated output and often has a demanding, intrusive, even quarrelsome conversational style. Spontaneous speech contains primarily semantic paraphasias and neologisms, whereas literal paraphasias may dominate attempts to respond to naming tests. Reading and writing are compromised. The production of logorrheic, paraphasic speech with multiple substitutions and in comprehensible gibberish is called *jargon aphasia,* a verbal output disorder that may also occur in conduction aphasia and transcortical sensory aphasia.[45,61] The relative preservation of comprehension in conduction aphasia and repetition in transcortical sensory aphasia distinguish these two disorders from Wernicke's aphasia.

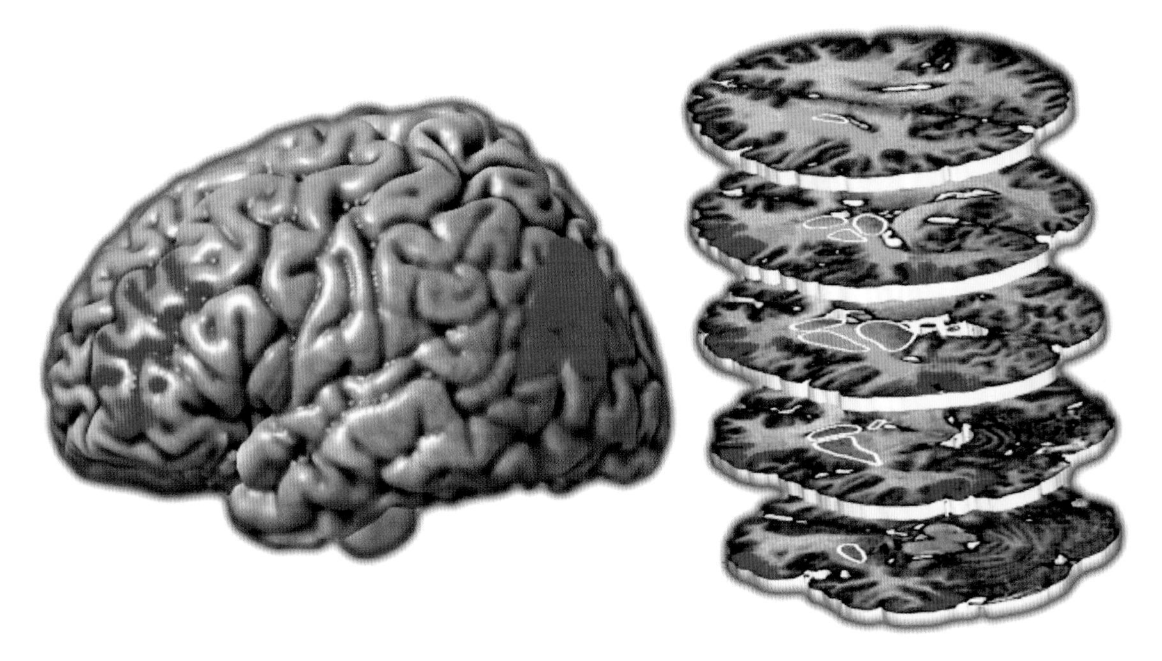

FIGURE 6.6 Mixed transcortical aphasia can occur when a lesion simultaneously affects posterior linguistic regions and the frontal lobes (red) and spares the arcuate fasiculus (green), thereby allowing normal repetition.

Although the cardinal features of Wernicke's aphasia (fluent output, poor comprehension, poor repetition) describe a basic syndrome, there are many variations in the clinical presentation. Comprehension may be mildly impaired with preserved ability to interpret moderately complex sentences, or it may be severely involved, sparing only simple midline and whole-body commands ("close your eyes," "open your mouth," "stand up," "sit down"). Comprehension of orally presented material may be relatively spared whereas written information is severely affected, or the reverse may occur. Greater involvement of auditory comprehension corresponds to more extensive involvement of temporal lobe structures, including primary auditory cortex, and greater compromise of reading comprehension may reflect extension of the lesion superiorly into inferior parietal lobe and angular gyrus.[15,62]

Pathologically, the lesion corresponding to Wernicke's aphasia involves the posterior third of the left superior temporal gyrus but rarely is limited to this region and frequently involves adjacent temporal and inferior parietal areas (Fig. 6.7).[15,46] The majority of patients with Wernicke's aphasia have had a cerebral infarction and, in most cases, the vascular occlusion was produced by an embolus arising from the heart.[63] Wernicke's aphasia is occasionally produced by a neoplasm or trauma and is rarely observed in patients with degenerative dementias.

A superior quadrantanopsia and cortical sensory loss in the face and the arm are common associated findings in patients with Wernicke's aphasia, and if the lesion extends into the posterior limb of the internal capsule, a hemiparesis will result.

Transcortical Sensory Aphasia Transcortical sensory aphasia shares many features with Wernicke's aphasia but is distinguished by the retained ability to repeat. The ease with which the patient repeats long sentences and phrases, but cannot comprehend them, is notable. Spontaneous speech is empty, circumlocutory, and paraphasic, and there is a mild to marked tendency to spontaneously repeat (echo) whatever the examiner says. The patient may be able to read aloud, but both reading and auditory comprehension are impaired.[15,45,48]

Transcortical sensory aphasia is produced by focal lesions involving the dominant angular gyrus, posterior middle temporal gyrus, and periventricular white matter pathways of the temporal isthmus underlying these cortical areas.[46,64,65] When it results from involvement of the angular gyrus, it is frequently accompanied by Gerstmann's syndrome, constructional disturbances, and other evidence of the angular gyrus syndrome (Fig. 6.8).[66] Transcortical sensory aphasia may also be seen during one stage of the evolution of Alzheimer's disease.[67]

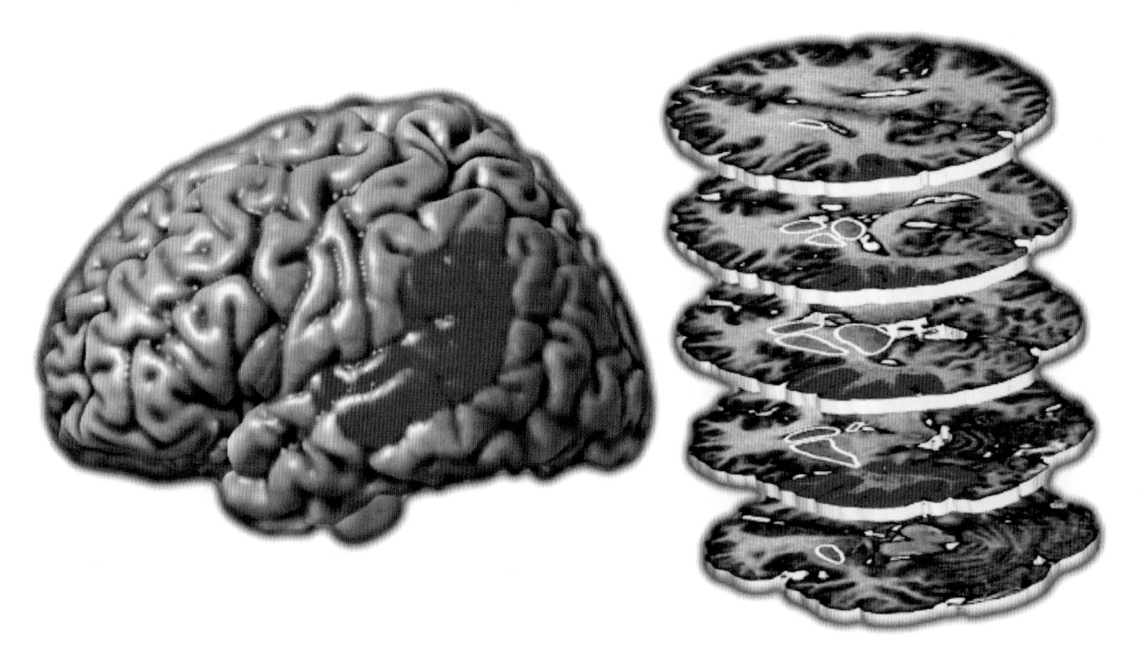

FIGURE 6.7 The lesion corresponding to Wernicke's asphasia involves the posterior third of the left superior tempral gyrus (red).

Conduction Aphasia *Conduction aphasia* is a unique fluent aphasic syndrome in which comprehension is relatively intact and repetition is disproportionately impaired. Spontaneous speech is characterized by word-finding pauses and a predominance of phonemic or literal paraphasias over semantic or neologistic paraphasias. Often the patient is aware of making errors and makes sequentially closer approximations to the intended word (conduit d'approche). Reading aloud is impaired, but reading comprehension is intact. Naming and writing are both abnormal and contain phonemic paraphasic substitutions.[15,45,48] Although compre-

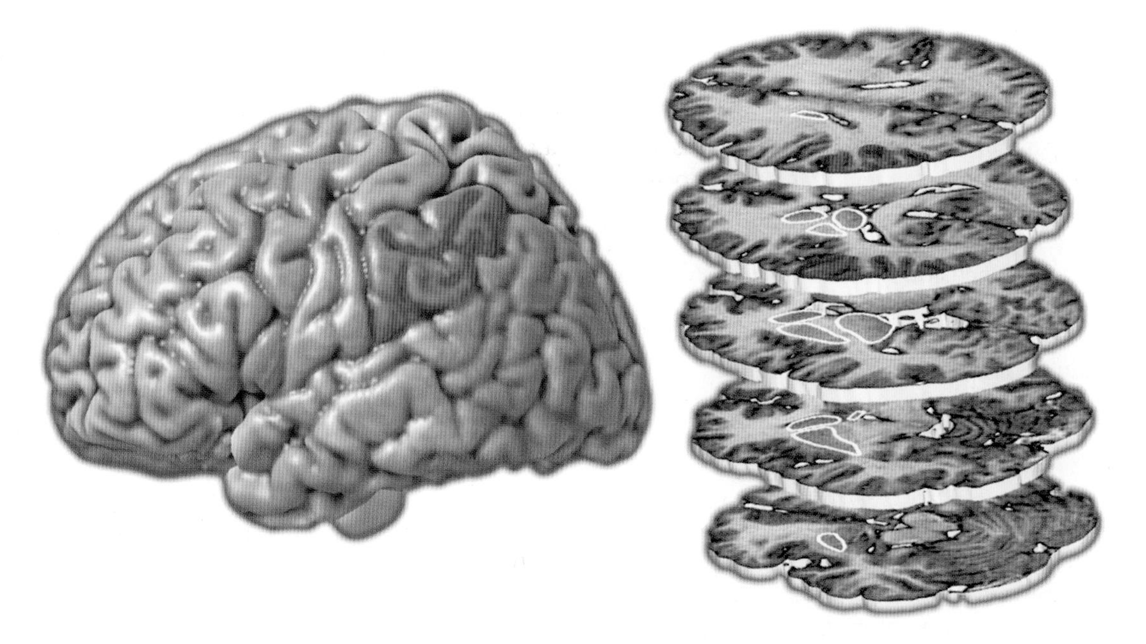

FIGURE 6.8 Transcortical sensory asphasia results from a lesion of the angular gyrus (red) and is similar to Wernicke's aphasia but with intact repetition.

hension is relatively preserved in conduction aphasia, some patients have syntactic comprehension defects similar to those described in Broca's aphasia.[68]

The lesion responsible for conduction aphasia typically involves the arcuate fasciculus in the left parietal operculum. Adjacent cortical areas mediating language comprehension are often involved.[46]

• Anomic Aphasia Anomia is a ubiquitous finding in disorders affecting the cerebral hemispheres and is present in all types of aphasia as well as in toxic-metabolic encephalopathies and with increased intracranial pressure. In the latter circumstances, anomia is a nonspecific indicator of cerebral dysfunction and has no localizing significance.[15,69] Three primary types of anomia occur in aphasic syndromes: word production, word selection, and semantic anomia.[69] *Word production anomia* is characterized by an inability to express the desired word. The primary problem is a disturbance of initiation of the word, and patients respond readily to phonemic cues (the first syllable or first sound of the word). Word production anomias are characteristic of nonfluent aphasias such as Broca's aphasia and transcortical motor aphasia. It is also the principal type of naming deficit in patients with subcortical dementias (Chapter 10).

Patients with *semantic anomia* have an impaired ability to name, do not respond to cues, and cannot recognize the word when it is said by the examiner. The sound of the word is bereft of meaning. Semantic anomia occurs in Wernicke's aphasia and transcortical sensory aphasia.

Word selection anomia features anomia, a failure to respond to phonemic cues, but an intact ability to recognize the word when provided. Word selection anomia is the principal feature of *anomic aphasia.*

Spontaneous speech has an empty, circumlocutory quality with frequent word-finding pauses, many words of indefinite reference ("it", "thing", etc.), and little paraphasia. Comprehension is relatively preserved, and repetition, reading aloud, and reading comprehension are spared. Anomia will be present on tests of confrontation naming and in spontaneous writing. The patient can usually, but not invariably, recognize the correct word when it is presented by the examiner.[15,45,48,69] Anomic aphasia usually indicates a lesion in the left angular gyrus or adjacent areas of the posterior second temporal gyrus. Some patients with anomic aphasia have had lesions of the left anterior temporal or temporal polar regions.[46] Anomic aphasia is frequently the residual deficit following recovery from more extensive aphasic syndromes (Wernicke's aphasia, conduction aphasia).

• Aphasia with Lesions of the Thalamus and Basal Ganglia Aphasia is regarded traditionally as a sign of cortical dysfunction. With the advent of computerized tomography (CT) and magnetic resonance imaging (MRI), however, it became apparent that subcortical lesions of the left hemisphere were sometimes associated with aphasic syndromes. The aphasic syndrome associated with hemorrhage in the dominant thalamus is the best known subcortical aphasia and consists of a fluent paraphasic output, variable compromise of comprehension (mild in most), good repetition, poor naming ability, impairment of reading aloud and writing, and relatively preserved reading comprehension. There may be rapid fluctuations in the degree of aphasia and marked fatigability of fluent output. The syndrome may closely resemble transcortical sensory aphasia, but there is often an initial mute period at the time of onset, and articulatory deficits may persist throughout the clinical course. The aphasia is often transient and is usually associated with attentional deficits, right-sided neglect, lack of appropriate concern, perseveration, and right hemiparesis.[10] A similar syndrome has been observed with infarctions of the dominant thalamus although in many cases there is no associated language disorder. Aphasia following circumscribed lesions of the thalamus is usually transient and studies of cerebral blood flow or cortical glucose metabolism indicate that reduced cortical perfusion or metabolism is present when a subcortical lesion is associated with an aphasia syndrome. These observations suggest that the thalamus has an important role in word production and cortical activation but thalamic dysfunction is not sufficient to produce a specific aphasia unless there is associated cortical dysfunction.[70]

Infarction of left-sided basal ganglia structures may also produce a syndrome of reduced generative language with prominent dysarthria and hypophonia[45] (Fig. 6.9). Syndromes associated with hemorrhage are similar but more severe. Nonhemorrhagic lesions may produce aphasic syndromes by disrupting subcortical white matter tracts and radiations or by extending to involve adjacent cortical regions.[43,71] The principal characteristics of the language syndrome associated with left basal ganglia dysfunction are word-finding deficits (lexical selection anomia), occasional semantic substitutions, and impaired comprehension of complex syntactically determined material. Generative language is reduced with deficient word list generation, poor sentence generation, increased latency and perseveration, and echolalia. These findings are nonspecific and compatible with loss of facilitating or activating influences exerted by subcortical structures on cortical activities. Metabolic studies reveal a reduc-

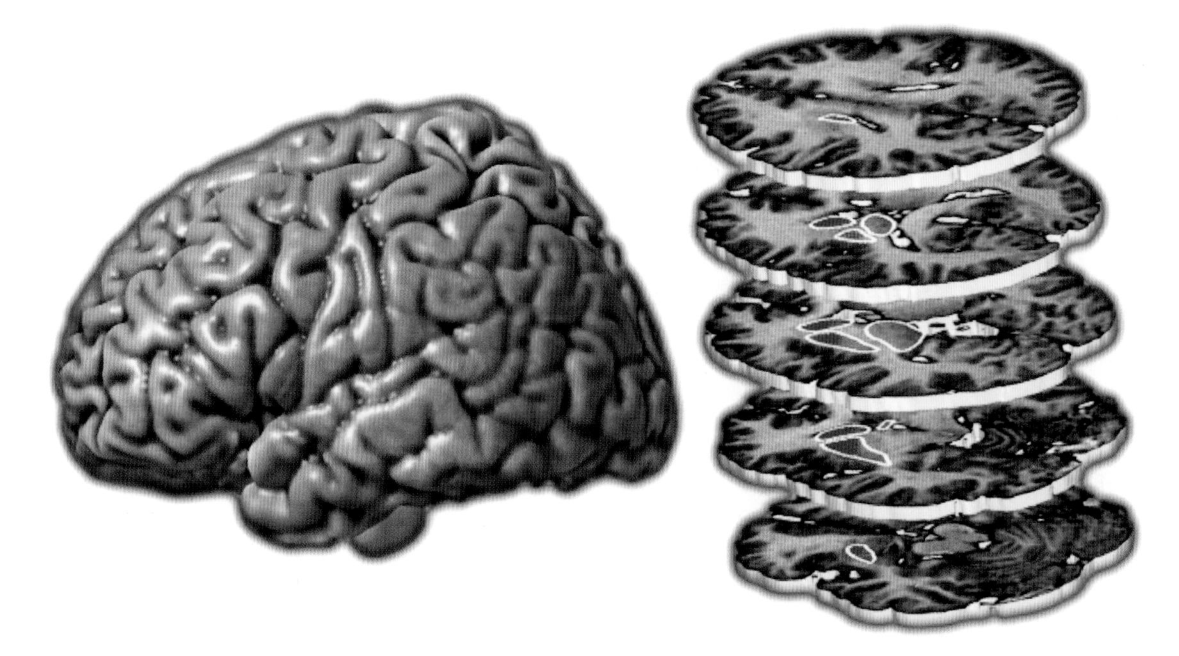

FIGURE 6.9 Infarction of left-sided basal ganglia structures (red) may produce a syndrome of reduced generative language with prominent dysarthria and hypophonia. The syndrome of subcortical aphasia occurs because of a disconnection of activating influences exerted by subcortical structures on cortical function.

tion in cortical activity proportional to the severity of the aphasia.[72,73]

Neuropsychiatric Complications, Prognosis, and Treatment of Aphasia Prognosis and treatment of the aphasic patient depend not only on the nature of the aphasia syndrome itself but also on the associated neuropsychiatric complications. Table 6.5 lists the neurological and neuropsychiatric disturbances commonly associated with anterior or posterior left hemisphere lesions. Catastrophic reactions may occur with either anterior or posterior lesions,[74,75] but they tend to be more common in patients with anterior subcortical lesions, particularly those with concomitant depression.[76] Patients with posterior lesions, particularly Wernicke's aphasics, may be unaware of their language deficits, and these patients exhibit the greatest tendency toward becoming suspicious and paranoid during the course of their illness.[77,78] Anxiety is most often found in patients with anomic aphasia and retrorolandic lesions,[74] although it also occurs with lesions of the left frontal cortex.[79] Depression is more common and more profound among patients with anterior than posterior lesions. Depression is not correlated with severity of physical or cognitive impairment but is related to the proximity of the lesion to the anterior pole of the frontal lobe (Chapter 14).[80,81] Moderate to severe depression in aphasics is associated with neurovegetative disturbances (sleep and appetite alterations) and an abnormal dexamethasone suppression test.[82]

Prognosis for language recovery varies with etiology of the aphasia and the type of linguistic deficit. The outcome of aphasias associated with neoplasms de-

TABLE 6.5. *Neurological and Neuropsychiatric Disturbances Associated with Anterior and Posterior Left Hemisphere Lesions*

Clinical Features	*Anterior Lesion*	*Posterior Lesion*
Aphasia	Nonfluent	Fluent
Neurological deficits		
Hemiparesis	+	+/−
Hemisensory loss	−	+/−
Hemianopsia	−	+/−
Behavioral alterations		
Depression	+	−
Denial and/or unawareness	−	+
Anxiety	+	+
Paranoia, suspiciousness	−	+
Catastrophic reactions	+	+/−

pends directly on the success of treating the tumor. Traumatic aphasias recover more completely than do aphasias produced by cerebrovascular disease, and among vascular aphasias, the greatest amount of recovery occurs within the first 3–6 months, although minor degrees of recovery may continue for 5 or more years.[83,84] Global aphasics have the worst prognosis for recovery of useful language skills; Broca's and Wernicke's aphasics have an overall fair prognosis for recovery with sizable variations from patient to patient; anomic, conduction, and transcortical aphasics have a relatively good prognosis, with some patients recovering completely.[83,85] Neuroimaging studies provide useful prognostic information. Lesions that directly involve the posterior superior temporal region of the left hemisphere suggest that there will be limited recovery of auditory comprehension, and large lesions affecting the rolandic area correlate with poor recovery of fluency.[86,87] In many cases, patients with more extensive linguistic deficits evolve into a stage of residual anomic aphasia. Younger patients with aphasia tend to recover more language skills than older patients, and left-handed patients have a better prognosis than dextrals.[83,88] In general, comprehension of language improves more than fluency of expressive output.[89,90]

Aphasia therapy may facilitate language recovery and should be offered to all interested patients.[91] In addition to traditional re-education techniques, recent efforts have been made to develop individualized techniques for specific types of aphasia, such as utilization of melodic intonation therapy in patients with Broca's aphasia, use of visual communication symbols by patients with global aphasic syndromes, or therapy for specific aspects of aphasic syndromes such as perseveration.[92–94]

Aphasia-Related Syndromes

• Alexia

Alexia refers to an acquired inability to read caused by brain damage and must be distinguished from dyslexia, a developmental abnormality in which the individual is unable to learn to read, and from illiteracy, which reflects a poor educational background.[15] Most aphasics are also alexic, but alexia may occur in the absence of aphasia and may occasionally be virtually the sole disability resulting from specific CNS lesions. The ability to read aloud and reading comprehension may be dissociated by some lesions and must be assessed independently. Table 6.6 presents a classification of alexia, and each alexic syndrome is discussed in the following paragraphs.

TABLE 6.6. *Classification of the Alexias*

Alexia without agraphia
Alexia with agraphia
Without aphasia
With fluent aphasia
With nonfluent aphasia (frontal alexia)
Deep dyslexia (deep alexia, paralexia)
Hemialexia (neglect syndrome)

Alexia Without Agraphia. Alexia without agraphia is a classic disconnection syndrome in which the usual lesion includes infarction of the left occipital cortex and the posterior aspect of the corpus callosum (Fig. 6.10). The occipital lesion produces a right homonymous hemianopsia, making reading in the right visual field impossible. The callosal lesion makes it impossible to transfer visual information from the left visual field (perceived by the intact right occipital region) across the corpus callosum to the left posterior hemispheric region, where graphic decoding can occur.[12,15,46,48]

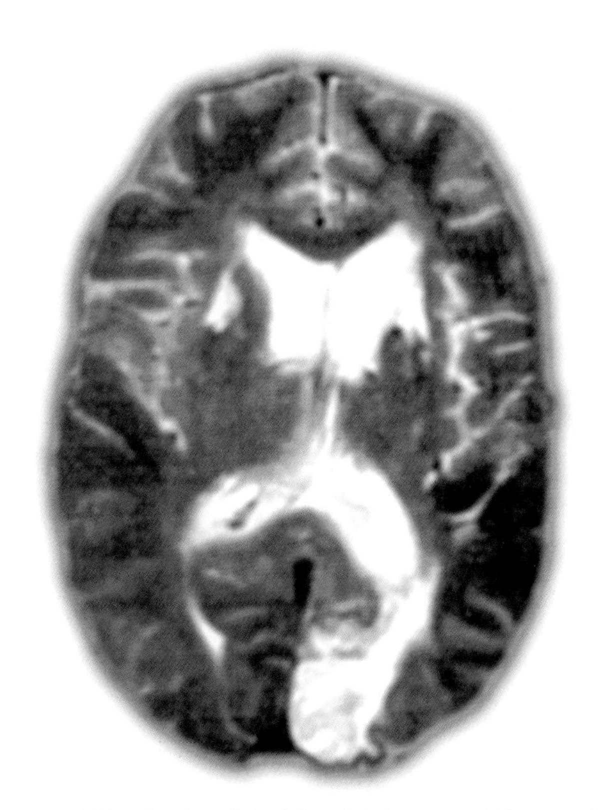

FIGURE 6.10 Infarction of the left occipital cortex and the posterior aspect of the corpus callosum produces alexia without agraphia.

Alexia without agraphia has rarely been associated with small lesions in the white matter beneath the angular gyrus[95] or with lesions of the left lateral geniculate plus the splenium of the corpus callosum.[96] In both cases, a disconnection similar to that occurring with the classic lesion occurs. In the alexia-without-agraphia syndrome, letter reading is superior to word reading. The patient retains the ability to spell and to recognize words spelled aloud, there is greater difficulty in copying words than in writing spontaneously, and a color anomia is frequently present.[15,48] In some cases a right hemiparesis, right hemisensory loss, and mild naming disturbance accompany the syndrome.[97]

Alexia with Agraphia. Alexia with agraphia may occur with no significant associated aphasia, with fluent aphasia, or with nonfluent aphasia. Alexia with agraphia in the absence of aphasia occurs with lesions in the region of the angular gyrus and often coexists with elements of Gerstmann's syndrome (discussed below). Both letter and word reading are impaired, and the patient cannot spell aloud or recognize spelled words—the syndrome is equivalent to an acquired illiteracy. Word copying is superior to spontaneous writing.[15,45] The syndrome is usually produced by occlusion of the angular branch of the middle cerebral artery but may occur as part of a border zone syndrome following carotid occlusion or with trauma or neoplasms.

Alexia with agraphia and fluent aphasia occurs with Wernicke's aphasia or transcortical sensory aphasia; in the latter, reading aloud may be preserved despite impaired reading comprehension. Alexia with agraphia and nonfluent aphasia is found in some patients with Broca's aphasia. Not all Broca aphasics are alexic, but when alexia is present, the reading disturbance has several distinctive characteristics. Word reading is superior to letter reading, and the words recognized are almost exclusively substantive nouns and verbs. The reading disability is comparable to other aspects of Broca's aphasia: syntactic comprehension and spontaneous production of the small grammatical functor words are impaired and comprehension and production of substantives is preserved. Spontaneous writing and copying of verbal material are also abnormal.[98]

Deep Dyslexia. This syndrome (deep alexia, paralexia) evolves in some aphasics with severe reading impairments in which semantically related paralexias are produced in response to written stimuli. The patient may read *automobile* as *car* or *infant* as *baby*. Such reading is thought to be mediated by the right hemisphere on the basis of iconic recognition.[98–100]

Hemialexia. Alexia may occur with hemispheric lesions that produce profound unilateral neglect. The syndrome usually occurs in patients with right hemispheric lesions and severe hemispatial inattention. The left half of words is ignored so that *northwest* is read as *west* or *baseball* as *ball*; or the left half may be misjudged so that *navigator* is read as *indicator*, *match* as *hatch*, or *alligator* as *narrator*.[98,101]

● **Agraphia** *Agraphia* indicates an acquired impairment of the ability to write.[15,102,103] It may reflect an aphasic disturbance with a writing deficit similar to that of oral language, or it may be a consequence of a motor system abnormality. Like alexia, agraphia must be distinguished from illiteracy, where writing skills were never developed. Table 6.7 presents the classification to be followed here. In the following paragraphs, aphasic agraphias are discussed first, and then nonaphasic agraphias are presented.

Aphasic Agraphia. As shown in Table 6.2, all aphasias are accompanied by writing disturbances. The type of writing disturbance usually closely parallels the disturbances of oral language, and in some cases the language abnormalities may be more marked in written than spoken language. In the nonfluent aphasias there is sparse graphic output, with clumsy calligraphy, agrammatism, and poor spelling. Fluent agraphias, on the other hand, have a normal quantity of well-formed letters, but with a lack of substantive words and insertion of literal, verbal, or neologistic paragraphias similar to oral paraphasias.[102]

Alexia with agraphia was discussed earlier with the alexias. The writing disturbance of alexia with agraphia is severe and has the characteristics of agraphias accompanying fluent aphasias. Similarly, the agraphia of Gerstmann's syndrome (discussed below) is a fluent form of agraphia but in its pure form lacks an accompanying disturbance of reading.

Pure agraphia is a controversial entity originally posited to occur with left frontal lobe lesions. No convincing cases with isolated pathologic involvement in this region have been described, but there are cases of relatively pure agraphia with left parietal lobe lesions.[104]

Chedru and Geschwind[105,106] observed that writing disturbances are among the most sensitive measures of confusional states associated with toxic and metabolic encephalopathies, and in some cases a relatively pure agraphia was the most prominent neuropsychological manifestation of the encephalopathy. The characteristics of the writing disturbance occurring in acute con-

TABLE 6.7. *Classification of the Agraphias*

Aphasic Agraphias	*Nonaphasic Agraphias*
Agraphia with fluent aphasia	Motor agraphia
Agraphia with nonfluent aphasia	Paretic agraphia
Alexia with agraphia	Hypokinetic agraphia
Gerstmann's syndrome agraphia	Micrographia with parkinsonism
Pure agraphia	Hyperkinetic agraphia
Agraphia in confusional states	Tremor
Deep agraphia	Chorea, athetosis, tics
Disconnection agraphia	Dystonia (writer's cramp)
Apraxic agraphia	Reiterative agraphia
	Perserveration
	Paligraphia
	Echographia
	Coprographia
	Visuospatial agraphia
	Hysterical agraphia

Source: Adapted from Benson DF, Cummings JL. Agraphia. In: Vinken PJ, Bruyn GW, eds. Disorders of Speech, Perception, and Symbolic Behavior, Vol. 4, 2nd ed., Handbook of Clinical Neurology. New York: American Elsevier Publishing Company, 1985, Table 34–1. With permission from Excerpta Medica, Elsevier Science Publishers.

fusional states include poor coordination and mild tremor, spatial misalignment, agrammatism, omission and substitution of letters (especially consonants), and reduplication of letters and words. Errors are concentrated at the endings of words.

Deep agraphia refers to a syndrome similar to deep dyslexia involving writing rather than reading. Concrete imageable words are written much better than abstract or nonsense words, and semantic paragraphias, similar to the semantic paralexias of deep dyslexia, are present.[102,107] The syndrome usually occurs in patients with severe alexia and left parietal lobe lesions.

Disconnection agraphia occurs with other aspects of callosal ideomotor apraxia (discussed below) in patients with lesions resulting in disconnection of the writing hand from the necessary input of the left hemisphere. The agraphia occurs in the left hand of right-handed patients with callosal lesions. The lesion prevents the transfer of linguistic information from the left to the right hemisphere controlling the left hand. Copying is usually superior to spontaneous writing or writing to dictation.[102,108]

Apractic agraphia is one manifestation of ideomotor apraxia affecting the limbs. The patient has difficulty forming letters when writing spontaneously and when copying. Performance may be improved by spelling with anagram letters. The associated lesion is typically located in the left superior parietal lobe.[72,103]

Nonaphasic Agraphia. Writing depends on a complex array of motor and visuospatial skills in addition to language abilities. Disruption of any aspect of the motor system—peripheral, corticospinal, extrapyramidal, cerebellar—will produce agraphia, and in each case the writing disturbance will have distinctive features. Lesions of the muscles, peripheral nerves, or corticospinal tracts produce a clumsy, uncoordinated agraphia secondary to limb paralysis. Micrographia is a common manifestation of parkinsonism and occurs in idiopathic, postencephalitic, and drug-induced parkinsonian disorders.[102] Extrapyramidal micrographia is characterized by a progressive diminution in the size of the letters, often accompanied by increased crowding. The micrographia may be most apparent in writing but eventually includes all written productions including constructions. Tests of micrographia that make the deficit apparent include obtaining the patient's current signature and comparing it with past signatures on licenses or legal documents, asking the patient to execute sequences such as the alphabet or consecutive digits, or having the patient draw repeated connected oval loops. While usually a manifestation of a degenerative parkinsonian disorder, tumors and other focal lesions of the basal ganglia can also produce micrographia.[109]

Action tremors (Chapter 18) of either the cerebellar or postural type produce disturbances in writing and may make written productions unintelligible. Postural tremor is a high-frequency (8–12 Hz), low-amplitude tremor that is precipitated by movement and disappears at rest. Postural tremors are very evident in written material, and in some cases writing is the only maneuver that elicits the tremor.[110] Cerebellar tremors are usually large-amplitude, intention tremors that are worsened by attempts to produce fine writing movements. Often the patient can make only a few sweeping marks on the page.

Chorea, athetosis, and tics are hyperkinetic movement disorders that influence writing in the same way that they affect all other volitional motor activity. In severe cases writing is impossible, and even in mild cases the output will be visibly distorted. The differential diagnosis and treatment of these disorders is discussed in Chapter 18.

"Writer's cramp" is among the most well-known and most misunderstood of all agraphias. The syndrome of progressive cramping of the hand and forearm among individuals in professions demanding fine finger movements, including writers, telegraphers, pianists, and violinists, was treated as a neurotic disorder by early psychoanalysts and as a learned disturbance by behavioral therapists. The progression of writer's cramp to a segmental dystonia involving the entire limb or even to generalized dystonia musculorum deformans, however, along with the absence of a consistent psychopathology among its victims, indicate that the disorder is a focal dystonia.[111] The cramping begins between ages 20 and 50, and there may be inconspicuous associated neurological deficits, including abnormal posturing or tremor of the affected limb, diminished arm swing, or increased limb tone.

Reiterative agraphias refer to the abnormal repetition of letters, words, or phrases in writing. Perseveration is a continuation of activity after the appropriate stimulus has stopped; paligraphia is the rewriting of phrases generated by the patient; and echographia is the rewriting of phrases produced by the examiner. These disorders occur in severely deteriorated patients, including those with advanced degenerative, vascular, or traumatic conditions, and in catatonic disturbances.[102] Coprographia occurs primarily in Gilles de la Tourette syndrome, where the patient occasionally has a compulsion to express coprolalic tendencies in writing.[112] (Chapter 19).

Visuospatial agraphia is manifested by a tendency to neglect one portion of the writing page; slanting of the lines upward or downward; and abnormal spacing between letters, syllables, or words. It is seen most often with right-sided lesions in the region of the temporoparietooccipital junction and is accompanied by other evidence of left-sided neglect.[103]

Agraphia may occasionally occur as a hysterical conversion symptom. The agraphia is usually part of a monoparesis in which the limb is weak throughout with slightly diminished tone and normal muscle stretch reflexes. Sensation may or may not be affected.[113] In some patients, the writing disturbance may be unaccompanied by other functional disturbances.[114] The disorder typically is short-lived, and the psychogenic cause seldom is subtle.

• Acalculia

There are three principal types of acalculia: (*1*) acalculia associated with language disturbances, including number paraphasia, number agraphia, or number alexia; (*2*) acalculia secondary to visuospatial dysfunction with malalignment of numbers and columns; and (*3*) a primary anarithmetria entailing disruption of the computation process (Table 6.8). A fourth type of acalculia, symbol agnosia, in which the patient loses the ability to understand the operational symbols that determine the mathematical process to be performed (+, ÷, ×, −), has occasionally been observed but has not been well studied and is rare.[115]

Aphasia-related disturbances of calculation include paraphasic errors in which the patient makes a verbal paraphasic error, substituting one number for another. Number alexia and number agraphia may also occur and, in some cases, may be disproportionately greater than letter reading and writing disturbances. Acalculia occurs with nearly all aphasias but is more severe in patients with lesions of the posterior aspect of the left hemisphere involving the parietal cortex.[116]

Visuospatial acalculia may occur with lesions of either hemisphere but is most common with right pari-

TABLE 6.8. *Classification of the Acalculias*

Language-related acalculias
Number paraphasia
Number alexia
Number agraphia
Visuospatial acalculia
Anarithmetria
Symbol agnosia

etal dysfunction. Spacing of multidigit numbers, place-holding values, and column alignment are disrupted.[116]

Primary anarithmetria occurs mainly in the context of Gerstmann's syndrome with lesions in the region of the dominant angular gyrus, but it may occasionally be seen as an isolated abnormality with disturbances of the same region. In this case there is no significant aphasic or visuospatial disturbance, but errors are made in the computation process.[117]

● **Gerstmann's Syndrome and Angular Gyrus Syndrome** In 1924, Josef Gerstmann described a syndrome occurring with discrete left angular gyrus lesions and consisting of a tetrad of clinical findings including dysgraphia, finger agnosia, inability to distinguish left from right, and acalculia. In 1940, Gerstmann reviewed the considerable literature that had evolved concerning the syndrome and concluded that the findings had clinical validity and localizing value.[118] The prominence of the different components varies in each individual case, and specific testing may be necessary to elicit subtle deficits.[119] When one or more of the elements of the syndrome is missing, the localizing implications of the remaining members are doubtful.[120]

In many cases a lesion of the dominant angular gyrus produces deficits in addition to Gerstmann's syndrome (Fig. 6.11). Some degree of aphasia is frequently present, alexia with agraphia may occur, and constructional disturbances often accompany the Gerstmann's syndrome elements. This combination of deficits may closely imitate the clinical findings of Alzheimer's disease.[66]

● **Apraxia** *Apraxia* refers to disorders of learned movement that cannot be accounted for on the basis of weakness, sensory loss, inattention, or failure to understand the requested action.[121] Two principal types of apraxia have been recognized: (*1*) ideational apraxia, in which the patient fails to correctly pantomime a mul-

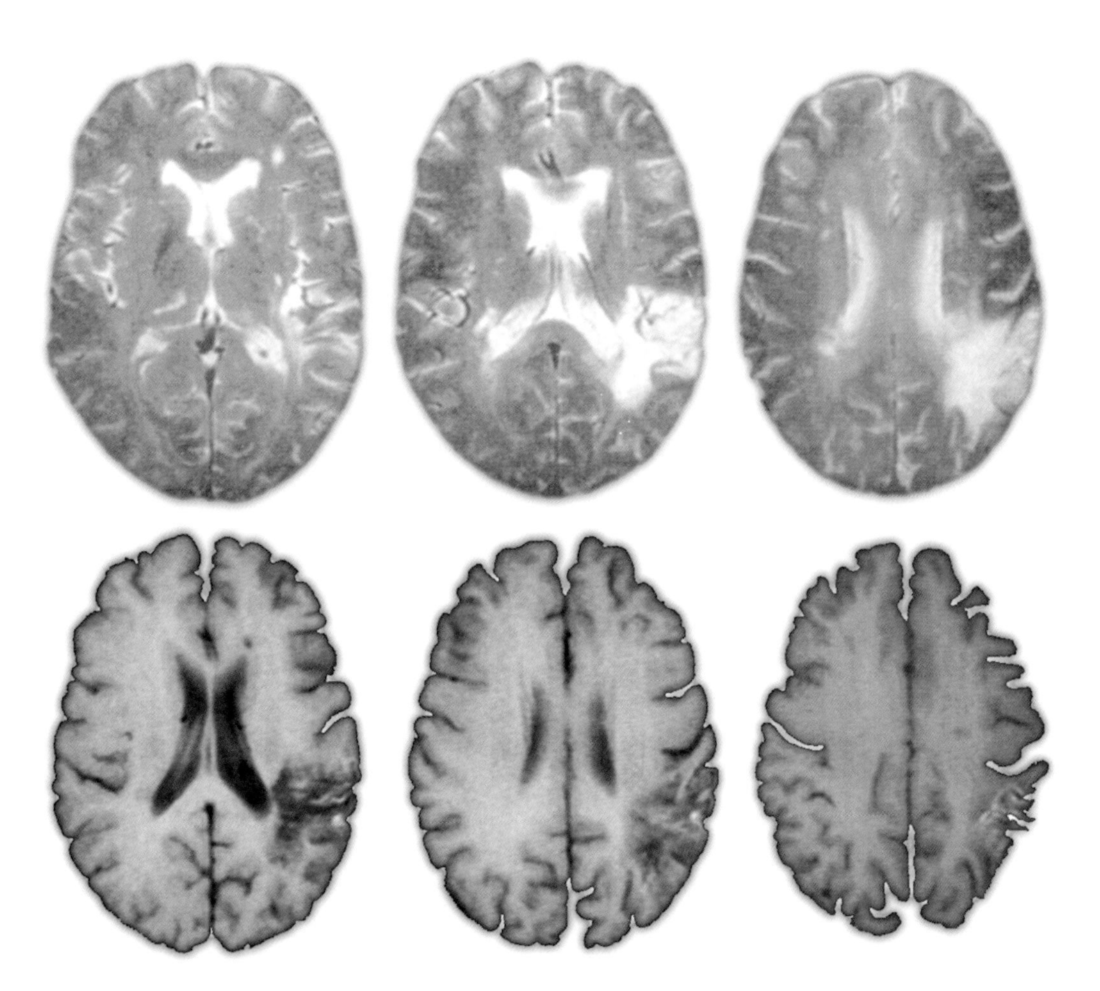

FIGURE 6.11 The Gerstmann's syndrome occurs with discrete left angular gyrus lesions and consists of a tetrad of clinical findings including dysgraphia, finger agnosia, inability to distinguish left from right, and acalculia.

TABLE 6.9. *Ideomotor Apraxias and Accompanying Clinical Findings*

	TYPE OF APRAXIA		
Characteristic	*Parietal*	*Sympathetic*	*Callosal*
Apraxia distribution	Bilateral limb, buccolingual	Left limbs, buccolingual	Left limbs
Aphasia	Conduction	Broca's	None or transcortical motor
Hemiparesis	± Right-sided weakness	Right hemiparesis	None
Lesion location	Inferior parietal lobule (arcuate fasciculus)	Left premotor cortex	Callosal fibers
Associated findings	± Right hemisensory loss	—	Commissurotomy syndrome*

*This involves left tactile anomia, left agraphia, poor crossed tactile matching, poor intermanual position matching, and right-hand constructional disturbance.

ticomponent sequence such as folding a letter, inserting it in an envelope, and stamping the envelope,[122] and (2) ideomotor apraxia, in which the patient fails to perform on command actions that can be done spontaneously, such as waving good-bye, hammering, thumbing a ride, sawing, sucking through a straw, or whistling. Ideational apraxias occur in dementias and in acute confusional states. Ideomotor apraxias occur with specific left hemisphere lesions.[121,123]

Table 6.9 presents the three principal types of ideomotor apraxia and their associated clinical findings. *Parietal apraxia* refers to the occurrence of apraxic movements in patients with lesions involving the inferior parietal lobule and the adjacent arcuate fasciculus. Thalamic lesions (left) have occasionally produced the syndrome.[124] The patients have fluent aphasia (usually conduction aphasia), may have a mild right hemiparesis and a hemisensory defect, and frequently fail to recognize that the apraxic movements are incorrectly performed.[125,126]

Sympathetic apraxia is the apraxia of the left limbs and buccolingual structures noted in patients with left frontal lesions. The apraxic limbs are "in sympathy" with the right hemiparesis produced by the frontal lesion. The patients also manifest a nonfluent Broca-type aphasia, have more prominent involvement of buccolingual than limb movements, and are likely to perceive that the apraxic movements are faulty.[121,127]

Callosal apraxia occurs when verbal directions mediated by the left hemisphere cannot cross the corpus callosum for execution of left-sided limb commands mediated by the right hemisphere (Fig. 6.12). The apraxia

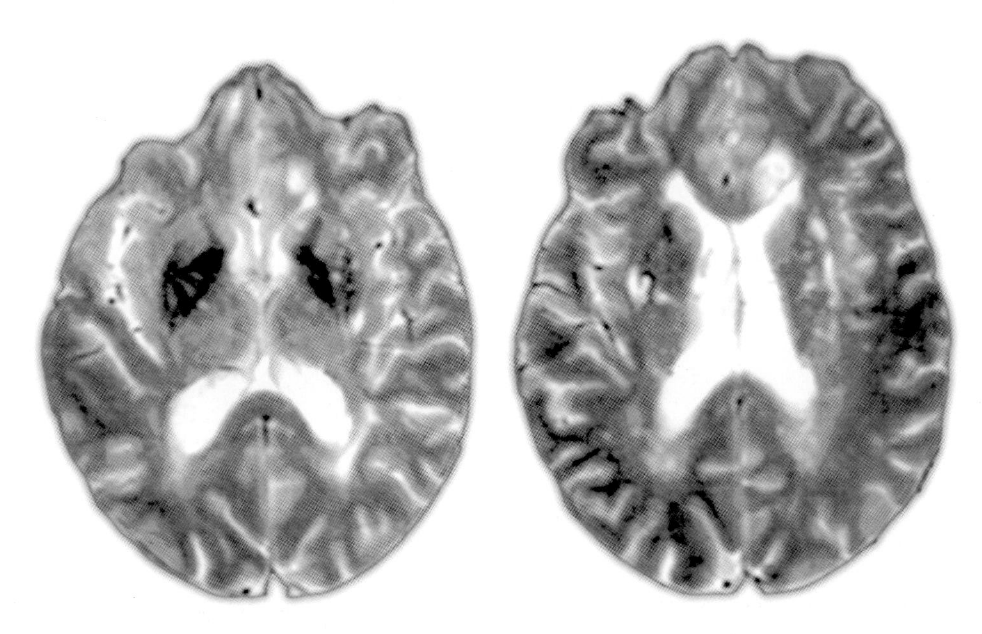

FIGURE 6.12 Callosal apraxia occurs when verbal directions mediated by the left hemisphere cannot cross the corpus callosum for execution of left-sided limb commands mediated by the right hemisphere.

involves only the left arm and leg, and in most cases there is no associated aphasia or hemiparesis.[128–131] Disruption of interhemispheric communication is manifested in a variety of disturbances in addition to the left limb apraxia, including left-hand tactile anomia, left-hand aphasic agraphia, right-hand constructional disturbances, and a variety of somesthetic disorders such as failure of intermanual tactile matching and intermanual matching of hand positions.[108,132] Corpus callosum injury may be produced by surgical sectioning for the control of intractable epilepsy, anterior cerebral artery occlusion, trauma, or neoplasm.[108,132,133]

• Auditory Agnosia and Pure Word Deafness

Auditory agnosia refers to the inability to recognize sounds (words or nonwords) despite the ability to hear. *Pure word deafness* is the inability to comprehend spoken language in the absence of aphasia; hearing and the ability to recognize nonspeech sounds are preserved. *Auditory agnosia for environmental sounds* is a syndrome in which speech but not environmental sounds can be recognized. All three groups of patients can read, write, and speak normally or with only minimal evidence of aphasia.[134] Patients with auditory agnosia cannot identify environmental sounds (ring of a telephone, honk of an automobile horn) and cannot understand spoken language. Patients with pure word deafness can recognize nonverbal noises but not spoken language. Patients recovering from auditory agnosia may go through a stage of pure word deafness.[135] The usual lesions producing these syndromes are in the primary auditory cortex of both hemispheres.[134]

• Aprosodia

Prosody consists of variations in sound pitch, stress, and rhythm that underlie speech melody and inflection.[136] Prosody imbues language with its emotional meaning (affective prosody) and contributes to semantic meaning (the difference between "we were in a *hothouse*" and "we were in a *hot* house") (prepositional prosody). *Aprosodia* or *dysprosodia* is the syndrome resulting from interruption of the normal prosodic contribution to spoken language. There are both *executive* (the ability to speak prosodically) and *receptive* (the ability to comprehend the prosodic elements of speech) aspects of prosody.

Prosody is particularly important in neuropsychiatric practice. The patient's emotional state is communicated primarily through prosodic influences on speech—*how* something is said is often of greater importance than *what* is said. Any impairment of prosodic executive skills will impair the patient's ability to communicate emotion and will compromise the ability of the clinician and of the patient's family to infer their emotional condition. Receptive prosodic abilities are critical to successful interpersonal function; it is on the basis of affective prosodic comprehension that the individual understands the emotions of others and can respond appropriately. Receptive dysprosody isolates the individual from the emotions of others. Individuals who sustain right brain injuries as children fail to develop interpersonal skills, cannot interpret social cues, have difficulty expressing themselves, and are at increased risk for adult psychopathology including depression, schizoid behavior, and episodic dyscontrol.[34,137–139]

Several brain regions contribute to the physiologic mechanisms underlying prosody and lesions in several areas will compromise executive or receptive prosody. Executive prosody is abnormal in patients with nonfluent aphasias when the premotor region is affected by the causative lesion.[50,140] These patients can often impart appropriate emotional inflection to their limited verbal output. Patients with right hemisphere damage involving the equivalent of Broca's area have an executive aprosodia.[31,140] Basal ganglia dysfunction also produces executive dysprosody with monotonic, uninflected, and often hypophonic output.[141] Cerebellar disorders with ataxic speech produce abnormalities of stress and pitch shift with a resulting characteristic change in prosody.[142]

Comprehension of the prosodic features of communication is also affected by lesions in different brain regions. Patients with right temporal–parietal lesions have a receptive aprosodia with impaired comprehension of emotion and affect in spoken language.[143–145] Comprehension of nonemotional prepositional prosody is compromised by both left and right posterior brain injury.[146,147] Thus, although not all studies report identical findings, most suggest that right temporoparietal lesions impair comprehension of both emotional and nonemotional prosody, whereas left posterior brain dysfunction affects primarily nonemotional prosodic comprehension. Emotional prosodic comprehension is also abnormal in patients with basal ganglia lesions.[140,141,145]

• Stuttering

Stuttering is a disturbance of speech rhythm with hesitations, prolongations, pauses, and repetitions of sounds within words. Stuttering may be of childhood type beginning insidiously between ages 2 and 10 or of acquired type beginning later in life in concert with a neurologic disorder. Childhood stuttering is often familial; 20% of sons of stutterers will stutter and 10% of daughters will stutter.[35] Stuttering remits before age 16 in most cases and is life-long in 20% of individuals. Conventional structural imaging is normal in stutterers, but studies of cerebral glucose metabolism with positron emission tomography (PET) reveal that stutters have reduced metabolism in the

TABLE 6.10. *Features That Distinguish Congenital from Acquired Stuttering*

Clinical Feature	*Congenital Stuttering*	*Acquired Stuttering*
Age at onset	Ages 2–10 years	Adulthood or with a brain disorder
Gender	More common in males	Either gender
Course	Insidious onset; remission	Abrupt or subacute onset; remission is uncommon*
Repetitions	Initial syllables; substantive word only	Any syllable; substantives and small grammatical words
Choral speech	Stuttering is rare	Stuttering is common
Emotional response	Anxious, frustrated	None or annoyed
Secondary symptoms (grimacing, fist clenching)	Present	Absent
Cause	Idiopathic; hereditary	Acquired brain disease: trauma, stroke, or basal ganglia degeneration

*Transient stuttering is common in patients recovering from aphasia.

anterior cingulate cortex and the superior and middle temporal gyri. The changes are more severe in the left than the right hemisphere.[148]

Acquired stuttering is associated with recovery from aphasia, traumatic or vascular injury to subcortical pyramidal and extrapyramidal systems of either hemisphere, or basal ganglia diseases such as Parkinson's disease and progressive supranuclear palsy.[149–151] Monoamine oxidase inhibitors (MAOIs; tranycypromine, phenelzine) have rarely been observed to induce stuttering.[152] Table 6.10 lists the features that distinguish childhood from acquired stuttering. Speech therapy and use of pacing techniques are of benefit to some patients. Relief of stuttering may also be provided by vocal cord injection with botulinum toxin.[153]

TABLE 6.11. *Causes of Echolalia*

Stroke with transcortical aphasia syndrome
Transcortical motor aphasia
Transcortical sensory aphasia
Mixed transcortical (isolation) aphasia
Frontotemporal dementias
Alzheimer's disease (advanced)
Creutzfeldt-Jakob disease
General paresis (syphilitic encephalitis)
Basal ganglia disorders
Huntington's disease
Gilles de la Tourette's syndrome
Postencephalitic parkinsonism
Hyperekplexias (e.g., jumping Frenchmen of Maine)
Neuroacanthocytosis
Autism
Mental retardation syndromes (including fragile X syndrome)
Developmental echolalia (normal between 1 month and 3 years of age)
Catatonia (schizophrenia, mood disorders, or due to a neurologic condition)
Schizophrenia

• Echolalia and Palilalia *Echolalia* refers to the repetition of words or phrases said by the clinician; *palilalia* is the repetition of words or phrases that were originally uttered by the patient. The two disorders occur in similar clinical circumstances although echolalia is more common and precedes the onset of palilalia in patients who eventually exhibit both disorders. Reported causes of echolalia are listed in Table 6.11. It occurs in the transcortical aphasias and in disorders that affect frontal lobe–basal ganglia circuits.[154] Echolalia may be a manifestation of stimulus boundedness and environmental dependency with echoing of perceived environmental stimuli.[60] Palilalia should be distinguished from stuttering (described above) and logoclonia (repetition of the final syllable of words).

DISORDERED VERBAL OUTPUT IN PSYCHOSIS

Psychosis is associated with alterations in thought content and thought form. The spontaneous verbalizations of psychotic patients with a thought disorder may include evidence of loosening of association, thought blocking, condensation, illogicality, neologisms, incoherence, and perseveration (Chapters 2, 12).[155,156] Analysis of the verbal output of psychotic patients may provide clues to the type of illness present and, in cases where the language is severely affected, may help distinguish between psychosis and aphasia.

Manic and schizophrenic psychoses have different effects on verbal output. Although both disorders have an adverse effect on discourse trajectory and coherence, manic speech has better preservation of hierarchical structures, more structural linkages, and is more successful in communicating complex ideas. Utterance length is shorter in schizophrenics than in manic patients. Loss of coherence in the two disorders reflects different underlying abnormalities: schizophrenic patients tend to have less structure in their discourse whereas manics shift more rapidly from topic to topic.[157,158] Schizophrenic patients are less able to anticipate the needs of the listener and structure conversation accordingly.[159]

The verbal output of a small percentage of schizophrenics, however, can sometimes closely resemble jargon aphasia and can be distinguished only with difficulty.[59,160–162] In some cases, aphasia may be mistaken for psychotic speech, resulting in delay of appropriate therapy and institution of misguided treatment.[163] The tendency for some fluent aphasics to develop paranoid syndromes resembling schizophrenia[75,77] makes the differential task even more challenging. The two disorders share many features, including fluent output, poverty of information content, relative preservation of syntax and phonology, paraphasia, perseveration, incoherence, and impairment of the pragmatic aspects of discourse.[163–169] Despite these similarities, there are differences between jargon aphasia and the disordered verbal output of schizophrenics (Table 6.12). In spon-

TABLE 6.12. *Clinical Features That Distinguish Fluent Aphasia from Verbal Output of Schizophrenia*

Clinical Characteristic	*Fluent Aphasia*	*Schizophrenia*
Spontaneous speech		
Length of response	Shorter	Extended
Awareness of deficit	Often present	Absent
Participation in conversation	Present	Absent
Content	Empty	Impoverished; bizarre; restricted themes
Neologisms and paraphasias	Common	Rare (stable meaning)
Prosody	Preserved	Impaired
Language testing		
Comprehension	Often impaired	Usually intact
Repetition	±	Intact
Naming	Impaired	Intact
Word list generation	Diminished	May be normal or bizarre
Reading	Impaired	Intact
Writing	Aphasic	Resembles spoken output
Associated characteristics		
Negative symptoms	−	+
Medical history	+	−
Psychiatric history	−	+
Age at onset	>50	<30
Family history	−	±
Neurological examination	± Focal findings (visual field defect; right-sided weakness)	± "Soft signs"

taneous speech, schizophrenics tend to have more extended replies to inquiries than do fluent aphasics. They are less aware of their communication deficit, are less engaged in conversational exchange, and care less about the listener's response. The ideational content of their output is more bizarre with a tendency to return to a few main themes and to use a restricted vocabulary.[162–164] Neologisms and paraphasia are common in fluent aphasia and rare in schizophrenia. When they occur in schiophrema, however, they may be highly distinctive, with the new word often acquiring a stable meaning within the schizophrenic's idiosyncratic vocabulary. For instance, one of the author's patients believed he had a "seisometer" behind his right eye that could receive and transmit instructions; Forrest reported a patient with a thesaurus of neologisms such as "semitiertology" or the study of half hundreds; one of Bleuler's patients used the word "snortse" to mean "to talk through the walls"; and a patient studied by Hamilton used a process of condensation to construct words such as "esamaxrider," meaning "he's a married man."[170–173] The stability of the paraphasic usage in schizophrenia differs markedly from fluent aphasia, where patients rarely make the same paraphasic substitution consistently. Language abnormalities are more evident in schizophrenic patients with prominent negative symptoms (Chapter 12) (affective flattening, avolition, alogia, anhedonia, disturbed attention), and these features are not characteristic of most aphasics.[174] Prosodic abnormalities are more characteristic of the spontaneous speech of schizophrenics than of fluent aphasics.[175]

Language testing in the course of the mental status examination can also be helpful in distinguishing aphasia from schizophrenic verbal output if the patient can be engaged in the testing process (Table 6.11).[59,161–163] Naming ability will invariably be impaired to some extent in aphasia and is usually preserved in schizophrenia. Likewise, comprehension and repetition may be impaired in aphasia, depending on the type of language deficit present, whereas they are preserved in schizophrenia. Generation of word lists such as the maximum number of animals one can name in a minute is diminished in aphasia and may be normal or contain bizarre entries in schizophrenia. For example, one patient examined by the author included "vaginal monster" and "Phoenician circus woman" among his animals. Reading and writing are impaired in aphasia; in schizophrenia reading is preserved, and writing may be normal or may resemble the disordered spoken output. For example, when one patient was asked to describe the weather, he wrote: "Rumors of a fiercer nature were let out about the phloral trumps of our Lord."

Finally, the clinical circumstances in which the verbal output disorder occurs may also facilitate differentiation of aphasia and schizophrenia (Table 6.11). Onset before age 30, history of psychosis, and absence of known medical illness all favor the diagnosis of schizophrenia; whereas onset after age 50 years, presence of a predisposing medical illness, absence of previous psychiatric disturbances, and focal findings on neurological examination all indicate a hemispheric insult and support a diagnosis of aphasia. These features are not infallible, however, and it must be remembered that schizophrenics are at the same risk as the general population for the development of cerebrovascular or neoplastic disease and may develop on aphasia-producing brain disorder.

REFERENCES

1. Davison C, Kelman H. Pathologic laughing and crying. Arch Neurol Psychiatry 1939;42:595–643.
2. Ironside R. Disorders of laughter due to brain lesions. Brain 1956;79:589–609.
3. Langworthy OR, Hesser FH. Syndrome of pseudobulbar palsy. Arch Intern Med 1940;65:106–121.
4. Lieberman A, Benson DF. Control of emotional espression in pseudobulbar palsy. Arch Neurol 1977;34:717–719.
5. Wilson SAK. Pathological laughing and crying. J Neurol Psychopathol 1924;4:299–333.
6. Cummings JL, Benson DF. Mutism: loss of neocortical and limbic vocalization. J Nerv Ment Dis 1983;171:255–259.
7. Crutchfield JS, Sawaya R, et al. Postoperative mutism in neurosurgery. J Neurosurg 1994;81:115–121.
8. Rekate HL, Grubb RL, et al. Muteness of cerebellar origin. Arch Neurol 1985;42:697–698.
9. Van Dongen HR, Catsman-Berrevoets CE, Van Mourik M. The syndrome of 'cerebellar' mutism and subsequent dysarthria. Neurology 1994;44:2040–2046.
10. Alexander MP, LoVerme SR Jr. Aphasia after left hemispheric intracerebral hemorrhage. Neurology 1980;30:1193–1202.
11. Alajouanine T, Lhermitte F. Acquired aphasia in childhood. Brain 1965;88:653.
12. Geschwind N. Disconnexion syndromes in animals and man. Brain 1965;88:237–294 and 585–644.
13. Levin HS, Madison CF, et al. Mutism after closed head injury. Arch Neurol 1983;40:601–606.
14. Bastian HC. On different kinds of aphasia, with special reference to their classification and ultimate pathology. BMJ 1887;2:931–936, 985–990.
15. Benson DF. Aphasia, Alexia, and Agraphia. New York: Churchill Livingston, 1979.
16. Blumstein SE, Alexander MP, et al. On the nature of the foreign accent syndrome: a case study. Brain Cogn 1987;31:215–244.
17. David AS, Bone I. Mutism following left hemisphere infarction. J Neurol Neurosurg Psychiatry 1984;47:1342–1344.
18. Takayama Y, Sugishita M, et al. A case of foreign accent syndrome without aphasia caused by a lesion of the left precentral gyrus. Neurology 1993;43:1361–1363.

19. Kushner M, Reivich M, et al. Regional cerebral glucose metabolism in aphemia: a case report. Brain Lang 1987;31:201–214.
20. Schiff HB, Alexander MP, Naeser MA. Aphemia: clinico-anatomic correlations. Arch Neurol 1983;40:720–727.
21. Alexander MP, Schmitt MA. The aphasia syndrome of stroke in the left anterior cerebral artery territory. Arch Neurol 1980; 37:97–100.
22. Cappa SF, Sterzi R. Infarction in the territory of the anterior choroidal artery: a cause of transcortical motor aphasia. Aphasiology 1990;4:213–217.
23. Freedman M, Alexander MP, Naeser MA. Anatomic basis of transcortical motor aphasia. Neurology 1984;34:409–417.
24. Laplane D, Talairach J, Meininger V. Clinical consequences of corticectomies involving the supplementary motor area in man. J Neurol Sci 1977;34:301–314.
25. McFarling D, Rothi LJG, Heilman KM. Transcortical aphasia from ischemic infarcts of the thalamus: a report of two cases. J Neurol Neurosurg Psychiatry 1982;45:107–112.
26. Tijssen CC, Tavy DLJ, et al. Aphasia with a left frontal interhemispheric hematoma. Neurology 1984;34:1261–1264.
27. Segarra JM. Cerebral vascular disease and behavior, 1: The syndrome of the mesencephalic artery (basilar artery bifurcation). Arch Neurol 1970;22:408–418.
28. Altshuler LL, Cummings JL, Mills MJ. Mutism: review, differential diagnosis, and report of 22 cases. Am J Psychiatry 1986; 143:1408–1414.
29. Castaigne P, Lhermitte F, et al. Paramedian thalamic and midbrain infarcts: clinical and neuropathological study. Ann Neurol 1981;10:127–148.
30. Crismon ML, Childs A, et al. The effect of bromocriptine on speech dysfunction in patients with diffuse brain injury (akinetic mutism). Clin Neuropharmacol 1988;11:462–466.
31. Ross ED, Stewart RM. Akinetic mutism from hypothalamic damage: successful treatment with dopamine agonists. Neurology 1981;31:1435–1439.
32. Scott TF, Lang D, et al. Prolonged akinetic mutism due to multiple sclerosis. J Neuropsychiatry Clin Neurosci 1995;7:90–92.
33. Stewart JT, Leadon M, Gonzalez-Rothi LJ. Treatment of a case of akinetic mutism with bromocriptine. J Neuropsychiatry Clin Neurosci 1990;2:462–463.
34. Weintraub S, Mesulam M-M. Developmental learning disabilities of the right hemisphere. Arch Neurol 1983;40:463–468.
35. American Psychiatric Association. Diagnostic and Statistical Manual of Mental Disorders, 4th ed. Washington DC: American Psychiatric Press, 1994.
36. Bishop DVM. Developmental disorders of speech and language. In: Rutter M, Taylor E, Hersov L, eds. Child and Adolescent Psychiatry. Boston: Blackwell Scientific Publications, 1994: 546–568.
37. Black B, Uhde T. Treatment of elective mutism with fluoxetine: a double-blind placebo-controlled study. J Am Acad Child Adolesc Psychiatry 1994;33:1000–1006.
38. Wijdicks EFM, Wiesner RH, et al. FK506-induced neurotoxicity in live transplantation. Ann Neurol 1994;35:498–501.
39. Love RJ. Motor Speech Disorders. New York: Marcel Dekker, 1995.
40. Goodglass H. Understanding Aphasia. New York: Academic Press, 1993.
41. Benton A. Aphasia, historical perspectives. In: Sarno MT, ed. Acquired Aphasia. 2nd ed. New York: Academic Press, 1991:1–24.
42. Benson DF. Fluency in aphasia: correlation with radioactive scan localization. Cortex 1967;3:373–394.
43. Damasio HD. Cerebral localization of the aphasias. In: Sarno MT, ed. Acquired Aphasia. New York: Academic Press, 1981: 27–50.
44. Aram DM. Acquired Aphasia in Children. New York: Academic Press, 1991.
45. Albert ML, Goodglass H, et al. Clinical Aspects of Dysphasia. New York: Springer-Verlag, 1981.
46. Damasio H. Neuroanatomical Correlates of the Aphasias. New York: Academic Press, 1991.
47. Goodglass H. Studies on the grammar of aphasics. In: Goodglass H, Blumstein S, eds. Psycholinguistics and Aphasia. Baltimore: Johns Hopkins University Press, 1973:183–215.
48. Damasio AR. Signs of aphasia. In: Sarno MT, ed. Acquired Aphasia, 2nd ed. New York: Academic Press, 1991:27–43.
49. Tonkonogy V, Goodglass H. Language function, foot of the third frontal gyrus, and rolandic operculum. Arch Neurol 1981;38:486–490.
50. Alexander MP, Naeser MA, Palumbo C. Broca's area aphasias: aphasia after lesions including the frontal operculum. Neurology 1990;40:353.
51. Henderson VW. Lesion localization in Broca's aphasia: implications from Broca's aphasia without hemiparesis. Arch Neurol 1985;42:1210–1212.
52. Masdeu JC, O'Hara RJ. Motor aphasia unaccompanied by facio-brachial weakness. Neurology 1983;33:519–521.
53. Mori E, Yamadori A, Furumoto M. Left precentral gyrus and Broca's aphasia: a clinicopathologic study. Neurology 1989;39: 51–54.
54. Kertesz A, Harlock W, Coates R. Computer tomographic localization, lesion size and prognosis in aphasia and nonverbal impairment. Brain Lang 1979;8:34–50.
55. Van Lancker D, Nicklay CKH. Comprehension of personally relevant (PERL) versus novel language in two globally aphasic patients. Aphasiology 1992;6:37–61.
56. Boller F, Green E. Comprehension in severe aphasics. Cortex 1972;8:382–394.
57. Van Horn G, Hawes A. Global aphasia without hemiparesis: a sign of embolic encephalopathy. Neurology 1982;31:403–406.
58. Geschwind N, Quadfasel F, Segarra J. Isolation of the speech area. Neuropsychologia 1968;6:327–340.
59. Ross ED. Left medial parietal lobe and receptive language functions: mixed transcortical sensory aphasia after left anterior cerebral artery infarction. Neurology 1980;30:144–151.
60. McPherson S, Kuratani J, et al. Mixed transcortical aphasia in Creutzfeldt-Jakob disease: insights into echolalia. Behav Neurol 1994;7:197–203.
61. Rapcsak SZ, Krupp LB, et al. Mixed transcortical aphasia without anatomic isolation of the speech area. Stroke 1990;21: 953–956.
62. Sevush S, Roeftgen DP, et al. Preserved oral reading in Wernicke's aphasia. Neurology 1983;33:916–920.
63. Knepper LE, Biller J, et al. Etiology of stroke in patients with Wernicke's aphasia. Stroke 1989;20:1730–1732.
64. Alexander MP, Hiltbrunner B, Fischer R. Distributed anatomy of transcortical sensory aphasia. Arch Neurol 1989;46:885–892.
65. Kertesz A, Sheppard A, Mackenzie R. Localization in transcortical sensory aphasia. Arch Neurol 1982;39:475–478.
66. Benson DF, Cummings JL, Tsai SY. Angular gyrus syndrome simulating Alzheimer's disease. Arch Neurol 1982;39:616–620.
67. Cummings JL, Benson DF. Dementia: A Clinical Approach. Boston: Butterworth-Heinemann, 1992.

68. Rothi LJ, Farling D, Heilman KM. Conduction aphasia, syntactic alexia and the anatomy of syntactic comprehension. Arch Neurol 1982;39:272–275.
69. Benson DF. Neurologic correlates of anomia. In: Whitaker H, Whitaker HA, eds. Studies in Neurolinguistics. New York: Academic Press, 1979:293–328.
70. Cappa SF, Vallar G. Neuropsychological disorders after subcortical lesions: implications for neural models of language and spatial attention. In: Vallar G, Cappa SF, Wallesch C-W, eds. Neuropsychological Disorders Associated with Subcortical Lesions. New York: Oxford University Press, 1992:7–41.
71. Naeser MA, Alexander MP, Helm-Estabrooks N. Aphasia with predominantly subcortical lesion sites: description of three capsular/putaminal aphasia syndromes. Arch Neurol 1982;39: 2–14.
72. Alexander MP, Fischer RS, Friedman R. Lesion localization in apractic agraphia. Arch Neurol 1992;49:246–251.
73. Mega MS, Alexander MP. Subcortical aphasia: the core profile of capsulostrital infarction. Neurology 1994;44:1824–1829.
74. Gainotti G. Emotional behavior and hemi-spheric side of the brain. Cortex 1972;8:41–55.
75. Ross ED. Acute agitation and other behaviors associated with Wernicke aphasia and their possible neurological bases. Neuropsychiatry Neuropsychol Behav Neurol 1993;6:9–18.
76. Starkstein SE, Federoff P, et al. Catastrophic reaction after cerebrovascular lesions: frequency, correlates, and validation of a scale. J Neuropsychiatry Clin Neurosci 1993;5:189–194.
77. Benson DF. Psychiatric aspects of aphasia. Br J Psychiatry 1973;123:555.
78. Signer S, Cummings JL, Benson DF. Delusions and mood disorders in patients with cerebrovascular injury. Arch Gen Psychiatry 1989;47:246–251.
79. Starkstein SE, Cohen BS, et al. Relationship between anxiety disorders and depressive disorders in patients with cerebrovascular injury. Arch Gen Psychiatry 1990;47:246–251.
80. Robinson RG, Benson DF. Depression in aphasic patients: frequency, severity and clinical pathological correlations. Brain Lang 1981;14:282–291.
81. Robinson RG, Szetela B. Mood change following left hemisphere brain injury. Ann Neurol 1981;9:447–453.
82. Finklestein S, Benowitz LI, et al. Mood, vegetative disturbance, and dexamethasone suppression test after stroke. Ann Neurol 1982;12:463–468.
83. Kertesz A, McCabe P. Recovery patterns and prognosis in aphasia. Brain 1977;100:1–18.
84. Levin HS, Grossman RG, et al. Linguistic recovery after closed head trauma. Brain Lang 1981;12:360–374.
85. Lomas J, Kertesz A. Patterns of spontaneous recovery in aphasic groups: a study of adult stroke patients. Cortex 1978;5:388–401.
86. Knopman DS, Selnes OA, et al. A longitudinal study of speech fluency in aphasia: CT correlates of recovery and persistent nonfluency. Neurology 1983;33:1170–1178.
87. Selnes OA, Knopman DS, et al. Computed tomographic scan correlates of auditory comprehension deficits in aphasia: a prospective recovery study. Ann Neurol 1983;13:558–566.
88. Subirana A. The prognosis in aphasia in relation to the factor of cerebral dominance and handedness. Brain 1958;8:415–425.
89. Cummings JL, Benson DF, et al. Left-to-right transfer of language dominance: a case study. Neurology 1979;29:1547–1550.
90. Prins RS, Snow CE, Wagenaar E. Recovery from aphasia: spontaneous speech versus language comprehension. Brain Lang 1978;6:192–211.
91. Wertz RT, Weiss DG, et al. Comparison of clinic, home, and deferred language treatment for aphasia: a Veterans Administration Cooperative Study. Arch Neurol 1986;43:653–658.
92. Gardner J, Zurif EB, et al. Visual communication in aphasia. Neuropsychologia 1976;14:275–292.
93. Helm-Estabrooks N, Emery P, Alber ML. Treatment of Aphasic Perseveration (TAP) Program: a new approach to aphasia therapy. Arch Neurol 1987;44:1253–1255.
94. Sparks R, Helm N, Albert ML. Aphasia rehabilitation resulting from melodic intonation therapy. Cortex 1974;10:303–316.
95. Greenblatt SH. Subangular alexia without agraphia or hemianopsia. Brain Lang 1976;3:229–245.
96. Stommel EW, Friedman RJ, Reeves AG. Alexia without agraphia associated with spleniogeniculate infarction. Neurology 1991;41:587–588.
97. Benson DF, Brown J, Tomlinson EB. Varieties of alexia. Neurology 1971;21:951.
98. Benson DF. Alexia. In: Frederiks JAM, ed. Handbook of Clinical Neurology. New York: Elsevier Science Publishers, 1985:433–455.
99. Coltheart M. Deep dyslexia: a review of the syndrome. In: Coltheart M, Patterson K, Marshall JC, eds. Deep Dyslexia. London: Routledge & Kegan Paul, 1980:22–47.
100. Marshall JC, Newcombe F. The conceptual status of deep dyslexia: a historical perspective. In: Coltheart M, Patterson K, Marshall JC, eds. Deep Dyslexia. Boston: Routledge and Kegan Paul, 1980:1–21.
101. Henderson VW, Alexander MP, Naeser MA. Right thalamic injury, impaired visuospatial perception, and alexia. Neurology 1982;32:235–240.
102. Benson DF, Cummings JL. Agraphia. In: Vinken J, Bruyn GW, eds. Handbook of Clinical Neurology. New York: American Elsevier Publishing Company, 1985:457–472.
103. Roeltgen DP. Agraphia. New York: Oxford University Press, 1993.
104. Auerbach SH, Alexander MP. Pure agraphia and unilateral optic ataxia associated with a left superior lobule lesion. J Neurol Neurosurg Psychiatry 1981;44:430.
105. Chedru F, Geschwind N. Disorders of higher cortical functions in acute confusional states. Cortex 1972;8:395–411.
106. Chedru F, Geschwind N. Writing disturbances in acute confusional states. Neuropsychologia 1972;10:343–353.
107. Bub D, Kertesz A. Deep agraphia. Brain Lang 1982;28:146–165.
108. Geschwind N, Kaplan E. A human cerebral de-connection syndrome. Neurology 1962;12:675–685.
109. LeWitt P. Micrographia as a focal sign of neurological disease. J Neurol Neurosurg Psychiatry 1983;46:1152–1157.
110. Klawans HL, Glantz R, et al. Primary writing tremor: selective action tremor. Neurology 1982;32:203–206.
111. Sheehy MP, Marsden CD. Writers cramp—a focal dystonia. Brain 1982;105:461–480.
112. Eriksson B, Persson T. Gilles de la Tourette's syndrome. Br J Psychiatry 1969;115:351–353.
113. Ziegler DK. Neurological disease and hysteria—the differential diagnosis. Int J Neuropsychiatry 1967;3:388–395.
114. Master DR, Lishman WA. Seizures, dyslexia, and dysgraphia of psychogenic origin. Arch Neurol 1984;41:889–890.
115. Grewel F. The acalculias. In: Vinken PJ, Bruyn GW, eds. Disorders of Speech, Perception, and Symbolic Behavior: Hand-

book of Clinical Neurology. New York: American Elsevier Publishing Company, 1969:181–194.
116. Rosselli M, Ardila A. Calculation deficits in patients with right and left hemisphere damage. Neuropsychologia 1989;27:607–617.
117. Takayama Y, Sugishita M, et al. Isolated acalculia due to left parietal lesion. Arch Neurol 1994;51:286–291.
118. Gerstmann J. Syndrome of finger agnosia, disorientation for right and left, agraphia, and acalculia. Arch Neurol Psychiatry 1940;44:398–408.
119. Roeltgen DP, Sevush S, Heilman KM. Pure Gerstmann's syndrome from a focal lesion. Arch Neurol 1983;40:46–47.
120. Benton AL. Gerstmann's syndrome. Arch Neurol 1992;49:445–447.
121. Geschwind N. The apraxias: neural mechanisms of disorders of learned movement. Am Sci 1975;63:188–195.
122. Denny-Brown D. The nature of apraxia. J Nerv Ment Dis 1958; 126:9–32.
123. Kertesz A, Ferro JM. Lesion size and location in ideomotor apraxia. Brain 1984;107:921–933.
124. Nadeau SE, Roeltgen DP, et al. Apraxia due to a pathologically documented thalamic infarction. Neurology 1994;44: 2133–2137.
125. Heilman K, Rothi LJG, Valenstein E. Two forms of ideomotor apraxia. Neurology 1982; 32:342–346.
126. Kertesz A, Hooper P. Praxis and language: the extent and variety of apraxia in aphasia. Neuropsychologica 1982;20:275–286.
127. Geschwind N. Sympathetic dyspraxia. Trans Am Neurol Assoc 1963;88:219–220.
128. Gersh F, Damasia AR. Praxis and writing of the left hand may be served by different callosal pathways. Arch Neurol 1981;38:634–636.
129. Graff-Radford NR, Welsh K, Godersky J. Callosal apraxia. Neurology 1987;37:100–105.
130. Volpe BT, Sidtis JJ, et al. Cortical mechanisms involved in praxis: observations following partial and complete section of the corpus callosum in man. Neurology 1982;32:645–651.
131. Watson RT, Heilman KM. Callosal apraxia. Brain 1983; 100:391–403.
132. Bogen JE. The callosal syndrome. In: Heilman KM, Valenstein E, eds. Clinical Neuropsychology. New York: Oxford University Press, 1979:308–359.
133. Rubens AB, Geschwind N, et al. Post-traumatic cerebral hemispheric disconnection syndrome. Arch Neurol 1977;34:750–755.
134. Coslett HB, Brashear HR, Heilman KM. Pure word deafness after bilateral primary auditory cortex infarcts. Neurology 1984;34:347–352.
135. Mendez MF, Geehan GR Jr. Cortical auditory disorders: clinical and psychoacoustic features. J Neurol Neurosurg Psychiatry 1988;51:1–9.
136. Monrad-Krohn GH. The third element of speech: prosody in the neuropsychiatric clinic. J Ment Sci 1957;103:326–331.
137. Bell WL, Davis DL, et al. Acquired aprosodia in children. J Child Neurol 1990;5:19–26.
138. Grace J, Malloy P. Neuropsychiatric aspects of right hemisphere learning disability. Neuropsychiatry Neuropsychol Behav Neurol 1992;5:194–204.
139. Voeller KKS. Right hemisphere deficit syndrome in children. Am J Psychiatry 1986;143:1004–1009.
140. Cancelliere AEB, Kertesz A. Lesion localization in acquired deficits of emotional expression and comprehension. Brain Cogn 1990;13:133–147.
141. Scott S, Caird FL, Williams BO. Evidence for an apparent sensory speech disorder in Parkinson's disease. J Neurol Neurosurg Psychiatry 1984;47:840–843.
142. Kent RD, Rosenbeck JC. Prosodic disturbance and neurologic lesion. Brain Lang 1982;15:259–291.
143. Darby DG. Sensory aprosodia: a clinical clue to lesions of the inferior division of the right middle cerebral artery? Neurology 1993;43:567–572.
144. Heilman KM, Scholes R, Watson RT. Auditory affective agnosia. J Neurol Neurosurg Psychiatry 1975;38:69–72.
145. Starkstein SE, Federoff P, et al. Neuropsychological and neuroradiologic correlates of emotional prosody comprehension. Neurology 1994;44:515–522.
146. Heilman KM, Bowers D, et al. Comprehension of affective and nonaffective prosody. Neurology 1984;34:917–921.
147. Weintraub S, Mesulam M-M, Kramer L. Disturbances of prosody. Arch Neurol 1981;38:742–744.
148. Pool KD, Devous MD Sr, et al. Regional cerebral blood flow in developmental stutterers. Arch Neurol 1991;48:509–512.
149. Helm NA, Butler RB, Benson DF. Acquired stuttering. Neurology 1978;28:1159–1165.
150. Koller WC. Disfluency (stuttering) in extrapyramidal disease. Arch Neurol 1983;40:175–177.
151. Ludlow CL, Rosenberg J, et al. Site of penetrating brain lesions causing chronic acquired stuttering. Ann Neurol 1987;22:60–66.
152. Duffy JD. Neurogenetic stuttering and lateralized motor deficits induced by tranylcypromine. Behav Neurol 1994;7:171–174.
153. Brin MF, Stewart C, et al. Laryngeal botulinum toxin injections for disabling stuttering in adults. Neurology 1994;44:262–266.
154. Ford RA. Neurobehavioral correlates of abnormal repetitive behavior. Behav Neurol 1991;4:113–119.
155. Andreasen NC. Thought, language, and communication disorders, 1: Clinical assessment, definition of terms, and evaluation of their reliability. Arch Gen Psychiatry 1979;36:1315–1321.
156. Andreasen NC. Thought, language, and communication disorders, II: Diagnostic significance. Arch Gen Psychiatry 1979;36: 1325–1330.
157. Hoffman RE, Stopek S, Andreasen NC. A comparative study of manic vs schizophrenic speech disorganization. Arch Gen Psychiatry 1986;43:831–838.
158. Wykes T, Leff J. Disordered speech: differences between manics and schizophrenics. Brain Lang 1982;15:117–124.
159. Rutter DR. Language in schizophrenia. Br J Psychiatry 1985; 146:399–404.
160. Faber R, Reichstein MB. Language dysfunction in schizophrenia. Br J Psychiatry 1981;139:519–522.
161. Faber R, Abrams R, et al. Comparison of schizophrenic patients with formal thought disorder and neurologically impaired patients with aphasia. Am J Psychiatry 1983;140:1348–1351.
162. Gerson SN, Benson DF, Frazier SH. Diagnosis: schizophrenia versus posterior aphasia. Am J Psychiatry 1977;134:966–969.
163. Sambunaris A, Hyde TM. Stroke-related aphasias mistaken for psychotic speech: two case reports. J Geriatr Psychiatry Neurol 1994;7:144–147.
164. Andreasen NC. The relationship between schizophrenic language and the aphasias. In: Henn FA, Nasrallah HA, eds. Schizophrenia as a Brain Disease. New York: Oxford University Press, 1982:99–111.

165. Chaika E. A linguist looks at "schizophrenic" language. Brain Lang 1974;1:257–276.
166. Critchley M. The neurology of psychotic speech. Br J Psychiatry 1964;110:353–364.
167. Lecours AR, Vanier-Clement M. Schizophrenia and jargonaphasia: a comparative description with comments on Chaika's and Fromkin's respective looks at "schizophrenic" language. Brain Lang 1976;3:516–565.
168. Maher B. The language of schizophrenia: a review and interpretation. Br J Psychiatry 1972;120:3–17.
169. Rochester SR, Martin JR, Thurston S. Thought-process disorder in schizophrenia: the listener's task. Brain Lang 1977;4:95–114.
170. Bleuler E. Dementia Praecox or the Group of Schizophrenias. New York: International Universities Press, 1950.
171. Forrest DV. New words and neologisms: with a thesaurus of coinages by a schizophrenic savant. Psychiatry 1969;32:44–73.
172. Heilman KM, Tucker DM, Valenstein E. A case of mixed transcortcal aphasia with intact naming. Brain 1976;99:415–426.
173. Kraeplin E. Dementia Praecox and Paraphrenia. New York: Krieger Publishing Company, 1971.
174. Landre NA, Taylor MA, Kearns KP. Language functioning in schizophrenic and aphasic patients. Neuropsychiatry Neuropsychol Behav Neurol 1992;5:7–14.
175. Fricchione G, Sedler MJ, Shukla S. Aprosodia in eight schizophrenic patients. Am J Psychiatry 1986;143:1457–1459.

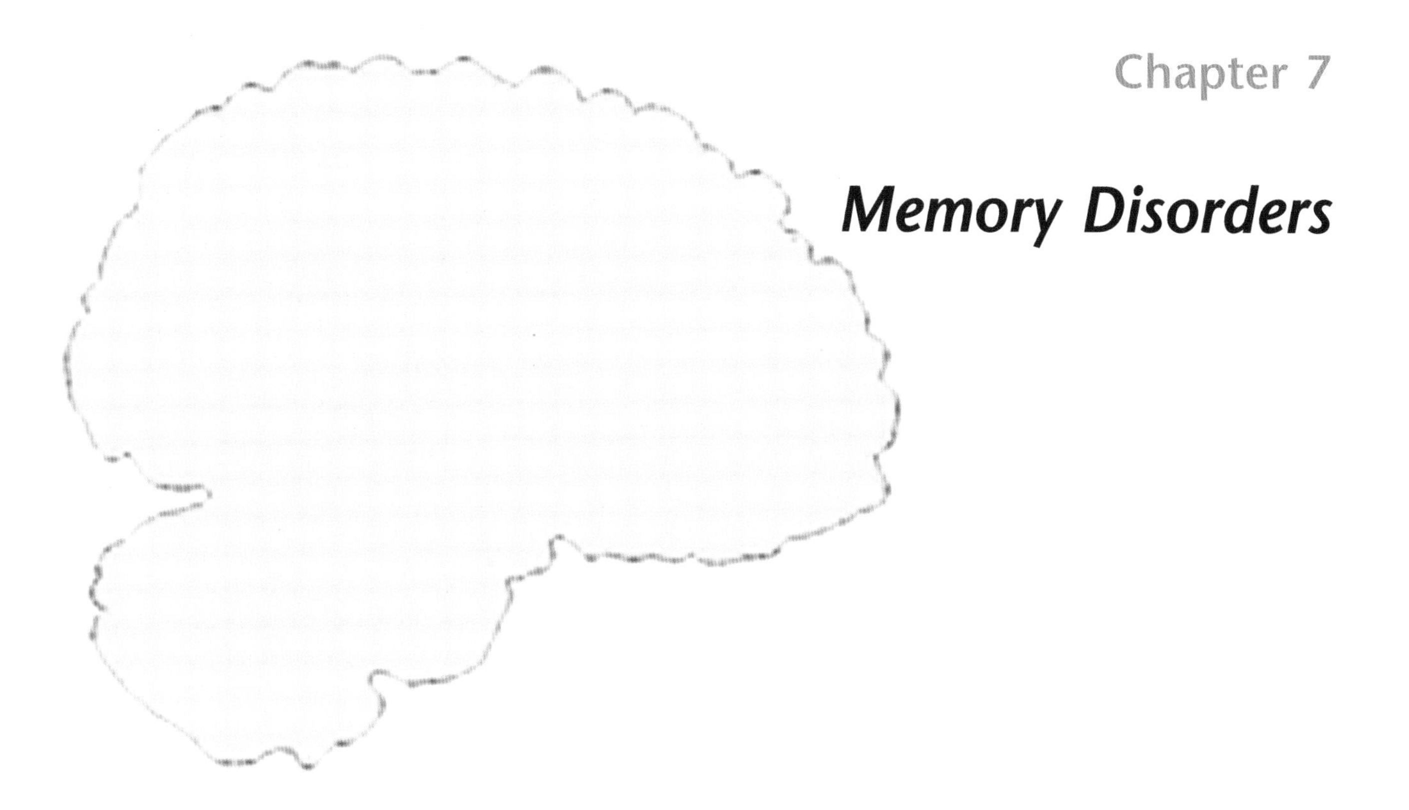

Chapter 7

Memory Disorders

Disturbed memory function is a ubiquitous finding in central nervous system (CNS) diseases, and identification of the presence and characteristics of memory impairment provides important information regarding the nature and location of brain dysfunction. Multiple subprocesses mediate different aspects of memory and all must be intact and well orchestrated for normal memory acquisition, retention, and recall to occur. Information must be registered, organized, and stored. To become functionally useful, information must be recalled within specific time requirements. Memory of verbal information, nonverbal information, and motor (procedural memory) skills are mediated by different subsystems in the brain and are subject to disruption by different brain diseases. Attentional disorders interfere with the registration of data and effectiveness of memory. Amnesias are conditions that prevent the adequate storage of information to be remembered. Frontal–subcortical circuit disorders disrupt organization of material that is to be remembered and impair timely recall of stored information.

This chapter discusses the memory disorders and associated conditions including amnesia, retrieval deficit syndromes, age-associated memory impairment, confabulation, and paramnesias. Chapter 22 describes memory disturbances occurring in dissociative states and fugues; Chapter 11 addresses memory loss in delirium; Chapter 10 discusses memory disorders in dementia; and Chapter 21 presents episodic memory disturbances in patients with epilepsy.

CLASSIFICATION OF MEMORY FUNCTIONS AND MEMORY DISORDERS

Several terminologies have evolved regarding the classification of memory functions and memory disorders. In this chapter memory processes will be divided between declarative (explicit) and nondeclarative (implicit) types (Fig. 7.1).[1] *Declarative memories* can be consciously recalled and include facts and events. Declarative memory consists of *semantic memory* (e.g., recall of facts independent of when or where the information was acquired) and *episodic memory* (e.g., recall of events occurring in one's life and integrally associated with specific times and places). Episodic memory is the autobiographic record of one's personal history and is the basis of autonoetic consciousness—the capacity to mentally represent personal subjective experiences in the past, present and future.[2] *Nondeclarative memory* is not available to consciousness and includes motor skill learning (procedural memory), habits, and

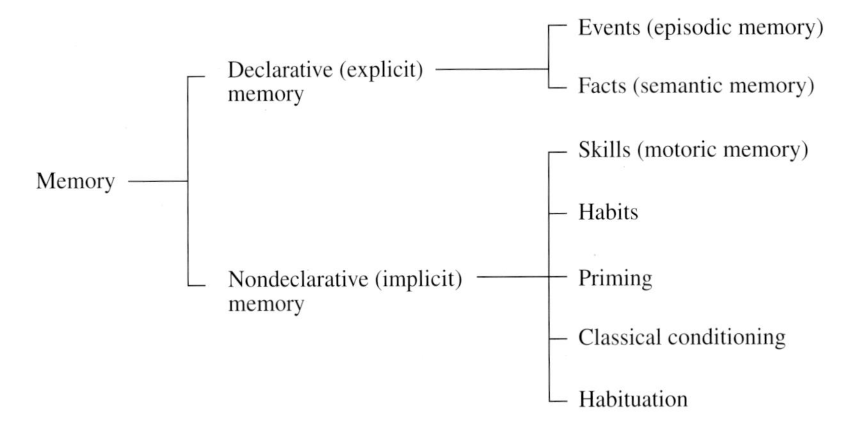

FIGURE 7.1 Conceptual organization and terminology of general memory function.

classical conditioning. Amnesias and retrieval deficit syndromes affect declarative memory; basal ganglia disorders impact nondeclarative memory. Working memory is an attentional system active for only brief periods of time (usually seconds); it allows utilization of acquired knowledge before responding to an environmental event. Working memory is mediated by frontal–subcortical systems and is discussed in Chapter 9.

AMNESIA

Clinical Features

Amnesia refers to a specific clinical condition in which there is an impairment in the ability to learn new information despite normal attention, preserved ability to recall remote information, and intact cognitive functions.[2] The amnesic patient has a normal or even supernormal digit span, indicating intact attention and immediate memory, and has normal linguistic and cognitive functions, including language comprehension, naming, reading, writing, calculating, drawing, and abstracting, but is unable to learn new information, such as the current day, month, or year or the current geographic location. The patient has impaired recall of three test words when asked to recall them after 3 minutes. Patients with amnesia involving nonverbal information cannot reproduce constructions that were drawn a few minutes previously. Amnesia results from the inability to store information and recall is not aided by clues (such as telling the patient that the word to be remembered was a vegetable when the to-be-remembered word was *cabbage*), by multiple-choice alternatives (telling the patient that the word to be remembered was either *red, blue,* or *green* when the to-be-remembered word was *green*), or by embedding the words to be remembered in a list of other words to determine if the patient can recognize previously seen material as evidenced by distinguishing between the target words (those seen earlier) and foils (those not seen previously). Amnesic disorders consist of anterograde amnesia and retrograde amnesia. *Anterograde amnesia* is difficulty in learning new information. It begins with the onset of the amnesia and continues indefinitely in permanent amnesic disorders or subsides in those amnesic syndromes that are transient. *Retrograde amnesia* refers to the failure to remember events that preceded the onset of the anterograde amnesia. It may extend for a few minutes (typical of post-traumatic amnesia) or a few years (common in Wernicke-Korsakoff's syndrome). Remote memory (beyond the period included in the retrograde amnesia) is normal. If recent memory recovers, the retrograde amnesia may progressively shrink to some finite period prior to the amnesia-inciting event, and the patient will be left with a period of permanent memory loss that extends from the beginning of the retrograde amnesia to the end of the anterograde amnesia.[3] Procedural memory (memory for motor skills) is typically spared in amnesic syndromes. Patients can learn new motor skills even though they may not be able to remember that they were taught the ability.[4,5] Table 7.1 lists the principal causes of amnesia and presents distinguishing findings commonly associated with each amnesic syndrome. *Confabulation* (discussed below) refers to the production of false answers in response to memory-related questions. Confabulation is common in the early stages of amnesia but rarely persists as a permanent phenomenon.

Amnesia is associated with dysfunction of a restricted set of brain structures. Neurologic conditions producing amnesia affect the hippocampus, fornix, mamillary bodies of the hypothalamus, mammillothalamic tract, or medial thalamic nuclei (Figs. 7.2 and 7.3). Hippocampi are affected in temporal lobectomy, head

TABLE 7.1. *Principal Amnesic Disorders*

Syndrome	*Clinical Features*
Wernicke-Korsakoff syndrome	Nystagmus, ataxia, peripheral neuropathy
Temporal lobectomy	Superior quadrantansopia (contralateral to lobectomy); history of surgery
Head trauma	Frontal lobe dysfunction; history of trauma
Hippocampal infarction	Homonymous hemianopsia (unilateral lesion) or cortical blindness (bilateral lesions)
Thalamic infarction or hemorrhage	Abrupt onset; vascular risk factors
Anoxia	History of cardiopulmonary arrest
Herpes simplex encephalitis	Kluver-Bucy syndrome, aphasia, seizures
Neoplasms	Homonymous hemianopsia, hemiparesis, headache
Basal frontal lesions (rupture of anterior communicating artery aneurysm)	Personality alterations, diabetes insipidus, hypothermia
Early Alzheimer's disease	Gradually progressive amnesia in an elderly person
Electroconvulsive therapy (ECT)	Depression with recent ECT
Transient global amnesia	Vascular etiology is most common; there may be associated evidence of cerebrovascular disease
Hypoglycemia	Insulin overdose
Psychogenic anmesia	Personal identity lost; may be amnesic for specific personal information

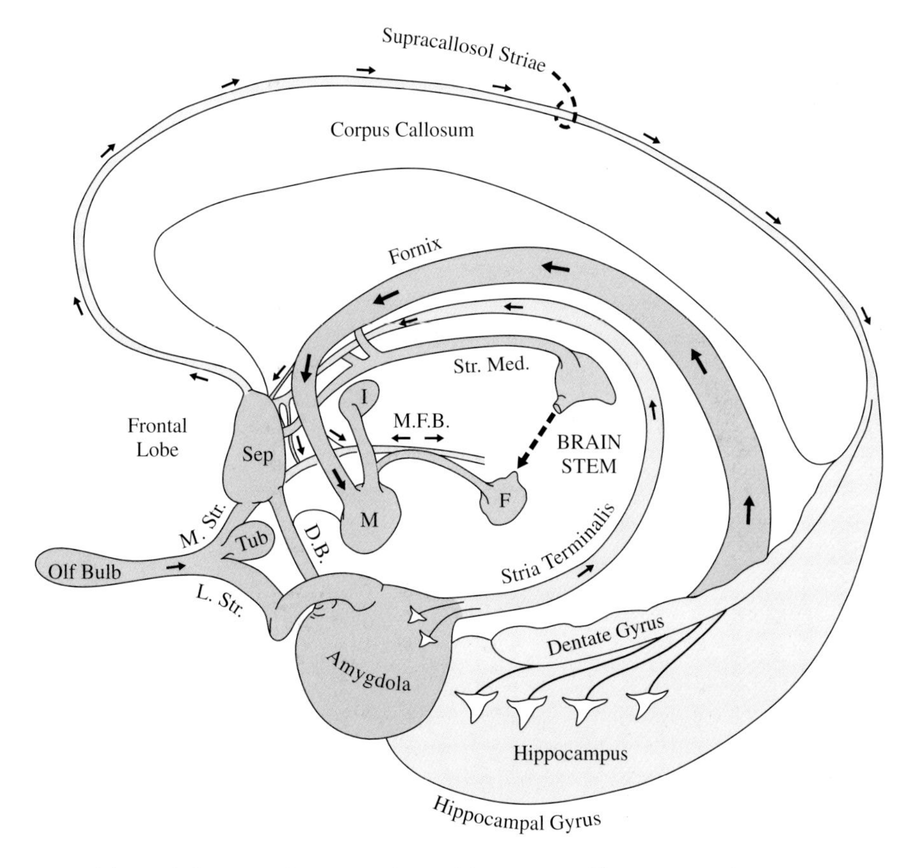

FIGURE 7.2 The medial circuit of Papez, as depicted by MacLean,[48] was proposed to support emotional processing in 1937.

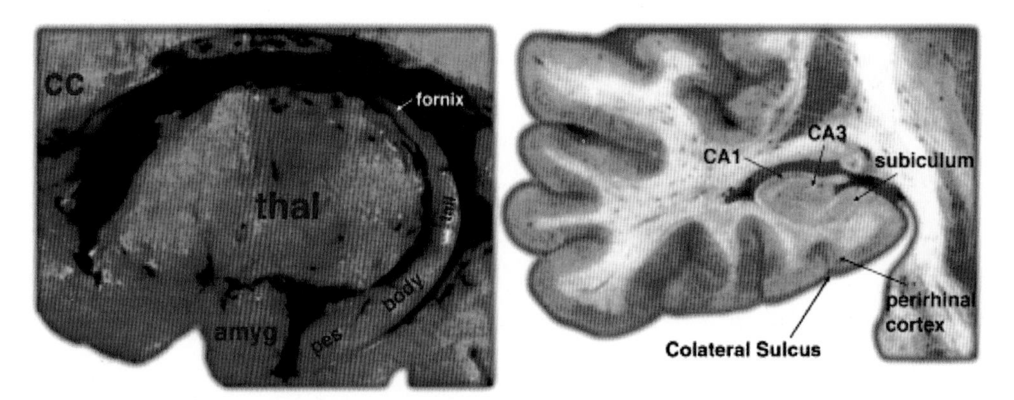

FIGURE 7.3 Anatomy of the hippocampus in relation to other brain regions (*left*) showing the anterior pes, body, and tail with extension into the fornix near the splenium of the corpus callosum (CC). Coronal section (*right*) shows the hippocampal fields transitioning from CA3, into CA2, and then CA1, which in turn extends into the subiculum and eventually into the entorhinal and perirhinal cortex within the banks of the collateral sulcus. amyg, amygdala; thal, thalamus.

trauma, posterior cerebral artery occlusion, hypoglycemia, and Alzheimer's disease; the fornix may be transected in the course of surgery or be affected in trauma and stroke, and can be infiltrated by neoplasms; the mamillary bodies are injured in the Wernicke-Korsakoff syndrome (thiamine deficiency syndrome) and are occasionally affected by neoplasms or during surgery to adjacent structures; the thalamus may be the site of tumors, infarction, hemorrhage, and damage in Wernicke-Korsakoff syndrome.

Amnesic disorders affecting primarily verbal information involve left hemisphere structures, whereas amnesias involving nonverbal information are associated with right hemisphere dysfunction. Severe amnesias occur with bilateral lesions and affect both verbal and nonverbal information. Procedural learning is mediated by the basal ganglia and cerebellum and is spared in amnesic disorders. Sparing of remote memory in amnesia suggests that recollection of remote information is independent of the hippocampal–thalamic circuit disrupted in the amnesias.

Clinical Syndromes with Amnesia

● Wernicke-Korsakoff Syndrome The Wernicke-Korsakoff syndrome is one of the earliest recognized amnesic syndromes and continues to be among the most common causes of isolated recent memory impairment. Wernicke described the acute phase of the illness, characterized classically by ophthalmoplegia, ataxia, and confusion, and included it in his three-volume handbook published in 1891–1883.[6] Korsakoff, a Russian neuropsychiatrist, called attention to the chronic amnesic phase of the illness and to the frequent coexistence of peripheral neuropathy. Consequently, in contemporary usage, *Wernicke's encephalopathy* denotes the ataxia, neuro-ophthalmologic abnormalities, autonomic disturbances, and confusional state. Horizontal nystagmus is present in 85% of patients, bilateral lateral rectus paralysis is in 54%, and conjugate gaze palsies are in 45%.[7] *Korsakoff's syndrome* refers to the more enduring syndrome with the characteristic memory deficit. A personality change, usually emotional indifference or apathy, frequently accompanies the amnesia. Neuroimaging changes are evident in some cases of Wernicke-Korsakoff's syndrome. Computerized tomography (CT) may reveal bilateral hypodense areas in the medial thalamus in patients with acute Wernicke's encephalopathy, and mamillary body atrophy may be demonstrated by magnetic resonance imaging (MRI) in some patients with the chronic Korsakoff's syndrome.[8]

Patients with Korsakoff syndrome have difficulty learning new information and usually have a retrograde amnesia that extends backward 3–20 years prior to the onset of the amnesia. Typically, patients remain amnesic for 1–3 months after onset and then begin to recover over a 1- to 10-month period. Of these patients, 25% recover completely, 50% show slight to moderate improvement, and 25% have no demonstrable recovery.[9] Confabulation is common during the early phases of the Korsakoff's syndrome but is unusual in the chronic phase of the condition. In some cases, administration of thiamine during the acute Wernicke phase prevents emergence of the chronic amnesic syndrome. Once the memory defect is established, however, thiamine has little effect except to prevent further deterioration.

Chronic nutritional deprivation associated with alcoholism is the most common cause of the thiamine deficiency producing Korsakoff syndrome, but other causes of thiamine deficiency may also cause the disorder. One of Korsakoff's original patients developed the syndrome from pyloric stenosis associated with intentional sulfuric acid ingestion, and other cases have been attributed to gastric carcinoma, hemodialysis, hyperemesis gravidarum, prolonged intravenous (IV) hyperalimentation, gastric plication, and dietary deprivation in prisoner of war (POW) camps.[9,10] An inherited abnormality of transketolase activity may render some patients vulnerable to the development of the Wernicke-Korsakoff syndrome under conditions in which dietary thiamine is marginal or inadequate.

Alcoholic Korsakoff syndrome must be distinguished from alcoholic dementia, in which deficits involve attention, word list generation, abstraction, and constructions as well as recent memory.

• Temporal Lobectomy, Fornicotomy, and Related Surgeries The search for an anatomy of memory was greatly stimulated when, in 1954, Scoville removed both temporal lobes of a patient with intractable seizures and produced a profound and lasting amnesia.[11] The epileptic patient, H. M., has become one of the most thoroughly studied cases in neuropsychology, and the results of extensive testing carried out on the patient have contributed significantly to understanding the role of temporal lobe structures in memory. Amnesia complicates temporal lobectomy when the hippocampal formations are removed bilaterally or when one hippocampus is removed and the other is dysfunctional. When one hippocampus is removed and the other is normal, the functioning member can compensate to a large extent for the missing structures, and the resulting deficits are relatively mild. In such cases the identifiable memory impairments are specific for the side of the lesion: left temporal lobectomy impairs learning and retention of verbal material, whereas right temporal lobe removal produces deficits in the recognition and recall of visual and auditory patterns that are not verbally coded (places, faces, melodies, nonsense patterns).[12]

Speculation that lesions outside the hippocampus proper might explain H.M.'s memory impairment and results from animal studies implicate other medial temporal structures subserving memory. The perirhinal cortex became a focus of the medial temporal memory system in monkey[13] because the delayed matching and non–matching-to-sample tasks (primarily recognition tests) were unimpaired if a hippocampal lesion did not also extend into the monkey's rhinal cortex. In H.M., the surgical lesion included the medial temporal pole, most of the amygdaloid complex, and all of the entorhinal cortex bilaterally (Fig. 7.4). In addition, the anterior 2 cm of the dentate gyrus, hippocampus, and subicular complex was removed. Given that the collateral sulcus and the cortex lining its banks are visible, then at least some of the posterioventral perirhinal cortex is intact in H.M. The posterior parahippocampal gyrus (areas TF and TH) was only slightly damaged rostrally. The lingual and fusiform gyri, lateral to the collateral sulcus, are also intact.

Another surgical case, P.B., had the entire left temporal lobe removed in a two-staged procedure and subsequently suffered a lifelong dense amnesia.[14] At autopsy, P.B. also had preservation of the posterior 22 mm of the hippocampus and parahippocampal cortex. Postoperative amnesia can now be avoided by preoperative conduction of a Wada test. In this procedure, the hemisphere ipsilateral to the temporal lobe to be removed is anesthetized with intracarotid injection of amyobarbital. If the contralateral hemisphere also has compromised memory function, memory deficits are evident during the test and surgery is avoided or modified to spare medial temporal lobe structures critical to memory function.[15]

In addition to temporal lobectomy, several other surgical procedures have resulted in amnesic syndromes. Bilateral injury to the fornix produced at the time of surgery or by traumatic brain injury has been associated with amnesic syndromes.[16] Surgical injury to the mamillary bodies at the time of removal of pituitary tumors has also produced amnesia, and a transient, Korsakoff-like syndrome affecting primarily orientation in time and learning of temporal sequences has been observed following cingulumotomy.

• Post-traumatic Amnesia Traumatic brain injuries are undoubtedly the most common cause of amnesic syndromes seen in clinical practice. The position of the temporal lobe in the middle cranial fossa, suspended between the petrous pyramid inferiorly and the greater wing of the sphenoid anteriorly and medially, renders it vulnerable to contusions with both coup and contrecoup injuries. Contusions following head injury are most commonly located in the anterior temporal lobes and the orbitofrontal cortex, and diffuse axonal injury and ischemic necrosis may contribute to the post-traumatic memory impairment.

The amnesia induced by closed head injury includes a period of retrograde memory loss extending for a few minutes to a few years prior to the injury, a variable period of unconsciousness caused by the injury, and a period of anterograde amnesia that lasts for a few hours to a few months or longer following recovery from coma. In cases with resolution of the anterograde am-

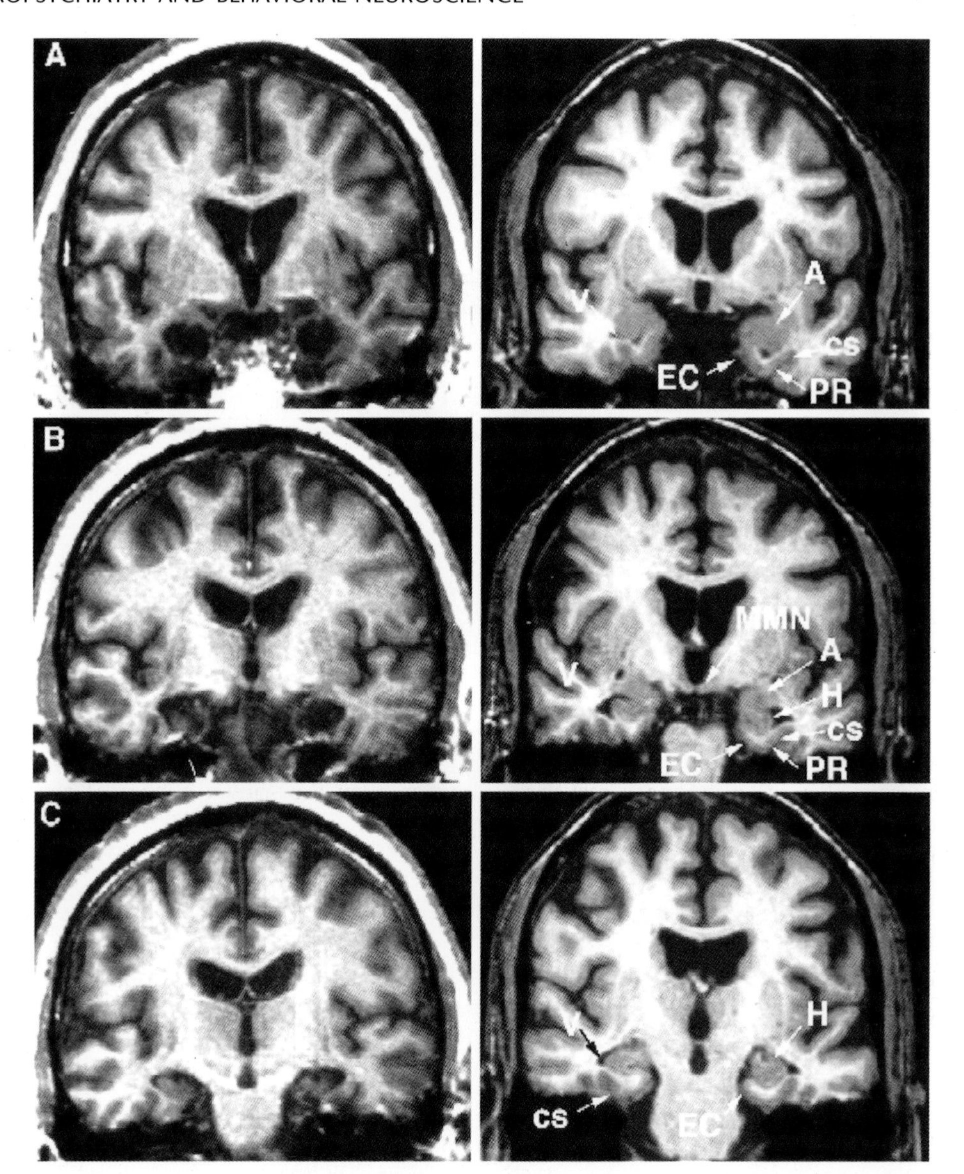

FIGURE 7.4 Coronal magnetic resonance images of H.M. from rostral (*A*) to caudal (C) showing the extent of bilateral hippocampal ablation (*left*) compared to a normal 66-year-old subject (*right*). Note the destruction of the amygdala (A), hippocampus (H), and entorhinal cortex (EC) anterior to the level of the mamillary bodies (MMN) with relative sparing of the posterior perirhinal cortex (PR) in the banks of the collateral sulcus (CS). V, temporal horn of lateral ventricle. Adapted from Corkin et al. (1997).[49]

nesia, there is often a concomitant shrinking of the period of retrograde amnesia to within a few minutes or hours of the trauma. Duration of the coma after head trauma is the best predictor of the severity of post-traumatic memory and cognitive deficits and duration of post-traumatic amnesia is correlated with post-trauma intellectual disorders.[17]

Hippocampal Infarction The hippocampus receives its blood supply from penetrating branches of the posterior cerebral artery, and occlusion of the basilar artery or proximal portions of both posterior cerebral arteries with bilateral hippocampal infarction will produce an amnesic syndrome.[18] The infarction is rarely limited to the hippocampal formation, and extension of the injury to include other occipital structures results in the frequent occurrence of visual field defects, prosopagnosia, environmental agnosia, central achromatopsia, alexia without agraphia, hemiparesis, or hemisensory loss with the amnesia. Occlusion of the posterior cerebral artery may be produced by embolism, thrombosis, or compression against the tentorium by an expanding hemispheric lesion or hemispheric edema.

Most cases of amnesia secondary to hippocampal infarction have had bilateral lesions, but a few cases appear to have damage limited to the hippocampal formation of the left hemisphere, which suggests that a unilateral infarction of the hippocampus of the language-dominant hemisphere can produce at least transient amnesia.

• **Thalamic Infarction and Hemorrhage** Lesions of the thalamus produce an amnesic disorder.[19] In most cases, the amnesia is accompanied by changes in personality, particularly apathy, and deficits in executive function may also be present. Both intrathalamic hemorrhages and ischemic infarctions have produced the syndrome. Involvement of the medial thalamic regions is required. Lesions affecting the right thalamus produce primarily a nonverbal amnesic disorder, whereas those involving the left thalamus cause an amnesia for verbal information.

• **Anoxia and Hypoglycemia** The widespread application of emergency cardiopulmonary resuscitation procedures has resulted in the emergence of a population of patients who have had profound but short-lived episodes of cerebral anoxia. Under such circumstances, only those areas of the brain that are most vulnerable to acute lack of oxygen become symptomatic, and in many cases it is the hippocampus that sustains the greatest injury.[20] The resulting clinical syndrome is characterized by profound amnesia with preserved performance in other realms of intellectual activity. The amnesia is often permanent, but a few cases have shown gradual resolution over a period of months to nearly normal levels of function. In addition to its occurrence following cardiopulmonary arrest, postanoxic amnesia has occurred following anesthetic accidents, carbon monoxide intoxication, and strangulation due to hanging. When the period of oxygen deprivation is more prolonged, widespread damage is incurred, and a postanoxic dementia ensues if the patient survives.

Acute hypoglycemia, like acute anoxia, has its greatest impact on hippocampal function, and subacute or chronic recurrent hypoglycemia often produces a memory disturbance. In many cases, however, other intellectual functions are also impaired and a dementia syndrome results.[21]

• **Herpes Simplex Encephalitis** Herpes encephalitis is the most common severe, nonepidemic encephalitis and is unique in its tendency to preferentially involve orbitofrontal and medial temporal areas of the brain (Chapter 26). The virus gains access to the brain by way of the olfactory nerves or through trigeminal nerves innervating the meninges of the anterior and middle cranial fossa. After entering the inferior frontal and medial temporal lobes, further spread of the virus is limited by immunologic control. Once in the brain, the virus initiates a destructive hemorrhagic process that may prove fatal. In most cases, the lesions are bilateral.

Among survivors of herpes encephalitis, the most common neuropsychological deficit is a profound amnesia.[22] The deficit in learning new information may be accompanied by other evidence of temporal lobe dysfunction, including aphasia, partial complex seizures, or the Klüver-Bucy syndrome.[23] Compared with patients with thiamine deficiency–induced Korsakoff syndrome, individuals rendered amnesic from herpes encephalitis appear to exhibit less confabulation, are more aware of their memory deficits, are more adept at learning logically ordered spatial arrangements and sequentially presented material, and benefit more from priming (are more likely to produce a word they have seen earlier when they are shown a part of the previously seen word).

• **Neoplasms** Neoplasms producing amnesic syndromes fall into two general categories: extracerebral tumors such as craniopharyngiomas that produce pressure on the base of the brain, and intracerebral tumors, particularly gliomas, that involve structures in the wall and floor of the third ventricle.[21] The extracerebral compressive lesions exert upward pressure on the mamillary bodies and adjacent tracts, and the gliomas invade the nuclei of the thalamus and hypothalamic structures to produce the disruption of memory function.

• **Basal Forebrain Lesions** Amnesia is not commonly associated with frontal lobe lesions. In some cases where memory disturbances are prominent, the patient is markedly distractible, and learning impairment is a product of impaired attention. In other cases, the patients manifest a retrieval deficit syndrome evidencing intact storage but difficulty accessing the information. A true amnesia, however, is sometimes observed in patients with frontal lobe disorders. In nearly all cases, the amnesia has been associated with rupture and/or surgical repair of an anterior communicating artery aneurysm located medially in the basal forebrain region.[24] In the acute phase, the patients are confused and tend to deny all deficits; as the confusion resolves they are left with amnesia and mild personality changes including disinhibition or apathy. Unilateral grasp reflexes, hemiparesis, diabetes insipidus, and hypothermia may coexist with the amnesia.

The pathological anatomy of the lesion responsible for the amnesia has not been fully determined. Penetrating branches from the anterior communicating artery supply the anterior hypothalamus, septal nucleus, lamina terminalis, portions of the head of the caudate nucleus, columns of the fornix, ventromedial corpus callosum, and anterior cingulate gyrus. Some of these structures (e.g., the columns of the fornix) are known to play a major role in anatomical circuits mediating memory functions. These vascular territories also contain the forebrain cholinergic nuclei that provide choline acetyltransferase to the cerebral cortex and the hippocampus. Injury to these nuclei may produce a cholinergic denervation of medial temporal structures, causing or contributing to the amnesic syndrome.

• Alzheimer's Disease Patients cannot meet diagnostic criteria for Alzheimer's disease (AD) if their cognitive deficits are limited to changes in memory (Chapter 10). Memory abnormalities, however, can be the initial manifestation of AD and in some cases may be the sole deficit for several years until other intellectual disturbances appear.[25] The syndrome of minimum cognitive impairment (MCI) consists of an isolated memory disorder, and patients with this syndrome progress to AD at the rate of 15% per year. The memory disorder of AD closely resembles amnesia, although at least subtle deficits in remote recall are typically present in addition to the severe alterations in recent memory. The most marked changes of AD involve the medial temporal regions, particularly entorhinal cortex, which serves as a major gateway for input to the hippocampus.

• Electroconvulsive Therapy Electroconvulsive therapy (ECT) produces both a retrograde amnesia for items learned a few minutes prior to the treatment and an anterograde amnesia for information introduced immediately after the treatment. The anterograde amnesia usually subsides within 4–6 hours of the most recent treatment, but mild memory deficits may persist for several weeks following termination of a full course of therapy (6–12 treatments).[26] Long-term follow-up studies show no objective evidence of memory impairment 6–9 months following a standard course of ECT. Use of bilateral electrodes induces impairment of both verbal and nonverbal learning; right unilateral ECT produces less memory disruption and preferentially affects memory for nonverbal material. The pathophysiology of the amnesia induced by ECT is unknown. The discharge is most likely mediated by nonspecific reticulothalamocortical circuits and may have a preferential effect on limbic mechanisms rendered vulnerable by their low seizure threshold. Electroconvulsive therapy also induces a variety of neurotransmitter, neuropeptide, and neuroendocrine changes that resemble those produced by antidepressant medications and may participate in the depression-resolving effects.

• Transient Global Amnesia *Transient global amnesia* (TGA) refers to a distinct clinical syndrome consisting of an acute period of amnesia of brief duration (<24 hours).[27] The amnesia includes an ongoing anterograde amnesia that usually persists for several hours and a retrograde amnesia of a few weeks' duration. As the anterograde learning deficit subsides, the retrograde amnesia shrinks to within a few minutes of the onset of the episode. During the amnesic period, patients usually recall their own identities and recognize familiar people but cannot remember recent occurrences or current circumstances and do not remember what they are told. Frequent repetition of the same question is one of the hallmarks of the syndrome and is the behavior that usually leads to recognition by family members or friends that something is wrong. These patients appear bewildered and recognize that a problem exists. Most patients have no neurological deficits during the episode although a few patients have evidence of minor brainstem dysfunction. The amnesia is disproportionately severe compared to other intellectual abnormalities, but passivity and difficulty copying complex constructions have been observed among patients carefully assessed while in the midst of an episode.

Most patients with TGA have a single attack and suffer no further amnesic episodes, but a few have repeated attacks and eventually have infarctions in the vertebrobasilar or posterior cerebral arterial territories. In some cases, specific circumstances such as intense emotional excitement, pain, sexual intercourse, or abrupt variations in temperature may acutely alter circulation and precipitate TGA in vulnerable individuals. In addition to cerebrovascular disease, a variety of other causes of TGA have been reported, including migraine, mild head trauma, valium overdose, seizures, and tumors (Table 7.2). Cerebral blood flow studies of patients with TGA reveal diminished blood flow in the posterior hemispheric or inferior temporal regions or the thalamus.[28]

• Psychogenic Amnesia There are several types of psychogenic memory disturbances including disso-

TABLE 7.2. *Characteristics That Distinguish Psychogenic Amnesia from Transient Global Amnesia*

Psychogenic Amnesia	*Transient Global Amnesia*
Personal identity lost	Personal identity retained
Ability to learn new information preserved	Unable to learn new information
Memory loss may be selective for specific personal information	Amnesia not selective
Temporal gradient absent	Temporal gradient present
Depression and anxiety common	Depression and anxiety infrequent
Indifference to amnesia	May be distressed by amnesia
Most common in younger patients (2nd to 4th decades)	More common in older patients (5th to 7th decades)

ciative amnesia (psychogenic amnesia), dissociative fugues (fugue states), and dissociative identity disorder (multiple personality disorder).[29] The latter two disorders entail the partial or total assumption of different identities for a finite period (usually days to weeks) that the patient has difficulty recalling later. These patients are not amnesic during the dissociated period and can learn and recall new information; later, however, they have no recall of the period during which they were in the dissociative state. Dissociation, fugues, and multiple personality are discussed in Chapter 22. Psychogenic amnesia or psychogenic loss of personal identity can, however, be confused with amnesias associated with dysfunction of medial limbic structures. Psychogenic amnesia is a hysterical conversion symptom in which patients suddenly forget their personal identities and life situations. They may not recall their names, addresses, families, or any other personal information. In some cases there is a selective loss of specific emotionally charged information such as whether one is married or the identity and whereabouts of one's parents. In other cases there is a failure to recall all autobiographical information. General information (how many inches in a foot) and skills (how to drive, etc.) are retained despite the loss of all personal memory. The amnesia is usually of short duration (24–48 hours) and either stops spontaneously or may be terminated by hypnosis, suggestion, or an amyobarbital (Amytal) interview. Unlike other types of conversion reaction, which are more common in women, there is an equal preponderance of men and women among patients exhibiting psychogenic amnesia. Psychogenic amnesia is often linked to anxiety and depression. Malingering is also common as a cause of psychogenic amnesia, particularly among "absconding treasurers, reluctant bridegrooms, and other criminals and wrongdoers seeking to evade the consequences of their actions."[30]

Psychogenic amnesia is most likely to be confused with TGA, but there are several characteristics that aid in the differentiation of psychogenic amnesia from TGA and from other types of amnesia associated with hippocampal, hypothalamic, or thalamic dysfunction (Table 7.2). Transient global amnesia almost never entails a loss of personal identity, whereas it is one of the hallmarks of psychogenic amnesia. By definition, patients with TGA have difficulty learning and retaining new information, whereas patients with psychogenic amnesia may be able to learn many details about their current situation at a time that they cannot recall information concerning their remote histories. The pattern of memory loss in TGA patients includes a temporal gradient with relative preservation of remote memory beyond the period of retrograde amnesia; patients with psychogenic amnesia do not exhibit a temporal gradient, and memory loss may be highly specific for selected personal information. Depression is common among patients with psychogenic amnesia, and they are usually indifferent to their memory losses, whereas TGA patients show no preponderance of any associated psychopathology and are distressed by their memory deficits. Psychogenic amnesia patients are usually in their teens, twenties, or thirties, whereas TGA is most common in patients in their fifth to seventh decade.

Psychogenic amnesia must also be distinguished from episodic disturbances of consciousness that may accompany seizures (Chapter 21). Partial complex seizures may occasionally produce twilight states lasting several hours, during which the patient is behaviorally active but later has no recall of the period. If they are ob-

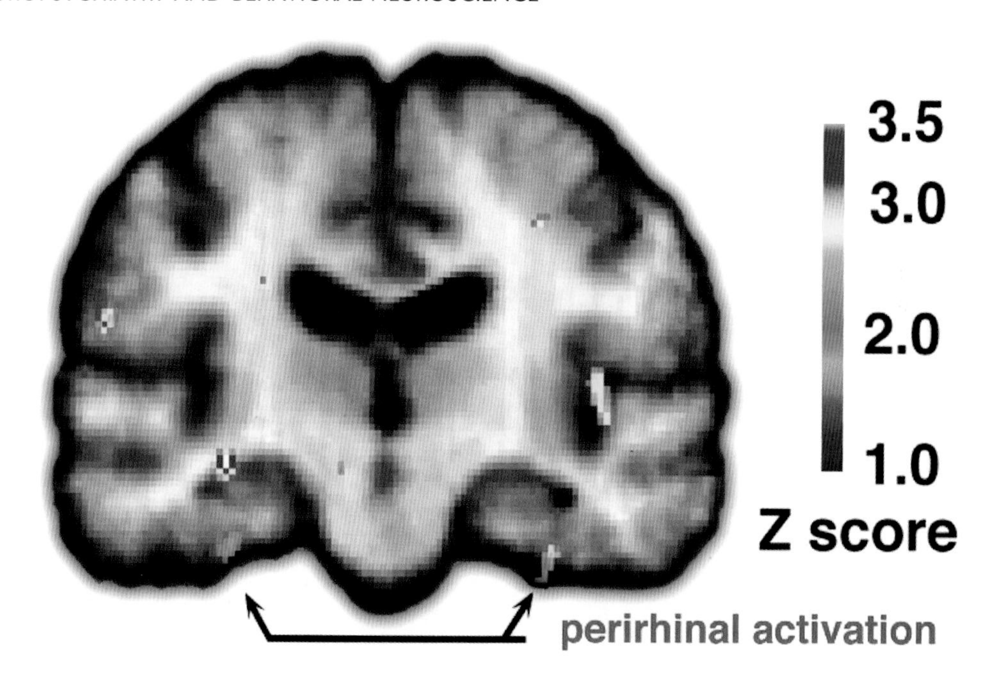

FIGURE 7.5 [^{18}F]-fluorodeoxy glucose positron emission tomography (FDG-PET) studies, registered to a probabilistic anatomic atlas of the average Alzheimer's disease (AD) brain,[50] were normalized across the groups' mean intensity levels, and subjected to a voxel-by-voxel subtraction of the post-minus pre-treatment studies using the cholinesterase inhibitor donepezil in 19 AD patients. Subvolume thresholding (SVT) corrected random lobar noise to produce a three-dimensional functional significance map. Note the bilateral perirhinal activation associated with cognitive improvement in the group after treatment.

served during such states, most patients are confused and exhibit difficulty learning and remembering. They may have a history of seizures or past head injury. Headaches can signal the presence of a brain tumor as the cause of the seizures. In some cases, evaluation with neuroimaging and electroencephalography may be necessary to aid in differentiating the two conditions.

Treatment of Amnesia

Amnesic disorders are treatment-resistant and no therapeutic intervention has produced marked improvement. Modest benefit in post-traumatic and post-encephalitic amnesias from treatment with physostigmine has been reported. Memory impairment in the Wernicke-Korsakoff syndrome has shown mild response to treatment with clonidine, methlysergide, or methylphenidate.[31]

Memory impairment in AD may respond to treatment with cholinesterase (ChE) inhibitors.[32] (Chapter 10). The cholinergic hypothesis of AD is supported by pharmacological studies that implicate the cholinergic system in cognitive and functional abilities. Scopolamine, for example, an antagonist of cholinergic transmission at muscarinic receptors, impairs neuropsychological function (attention and memory performance) in normal individuals, and exacerbates behavioral and cognitive symptoms in AD patients. Conversely, physostigmine, a ChE inhibitor, reverses the effects of scopolamine and improves cognitive performance in animals, normal subjects, and some AD patients. If ChE inhibitors influence memory function in AD, then augmentation of medial temporal activity should occur with treatment. Preliminary evidence using [^{18}F]-fluorodeoxyglucose positron emission tomography (FDG-PET) before and after 8 weeks of ChE inhibitor (donepezil) treatment in 19 AD patients showed an increase in parahippocampal and prefrontal activity with increased cholinergic tone (see Fig. 7.5).[33] These increases were associated with significant post-treatment improvement in Mini-Mental State exam (MMSE) scores. Cholinergic therapy may exert effects on memory, through increasing attentional function.

RETRIEVAL DEFICIT SYNDROME

Amnesia is not the most common form of memory impairment; difficulty retrieving information that has been learned and stored is more frequently encountered. Impaired retrieval is demonstrated by recognizing information as correct that one cannot spontaneously recall. The superiority of recognition memory over recall is evident in normals who perform better on multiple-choice examinations (recognition tests) than on tests that require the individual to recall the material without the aid of clues. Likewise, the ubiquitous difficulty among normals of recalling names is almost never accompanied by difficulty recognizing the correct name when it eventually comes to mind or is offered by someone else. This normal retrieval limitation is increased by fatigue and stress.

Age-associated memory impairment (AAMI) is the change in memory that occurs in the course of normal aging. There are two principal components to AAMI—difficulty acquiring new information and diminished retrieval—the latter represents an exaggeration of the normal difficulty with spontaneous recall of information. Ten to 20 percent of individuals over age 60 exhibit an identifiable AAMI syndrome;[34] they may be a separate subgroup of aged individuals or they may represent the low end of a normal distribution of memory function in the elderly. Complaints of memory loss in the elderly more often signal the presence of depression than dementia, and the syndrome is not the harbinger of AD. Age-associated memory impairment must be distinguished from type–type memory abnormalities that commonly presage the emergence of diagnosable AD. The pathologic basis for AAMI has not been determined. Cell loss, biochemical changes, and atrophy of medial temporal lobe structures have all been suggested as responsible for the memory alterations of AAMI.

Disorders that affect the *frontal–subcortical circuits* also produce a retrieval deficit syndrome. Diseases affecting the frontal lobes and the basal ganglia produce a deficit in executive function that includes poor organization of material to be remembered, diminished recall of stored information, and preserved recognition of learned data.[35] The retrieval disorder is evident in attempts to recall remote as well as recently learned information; there is no temporal gradient of the type seen in amnesias. In all these situations, the patient is better able to recognize than recall the information. The member structures of the frontal–subcortical circuit mediating executive function and attention include the dorsolateral prefrontal cortex and anterior cingulate as well as their subcortical connections in the caudate nucleus, globus pallidus, and medial thalamus (Fig. 7.6). As noted above, thalamic dysfunction produces a true amnesia, but disorders affecting the other structures of the dorsolateral prefrontal–subcortical circuits manifest executive dysfunction and a retrieval deficit syndrome. Retrieval abnormalities are present in patients with frontotemporal dementias, Huntington's disease, progressive supranuclear palsy, and subcortical vascular dementia (Chapter 10). Table 7.3 summarizes the principal differences between amnesia and the retrieval deficit syndrome.

Functional Imaging Studies

● **The Encoding System** Functional imaging studies of normal subjects performing tasks of episodic encoding, retrieval, and recognition also support the importance of Papez' circuit and frontal lobe function in the memory process.[36] Functional imaging studies have confirmed the involvement of medial temporal

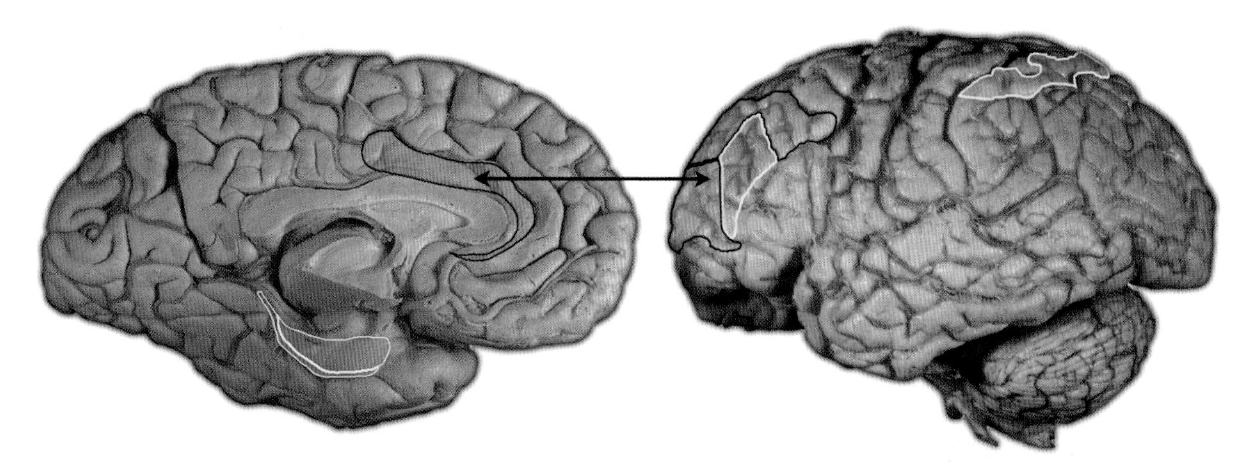

FIGURE 7.6 The dorsolateral frontal cortex is linked by reciprocal connections with the anterior cingulate and the posterior association cortex of the parietal lobe (outlined in yellow on right). The cingulate, in turn, has strong reciprocal connections with the anterior perirhinal and parahippocampal cortex (outlined in yellow on left). This anatomic circuit supports the coordinated interaction of attentional-executive functions with encoding and retrieval.

TABLE 7.3. *Principal Differences between Amnesia and Retrieval Deficit Syndrome*

Feature	*Amnesia*	*Retrieval Deficit Syndrome*
Registration	Intact	Intact (if attention intact)
Recall	Impaired	Impaired
Recognition	Impaired	Intact
Response to clues to aid recall	Impaired	Intact
Anatomy	Hippocampal–mammillary body– thalamic circuit	Frontal–subcortical circuit

structures in the encoding process (see Fig. 7.7). Application of either distracter tasks or varying the level of cognitive processing during the presentation of items to be learned can affect encoding success.[37] By judging the abstract quality, or deeper associations, of words, as opposed to their surface orthographic features, subsequent recall is significantly enhanced. Such strategies of leveraging the associations of items to be remembered has been used since the ancient Greek orators. Modern imaging has revealed that this deeper processing, compared to the shallow inspection of individual letters, recruits dorsolateral prefrontal regions during encoding that are also engaged during retrieval (see Fig. 7.8).

Most functional imaging studies are focused on the assessment of activity occurring on the same day of testing. Long-term dynamic changes occur with the eventual consolidation of learned information. Functional imaging studies have just begun to probe this dynamic consolidation process. After hippocampal and medial temporal regions are engaged with initial encoding, anterior cingulate and temporal cortices appear to mediate the retrieval of learned information. The anterior cingulate has been consistently activated in paradigms

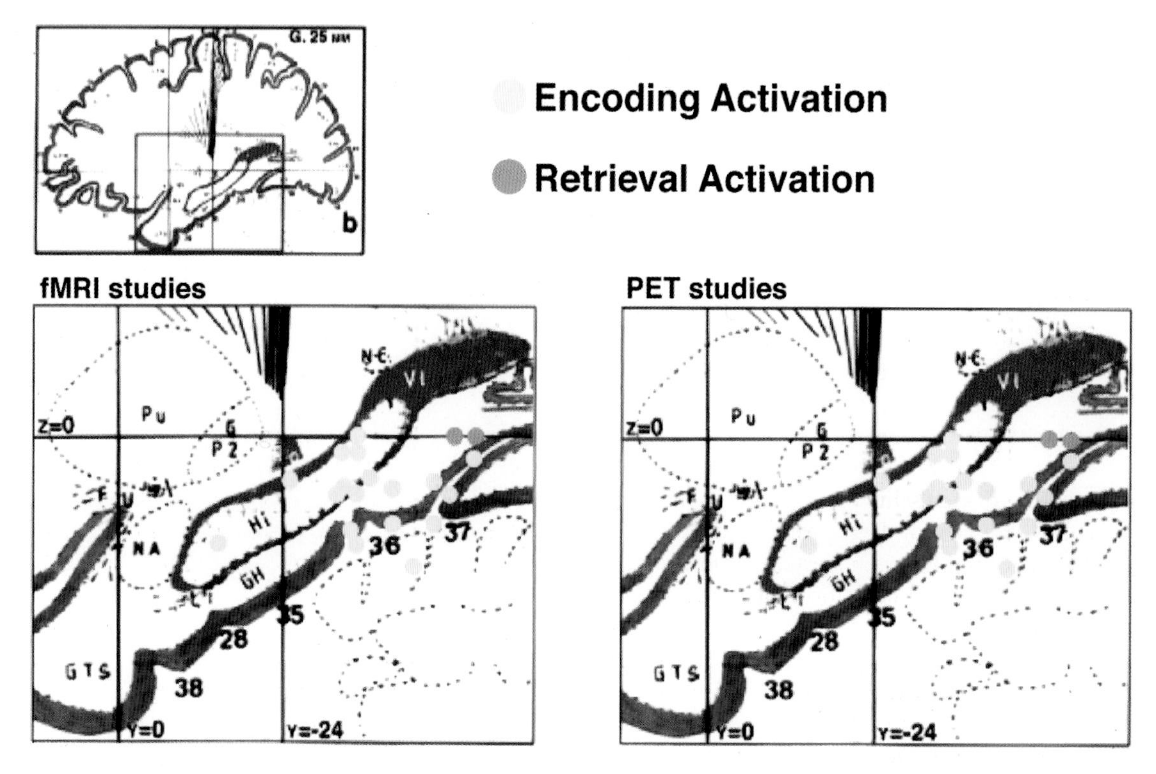

FIGURE 7.7 Regional mapping of the medial temporal activations found in encoding and retrieval tasks of episodic memory for both fMRI and PET studies of normals. Adapted from Schacter et al. (1999).[51]

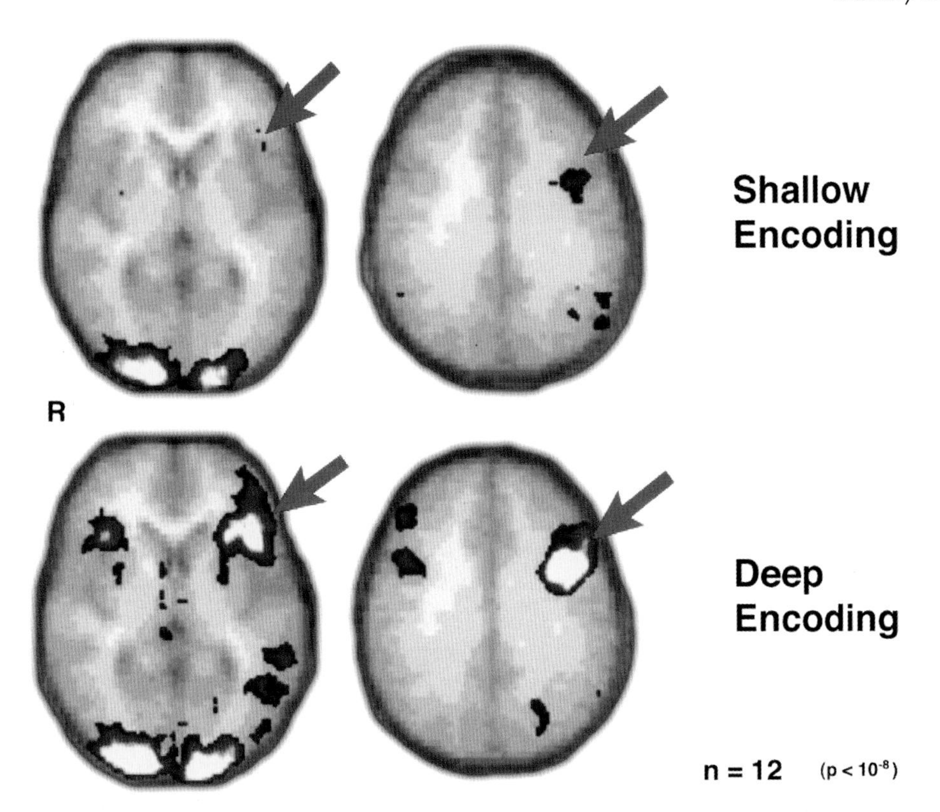

FIGURE 7.8 Functional magnetic resonance imaging (fMRI) activation maps for "shallow" and "deep" encoding tasks, both contrasted with fixation. Both activate posterior visual areas whereas only the deep encoding task shows increased activation of left inferior and dorsolateral frontal areas (arrows). These activations are at peak Talairach[52] coordinates (x, y, z) of −40, 9, 34 and −46, 6, 28 for the more dorsal activations and −40, 19, 3 and −43, 19, 12 for the more ventral prefrontal activations. Adapted from Buckner and Koutstaal (1998).[53]

that require sustained attention to *novel* tasks. In a subtraction-based paradigm of memory encoding combined with a motor task demanding sustained divided attention,[38] the anterior cingulate was singularly activated by the sustained vigilance demanded with dividing effort between the two tasks. PET activation studies using varied designs[39,40] consistently activate the anterior cingulate when subjects are motivated to succeed in whatever task is given them. When motivation to master a task is no longer required, and accurate performance of a task becomes routine, the anterior cingulate returns to a baseline activity level. In addition to its role in the consolidation of declarative memory, the posterior cingulate is also active during associative learning in classical conditioning paradigms.

• **The Retrieval System** Dynamic hippocampal–cortical interactions occur during the memory consolidation process with a gradual reorganization of the neural substrates underlying long-term memory storage. Retrieval of remotely acquired information then involves the anterior cingulate and other neocortical regions to access the previously stored representations.[41] This anterior cingulate activation may be related to increased attention and internal search strategies. The acquisition of novel cognitive strategies requires the "dynamic vigilance" of the anterior cingulate, but with practice the motivation required to entrain new cognitive networks to a novel task is no longer necessary. In humans, cingulate and temporal cortices are enlisted during the retrieval of autobiographic memory.[42] Prefrontal activation, right greater than left, is typically observed in brain imaging studies of explicit retrieval (see Fig. 7.9), and is perhaps related to the temporary sequencing and organization of information and to the search strategies employed to recall previously studied items.

DIFFERENTIAL DIAGNOSIS OF MEMORY DISORDERS

Table 7.4 presents the differential diagnosis of memory disorders. For clinical purposes, memory distur-

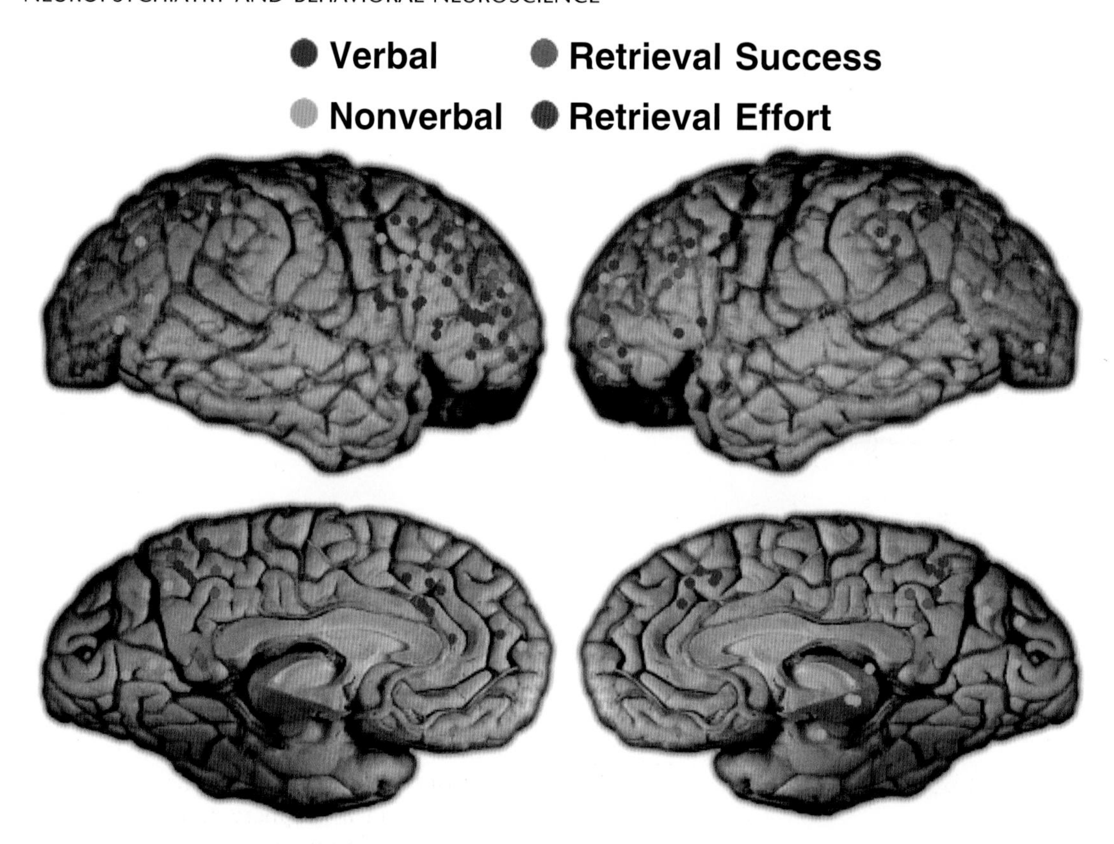

FIGURE 7.9 Summary of the peak regions of significance in functional imaging studies mapping the success and effort in the retrieval of verbal and nonverbal material. Adapted from Carbeza and Nyberg (2000).[54]

bances can be divided into those that are short-lived (usually less than 24–48 hours), those that are either more prolonged or are stable (lasting for more than 48 hours), and those that are progressive.

Transient, short-duration episodes of memory loss include psychogenic amnesia, some cases of post-traumatic amnesia, and TGA. These must be distinguished from other brief interruptions of consciousness with memory lapses, including complex partial seizures, alcoholic blackout spells, migraine, and toxic-metabolic confusional states. Seizures as a cause of memory lapses should be considered in individuals with a history of head trauma or other predisposing circumstances, a known history of epilepsy, or other symptoms suggestive of an epileptic ictus such as an aura, incontinence, or postictal confusion. Most seizure-related episodes are short-lived, lasting for only minutes, but automatic behavior may occasionally persist for hours or even several days. Integrated, purposeful behavior is rare during seizures. Alcoholic blackouts are periods of unrecalled behavior during which the intoxicated individual's behavior seems normal but the episode cannot be recalled later. Amnesia may occur with relatively low blood alcohol levels and becomes more common as the concentration rises. The amnesia of alcoholic blackouts is a product of impaired information storage. Memory lapses also occur with benzodiazepine ingestion. Migraine as a cause of confusional states with memory lapses is usually accompanied by other migrainous symptoms such as photophobia, visual hallucinations, nausea, and headache, although the latter need not be prominent. Toxic and metabolic encephalopathies, particularly those associated with drug ingestion and transient hypoglycemia, must also be considered in the differential diagnosis of transient memory alterations.

Memory disturbances lasting for more prolonged periods include the amnesia syndromes discussed earlier (Table 7.1) and the more long-lasting dissociative states, including fugues and multiple personality. The latter are not amnesias as defined here, in that there is no disturbance of recent memory during the episode. Rather, on recovering from the dissociative state, the patient is unable to recall all or most of the events transpiring during the dissociated period. Progressive memory loss is characteristic of dementia syndromes (Chap-

TABLE 7.4. *Differential Diagnosis of Memory Disturbances*

Transient episodes of memory loss (<48 hours)
Amnesias (nonmemory functions intact)
Transient global amnesia
Psychogenic amnesia
Post-traumatic amnesia
Memory lapses with altered attention
Seizures
Alcoholic blackouts
Benzodiazepine-related amnesia
Migraine
Toxic-metabolic confusional states
Prolonged periods of memory loss syndromes (>48 hours)
Amnestic syndromes
Dissociative states
Fugues
Multiple personality
Progressive memory dysfunction
Cortical dementias
Frontal–subcortical dementias
Minimal cognitive impairment (MCI)

ter 10). Two principal types of memory disorders occur in dementing illnesses: a type–type storage disorder characterized by impaired recall and recognition and a retrieval deficit syndrome featuring diminished recall with preserved recognition. The amnesic type disorder occurs in AD and other conditions in which the medial temporal structures are affected. The retrieval deficit syndrome occurs in diseases that involve the frontal lobes and the basal ganglia. It is essential when attempting to differentiate these two disorders that the patient have sufficient attention to ensure information registration. Distractible, inattentive individuals will have reduced storage and an amnesic pattern of performance regardless of the cause or the site of any underlying pathology. Information registration is a prerequisite for demonstrating a distinction between storage and retrieval deficits. Dementia syndromes have impairments in other cognitive domains as well as memory. Patients with AD evidence language and visuospatial deficits, and patients with frontal–subcortical dementias typically manifest executive function deficits and prominent mood and behavior changes.[18]

Progressive memory abnormalities must be distinguished from AAMI. The latter is confined to memory performance and does not involve other major cognitive domains. It is not incapacitating. Individuals with AAMI do not meet criteria for the definition of dementia (Chapter 10).

REDUPLICATIVE PARAMNESIA

Reduplicative paramnesia or *double orientation* refers to a peculiar disorientation syndrome in which the patient claims to be present simultaneously in two or more locations. The reduplicative paramnesia may take the form of believing that the location has two names, that the two locations are contiguous, or that the patient has recently made a journey from one to the other. For example, one of the author's patients insisted that he was in the Rose Garden Hotel, which was a branch of UCLA Medical Center (his actual location). He explained that the hospital had been built around the hotel. Reduplicative paramnesia most commonly occurs during the recovery phase of post-traumatic encephalopathy, but it has been observed in patients with tumors, infarctions, and arteriovenous malformations as well as in metabolic and toxic encephalopathies.[43]

In most cases with focal lesions, there has been damage to the right hemisphere and to both frontal lobes. A right frontal lesion may be sufficient to produce the syndrome. A possible explanation for the unique phenomenon is that the right hemispheric lesion makes it difficult for patients to integrate spatially significant information, whereas concomitant frontal lobe dysfunction or the confusional state impairs patients' abilities to appreciate and correct the discrepancy in their beliefs.

Alexander et al.[44] have suggested that the *Capgras syndrome*, the delusional belief that people have been replaced by identical-appearing imposters, shares many features with reduplicative paramnesia, including the combination of right hemispheric and frontal dysfunction, and might be regarded as a similar paramnesia syndrome. The Capgras syndrome is presented more thoroughly in Chapter 12.

CONFABULATION

Confabulation refers to the production of erroneous answers by patients with memory defects and represents a failure of error recognition rather than a desire to deliberately mislead. Two basic forms of confabulation have been distinguished: (*1*) confabulation of embarrassment, in which the amnesic patient provides incorrect answers based on personal past experience, and (*2*) fantastic confabulation, in which patients with impaired judgment and current or recent amnesia spontaneously describe impossible, adventurous, and often gruesome experiences.[45] Confabulation of embarrassment typically occurs in the acute stages of Wernicke-Korsakoff syndrome. Patients respond to questions regarding the date, location, and their employment with answers derived from their past. The answers are usually coherent or possible, but incorrect. If the syndrome enters its more chronic phases, confabulation diminishes and may ultimately disappear altogether.

Studies of confabulation demonstrate that it represents a failure of error recognition and self-monitoring and most likely reflects impairment of executive function.[46] Confabulation frequently coexists with other evidence of executive dysfunction such as perseveration and apathy. Functional imaging demonstrates abnormalities of orbital and medial frontal regions that resolve in concert with resolution of the confabulatory state.[47] Confabulation of embarrassment may occur in other amnesic states and in degenerative dementias as well as in Korsakoff's syndrome, but appears to be most common in patients with lesions in the mamillary bodies or thalamus.

Fantastic confabulation is a more rare and more colorful syndrome. One of the present authors' patients described how he had been taken into a spaceship by extraterrestrial visitors and taught how to drive their ship; on another occasion he told how he had single-handedly decapitated 39 enemy soldiers as they were lined up shooting from behind a log. Most patients with fantastic confabulation have obvious frontal lobe syndromes, but their amnesia may be mild. The syndrome has been observed in post-traumatic encephalopathy, the Wernicke-Korsakoff syndrome, and degenerative dementias. If the patient consistently tells the same story and appears to endorse the story as true, the syndrome may be better viewed as a delusional disorder.

Confabulation must be distinguished from *prevarication*, where the patient attempts to deliberately mislead the examiner. *Pseudologia fantastica* is a distinctive form of lying often associated with Munchausen's syndrome.

REFERENCES

1. Squire LR, Zola-Morgan S. The medial temporal lobe memory system. Science 1991;253:1380–1386.
2. Bauer RM, Tobias B, et al. Amnestic disorders. In: Heilman KM, Valenstein E, eds. Clinical Neuropsychology, 3rd ed. New York: Oxford University Press; 1993:523–602.
3. Benson DF, Geschwind N. Shrinking retrograde amnesia. J Neurol Neurosurg Psychiatry 1967;30:539–544.
4. Cohen NJ, Eichenbaum H, et al. Different memory systems underlying acquisition of procedural and declarative knowledge. Ann NY Acad Sci 1985;444:54–71.
5. Parkin AJ. Residual learning capability in organic amnesia. Cortex 1982;18:417–440.
6. Brody IA, Wilkins RH. Wernicke's encephalopathy. Arch Neurol 1968;19:228–232.
7. Reuler JB, Girard DE, et al. Wernicke's encephalopathy. N Engl J Med 1985;312:1035–1039.
8. Charness ME, De La Paz RL. Mamillary body atrophy in Wernicke's encephalopathy: antemortem identification using magnetic resonance imaging. Ann Neurol 1987;22:595–600.
9. Victor M, Adams RD, et al. The Wernicke-Korsakoff Syndrome and Related Neurologic Disorders Due to Alcoholism and Malnutrition, 2nd ed. Philadelphia: F.A. Davis, 1989.
10. Tan GH, Farnell GF, et al. Acute Wernicke's encephalopathy attributable to pure dietary thiamine deficiency. Mayo Clin Proc 1994;69:849–850.
11. Scoville WB, Milner B. Loss of recent memory after bilateral hippocampal lesions. J Neurol Neurosurg Psychiatry 1957;20:11–21.
12. Gabrieli JDE, Cohen NJ, et al. The impaired learning of semantic knowledge following bilateral medial temporal-lobe resection. Brain Cogn 1988;7:157–177.
13. Zola-Morgan S, Squire LR, et al. Lesions of perirhinal and parahippocampal cortex that spare the amygdala and hippocampal formation produce severe memory impairment. J Neurosci 1989;9:4355–4370.
14. Penfield W, Milner B. Memory deficit produced by bilateral lesions in the hippocampal zone. Arch Neurol Psychiatry 1958; 79:475–497.
15. Sass KJ, Lencz T, et al. The neural substrate of memory impairment demonstrated by the intracarotid amobarbital procedure. Arch Neurol 1991;48:48–52.
16. Gaffan EA, Gaffan D, et al. Amnesia following damage to the left fornix and to other sites. Brain 1991;114:1297–1313.
17. Levin HS. Memory deficit after closed head injury. In: Boller F, Grafman J, eds. Handbook of Neuropsychology. Amsterdam: Elsevier; 1989:183–207.
18. Schnider A, Regard M, et al. Anterograde and retrograde amnesia following bitemporal infarction. Behav Neurol 1994;7:87–92.
19. Markowitsch HJ, von Cramon DY, et al. Amnestic performance profile of a bilateral diencephalic infarct patient with preserved intelligence and severe amnesic disturbances. J Clin Exp Neuropsychol 1993;15:627–652.
20. Cummings JL, Tomiyasu U, et al. Amnesia with hippocampal lesions after cardiopulmonary arrest. Neurology 1984;34:679–681.
21. Lishman WA. Organic Psychiatry, 3rd ed. Oxford: Blackwell Scientific, 1998.
22. Elsinger PJ, Damasio H, et al. Nonverbal amnesia and asymmetric cerebral lesions following encephalitis. Brain Cogn 1993; 21:140–152.
23. Lilly R, Cummings JL, et al. The human Klüver-Bucy syndrome. Neurology 1983;33:1141–1145.
24. Irle E, Wowra B, et al. Memory disturbances following anterior communicating artery rupture. Ann Neurol 1992;31:473–480.
25. McGlone J, Gupta S, et al. Screening for early dementia using memory complaints from patients and relatives. Arch Neurol 1990;47:1189–1193.
26. Coffey CE, Weiner RD. Electroconvulsive therapy: an update. Hosp Commun Psychiatry 1990;41:515–521.
27. Kritchevsky M, Squire LR. Transient global amnesia: evidence

of extensive, temporally graded retrograde amnesia. Neurology 1989;39:213–218.
28. Goldenberg G, Podrecka I, et al. Thalamic ischemia in transient global amnesia: a SPECT study. Neurology 1991;41:1748–1752.
29. American Psychiatric Association. Diagnostic and Statistical Manual of Mental Disorders, 4th ed. Washington, DC: American Psychiatric Press, 1994.
30. Kennedy A, Neville J. Sudden loss of memory. BMJ 1957;2: 428–433.
31. McEntee WJ, Crook TH. Age-associated memory impairment: a role for catecholamines. Neurology 1990;40:526–530.
32. Rogers SL, Friedhoff LT, et al. The efficacy and safety of donepezil in patients with Alzheimer's disease: results of a US multicentre, randomized, double-blind, placebo-controlled trial. Dementia 1996;7:293–303.
33. Mega MS, Cummings JL, et al. Metabolic response to donepezil therapy in Alzheimer's disease. Neurology 2000;54:A416.
34. Larrabee GJ, Levin HS, et al. Senescent forgetfulness: a quantitative study. Dev Neuropsychol 1986;2:373–385.
35. Cummings JL. Frontal–subcortical circuits and human behavior. Arch Neurol 1993;50:873–880.
36. Cohen NJ, Ryan J, et al. Hippocampal system and declarative (relational) memory: summarizing the data from functional neuroimaging studies. Hippocampus 1999;9:83–98.
37. Gabrieli JDE. Cognitive neuroscience of human memory. Annu Rev Psychol 1998;49:87–115.
38. Fletcher PC, Firth CD, et al. Brain systems for encoding and retrieval of auditory-verbal memory. An in vivo study in humans. Brain 1995;118:401–416.
39. Petersen SE, Fox PT, et al. Positron emission tomographic studies of the cortical anatomy of single word processing. Nature 1988;331:585–589.
40. Corbetta M, Miezin FM, et al. Selective and divided attention during visual discriminations of shape, color and speed: functional anatomy by positron emission tomography. J Neurosci 1991:2383–2402.
41. Squire LR, Alvarez P. Retrograde amnesia and memory consolidation: a neurobiological perspective. Curr Opin Neurobiol 1995;5:169–177.
42. Fink GR. Cerebral representation of one's own past: neural networks involved in autobiographical memory. J Neurosci 1996; 16:4275–4282.
43. Kapur N, Turner A, et al. Reduplicative paramnesia: possible anatomical and neuropsychological mechanisms. J Neurol Neurosurg Psychiatry 1988;51:579–581.
44. Alexander MP, Stuss DT, et al. Capgras syndrome: a reduplicative phenomenon. Neurology 1979;29:334–339.
45. Berlyne N. Confabulation. Br J Psychiatry 1972;120:31–39.
46. Shapiro BE, Alexander MP, et al. Mechanisms of confabulation. Neurology 1981;31:1070–1076.
47. Benson DF, Djenderendjian A, et al. Neural basis of confabulation. Neurology 1996;46:1239–1243.
48. MacLean PD. Psychosomatic disease and the "visceral brain." Recent developments bearing on the Papez theory of emotion. Psychosom Med 1949;11:338–353.
49. Corkin S, Amaral DG, et al. H. M.'s medial temporal lobe lesion: findings from magnetic resonance imaging. J Neurosci 1997;17:3964–3979.
50. Thompson PM, Woods RP, et al. Mathematical and computational challenges in creating deformable and probabilistic atlases of the human brain. Hum Brain Mapping 2000;9:81–92.
51. Schacter DL, Wagner AD. Medial temporal lobe activations in fMRI and PET studies of episodic encoding and retrieval. Hippocampus 1999;9:7–24.
52. Talairach J, Tournoux P. Principe et Technique des Etudes Anatomiques. New York: Thieme Medical Publishers, 1988.
53. Buckner R, Koutstaal W. Functional neuroimaging studies of encoding, priming, and explicit memory retrieval. Proc Natl Acad Sci USA 1998;95:891–898.
54. Carbeza R, Nyberg L. Imaging cognition II: An empirical review of 275 PET and fMRI studies. J Cogn Neurosci 2000;12:1–47.

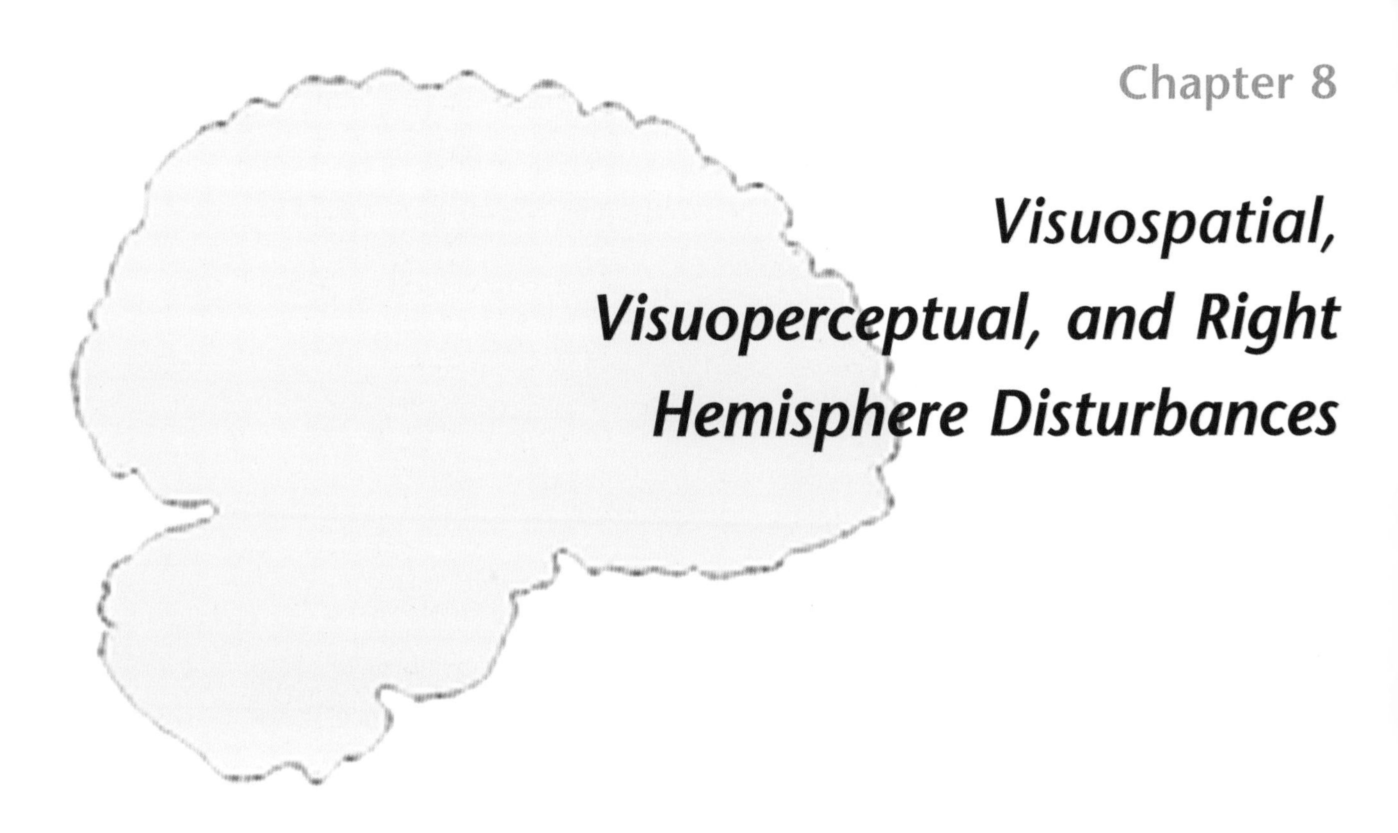

Chapter 8

Visuospatial, Visuoperceptual, and Right Hemisphere Disturbances

Humans are heavily reliant on vision for spatial orientation and environmental interaction. When portions of the brain mediating visuospatial and visuoperceptual function are injured, significant disability results. Visuospatial abilities are represented in both cerebral hemispheres, but the right brain is more involved than the left in this mental domain. Functions with greater personal relevance (e.g., facial and voice recognition) are also more strongly linked to the right than the left hemisphere. This linkage between visuospatial skills and interpersonal relevance may reflect the dependence of successful social interactions on visual recognition, visual social signaling (e.g., smiling, crying), and territorial boundaries (e.g., interpersonal space issues) on visual and spatial abilities. Thus, right brain disorders produce visual, spatial, and interpersonal dysfunction.

In this chapter, a general classification of visuospatial functions and the corresponding clinical deficits are presented. Then, three types of visuospatial disability are described: (*1*) visual agnosias, (*2*) disorders of spatial attention including neglect, denial, and anosognosia, and (*3*) visuoconstructive deficits. In the final sections, a variety of miscellaneous visuospatial disorders are discussed and an overview of the role of the right hemisphere in mediating emotions is presented.

CLASSIFICATION OF VISUOSPATIAL FUNCTION

A wide variety of visuospatial functions have been identified (Table 8.1). Most of these have been discovered by observing patients with functional deficits, analyzing the failures, and identifying the corresponding components of visuospatial processing. Visual recognition requires a multistep chain of events that begins with elementary visual perception and gradually results in a central nervous system (CNS) "representation" of the stimulus. Memory is invoked to compare the representation with earlier experiences and allow recognition as a novel or familiar stimulus. Clinically observed deficits correspond to impairments at different levels of visuocognitive processing.

VISUAL AGNOSIAS AND RELATED PHENOMENA

Agnosia refers to a clinical syndrome in which the patient is able to perceive sensory stimuli normally and has sufficient language capacity to name stimuli, but recognition of perceived material is impaired. The patient has a percept stripped of its meaning.[1] The spe-

TABLE 8.1. *Classification of Visuospatial Functions and Corresponding Deficits Observed Clinically*

Visuospatial Function	*Clinical Deficit*
Visual sensory ability	
Visual acuity	Impaired acuity
Visual field perception	Homonymous hemianopia
Color vision	Achromatopsia
Depth perception	Astereopsis
Motion perception	Cerebral akinetopsia
Visual perception	Cortical blindness
Visual discrimination ability	
Facial matching	Inability to match similar faces seen at different angles
Visual recognition	
Familiar faces	Prosopagnosia
Familiar places	Environmental agnosia (topographagnosia)
Color	Color agnosia
Object features	Apperceptive visual agnosia
Object identity	Associative visual agnosia
Visuospatial attention	
Spatial attention	Unilateral neglect
Simultanagnosia	Narrowing of effective visual fields
Awareness of deficit	Anosognosia
Visuomotor ability	
Figure copying and drawing	Constructional disturbances
Visually guided movements	Optic ataxia (Balint's syndrome)
Volitional redirection of gaze	Sticky fixation (Balint's syndrome)
Visuospatial cognition	
Revisualization	Charcot-Wilbrand syndrome
Visuospatial imagination	Neglect of hemi-imaginative space
Mental figure rotation	Impaired manipulation of mental images
Visual organization	Difficulty with mental reconstruction of partial figures into wholes
Recognition of embedded figures	Difficulty with figural disambiguation
Design generation	Impairment of design fluency
Visuospatial memory	
Storage of visuospatial information	Visuospatial amnesia
Recall of visuospatial information	Visuospatial recall deficit
Body-spatial orientation	
Finger recognition	Finger agnosia
Right–left orientation	Right–left disorientation
Dressing	Dressing disturbance
Map reading	Planotopokinesia

cific characteristics of agnosia reflect the stage of recognition disrupted and the class of stimulus object most involved.

Visual Object Agnosia

The existence of visual object agnosia has been challenged,[2,3] but a sufficient number of credible cases have been reported to establish the syndrome as an identifiable, if rare, clinical entity. Following Lissauer's 1889 suggestion, the visual object agnosias are divided into two principal types: apperceptive visual agnosias and associative visual agnosias.[4] The former involve deficits at a lower level of visual analysis at the boundary between perception and recognition; the latter involve higher levels of visual processing and are almost pure recognition deficits. In *apperceptive visual agnosia*, the patient is able to see to the extent of identifying color, movement and direction of movement, line direction and dimension, and light intensity. The patient cannot, however, distinguish one form from another (e.g., a cross from a circle) and cannot draw presented objects. Identification through other sensory modalities—auditory, tactile, and olfactory—is intact. Most patients have had extensive bilateral posterior cerebral insults.[4–6]

Associative visual agnosia is a disturbance of visual recognition with intact visual perception. The integrity of perceptual processes is attested to by the ability of the patient to make drawings or copies of objects that cannot be visually identified. The preservation of the ability to copy and match similar visual stimuli is the primary clinical characteristic that distinguishes associative from apperceptive visual agnosia. Impairment of facial recognition (prosopagnosia), color recognition (color agnosia), and reading (alexia) usually accompanies the associative visual agnosias, although there have been rare cases of agnosia for nonverbal material with preserved reading abilities.[7,8] Associative visual agnosia can be distinguished from anomia by the intact ability of anomic patients to select the correct name from a list of choices or to describe the correct use of the object. Patients with associative agnosia cannot name a visual stimulus but demonstrate intact naming abilities when auditory or tactile information is available. Pathologically, nearly all autopsied cases of associative visual agnosia have had bilateral medial occipitotemporal lesions involving the inferior longitudinal fasciculi connecting occipital and temporal lobes (Fig. 8.1).[9–11]

Prosopagnosia

Prosopagnosia is a unique syndrome in which the patient's major difficulty is an inability to recognize fa-

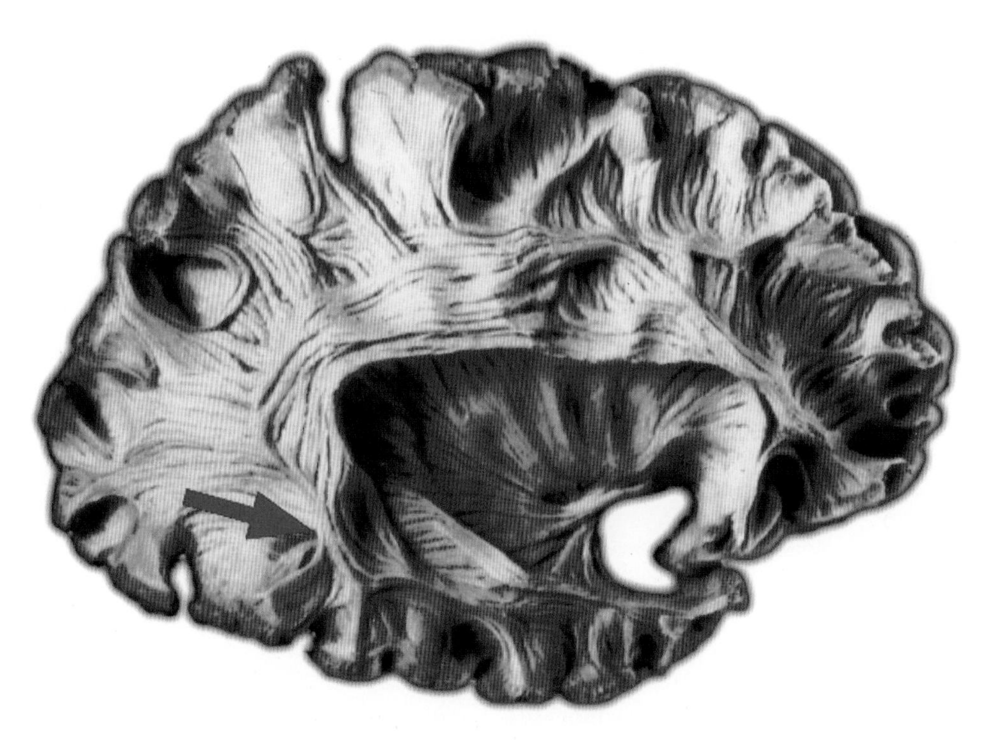

FIGURE 8.1 Transaxial magnetic resonance imaging shows the location of the inferior longitudinal fasciculi involved in associative visual agnosia.

miliar faces. Although able to see normally and demonstrating no visual agnosia for most other classes of objects, patients with prosopagnosia are unable to recognize spouses, friends, or relatives. They may be able to correctly describe who they see, but no sense of recognition ensues, and they must develop alternate strategies to compensate for the visual recognition deficit (e.g., the sound of the unrecognized person's voice or specific identifying features such as mustache, hair color, etc., are used to aid identification).[12–14]

Although the defect in prosopagnosia involves primarily recognition of familiar faces, some patients have deficits in recognizing individual members of other classes of objects, such as specific animals (e.g., individual cows) or specific automobiles, with which the patient was previously familiar.[12,15] Patients with prosopagnosia may be able to discriminate and match unfamiliar faces normally, but they cannot learn new faces and recognize them at a later time.[16] Difficulty with discriminating and matching unfamiliar faces is a ubiquitous abnormality among patients with posterior right hemisphere lesions, and such patients seldom have concomitant prosopagnosia.[17–19] Patients with prosopagnosia have electrodermal responses to familiar faces, which suggests that they have preserved psychophysiological recognition even though they have no conscious or declarative recognition.[20,21]

The anatomic basis of prosopagnosia is controversial. Most cases have had bilateral medial occipital lesions,[12,22] but several cases studied have had unilateral right-sided posteromedial lesions, suggesting that unilateral right-sided lesions may be sufficient to interrupt recognition of familiar faces.[23,24]

Environmental Agnosia

Environmental agnosia (also called *topographagnosia, topagnosia,* and *topographic memory loss*), like prosopagnosia, involves loss of recognition of a specific class of familiar visual stimuli—in this case, one's own environment. The patient can see adequately, can usually describe the environment correctly, can often correctly utilize maps to find directions, and may even be able to draw an accurate map. When faced with the actual environment, however, the patient has no sense of familiarity or recognition.[25,26] Most patients capitalize on their intact linguistic capacities and adopt verbal strategies to help compensate for their deficits. For example, they use street names and house numbers to find their way around, and they may not be able to find their own houses, except in this way! Environmental agnosia frequently occurs as part of a clinical triad with prosopagnosia and central achromatopsia (described below), but a variety of concomitant abnormalities have been described, including palinopsia, absent mental revisualization (Charcot-Wilbrand syndrome), dressing disturbances, visual allesthesia, and disturbances of brightness modulation.[26] The underlying defect in environmental agnosia appears to be an inability to relate current perceptions to stored memories that would allow environmental recognition to occur.

Color Agnosia

Color agnosia refers to the specific inability to recognize colors.[27] Color agnosia is difficult to assess, since color is exclusively a visual phenomenon and it is not possible to demonstrate that patients who are unable to name colors or point to named colors could recognize them when presented in another modality (e.g., colors have no auditory or tactile dimension). Disturbances of color-related cognition from which color agnosia should be distinguished include achromatopsia (discussed below), color-specific naming defects that occur in patients with alexia without agraphia, and color aphasia (aphasic syndrome with disproportionate severity for color naming and understanding color words).[1,28] Patients with color agnosia can sort by color category or arrange colors by shade, whereas patients with achromatopsia cannot perform these tasks.

Finger Agnosia

Finger agnosia occurs as one element of Gerstmann's syndrome along with dysgraphia, right–left disorientation, and acalculia. When all four elements of the syndrome are simultaneously present in the absence of dementia or delirium, it reliably indicates the presence of a lesion of the left angular gyrus.[29] Finger agnosia is a form of autotopagnosia reflecting a deficit in the ability to localize stimuli applied to one's body (described below). When the syndrome is present in its most severe form, the patient cannot point to named fingers. In its more mild forms, the syndrome can be demonstrated by asking the patient to show the finger on his or her hand that corresponds to a finger on the other hand touched by the examiner when the patient's eyes are closed, or by asking the patient to state how many fingers are between two touched by the examiner while the patient's eyes are closed.

Simultanagnosia

Simultanagnosia refers to a curious neuropsychological deficit in which the patient is unable to simultaneously perceive more than one stimulus item or more than one part of a complex pattern. For example, if shown a piece of paper with a circle and a cross on it, the patient will see only one of the items; or, if shown a complex picture, only one portion of it is visible to the patient.[30,31] The disorder cannot be attributed to elementary visual abnormalities or to hemispatial neglect. Simultanagnosia does not meet the usual criteria for an agnosia since the patient has a narrowing of effective visual field excluding more than one element and allowing recognition of only one element at a time. The patient cannot see the other elements, thus simultanagnosia is a specific type of visual attentional failure rather than an agnosia.

Simultanagnosia is thought to be due to an impaired ability to integrate more than one item at a time or an inability to use visual cues that allow rapid analysis of complex figures.[31] Most patients with simultanagnosia have had bilateral parietooccipital lesions, although simultanagnosia due to a lesion of the left occipital lobe has been described.[1] Simultanagnosia often occurs as one element of Balint's syndrome (described below).

NEGLECT, DENIAL, AND ANOSOGNOSIA

Hemispatial Neglect

Unilateral neglect refers to a hemi-inattention syndrome in which the patient fails to notice, report, or respond to stimuli in one-half of space. All sensory modalities can be included in a neglect syndrome, and the severity of the attentional deficit varies considerably among patients. In severe cases, the patient notices nothing in the neglected half-field; will draw only one-half of constructions; may dress, shave, or apply make-up to only one-half of the body; ignores visual, auditory, and tactile stimuli on the neglected side; and will frankly deny the presence of any motor or sensory deficit on the side of the body in the affected hemispace. In milder cases, there will be a less pronounced tendency to ignore stimuli on the involved side, the patient's attention can be directed to the affected hemispace, and the patient will not overtly deny the deficits. In its least severe form, the neglect will be revealed only by extinguishing one of a pair of stimuli during double simultaneous sensory stimulation.[32,33] The severity of neglect tends to be greater in patients with larger lesions and in those with coexisting cerebral atrophy.[34]

Right parietal lesions produce the most marked neglect syndrome, but hemispatial neglect is not uniquely related to damage to this brain region. Neglect has also been observed in patients with left parietal lesions, damage to the left and right dorsolateral frontal areas, left and right medial frontal regions, left and right striatum, right thalamus, and the white matter of the internal capsule and hemispheric white matter.[32,35–43] These structures receive projections from the ascending reticular activating system and comprise a reticulo-limbic-neocortical network responsible for directed attention.[44] Thalamoparietal dysfunction produces primarily sensory neglect; dorsolateral frontal and striatal lesions produce motor neglect; and anterior cingulate injury results in diminished motivation[45–47] (Table 8.2)

The right hemisphere is dominant for spatial attention and mediates aspects of attention directed to both right and left hemispace, whereas the left hemisphere

TABLE 8.2. *Anatomic Location of Lesions Producing Hemispatial Neglect and Clinical Features of the Corresponding Syndrome*

Location	*Clinical Features*
Parietal lobe	Sensory neglect with inattention to contralateral stimuli Failure to detect contralateral stimuli and extinguishing contralateral stimuli during bilateral simultaneous stimulation Anosognosia
Medial thalamus	Sensory neglect similar to that observed in patients with parietal lesions
Dorsolateral prefrontal lobe	Contralateral hypokinesia, hypometric limb movements, and reduced action directed toward contralateral space
Striatum	Motor neglect similar to that observed in patients with dorsolateral prefrontal lesions
Anterior cingulate cortex	Reduced motivation to perform sensory and motor tasks in the contralateral hemispace

FIGURE 8.2 On this line-crossing test, a patient with hemispatial neglect was instructed to cross each line in the middle. He crossed only the lines on the right side of the page, neglecting the left hemispace.

mediates only contralateral attention. Thus, right brain dysfunction produces contralateral as well as ipsilateral attentional deficits, while left brain lesions produce only contralateral neglect.[44,48] Compensation for left hemisphere dysfunction by the right brain leads to early resolution of right-sided neglect following left brain injury, but the left hemisphere is less able to compensate for right brain dysfunction and there is more sustained left-sided neglect after right brain injury. Cortical dysfunction may be the essential element in all forms of neglect. Subcortical lesions with neglect have associated functional disruption of cortical activity when studied with single photon emission computed tomography (SPECT), and recovery of neglect corresponds to the recovery of cortical function.[35] Subcortical lesions may cause cortical dysfunction through disconnection of cortical afferent connections, concomitant hypoperfusion, or diaschesis.

Tests for unilateral neglect should include both sensory and motor function. The line-crossing test (Fig. 8.2) is one simple and effective way of demonstrating visual neglect.[49] The patient is presented with a sheet of paper with random lines on it and asked to mark the center of each line. If neglect is present, all or a portion of the lines in the neglected field will not be marked. Drawing and copying tasks may also reveal hemispatial neglect. The patient will reproduce only the portion of the model figure in the non-neglected field or will omit details of the figure that appear in the neglected hemifield (Fig. 8.3). Likewise, when drawing new figures (rather than copying model figures), the patient will make errors of omission or placement on the side of the figure in the neglected hemispace (Fig. 8.4). There is a tendency for patients with lesions of the parietal cortex to make more errors when copying than when drawing and for patients with frontal lobe lesions to make more errors when drawing than when copying. Neglect can also be demonstrated by cancellation tasks (requiring the patient to search a visual space and mark or "cancel" specific target letters or objects.[50] Hemialexia is best demonstrated by asking the patient to read compound words such as *baseball* and *northwest*; the patient will read only the word in the non-neglected hemifield.[51] Hemiacalculia is observed when patients are asked to add figures arranged in columns; they will add only the numbers in the non-neglected field. Patients with motor neglect often appear to have a hemiplegia, since they fail to use the neglected limbs. When each limb is tested individually, the strength on the neglected side is found to be normal or only partially diminished. Motor neglect is also evidenced by motor impersistence and hypokinetic movements of the neglected side (for example, when asked to draw large circles in the air with arms extended, the patient will draw smaller circles with the neglected limb).[47,52]

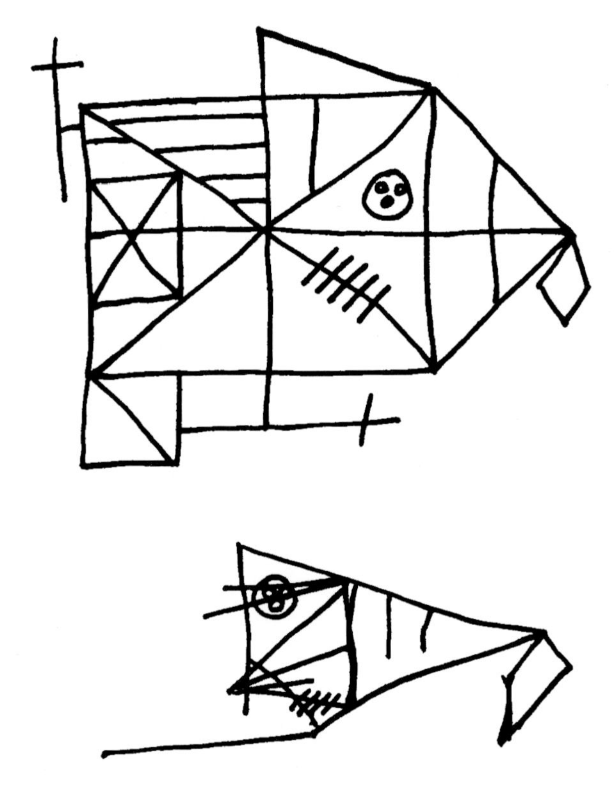

FIGURE 8.3 When asked to copy the Rey-Osterrieth Complex Figure (*top*), the patient neglected the left half of the figure and copied only the portion appearing in the right hemifield (*bottom*).

Neglect is not a product of any coexisting visual field defect; many patients with homonymous hemianopia do not have neglect syndromes, and many with neglect syndromes do not have hemianopias.

Extinction of one stimulus during double simultaneous sensory stimulations is a mild form of neglect and can be demonstrated in the visual sphere by stimulating homonymous portions of the visual fields during confrontation testing (a visual field defect will, if present, also obliterate perception of one stimulus and invalidate this test), in the auditory sphere by snapping one's fingers behind the ears on each side, and in the somatosensory sphere by simultaneously stimulating both sides of the body.[32]

Neglect is most pronounced during the acute phases of an acquired cerebral insult and, in the case of static lesions, gradually improves. Subtle evidence of neglect may persist for months or even longer, however, and may be one of the factors most limiting to the rehabilitation of the brain-injured patients.[53] Improvement in neglect of hemispace can be encouraged by tasks that require the patient to direct their attention toward the neglected side, such as extending a line in that direction or producing a series of numbers that extend into the neglected space.[54,55]

Anosognosia

The term *anosognosia* was originally used by Babinski to describe patients with denial of hemiparesis. A variety of types of behavioral disturbances have been observed in patients with unilateral weakness, ranging from indifference to the deficit (anosodiaphoria), to denial of weakness (anosognosia), hatred of the paralyzed limbs (misoplegia) or denial of ownership of the weak extremities (somatoparaphrenia) (Table 8.3).[36,56–61]

Most patients with anosognosia have right brain lesions with left hemisensory loss of left-sided neglect. In

FIGURE 8.4 The patient was asked to draw a man. The left side of the figure is less developed (note the small ear of the left side of the drawing) and has several errors (discontinuity of the arm and leg on the left side of the drawing).

TABLE 8.3. *Varieties of Anosognosia and Related Phenomena*

Syndrome	*Clinical Abnormality*
Asomatognosia	Distorted awareness of a part of the body
Extinguishing	Nonperception of stimulus on affected side during double simultaneous stimulation to analogous areas on two sides of the body
Anosognosia	Unawareness of hemiparesis
Denial of illness	Denial of importance of weakness; undue optimism regarding prognosis
Anosodiaphoria	Indifference to hemiparesis
Misoplegia	Hatred of paralyzed limbs
Personification	Naming of the paralyzed limb, addressing it as one would a pet or a person
Somatoparaphrenia	Denial of ownership of paralyzed limbs and belief that they belong to someone else
Conscious hemiasomatognosia	Conscious experience of having lost perception of one-half of the body
Nosagnosic overestimation	Exaggeration of the strength of the unaffected limbs
Hyperschematia	Experience of undue heaviness or lifelessness of the affected half of the body
Autotopagnosia	Inability to localize stimuli accurately on the affected side of the body
Allochiria	Mislocation of sensory stimuli to the corresponding point on the opposite side of the body
Allesthesia	Mislocation of sensory stimuli to different points on the same extremity
Reduplication or phantom body parts	Experience of additional body parts on paralyzed side
Kinesthetic hallucinations	False experience of movements of paralyzed limbs
Macrosomatognosia	Parts of the body are perceived as unusually large
Microsomatognosia	Parts of the body are perceived as unusually small

addition, most patients with the more marked forms of the syndrome are apathetic and have mild to moderate cognitive impairment.[62,63]

Anosognosia is most common in patients with acute stroke involving the supramarginal gyrus of the inferior parietal lobule of the right hemisphere or the underlying white matter of the thalamoparietal peduncle.[62,64] Cerebral atrophy is commonly present in anosognosic patients. Up to 80% of patients with acute left hemiparesis will have anosognosia or anosognosic phenomena, while approximately 20% of patients with left brain lesions and right-sided weakness will exhibit such behaviors.[57] Patients with Wernicke's aphasia or jargon aphasia (Chapter 6) often have a language-related form of anosognosia, denying the presence of language disturbances in the acute period immediately following the onset of the disorder.[65,66]

Anton's Syndrome

Anton's syndrome features blindness and denial of blindness.[67] The patient is blind but denies sightlessness and readily confabulates answers to questions concerning visual information. If pressed, the patient may admit to having vision that is "slightly blurred" or that "the light is dim" but gives no other hint concerning the presence or severity of the deficit. The syndrome occurs most commonly with cortical blindness produced by bilateral lesions of the occipital cortex or retrogeniculate visual radiations, but it also has been described in patients with blindness from ocular or optic tract disease and who have an associated dementia or confusional state.[68–70]

CONSTRUCTIONAL DISTURBANCES

Constructional disturbances have traditionally been called *apraxias* in the neurological and neuropsychological literature, but they do not fit the definition of apraxia adopted in Chapter 6 and are referred to simply as constructional disturbances in this discussion. The evaluation of constructional abilities is a rapid and discriminating technique for assessment of the integrity of visuospatial skills and should be included in all neuropsychiatric assessments. Constructional abilities are disturbed by a variety of brain disorders and by lesions in several different brain regions. Most idiopathic psychiatric disorders spare constructional abilities and the occurrence of constructional abnormalities usually indicates the presence of a neurologic disease.[71]

Many tasks can be used to evaluate constructional abilities, including copying model figures such as a circle, cross and cube; copying complex scorable figures such as the Rey-Osterrieth Figure (Fig. 8.2); drawing figures without a model such as a flower, house, clock, or person (Fig. 8.3); or reproducing model figures using matchsticks. Deficits are more evident with tests requiring reproduction of three-dimensional perspective or complex figures than with less demanding tasks.[72] Constructional disturbances are one manifestation of the disruption of visuospatial skills produced by a brain disorder, and other types of visuospatial abnormalities—dressing disorders, environmental disorientation—are present in most patients with constructional deficits.

Successful completion of a constructional task depends on integrating a variety of contributing neuropsychological abilities including spatially distributed attention, accurate perception, development of a drawing or copying strategy, execution of a sequence of motor acts, and use of feedback to modify the process as the construction evolves. Many brain regions contribute to this process, and the presence of a constructional deficit has little localizing value. Isolated lesions of the temporal lobes have few effects on drawing, but dysfunction of the frontal lobes, parietal lobes, occipital lobes, and basal ganglia of either hemisphere can adversely affect constructional abilities. Constructional deficits are most marked with right posterior lesions.[73]

The features of the patient's drawing or copying may provide some insight into the location of brain lesions, although additional information from the clinical history, neurologic examination, and neuroimaging is usually necessary to confirm the localization suggested by the characteristics of the constructions. Lesions of the right and left hemispheres tend to have different effects on construction (Table 8.4). Right brain lesions are as-

TABLE 8.4. *Contrasting Features of Constructions Produced by Right or Left Brain Dysfunction**

Right Brain Disorder	*Left Brain Disorder*
Left-sided neglect	Right-sided neglect
Right-to-left sequence	Left-to-right sequence
Segmented approach	More organized approach
Tendency to add extraneous detail	Copying or drawing only what is requested
Greater effect on external configuration	Greater effect on internal details
Faulty orientation	Less effect on orientation
Copies are larger than model	Copies are smaller than model or are same size
Elaboration of details	Omissions and simplifications
Overscoring of lines	Single lines
Faster execution	Slower execution
More severe disruption	Less severe disruption
More limited recovery after insult	Better recovery

*None of these features alone distinguishes the laterality of the lesion; taken together, they provide tentative evidence of the side of the disorder.

sociated with left-sided neglect, a right-to-left drawing strategy, loss of perspective, abnormalities of the external configuration, and inappropriate relationships of the constituent parts. Left brain lesions produce less marked abnormalities, with a tendency toward right-sided neglect and omission or simplification of the internal detail of the design.[73,74] Copies made by patients with right brain damage tend to be larger than the model, while copies from patients with left brain lesions are more likely to be smaller than, or the same size as, the model.[75] Patients with right brain damage have more difficulty with judgment of line orientation and figure matching, which suggests that their constructional deficits are a product of an inability to accurately perceive the figures.[76] Frontal lobe lesions, particularly those on the right, disrupt planning and result in segmented drawings developed with a piecemeal approach. Observing exactly how the patient performs the task is critical to detecting this type of disturbance.[77] Frontal lobe lesions tend to disrupt drawing more than copying, while parietal lobe lesions have a greater effect on copying than drawing. Patients with frontal lobe disturbances are also likely to evidence perseveration and to overemphasize high-stimulus areas of a drawing at the expense of other details. They may convert emotionally neutral stimulus items of a construction into "happy faces."[78]

Constructional deficits are etiologically nonspecific. They may be observed in patients with stroke, tumors, multiple sclerosis, or trauma, as well as in patients with delirium and dementia (Chapters 10, 11, and 26).[79,80]

RELATED VISUOSPATIAL DISORDERS

Cortical Blindness

Cortical blindness refers to loss of vision secondary to injury to the occipital cortices or genicocalcarine radiations bilaterally. The most frequent cause of cortical blindness is vertebrobasilar artery disease and bilateral occlusion of the posterior cerebral arteries. It also has been reported following surgery, particularly cardiac surgery, following cerebral angiography, and in patients with trauma and carbon monoxide intoxication.[81] There may be an accompanying amnesia secondary to anoxic injury or infarction of the hippocampi. Recovery is most limited in older patients who have sustained a stroke and who have visible lesions on cerebral.[81]

Patients with cortical blindness may exhibit the phenomenon of *blindsight*, an apparently paradoxical syndrome that reflects the capacity for visual processing in the extrageniculocortical system. Twenty to 30 percent of fibers of the optic tract are directed to nongeniculate destinations, particularly the superior colliculi and pretectal region of the brainstem. The nongeniculate system allows cortically "blind" patients to orient toward visual stimuli and to detect object movement while experiencing no conscious visual perception.[82] These phenomena are not demonstrable if the blindness is the result of preginuculate lesions.[83] Blindsight must be distinguished from residual rudimentary vision mediated by small islands of preserved cortex.[84]

Blindness from cortical lesions or ocular pathology must occasionally be differentiated from hysterical blindness. Patients with cortical blindness have retained pupillary responses but lose optokinetic nystagmus (produced by moving a striped target in front of the patient's eyes); patients with anterior blindness typically lose both pupillary responses and optokinetic responses; and patients with blindness as a conversion reaction retain both pupillary responses and optokinetic nystagmus.[85]

Balint's Syndrome

Balint's syndrome is a complex disturbance including (*1*) "psychic paralysis of visual fixation" (or "sticky fixation") in which the patient cannot volitionally shift gaze from one object to another; (*2*) "optic ataxia," evidenced by an inability to execute visually guided manual movements; and (*3*) simultanagnosia, characterized by an inability to see any but the most prominent visual stimuli[29] (described below).

Pathologically, Balint's syndrome is produced by bilateral parietooccipital lesions involving the lateral aspects of the hemisphere.[86–88]

Charcot-Wilbrand Syndrome

The Charcot-Wilbrand syndrome, or "irreminiscence," is characterized by the inability to generate an internal mental image or "revisualize" an object.[36] When asked to imagine an elephant, flag, bicycle, or any other object, the person is unable to generate an internal mental image. One of the authors' patients said of the experience, "Doc, it's like having your picture tube go out." Patients with the syndrome typically have much more difficulty generating objects through drawing than copying model figures. Some patients have reported a loss of dream imagery.[36,89,90]

Most patients with Charcot-Wilbrand syndrome have bilateral parietal lobe lesions and in some cases the patients have both Charcot-Wilbrand and Balint's syndrome.[36,90,91]

Central Achromatopsia

Central achromatopsia refers to loss of color vision produced by occipital lobe lesions. It must be distinguished from inherited abnormalities of color vision, acquired disorders of the optic nerves that impair color vision, color agnosia, and color anomia. Central processing of color perception in the occipital lobes occurs inferior to the calcarine sulcus and anterior to the region mediating visual field information. Bilateral injury to this area produces complete color blindness, whereas unilateral damage produces contralateral hemiachromatopsia.[92]

Dressing Disturbances

Dressing disturbances may be produced by a variety of CNS lesions and occur in the context of several different clinical syndromes: (*1*) in disorders with profound unilateral neglect, only the non-neglected side of the body may be bathed, toileted, and dressed; (*2*) in acute confusional states, dementing disorders, and schizophrenia, patients may don multiple layers of clothing when such bundling is inappropriate for the weather; and (*3*) a disturbance of true body–garment orientation may occur. In the third case, the patient may be unable to correctly orient the arm to the sleeve, may try to wear a shirt on the legs, or may put pants on backwards. The syndrome appears to be uniquely associated with right parietal lesions.[93]

Dazzle

Central dazzle is a syndrome of painless photophobia.[94] It is associated with lateral geniculate lesions and may be analogous to a thalamic pain syndrome in the visual domain.

Planotopokinesia

Planotopokinesia refers to the inability to use a map. Patients cannot draw a map or use a map to orient themselves in the environment.[95] This syndrome is commonly accompanied by left neglect and constructional disturbances.

RIGHT HEMISPHERE DYSFUNCTION AND NEUROPSYCHIATRIC DISORDERS

The left hemisphere mediates propositional linguistic capacities, arithmetic skills, and praxis (Chapter 6). Verbal memory is also subserved primarily by left hemispheric mechanisms and structures (Chapter 7). Visuospatial skills and nonverbal memory are preferentially mediated by the right hemisphere. Interpersonal and some aspects of emotional functions also are mediated by the right brain.

Patients with right brain lesions are consistently impaired in a variety of emotionally relevant abilities. Van Lancker[96] proposed that personal relevance represents an organizational principle that subsumes many observations of patients with right brain dysfunction. Recognizing familiar faces and discriminating among facial expressions, for example, are extremely important to successful emotional–social engagement, and these are compromised in patients with right hemispheric dysfunction.[97] Recognition of familiar environments, likewise, is a personally relevant ability and may be impaired in patients with right occipital–temporal lesions in the syndrome of environmental agnosia (discussed above). Display of mood-congruent facial expression, important for telegraphing one's emotional state, is more abnormal in patients with right brain lesions than in those with left brain lesions.[98] Recognition of familiar voices and deduction of the speaker's emotional state are also primarily dependent on right hemisphere function and may be abnormal in patients with right parietal damage.[99,100] Execution of prosodic speech qualities that allow listeners to perceive the speaker's emotion is disturbed by right brain lesions.[101]

Table 8.5 lists the clinical disorders observed in patients with right brain lesions that are highly relevant to personal emotional–social function. When these dis-

TABLE 8.5. *Disorders of Personal Relevance and Social–Emotional Function That Are Associated with Damage to the Right Hemisphere*

Recognition of familiar faces (prosopagnosia)
Recognition of familiar environments (topographagnosia)
Discrimination of unfamiliar faces
Discrimination of emotional facial expressions
Execution of facial expression
Recognition of familiar voices (phonagnosia)
Recognition of vocal emotion (receptive aprosodia)
Execution of vocal emotion (executive aprosodia)
Recognition of familiar environment (environmental agnosia)
Recognition of familiar handwriting
Integration of personal memory (reduplicative paramnesia)
Acceptance of the identity of a familiar-looking individual (Capgras syndrome and related misidentification disorders)

orders occur in adults, they produce emotional isolation along with an inability to communicate one's emotions or to comprehend the emotions of others; when they occur in children, they may permanently disrupt the ability to develop interpersonal relationships and result in isolating, shy, aggressive, and schizoid behaviors.[85,102] Of the three major axes of emotion—perception of emotionally relevant stimuli, execution of emotionally relevant behaviors, and subjective emotion experience—the right hemisphere is vitally concerned with mediation of perception of emotions and execution of emotional behaviors. Emotional experience is mediated primarily by limbic system structures.

In addition to the disorders of personal relevance and abnormalities of emotional perception and expression, a variety of other neuropsychiatric disorders have been linked to right hemisphere dysfunction. Secondary mania occurs nearly exclusively with lesions of the right hemisphere (Chapter 14), visual hallucinations are common with posterior right brain dysfunction (Chapter 13), and psychosis may also occur with right hemisphere disorders. Misidentification disorders—Capgras syndrome, Frigoli syndrome, intermetamorphosis syndrome, and related conditions—are more common with right than with left hemisphere lesions (Chapter 12). Anxiety has been linked to lesions of the right medial temporal lobe (Chapter 17). Reduced sexual activity has been observed more commonly after right than left brain lesions (Chapter 23). Frontotemporal dementia with disproportionate right frontal involvement produces more marked disruption of social behavior than does frontal degenerative syndrome affecting primarily the left hemisphere.

REFERENCES

1. Bauer RM. Agnosia: Visuoperceptual, Visuospatial, and Visuoconstructive Disorders. New York: Oxford University Press, 1993.
2. Bay E. Disturbances of visual perception and their examination. Brain 1953;76:515–550.
3. Critchley M. Psychiatric symptoms and parietal disease: differential diagnosis. Proc R Soc Med 1964;57:422–428.
4. Benson DF, Greenberg JP. Visual form agnosia. Arch Neurol 1969;20:82–89.
5. Shelton PA, Bowers D, et al. Apperceptive agnosia: a case study. Brain Cogn 1994;25:1–23.
6. Warrington EK, James M. Visual apperceptive agnosia: a clinico-anatomical study of three cases. Cortex 1988;24:13–32.
7. Gomori AJ, Hawryluk GA. Visual agnosia without alexia. Neurology 1984;34:947–980.
8. Mendez MF. Visuoperceptual function in visual agnosia. Neurology 1988;38:1754–1759.
9. Albert ML, Soffer D, et al. The anatomic basis of visual agnosia. Neurology 1979;28:876–879.
10. Benson DF, Segarra J, Albert ML. Visual agnosia-prosopagnosia: a clinicopathologic correlation. Arch Neurol 1974;30: 307–310.
11. Kawata N, Nagata K. A case of associative visual agnosia: neuropsychological findings and theoretical considerations. J Clin Exp Neuropsychol 1989;11:645–664.
12. Damasio AR, Damasio H, Van Hoesen GW. Prosopagnosia: anatomic basis and behavioral mechanisms. Neurology 1982; 32:331–341.
13. Shuttleworth ECJ, Syring V, Allen N. Further observations on the nature of prosopagnosia. Brain Cogn 1982;1:307–322.
14. Tranel D, Damasio A, Damasio H. Intact recognition of facial expression, gender, and age in patients with impaired recognition of face identity. Neurology 1988;38:690–696.
15. Bronstein B, Sroka H, Munitz H. Prosopagnosia with animal face agnosia. Cortex 1969;5:164–169.
16. Malone DR, Morris HH, et al. Prosopagnosia: a double dissociation between the recognition of familiar and unfamiliar faces. J Neurol Neurosurg Psychiatry 1982;45:820–822.
17. De Renzi E, Spinnler H. Facial recognition in brain-damaged patients. Neurology 1966;16:145–152.
18. De Renzi E, Scotti G, Spinnler H. Perceptual and associative disorders of visual recognition. Neurology 1969;19:634–642.
19. Hamsher KDS, Levin HS, Benton AL. Facial recognition in patients with focal brain lesions. Arch Neurol 1979;36:837–839.
20. Bauer RM. Autonomic recognition of names and faces in prosopagnosia: a neuropsychological application of the Guilty Knowledge Test. Neuropsychologia 1984;22:457–469.
21. Bauer RM, Verfaellie M. Electrodermal discrimination of familiar but not unfamiliar faces in prosopagnosia. Brain Cogn 1988;8:240–252.
22. Nardelli E, Buonanno F, et al. Prosopragnosia: report of four cases. Eur Neurol 1982;21:289–297.
23. Campbell R, Landis T, Regard M. Face recognition and lipreading. Brain 1986;109:509–521.
24. Landis T, Cummings JL, et al. Are unilateral right posterior cerebral lesions sufficient to cause prosopagnosia? Clinical and radiological findings in six additional patients. Cortex 1986;22: 243–252.
25. Hecaen H, Tzortzis C, Rondot P. Loss of topographic memory learning deficits. Cortex 1980;16:525–542.
26. Landis T, Cummings JL, et al. Loss of topographic familiarity: an environmental agnosia. Arch Neurol 1986;43:132–136.
27. Kinsbourne M, Warrington EK. Observations on color agnosia. J Neurol Neurosurg Psychiatry 1964;27:296–299.
28. McCarthy RA, Warrington EK. Cognitive Neuropsychology: A Clinical Introduction. New York: Academic Press, 1990.
29. Kirshner HS. Behavioral Neurology: A Practical Approach. New York: Churchill Livingstone, 1986.
30. Hecaen H, Albert ML. Human Neuropsychology. New York: John Wiley & Sons, 1978.
31. Luria AR. Disorders of "simultaneous perception" in a case of bilateral occipito-parietal brain injury. Brain 1959;82:437–449.
32. Heilman KM, Watson RT, Valenstein E. Neglect and related disorders. In: M. HK, Valenstein E, eds. Clinical Neuropsychology, 3rd ed. New York: Oxford University Press, 1993: 279–336.
33. Stone SP, Wilson B, et al. The assessment of visuo-spatial neglect after acute stroke. J Neurol Neurosurg Psychiatry 1991; 54:345–350.
34. Levine DN, Warach JD, et al. Left spatial neglect: effects of lesion size and premorbid brain atrophy on severity and recovery following right cerebral infarction. Neurology 1986;36: 362–366.

35. Bogousslavsky J, Miklossy J, et al. Subcortical neglect: neuropsychological, SPECT, and neuropathological correlations with anterior choroidal artery territory infarction. Ann Neurol 1988;23:448–452.
36. Critchley M. The Parietal Lobes. New York: Hafner Press, 1953.
37. Ferro JM, Kertesz A, Black SE. Subcortical neglect: quantitation, anatomy, and recovery. Neurology 1987;37:1487–1492.
38. Ferro JM, Kertesz A, Black SE. Posterior internal capsule infarction associated with neglect. Arch Neurol 1984;41:422–424.
39. Healton EB, Navarro C, et al. Subcortical neglect. Neurology 1982;32:776–778.
40. Heilman KM, Valenstein E. Frontal lobe neglect in man. Neurology 1972;22:660–664.
41. Stein S, Volpe BT. Classical "parietal" neglect syndrome after right frontal lobe infarction. Neurology 1983;33:797–799.
42. Watson RT, Heilman KM. Thalamic neglect. Neurology 1979;29:690–694.
43. Watson RT, Valentein E, Heilman KM. Thalamic neglect: possible role of the medial thalamus and nucleus reticularis in behavior. Arch Neurol 1981;38:501–506.
44. Mesulam M-M. A cortical network for directed attention and unilateral neglect. Ann Neurol 1981;10:309–325.
45. Daffner KR, Ahern GL, et al. Dissociated neglect behavior following sequential strokes in the right hemisphere. Ann Neurol 1990;28:97–101.
46. Coslett HB, Heilman KM. Hemihypokinesia after right hemisphere stroke. Brain Cogn 1989;9:267–278.
47. Meador KJ, Watson RT, et al. Hypometric with hemispatial and limb motor neglect. Brain 1986;109:293–305.
48. Weintraub S, Mesulam M-M. Right cerebral dominance in spatial attention. Arch Neurol 1987;44:621–625.
49. Albert ML. A simple test of visual neglect. Neurology 1973;23: 658–664.
50. Binder J, Marshall R, et al. Distinct syndromes of hemineglect. Arch Neurol 1992;49:1187–1194.
51. Behrmann M, Moscovitch M, et al. Perceptual and conceptual mechanisms in neglect dyslexia. Brain 1990;113:1163–1183.
52. Roeltgen MG, Roeltgen DP, Heilman KM. Unilateral motor impersistence and hemispatial neglect from a right striatal lesion. Neuropsychiatry Neuropsychol Behav Neurol 1989;2: 125–135.
53. Columbo A, DeRenzi E, Gentilini M. The time course of the visual hemi-inattention. Acta Psychiatr Nervenkr 1982;231: 539–546.
54. Ishiai S, Sugishita M, et al. Improvement of unilateral spatial neglect with numbering. Neurology 1990;40:1395–1398.
55. Pizzamiglio L, Antonucci G, et al. Cognitive rehabilitation of the hemineglect disorder in chronic patients with unilateral right brain damage. J Clin Exp Neuropsychol 1992;14:901–923.
56. Ames D. Self shooting of a phantom head. Br J Psychiatry 1984;145:193–194.
57. Cutting J. Study of anosognosia. J Neurol Neurosurg Psychiatry 1978;41:548–555.
58. Frederiks JAM. Disorders of the body schema. In: Frederiks JAM, ed. Clinical Neuropsychology. New York: Elsevier Science Publishers, 1985:373–393.
59. Meador KJ, Allen ME, et al. Allochiria or allesthesia: is there a misperception? Arch Neurol 1991;48:546–549.
60. Nightingale S. Somatophrenia: a case report. Cortex 1982;18: 463–467.
61. Weinstein EA, Kahn RL, et al. Delusional reduplication of parts of the body. Brain 1954;77:45–60.
62. Levine DN, Calvanio R, Rinn WE. The pathogenesis of anosognosia for hemiplegia. Neurology 1991;41:1770–1781.
63. Starkstein SE, Federoff JP, et al. Neuropsychological deficits in patients with anosognosia. Neuropsychiatry Neuropsychol Behav Neurol 1993;6:43–48.
64. Feinberg TE, Haber LD, Leeds NE. Verbal asomatagnosia. Neurology 1990;40:1391–1394.
65. Breier JI, Adair JC, et al. Dissociation of anosognosia for hemiplegia and aphasia during left-hemisphere anesthesia. Neurology 1995;45:65–67.
66. Shuren JE, Hammond CS, et al. Attention and anosognosia: the case of a jargonaphasic patient with unawareness of language deficit. Neurology 1995;45:376–378.
67. Forstl H, Owen AM, David AS. Gabriel Anton and "Anton's symptom": on focal diseases of the brain which are not perceived by the patient (1898). Neuropsychiatry Neuropsychol Behav Neurol 1993;6:1–8.
68. Bergman PS. Cerebral blindness. Arch Neurol Psychiatry 1957; 78:568–584.
69. Redlich FC, Dorsey JF. Denial of blindness by patients with cerebral disease. Arch Nuerol Psychiat 1945;53:407–417.
70. Symonds C, MacKenzie I. Bilateral loss of vision from cerebral infarction. Brain 1957;80:415–455.
71. Nahor A, Benson DF. A screening test for organic brain disease in emergency psychiatric evaluation. Behav Psychiatry 1970;2:23–26.
72. Griffiths KM, Cook ML, Newcombe RLG. Cube copying after cerebral damage. Neuropsychology 1988;10:800–812.
73. Swindell CS, Holland AL, et al. Characteristics of recovery of drawing ability in left and right brain-damaged patients. Brain Cogn 1988;7:16–30.
74. Gainotti G, Tiacci C. Patterns of drawing disability in right and left hemispheric patients. Neuropsychologia 1970;8:379–384.
75. Larrabee GJ, Kane RL. Differential drawing size associated with unilateral brain damage. Neuropsychologia 1983;21:173–177.
76. Carlesimo GA, Fadda L, Caltagirone C. Basic mechanisms of constructional apraxia in unilateral brain-damaged patients: role of visuo-perceptual and executive disorders. J Clin Exp Neuropsychol 1993;15:342–358.
77. Albert MS, Kaplan E. Organic implications of neuropsychological deficits in the elderly. In: Poon LW, Foard JL, Cermak LS, Arenberg D, Thompson LW, eds. New Directions in Memory and Aging. Hillsdale, NJ: Lawrence Erlbaum Associates, 1980:403–432.
78. Regard M, Landis T. The "smiley": a graphical expression of mood in right anterior cerebral lesions. Neuropsychiatry Neuropsychol Behav Neurol 1994;7:303–307.
79. Henderson VW, Mack W, Williams BW. Spatial disorientation in Alzheimer's disease. Arch Neurol 1989;46:391–394.
80. Watson YI, Arfken CL, Birge SJ. Clock completion: an objective screening test for dementia. J Am Geriatr Soc 1993;41: 1235–1240.
81. Aldrich MS, Alessi AG, et al. Cortical blindness: etiology, diagnosis, and prognosis. Ann Neurol 1987;21:149–158.
82. Cowey A, Stoerig P. The neurobiology of blindsight. Trends Neurosci 1991;14:140–145.
83. Perenin MT, Jeannerod M. Residual vision in cortically blind hemiphields. Neuropsychologia 1975;13:1–7.
84. Celesia GG, Bushnell D, et al. Cortical blindness and residual vision: is the "second" visual system in humans capable of more than rudimentary visual perception? Neurology 1991;41:862–869.
85. Weintraub S, Mesulam M-M. Developmental learning disabilities of the right hemisphere. Arch Neurol 1983;40:463–468.

86. Pierrot-Deseilligny CH, Gray F, Brunit P. Infarcts of both inferior parietal lobules with impairment of visually guided eye movements, peripheral visual inattention and optic ataxia. Brain 1986;109:81–97.
87. Verfaellie M, Rapcsak SZ, Heilman KM. Impaired shifting of attention in Balint's syndrome. Brain Cogn 1990;12:195–204.
88. Damasio AR, Benton AL. Impairment of hand movements under visual guidance. Neurology 1979;29:170–178.
89. Botez MI, Olivier M, et al. Defective revisualization: dissociation between cognitive and imagistic thought: case report and short review of the literature. Cortex 1985;21:375–389.
90. Farah MJ, Levine DN, Calvanio R. A case study of mental imagery deficit. Brain Cogn 1988;8:147–164.
91. Trojano L, Grossi D. A critical review of mental imagery defects. Brain Cogn 1994;24:213–243.
92. Rizzo M, Smith V, et al. Color perception profiles in central achromatopsia. Neurology 1993;43:995–1001.
93. Hemphill RE, Klein R. Contribution to the dressing disability as a focal sign and to the imperception phenomenon. J Ment Sci 1948;94:611–622.
94. Cummings JL, Gittinger WJ. Central dazzle: a thalamic syndrome? Arch Neurol 1981;38:372–374.
95. Hecaen H, Penfield W, et al. The syndrome of apractogosia due to lesions of the minor cerebral hemisphere. Arch Neurol Psychiatry 1956;75:400–434.
96. Van Lancker D. Personal relevance and the human right hemisphere. Brain Cogn 1991;17:64–92.
97. DeKosky ST, Heilman KM, et al. Recognition and discrimination of emotional faces and pictures. Brain Lang 1980;9:206–214.
98. Borod JC, Koff E, et al. The expression and perception of facial emotion in brain-damaged patients. Neuropsychologia 1986;24:169–180.
99. Heilman KM, Bowers D, et al. Comprehension of affective and nonaffective prosody. Neurology 1984:917–921.
100. Van Lancker DR, Canter GJ. Impairment of voice and face recognition in patients with hemispheric damage. Brain Cogn 1982;1:185–195.
101. Ross ED, Mesulam MM. Dominant language function of the right hemisphere? Prosody and emotional gesturing. Arch Neurol 1979;36:144–148.
102. Grace J, Malloy P. Neuropsychiatric aspects of right hemisphere learning disability. Neuropsychiatry Neuropsychol Behav Neurol 1992;5:194–204.

Frontal Lobe Dysfunction

The frontal lobe is the largest lobe of the human brain, comprising approximately one-third of the total cortical volume. It is among the most recent phylogenetic acquisitions and is one of the last regions to mature and myelinate in ontogenetic development.[1] The frontal lobes mediate behaviors that are distinctively human. They are the focal point for the integration of information from the environment, the internal milieu of the body, and the emotional state of the individual. The frontal lobes generate behavior and mediate action on the environment. Frontal lobe dysfunction produces some of the most extravagant syndromes encountered in neuropsychiatry. Disorders of cognition, mood, motivation, and behavioral control emerge in patients with frontal lobe disorders.

Organized volitional activity requires an environmental assessment, integration of perception with past historical information and one's emotional state, generation of action or response contingencies, choice of an action plan that both accomplishes one's goal and anticipates environmental consequences, implementation of the plan through appropriate motor activity, monitoring the action, and adjustment based on feedback. Accomplishment of this complex series of activities is dependent primarily on frontal lobe function. Motivation to action, anticipation of the social and interpersonal consequences of one's actions, and ongoing surveillance and modification of one's behavior are accomplished with seamless action plans when the frontal lobe is intact. With frontal lobe dysfunction, a variety of clinical syndromes representing breakdown in the components of the process emerge.

The frontal lobes are not homogenous entities but are divided into functionally specialized subregions, and injury to different areas produces clinically distinct psychosyndromes and behavioral alterations. This chapter presents the recognized syndromes resulting from restricted frontal damage and describes additional behavioral symptoms to which frontal dysfunction contributes. The anatomic correlates of the principal syndromes and the common etiologies of frontal damage are discussed. The anatomic relationships between the frontal lobes and subcortical structures as well as the similarity of behavioral changes associated with frontal lobe, basal ganglionic, and thalamic dysfunction are described. Finally, the treatment of behavioral disorders associated with frontal lobe dysfunction is presented.

FRONTAL LOBE SYNDROMES AND SYMPTOMS

Sensory and Motor Abnormalities

The frontal lobe mediates a wide range of behaviors, and frontal lobe disorders produce abnormalities of cranial nerve function, motor ability, language, and memory, as well as overt behavior and cognition (Table 9.1). The olfactory nerves lie on the inferior surface of the frontal lobes between the bony roof of the orbit and the orbitofrontal cortex. Injury to these nerves with resultant anosmia may occur with head trauma and from compression by tumors arising from the floor of the anterior cranial fossa.

The motor cortex of the frontal lobe is the origin of the pyramidal tract and injury to this central brain region produces contralateral weakness, spastic tone, a flexion posture in the upper limb and extension of lower limb, and an extensor response to plantar stimulation (Babinski sign) (Chapter 2).[2]

The frontal eye field is anterior to the motor strip and mediates volitional (saccadic) eye movements. Acute damage to the eye field results in a gaze preference toward the side of the lesion, and chronic lesions often result in gaze impersistence when the patient is asked to maintain gaze away from the side of the lesions. Patients with frontal lesions have particular difficulty with antisaccadic tasks requiring them to look away from a visual stimulus.

Primitive Reflexes

Primitive reflexes are characteristic of frontal lobe disorders (Chapter 2). These represent the re-emergence of motor acts that were appropriate at developmentally earlier stages, were suppressed by the maturation of the frontal lobes, and reappear with frontal lobe dysfunction. The grasp and suck reflexes are the two most prominent primitive reflexes encountered in patients with frontal lobe disorders (Chapter 2).[3,4]

Praxis and Language Disturbances

Apraxia occurs with lesions of the frontal lobes (Chapter 6). Sympathetic apraxia is associated with lesions of the left frontal lobe. Patients who have Broca's aphasia exhibit right hemiparesis and apraxia of the left limbs. Callosal apraxia is evident in patients with medial frontal lesions affecting the anterior fibers of the corpus callosum.[5] Patients with left medial frontal lesions may have a transcortical motor type of aphasia as well as apraxia of the left limbs.

Several types of language disorders are associated with frontal lobe dysfunction (Chapter 6). Broca's aphasia occurs with lesions of the inferior premotor and prefrontal regions of the left frontal lobe, and executive aprosodia is evident in patients with lesions of the corresponding region of the right frontal lobe. Transcortical motor aphasia is produced by lesions superior to Broca's area, on the medial surface of the left frontal lobe, or in the white matter connecting the supplementary motor area and Broca's area (Fig. 9.1).[6,7] Lesions causing global aphasia include frontal as well as temporal and parietal regions of the left hemisphere. Aphemia is a unique language disorder beginning with mutism and evolving into the "foreign accent syndrome"; it is observed in patients with small lesions confined to or undercutting Broca's area.

- **Memory Disturbances** Memory deficits may also occur in patients with frontal lobe disorders (Chapter 7). Retrieval deficit syndromes are common in patients with dorsolateral prefrontal dysfunction and true amnesias may occur in patients with inferomedial lesions affecting the nucleus basalis of Meynert or the fibers of the fornix.[8]

PREFRONTAL SYNDROMES

Three principal behavioral syndromes are associated with frontal lobe dysfunction (Table 9.2): an orbitofrontal syndrome, characterized by disinhibition and impulsiveness; a dorsolateral prefrontal syndrome, manifested primarily by executive dysfunction; and a medial frontal syndrome featuring apathy and akinesia (Fig. 9.1). Each of these syndromes is associated with abnormalities of a distinct region within the frontal lobe. Similar behaviors may be seen with disorders of the caudate nucleus, globus pallidus, and thalamus, to which the frontal lobes are connected in frontal–subcortical circuits (described below).[9] Thus, these behaviors are markers of frontal–subcortical circuit dysfunction and are not uniquely "frontal lobe" syndromes.

Frontal lobe syndromes are rarely seen in isolation. Few diseases affect only one region of the frontal lobe, and mixed syndromes with elements of dorsolateral, orbitofrontal, and medial frontal dysfunction are common. Moreover, because the two frontal lobes are juxtaposed, bilateral frontal lobe involvement is common.

Orbitofrontal Syndrome

Behaviorally, the outstanding feature of the orbitofrontal syndrome is disinhibition. These patients

TABLE 9.1. *Principal Behavioral Symptoms and Syndromes Associated with Frontal Lobe Dysfunction*

Behavioral Alteration	*Region of Frontal Dysfunction*
Executive deficits	Dorsolateral prefrontal
Memory disorders	
Retrieval deficit syndrome	Dorsolateral prefrontal
Amnesia	Fornix or medial forebrain
Confabulation	Dorsolateral prefrontal
Reduplicative paramnesia	Right frontal
Language and speech disorders	
Broca's aphasia	Left inferior premotor and prefrontal cortex (large lesions)
Transcortical motor aphasia	Left medial frontal or region superior to Broca's area
Aphemia	Broca's area (small lesion)
Dysprosody	Right premotor and prefrontal area
Ideomotor apraxia	
Sympathetic apraxia	Left premotor area
Callosal apraxia	Corpus callosum
Neuropsychiatric disorders	
Disinhibition	Orbitofrontal
Apathy	Medial frontal
Mania	Right inferomedial region
Depression	Dorsolateral prefrontal, orbitofrontal caudate
Obsessive-compulsive disorder	Orbitofrontal (increased metabolism)
Schizophrenia	Dorsolateral prefrontal
Catatonia	Multiple frontal areas
Miscellaneous	
Perseveration	Multiple frontal areas
Motor neglect	Dorsolateral and medial prefrontal
Elementary neurological deficits	
Hemiparesis	Motor cortex
Gaze abnormalities	Frontal eye fields
Primitive reflexes	Multiple frontal areas

lack social judgment, make tactless and socially inappropriate comments, may commit antisocial acts, and exhibit a general coarsening of interpersonal style. They may manifest an inane euphoria and inappropriate jocularity (*Witzelsucht*). Sexual preoccupations, inappropriate sexual jesting, and improper sexual comments are frequent, but overt sexual aggression is rare.[10–12] Orbitofrontal dysfunction has also been associated with imitation and utilization behavior.[13–15] *Utilization behavior* is manifested by the drive to pick up available objects (pencils, stethoscope, tongue blades) and use them or pantomime their use; *imitation behavior* refers

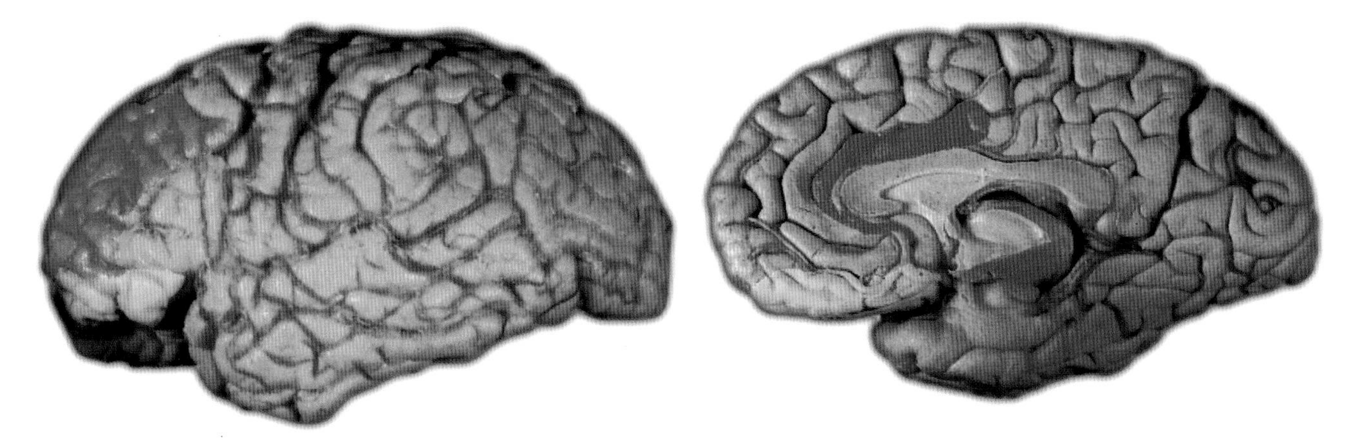

FIGURE 9.1 Anatomy of three prefrontal regions corresponding to orbitofrontal (green), dorsolateral (blue), and medial frontal (pink) syndromes.

to the drive to imitate the behavior of the clinician or others in the environment. The patients' insight into their own behavior is limited. Empathy, the ability to appreciate another's feelings, is also compromised in patients with orbitofrontal lesions, often dramatically so.[16]

A variety of mood changes have been associated with orbitofrontal dysfunction. Emotional lability with rapid shifts among affective states—happiness, anger, frustration, sadness—is common. Mania or hypomania may occur in a minority of patients with lesions in this area.[17] Depression has also been associated with

TABLE 9.2. *Behavioral Characteristics of Prefrontal Syndromes*

Prefrontal Region	*Clinical Disorder*
Orbitofrontal cortex	Disinhibition and diminished self-supervision of behavior
	Impulsiveness
	Tactlessness and loss of interpersonal sensitivity
	Poor social judgment
	Limited insight
	Irritability
	Mood lability, hypomania, depression
	Poor hygiene, neglect of personal core
Dorsolateral convexity	Retrieval deficit syndrome
	Difficulty altering set in response to changing contingencies
	Impaired strategy generation for solving complex problems
	Poor abstraction
	Reduced mental control
	Depression
Medial frontal cortex	Reduced interest
	Poor motivation
	Impaired initiation
	Reduced activity
	Impaired task maintenance
	Decreased emotional concern

dysfunction of orbitofrontal cortex: depressed patients with Parkinson's disease and Huntington's disease have decreased metabolism of the orbitofrontal area, compared to nondepressed patients with these diseases.[18,19]

Neuropsychological deficits are minimal in patients with orbitofrontal lesions. Impulsiveness, distractibility, and lack of concern for correct performance may interfere with intellectual assessment; however, when these can be contained, basic language, memory, and cognitive skills are usually found to be intact.[11,20] When the lesion affects the descending columns of the fornix or the medial forebrain nuclei, an amnesic-type memory disorder is present.[21,22]

Elementary neurological deficits are not prominent in patients with lesions limited to the orbitofrontal cortex. Primary motor, somatosensory, and visual functions are normal if the lesion is limited to the orbitofrontal region. Anosmia may be present in patients with orbitofrontal trauma or compressive neoplasms.

Frontal Convexity Syndrome

Dysfunction of the dorsolateral prefrontal convexity is associated with a variety of neuropsychological deficits. Table 9.3 lists the major categories of cognitive functions mediated by the dorsolateral convexity, the types of deficits seen with dysfunction of this area, and typical tests used to assess these functions. The prefrontal cortex mediates working memory, the memorandum that is active for a short period of time and allows information to be held "on-line." It allows the guidance of behavior on the basis of representations of the world, independent of immediate environmental stimulation.[23] Actions can be based on ideas, thoughts, environmental circumstances, and multisource information. Thus, the prefrontal cortex mediates generation of response contingencies that free the individual from the immediate demands of the environment.

Patients with dorsolateral lesions exhibit a retrieval deficit syndrome affecting spontaneous recall of previously learned information.[24,25] This applies to tests of recent memory where there is decreased recall with preserved recognition memory. Assessment of recall of remote information reveals diminished spontaneous recall and preserved performance on multiple choice tests. Classical delayed-response tasks used to assess frontal dysfunction in nonhuman primates and in humans with frontal disorders also depend on application of memory to produce an accurate response after a delay;[26] these are performed poorly by patients with prefrontal syndromes.

Word list generation is compromised; patients have decreased performance on tests such as the number of animals that can be named in 1 minute (normal, 18/minute; abnormal, <12/minute) or the number of words beginning with *F*, *A*, or *S* in 1 minute (normal, 15/minute for each letter; abnormal, <10/minute).[27] Verbal fluency tasks are more sensitive to left frontal lobe dysfunction, whereas design fluency (described below) tests are more affected in right frontal lobe disorders.

Difficulty altering set in response to changing contingencies is typical of patients with prefrontal convexity lesions.[16,24,28] On the Wisconsin Card Sort Test (WCST), for example, they may exhibit loss of set or they may be unable to change set when necessary, exhibiting perseveration.[25,29] Delayed-alternation tasks, requiring patients to respond to the last unrewarded stimulus and ignoring the last rewarded stimulus, are another means of testing the patient's ability to change sets in response to changing circumstances.[26]

Impaired planning with development of poor strategies for solving complex problems is typical of patients with dorsolateral prefrontal dysfunction and is evident in memory tests, construction tasks, and cognitive assessment.[1,24] The California Verbal Learning Test (CVLT), for example, assesses learning strategies by determining if the patient exhibits the clustering of semantically related words observed in normal individuals. The CVLT requires patients to learn a list of 16 words consisting of four groups of four related words (e.g., *tools, fruit*). Patients with intact frontal function cluster related words together to facilitate learning, whereas patients with frontal dysfunction fail to generate this learning strategy. Likewise, when confronted with a complex construction such as the Rey-Osterrieth Figure or Taylor Figure, patients with frontal lobe syndromes do not generate a coherent integrated structure but exhibit a piecemeal segmented approach (Fig. 9.2).[30] They may also exhibit stimulus boundedness, emphasizing elements that attract their attention, and they may assign affective value to emotionally neutral stimuli such as putting a "smiley" face on the figures.[31]

Another visuospatial task requiring generative capacity, which is frequently compromised in frontal lobe disorders, is the test of design fluency.[27,32] Several versions of this test exist. The form shown in Figure 9.3 requires the patients to draw as many different figures as possible in 1 minute. Each figure is to consist of four lines, and every line should touch at least one other line. The lines are to be approximately the same length, and no figure should be repeated. Patients with frontal lobe syndrome have difficulty generating novel figures

TABLE 9.3. *Assessment of Neuropsychological Deficits Associated with Dorsolateral Prefrontal Dysfunction*

Neuropsychological Function	*Clinical Test*
Retrieval of learned information	
Recall and recognition	Word list memory
Remote memory	Historical facts
Word list generation	Animals named/minute; FAS test
Use of cross-temporal stimulus information	Delayed-response task
Effort-demanding memory	CVLT
Altering set in response to changing contingencies	
Set maintenance	WCST (categories)
Ability to change set	WCST (perseverative errors)
	Alternating programs*
	Reciprocal programs
	Go–no go test
Use of cross-temporal stimulus information	Delayed-alternation task
Strategy generation for solving complex problems	
Complex constructions	Rey-Osterrieth Complex Figure (copy)
Design fluency	Novel figure generation
Advanced planning	Tower tests
Programming sequential motor acts	Alternating programs
	Reciprocal programs
	Go–no go test
	Serial hand sequences
Abstraction	Proverbs
	Similarities and differences
Mental control	
Dual track tasks	Consonant trigrams
Reversal of automatic tasks	Reciting months of the year in reverse order
Suppression of habitual in favor of novel responses	Stroop Color-Word Test
	Trails B

CVLT, California Verbal Learning Test; WCST, Wisconsin Card Sort Test.
*Some tests appear on this list more than once because more than one type of deficit associated with frontal lobe dysfunction contributes to failures on the test.

(8 to 10/minute is normal) and tend to reproduce the same figure or closely related variants several times.

Planning and strategy generation are assessed with tower tests (e.g., the Tower of London, Tower of Toronto, Tower of Hanoi) that require the patient to rearrange blocks with the fewest possible moves to achieve a specified end configuration.[29,30] Patients with frontal disorders have impoverished strategies and use more than the minimum number of moves to solve the problem.

Patients with frontal dysfunction tend to make more bizarre responses on tests of cognitive estimation (e.g.,

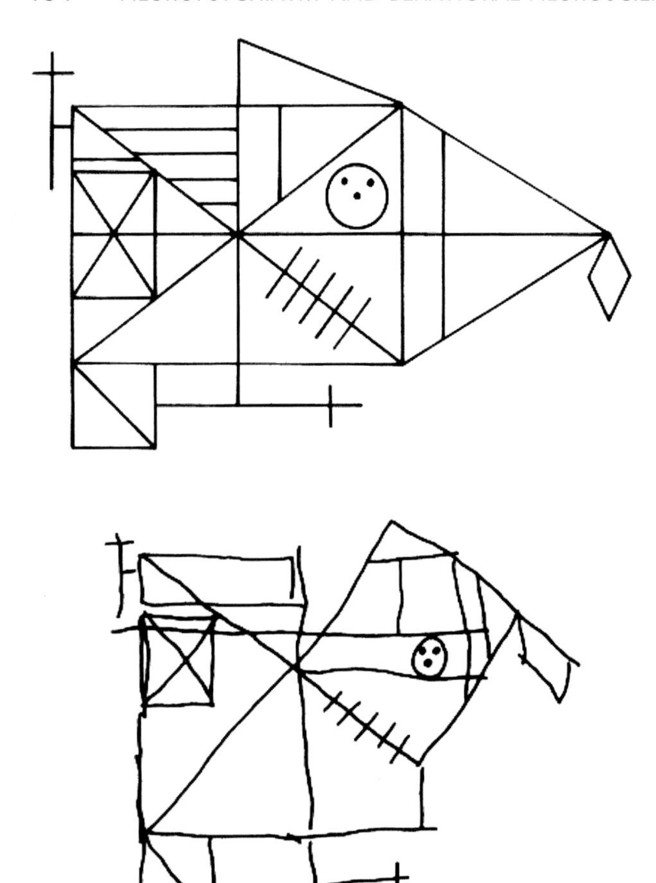

FIGURE 9.2 Rey-Osterrieth Figure drawn by a patient with a frontal lobe syndrome (model *above*, copy *below*). The figure was drawn in a segmented fashion and exaggerates areas with high stimulus value.

"How long is an average man's spine?"), a deficit interpreted in terms of poor use of available information and absence of response monitoring.[33]

Sequential motor acts also depend on intact dorsolateral prefrontal function. Alternating programs, multiple loops, reciprocal programs, go–no go tests, and serial hand sequences provide insight into the ability of the frontal lobe to guide sequential behaviors.[20,34] Alternating programs require the patients to alternate between two shapes such as squares and points or *m*'s and *n*'s (Fig. 9.4). Patients with dorsolateral prefrontal dysfunction frequently perseverate, failing to alternate regularly between figures. A test copying multiple loops (Fig. 9.5) can reveal a variety of types of frontal lobe–related abnormalities. The figures may be misinterpreted concretely and copied as 3's, or the patient may perseverate, adding additional loops beyond those of the model. It is important when administering both the alternating programs and multiple loops that the patient be encouraged to make more figures than provided in the clinician's model. Patients with frontal lobe dysfunction are frequently stimulus-bound, can slavishly copy models, and may reveal their perseverative tendencies only when they are required to personally generate the features of the figure beyond the model provided by the examiner.

Reciprocal program tasks require the patient to perform different movements with each hand or to respond reciprocally to a movement made by the clinician. For example, the patient may be asked to make a fist with one hand and to extend the other and to rapidly alternate between the two positions. When executing reciprocal programs, if the examiner taps once, the patient must tap twice, if the examiner taps twice, the patient taps once. Patients with frontal lobe lesions are likely to lose set and echo the taps of the clinician.

The go–no go test requires the patient to withhold a response. The patient is asked to tap twice if the examiner taps once and to do nothing if the examiner taps twice. Response inhibition is difficult for many patients with frontal lobe disorders and they may echo the examiner's movements.

Serial hand sequence tasks require the patient to move through a series of hand positions from fist, to slap, and to cut (Fig. 3.1). The examiner begins by demonstrating the sequence and asking patients to produce it on their own. If the patients fail, they are asked to say aloud "fist, slap, side" to determine if they can use verbalization to guide their motor behavior. Patients with frontal lobe disorders frequently cannot take advantage of verbal input and have a verbal–manual dissociation, saying one sequence while executing another. Verbal–manual dissociation is an indication of executive dysfunction consistent with a frontal lobe disorder. Other types of failures observed on the serial hand sequences (e.g., difficulty learning the sequences) have less localizing value.

Abstraction abilities also depend on frontal lobe integrity.[30] Interpretation of proverbs requires the patient to make generalizations based on an example. For instance, they must derive a generalizable meaning from a saying such as "People who live in glass houses should not throw stones." Abstract interpretation of similarities requires that the patient derive a common theme from two examples—for instance, "How are a tulip and a rose alike?" or "How are a watch and ruler alike?" Interpretation of differences requires the patient to infer the essential difference between two items that are superficially similar—for example, "What is the difference between a child and a midget?" or "What is the difference between a lie and a mistake?" Patients

FIGURE 9.3 In a test of design fluency, the patient is asked to make as many figures as possible consisting of four lines, each of which touch at least one other line. The figures should not be repeated. The test is terminated after 1 minute. The first two figures are examples provided by the examiner. The patient produced few figures and they are all similar.

with frontal lobe dysfunction are stimulus-bound, concrete, and environmentally dependent. These qualities interfere with their ability to abstract meaning beyond the surface content of these tests. Thus, they find little that is similar between a watch and a ruler (they don't look alike) and little that is different between a lie and a mistake (they seem similar). They cannot derive a second meaning or moral from the superficial features of a proverb (e.g., "They will break their windows" for the proverb given above). Proverb interpretation is conditioned by education and cultural background and cannot be used to test poorly educated persons or individuals unfamiliar with the proverb.

Mental control tests are tasks assessing complex attentional mechanisms mediated in part by the frontal lobes. Patients with frontal lobe dysfunction may do well on elementary attentional tests such as digit span. They may perform normally on continuous performance tasks (for example, requiring patients to raise their hand each time the examiner reads an *A* among

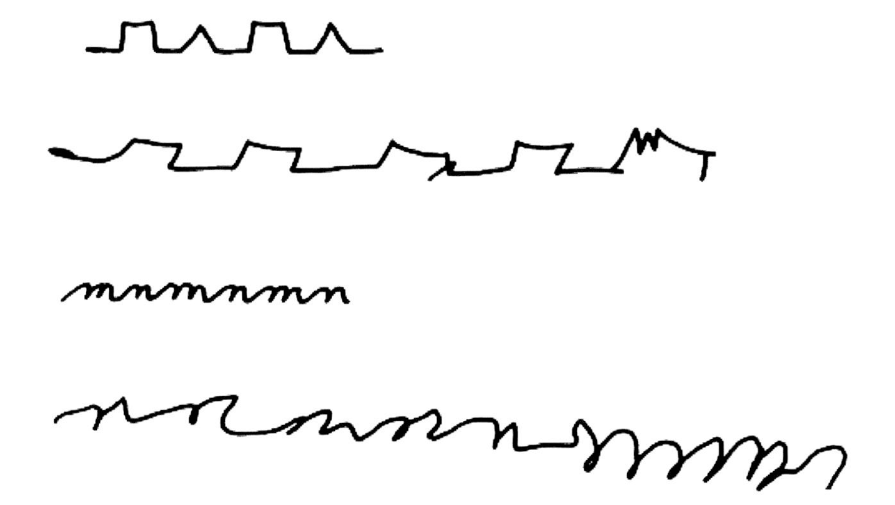

FIGURE 9.4 In the alternating program test; the patient is provided with a sample (*above*) and asked to copy the sample and extend the pattern (*below*). Perseveration is evident.

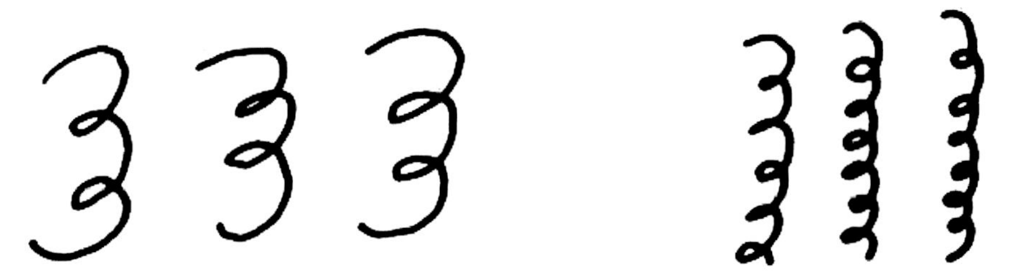

FIGURE 9.5 In the multiple loops test, the patient is provided with a sample (three figures on the left) and asked to make multiple copies. Perseveration is demonstrated here.

a long list of letters) or may be distractible and exhibit errors of omission.[29] More challenging for patients with frontal lobe disorders are mental control tasks, such as repeating the months of the year in reverse order. If patients fail this task, less demanding tests may be tried—e.g., serial subtraction or reverse spelling—to explore the severity of the deficit. Consonant trigrams are another type of mental control task that is difficult for patients with frontal lobe disorders to perform.[25] The patient is given three letters to remember and then immediately asked to count backward by 3's for 15 or 30 seconds. This requires the patient to perform two tasks at the same time (rehearse the letters and count backwards), and this may exceed the attentional resources of the patient.

Suppression of routine responses in favor of novel responses is another type of mental control task that is difficult for patients with dorsolateral prefrontal dysfunction. On the Trails B test, the patient is required to alternate sequentially between numbers and letters (1–A–2–B–3–C . . .), and the patients tend to fall out of set, executing the conventional sequences (1, 2, 3 or A, B, C). Similarly, in the Stroop Test (Part C), the patient must suppress the usual response (read the color name) in favor of a novel response (state the color in which the word is written). Patients cannot suppress the habitual response in favor of the novel response and fail the tests.[25]

The emotional concomitants of dorsolateral prefrontal dysfunction have not been fully explored. Depression is the most commonly described neuropsychiatric disorder observed in patients with lesions in the region, and anxiety frequently accompanies the depressive syndrome occurring with dorsolateral prefrontal injury.[35]

A few generalizations about dorsolateral prefrontal dysfunction are warranted. First, inferences about prefrontal functions are justified only when other more basic instrumental abilities are intact. The function of the frontal lobes is to synthesize information, create a strategy for problem resolution, and implement the action plan. Deficits in the elementary component of these functions make it impossible to assess the integrity of the frontal lobe. For example, aphasia will adversely affect word list generation and make if difficult to draw conclusions about frontal lobe function on the basis of verbal fluency tests. Likewise, amnesia or aphasia will affect the patient's performance on the WCST, and right parietal lobe disorders will disrupt visuospatial skills and affect deductions concerning frontal dysfunction that are based on copying or drawing performances.[36] Similarly, attentional disorders will disrupt mental control abilities, and motor system disorders will adversely affect sequential motor tasks. Inferences about frontal lobe function can be drawn only when other relevant cognitive and motor abilities are known to be intact.

Second, patients rarely fail all tests of functions ascribed to the dorsolateral prefrontal cortex. Most patients fail some tasks and pass others. It is the overall performance that permits the conclusion that dorsolateral prefrontal dysfunction is present, rather than performance on any single test. Moreover, patients with prefrontal lesions exhibit more than usual day-to-day variability in cognitive performance.

Third, a few basic types of dysfunction appear to underlie most of the clinical phenomena observed in patients with prefrontal disorders. Loss of generative and strategic capacities with a consequent stimulus-boundedness and environmental dependence is the principal determinant of the retrieval deficit syndrome, reduced verbal fluency, impaired nonverbal fluency, abnormalities in set shifting, poor planning and problem solving, poor abstraction, and abnormalities of mental control. Another major deficit is the failure to adjust behavior on the basis of ongoing information collection. This contributes to loss of set and perseveration and the observed lack of flexibility in the behavior of patients with frontal lobe disorders. This suggests that the critical contributions of the dorsolateral frontal region to human behavior are the uncoupling of the perception of the immediate environmental stimulus from a considered action plan based on rule-guided behavior, reflection and judgment, synthesis of historical information, and abstraction of common themes from one setting to another.

Medial Frontal Syndrome

The medial frontal syndrome is the third major behavioral disorder observed in patients with frontal lobe dysfunction. Apathy is the marker behavior of the medial frontal syndrome and the critical lesions involve the anterior cingulate region. The central features of apathy are loss of motivation and reduced goal-directed activities. Apathy has affective, emotional, cognitive, and motoric dimensions.[37,38] Affective aspects of apathy include flat and unchanging expression. Emotional dimensions of apathy include absence of interest, excitement, or emotional intensity, and lack of emotional responsiveness to events (positive or negative). The cognitive aspect of apathy includes reduced generative thinking, diminished curiosity, decreased engagement with usual activities (social, occupational, recreational), lack of interest in learning and new experiences, and lack of concern (for one's health, family, or future). The motor expression of apathy includes lack of effort, reduced productivity, decreased ability to sustain activities, decreased initiation of new activities, and increased dependence on others to structure activities.

When the supplementary motor area of the medial frontal region is affected, a marked akinesia may accompany the syndrome. Akinesia is characterized by difficulty initiating motor acts, hesitation when starting a new movement, and a tendency not to gesture when speaking. The anterior cingulate gyrus and the supplementary motor area are both involved in many patients with neurologic disorders and the combination of apathy and akinesia is common. Acute unilateral lesions usually produce marked apathy in the initial stage of the disorder with gradual improvement to a state of reduced motivation as the patient recovers. Treatment with psychostimulants or dopaminergic agents (discussed below) may improve motivation and reduce apathy.

The most marked apathetic syndrome accompanies bilateral anterior cingulate lesions and is characterized by *akinetic mutism*. In this remarkable syndrome, the patients are typically awake and may follow environmental activities with their eyes. They are not paralyzed and will eat if fed, urinate or defecate if taken to toilet, and may occasionally make voluntary postural adjustments or utter brief statements. They show no response to pain and no concern about their dramatic impairment. The state often has a fatal outcome. Akinetic mutism must be distinguished from lock-in syndrome, where the patient is paralyzed and from persistent vegetative state in which widespread brain injury prohibits meaningful behavior. The syndrome may be mistaken for catatonic unresponsiveness. Akinetic mutism is produced by lesions that involve both anterior cingulate gyri or interrupt cingulate connections.[39–41]

Neuropsychological deficits in patients with the medial frontal syndrome are relatively modest. In testable patients, general intellectual functions are not affected. As noted, patients with left medial lesions may have transcortical motor aphasia, and those with damage to the adjacent corpus callosum will manifest a callosal apraxia with apraxic movements of the left limbs (Chapter 6). In addition, patients with anterior cingulate lesions produce excessive failures on the WCST and on go–no go tests.[16,42,43]

Elementary neurological deficits will be present if the lesion extends sufficiently far posteriorly to affect the medial aspect of the motor and sensory cortex. The leg and sphincters are represented medially and lesions of this region will produce leg weakness, sensory deficits in the leg, and incontinence.[44] Grasp and suck reflexes are frequently present in patients with medial frontal lesions.

Shared Behavior Disturbance of the Prefrontal Syndrome

While the three major syndromes associated with prefrontal dysfunction have distinct behavioral profiles, they share at least one common feature. A major theme of each syndrome is the loss of internally generated behavioral schemata and an ensuing enslavement to external influences. The impulsiveness of the orbitofrontal syndrome occurs in response to external contingencies without reflection on the consequences. The transcortical motor aphasia of the medial frontal syndrome is characterized by markedly reduced spontaneous (internally generated) speech but preservation of repetition and echolalia (repetition of external stimuli). Patients with dorsolateral prefrontal dysfunction exhibit poor recall of information (self-generated) but preserved recognition memory (externally structured) and poor word list generation (internally generated) but preserved confrontation naming (externally structured). Thus, all three of the prefrontal syndromes share a basic principle of loss of volitional control of behavior and generative function with ensuing dependency on the environment for response options.

BEHAVIORAL SYMPTOMS ASSOCIATED WITH FRONTAL LOBE DYSFUNCTION

In addition to the three primary frontal lobe syndromes, there are a variety of individual behavioral symptoms and syndromes that occur with frontal lobe dysfunction (Table 10.1). The major language, memory, executive, visuospatial, and elementary neurologic disturbances associated with frontal lobe disorders have been described above or in previous chapters (Chapters 2, 6, 7, and 8). This section will address behavioral disorders in which frontal lobe dysfunction has been identified, although other types of brain dysfunction are usually simultaneously present.

Confabulation and reduplicative paramnesia are memory-related abnormalities that occur with frontal lobe dysfunction. *Confabulation* may be of either the "embarrassment" or "momentary" type, in which the patient fabricates answers based on past experience, or it may be the "fantastic" type, in which the patient elaborates bizarre, macabre, and impossible stories.[45] Experimental studies demonstrate that confabulation correlates better with lack of self-monitoring than with memory impairment, and clinicopathological correlations suggest that confabulation occurs when memory disturbances are present in patients with concomitant frontal lobe dysfunction (Chapter 7),[46,47]

Reduplicative paramnesia is a disorder of spatial orientation in which patients insist that their current environments are relocated elsewhere, usually nearer to their homes (Chapter 7). In most cases the patients are recovering from post-traumatic encephalopathy or other acute cerebral insults and have combined frontal and right parietal lesions.[48,49]

Mood disorders are commonly associated with frontal lobe dysfunction. As noted above, patients with orbitofrontal lesions may exhibit euphoria, irritability, mood and affect lability, and inappropriate jocularity. *Mania and hypomania* have also been associated with frontal lobe dysfunction. Lesions producing mania usually affect the inferior medial frontal regions of the right hemisphere and are produced by stroke or trauma.[50–52]

Both idiopathic and symptomatic *depression* are associated with frontal lobe dysfunction. Fluorodeoxyglucose positron emission tomography (FDG-PET) in patients with idiopathic depression reveals diminished metabolism in the dorsolateral prefrontal regions. The reductions are more marked on the left than on the right.[53,54] Depression is common in the acute post-stroke period in patients with lesions affecting the left dorsolateral prefrontal cortex,[35] and patients with depression accompanying Parkinson's disease and Huntington's disease have reduced metabolism in the orbitofrontal area.[18,19]

Frontal lobe function has been studied in *obsessive-compulsive disorder* (OCD) using PET. Patients with OCD have increased metabolism in the orbitofrontal cortex[55] and increased blood flow in medial prefrontal regions.

Agitation is a frontal lobe syndrome that results from a loss of behavioral modulation interacting with other types of psychopathology. Psychosis, depression, and mania are all associated with agitation; the features of the syndrome reflect the comorbid syndrome. Agitation results from frontal lobe dysfunction and reduction of the threshold for behavioral disturbances relating to the associated psychopathological process.

Schizophrenia is consistently associated with both neuropsychological and neuroimaging evidence of prefrontal dysfunction. Despite normal intelligence, patients with schizophrenia perform poorly on tests of frontal lobe function, including the WCST, verbal fluency, design fluency, and tower tests.[56,57] Many schizophrenic patients have reduced blood flow or metabolism in the frontal lobes while in the resting state and they fail to activate the frontal lobe normally during tests that increase frontal function in nonschizophrenic individuals.[58]

Catatonia is a clinical syndrome characterized by immobility, mutism, and withdrawal. Associated symptoms include staring, rigidity, posturing, negativism, waxy flexibility, echophenomena, mannerisms, stereotypy, and verbigeration (Chapter 20).[59] Catatonia is etiologically nonspecific and frontal lobe disorders are among the conditions that may induce catatonic states. Frontal lobe disorders that have produced catatonia include anterior communicating artery aneurysms, frontal lobe trauma and neoplasms, arteriovenous malformations, and frontal infections.[60]

Perseveration refers to the continuation or recurrence of an experience or activity without appropriate stimulus.[61] Three types of perseveration are recognized: (*1*) recurrence of a previous response to a subsequent stimulus (recurrent perseveration), (*2*) inappropriate maintenance of a category or framework of activity (stuck-in-set perseveration), and (*3*) abnormal prolongation or continuation without cessation of a current behavior (continuous perseveration).[62] Recurrent perseveration is observed in tasks such as naming, word list generation, drawing, and design fluency in which a previous response recurs at a later time in the testing. Stuck-in-set perseveration occurs on tasks such as the WCST, in which the patient is required to change sets. Continuous perseveration is evident on graphic tests

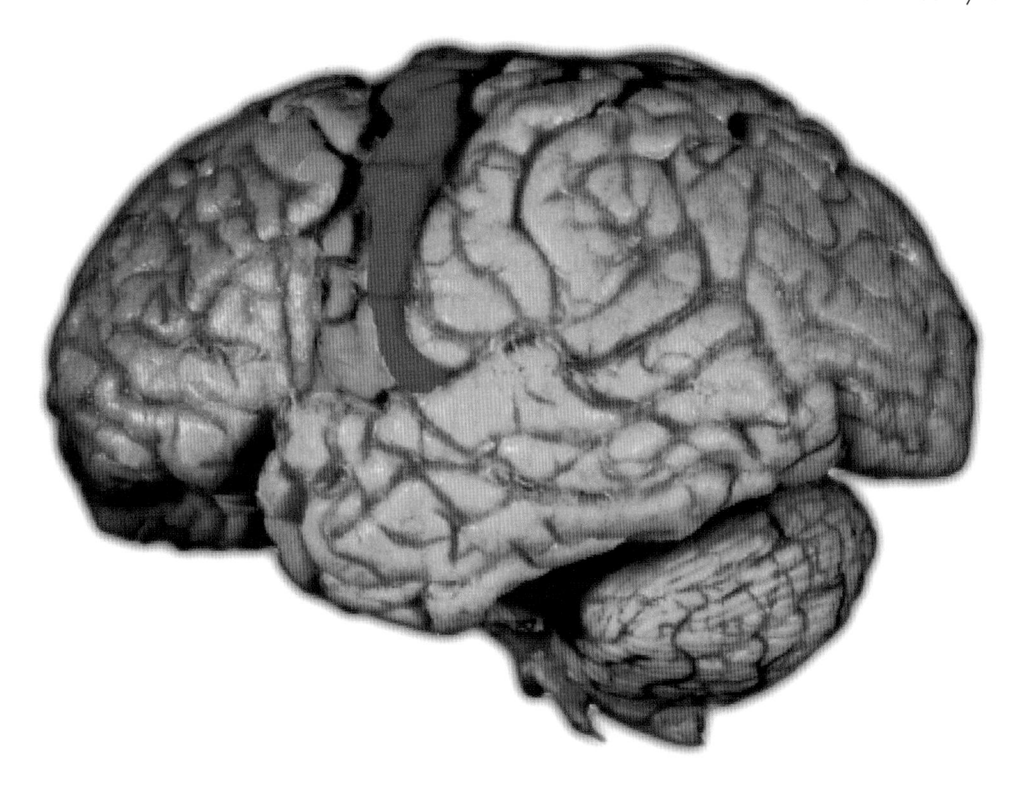

FIGURE 9.6 Anatomy of the frontal cortex Motor strip is depicted in red; premotor cortex, green; frontal eye fields, purple; prefrontal cortex, blue.

such as alternating programs (Fig. 9.5) and multiple loops (Fig. 9.6), the patient continues an ongoing behavior inappropriately. Perseveration is common in frontal lobe disorders but is not unique to frontal dysfunction and is seen in patients with lesions in other brain regions.[61,62]

ANATOMIC CORRELATES OF FRONTAL LOBE SYNDROMES

There are three principal subdivisions of the frontal lobes: (*1*) the precentral cortex, comprising the motor strip (Brodmann area 4); (*2*) premotor cortex (Brodmann area 6), including the frontal eye fields (Brodmann area 8) and supplementary motor area (extension of premotor area 6 on the medial aspect of the hemisphere); and (*3*) prefrontal cortex (Fig. 9.6).[63] Prefrontal cortex includes the dorsolateral prefrontal area, the orbitofrontal area, and the medial frontal/anterior cingulate regions, corresponding to the three prefrontal behavior syndromes described above (Fig. 9.2). Dorsolateral prefrontal cortex (Brodmann areas 9, 10, and 46) is composed of six-layered neocortex; orbitofrontal and anterior cingulate areas are paralimbic areas composed of transitional cortex (Chapter 2).[64] Each of these areas has different afferent and efferent relationships with other cortical and subcortical structures that reflect the different role each plays in human behavior. Likewise, the behavioral syndromes associated with each area are determined by their contrasting structures and functions.

The *dorsolateral prefrontal cortex* receives remarkably diverse input. It has afferent connections from posterior sensory association cortex via the long hemispheric white matter tracts. It also receives input from the limbic cortical regions via short association fibers from the adjacent cingulate and orbital cortices. From subcortical structures, it receives a robust projection from the dorsomedial nucleus of the thalamus and more limited input from the hypothalamus.[64,65] Thus, the dorsolateral prefrontal cortex is poised to integrate information regarding the external environment through sensory regions, the internal milieu from the hypothalamus, and the emotional state of the organism from limbic regions. Historical information is available through thalamic and hippocampal–limbic projections. The frontal lobes initiate activity, monitor the effect of the action (through its widespread afferent connections), and modify actions accordingly. They modulate their own input to allow focusing of attention on relevant information and exclude irrelevant information. Deficits corresponding to disruption of these activities include distractibility, impersistence, perseveration, and failure to adjust behavior on the basis of feedback. The frontal lobes are the only anatomical areas with information sufficient to form a global view of the person, the environment, and the individual's history. They are positioned to defer immediate environmentally determined responses and construct a response informed by the past history and future goals of the individual. Lesions of the frontal lobes return the individual to a state of environmental dependency and stimulus-boundedness.

The dorsolateral prefrontal cortex also has extensive efferent connections. It has reciprocal connections with most of the areas from which it receives input. In addition, it has a large projection to the head of the caudate nucleus, the first link of the dorsolateral prefrontal–subcortical circuit that includes the frontal convexity, caudate nucleus, globus pallidus and substantia nigra, and the dorsomedial nucleus of the thalamus (Fig. 9.7).[9,66]

The *orbitofrontal cortex* receives projections from the temporal lobe via the uncinate fasciculus, dorsomedial nucleus of the thalamus, hypothalamus, and amygdala.[63,64] There is also input from the sensory association areas and the frontal association cortex. In addition to reciprocal connections with these areas, the orbitofrontal cortex serves as the origin of the orbitofrontal–subcortical circuit projecting to ventral caudate nucleus, globus pallidus and substantia nigra, and dorsomedial thalamus. This circuit is parallel to, but discrete from, the dorsolateral prefrontal–subcortical circuit.[66] Lesions of this area interrupt the monitoring function of the orbitofrontal cortex, release limbically mediated behaviors, and produce impulsive, disinhibited behaviors. Social rules of conduct, consideration of consequences of one's behavior, and concern for the feelings of others are abandoned in favor of ungoverned behavior responsive to immediate environmental stimuli and bodily urges.

The *anterior cingulate cortex* receives major input from the hippocampus, dorsomedial nucleus of the thalamus, hypothalamus, amygdala, sensory association cortex, and dorsolateral prefrontal cortex.[63,64] The anterior cingulate–subcortical circuit projects to the nucleus accumbens, ventral globus pallidus and substantia nigra, and the dorsomedial nucleus of the thalamus.[66] Several member structures of this circuit—anterior cingulate, nucleus accumbens, ventral pallidum—receive dopaminergic projections from the

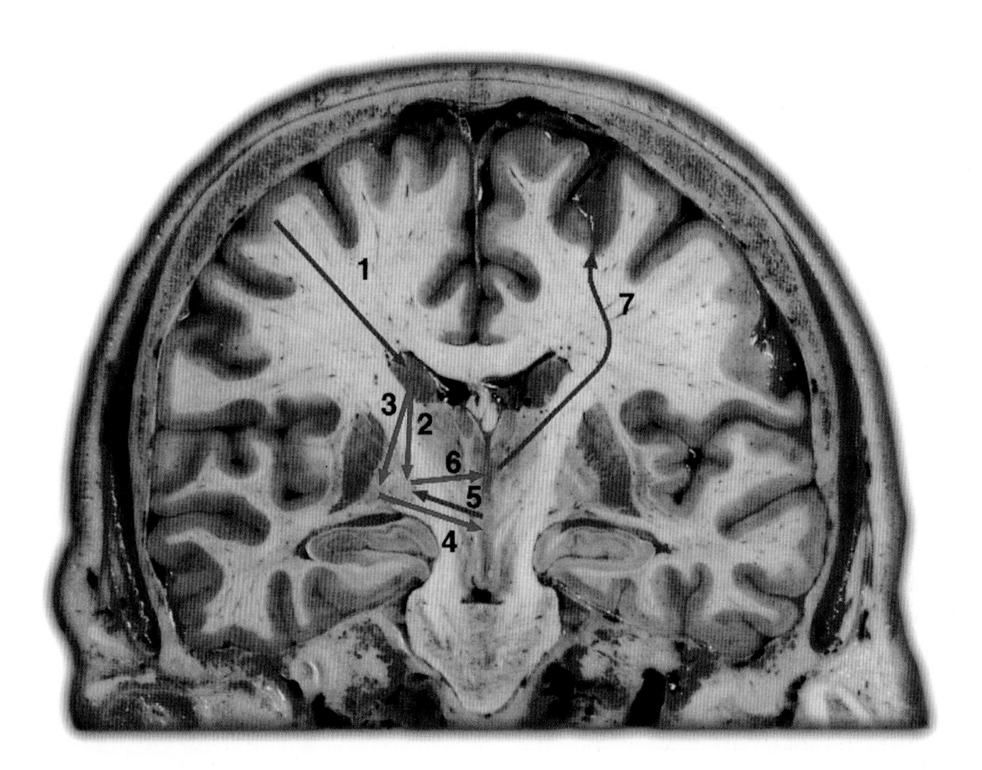

FIGURE 9.7 Frontal-subcortical circuits. There are five circuits, each consisting of projections from the frontal cortex to striatum (1); striatum to globus pallidus/substantia nigra (2,3); pallidum to and from subthalamic nucleus (4,5); pallidum to thalamus (6); and thalamus to frontoparietal cortex (7).

ventral tegmental area.[67] The anterior cingulate cortex mediates motivation, and dysfunction of this region results in apathy.[9]

ETIOLOGIES OF FRONTAL LOBE SYNDROMES

Most conditions affecting the frontal lobes, including trauma, neoplasms, infections, and demyelinating disorders, affect both lobes, and most frontal lobe syndromes reflect bilateral frontal dysfunction. Vascular occlusions involving the anterior or middle cerebral arteries are capable of producing unilateral frontal lesions.

The orbitofrontal syndrome arises from damage to the inferior frontal lobe overlying the bony roofs of the orbits. The orbital roofs comprise the floor of the anterior cranial fossa. The most common injury to the orbitofrontal cortex results from blunt head trauma with contusion of the inferior frontal cortex and adjacent white matter connections by the irregular bony surface (Fig. 9.8).[68,69] A second common etiology is compression by a subfrontal neoplasm arising from the pituitary fossa, olfactory groove, or sphenoid ridge. Meningiomas and chromophobe adenomas are particularly frequent in these locations.[11] Aneurysms arising from the anterior communicating artery can act as mass lesions or rupture into the orbitofrontal and medial frontal areas.[12,70] Frontotemporal dementias may present with an orbitofrontal syndrome and the gradual occurrence of such changes for the first time late in life nearly always signals the presence of a degenerative or neoplastic brain disease (Chapter 21).[71] Intrinsic brain tumors, vascular infarction, infections, psychosurgery, and demyelinating diseases are less frequent causes of the orbitofrontal syndrome.[10]

Frontal convexity lesions result from disorders similar to those producing the orbitofrontal syndrome, accounting for the frequent simultaneous occurrence of orbitofrontal and convexity symptoms. The differential diagnosis of etiologies of the frontal convexity syndrome includes trauma, stroke, hydrocephalus, external compressive neoplasms (meningiomas) and intrinsic tumors (gliomas), infectious disorders, demyelinating disorders, and degenerative diseases.[10,71,72]

The akinetic frontal lobe syndrome and its variants are associated with bilateral medial lesions, usually of a vascular nature.[39] The syndrome has also been produced by midline tumors, hydrocephalus, and trauma.[40,73,74] Apathetic states less marked than akinetic mutism may occur with unilateral lesions affecting the anterior cingulate region. Occlusion of an an-

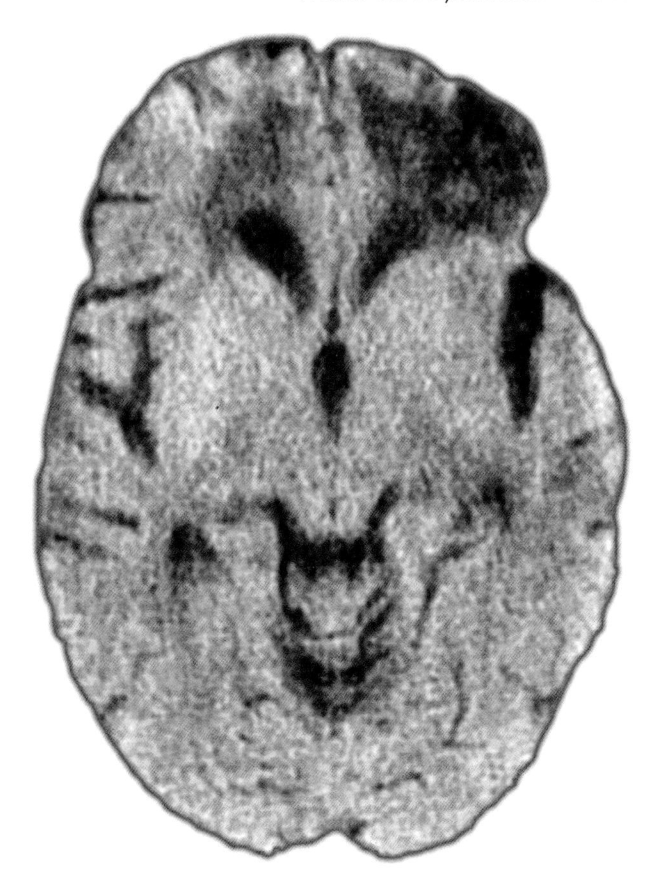

FIGURE 9.8 Computerized tomogram showing orbitofrontal injury associated with closed head trauma.

terior cerebral artery is the most common cause of this disorder (Fig. 9.9), but it may also be seen with falcine meningiomas, lateralized gliomas, and other focal pathology. Apathy may be prominent early in the course of frontotemporal dementias and is the rule late in their course.[72]

Behavioral neurosurgery may produce any of the frontal syndromes, depending on the site and extent of the surgical lesions. Orbitofrontal lesions occasionally produce marked personality changes manifested by euphoria, increased motor activity, and impulsiveness. Neuropsychological testing of patients who have undergone traditional psychosurgery reveals deficits on the WCST reflecting abnormalities in abstracting and maintaining and shifting mental sets appropriately.[75–79] More laterally placed lesions (Fig. 9.10) produce different personality alterations and more extensive neuropsychological deficits. The personality changes include loss of drive, stimulus-boundedness, and impaired ability to elaborate experiences.[80] Neuropsychological assessment reveals difficulty with abstraction as well as deteriorated performance on tests of intelligence.[81]

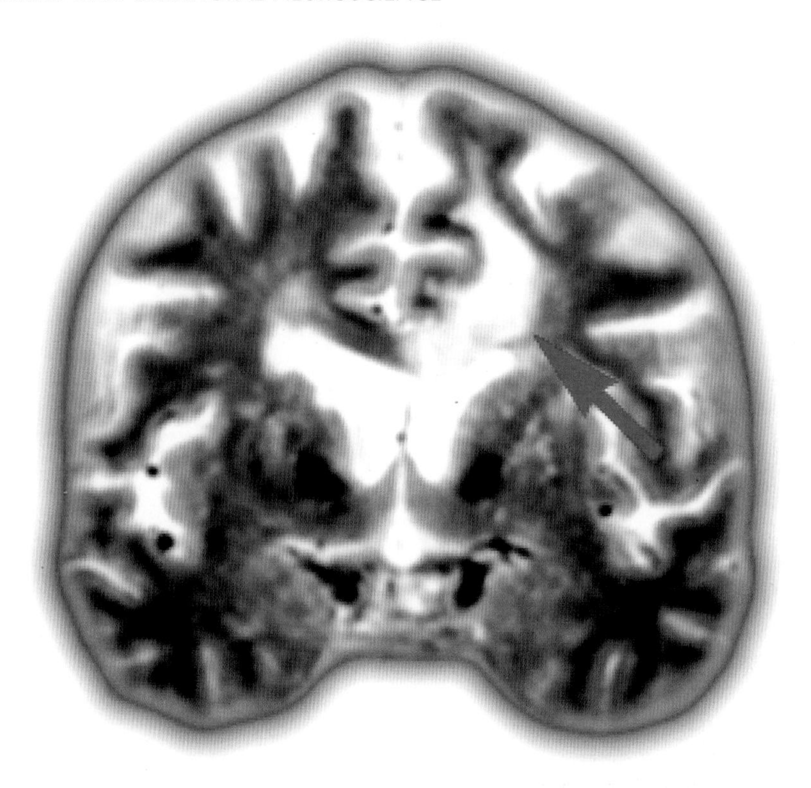

FIGURE 9.9 Magnetic resonance imaging (coronal section) reveals a left anterior cerebral artery territory infarction (arrow) in a patient who presented with mutism, apathy, and leg weakness.

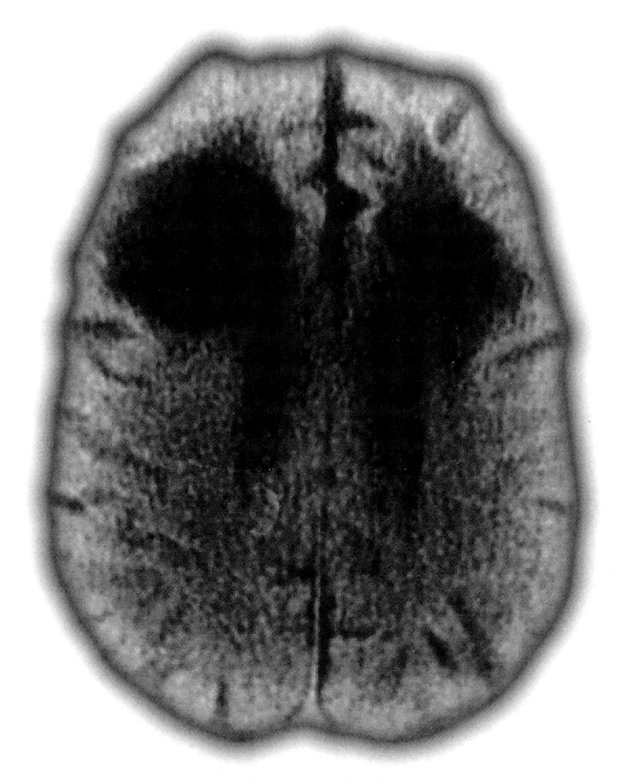

FIGURE 9.10 Computerized tomography (horizontal section) demonstrates large bilateral lucencies in the frontal lobes of a patient who had undergone a conventional frontal leukotomy.

Frontal lobe syndromes are usually the result of overt structural changes, but some chronic toxic encephalopathies appear to involve primarily frontal lobe functions. Chronic alcoholic dementia has a predilection for affecting neuropsychological abilities mediated by the frontal lobes or related subcortical structures. Typical personality changes include an irritable apathy, and cognitive deficits include impaired word list generation, poor attention, visuospatial abnormalities, impaired memory, and poor abstraction.[82–84] The deficits may partially or completely remit if the patient remains abstinent.

TREATMENT OF FRONTAL LOBE SYNDROMES

No specific treatments have been devised for the management of either the orbitofrontal syndrome or the frontal convexity syndrome. In all cases, the preliminary management is directed at discovering the underlying etiology and treating the causative process. Behavioral improvement may follow resection of a neoplasm or shunting of hydrocephalus, and considerable spontaneous resolution may follow vascular or traumatic insults to the frontal lobes. Psychostimulants have been used with some success in the attentional disorders that accompany these syndromes. Post-traumatic encephalopathy, vascular syndromes, or other frontal disorders with attentional deficits may be treated with methylphenidate, dextroamphetamine, or pemoline.[69]

A variety of pharmacological agents have been tried in an attempt to modify the disinhibited behaviors of patients with orbitofrontal injuries. Antipsychotic agents, benzodiazepines, buspirone, carbamazepine, trazodone, propranolol, valproic acid, antidepressants, and lithium have all been used with some success in individual patients, but none has proved to be uniformly reliable (Chapter 4).[85]

The apathy accompanying the medial frontal syndrome may respond to pharmacotherapy. Some patients improve with psychostimulants while others respond to treatment with dopamine receptor agonists.[40,74,86] Apathetic patients should be treated first with dopaminergic agents, since these drugs have fewer adverse long-term side effects. In those who fail to respond, a trial of psychostimulants is warranted.

Pharmacotherapy should be combined with education, family therapy, and individual therapy as appropriate for the underlying condition (Chapter 4).[87]

REFERENCES

1. Fuster JM. The Prefrontal Cortex. New York: Raven Press, 1989.
2. Estanol B. Temporal course of the threshold and size of the receptive field of the Babinski sign. J Neurol Neurosurg Psychiatry 1983;46:1055–1057.
3. DeRenzi E, Barbieri C. The incidence of the grasp reflex following hemispheric lesion and its relation to frontal damage. Brain 1992;115:293–313.
4. Jenkyn LR, Walsh DB, et al. Clinical signs in diffuse cerebral dysfunction. J Neurol Neurosurg Psychiatry 1977;40:956–966.
5. Watson RT, Heilman KM. Callosal apraxia. Brain 1983;106: 391–403.
6. Alexander MP, Schmitt MA. The aphasia syndrome of stroke in the left anterior cerebral artery territory. Arch Neurol 1980;37: 97.
7. Freedman M, Alexander MP, Naeser MA. Anatomic basis of transcortical motor aphasia. Neurology 1984;34:409–417.
8. Alexander MP, Freedman M. Amnesia after anterior communicating artery aneurysm rupture. Neurology 1984;34:752–757.
9. Cummings JL. Frontal–subcortical circuits and human behavior. Arch Neurol 1993;50:873–880.
10. Blumer D, Benson DF. Personality changes with frontal and temporal lobe lesions. In: Benson DF, Blumer D, eds. Psychiatric Aspects of Neurologic Disease. New York: Grune & Stratton, 1975:151–169.
11. Eslinger PJ, Damasio AR. Severe disturbance of higher cognition after bilateral frontal lobe ablation: patient EVR. Neurology 1985;35:1731–1741.
12. Logue V, Durward M, et al. The quality of survival after rupture of an anterior cerebral aneurysm. Br J Psychiatry 1968;114: 137–160.
13. Lhermitte F. "Utilization behavior" and its relation to lesions of the frontal lobes. Brain 1983;106:237–255.
14. Lhermitte F. Human autonomy and the frontal lobes, Part I: imitation and utilization behavior: a neurpsychological study. Ann Neurol 1986;19:326–334.
15. Lhermitte F. Human autonomy and the frontal lobes, Part II: Patient behavior in complex and social situations: The "environmental dependency syndrome". Ann Neurol 1986;19:335–343.
16. Grattan LM, Bloomer RH, et al. Cognitive flexibility and empathy after frontal lobe lesion. Neuropsychiatry Neuropsychol Behav Neurol 1994;7:251–259.
17. Starkstein SE, Boston JD, Robinson RG. Mechanisms of mania after brain injury: 12 case reports and review of the literature. J Nerv Ment Dis 1988;176:87–100.
18. Mayberg HS, Starkstein SE, et al. Selective hypometabolism in inferior frontal lobe in depressed patients with Parkinson's disease. Ann Neurol 1990;28:57–64.
19. Mayberg HS, Starkstein SE, et al. Paralimbic frontal lobe hypometabolism in depression associated with Huntington's disease. Neurology 1992;42:1791–1797.
20. Luria AR. Higher Cortical Functions in Man. New York: Basic Books, 1980.
21. DeLuca J. Cognitive dysfunction after aneurysm of the anterior communicating aneurysm. J Clin Exp Neuropsychol 1992;14: 924–934.
22. Van der Linden M, Bruyer R, et al. Proactive interference in patients with amnesia resulting from anterior communicating artery aneurysm. J Clin Exp Neuropsychol 1993;15:525–536.
23. Goldman-Rakic PS, Friedman HR. The circuitry of working memory revealed by anatomy and metabolic imaging. In: Levin HS, Eisenberg HM, Benton AL, eds. Frontal Lobe Function and Dysfunction. New York: Oxford University Press, 1991:72–91.
24. Stuss DT, Benson DF. The Frontal Lobes. New York: Raven Press, 1986.
25. Stuss DT, Ely P, et al. Subtle neuropsychological deficits in patients wtih good recovery after closed head injury. Neurosurgery 1985;17:41–47.
26. Oscar-Berman M, McNamara P, Freedman M. Delayed-response tasks: parallels between experimental ablation studies and findings in patients with frontal lesions. In: Goldman-Rakic PS, ed. Frontal Lobe Function and Dysfunction. New York: Oxford University Press, 1991:230–255.
27. Malloy PF, Richardson ED. Assessment of frontal lobe functions. J Neuropsychiatry Clin Neurosci 1994;6:399–410.
28. Della Malva CL, Stuss DT, et al. Capture errors and sequencing after frontal brain lesions. Neuropsychologia 1993;31:362–372.
29. Rezai K, Andreasen NC, et al. The neuropsychology of the prefrontal cortex. Arch Neurol 1993;50:636–642.
30. Walsh K. Neuropsychology: A Clinical Approach. New York: Churchill Livingstone, 1994.
31. Regard M, Landis T. The "smiley": a graphical expression of mood in right anterior cerebral lesions. Neuropsychiatry Neuropsychol Behav Neurol 1994;7:303–307.
32. Jones-Gotman M, Milner B. Design fluency: the invention of nonsense drawings after focal cortical lesions. Neuropsychologia 1977;15:653–674.
33. Shallice T, Evans ME. The involvement of the frontal lobes in cognitive estimation. Cortex 1978;14:294–303.
34. Glosser G, Goodglass H. Disorders in executive control functions among aphasic and other brain-damaged persons. J Clin Exp Neuropsychol 1990;12:485–501.

35. Starkstein SE, Robinson RG. Neuropsychiatric aspects of stroke. In: Coffey CE, Cummings JL, eds. Textbook of Geriatric Neuropsychiatry. Washington, DC: American Psychiatric Press, 1994:457–475.
36. Anderson SW, Damasio H, et al. Wisconsin Card Sorting Test performance as a measure of frontal lobe damage. J Clin Exp Neuropsychol 1991;13:909–922.
37. Marin RS. Differential diagnosis and classification of apathy. Am J Psychiatry 1990;147:22–30.
38. Marin RS. Apathy: a neuropsychiatric syndrome. J Neuropsychiatry Clin Neurosci 1991;3:243–254.
39. Barris RW, Schuman HR. Bilateral anterior cingulate gyrus lesions. Neurology 1953;3:44–52.
40. Ross ED, Stewart RM. Akinetic mutism from hypothalamic damage: successful treatment with dopamine agonists. Neurology 1981;31:1435–1439.
41. Segarra JM. Cerebral vascular disease and behavior: the syndrome of the mesencephalic artery (basilar artery bifurcation). Arch Neurol 1970;22:408–418.
42. Drewe EA. Go–no go learning after frontal lobe lesions in humans. Cortex 1975;11:8–16.
43. Leimkuhler ME, Mesulam M-M. Reversible go–no go deficits in a case of frontal lobe tumor. Ann Neurol 1985;18:617–619.
44. Andrew J, Nathan PW. Lesions of the anterior frontal lobes and disturbances of micturition and defecation. Brain 1964;87:233–262.
45. Berlyne N. Confabulation. Br J Psychiatry 1972;120:130–139.
46. Shapiro BE, Alexander MP, et al. Mechanisms of confabulation. Neurology 1981;31:1070–1076.
47. Stuss DT, Alexander MP, et al. An extraordinary form of confabulation. Neurology 1978;28:1166–1172.
48. Hakim H, Verma NP, Greiffenstein MF. Pathogenesis of reduplicative paramnesia. J Neurol Neurosurg Psychiatry 1988;51:839–841.
49. Kapur N, Turner A, King C. Reduplicative paramnesia: possible anatomical and neuropsychological mechanisms. J Neurol Neurosurg Psychiatry 1988;51:579–581.
50. Shukla S, Cook BL, et al. Mania after head trauma. Am J Psychiatry 1987;144:93–96.
51. Starkstein SE, Pearlson GD, et al. Mania after brain injury: a controlled study of causative factors. Arch Neurol 1987;44:1069–1073.
52. Starkstein SE, Mayberg HS, et al. Mania after brain injury: neuroradiological and metabolic findings. Ann Neurol 1990;27:652–659.
53. Baxter LR Jr, Phelps ME, et al. Cerebral metabolic rates for glucose in mood disorders. Arch Gen Psychiatry 1985;42:441–447.
54. Baxter LR, Schwartz JM, et al. Reduction of prefrontal cortex glucose metabolism common to three types of depression. Arch Gen Psychiatry 1989;46:243–250.
55. Baxter LR, Phelps ME, et al. Local cerebral glucose metabolic rates in obsessive-compulsive disorder: a comparison with rates in unipolar depression and normal controls. Arch Gen Psychiatry 1987;44:211–218.
56. Goldberg TE, Saint-Cyr JA, Weinberger DR. Assessment of procedural learning and problem solving in schizophrenic patients by Tower of Hanoi type tests. J Neuropsychiatry Clin Neurosci 1990;2:165–173.
57. Kolb B, Whishaw. Performance of schizophrenic patients on tests sensitive to left or right frontal, temporal, or parietal functions in neurological patients. J Nerv Ment Dis 1983;171:435–443.
58. Holcomb HH, Links J, et al. Positron emission tomography: measuring the metabolic and neurochemical characteristics of the living human nervous system. In: Andreasen NC, ed. Brain Imaging: Applications in Psychiatry. Washington, DC: American Psychiatric Press, 1989:235–270.
59. Carroll BT, Anfinson TJ, et al. Catatonic disorder due to general medical conditions. J Neuropsychiatry Clin Neurosci 1994;6:122–133.
60. Taylor MA. Catatonia. A review of a behavioral neurologic syndrome. Neuropsychiatry Neuropsychol Behav Neurol 1990;3:48–72.
61. Allison RS. Perseveration as a sign of diffuse and focal brain damage. BMJ 1966;2:1027–1032 (I), 1095–1101 (II).
62. Sandson J, Albert ML. Perseveration in behavioral neurology. Neurology 1987;37:1736–1741.
63. Carpenter MB. Core Text of Neuroanatomy. Baltimore: Williams and Wilkins, 1991.
64. Mesulam M-M. Patterns of behavioral neuroanatomy: association areas, the limbic system, and hemispheric specialization. In: Mesulam M-M, ed. Principles of Behavioral Neurology. Philadelphia: F.A. Davis, 1985:1–70.
65. Weinberger DR. A connectionist approach to the frontal cortex. J Neuropsychiatry Clin Neurosci 1993;5:241–253.
66. Alexander GE, DeLong MR, L. Stritch P. Parallel organization of functionally segregated circuits linking basal ganglia and cortex. Annu Rev Neurosci 1986;9:357–381.
67. Domesick VB. Neuroanatomical organization of dopamine neurons in the ventral tegmental area. In: Kalivas PW, Nemeroff CB, eds. The Mesocorticolimbic Dopamine System. New York: New York Academy of Sciences, 1988:10–26.
68. Mattson AJ, Levin HS. Frontal lobe dysfunction following closed head injury: a review of the literature. J Nerv Ment Dis 1990;178:282–291.
69. O'Shanick GJ, O'Shanick AM. Personality and intellectual changes. In: Silver JM, Yuclofsky SC, Hales RE, eds. Neuropsychiatry of Traumatic Brain Injury. Washington, DC: American Psychiatric Press, 1994:163–188.
70. Weisberg LS. Ruptured aneurysms of anterior cerebral or anterior communicating arteries: CT patterns. Neurology 1985;35:1562–1566.
71. Neary D, Snowden JS, et al. Dementia of frontal lobe type. J Neurol Neurosurg Psychiatry 1988;51:353–361.
72. Cummings JL, Benson DF. Dementia: A Clinical Approach. Boston: Butterworth-Heinemann, 1992.
73. Messert B, Baker NH. Syndrome of progressive spastic ataxia and apraxia associated with occult hydrocephalus. Neurology 1966;16:440–452.
74. Stewart JT, Leadon M, Gonzalez-Rothi LJ. Treatment of a case of akinetic mutism with bromocriptine. J Neuropsychiatry Clin Neurosci 1990;2:462–463.
75. Benson DF, Stuss DT. Motor abilities after frontal leukotomy. Neurology 1982;32:1353–1357.
76. Benson DF, Stuss DT, et al. The long-term effects of prefrontal leukotomy. Arch Neurol 1981;38:165–169.
77. Stuss DT, Benson DF, et al. Leucotomized and nonleucotomized schizophrenics: comparison tests of attention. Biol Psychiatry 1981;16:1085–1100.
78. Stuss DT, Kaplan EF, et al. Long-term effects of prefrontal leucotomy—an overview of neuropsychologic residuals. J Clin Neuropsychol 1981;3:13–32.
79. Stuss DT, Benson DF, et al. The involvement of orbitofrontal cerebrum in cognitive tasks. Neuropsychologia 1983;21:235–248.
80. Greenblatt M, Solomon HC. Studies of lobotomy. Assn Res New Ment Dis 1958;36:19–34.

81. Hamlin RM. Intellectual function 14 years after frontal lobe surgery. Cortex 1970;6:299–307.
82. Brandt J, Butters N, et al. Cognitive loss and recovery in long-term alcohol abusers. Arch Gen Psychiatry 1983;40:435–442.
83. Lishman WA. Cerebral disorder in alcoholism: syndromes of impairment. Brain 1981;104:1–20.
84. Ron MA. Brain damage in chronic alcoholism: a neuropathological, neuroradiological, and psychological review. Psychol Med 1977;7:103–112.
85. Silver JM, Yudofsky SC. Psychopharmacology. In: Silver JM, Yudofsky SC, Hales RE, eds. Neuropsychiatry of Traumatic Brain Injury. Washington, DC: American Psychiatric Press, 1994:631–670.
86. Parks RW, Crockett DJ, et al. Assessment of bromocriptine intervention for the treatment of frontal lobe syndrome: a case study. J Neuropsychiatry Clin Neurosci 1992;4:109–111.
87. Lewis L, Athey GI Jr, et al. Psychological treatment of adult psychiatric patients with traumatic frontal lobe injury. J Neurpsychiatry Clin Neurosci 1992;4:323–330.

Chapter 10

Dementia

Dementia is a syndrome of acquired intellectual impairment characterized by persistent deficits in at least three of the following areas of mental activity: memory, language, visuospatial skills, personality or emotional state, and cognition (abstraction, mathematics, judgment).[1] Its acquired nature distinguishes dementia from mental retardation, whereas its persistence differentiates it from the delirium of acute confusional states (ACSs). The requirement, in the above criteria, that the intellectual alterations include multiple areas of mental function distinguishes dementia from aphasic, amnesic, and other monosymptomatic cognitive deficits. A second, more constrained definition of dementia is found in the *Diagnostic and Statistical Manual of Mental Disorders, 4th ed.* (DSM-IV) criteria,[2] which require acquired impairment in memory with additional decline in at least one other domain (e.g., language, praxis, gnosis, executive skills) that interfere with either occupational or social functioning or interpersonal relationships. Abnormalities cannot be secondary to a delirium and should not be based on depression, schizophrenia, or other psychiatric illness). The application of the DSM-IV criteria is often hampered in the nursing home setting where no social or occupational challenges exist but where patients frequently manifest an acquired, persistent decline in multiple domains. For either set of criteria, dementia is a clinical syndrome with many etiologies that may be reversible or irreversible.

Although dementia prevalence estimates are affected by the dementia criteria applied,[3] 75% of patients with moderate to severe dementia and over 95% with mild impairment escape diagnosis in the primary care setting.[4] Dementia is a rapidly growing public health concern. It is largely a problem of elderly individuals, and the size of the aged population is expanding more rapidly than any other segment of the population. Approximately 5% of individuals over age 65 are severely demented, and an additional 10%–15% are mildly to moderately intellectually impaired.[5] In 1950, 8% of the population was over age 65 (12.3 million individuals), and by 1978 the proportion had increased to 11% (22 million); it is estimated that by the year 2030 it will represent 20% of the U.S. population (51 million persons). As the number of aged persons increases, dementia will demand an increasing share of the health-care budget, health-care work-hours, and hospital and nursing home beds. Responding to this challenge demands an integrated, comprehensive, and systematic approach to identification of dementing illnesses, differentiation of the many etiologies of dementia, and development of management programs.

TABLE 10.1. *Differential Diagnosis of the Dementia Syndrome*

Alzheimer's disease
Frontotemporal dementia
Vascular dementia
Dementia with Lewy bodies
Parkinsonian syndromes with dementia
Parkinson's disease
Progressive supranuclear palsy
Corticobasal degeneration
Rare parkinsonian syndromes
Huntington's disease
Prion diseases
Viral and other infectious dementias
HIV dementia
Syphilis
Chronic meningitis
Miscellaneous CNS infections
Toxic and metabolic dementias
Hydrocephalic dementias
Traumatic dementias
Neoplastic dementias
Myelin diseases with dementia
Dementias associated with psychiatric disorders
Depression
Schizophrenia
Miscellaneous psychiatric disorders

ETIOLOGIES OF DEMENTIA

Table 10.1 presents the differential diagnosis of the dementia syndrome. Each major etiology of dementia is discussed and the distinguishing clinical features are described. The relationship between the topography of central nervous system (CNS) involvement and the pattern of intellectual alteration and profile of neuropsychiatric symptoms is emphasized.

Alzheimer's Disease

Alzheimer's disease (AD) almost invariably begins after the age of 50 and becomes increasingly common with advancing age. In 3% of cases the disease is inherited as an autosomal dominant condition, and in the remaining 97% there is an increased incidence of AD among family members.[5] Clinical diagnostic criteria for probable AD[6] (see Table 10.2) have helped standardize its assessment and diagnosis. The accuracy of the clinical diagnosis of AD, compared with pathological assessment, has increased from above 80% in the 1980s to above 90% in the 1990s, mainly because of the application of standardized diagnostic criteria.[7,8]

Alzheimer's disease progresses through three stages in a relatively orderly and consistent manner (Table 10.3).[5] In the first stage the patient has empty speech with few substantive words and a paucity of ideas. There may be an anomia on tests of naming, and the patient will have difficulty generating word lists (e.g., number of animals names produced in 1 minute). Memory, cognition, and visuospatial skills are also compromised in the early phases of the disease, but speech articulation and other motor functions remain normal. Patients usually are indifferent to their deficits, and an electroencephalogram (EEG) may be normal while computerized tomographic (CT) or magnetic resonance imaging (MRI) scans are remarkable only for medial temporal atrophy.[9] Functional imaging using single photon emission computed tomography (SPECT) or positron emission tomography (PET) may be the best diagnostic tools for distinguishing early AD from early frontotemporal dementia (Fig. 10.1).

In the second stage of the disease all intellectual functions continue to deteriorate. Language is characterized by a fluent paraphasic output, impaired comprehension, and relatively preserved repetition.[10] Memory, for both recent and remote information, is severely impaired; visuospatial abilities are further compromised, and patients cannot find their way about or copy constructions; cognitive skills, including calculation and abstraction, are severely impaired. Apraxia and agnosia are present but difficult to demonstrate because of the patients' limited language and memory. Motor strength and coordination are normal, but the patients become

TABLE 10.2. *Criteria for Diagnosis of Probable Alzheimer's Disease*[6]

1. Dementia present
2. Onset between 40 and 90 years of age
3. Deficits in two or more cognitive areas
4. Progression of deficits >6 months
5. Consciousness undisturbed
6. Absence of other potential etiology

The best that can be achieved in the premorbid diagnosis for these and most other degenerative disorders is a "probable" diagnosis. The diagnosis shifts to "possible Alzheimer's disease" if it is very early in its course, or another brain-based disorder is present but does not significantly contribute to the clinical picture, or if there is an atypical presentation. "Definite" diagnosis requires a clinically probable diagnosis and autopsy or biopsy confirmation.

TABLE 10.3. *Characteristics of the Three Stages of Alzheimer's Disease and Frontotemporal Dementia*

Stage	*Alzheimer's Disease*	*Frontotemporal Dementia*
Stage I		
Language	Anomia, empty speech	Anomia
Memory	Defective	Relatively spared
Visuospatial skills	Impaired	Relatively spared
Calculation	Impaired	Relatively spared
Personality	Indifferent	Disinhibited
Klüver-Bucy syndrome	Absent	Present
Motor system	Normal	Normal
EEG	Normal	Normal
Structural scan	Medial temporal atrophy	Normal
Stage II		
Language	Fluent aphasia	Aphasia, stereotyped output
Memory	Severely impaired	Impaired
Visuospatial skills	Severely impaired	Impaired
Personality	Indifferent	Disinhibited
Motor system	Restlessness	Restless, stereotyped behavior
EEG	Background slowing	Background or frontotemporal slowing
Structural scan	Temporal–parietal atrophy	Frontal or temporal atrophy
Stage III		
Intellectual function	Severely impaired	Severely impaired
Language	Palilalia, echolalia, or mutism	Echolalia, mutism
Sphincter control	Incontinence	Incontinence
EEG	Diffuse slowing	Diffuse or frontotemporal slowing
Structural scan	Diffuse atrophy	Frontal or temporal atrophy

EEG, electroencehalogram.

markedly restless, wandering and pacing incessantly. There is usually theta-range slowing of the EEG, and structural imaging shows greater medial temporal atrophy with moderate parietal cortical atrophy as well.

In the final stage all intellectual functions are severely impaired, and the patients' cognitive abilities are largely untestable. Verbal output is reduced to echolalia, palilalia, or mutism. Sphincter control is lost, and the patient's limbs assume a rigid, flexed position. The EEG reveals delta-range slowing, and structural imaging demonstrates diffuse cerebral atrophy with ventricular dilatation and general sulcal enlargement. Death results from aspiration pneumonia or urinary tract infection with sepsis.

The diagnosis of AD is conventionally approached as a matter of exclusion based on the elimination of other more easily diagnosed causes of dementia. Negative laboratory evaluations and nonspecific physical findings, however, also characterize other types of dementia, such as the dementia syndrome of depression, dementia with Lewy bodies (DLB), some insidiously progressive toxic and metabolic processes, and some types of vascular dementia. Diagnosis by exclusion results in the inclusion of all such unrecognized dementias as AD and thus renders AD a nonspecific diagnosis. To ensure that the diagnosis of AD is applied to a homogeneous group of patients suffering from the same illness, all patients identified as suffering from AD

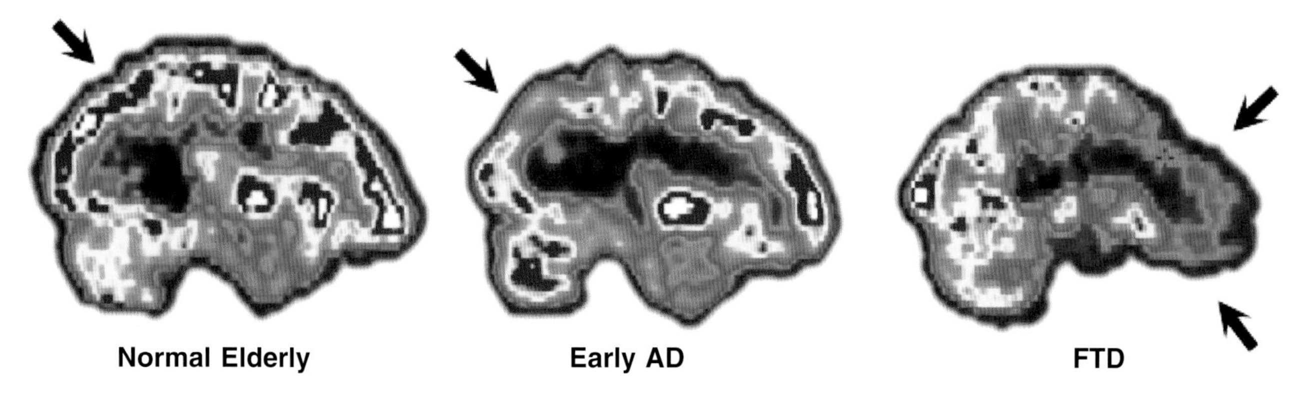

FIGURE 10.1 [15]Fluorodeoxyglucose positron emission tomography (FDG-PET) scans of a normal elderly control (*left*), a patient with early Alzheimer's disease (AD) (*middle*), and a patient with frontotemporal dementia (FTD) (*right*). Note the presence of a metabolic defect in the posterior parietal cortex in the AD patient, but realtively preserved frontal metabolism as noted by absence of "cooler" colors.

should have clinical features and course similar to that described here and thus be called AD through a diagnostic process of *inclusion*. They should have an insidiously progressive clinical syndrome, including aphasia, amnesia, and visuospatial abnormalities, and motor, reflex, and sensory function should be normal throughout most of the clinical course. The diagnosis of AD should be viewed skeptically if the patient's clinical features or course deviates substantially from this pattern.

The neuropsychiatric features of AD include prominent apathy, agitation, anxiety, irritability, and depression.[10a] Delusions are present in approximately 25% of patients in cross-sectional studies and the cumulative prevalence approaches 50%. Hallucinations are present in approximately 10%. Most neuropsychiatric symptoms are increasingly common with disease progression. Agitation and psychosis have been linked to neurofibrillary tangles in the frontal lobes,[10b] and the cholinergic abnormalities also contribute to the psychopathology.[10c]

Pathologically, the brains of AD patients are atrophic with ventricular and sulcal enlargement. Histological investigation reveals progressive neuronal loss, amyloid deposition within senile and neuritic plaques, and neurofibrillary tangles.[11,12] Early changes are most abundant in the hippocampus, then parietal, and frontal association areas, and finally in primary sensory-motor cortex at the end stage of the disease. Neurons in the hippocampus also exhibit numerous intracellular granulovacuolar changes. Neurochemical analyses demonstrate a preferential loss of presynaptic cholinergic neurons in late-stage patients.[13] Loss of cholinergic neurons from the nucleus basalis in the inferior medial forebrain area correlates with the loss of cholinergic innervation of the cerebral cortex.[14] Cortical cholinergic markers may be normal in early and mid-stage patients.[15]

The overproduction of beta-amyloid protein (in hereditary cases of AD) or the abnormal accumulation of beta-amyloid protein (in sporadic cases) is increasingly viewed as the central pathophysiologic event in AD. Beta-amyloid is released from amyloid precursor protein by beta- and gamma-secretases. The amyloid protein oligomerizes to protofibrils which are neurotoxic and initiate a cascade of events including oxidation, inflammation, apoptosis-like neurodegenerative changes, and formation of neurofibrillary tangles.[16] The protein accumulates in plaques, one of the key histopathologic hallmarks of the disease.

- **Treatment** Although the etiology of AD is unknown, treatment is directed toward symptomatic improvement with cholinergic agents, use of psychotropic agents to reduce behavioral disturbances, and disease modification.[17,18] Anti-inflammatory agents, and estrogen replacement in postmenopausal woman,[19,20] may reduce the rate of development of AD.

Treatment with a cholinesterase inhibitor such as donepezil,[21] rivastigmine,[22] or galantamine[23] is standard therapy for AD (Table 10-4a). Patients may exhibit cognition improvement or slowing of cognitive decline, greater preservation of daily function, or behavioral benefit. Behavioral improvement with cholinergic treatment may be more robust than cognitive response.[24] Behavioral response to therapy may be predicted by a baseline orbital frontal paralimbic functional defect in the brains of AD patients as assessed with functional imaging.[25] After treatment with a cholinesterase inhibitor, metabolic increases in the an-

TABLE 10.4A. *Cholinesterase Inhibitors Commonly Used in the Treatment of AD*

Agent	*Initial Dose*	*Final Dose*	*Comment*
Donepezil (Aricept)	5 mg/d	10 mg/d	Single daily dose; few side effects
Galantamine (Reminyl)	4 mg b.i.d.	8–12 b.i.d.	Twice per day dosing; allosteric nicotine modulation plus acetyl-cholinesterase inhibition
Rivastigmine (Exelon)	1.5 mg b.i.d.	3–6 mg b.i.d.	Twice daily dosing; butyrylcholinesterase inhibition; non-hepatic metabolism; monitor for weight loss

terior cingulate and dorsolateral association cortex have been found in patients who evidenced a cognitive response (Fig. 10.2).[26]

Frontotemporal Dementia

Frontotemporal dementia (FTD), like AD, affects primarily cortical structures. The two diseases are clinically similar, beginning in late middle life or later and progressing through a series of stages (Table 10.3).[5] It may be difficult to distinguish the two disorders in the early stages, but several clinical, imaging, and EEG features have differentiating value, particularly when the patient is observed in the early phases of the illness. Progressive nonfluent aphasia, semantic dementia, and a disinhibition syndrome are the three recognized forms of FTD.[27] Table 10.4b provides one approach to the diagnosis of FTD.[28] Compared with AD patients, FTD victims have less memory, calculation, and visuospatial impairment and more extravagant personality alterations. Both diseases produce aphasia, but FTD patients have a greater tendency to produce a stereotyped verbal output, repeating the same story or joke again and again. Those with progressive nonfluent aphasia or semantic dementia have aphasia syndromes distinct from the transcortical sensory aphasia of AD. Features of the Klüver-Bucy syndrome (hyperorality, dietary changes, hypermetamorphosis, placidity, hypersexuality, sensory agnosia) may appear early in the course of FTD, whereas they are confined to the late stages of AD.[29] The EEG and structural imaging scans are normal in the initial stages of FTD, but as the disease progresses, frontotemporal slowing may appear on the EEG, and focal frontal and/or temporal abnormalities may be evident on structural and functional imaging (Fig. 10.1); an exclusionary criterion for FTD is prominent post-central imaging defects.[27]

The most prominent neuropsychiatric features of FTD are apathy and disinhibition.[29a] Mood changes including irritability, depression, and a fatuous euphoria may be present. Some patients develop a distant or bizarre affect with unusual laughter. Tactlessness, loss of concern for the feelings of others, lack of empathy, and reduced emotional engagement are common.

Pathologically, the brains of FTD patients have focal atrophy involving the frontal or anterior temporal lobes.[5] Histologically there is gliosis and neuronal loss in the frontal lobe and temporal lobe. Subcortical white matter may be affected. Occasionally the presence of inflated neurons and neurons containing highly argy-

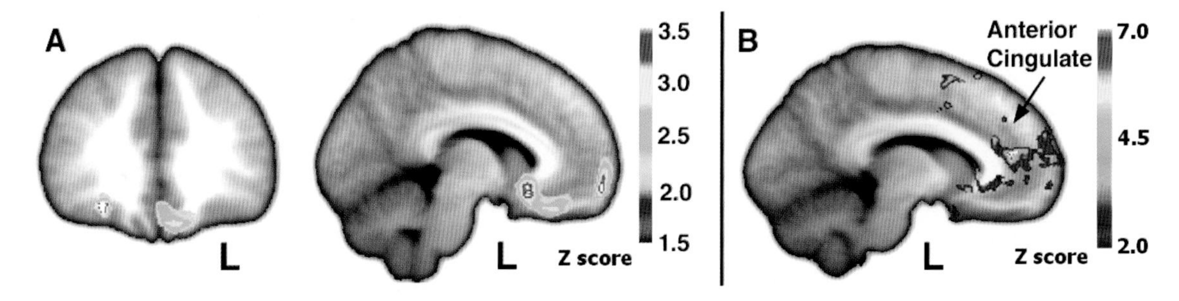

FIGURE 10.2 (*A*) Behavioral response to cholinesterase inhibitor therapy may be predicted by a baseline orbital frontal paralimbic functional defect in the brains of Alzheimer's disease (AD) patients as assessed with technetium-99m D,L hexamethylpropeleneamine oxime (^{99m}Tc-HMPAO) single photon emission computed tomography (SPECT).[50] (*B*) After treatment with metrifonate (a cholinesterase inhibitor), metabolic increases in the anterior cingulate and dorsolateral association cortex were found in a separate group of AD patients who achieved criteria for a cognitive response.[51]

TABLE 10.4B. *Clinical Criteria for Frontotemporal Dementia*[28]

1. Development of a behavioral or cognitive decline manifested by:
a. Progressive personality or behavioral change, or
b. Progressive expressive language or naming impairment
2. Deficits significantly impair social or occupational function
3. Course is manifested by gradual and significant decline
4. No other CNS or systemic cause is responsible
5. Deficits do not occur exclusively during a delirium
6. Deficits are not due to a psychiatric diagnosis

TABLE 10.5. *Criteria for Diagnosis of Probable Vascular Dementia*[68]

1. Dementia is present
2a. Focal neurologic signs
2b. Vascular lesion(s) on brain imaging
3. A relationship between points 1 and 2 with either
Onset within 3 months of a stroke
Abrupt cognitive deterioration
Fluctuating stepwise progression

The criterion for the diagnosis of possible vascular dementia is met if the 3-month time requirement is not achieved.

rophilic Pick bodies are found in the Pick's disease subtype of FTD. Familial FTD resulting from mutations within chromosome 17, some of which are located within the *tau* gene, may manifest a clinical–pathological spectrum within affected families that includes frontal lobe dementia, progressive aphasia, lobar atrophy without distinctive histopathology, and Pick-like inclusions.[30–32] Immunolabeling with antibodies to individual tau epitopes has revealed neuronal and glial tauopathy in these families and several related sporadic degenerative brain diseases with tau inclusions (e.g., progressive supranuclear palsy, and corticobasal degeneration) are regarded as *tauopathies*.[33]

Treatment of FTD is directed at the control of aberrant behavior and prevention of secondary complications.

Vascular Dementia

Vascular occlusions lead to tissue infarction with progressive disruption of brain function and may produce dementia. The characteristics of the dementia are highly variable and may include predominantly cortical features (aphasia, amnesia, agnosia, apraxia) or mainly subcortical symptoms (slowness, depression, forgetfulness, cognitive dilapidation).[5] Combinations of cortical and subcortical characteristics are common, but some neurobehavioral features, including psychomotor retardation and emotional lability are present in most cases. Table 10.5 lists the clinical criteria for vascular dementia (VaD).[34]

Sustained hypertension leading to fibrinoid necrosis and occlusion of cerebral arterioles is the most common cause of VaD. Infarctions (lacunes) are concentrated in the thalamus, basal ganglia, and internal capsule near the base of the brain. In Binswanger's disease, the ischemic injury preferentially involves the hemispheric white matter (Fig. 10.3) and functional imaging shows multifocal defects (Fig. 10.4).[34] The most common form of VaD includes multiple lacunaes and white matter ischemia.

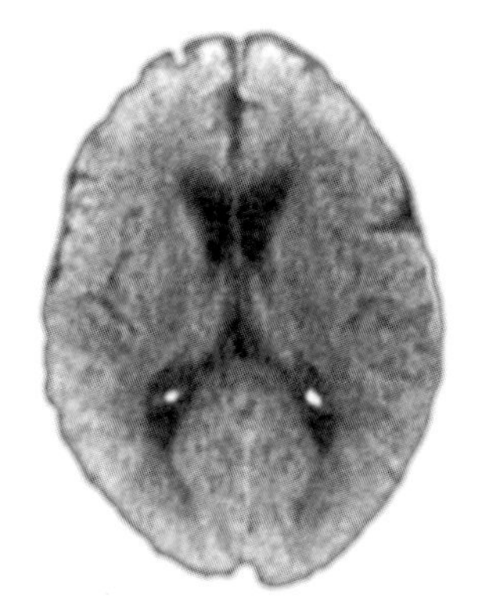
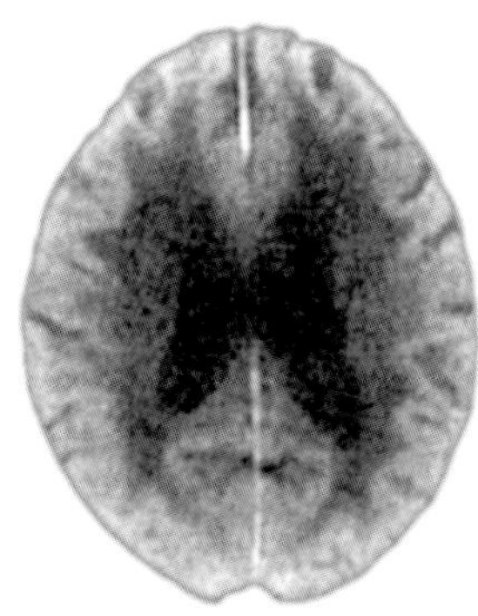
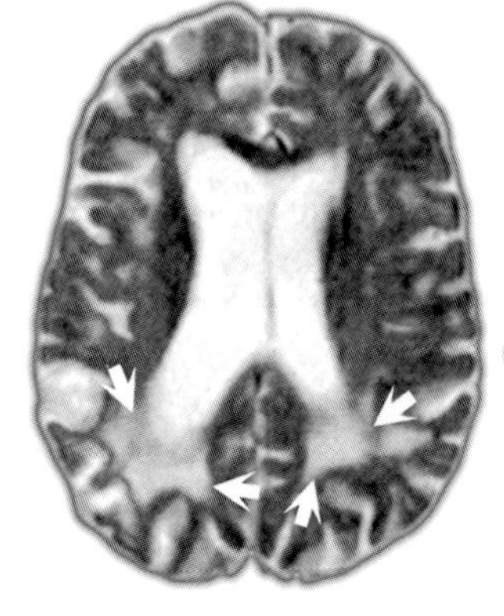
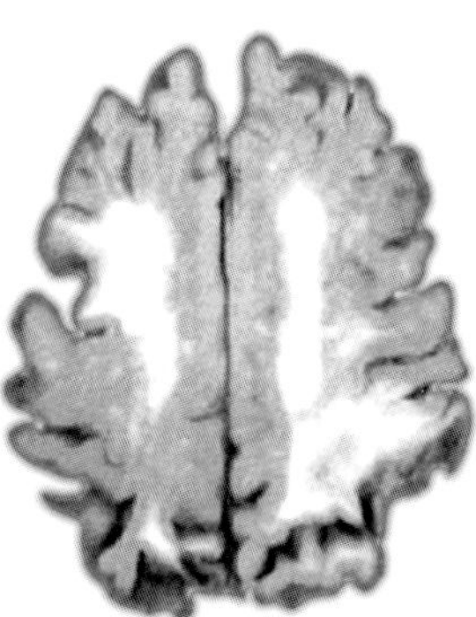

FIGURE 10.3 Differing manifestations of periventricular subcortical white matter disease on structural imaging studies, computed assisted tomography, T_1-weighted magnetic resonance imaging (MRI), T_2-weighted MRI, and FLAIR MRI, from left to right.

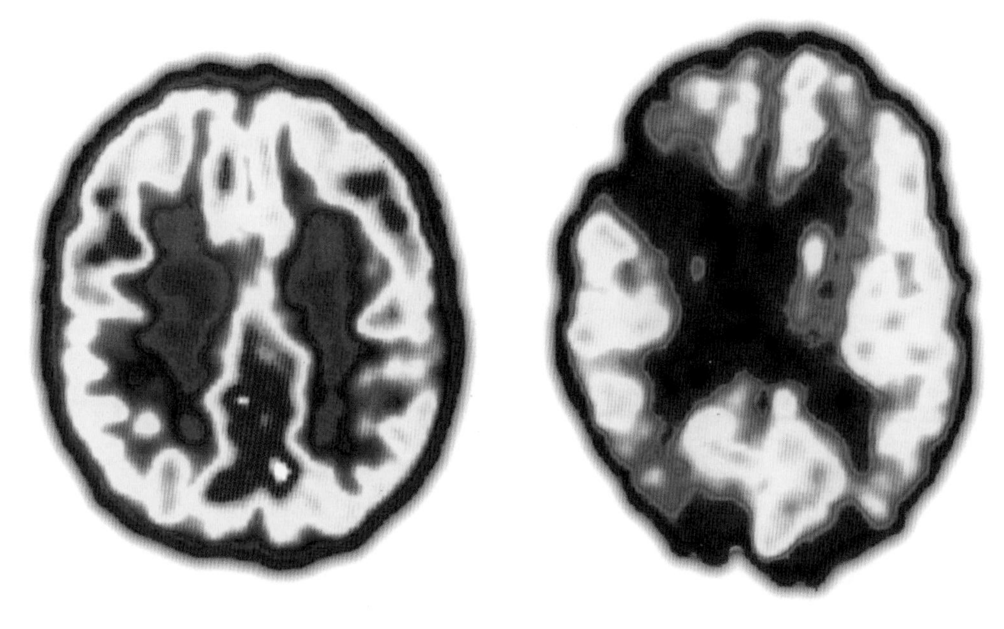

FIGURE 10.4 $^{[15]}$Fluorodeoxyglucose positron emission tomography (FDG-PET) scans of a normal elderly control (*left*), and a patient with multiple cerebral vascular lesions (*right*).

In addition to hypertension-related arteriosclerosis, a variety of other vascular disorders can produce multiple cerebral infarctions, including arteriosclerosis of larger vessels, emboli from the heart or vessels of the neck, inflammatory conditions, and hematologic disorders.[5]

Infarctions are frequently visible on brain imaging, but in some cases they may be too small to be visualized on CT or MRI scans with thick slices (≥10 mm). A scan similar to that seen in obstructive hydrocephalus may be produced in VaD when concentration of the infarctions in the deep periventricular structures leads to ventricular dilation with little enlargement of the cortical sulci. Electroencephalography may reveal multifocal slowing in cases of VaD.

Treatment of VaD is directed at resolving the underlying condition and prevention of future ischemic disease. Dysarthria and aphasia may improve with speech and language therapy, while cholinesterase inhibitor therapy may also be of benefit.[35]

Prion Diseases

Creutzfeldt-Jakob disease (CJD) is one of the many transmissible spongiform encephalopathies that are caused by prions affecting humans and animals. Prion diseases are fatal neurodegenerative disorders caused by protease-resistant isoforms of the prion protein (PrPres). An epidemic of bovine spongiform encephalopathy ("mad cow" disease) and cross-species transmission to humans in the form of variant Creutzfeldt-Jakob disease (vCJD) has heightened public awareness of prion diseases.[36,37] Scrapie is a naturally occurring form of prion diseases first recognized 200 years ago in sheep and goats. In humans, the phenotypic spectrum of prion disease now encompasses CJD, vCJD, the familial Gerstmann-Sträussler-Scheinker syndrome, fatal familial insomnia, kuru, and other less distinct neuropsychiatric disorders. In contrast to kuru, now almost eradicated since the outlawing of ritualistic endocannibalism, sporadic CJD occurs in 85%–90% of patients suffering from prion diseases while the genetically determined forms occur in 8%–13% of patients;[38,39] the remaining transmitted cases are usually caused by invasive medical procedures such as corneal transplantations and surgical or dental procedures from an infected to an uninfected host and in cases where contaminated depth electrodes were used for recording in uninfected patients.[40] Infectivity of prions is due to an abnormal structural conformation of the cell surface protein known as the prion protein (PrP).[40] The wild-type isoform (PrPC) in humans is encoded by a single copy gene (*PRNP*) on chromosome 20, and detailed analysis in sporadic cases has failed to show any primary sequence or post-translational differences between PrPC and the abnormal, protease-resistant, disease-associated isoform, PrPres. Prion replication and infectivity have an absolute dependence on the simultaneous presence of PrPres and ongoing de novo synthesis of PrPC. PrPres is unusually resis-

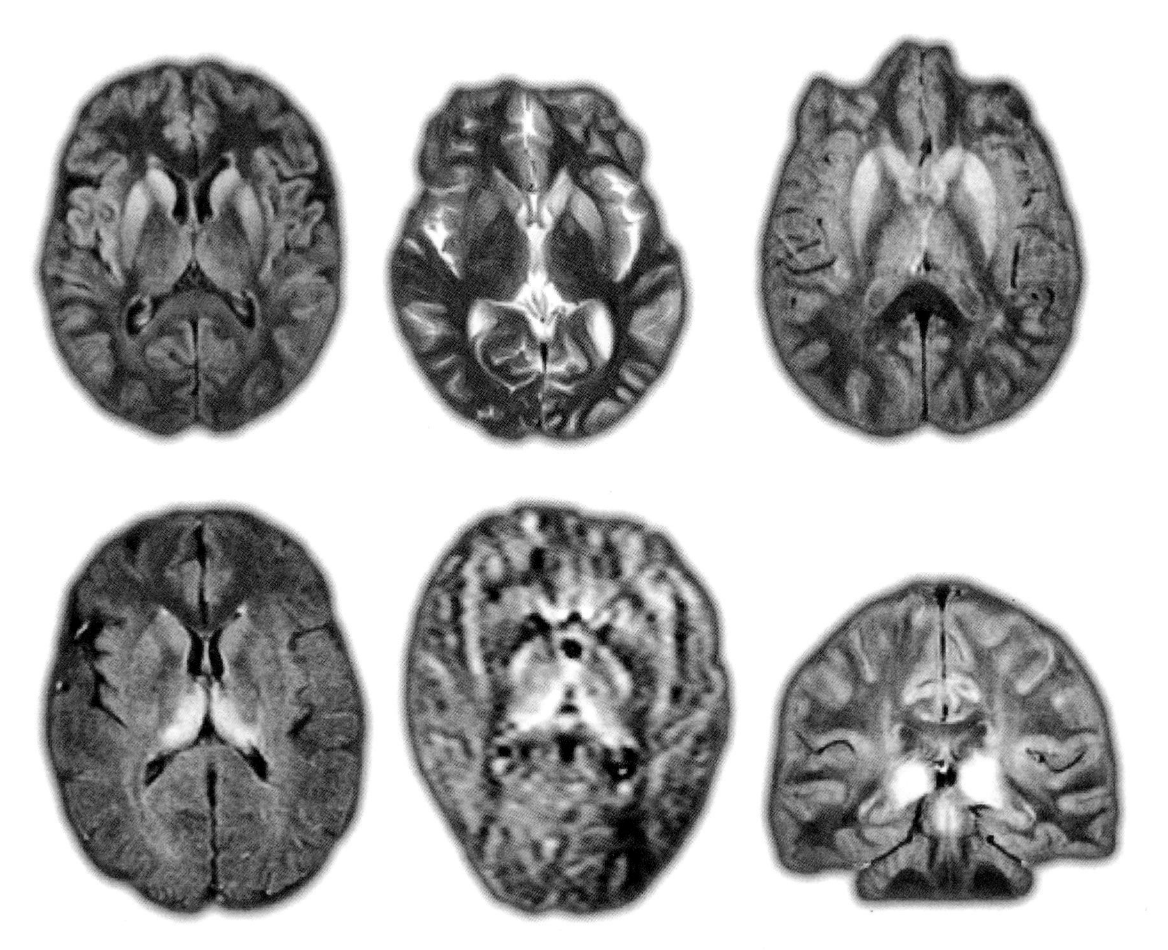

FIGURE 10.5 Differing manifestations of sporadic Creutzfeldt-Jakob disease (CJD) on structural imaging studies demonstrating striatal pathology (*top row*) on T_1-weighted magnetic resonance imaging (MRI), T_2-weighted MRI, and proton-dense MRI, from left to right. Adapted from Finkenstaedt et al. (1996).[87] Thalamic involvement can also be seen in the variant form of CJD (vCJD) on MRI; prion disease is shown (*bottom row*) on computed assisted tomography and T_2 MRI. Adapted from Zeidler et al. (1997).[88]

tant to traditional physical and chemical disinfectant techniques, thus continued surgical or dental transmissions are of great concern when a suspected case is identified.

Creutzfeldt-Jakob disease can be caused by six clinicopathological subtypes that comprise the entire phenotypic range. Seventy percent of cases present with the classic rapidly progressive dementia, periodic sharp wave complexes (PSWC) on EEG, and myoclonus, while the remaining 30% of cases are due to five less common subtypes presenting either with a slower dementia greater than a year without the typical EEG findings, initial ataxia, prominent thalamic degeneration, or kuru plaques.[41,42] Detection of the 14-3-3 protein in cerebrospinal fluid (CSF) and the finding of hyperintensity of the basal ganglia (sporadic CJD)[43] or thalamus (vCJD)[44] on MRI (Fig. 10.5) are additional useful diagnostic signs of CJD, since the EEG findings of PSWC are only reliable in a subset of patients. The CSF identification for 14-3-3 protein shows high sensitivity in most subgroups.[45] The sensitivity of MRI can be 70% across subgroup and is highest for Kuru patients.[45] The diagnostic criteria for definite, probable, and possible CJD are shown in Table 10.6.

Pathologically all prion diseases show widespread vacuolation of gray matter (spongiform change), hypertrophy of large fibrous astrocytes, neuronal loss, and deposition of PrPres that stain similar to amyloid deposits.[462] No effective treatment is available.

Dementias Produced by Viruses and Other Infectious Agents

Dementias may be produced by viruses, bacterial encephalitis (syphilis, Whipple's disease), or chronic meningitis (tuberculous, fungal, parasitic) (Table 10.7).

TABLE 10.6. *Criteria for the Diagnosis of Creutzfeldt-Jakob Disease*[41,46]

Definite CJD
Neuropathologically confirmed PrPres in brain
Probable CJD
Progressive dementia of less than 2 years duration plus two of following: Myoclonus Visual or cerebellar signs Pyramidal or extrapyramidal signs Akinetic mutism
Plus one of the following: Typical PSWCs in the EEG 14–3–3 proteins clearly detectable in the CSF
Possible CJD
Clinical features as for probable CJD, but no PSWCs, and CSF negative for 14–3–3 proteins

CJD, Creutzfeldt-Jakob disease; CSF, cerebrospinal fluid; EEG, electroencephal electroencephalogram; PrPres, protein-resistant isoforms of the prion protein; PSWC, periodic sharp wave complex.

• Human Immunodeficiency Dementia All lentiviruses, a group of retroviruses that includes human immunodeficiency virus-1 (HIV-1), are neurovirulent and can cause a subcortical dementia in infected patients.[2] Memory impairment typically is of a retrieval type deficit and is found in association with apathy and psychomotor slowing. Motor symptoms are significant for poor coordination. Nearly half of patients with acquired immunodeficiency syndrome (AIDS) may suffer from dementia,[47] which usually occurs later in the course of the disease.

Neuropathological changes include pallor of the myelin sheaths, astrocytic proliferation, and multinucleated giant cells (syncytia). Microglia and brain macrophages are the main cellular targets for HIV infection in the brain, and syncytium formation is a signature finding for infection. The syncytium results from viral glycoproteins gp120 and gp41 interaction with the principal cellular HIV-1 receptors in microglia: CD4 and CCR5. The presence of multinucleated giant cells in the CNS is the most specific finding in HIV infection and is a better correlate of the dementia than the extent of brain viral load.[48] Similar findings in the simian immunodeficiency virus (SIV) animal model suggest that in late-stage HIV infection, an increase in trafficking of monocytes (the precursors of macrophages) to the brain may be associated with the development of HIV-related neurological disease.[49] HIV replication in bone marrow, along with chronic systemic inflammation, could further increase activated monocytes. Thus, the best proposed treatment for HIV dementia would be effective control of systemic HIV replication and decreasing chronic macrophage activation.

TABLE 10.7. *Infectious Conditions Producing Dementia*

Condition	*Specific Disorder(s)*
Viral diseases	Human immunodeficiency dementia
	Progressive multifocal leukoencephalopathy
	Paraneoplastic limbic encephalitis
	Subacute sclerosing panencephalitis
	Progressive rubella panencephalitis
Bacterial encephalitis	Syphilis (general paresis)
	Whipple's disease
Chronic meningitis	Bacterial
	Tuberculosis
	Syphilis (meningovascular form)
Fungal	
Parasitic	

• Non-HIV Viral Dementias Other viral dementias include progressive multifocal leukoencephalopathy, paraneoplastic limbic encephalitis, subacute sclerosing panencephalitis, and progressive rubella panencephalitis. The latter two disorders are confined to children, adolescents, and young adults infected prenatally or in early life by rubeola or rubella viruses, respectively.[5]

Progressive multifocal leukoencephalopathy (PML) is a slowly progressive papovavirus infection occurring almost exclusively in patients with chronic lymphoproliferative, myeloproliferative, or granulomatous diseases.[5] There is a gradual accrual of focal neurological deficits, and dementia supervenes in the final stages. Pathologically, the brains have multiple demyelinated lesions with abnormal astrocytes, inclusion-containing oligodendrocytes, and inflammatory cells.

Limbic encephalitis is a syndrome of progressive intellectual impairment occurring in patients with systemic neoplasms, particularly oat cell carcinoma of the lung.[50] Amnesia is the most prominent neurological deficit; bizarre behavior and Kluver-Bucy type symptoms have been reported. Maximal pathological changes are found in the hippocampi and cingulate. At autopsy, alterations include inflammatory changes most consistent with a viral infection, although no agent has been consistently isolated.

Bacterial Infections Producing Dementia

Syphilitic general paresis is perhaps the best example of a dementia syndrome produced by a bacterial infection of the brain. Although currently rare, general paresis once accounted for 10%–30% of all mental hospital admissions. The disorder typically becomes manifest 15–30 years after the initial infection and is characterized by progressive intellectual impairment combined, in some cases, with psychosis.[5] Facial and lingual tremors are common, and two-thirds of patients have pupillary abnormalities. Pathologically, the *Treponema pallidum* organisms are most abundant in the frontal cortex, and 75% of patients improve with penicillin treatment.

Whipple's disease is a rare bacterial meningoencephalitis manifest by diarrhea and gastrointestinal malabsorption symptoms, lymphadenopathy, arthritis, anemia, and fever as well as neurological abnormalities, including dementia. If initiated early, antibiotic therapy may halt progression of the disease.[5]

• Chronic Meningitis Syphilis, tuberculosis, fungi, or parasites may produce a chronic meningitis that causes a dementia. In addition to intellectual deterioration, affected patients manifest cranial nerve palsies, stiff neck, and headache. Focal neurological deficits may occur, and hydrocephalus may be an acute or late complication. Early recognition and treatment reverses the deficits in most cases.[5]

Toxic and Metabolic Dementias

Adequate intellectual and emotional function demands normal cellular metabolism with cells supplied sufficient amounts of appropriate metabolic ingredients free of toxins. Metabolic and toxic disturbances frequently produce intellectual disturbances, and in some cases the neurobehavioral alterations may overshadow systemic manifestations. When the mental state disturbance has an abrupt onset and short course and causes attentional disturbance, a delirium is diagnosed (Chapter 11); when the intellectual deterioration has a gradual onset and is insidiously progressive, persisting for weeks, months, or longer, the disorder is a dementia syndrome.

Like other dementias, toxic and metabolic dementias are most prevalent in the elderly population; 86% of individuals over age 65 suffer from one chronic illness, and 50% have two or more.[51] Many of these disorders are capable of compromising intellectual function. The high frequency of systemic illnesses also results in a high rate of drug consumption in the elderly. Those over age 65 comprise 11% of the population and use 25% of all prescribed pharmaceuticals. Almost 70% of elderly patients regularly use over-the-counter medication, as compared with 10% of the general adult population.[52] Furthermore, alterations in body mass, serum-binding proteins, and drug metabolism render the elderly vulnerable to drug intoxication even when conventional doses are administered. The prevalence of chronic physical illness in the aged and the consumption of the drugs used to treat these disorders thus converge to produce an increased prevalence of metabolic and toxic dementias in older individuals. Recognition of these dementia syndromes is particularly important since most are treatable with partial or complete reversal of neuropsychological dysfunction.

Metabolic Disorders Causing Dementia

Table 10.8 lists the principal etiologies of chronic toxic and metabolic encephalopathies. Cerebral anoxia compromises cerebral metabolism and affects intellectual function. The anoxia may be a product of cardiac failure, pulmonary disease, or severe anemia.[5] Severe pulmonary disease with continuous oxygen deprivation and carbon dioxide retention produces a syndrome of confusion, headache, papilledema, tremor and twitching of the extremities, and evidence of cardiopulmonary decompensation.[5] Acute profound anoxia, as occurs with cardiopulmonary arrest, may produce a prominent post-anoxic dementia syndrome.[5]

Patients with chronic renal failure may develop a uremic encephalopathy, and patients on dialysis are vulnerable to dialysis dementia. Uremia-related dementias are characterized by mental status changes, tremor, asterixis, and slowing of the EEG.[5] Dialysis dementia is a progressive fatal encephalopathy with intellectual deterioration, prominent dysarthria, and myoclonus.[5] The EEG in dialysis dementia reveals generalized slowing with intermittent bursts of polyspike complexes. The occurrence of dialysis dementia appears to be related to the amount of aluminum in the dialysis solution, and the incidence has diminished with attention to the amount of parenteral aluminum received by dialysis patients.

Hepatic (portosystemic) encephalopathy is produced by advanced hepatic disease and shunting of portal venous blood into the general circulation. Clinically, the syndrome includes a chronic confusional state, tremor, asterixis, and evidence of hepatic dysfunction. The EEG is generally slow, with bursts of delta frequency triphasic waves. Strict limitation of dietary protein and prevention of ammonia production by gastrointestinal

TABLE 10.8. *Metabolic and Toxic Causes of Dementia*

Metabolic Disorders	*Toxic Disorders*
Hypoxia	Drugs
Cardiopulmonary disease or failure	Psychotropic agents
Anemia	Anticholinergic drugs
Uremia and dialysis dementia	Antihypertensive agents
Hepatic encephalopathy	Anticonvulsants
Vitamin deficiencies	Antineoplastic drugs
B_{12}	Antibiotics
Folate	Miscellaneous therapeutic agents
Niacin	Alcohol and drug abuse (glue sniffing, etc.)
Endocrine diseases	Metals
Thyroid and parathyroid disorders	Lead
Adrenal abnormalities	Mercury
Electrolyte disturbances	Manganese
Porphyria	Arsenic
	Thallium
	Industrial agents
	Organic solvents
	Organophosphate insecticides
	Carbon monoxide

flora through the use of neomycin and lactulose frequently result in an improved mental state. A few patients with persistent hepatic encephalopathy develop an irreversible dementia syndrome and a choreiform movement disorder.

Vitamin deficiencies, including lack of B_{12}, folate, or niacin, produce dementia syndromes. Thiamine deficiency leads to the Wernicke-Korsakoff syndrome, an amnesic state with little impairment of other intellectual functions (Chapter 7). Vitamin B_{12} deficiency leads to dementia, myelopathy, peripheral neuropathy, optic neuropathy, and anemia. The dementia may be the predominant neurological abnormality and may precede blood and marrow changes.[5] Folate deficiency may exist without concomitant neurological deficits but in some cases leads to a clinical syndrome, including a dementia closely resembling that of B_{12} deficiency. Lack of niacin produces a syndrome manifested by gastrointestinal tract lesions (gingivitis, glossitis, enteritis), diarrhea, dermatitis, and dementia.

Endocrine disturbances are well-known causes of dementia.[5] Hypothyroidism produces dementia, psychosis, neuropathy, and myopathy. Hyperthyroidism classically produces anxiety, restlessness, tachycardia, palpitations, and heat intolerance, but in the elderly it may present as a "simple" dementia syndrome with little systemic evidence of excessive thyroid activity. Hyperparathyroidism causes elevated serum calcium levels, dementia, weight loss, renal colic, abdominal pain, and bone and joint pain. In some cases, dementia may be the presenting manifestation. Hypoparathyroidism leads to basal ganglia calcification, dementia, and parkinsonism or choreoathetosis. In both Cushing's and Addison's disease, reflecting excessive and inadequate adrenal function respectively, dementia is one manifestation of the clinical symptomatology.

Toxic Conditions Causing Dementia

A wide variety of drugs have been associated with toxic dementia syndromes. Nearly any drug can produce intellectual compromise if administered in excessive amounts, but some can alter CNS function even in conventional dosages. The agents that have most often been implicated in drug-induced chronic confusional states include neuroleptic agents, antidepressants, lithium, and minor tranquilizers.[5] Anticholinergic agents used in the treatment of extrapyramidal and cardiac disorders and psychotropic agents with prominent anticholinergic activity impair central cholinergic activity and disturb intellectual function. Antihypertensive agents and anticonvulsants produce dementia syndromes when present in toxic amounts, and in some patients, dementia may be the only toxic manifestation. Antibiotics, antineoplastic agents, digitalis, disulfiram, bromides, levodopa, and many other agents may produce dementia syndromes in specific circumstances.[5]

In addition to iatrogenic dementias, chronic toxic encephalopathies occur commonly in multidrug abusers. In some cases the neuropsychological deficits remit with abstinence, but in others the dementia may be permanent. Electroencephalographic abnormalities often accompany the intellectual impairment.[53]

Clinically significant dementia also occurs in approximately 3% of alcoholic patients, and careful neuropsychological testing indicates subtle intellectual deficits in up to 50%. The mental status impairments include disturbances of attention and abstraction, visuospatial alterations, and memory deficits.[54] The dementia syndrome includes many deficits beyond the restricted memory disturbance occurring in alcoholics with the Wernicke-Korsakoff syndrome, and alcoholic dementia is far more common than the Korsakoff state. The dementia is at least partially reversible with abstinence, and the atrophy visible on structural imaging of some alcoholics also reverses in many who remain abstinent.

Metals (lead, mercury, manganese, arsenic, thallium) are highly toxic to the nervous system, and excessive exposure produces dementia and peripheral neuropathy.[5,53,54] Industrial and agricultural agents, including organic solvents (trichloroethylene, toluene, carbon tetrachloride, carbon disulfide), organophosphate insecticides, and carbon monoxide, produce dementia in circumstances involving chronic excessive exposure. As in other toxic and metabolic conditions, elimination of the exposure leads to symptomatic improvement in a majority of cases.

Hydrocephalic Dementias

Hydrocephalus refers to the presence of excessive CSF in the head and in essentially all cases entails ventricular enlargement with increased fluid within the ventricular cavities. Hydrocephalus can be the end result of several processes as shown in Table 10.9. In hydrocephalus ex vacuo, the ventricular dilatation is a product of tissue loss with no change in the dynamics of CSF flow (nonobstructive hydrocephalus). Obstructive hydrocephalus, by contrast, occurs when there is a blockage of CSF pathways. The obstruction may be within the ventricular system or at the level of the outlet foramina, preventing the fluid from moving from within the ventricular system to the subarachnoid space (noncommunicating hydrocephalus), or it may be within the subarachnoid space, preventing absorption of the fluid by the pacchionian villi of the arachnoid granulations (communicating hydrocephalus). In the former, ventricular pressure is usually increased, whereas in communicating hydrocephalus the intracranial pressure often remains normal (normal-pressure hydrocephalus).[5]

TABLE 10.9. *Differential Diagnosis of Conditions Causing Hydrocephalus*

Nonobstructive hydrocephalus
Hydrocephalus ex vacuo (secondary to loss of cerebral tissue)
Obstructive hydrocephalus
Noncommunicating
Intraventricular blockade
Aqueductal stenosis
Ventricular masses
Obstruction of ventricular outlet foramina
Posterior fossa neoplasms
Basilar meningitis
Congenital malformations
Communicating (normal-pressure hydrocephalus)
Post-traumatic
Post-hemorrhagic
Post-infectious
Idiopathic
Ectatic basilar artery

The clinical characteristics of obstructive hydrocephalus include dementia, gait disturbance, and incontinence. Apathy, inattention, poor memory, and impaired judgment and abstraction are typical of hydrocephalic dementia. The gait disturbance is varied and may present as ataxic, apraxic, or spastic; the incontinence is usually a late feature.

Intraventricular blockade in noncommunicating obstructive hydrocephalus may be produced by aqueductal stenosis or ventricular masses (neoplasms, hematomas, colloid cysts). Foraminal obstruction can occur with posterior fossa neoplasms, basilar meningitis, or congenital malformations. Absorption blockade in communicating hydrocephalus usually follows trauma with subarachnoid bleeding, subarachnoid hemorrhage from aneurysms or vascular malformations, or CNS infections.

Determination of the type of hydrocephalus is made by a combination of structural imaging and cisternographic findings, although cisternography does not aid in prediction of treatment response. Structural imaging reveals large ventricles in all cases of hydrocephalus. Patients with hydrocephalus ex vacuo usually have symmetrically dilated ventricles and enlarged cerebral sulci, whereas patients with obstructive hydrocephalus have ventricular enlargement out of proportion to the sulcal enlargement, the anterior portions of the ventricles are more enlarged than the posterior portions, and there may be periventricular edema. Radioisotope flow studies in hydrocephalus ex vacuo show ventricular reflux of the tracer substance and normal flow over the cerebral convexities; in normal-pressure hydrocephalus (obstructive communicating hydrocephalus) there is ventricular reflux and blockage of flow over the convexities, and in noncommunicating hydrocephalus there is no ventricular reflux and normal flow over the convexities.[5] Differentiating AD with ventricular enlargement from the dementia associated with normal pressure hydrocephalus may be aided by functional imaging abnormalities in the high parietal region and preferential medial temporal atrophy on structural imaging characteristic of AD.[55–58]

Patients with normal pressure hydrocephalus who benefit from shunting may have a greater CSF resistance to outflow[59] and those who receive a low pressure shunt appear to do better than those who receive a medium pressure shunt.[60]

Traumatic Dementias

Cerebral trauma is the most common cause of dementia in young individuals and in Western countries is usually a product of motor vehicle accidents. The lesions may be primarily contusions of the cerebral gray matter or shearing lesions of the subcortical white matter.[5,53] Amnesia and personality changes are common and reflect medial temporal and orbitofrontal damage, respectively. Aphasia, impaired concentration, poor abstracting abilities, and apraxia occur in some cases. The long-term prognosis for recovery is good in most cases, but intellectual restitution may take several years and may never be complete.

Subdural hematomas should be considered in any patient with mental status changes following trauma, and in the elderly the inciting traumatic event may be minimal.[54] Mental status alterations include fluctuating arousal, irritability, poor attention, and impaired memory. Focal neurological signs may be present. In most cases, subdural blood collections are visible on CT or MRI scans, but they may become isodense on CT with brain tissue and thus be difficult to detect.

Dementia pugilistica is an uncommon dementia syndrome occurring in boxers who have sustained multiple episodes of cerebral trauma. The dementia begins late in the boxer's career or after retirement and gradually progresses. The intellectual impairment is combined with ataxia and extrapyramidal disturbances, and autopsy studies reveal neuronal loss, astrocytic proliferation, and prominent neurofibrillary tangles.[5]

Neoplastic Dementias

Brain tumors produce dementia syndromes by causing local tissue destruction or compression, by compromising cerebral blood flow, by increasing intracranial pressure, and, in some cases, by obstructing CSF flow and producing hydrocephalus. Dementia is most common with tumors of the frontal lobe. Such tumors may impair judgment and abstraction and increase intracranial pressure without producing focal neurological disturbances.[50] Temporal lobe tumors, tumors of subcortical structures, and neoplastic meningitis may also produce dementia syndromes.

Myelin Diseases with Dementia

Multiple sclerosis may have a relatively benign clinical course with little intellectual impairment, or it may be an aggressive, remitting, and relapsing or chronically progressive disorder with profound neuropsychological deterioration.[53] Nearly all patients with multiple sclerosis have eye movement disturbances, as well as motor, sensory, and reflex abnormalities. Structural imaging may reveal periventricular cerebral lesions, and the spinal fluid may contain an increased number of lymphocytes, elevated protein, or excessive gamma globulin content.

Metachromatic leukodystrophy and adrenoleukodystrophy are two rare inherited diseases of myelin metabolism that cause dementia in adults. Metachromatic leukodystrophy is a recessively inherited disorder manifesting dementia and peripheral neuropathy.[53] Adrenoleukodystrophy is an X-linked recessive disorder producing dementia and adrenal failure.[53]

Dementias Associated with Psychiatric Disorders

Dementia syndromes associated with psychiatric disorders have often been called *pseudodementias*, but the

TABLE 10.10. *Psychiatric Disorders Associated with Dementia Syndromes*

Affective disorders
Depression
Mania
Schizophrenia
Hysteria
Conversion symptoms
Ganser syndrome
Miscellaneous
Anxiety
Obsessive-compulsive disorders
Malingering

prefix *pseudo-* is inappropriate when dementia is defined as a clinical syndrome produced by a variety of diverse disorders. Patients with intellectual impairment associated with psychiatric disturbances meet the syndromic definition of dementia used here. Their potential treatability makes recognition of the dementia syndromes occurring with psychiatric disorders particularly important. Table 10.10 lists the psychiatric conditions that may compromise intellectual performance.

- **Depression** Depression is the most common psychiatric disorder that produces a syndrome of intellectual impairment. Follow-up studies of patients diagnosed as suffering from degenerative dementia have revealed that 30%–50% do not undergo the expected neuropsychological deterioration and are eventually rediagnosed. The disorder most frequently misidentified as degenerative dementia is depression and can be difficult to distinguish from a degenerative process without a comprehensive evaluation.[61] New onset late life depression increases the risk for developing AD and thus suggests that personality and mood changes may predate the cognitive dysfunction in AD.[62,63]

The dementia syndrome of depression occurs primarily in elderly individuals with manic-depressive illness, recurrent unipolar depression, or late-onset endogenous depression. Attention and memory deficits are ubiquitous in depressed patients of all ages, but intellectual impairment sufficient to produce a dementia syndrome is rare in young patients and occurs in at least 10% of aged depressed patients.[5]

The dementia usually occurs in patients with retarded depressions manifesting both neurovegetative and motor disturbances. They have a parkinsonian-like appearance with psychomotor retardation, bowed posture, and hypophonic speech. In addition, they suffer from insomnia, loss of appetite, constipation, and diminished libido. Mental status alterations characteristic of the dementia syndrome of depression include slowness of responses, lack of attention, poor memory, disorientation, impaired motivation, and disturbed ability to abstract and grasp the meaning of situations. Poor word list generation and simplification of constructions are also common, and incomplete performances and "I don't know" and "I can't" responses are frequent. These patients may have mood-congruent hallucinations and delusions and ideas of reference.

Laboratory studies of patients with the dementia syndrome of depression usually reveal cerebral atrophy on structural imaging, a positive dexamethasone suppression test (failure of suppression of endogenous cortisol secretion by administration of dexamethasone), and a normal EEG. Unfortunately, enlarged ventricles and sulci are common in normal elderly individuals as well as those with dementia, and abnormal dexamethasone suppression tests occur in many types of dementia without depression. These tests do not effectively distinguish depressed patients from those with other causes of dementia; focal left hippocampal atrophy may suggest underlying AD.[64]

In some cases, the diagnosis of depression-related dementia may depend on treatment responsiveness, and antidepressant therapy should be considered in any patient in whom depression may be producing or exacerbating a dementia syndrome. Serotonin reuptake inhibitors (SSRIs) heterocyclic antidepressants, monoamine oxidase (MAO) inhibitors, lithium, and electroconvulsive therapy (ECT) have all been used successfully to treat the depression of patients with an associated dementia syndrome and can stabilize cognition for many years.[65] Even if the cognitive dysfunction normalizes with successful treatment of the mood disorder these individuals remain at a higher risk for ensuing AD than depressed patients without associated cognitive difficulty.[66]

- **Mania** Mania is a rare cause of dementia, but a few cases have been reported in which the disorganized, disinhibited behavior led to a dementia syndrome.[5] Memory disturbances and disorientation are common in advanced stages of mania, but in most cases the associated hyperactivity, flight of ideas, pressured speech, and expansive grandiosity make the diagnosis obvious. Improvement usually follows treatment with lithium, carbamazapine or valproate, and many patients require an antipsychotic to control the acute episode. Elderly manic patients may develop confusional states in the course of manic episodes and respond more slowly to treatment.

Schizophrenia Intellectual impairment may occur in schizophrenia as part of an acute psychotic episode in the buffoonery syndrome when the patient manifests clowning, jocularity, and facetious responses or as an integral part of the schizophrenic disorder in a specific subpopulation of schizophrenic patients.[5] The latter group is distinguished from schizophrenia without neuropsychological deterioration by ventricular enlargement on structural imaging, a preponderance of negative schizophrenic symptoms (apathy, withdrawal, flat affect, anhedonia), poor premorbid adjustment, and poor response to treatment (Chapter 12).

Hysteria Dementia as a manifestation of an hysterical conversion reaction is rare. The hallmark of the syndrome is the marked contrast between the patient's relatively normal performance in unstructured circumstances and markedly impoverished performance in the testing situation. Whenever an hysterical conversion syndrome is identified, it must be borne in mind that the symptom complex is usually the harbinger of a neurological or major psychiatric disorder (Chapter 15).

Ganser Syndrome The Ganser syndrome is considered by many to be a variant of hysterical dementia but has separate and unique clinical features.[5] The most unusual and characteristic feature is the patient's penchant for replying to simple questions (e.g., "How many legs does a dog have?" "Where was the battle of Waterloo fought?") with ridiculous or approximate answers. In addition, the typical Ganser syndrome includes disturbances of consciousness, amnesia for the episode, hallucinations, and motor or sensory deficits similar to those found in conversion reactions. The disorder usually improves spontaneously, but in many cases an underlying metabolic or neurological disorder contributes to the symptomatology.

Miscellaneous Psychosyndromes with Dementia In rare cases, severe anxiety, disabling obsessive-compulsive symptoms, or malingering may interrupt the patient's performance and create an appearance of dementia. Observation and response to treatment usually clarify the diagnosis.

DEMENTIA EVALUATION

It is impossible to construct a laboratory battery that would adequately screen for all the causes of dementia. In addition, many syndromes (degenerative dementias, depression) lack pathognomonic laboratory features that would allow such identification. Instead, recognition of the different causes of dementia depends on integration of information from the clinical history, neurological and general physical examinations, and mental status assessment as well as from selected laboratory tests (Table 10.11). A thorough history will reveal any evidence of industrial exposure, drug inges-

TABLE 10.11. *Laboratory Evaluation of the Demented Patient*

Required Laboratory Tests	*Optional Laboratory Tests*
Complete blood count	Syphilis serology
Serum electrolytes, calcium, glucose, blood urea nitrogen/creatinine, liver function tests	Sedimentation rate
Thyroid-stimulating hormone	HIV testing
Serum vitamin B_{12}	Chest X-ray
Structural imaging study	Urinalysis, 24-hour urine for heavy metals, toxicology screen
	Neuropsychological testing
	Lumbar puncture
	Apo-E genotyping, $A\beta_{42}$/tau CSF analysis
	Electroencephalography
	Single-photon emission computed tomography
	Positron emission tomography

tion, past physical illnesses or psychiatric disturbances, and any family history suggestive of an inherited systemic, neurologic, or psychiatric disability. A general physical examination may uncover evidence of a systemic or toxic disturbance that may be compromising intellectual function, and a neurological examination will provide evidence for focal, multifocal, or diffuse involvement of the CNS. The mental status assessment may be of great value in determining whether the pattern of intellectual deficits is most consistent with predominantly cortical dysfunction (AD, FTD), subcortical dysfunction (extrapyramidal syndromes, lacunar state), or mixed cortical and subcortical involvement (VaD, DLB, CNS infections, etc.). Mental status examination will also provide evidence for any psychiatric disturbance (depression, mania, schizophrenia) that may be etiologically relevant to the dementia syndrome.

Laboratory assessment of patients with a question of dementia is targeted at identifying reversible causes. With the improvement in primary care screening the percentage of patients with a reversible dementia presenting to specialty memory disorder clinics has declined from 11% in 1972 to 1% in 1994.[67] A core group of laboratory studies should be obtained on all demented patients to evaluate the most common systemic disorders responsible for dementia syndromes (Table 10.11). This required laboratory assessment includes a complete blood count, serum electrolytes, blood glucose, blood urea nitrogen, serum calcium and phosphorous levels, liver function and thyroid function tests, and analyses of serum vitamin B_{12};[68] ancillary determination of erythrocyte sedimentation rate (ESR) and a serologic test for syphilis is recommended if suspicion is high. A lumbar puncture should be performed whenever there is question of an infectious, inflammatory, or demyelinating disorder involving the CNS. Structural imaging is also required in the initial assessment to identify reversible causes as well as focal lesions, significant white matter disease implicating VaD and preventative treatment, or significant medial temporal atrophy consistent with AD.[68] Enlarged ventricles and cortical sulci are evident on the scans of many demented patients, but this atrophy correlates best with the patient's age, is a poor index of intellectual function, and cannot be used as evidence for the existence of a cortical degenerative process such as AD. Repeated imaging studies after a diagnosis is made are unnecessary unless an acute process is suspected. In AD, functional studies such as PET or SPECT typically show abnormalities most marked in the parietal and temporal lobes bilaterally;[9] they are most useful in differentiating early AD from normal aging or FTD. Apolipoprotein E genotyping is not useful in isolation from applying the clinical criteria of AD but may increase specificity of the diagnosis (correctly identifying those without the disease), when patients do not have the ϵ-4 allele, if the diagnosis is in question.[69] Another potential biomarker is the combined assessment of CSF amyloid $\beta_{(1-42)}$ protein ($A\beta_{42}$) and tau concentrations (low $A\beta_{42}$ protein with high tau), which have a positive predictive value of 90% and negative predictive value of 95% based on the *clinical diagnosis* of probable AD.[70] The best current biomarker based on the *pathologic diagnosis* of definite AD, which is superior to the clinical diagnosis of probable AD, is the combination of medial temporal atrophy on structural imaging with parietal impairment on functional imaging.[71]

The EEG may aid in the evaluation of the dementia patient. Focal abnormalities are most consistent with localized disorders such as tumors, abscesses, subdural hematomas, or cerebral infarctions; diffuse slowing occurs in toxic and metabolic disorders and in advanced degenerative diseases; and normal records suggest a dementia syndrome of depression or an early degenerative disorder.

MANAGEMENT OF THE DEMENTIA PATIENT

Management of demented patients involves four separate objectives: (*1*) etiologic diagnosis and disease-specific management, (*2*) management of behaviors produced by the dementia, (*3*) prevention of secondary complications, and (*4*) support of the patient's family. This chapter has emphasized the need to identify the cause of the dementia syndrome in any patient with acquired intellectual impairment. Dementia cannot be appropriately managed when considered as a unitary syndrome of brain failure. Rather, dementia must be recognized as a complex clinical syndrome produced by a multitude of different disease processes. Proper management depends on identifying the etiologic disorder and instituting disease-specific treatment. Thus control of hypertension or elimination of a source of emboli can halt the progression of VaD, shunting may reverse the deficits in hydrocephalic dementia, some infectious dementias can be treated with antibiotics, metabolic dementias respond to treatment of the underlying condition, toxic dementias usually resolve when harmful exposure is eliminated, and the dementia syndrome of depression responds to pharmacotherapy or ECT. The dementia of Parkinson's disease may show limited improvement with levodopa therapy, and the dementia of Wilson's disease may be prevented or reversed by the timely administration of penicillamine (Chapter 18). Cholinesterase inhibitors are useful in AD, VaD, dementia with Lewy bodies and Parkinson's disease with dementia.

For many demented patients, however, no specific treatment is available (e.g., FTD and other degenerative dementias), and it is not uncommon for patients with treatable illnesses to remain partially impaired even after appropriate therapy has been initiated. Treatment of these patients is directed at minimizing the disabling effects of the dementia and controlling unacceptable behaviors. A safe, contained environment is necessary for AD and FTD patients whose restless wandering and hyperoral behavior may lead to their getting lost or ingesting inappropriate items. Adequate hydration and nutrition must be maintained and social and sensory deprivation eliminated. Nocturnal confusion can he minimized by a nightlight, and a soft restraint may be necessary to keep the patient in bed; however, restraints should be used as sparingly as possible. Urinary infections and aspiration pneumonia must be guarded against, and, in the final stages when the patient is bed-bound, decubiti must be avoided by frequent turning and protective cushioning.

When drugs are necessary to gain control of unacceptable behavior atypical antipsychotic, given in small doses, should be utilized. Minor tranquilizers and soporific agents should be avoided, since they tend to increase confusion in the intellectually compromised patient. Depression may accompany many dementia syndromes, particularly extrapyramidal disorders and VaD, and can exacerbate any preexisting neuropsychological impairment. Treatment of the depression may reverse at least a portion of the mental status deficits. Anticoagulants, cerebral stimulants, and vasodilators have all been used in the treatment of dementia patients, but with limited success, and their role in the management of dementia is not established.

Attention must also be directed toward the family of the demented patient. Education is a primary goal: family members should be informed about the cause of the patient's changed behavior, any available treatments, and the patient's prognosis. In addition, provision of social supports such as home health aids and visiting nurses may allow the patient to be maintained in the home for an extended period of time. Legal advice is necessary to aid the family regarding estate disposition and conservatorship. Finally, psychotherapy, either with an individual therapist or through disease-oriented support groups, may provide insight into feelings of loss, grief, and guilt that are common among family members.

REFERENCES

1. Cummings JL, Benson DF, LoVerme SJ. Reversible dementia. JAMA 1980;243:2434–2439.
2. American Psychiatric Association. Diagnostic and Statistical Manual of Mental Disorders, 4th ed. Washington, DC: American Psychiatric Press, 1994.
3. Erkinjuntti T, Østbye T, et al. The effect of different diagnostic criteria on the prevalence of dementia. N Engl J Med 1997;337: 1667–1674.
4. Gifford DR, Cummings JL. Rating dementia screening tests: methodologic standards to rate their performance. Neurology 1999;52:224–227.
5. Cummings JL, Benson DF. Dementia: A Clinical Approach, 2nd ed. Boston: Butterworths, 1992.
6. McKhann G, Drachman D, et al. Clinical diagnosis of Alzheimer's disease: report of the NINCDS-ADRDA Work Group, Department of Health and Human Services Task Force on Alzheimer's Disease. Neurology 1984;34:939–944.
7. Lopez OL, Becker JT, et al. Research evaluation and diagnosis of probable Alzheimer's disease over the last two decades: I. Neurology 2000;55:1854–1862.
8. Victorhoff J, Mack WJ, et al. Multicenter clinicopathological correlation in dementia. Am J Psychiatry 1995;152:1476–1484.
9. Mega MS, Thompson PM, et al. Neuroimaging in dementia. In: Mazziotta JC, Toga AW, Frackowiak R, eds. Brain Mapping: The Disorders. San Diego: Academic Press, 2000:217–293.
10. Cummings JL, Benson DF, et al. Aphasia in dementia of the Alzheimer type. Neurology 1985;35:394–397.
10a. Cummings J. Neuropsychiatric assessment and intervention in Alzheimer's disease. Int Psychogeriatr 1996;8:25–30.
10b. Tekin S, Mega MS, Masterman DL, et al. Orbitofrontal and anterior cingulate cortex: neurofibrillary tangle burden is associated with agitation in Alzheimer's disease. Ann Neurol 2001; 49:355–361.
10c. Cummings JL, Kaufer D. Neuropsychiatric aspects of Alzheimer's disease: the cholinergic hypothesis revisited. Neurology 1996;47:876–883.
11. Braak H, Braak E. Neuropathological staging of Alzheimer-related changes. Acta Neuropathol 1991;82:239–259.
12. Price JL, Davis PB, et al. The distribution of tangles, plaques and related immunohistochemical markers in healthy aging and Alzheimer's disease. Neurobiol Aging 1991;12:295–312.
13. Davies P. Studies on the neurochemistry of central cholinergic systems in Alzheimer's disease. In: Katzman R, Terry RIY, eds. Alzheimer's Disease: Senile Dementia Disorders. New York: Raven Press, 1978:453–468.
14. Whitehouse PJ, Price DL, et al. Alzheimer's disease and senile dementia: loss of neurons in the basal forebrain. Science 1982; 215:1237–1239.
15. Davis KL, Mohs RC, et al. Cholinergic markers in elderly patients with early signs of Alzheimer disease. JAMA 1999;281: 1401–1406.
16. Cummings J, Cole GM. Alzheimer disesae. JAMA 2002;287: 2335–2338.
17. Cummings JL. Treatment of Alzheimer's disease. Clin Cornerstone 2001;3:27–39.
18. Sano M, Ernesto C, et al. A controlled trial of selegiline, alphatocopherol, or both as treatment for Alzheimer's disease. The Alzheimer's disease cooperative study. N Engl J Med 1997;336: 1216–1222.
19. in 't Veld BA, Ruitenberg A, et al. Nonsteroidal antiinflamma-

tory drugs and the risk of Alzheimer's disease. N Engl J Med 2001;345:1515–1521.
20. Kawas C, Resnick S, et al. A prospective study of estrogen replacement therapy and the risk of developing Alzheimer's disease: the Baltimore Longitudinal Study of Aging. Neurology 1997;48:1517–1521.
21. Rogers SL, Farlow MR, et al. A 24-week, double-blind, placebo-controlled trial of Aricept® in patients with Alzheimer's disease. Neurology 1998;50:136–145.
22. Corey-Bloom J, Anand R, et al. A randomized trial evaluating the efficacy and safety of ENA 713 (rivastigmine tartrate), a new acetylcholinesterase inhibitor, in patients with mild to moderately severe Alzheimer's disease. Int J Geriatr Psychopharmacol 1998;1:55–65.
23. Tariot PN, Solomon PR, et al. A 5-month, randomized, placebo-controlled trial of galantamine in AD. Neurology 2000;54: 2269–2276.
24. Mega MS, Masterman DM, et al. The spectrum of behavioral responses with cholinesterase inhibitor therapy in Alzheimer's disease. Arch Neurol 1999;56:1388–1393.
25. Mega MS, Dinov ID, et al. Orbital and dorsolateral frontal perfusion defects associated with behavioral response to cholinesterase inhibitor therapy in Alzheimer's disease. J Neuropsychiatry Clin Neurosci 2000;12:209–218.
26. Mega MS, Cummings JL, et al. Cognitive and metabolic responses to metrifonate therapy in Alzheimer's disease. Neuropsychiatry Neuropsychol Behav Neurol 2001;14:63–68.
27. Neary D, Snowden JS, et al. Frontotemporal lobar degeneration: a consensus on clinical diagnostic criteria. Neurology 1998;51: 1546–1554.
28. McKhann GM, Albert MS, et al. Clinical and pathological diagnosis of frontotemporal dementia. Arch Neurol 2001;58: 1803–1809.
29. Cummings JL, Duchen LW. The Klüver-Bucy syndrome in Pick disease. Neurology 1981;31:1415–1422.
29a. Levy ML, Miller BL, Cummings JL, et al. Alzheimer disease and frontotemporal dementias: behavioral distinctions. Arch Neurol 1996;53:687–690.
30. Murrell JR, Spillantini MG, et al. Tau gene mutation G398R causes a tauopathy with abundant Pick body–like inclusions and axonal deposits. J Neuropathol Exp Neurol 1999;58:1207–1226.
31. Lendon CL, Lynch T, et al. Hereditary dysphasic disinhibition dementia: a frontotemporal dementia linked to 17q21-22. Neurology 1998;50:1546–1555.
32. Bird TD, Wijsman EM, et al. Chromosome 17 and hereditary dementia: linkage studies in three non-Alzheimer families and kindreds with late-onset FAD. Neurology 1997;48:949–954.
33. Kertesz A, Martinez-Lage P, et al. The corticobasal degeneration syndrome overlaps progressive aphasia and frontotemporal dementia. Neurology 2000;55:1368–1375.
34. Roman GC, Tatemichi TK, et al. Vascular dementia: diagnostic criteria for research studies: Report of the NINDS-AIREN International Workshop. Neurology 1993;43:250–260.
35. Erkinjuntti T, Kurz A, et al. Efficacy of galantamine in probable vascular dementia and Alzheimer's disease combined with cerebrovascular disease: a randomized trial. Lancet 2002;359: 1283–1290.
36. Will RG, Ironside JW, et al. A new variant of Creutzfeldt-Jakob disease in the UK. Lancet 1996;347:921–925.
37. Almond J, Pattison J. Human BSE. Nature 1997;389:437–438.
38. Will RG, Alperovitch A, et al. Descriptive epidemiology of Creutzfeldt-Jakob disease in six European countries, 1993–1995. EU Collaborative Study Group for CJD. Ann Neurol 1998;43: 763–767.
39. Windl O, Dempster M, et al. Genetic basis of Creutzfeldt-Jakob disease in the United Kingdom: a systematic analysis of predisposing mutations and allelic variation in the *PRNP* gene. Hum Genet 1996;98:259–264.
40. Prusiner S. Natural and experimental prion diseases of humans and animals. Curr Opin Neurobiol 1992;2:638–647.
41. Collins S, Boyd A, et al. Recent advances in the pre-mortem diagnosis of Creutzfeldt-Jakob disease. J Clin Neurosci 2000;7: 195–202.
42. Parchi P, Giese A, et al. Classification of sporadic Creutzfeldt-Jakob disease based on molecular and phenotypic analysis of 300 subjects. Ann Neurol 1999;46:224–233.
43. Finkenstaedt M, Szudra A, et al. MR imaging of Creutzfeldt-Jakob disease. Radiology 1996;199:793–798.
44. Zeidler M, Stewart GE, et al. New variant Creutzfeldt-Jakob disease: neurological features and diagnostic tests. Lancet 1997;350: 903–907.
45. Zerr I, Schulz-Schaeffer WJ, et al. Current clinical diagnosis in Creutzfeldt-Jakob disease: identification of uncommon variants. Ann Neurol 2000;48:323–329.
46. Kretzschmar HA, Ironside JW, et al. Diagnostic criteria for sporadic Creutzfeldt-Jakob disease. Arch Neurol 1996;53:913–920.
47. Goplen AK, Liestol K, et al. Dementia in AIDS patients in Oslo: the role of HIV encephalitis and CMV encephalitis. Scand J Infect Dis 2001;33:755–758.
48. Glass JD, Fedor H, et al. Immunocytochemical quantitation of human immunodeficiency virus in the brain: correlations with dementia. Ann Neurol 1995;38:755–762.
49. Gartner S. HIV infection and dementia. Science 2000;287:602–604.
50. Posner JB. Neurologic Complications of Cancer. Philadelphia: F.A. Davis Company; 1995.
51. Jarvik LF, Perl M. Overview of physiologic dysfunctions related to psychiatric problems in the elderly. In: Levenson AJ, Hall RCW, eds. Neuropsychiatric Manifestations of Physical Disease in the Elderly. New York: Raven Press, 1981:1–15.
52. Beyth RJ, Shorr RI. Medication use. In: Practice of Geriatrics. Duthie EHJ, Katz PR (eds). Philadelphia: W. B. Saunders Company; 1998; pp. 38–47.
53. Filley CM. The behavioral neurology of white matter. New York: University Press; 2001.
54. Lishman WA. Organic Psychiatry, 3rd ed. Blackwell, London, 1998.
55. George AE, Holodny A, et al. The differential diagnosis of Alzheimer's disease. Cerebral atrophy versus normal pressure hydrocephalus. Neuroimaging Clin N Am 1995;5:19–31.
56. Granado JM, Diaz F, et al. Evaluation of brain SPECT in the diagnosis and prognosis of the normal pressure hydrocephalus syndrome. Acta Neurochir 1991;112:88–91.
57. Holodny AI, Waxman R, et al. MR differential diagnosis of normal-pressure hydrocephalus and Alzheimer disease: significance of perihippocampal fissures. Am J Neuroradiol 1998;19:813–819.
58. Savolainen S, Laakso MP, et al. MR imaging of the hippocampus in normal pressure hydrocephalus: correlations with cortical Alzheimer's disease confirmed by pathologic analysis. Am J Neuroradiol 2000;21:409–414.
59. Boon AJ, Tans JT, et al. Dutch normal-pressure hydrocephalus study: prediction of outcome after shunting by resistance to outflow of cerebrospinal fluid. J Neurosurg 1997;87:687–693.

60. Boon AJ, Tans JT, et al. Dutch Normal-Pressure Hydrocephalus Study: randomized comparison of low- and medium-pressure shunts. J Neurosurg 1998;88:490–495.
61. Rubin EH, Kinscherf DA, et al. The influence of major depression on clinical and psychometric assessment of senile dementia of the Alzheimer type. Am J Psychiatry 1991;148:1164–1171.
62. van Reekum R, Simard M, et al. Late-life depression as a possible predictor of dementia: cross-sectional and short-term follow-up results. Am J Geriatr Psychiatry 1999;7:151–159.
63. Buntinx F, Kester A, et al. Is depression in elderly people followed by dementia? A retrospective cohort study based in general practice. Age Ageing 1996;25:231–233.
64. Steffens DC, Payne ME, et al. Hippocampal volume and incident dementia in geriatric depression. Am J Geriatr Psychiatry 2002;10:62–71.
65. McNeil JK. Neuropsychological characteristics of the dementia syndrome of depression: onset, resolution, and three-year follow-up. Clin Neuropsychol 1999;13:136–146.
66. Alexopoulos GS, Meyers BS, et al. The course of geriatric depression with "reversible dementia": a controlled study. Am J Psychiatry 1993;150:1693–1699.
67. Weytingh MD, Bossuyt PMM, Crevel Hvan. Reversible dementia: more than 10% or less than 1%? A quantitative review. J Neurol 1995;242:466–471.
68. Knopman DS, DeKosky ST, Cummings JL, et al. Practice parameter: Diagnosis of dementia (an evidence-based review). Report of the Quality Standards Subcommittee of the American Academy of Neurology. Neurology 2001;56:1143–1153.
69. Mayeux R, Saunders AM, et al. Utility of the apolipoprotein E genotype in the diagnosis of Alzheimer's disease. N Engl J Med 1999;338:506–511.
70. Andreasen N, Minthon L, et al. Evaluation of CSF-tau and CSF-A42 as diagnostic markers for Alzheimer disease in clinical practice. Arch Neurol 2001;58:373–379.
71. Jobst KA, Barnetson LPD, et al. Accurate prediction of histologically confirmed Alzheimer's disease and the differential diagnosis of dementia: the use of NINCDS-ADRDA and DSM-III-R criteria, SPECT, X-ray CT, and Apo E4 in medial temporal lobe dementias. Int Psychogeriatr 1998;10:271–302.

Chapter 11

Delirium

Delirium, synonymous with the acute confusional state, is a condition of relatively abrupt onset and short duration whose major behavioral characteristic is altered attention.[1–4] The *Diagnostic and Statistical Manual of Mental Disorders, 4th ed.* (DSM-IV) criteria for delirium[5] are outlined in Table 11.1. Other behavioral abnormalities frequently coexist with the clouded, reduced, or shifting attention, including other cognitive disturbances, hallucinations and delusions, sleep cycle abnormalities, and autonomic dysfunction. In most cases the EEG reveals diffuse slowing.[6,7] Standardized assessments for delirium may also be useful in the clinical and research settings and are of benefit in diagnosis as well as following treatment response.[8,9]

Confusion is among the most misused of all behavioral terms. It is often applied vaguely and indiscriminately to patients whose behavior is erratic, incoherent, or psychotic. Confusion is often linked to disorientation, dementia, delirium, or psychosis without specifying the behavior to which it refers, When used in this vague and undefined way, *confusion* loses all meaning as a behavioral descriptor and fails to communicate any relevant information about the state of the patient. In this chapter, *confusion* is used to refer only to states characterized predominantly by attentional alterations. The terms *acute confusional state* and *delirium* are used synonymously; *chronic confusional state* is one of the etiologies of the dementia syndrome and is considered in Chapter 10.

CLINICAL CHARACTERISTICS

Table 11.2 lists the principal behavioral manifestations of delirium. The behavioral change that defines the syndrome and determines many of the other clinical alterations is an alteration of attention.[2] The level of consciousness may be reduced or may fluctuate between drowsiness and hypervigilance, but the patient is unable to maintain attention for any substantial period of time. Even when arousal and level of consciousness per se are not abnormal, subtle attentional deficits can usually be elicited on examination. Digit span (number of digits the patient can repeat forward after presentation by the examiner; normal performance is 7 ± 2) and continuous-performance tests (asking patients to raise their hand each time they hear an *A* within a sequence of spoken letters) are among the best mental status tests for detecting attentional deficits and should be performed on all patients in whom delirium is a consideration.

TABLE 11.1. *DSM-IV Criteria for Delirium*[5]

Disturbance of consciousness
Reduced awareness of environment
Reduced ability to focus attention
Reduced ability to sustain attention
Reduced ability to shift attention
Cognitive change not resulting from dementia
Quickly evolving (hours or days) with fluctuations
There is evidence that the disturbance is caused by the direct physiological consequences of a general medical condition

The attentional impairment is associated with and contributes to alterations in language, memory, constructions, perceptions, and mood.[10] Language abnormalities noted during delirium include abnormal spontaneous speech, anomia, the syndrome of nonaphasic misnaming, and agraphia. Spontaneous speech is often incoherent, rambling, and shifts from topic to topic. Intelligibility is further limited by the hypophonia and slurring of speech that frequently coexist with the language changes. Anomia may be noted in the course of spontaneous speech or on tests of confrontation naming. Paraphasia is rare, and the anomia is usually manifested as a simple failure to recall the correct name.[2,11] In the syndrome of nonaphasic misnaming, naming errors are most pronounced for illness-related items, the anomia propagates (i.e., all misnamed items are named according to a specific theme), the speech style is pedan-

TABLE 11.2. *Principal Clinical Characteristics of Acute Confusional States*

Factor	*Characteristics*
Alertness	Clouded or fluctuating
Attention	Impaired attention, distractible
Language	Incoherent spontaneous speech
	Anomia
	Agraphia
	Variable comprehension defect
	Nonaphasic misnaming
Memory	Disoriented, poor recent memory
Constructions	Visuospatial deficits
Cognition	Incoherent thought
	Concrete thinking
	Dyscalculia
Other behavioral alterations	Perseveration and/or impersistence
	Occupational pantomime
Neuropsychiatric disorders	Hallucinations
	Delusions
	Mood alterations
Motor system abnormalities	Psychomotor retardation and/or hyperactivity
	Action tremor
	Asterixis
	Myoclonus
	Dysarthria
	Tone and reflex abnormalities
Miscellaneous	Sleep disturbances
	Autonomic dysfunction
Electroencephalogram	Diffuse slowing

tic or bombastic, and there is a tendency to make facetious and bizarre responses.[12–14] In almost all reported cases, patients manifesting aphasic misnaming have experienced delirium and have an anomia in association with their "nonaphasic" abnormalities.

Agraphia may be marked in delirium, and writing errors may be out of proportion to other behavioral and linguistic alterations. Writing abnormalities include illegibility and abnormal spatial alignment, abbreviated agrammatical sentences, and spelling errors. The latter tend to involve small grammatical words and ends of words, and omissions, substitutions, and duplication errors are particularly common.[11,15]

Memory and learning abnormalities are virtually always present in delirium, although their severity may vary. The leaning deficits are attributed to abnormalities of registration produced by the attentional limitations. Memory abnormalities are evident in the patient's inability to recall three words after 3 minutes, and the patient may even be unable to repeat the three words immediately after hearing them. The patients are usually disoriented with regard to time and place and may exhibit reduplicative paramnesia (the belief that the hospital has been relocated closer to one's home) (Chapter 12).[8] Disorientation, however, is a nonspecific finding occurring in amnesia, dementia, and a variety of other disorders, and its occurrence does not necessarily imply the presence of delirium.

Constructional tasks, like writing, are also abnormal in delirium. The drawings may be distorted or unrecognizable. Three-dimensional aspects are lost, and lines and angles are omitted.[8]

Executive function abnormalities are also pervasive in delirium. Errors in calculation occur, particularly when the problem requires the patient to "carry" one sum to the next column. There is a lack of coherent thought with loss of normal associations and intrusion of abnormal associative connections. Discrepancies do not trouble the patient, who is unlikely to recognize incongruities when they are pointed out.[2,3,8] Thinking is concrete, and abstraction and categorization skills are limited.

Perseveration is common in all aspects of behavior of patients in delirium. It contributes to the duplication errors in writing and speaking, is apparent in recurrent but irrelevant themes of conversation, and contaminates the motor system examination.[2,3,8] Paradoxically, impersistence may occur in the same patient, with some tasks completed and perseverated while others go unfinished.

Occupational delirium refers to elaborate pantomimes performed by the patient in delirium.[16] Patients act as if they are continuing their usual occupations of sweeping, driving, or working, despite being in a hospital bed. Patients in delirium may remove nonexistent glasses, take nonexistent pills, and pantomime other activities of everyday life.

Neuropsychiatric abnormalities, including hallucinations, delusions, and mood alterations, are common in delirium. In some cases, these disturbances dominate the clinical picture to such an extent that the patient is thought to be suffering from an idiopathic psychiatric disorder and is referred for psychiatric care. The hallucinations of delirium tend to be silent visual images that may be fully formed, such as dogs walking through the room or people standing at the bedside or peering in through the window.[3,7] Tactile (formication) hallucinations are not unusual in delirium, particularly in alcohol and cocaine withdrawal syndromes. Auditory hallucinations (voices, sounds) may occur but are less common than visual hallucinations.

The delusions occurring in delirium may be simple, transient, and loosely held or may be complex, intricately structured, and rigidly endorsed (Chapter 12).[17,18] Occasionally, specific delusional beliefs such as the Capgras syndrome (the belief that significant others have been replaced by identical-appearing impostors) may be the principal manifestation of delirium.[17,19,20] Delusions can motivate combative, self-destructive, or paranoid behavior and can be among the most difficult aspects of delirium to manage. Improvement in the underlying condition usually leads to resolution of the false beliefs, but in some cases small doses of a tranquilizer may have to be used to treat the delusions, limit abnormal behavior related to the delusions, and facilitate management of the etiologic disorder.

A diverse array of mood alterations has also been observed in patients in delirium. The most common is a labile, perplexed, excitable state. Affective alterations ranging from euphoria to depression and fearful paranoia to indifference and apathy also occur.[2,3,8] Mood changes are often congruent with the belief content of delusions in delusional patients.

Patients in delirium may be entirely devoid of motor system abnormalities, but more commonly they manifest alterations in general activity level, tremor, asterixis, myoclonus, or tone and reflex alterations. Changes in activity level may be in the direction of hypo- or hyperactivity.[8] Myxedema is most likely to lead to diminished psychomotor activity, whereas hyperthyroidism and delirium tremens are characterized by increased activity and motor restlessness. Patients in delirium may not have such predictable alterations and many factors such as rate of change in metabolic status, age of the patient, and severity of the encephalopathy influence the behavioral manifestations.

In some cases, patients may have alternating periods of hypo- and hyperactivity.

Tremor is one of the most frequent concomitants of toxic-metabolic disturbances. The typical tremor is a slightly irregular, oscillating, distal movement that is absent at rest and precipitated by action (Chapter 18). It is visible in the outstretched hands and may also be apparent in the neck, lids, tongue, or jaw.

Asterixis is the sudden jerk produced by brief interruptions of the muscular activity involved in sustaining a fixed posture. Although it is common in hepatic encephalopathy and has been called "liver flap," it occurs in many toxic and metabolic derangements and is an etiologically nonspecific sign of encephalopathy.[21] Focal lesions involving the midbrain, thalamus, and parietal cortex may produce unilateral asterixis affecting the contralateral limb and must be excluded before asterixis can be attributed to a metabolic disorder.[21]

Myoclonus is common in uremic and postanoxic encephalopathy but, like asterixis, is a nonspecific finding and occurs in a large number of toxic and metabolic disturbances.[22] Alterations in muscle tone and reflexes are also frequent in metabolic encephalopathies. The tone is symmetrically increased, imparting a plastic resistance to passive movement, and reflex changes include generalized hyperreflexia and extensor plantar responses.

Sleep disturbances comprise an important part of delirium. There may be a disruption of the normal circadian cycle with excessive drowsiness during the day and restless wakefulness at night.[3,7,8] When delirium results from withdrawal from alcohol or other agents that suppress rapid eye movement (REM) sleep, REM rebound may occur and the nocturnal sleep pattern will be dominated by REM sleep.

Autonomic disturbances, including tachycardia, diaphoresis, and pupillary dilatation, are common in delirium, particularly those associated with alcohol and drug withdrawal.[3,8]

The electroencephalogram (EEG) is the most useful laboratory tool for the identification of metabolic encephalopathies. The tracing reveals a generalized symmetrical slowing in the theta or delta range.[7,8]

TABLE 11.3. *Etiologies of Acute Confusional States*

Systemic conditions	Withdrawal syndromes
Cardiac failure	Drugs
Pulmonary disease	Alcohol
Uremia	
Hepatic encephalopathy	Infections
Electrolyte disturbances	Systemic infections with fever
Hypoglycemia	Meningitis
Inflammatory disorders	Encephalitis
Anemia	
Porphyria	Intracranial disorders
Carcinoid syndrome	Head trauma
	Cerebral edema
Endocrinopathies	Epilepsy (ictal and postictal confusion)
Thyroid dysfunction	Hypertensive encephalopathy
Parathyroid dysfunction	Intracranial inflammatory disease
Adrenal dysfunction	Cerebrovascular accident (acute phase)
Pituitary dysfunction	Migraine
	Subdural hematoma
Nutritional deficiencies	
Thiamine (Wernicke's encephalopathy)	Focal cerebral lesions
Niacin	Right parietal lesions
B_{12}	Bilateral occipitotemporal lesions
Folic acid	
	Miscellaneous conditions
Intoxications	Heatstroke
Drugs	Radiation
Iatrogenic	Electrocution
Self-administered	Hypersensitivity reaction
Alcohol	Sleep deprivation
Metals	Postoperative confusion
Industrial agents	
Biocides	Idiopathic psychiatric disorders
	Mania (particularly in elderly individuals)
	Schizophrenia
	Depression

ETIOLOGIES

Delirium reflects an acute interruption of cerebral function and as such can be produced by a large number of metabolic, toxic, and intracranial conditions (Table 11.3). It is particularly likely to occur in patients with preexisting intellectual impairment and in the elderly.[7,23,24] The prevalence of delirium in the hospitalized elderly is 10% to 40%, while 51% of postoperative patients develop delirium and up to 80% of terminally ill patients will become delirious.[8,25]

Intracranial disorders presenting as delirium include head trauma, cerebral edema, hypertensive encephalopathy, intracranial inflammatory diseases, acute cerebrovascular accidents, meningitis, encephalitis, epilepsy, and migraine. Confusional behavior in epilepsy occurs in the ictal and postictal stages and may last for hours or (rarely) days in complex partial status epilepticus or petit mal status epilepticus.[3,26] Acute confusional migraine occurs almost exclusively in children and adolescents and is characterized by confused behavior occurring as a prodrome to the migraine headache.[27]

The metabolic and toxic disorders producing delirium include systemic disturbances, endocrinopathies, nutritional deficiencies, drug intoxications, withdrawal syndromes, and infections. Among the most common metabolic conditions producing delirium are infections, dehydration and electrolyte abnormalities, cardiopulmonary failure, uremia, and hepatic encephalopathy.[3,7]

Drug-induced delirium is also common. Encephalopathies are particularly likely to be produced by anticholinergic agents but may occur with virtually any drug reaching high serum concentrations. Altered drug metabolism and disposition in the elderly render them vulnerable to developing an iatrogenic delirium even when conventional dosages of medications are prescribed.[28]

Postoperative confusion deserves special consideration because of its frequency and because its etiology is often perplexing. Acute confusional states presenting immediately after surgery are usually due to anoxia or persistent medication effects, particularly the effects of anticholinergic medication.[3] When delirium appears later in the postoperative course, it is likely to be a product of multiple factors, including metabolic abnormalities, sleep deprivation, pain, and sensory isolation.[3]

In addition to the acute disruption of cellular function produced by metabolic and toxic encephalopathies and the disturbances of arousal resulting from epilepsy, migraine, and sleep deprivation, delirium can also be produced by specific focal central nervous system (CNS) lesions and, rarely, by idiopathic psychiatric disorders.

The common manifestation of all these conditions is a disturbance of attention, the hallmark of delirium. Two focal lesions that have been associated with acute confusional behavior are right parietal lesions and bilateral medial occipitotemporal lesions.[29–31]

Idiopathic psychiatric disorders usually present with distinctive behavioral alterations indicative of schizophrenia, mania, or depression. Rarely, however, such patients appear to have delirium as a major feature of their psychiatric disorder. The most frequent circumstance in which this is noted is in the course of a manic or depressive episode in an elderly individual.[3] The patients manifest a significant attentional impairment in addition to the typical symptoms of mania or depression, and the attentional deficits resolve with successful treatment of the psychiatric disorder. Other medical and drug-induced causes of delirium must be carefully excluded in these patients, but the evaluation will frequently be unrevealing, and delirium will be determined to be a product of the psychiatric illness.

DIFFERENTIAL DIAGNOSIS

The differential diagnosis of delirium includes dementia, amnesia, catatonic stupor, and hysterical unresponsiveness.[3,7,28] The only definitive criterion distinguishing delirium from dementia is duration: delirium persists for hours, days, or rarely weeks, whereas dementia usually implies persistence of intellectual deficits for months or years. Other features that may facilitate differentiation between the two syndromes include greater attentional impairment in delirium, along with more frequent delusions and hallucinations. The EEG is also more abnormal in delirium than in most dementias (Chapter 10).

Amnesia enters the differential diagnosis of delirium because disorientation is a prominent feature of both.[8] *Amnesia*, however, refers to an impairment of new leaning with intact attention and intellect (Chapter 7). *Delirium*, by contrast, has prominent attentional deficits along with impairment of language, memory, cognition, visuospatial skills, and personality. Disorientation in delirium is a product of inattention and is one among a host of deficits. The amnesic patient is not "confused" when "confusion" implies an attentional disturbance, and the disorientation accompanying amnesia is a product of the failure to retain spatial and temporal information.

Catatonia, including catatonic stupor, can occur in affective disorders and schizophrenia as well as in a variety of neurological and metabolic disturbances (Chapter 18).[32] When stupor is a manifestation of an idiopathic psychiatric disturbance, it usually persists for less than 1 week and is distinguished from other causes of stupor by normal reflex function and a normal EEG. Similarly, hysterical unresponsiveness is characterized by normal reflex responses and a normal EEG.

MANAGEMENT

The principal effort in the management of the patient in delirium is directed at identifying and treating the underlying disease process. Once the presence of delirium is recognized by careful mental status testing and

TABLE 11.4. *Pharmacologic Approaches in the Treatment of Delirium with Agitation*

Agent	*Dose*
Antipsychotics	
Droperidol	5 mg (intramuscular)
Haloperidol	0.25– 2 mg every 4 hr
Benzodiazepines	
Lorazepam	0.5–1 mg
Cholinergics	
Physostigmine	0.16–3 mg/hr (intravenous)
Donepezil	5 mg p.o qd

identification of the attentional deficits, the clinician must immediately search for the etiology of the encephalopathy. In most cases a reversible metabolic or toxic condition will be discovered. A careful history may suggest the presence of drug intoxication, a medical illness resulting from exposure to industrial toxins, or alcoholism. Hypoxia, uremia, hepatic encephalopathy, electrolyte disorders, and endocrine disturbances can be identified by the appropriate laboratory studies. In addition to these measures, there are general management strategies that apply to most patients in delirium, including maintaining proper nutrition, hydration, and electrolyte balance; ensuring adequate sleep; providing an appropriate amount of sensory and social stimulation; and sedating patients whose agitation prevents evaluation and management of the underlying condition.[33] Drugs should be used sparingly since they may exaggerate delirium and are avoided unless agitation becomes severe. Table 11.4 lists those agents commonly used for the treatment of delirium with agitation.

When drug management is necessary to control agitation, small doses of a major tranquilizer should be utilized.[33,34] Droperidol has been shown to be superior over haloperidol in delirium[35] and is preferred over benzodiazepines (except in alcohol or benzodiazepine withdrawal). Cholinergic agents may also be useful; the most refractory agitation is best treated with intubation, morphine, and paralysis. Restoration of the metabolic milieu of the CNS lags behind normalization of the peripheral blood and serum values, and delirium may persist for several days or more after appropriate treatment of the etiologic condition. In the elderly, delirium may endure for several weeks after improvement of the underlying disorder. Death is not an uncommon outcome of delirium because of the seriousness of many of the etiologic conditions.

REFERENCES

1. Berrios GE. Delirium and confusion in the 19th century: a conceptual history. Br J Psychiatry 1981;139:439–449.
2. Chedru F, Geschwind N. Disorders of higher cortical functions in acute confusional states. Cortex 1972;10:395–411.
3. Lipowski ZJ. Delirium: Acute Confusoinal States. New York: Oxford University Press, 1990.
4. Liptzin B, Levkoff SE, Cleary PD, et al. An empirical study of diagnostic criteria for delirium. Am J Psychiatry 1991;148:454–457.
5. American Psychiatric Association. Diagnostic and Statistical Manual of Mental Disorders, 4th ed. Washington, DC: American Psychiatric Press, 1994.
6. Liptzin B, Levkoff SE, Gottlieb GL, et al. Delirium. J Neuropsychiatry Clin Neurosci 1993;5:154–160.
7. Rummans T, Evans JM, Krahn LE, et al. Delirium in elderly patients: evaluation and management. Mayo Clin Proc 1995;70: 989–998.
8. Trzepacz PT. The neuropathogenesis of delirium. Psychosomatics 1994;35:374–391.
9. Bettin KM, Gabe JM, et al. Measuring delirium severity in older general hospital inpatients without dementia. Am J Geriatr Psychiatry 1998;6:296–307.
10. Trzepacz PT, Mittal D, et al. Validation of the Delirium Rating Scale-Revised-98: Comparison with the Delirium Rating Scale and the Cognitive Test for Delirium. J Neuropsychiatry Clin Neurosci 2001;13:229–242.
11. Benson DF. Aphasia, Alexia, and Agraphia. New York: Churchill Livingston, 1979.
12. Cummings JL, Hebben NA, et al. Nonaphasic misnaming and other neurobehavioral features of an unusual toxic encephalopathy: case study. Cortex 1980;16:315–323.
13. Geschwind N. Non-]aphasic disorders of speech. lnt J Neurol 1964;4:207–214.
14. Weinstein EA, Kahn RL. Nonaphasic misnaming (paraphasia) in organic brain disease. Arch Neurol Psychiatry 1952;67:72–79.
15. Chedru F, Geschwind N. Writing disturbances in acute confusional states. Neuropsychologia 1972;10:343–353.
16. Wolff HG, Curran D. Nature of delirium and allied states. Arch Neurol Psychiatry 1935;33:1175–1215.
17. Cummings JL. Organic delusions: phenomenology, anatomic correlations, and review. Br J Psychiatry 1985;46:184–197.
18. Nash JL. Delusions. Philadelphia: J.B. Lippincott, 1983.
19. Hay GG, Jolley DJ, Jones RG. A case of the Capgras syndrome in association with pseudo-hypoparathyroidism. Psychiatr Scand 1974;50:73–77.
20. Madakasira S, Hall TB. Capgras syndrome in a patient with myxedema. Am J Psychiatry 1981;138:1506–1508.
21. Weiner WJ, Lang AE. Mov Disord: A comprehensive survey. Mount Kisco, New York: Future Publishing Company, 1989.
22. Lang AE. Movement disorders: approach, definitions, and differential diagnosis. In: Drug-induced Movement Disorders. Lang

AE, Weiner WJ (eds). Mount Kisco, New York: Future Publishing Co., 1992; pp. 1–20.
23. O'Keeffe S, Lavan J. The prognostic significance of delirium in older hospital patients. J Am Geriatr Soc 1997;45:174–178.
24. Pompeii P, Foreman M, Rudberg MA, et al. Delirium in hospitalized older persons: outcomes and predictors. J Am Geriatr Soc 1994;42:809–815.
25. Tune LE. Post-operative delirium. Psychogeriatrics 1991;3:325–332.
26. Ellis JM, Lee SI. Acute prolonged confusion in later life as an ictal state. Epilepsia 1978;19:119–128.
27. Ehyai A, Fenichel GM. The natural history of acute confusional migraine. Arch Neurol 1978;35:368–369.
28. Lindesay J, Macdonald A, Starke I. Delirium in the elderly. Oxford: Oxford University Press, 1990.
29. Horenstein S, Chamberlain W, Conomy J. Infarction of the fusiform and calcarine regions: agitated delirium and hemianopsia. Trans Am Neurol Assoc 1967;92:85–87.
30. Medina JL, Chokroverty S, Rubino FA. Syndrome of agitated delirium and visual impairment: a manifestation of medial temporal-occipital infarction. J Neurol Neurosurg Psychiatry 1977;40:861–864.
31. Mesulam MM, Waxman SG, et al. Acute confusional states with right middle cerebral artery infarctions. J Neurol Neurosurg Psychiatry 1976;39:84–89.
32. Carroll BT, Anfinson Tj, Kennedy JC, et al. Catatonic disorder due to general medical conditions. J Neuropsychiatry Clin Neurosci 1994;6:122–133.
33. American Psychiatric Association. Practice Guideline for the Treatment of Patients with Delirium. Am J Psychiatry 1999;156: (Supplement): 1–20.
34. Breitbart W, Marotta R, Platt MM, et al. A double-blind trial of haloperidol, chlorpromazine, and lorazepam in the treatment of delirium in hospitalized AIDS patients. Am J Psychiatry 1996;153:231–237.
35. Thomas H, Schwartz E, Petrilli R. Droperidol versus haloperidol for chemical restraint of agitated and combative patients. Ann Emerg Med 1992;21:407–413.
36. Francis J, Martin D, Kapoor WN. A prospective study of delirium in hospitalized elderly. JAMA 1990;263:1097–1101.

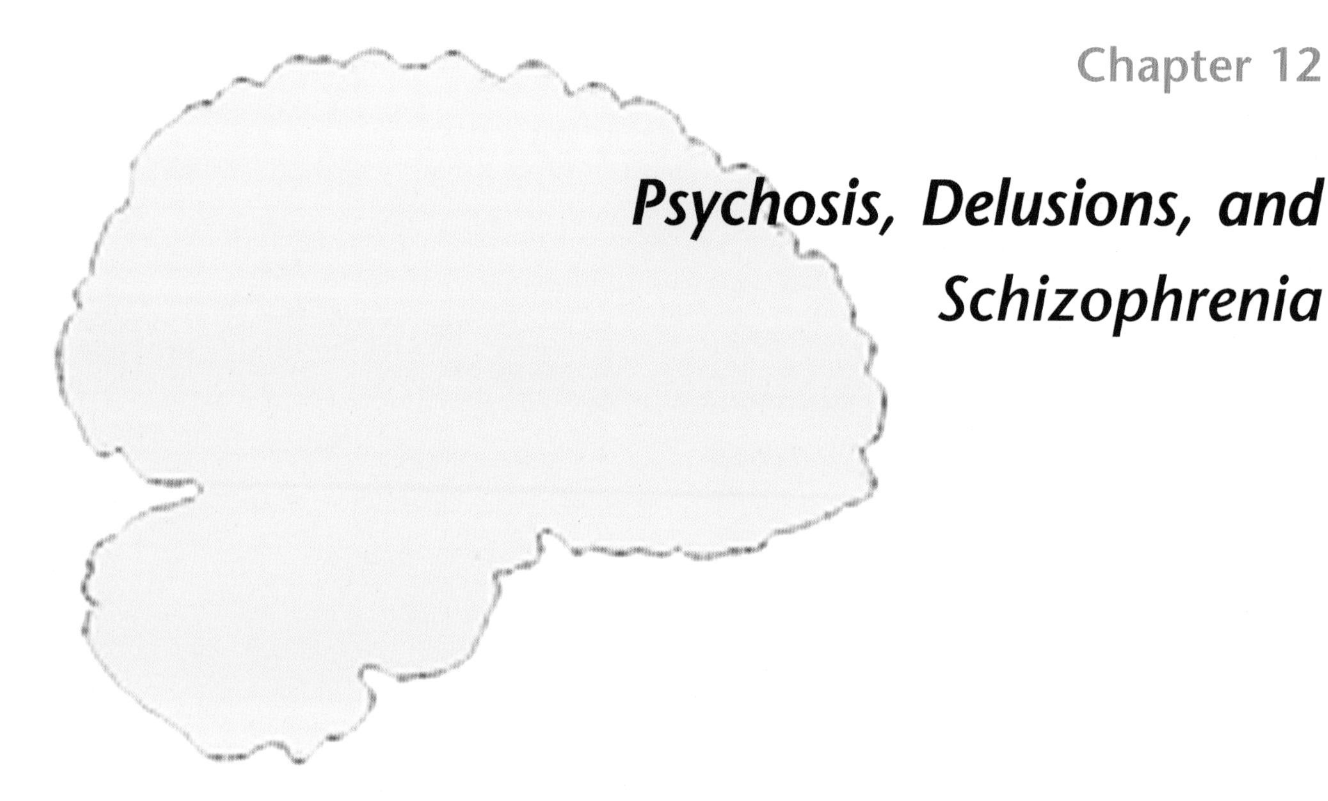

Chapter 12

Psychosis, Delusions, and Schizophrenia

Psychosis is defined as a loss of reality testing such that affected individuals cannot evaluate the accuracy of their perceptions or thoughts and draw incorrect inferences about external reality.[1] *Delusions* are specific false beliefs that are firmly held despite evidence to the contrary and that are not endorsed by members of patients' cultures or subcultures.[1,2] Delusions must be distinguished from hallucinations and from confabulation. *Hallucinations* are false perceptions and have a delusional aspect if patients endorse them as reality. *Confabulations* are spontaneous untruths occurring in patients with amnesia and frequently changing with each interview (Chapter 7).

This chapter reviews secondary psychoses and symptomatic delusions, describes the neurologic aspects of schizophrenia, and briefly reviews the pharmacological treatment of psychotic disorders.

SECONDARY PSYCHOSES

A wide variety of neurological, toxic, and metabolic disorders can have secondary psychosis as their presenting manifestation or as one aspect that emerges during the course of the disorder. The principal etiologies of secondary psychoses are presented in Tables 12.1, 12.2, and 12.3.

Neurological Diseases with Secondary Psychoses

Table 12.1 lists the neurological diseases that can produce secondary psychoses. Psychoses are particularly common among degenerative, traumatic, infectious, neoplastic, and vascular disorders affecting the limbic system.[3,4]

Extrapyramidal Disorders

Psychoses are not uncommon in von Economo's encephalitis and postencephalitic Parkinson's disease. Delusions were the principal neuropsychiatric manifestation of the disease in up to 25% of Fairweather's large sample,[5] and similar frequencies of schizophrenia-like disorders and paranoia have been noted by others.[6–8] Among patients with idiopathic Parkinson's disease (paralysis agitans), psychosis is much less common. Individuals with paralysis agitans and schizophrenia-like disorders have rarely been reported and most psychoses unrelated to depression are induced by the drugs required to treat the Parkinson's disease.[9–13]

Choreiform disorders may also be associated with psychosis, and psychosis is more frequent among hyperkinetic diseases (choreic conditions) than among most parkinsonian disorders. Psychosis may occur in

TABLE 12.1. *Neurological Causes of Secondary Psychoses*

Disorder Type	*Specific Disorder*
Extrapyramidal disturbances	Postencephalitic Parkinson's disease
	Huntington's disease
	Sydenham's chorea
	Wilson's disease
	Idiopathic basal ganglia calcification
	Spinocerebellar degeneration
Central nervous system infections	Herpes
	Rabies
	Mumps
	Asian influenza
	HIV encephalopathy
	Subacute sclerosing panencephalitis
	Creutzfeldt-Jakob disease
	Malaria
	Syphilis
	Cysticercosis
	Trypanosomiasis
	Schistosomiasis
Diseases affecting myelin	Multiple sclerosis
	Metachromatic leukodystrophy
	Adrenoleukodystrophy
	Marchiafava-Bignami disease
	Cerebrotendinous xanthomatosis
	Ischemic demyelination
Primary degenerative dementias	Alzheimer's disease
	Parkinson's disease with dementia
	Dementia with Lewy bodies
Other acquired CNS disturbances	Epilepsy
	Primary generalized seizures
	Complex-partial seizures
	Stroke
	Vascular dementia
	Post-traumatic encephalopathy
	Postanoxic encephalopathy
	Neoplasms
	Hydrocephalus
Miscellaneous disorders	Leber's hereditary aptic atrophy
	Cerebral lipidoses
	Niemann-Pick disease
	Narcolepsy
	Agenesis of corpus callosum
	Intracranial cysts
	GM2 gangliosidosis
	Neuronal ceroid lipofuscinosis
	Leigh's syndrome
	Mitochondrial encephalopathy

TABLE 12.2. *Metabolic Disorders Associated with Psychosis*

Disorder Type	*Specific Disorders*
Systemic illnesses	Uremia and dialysis dementia
	Hepatic encephalopathy
	Pancreatic encephalopathy
	Anoxia (cardiopulmonary insufficiency)
	Hypoxia
	Subacute bacterial endocarditis
	Hyponatremia
	Hypercalcemia
	Hypoglycemia
	Poryphria
	Postoperative and intensive care unit psychoses
Endocrine disturbances	Addison's disease (adrenal insufficiency
	Cushing's disease (hyperadrenalism)
	Hypothyroidism
	Hyperthyroidism
	Hypoparathyroidism
	Hyperparathyroidism
	Panhypopituitarism
	Recurrent menstrual psychosis
	Postpartum psychosis
Deficiency states	Thiamine (Wernicke-Korsakoff syndrome)
	Vitamin B_{12}
	Folate
	Niacin
Inflammatory disorders	Systemic lupus erythematosus
	Temporal arteritis
	Sarcoidosis
	Antiphospholipid antibody syndrome

Huntington's diseases.[14] Among 186 patients assessed by Folstein,[15] a schizophrenia-like disorder was present in 6%; most studies report psychosis in 4% to 12% of patients.[16] Among patients with Sydenham's chorea, which is chorea associated with rheumatic fever in childhood, psychosis may occur concomitantly with the movement disorder.

Psychoses also occur in patients with Wilson's disease, idiopathic basal ganglia calcification, and the spinocerebellar degenerations. Patients with Wilson's disease may manifest schizophrenia-like illnesses with paranoid delusions and auditory hallucinations indistinguishable from those occurring in idiopathic schizophrenia.[17,18] Idiopathic basal ganglia calcification

TABLE 12.3. *Drugs and Toxins Reported to Produce Psychosis*

Anticholinergic agents
Dopaminergic drugs
Endocrine agents
Anticonvulsants
Antidepressants
Sedative-hypnotics/anxiolytics
Hallucinogens
Psychostimulants
Appetite suppressants
Antihypertensive agents
Cardiac agents
Respiratory and pulmonary drugs
Gastrointestinal agents (also see anticholinergic agents)
Antibiotics
Antineoplastic agents
Analgesics
Nonsteroidal anti-inflammatory agents
Toxins

commonly presents in the third or fourth decade of life with a schizophrenia-like illness.[19,20] Psychosis is less common in patients with spinocerebellar degenerations but has been observed in association with Friedreich's ataxia and olivopontocerebellar degenerations.[21–23]

Central Nervous System Infections

A variety of central nervous system (CNS) infections produce psychosis as a prominent clinical manifestation. Herpes simplex encephalitis preferentially involves the medial temporal lobes and inferior frontal lobes, cerebral cortical areas that are included in the limbic system, and frequently presents with psychosis, delusions, and/or auditory hallucinations as the earliest expressions of the infection.[24,25] Nonherpetic viral encephalitides producing psychosis include rabies, mumps, measles, cysticerosis, infectious mononucleosis, and encephalitis associated with Asian influenza as well as a number of cases of encephalitis in which the causal organism was not identified.[26–31]

Two slow virus diseases, human immunodeficiency virus (HIV) encephalopathy and subacute sclerosing panencephalitis, may have psychosis as a premonitory manifestation prior to the emergence of the progressive dementia.[32–34]

Nonviral causes of infectious psychoses have included cerebral malaria, CNS syphilis, trypanosomiasis, and schistosomiasis.[35–39] Creutzfeldt-Jakob disease, a prion disorder, may also produce a delusional disorder.[40]

Diseases Affecting Myelin

Multiple sclerosis is more likely to be associated with depression than with psychosis, but in some cases progressive demyelination has resulted in a psychotic disorder.[41,42] Marchiafava-Bignami disease, an acquired demyelinating disorder affecting the corpus callosum and white matter of the frontal lobes, is a rare condition occurring primarily in chronic alcoholics. It has been associated with a symptomatic psychosis.[43] Inherited diseases of myelin-producing psychoses include metachromatic leukodystrophy, adrenoleukodystrophy, and cerebrotendinous xanthomatosis.[44–48]

Late-onset psychosis has a variety of causes, but magnetic resonance imaging (MRI) demonstrates that older individuals with new-onset psychotic disorders and no specific identifiable cause have an increased frequency of ischemic lesions in the cerebral white matter compared to nonpsychotic elderly individuals.[49,50] These observations suggest that ischemic demyelination may contribute to some cases of late-onset psychosis.

Primary Degenerative Dementia

Primary degenerative diseases of the cerebral cortex such as Alzheimer's disease, dementia with Lewy bodies and frontotemporal dementia are frequently accompanied by psychoses.[51] Between 40% and 70% of patients with Alzheimer's disease manifest delusions, making it the most common cause of psychosis after schizophrenia. Cummings[3] noted that the delusions in Alzheimer's disease are simple, loosely held, and often transient. Typically, Alzheimer's patients with delusions believe that family members or others are trying to steal their money or home or that uninvited and unwelcome strangers are coming into the house. Pick's disease and other frontotemporal dementias may produce delusions, but psychosis is less common than in Alzheimer's diseases.

Dementia with Lewy bodies (Chapters 10, 18) is frequently accompanied by delusions. Patients have prominent visual hallucinations and endorse the hallucinatory experiences as veridical.[52,53]

Other Acquired Central Nervous System Disturbances

Psychosis is associated with both primary generalized epilepsy and with epilepsy manifested by complex partial seizures (Chapter 21). Psychosis has its highest association with temporal lobe epilepsy and is more common with seizures originating from left-sided epileptic foci.[54–56]

Cerebrovascular disorders such as infarctions, aneurysms, and arteriovenous malformations have produced psychoses. Psychotic manifestations are particularly common with bilateral lesions affecting limbic structures in the subcortical regions, temporal lobes, or temporoparietal areas.[3,57] Left hemispheric lesions manifest themselves primarily as ideas of reference and persecution,[58,59] whereas right-sided lesions tend to produce delusions with visual hallucinations.[60–62]

Traumatic brain injuries are also followed by an increased incidence of schizophrenia-like psychoses.[63,64] The traumatic lesions are concentrated in the temporal lobes and in most series the psychoses have been more commonly associated with left-sided damage.

Cerebral neoplasms manifesting psychosis share the anatomic characteristics noted for epileptic, traumatic, and vascular lesions. Tumor-related psychoses occur with masses affecting the brainstem or temporal lobes.[65,66]

Other acquired conditions associated with symptomatic psychoses include hydrocephalus and postanoxic encephalopathy.[4,67–69] An unusually frequent association between aqueductal stenosis, hydrocephalus, and a schizophrenia-like illness has been noted.[70–72]

Miscellaneous Secondary Psychoses

A number of other neurological conditions with secondary psychoses have been described, including Leber's hereditary optic atrophy, Niemann-Pick disease, agenesis of the corpus callosum, lipoid proteinosis, intracranial cysts, and inherited metabolic disorder.[4,73–77] A small number of cases of narcolepsy with psychosis have been reported. Most often, psychosis in a narcoleptic patient will be a product of treatment with psychostimulants, but in a few instances psychosis not attributable to medications have been described.[78,79]

Systemic Disorders with Secondary Psychoses

A large number of endogenous metabolic disturbances and systemic illnesses have produced secondary psychoses as part of their clinical presentation. In most cases, the psychosis is accompanied by evidence of an acute metabolic encephalopathy (fluctuating arousal, impaired attention, etc.), but the psychosis is occasionally the sole behavioral expression of the disorder. Table 12.2 outlines the metabolic causes of secondary psychoses.

Kidney diseases leading to uremia may produce psychotic states, and psychosis is also common among patients with dialysis dementia.[80,81] Similarly, hepatic encephalopathy and pancreatic failure may each produce secondary psychosis as one manifestation of their clinical symptomatology.[82,83] Cerebral anoxia or hypoxia associated with pulmonary or cardiac insufficiency can cause psychosis or exacerbate an existing idiopathic psychosis. In a study of patients with subacute bacterial endocarditis, 60% had behavioral abnormalities, including some with paranoid psychoses. Serum abnormalities, including hyponatremia, hypoglycemia, and hypercalcemia, can also cause symptomatic cerebral dysfunction with psychosis.[84,85] Postoperative and intensive care unit psychoses appear to be determined by multiple factors, including sensory deprivation, sleep deprivation, electrolyte imbalance, drug administration, and organ failure.[86–88]

Acute intermittent porphyria produces encephalopathy with psychosis along with abdominal pain and peripheral neuropathy.[89,90] Magnetic resonance imaging demonstrates areas of increased signal intensity during acute attacks; the lesions resolve following the episode.[91]

Endocrine disturbances are commonly associated with secondary psychoses. Adrenal insufficiency and excess, hypo- and hyperthyroidism, hypo- and hyperparathyroidism, and panhypopituitarism have all produced secondary psychoses.[92–96] Recurrent psychoses have been associated with menstrual disorders, and endocrinologic factors are suspected to contribute to postpartum psychoses.[97–99]

Deficiency states associated with psychosis include thiamine deficiency, producing Korsakoff's psychosis, vitamin B_{12} deficiency, folate deficiency, and niacin deficiency.[100–102]

Inflammatory diseases causing secondary psychoses include systemic lupus erythematosus, antiphospholipid antibody syndrome, temporal arteritis, and sarcoidosis.[103–107] Antiribosomal P-protein antibodies are increased in the cerebrospinal fluid (CSF) of patients with systemic lupus erythematosus and psychosis.[108]

Toxic Encephalopathies

Toxic encephalopathies induced by drugs and metals are associated with secondary psychoses in some cases (Table 12.3). Essentially every drug can produce psychosis when administered in sufficiently large doses; in most cases the delusions are part of a delirious syndrome. The does required to cause psychosis differs substantially among individuals. Some agents, however, are reported to produce delusions without delirium or are associated with psychosis with an unusually high frequency. Drugs with a propensity to cause psychosis include anticholinergic agents, dopaminergic drugs, antituberculosis agents, anticonvulsants, endocrine agents, antimalarials, appetite suppressants, antidepressants, antihypertensive agents, hallucinogens, and psychostimulants.[11,13,109–116]

Abrupt withdrawal of drugs can precipitate a toxic psychosis with delusions. The principal agents capable of producing withdrawal syndromes when abruptly discontinued include alcohol, sedative-hypnotics, and psychostimulants such as amphetamines, cocaine, and even sympathomimetics.[117] Psychosis has also been associated with withdrawal of propranolol or baclofen in occasional patients.

Metal intoxications associated with secondary psychosis include the encephalopathies induced by excess mercury, arsenic, manganese, thallium, and bismuth.[113,114]

CONTENT-SPECIFIC DELUSIONS

Most of the secondary psychoses are manifested by paranoid delusions, ideas of reference, and persecutory thoughts. In some cases, however, delusions have a specific theme or are confined to a single topic. Table 12.4 presents the principal content-specific delusions. These specific types of delusion have been associated with both idiopathic psychoses (mania, depression, schizophrenia, delusional disorder) and neurological and toxic-metabolic conditions. None of these delusions is disease-specific; any of them may be the sole manifestation of psychosis or may occur in conjunction with other delusional beliefs.

Schneiderian first-rank symptoms are specific psychotic symptoms that occur primarily in schizophrenia but have also been noted in a small number of patients with manic, depressive, and neurological psychoses.[118,119] First-rank symptoms include aberrations of thought such as thought insertion, thought withdrawal, and thought broadcasting, as well as certain types of auditory hallucination, delusional perception, and passivity experiences involving the feeling that bodily sensations or one's emotions, impulses, or actions are imposed from the outside (Table 12.4). Neurological and toxic-metabolic disorders producing first-rank symptoms include idiopathic basal ganglia calcification, post-traumatic encephalopathy, temporal lobe neoplasms, postencephalitic parkinsonism, temporal lobe epilepsy, viral encephalitis, cerebrovascular disease, hydrocephalus, hypothyroidism, Addison's disease, isosafrol, LSD, amphetamines, diethylproprion, clonazepam, podophyllin, Actifed, Inderal, and metrizamide encephalopathy.[19,27,55,56,64,71,105]

The *Capgras syndrome* is a specific delusional belief in which the patient is convinced that some important person (usually the spouse) has been replaced by an identical-appearing impostor. This syndrome occurs most commonly in schizophrenia but has also been described in manic-depressive psychosis, paraphrenia, and postpartum psychosis.[120–122] Neurological disorders producing the Capgras syndrome include intracerebral hemorrhage, HIV encephalopathy, post-traumatic encephalopathy, temporal lobe epilepsy, postencephalitic parkinsonism, varicella encephalitis, migraine, neurocysticercosis, and tuberous sclerosis.[123–126] Among reported cases with structural lesions, there is a preponderance of right hemispheric lesions, suggesting that right-sided dysfunction may predispose to this particular delusional misinterpretation.[123,127,128] Metabolic disorders reported to produce the Capgras syndrome include vitamin B_{12} deficiency, hepatic encephalopathy, pneumonia, malnutrition, metrizamide myelography, diabetic encephalopathy, hypothyroidism, and pseudohypoparathyroidism.[129–133] Capgras syndrome is sometimes associated with violence directed at the presumed impostor and the clinician should evaluate the threat of aggression in patients with this disorder.[134]

Two syndromes that resemble the Capgras syndrome are the Fregoli syndrome and the intermetamorphosis syndrome. The *Fregoli syndrome* refers to a delusion in which the patient believes that a persecutor is able to take on the appearance of others in the patient's environment, changing faces like an actor.[135,136] In the *intermetamorphosis syndrome* one believes that those in one's environment begin to look like a persecutor or other object of the delusion. Both of these syndromes have been noted in schizophrenia and in behavioral syndromes associated with epilepsy.[137]

Heutoscopy (the syndrome of doubles, the doppelgänger) is the delusion that one has an exact double. The double may or may not be visible. It occurs primarily in schizophrenia but is also observed in migraine, toxic psychoses, encephalitis, post-traumatic encephalopathy, epilepsy, and intracranial hemorrhage.[138] The syndrome of doubles must be distinguished from autoscopy, where one has an hallucination of oneself but recognizes that the experience is hallucinatory.

De Clerambault syndrome, or *erotomania*, is a delusional belief, most common in women, that an older, more influential male is in love with her despite outward evidence to the contrary. The patient may pursue

TABLE 12.4. *Content-Specific Delusions*

Delusion Name	*Delusion Content*
Schneiderian first-rank symptoms	Thought insertion
	Thought withdrawal
	Thought broadcasting
	Hearing one's thought spoken aloud
	Hearing voices arguing about or discussing oneself
	Hearing voices commenting on one's actions
	Delusional interpretation of one's perceptions
	Imposition of sensations
	Imposition of emotions
	Imposition of impulses
	Imposition of actions
Capgras syndrome	Someone (usually a family member) has been replaced by an identical-appearing impostor (in dementias this often takes a more elementary form of denying that someone is who they claim to be without necessarily claiming replacement by an impostor)
Fregoli syndrome	A persecutor takes on the form of others in the environment (i.e., a persecutor takes on the form of a patient's nurse or physician)
Intermetamorphosis syndrome	Individuals in the environment take on the appearance of others significant to the patient (i.e., one's nurse begins to look like one's mother or sister)
Double or doppelgänger	One has a twin or second self
Othello syndrome	One's mate is unfaithful
Parasitosis, infestation, or Ekbom syndrome	One is infested by insects or vermin
Lycanthropy	One is periodically transformed into a wolf or other animal
de Clerambault's syndrome	One is loved by an individual of higher socioeconomic status
Incubus syndrome	One is visited by a demon lover or phantom lover
Phantom boarder	Unwelcome guests are living in the home
Picture sign	Individuals seen on television or in magazines are present in the home
Koro	One's penis is shrinking and retracting into the abdomen
Dorian Gray syndrome	One is not aging
Cotard's syndrome	One is dead

her victim relentlessly, trying to establish contact and allow him to demonstrate his love. The syndrome has occurred in schizophrenia and in toxic psychoses, epilepsy, Alzheimer's disease, post-traumatic encephalopathy, and CNS tumors.[139–141]

Incubus syndrome is one of the most common monosymptomatic delusions. Also known as the *Othello syndrome*, delusional jealousy is manifested by an unjustified conviction of the spouse's infidelity. Delusional jealousy occurs in idiopathic psychoses as well as in Huntington's disease, encephalitis, CNS neoplasms, Alzheimer's disease, multiple sclerosis, epilepsy, Parkinson's disease, general paresis, and drug intoxication.[26,142]

Delusions of infestation, *Ekbom syndrome* (acrophobia, parasitophobia), or the delusional belief that one's body is inhabited by worms or insects, habr been observed in patients with vitamin B_{12} deficiency, iron deficiency, and toxic psychoses.[143,144] In *lycanthropy*, or werewolfism, one believes that one have been turned into a wolf. The syndrome has been produced by LSD use and an undiagnosed primary dementing illness.[145] *Phantom boarder*, *picture sign*, and the *Dorian Gray delusion* are all delusional syndromes that have been observed in patients with Alzheimer's disease and occasionally occur with other neurological disorders.[146,147] *Koro* is the unusual delusion that one's penis is shrinking and retracting into the abdomen. While usually occurring as a culture-bound manifestation of anxiety and depersonalization among Asian individuals.[148] it has been observed among non-Asians with corpus callosum tumors, frontotemporal tumors, or right brain strokes.[149,150] *Cotard's syndrome* is the delusional belief that one is dead. It has been observed in patients with multifocal post-traumatic contusions and with frontal lobe atrophy.[151,152]

A few delusions that have more closely determined associations with specific CNS lesions have been mentioned in previous chapters. Denial of illness must be regarded as a delusional belief in one's well-being; thus anosognosia syndromes, such as the denial of hemiparesis and denial of blindness (Anton's syndrome), are delusional disorders (Chapter 8).[153] When the patient claims that the neglected limb belongs to someone else, the term *somatoparaphrenia* may be used. Occasional patients with anosognosia develop a delusional conviction that a third limb exists on the paretic side.[154] Anosognosic syndromes are commonly associated with unilateral neglect and occur with posterior hemispheric lesions (Chapter 8). *Reduplicative paramnesia*, another delusion closely correlated with specific CNS lesions, is the belief that one has been relocated, usually to a position closer to one's home. It occurs in patients with right hemispheric lesions in conjunction with frontal lesions or during recovery from acute confusional states (Chapter 7).[155,156]

Relationship of Brain Disorders to Psychosis

Many types of brain disorders manifest psychosis (Table 12.1). The principal lesions in patients with neurological disorders and psychosis involve the limbic system, but the limbic system is widely represented in the brain with parts of the temporal lobes, frontal lobes, basal ganglia, thalamus, and connecting white matter tracts participating. Thus, the presence of psychosis suggests that the patient has limbic system dysfunction but does not predict the exact location of an associated brain disorder. Many types of disease affect the limbic system and psychoses do not suggest a specific etiology for the associated brain disorder. Moreover, since many limbic system lesions are not accompanied by psychotic symptoms, additional factors must be present for psychosis to occur. Bilateral brain involvement (bilateral structural lesions, one lesion plus cerebral atrophy, or one lesion and evidence of bilateral metabolic brain dysfunction) increases the likelihood of occurrence of psychosis.[57] Onset of the disorder in adolescence or young adulthood increases the risk of psychosis, compared to disorders that have their onset in midlife.[46] A second peak of psychoses associated with brain disorders occurs in late life with the increased frequency of Alzheimer's disease, dementia with Lewy bodies, frontotemporal degenerations, and stroke. Most psychoses do not begin immediately at the time of brain injury; elapsed time since injury may affect the frequency of delusions.[60] The occurrence of seizures with a brain disorder also appears to increase the likelihood of an associated psychosis.[60]

Many types of drugs are also capable of inducing psychosis (Table 12.3), but drugs that increase monoamine activity and particularly those that increase dopamine function are most likely to induce delusions. Dopaminergic agents used in the treatment of Parkinson's disease produce psychosis in 10% of patients.[11] Neurologic conditions associated with reduced brain dopamine, however, are rarely associated with psychotic manifestations (such as untreated Parkinson's disease), which suggests that dopamine plays a critical role in the pathopsychology of psychosis. There may also be critical balances between dopamine and other transmitters such as acetylcholine or serotonin that, when affected, result in psychosis. Delusions are common in Alzheimer's disease when dopamine levels are preserved and cholinergic activity is reduced,[157] and delusions with hallucinations are characteristic features of dementia with Lewy bodies; in this disorder, the ratio of acetylcholine to serotonin and to dopamine found at autopsy has the highest correlation with psychotic features.[158] Dopaminergic hyperactivity has been posited to play a central role in schizophrenia (discussed below). Dopamine is a major neurotransmitter within the limbic system and thus both anatomic and neurochemical observations suggest that limbic system dysfunction is a key pathogenetic feature in psychosis.

Psychotic disorders produced by neurologic conditions are typically manifested by paranoia with persecutory delusions.[159,160] Mood changes are common in neurologic disorders and grandiose or nihilistic delusions may occur when there is an associated major mood change.[161,162] Content-specific delusions may also be the principal manifestation of a brain disorder with psychosis. Auditory hallucinations are common, and visual hallucinations are more frequent than in idiopathic psychiatric disorders.[163] Lack of insight, neuropsychological deficits, sleep disorders, self-neglect, social withdrawal, and appetite disorders are common among patients with neurologic disorders and psychosis.[159,161] Aggressive behavior occurs in approximately one-third of patients with neurologic disease and psychosis.[147,161] Thought disorganization and fragmentation may occur but formal thought disorders of the type seen in schizophrenia are not common.[161,164] In patients with dementia, there is a limited relationship between the psychosis and the specific neuropsychological deficits, although psychosis tends to be more common in the middle and later phases of dementing diseases.[147] The intellect is in the service of the psychotic process; patients with cognitive impairment associated with their neurologic disorder (such as patients with dementing disorders) tend to manifest loosely structured delusions, whereas those with greater cognitive integrity (such as epilepsy patients) have more complex delusional content. Schneiderian first-rank symptoms are more indicative of left hemisphere dysfunction,[165] while misidentification syndromes and prominent visual hallucinations are more typical of right hemisphere disorders.[166]

Treatment of Psychosis in Neurologic Disorders

Treatment of neurologic disorders with psychosis (discussed in Chapter 4) must address both the underlying disease (e.g., epilepsy, multiple sclerosis) and the psychosis. The latter is treated with conventional neuroleptics or novel antipsychotic agents except in special circumstances such as peri-ictal psychoses, which are treated with anticonvulsants (Chapter 21). The effects of drugs used in the treatment of neuromedical illnesses (i.e., dopaminergic agents used in Parkinson's disease) must be reviewed for their potential role in the genesis of delusions.

Neurologic Assessment of Patients with Psychosis

A critical and practical clinical question is, "When should a patient presenting with psychosis undergo a neurologic evaluation and what tests should be pursued if a neurologic assessment is warranted?". All patients presenting with a psychiatric disorder should have a careful mental status and neurologic examination; the elements of an examination most relevant to psychiatry are presented in Chapter 3. Basic laboratory tests are also indicated in all patients with psychiatric presentations. These should include a complete blood count, erythrocyte sedimentation rate, electrolytes, blood sugar, blood urea nitrogen, liver function tests, and thyroid stimulating hormone. Patients with atypical features should have more extensive testing.

A structural imaging procedure such as computerized tomography (CT) or MRI is the next step in neurodiagnostic assessment and should be obtained in all patients in whom a comprehensive evaluation is pursued. Magnetic resonance imaging is superior to CT in terms of sensitivity for detection of most types of cerebral pathology. Magnetic resonance imaging is contraindicated in patients who have pacemakers or who have metal clips in their head. Patients with symptoms suggestive of epilepsy and those with possible toxic encephalopathies should have an electroencephalogram (EEG), and patients with suspected or documented cognitive deficits should have neuropsychological assessment. More advanced imaging studies such as positron emission tomography (PET) or single photon emission computed tomography (SPECT) may be useful in selected circumstances. Specific additional laboratory tests are chosen on the basis of a careful history and examination and might include HIV antibodies, antiphospholipid antibodies, blood and urine drug screens, and other specialized measures. Cerebrospinal fluid examination is indicated in patients with evidence of demyelinating, inflammatory, or infectious brain disorders and should usually follow imaging procedures.

Clinical features that should trigger consideration of a neurodiagnostic assessment of psychotic patients include atypical age of onset, especially after age 45; absence of a family history of psychiatric illness; absence of any past psychiatric disturbances or premorbid behaviors characteristic of idiopathic psychiatric disorders; presence of family history of a neurologic disorder such as Huntington's disease; presence of focal neurological signs; presence of mental status deficits suggestive of focal or degenerative brain disorder; presence of unusual psychiatric syndromes or atypical mixed states (such as prominent mood changes with mood-incongruent delusion); history of a medical disorder or neurologic condition, even if remote; presence of unusual temporal features such as abrupt onset, quick resolution, or rapid fluctuation; and treatment resistance or unusual treatment responses.

NEUROLOGICAL ASPECTS OF SCHIZOPHRENIA

Schizophrenia is an idiopathic neuropsychiatric disorder characterized by bizarre, persecutory, or religious delusions; auditory hallucinations; illogical thinking with loosening of associations and poverty of information content; and deterioration from a previous level of functioning. The disease develops in adolescence or early adulthood, and the active phase is often preceded by a prodrome of deteriorating abilities and social withdrawal. The course of the disease is lifelong, although periods of active illness are followed by remissions or a residual phase similar to the prodromal

TABLE 12.5. *Positive and Negative Features in Schizophrenia*

	Positive Symptoms Predominate	*Negative Symptoms Predominate*
Clinical features	Hallucinations	Blunted affect
	Delusions	Anhedonia
	Bizarre behavior	Avolition
	Positive thought disorder (loose associations, verbigeration, etc.)	Negative thought disorder (alogia, poverty of thought)
		Attentional disturbance
Neurological abnormalities	Less commonly associated	More commonly associated
Neuropsychological deficits	Tasks requiring auditory processing and involving linguistic stimuli	Tasks involving visual information processing and visuomotor integration
Educational attainment	Higher	Lower
Premorbid social function	Higher	Lower
Genetic influence	Less evident	More evident
Prognosis	Better	Worse
Vulnerability to EPS	Lower	Higher
Response to anticholinergic treatment	Worsened or no effect	Improved or no effect
Response to levodopa treatment	Worsened or no effect	Improved or no effect
Response to neuroleptic treatment	Improved or no effect	Worsened or no effect
Response to atypical antipsychotic treatment	Improved	Improved
Abnormal CT or MRI	Less commonly associated	More commonly associated

CT, computerized tomography; EPS, extrapyramidal syndrome, including both parkinsonism and tardive dyskinesia; MRI, magnetic resonance imaging.

period.[1] The abnormalities of verbal output characterizing the speech of some schizophrenics are described in Chapter 6.

Attempts to subdivide schizophrenia into clinical types that have clinical or prognostic significance have generally met with limited success. Subtypes that are currently recognized include the paranoid, disorganized, catatonic, undifferentiated, and residual types. Subtypes are based on the symptom complex observed during the treatment episode and may change over time.

Identification of positive and negative features of schizophrenia offers another avenue of insight into the disorder (Table 12.5). These characteristics do not distinguish two types of schizophrenia; mixtures of positive and negative symptoms are present in most schizophrenic patients; and patients may vary in the proportion of positive and negative symptoms they display during different periods of their illness. The two sets of clinical characteristics, however, have differing clinical and prognostic correlates, and they may provide useful information when one set of symptoms predominates. Generally, negative symptoms are found in patients who have a poor premorbid adjustment, exhibit more neuropsychological deficits and more neurological abnormalities, are more likely to have a positive family history of psychosis, and tend to have more evidence of atrophy on brain imaging studies. Negative symptoms tend to worsen with neuroleptic therapy and improve with levodopa treatment, whereas positive symptoms exhibit the reverse treatment response.[167–169]

Neurological abnormalities are commonly present when schizophrenic patients are examined. These abnormalities are more common in those with negative symptoms but are not limited to any schizophrenic subtype.[169,170] Abnormalities are most evident on examination of cranial nerves, motor system, and sensory function. Neuro-ophthalmologic abnormalities commonly observed include increased blinking, difficulty moving the eyes without moving the head, and eye-tracking dysfunction (volitional eye movements) during smooth pursuit tasks.[170,171] "Soft" (or subtle) signs of neurologic dysfunction are present in 30% to 80% of patients, including poor coordination, clumsiness, and impaired graphesthesia (Chapter 3).[172–174]

Neuropsychological abnormalities are common in schizophrenia. While some of the deficits may be attributable to poor motivation, intrusion of psychotic thoughts, formal though disorder, or medication effects, cognitive abnormalities are identified even when these factors are absent or controlled. Abnormalities involve most cognitive domains but affect frontally mediated tasks such as the Wisconsin Card Sort Test, verbal fluency, and spontaneous recall disproportionately.[175–177]

Computerized tomography of the brain commonly reveals abnormalities in schizophrenia. Enlarged lateral ventricles with an abnormal ventriculo-brain ratio (VBR) is the most frequently reported abnormality. The abnormal VBR has been found to correlate with poor premorbid adjustment, poor prognosis, negative symptoms, less benefit from antipsychotic medication, and higher prevalence of extrapyramidal drug-induced side effects.[178] An enlarged third ventricle and increased cortical sulcal size, particularly in the frontal lobe, have also been found in schizophrenics, compared to normal controls.[178,179]

Studies using MRI have confirmed and extended observations made with CT.[180,181] In a study of identical twins discordant for schizophrenia, it was demonstrated that the affected twin had larger ventricles in all but one of the twin pairs.[182] The ventricles in the affected twin were not necessarily abnormal in size compared to age-matched controls but were enlarged when compared to the ventricles of their genetically identical sibling. This suggests that ventricular enlargement is present in most schizophrenic patients and the frequency of ventriculomegaly is underestimated in group studies. The temporal lobe and thalamus have been shown to be smaller in schizophrenia than in normal controls,[183,184] and some studies have reported that the frontal lobes are smaller.[185] Verbal memory, abstraction, and categorization deficits correlate with the reduced temporal lobe volume.[186] Moreover, the severity of the schizophrenic thought disorder is also related to the size of the temporal lobe.[187]

Functional brain imaging also demonstrates abnormalities in schizophrenia, although there is less consistency among these studies. Reduced blood flow in anterior brain regions, particularly in patients with chronic schizophrenia, has been observed commonly,[188] and patients fail to exhibit an increase in cerebral blood flow in response to neuropsychological tests that normally activate brain function.[189] Cerebral glucose metabolism measured by PET shows a reduced ratio of anterior-to-posterior metabolic activity, consistent with underactivation of frontal lobe structures.[190]

Postmortem studies have failed to identify a consistent pathognomonic abnormality in the brains of schizophrenic patients. Several changes, however, have been observed. Grossly, the brains of schizophrenic patients tend to have reduced weight, cortical atrophy, and ventricular enlargement.[191] Neuronal density is reduced in the prefrontal cortex and cingulate gyrus.[192] Temporal lobe volume is reduced, hippocampal neuronal density is decreased, and hippocampal pyramidal cells exhibit excessive architectural disarray.[193,194] Thus, limbic and frontal cortical abnormalities, although subtle, are present in the brains of many schizophrenic patients studied at autopsy.

The evidence robustly supports the proposition that schizophrenia is a neurologic disorder manifested by behavioral, neuropsychological, neuroimaging, and neuroanatomic alterations. Dopamine hyperactivity is implicated in the pathogenesis of schizophrenia because of the fidelity with which amphetamines and levodopa can reproduce the positive features of the syndrome and the readiness with which these symptoms respond to dopamine blockade with neuroleptics. The discovery of elevated levels of dopamine D_4 receptors in schizophrenics supports the dopamine theory of schizophrenic psychosis.[195] Weinberger[196] has proposed a neurodevelopmental theory of schizophrenia placing dopamine dysfunction in a developmental context. It is proposed that there is a congenital brain insult that becomes symptomatic when developmental demands exceed the capacity of the system, usually in early adulthood. Dorsolateral prefrontal dysfunction produces the negative symptom complex, whereas dopamine hyperactivity in limbic system structures causes positive symptoms. Individuals vary in the amount of prefrontal and limbic system dysfunction, accounting for the clinical heterogeneity of the disorder. The congenital damage could be produced by a variety of conditions (traumatic, infectious, vascular, etc.), allowing etiologic heterogeneity for schizophrenia.

Paraphrenia, Late-Onset Psychosis, and Late-Onset Schizophrenia

Schizophrenia is a lifelong disorder. As schizophrenic patients age, their symptoms change somewhat; they continue to experience psychotic symptoms but positive symptoms tend to become less severe, while negative symptoms worsen. Cognitive decline occurs in many elderly schizophrenics.[197] Thus, the late stage of schizophrenia is characterized primarily by an apathetic state with impaired cognition and a reduction in delusions, hallucinations, and thought disorder.

Schizophrenia may begin late in life (e.g., late-onset "paraphrenia"). Although the majority of schizophrenic patients experience the onset of symptoms before age 30, in a few patients symptoms begin after age 40, and some have the onset of the disorder as late as age 60. Late-onset schizophrenia is characterized by paranoid symptoms, auditory hallucinations, a predominance in women, and deterioration of personal–social function after onset.[198,199] A family history of schizophrenia is equally common in early- and late-onset schizophrenia.[200] Magnetic resonance imaging reveals that thalamic volumes are larger in those with late-onset symptoms than in those with early-onset disorders.[201]

Most patients with late-onset psychosis do not have schizophrenia; they have a neurologic, medical, or toxic cause of their psychotic disorder. Only 25% to 40% of patients with late-onset psychosis will be found to have schizophrenia, while vascular disease, degenerative disorders such as Alzheimer's disease, dementia with Lewy bodies, tumors, and toxic-metabolic encephalopathies will account for the psychosis in 60% to 75%.[202,203] Patients over age 45 presenting with the first onset of psychotic symptoms should have a thorough neuromedical assessment.

REFERENCES

1. American Psychiatric Association. Diagnostic and Statistical Manual of Mental Disorders, 4th ed. Washington, DC: American Psychiatric Press, 1994.
2. Jaspers K. General Psychopathology. Chicago: University of Chicago Press, 1963.
3. Cummings JL. Organic delusions: phenomenology, anatomic correlates and review. Br J Psychiatry 1985;146:184–187.
4. Davison K, Bagley OR. Schizophrenia-like psychoses associated with organic disorders of the central nervous system: a review of the literature. Br J Med Psychol 1980;53:75–83.
5. Fairweather DS. Psychiatric aspects of the postencephalitic syndrome. J Ment Sci 1947;93:201–254.
6. Bromberg W. Mental states in chronic encephalitis. Psychiatr Q 1930;4:537–566.
7. Kirby GH, Davis TK. Psychiatric aspects of epidemic encephalitis. Arch Neurol Psychiatry 1921;5:491–551.
8. Sands IJ. The acute psychiatric type of epidemic encephalitis. Am J Psychiatry 1928;84:975–987.
9. Celesia GG, Barr AN. Psychosis and other psychiatric manifestations of levodopa therapy. Adv Neurol 1970;23:193–200.
10. Crow TJ, Johnstone EG, McClelland HA. The coincidence of schizophrenia and parkinsonism: some neurochemical implications. Psychol Med 1976;6:227–233.
11. Cummings JL. Behavioral complications of drug treatment of Parkinson's disease. J Am Geriatr Soc 1991;39:708–716.
12. Mindham RHS. Psychiatric symptoms in parkinsonism. J Neurol Neurosurg Psychiatry 1970;33:188–191.
13. Moskovitz C, Moses H III, Klawans HL. Levodopa induced psychosis: a kindling phenomenon. Am J Psychiatry 1978;135:669–675.
14. Caine ED, Shoulson I. Psychiatric syndromes in Huntington's disease. Am J Psychiatry 1983;140:728–733.
15. Folstein SE. Huntington's Disease: A Disorder of Families. Baltimore: Johns Hopkins University Press, 1989.
16. Morris M, Scourfield J. Psychiatric aspects of Huntington's disease. In: Harper PS, ed. Huntington's Disease. Philadelphia: W.B. Saunders, 2nd edition. 1996:73–121.
17. Medalia A, Scheinberg IH. Psychopathology in patients with Wilson's disease. Am J Psychiatry 1989;146:662–664.
18. Scheinberg IH, Sternlieb I. Wilson's Disease. Philadelphia: W.B. Saunders, 1984.
19. Cummings JL, Gosenfeld LF, et al. Neuropsychiatric disturbances associated with idiopathic calcification of the basal ganglia. Biol Psychiatry 1983;18:591–601.
20. Francis AF. Familial basal ganglial calcification and schizophreniform psychosis. Br J Psychiatry 1979;135:360–362.
21. Chandler JH, Bebin J. Hereditary cerebellar ataxia. Neurology 1956;6:187–195.
22. Davies DL. Psychiatric changes associated with Friedreich's ataxia. J Neurol Neurosurg Psychiatry 1949;12:246–250.
23. Hamilton NG, Frick RB, et al. Psychiatric symptoms and cerebellar pathology. Am J Psychiatry 1983;140:1322–1326.
24. Johnson KP, Rosenthal MS, Lerner PI. Herpes simplex encephalitis. Arch Neurol 1971;27:103–108.
25. Williams BB, Lerner AM. Some previously unrecognized features of herpes simplex virus encephalitis. Neurology 1978;28:1193–1196.
26. Keddie KMG. Toxic psychosis following mumps. Br J Psychiatry 1965;111:691–696.
27. Misra PC, Hay GG. Encephalitis presenting as acute schizophrenia. BMJ 1971;1:532–533.
28. Raymond RW, Williams RL. Infectious mononucleosis with psychosis. N Engl J Med 1948;239:542–544.
29. Still RML. Psychosis following Asian influenza in Barbados. Lancet 1958;3:20–21.
30. Signore RJ, Laymeyer HW. Acute psychosis in a patient with cerebral cysticercosis. Psychosomatics 1988;29:106–108.
31. Stoler M, Meshulam B, et al. Schizophreniform episode following measles infection. Br J Psychiatry 1987;150:861–862.
32. Halstead S, Riccio M, et al. Psychosis associated with HIV infection. Br J Psychiatry 1988;153:618–623.
33. Harris MJ, Jeste DV, et al. New-onset psychosis in HIV-infected patients. J Clin Psychiatry 1991;52:369–376.
34. Risk WS, Hadded FS. The variable natural history of subacute sclerosing panencephalitis. Arch Neurol 1979;86:510–614.
35. Arieti S. Histopathologic changes in cerebral malaria and their relation to psychotic sequels. Arch Neurol Psychiatry 1946;56:79–104.
36. Blankfein RJ, Chirico A-M. Cerebral schistosomiasis. Neurology 1965;15:957–967.
37. Blocker WWJ, Kastl AJJ, Daroff RB. The psychiatric manifestations of cerebral malaria. Am J Psychiatry 1968;125:192–196.
38. Rothschild D. Dementia paralytica accompanied by manic-depressive and schizophrenic psychoses. Am J Psychiatry 1940;96:1043–1060.
39. Schube PG. Emotional states of general paresis. Am J Psychiatry 1934;91:625–638.
40. Goldhammer Y, Bubis JJ, et al. Subacute spongiform en-

cephalopathy and its relation to Jakob-Creutzfeldt disease: report of six cases. J Neurol Neurosurg Psychiatry 1972;35:1–10.
41. Honer WG, Hurwitz T, et al. Temporal lobe involvement in multiple sclerosis patients with psychiatric disorders. Arch Neurol 1987;44:187–190.
42. Kohler J, Heilmeyer H, Volk B. Multiple sclerosis as chronic atypical psychosis. J Neurol Neurosurg Psychiatry 1988;51: 281–284.
43. Freeman AMI. Delusions, depersonalization and unusual psychopathological symptoms. In: Hall RCW, ed. Psychiatric Presentations of Medical Illness. New York: SP Medical Scientific Books, 1980:75–89.
44. Berginer VM, Foster NL, et al. Psychiatric disorders in patients with cerebrotendinous xanthomatosis. Am J Psychiatry 1988; 145:354–357.
45. Finelli PF. Metachromatic leukodystrophy manifesting as a schizophrenic disorder: computed tomographic correlation. Ann Neurol 1985;18:94–95.
46. Hyde TM, Ziegler JC, Weinberger DR. Psychiatric disturbances in metachromatic leukodystrophy. Arch Neurol 1992;49:401–406.
47. James ACD, Kaplan P, et al. Schizophreniform psychosis and adrenomyeloneuropathy. J R Soc Med 1984;77:882–884.
48. Kitchin W, Cohen-Cole SA, Mickel SF. Adrenoleukodystrophy: frequency of presentation as a psychiatric disorder. Biol Psychiatry 1987;22:1375–1387.
49. Breitner JCS, Husain MM, et al. Cerebral white matter disease in late-onset paranoid psychosis. Biol Psychiatry 1990;28:266–274.
50. Miller BL, Lesser IM, et al. Brain white-matter lesions and psychosis. Br J Psychiatry 1989;155:73–78.
51. Wragg RE, Jeste DV. Overview of depression and psychosis in Alzheimer's disease. Am J Psychiatry 1989;146:577–587.
52. Chacko RC, Hurley RA, Jankovic J. Clozapine use in diffuse Lewy body disease. J Neuropsychiatry Clin Neurosci 1993;5: 206–208.
53. McKeith IG, Perry RH, et al. Operational criteria for senile dementia of Lewy body type (SDLT). Psychol Med 1992;22:911–922.
54. Perez MM, Trimble MR, et al. Epileptic psychosis: an evaluation of PSE profiles. Br J Psychiatry 1985;46:155–163.
55. Slater E, Beard AW, Glithero E. This schizophrenia-like psychoses of epilepsy. Br J Psychiatry 1963;109:95–150.
56. Trimble MR, Cummings JL. Neuropsychiatric disturbances following brainstem lesions. Br J Psychiatry 1981;138:56–59.
57. Rabins PV, Starkstein SE, Robinson RG. Risk factors for developing atypical (schizophreniform) psychoses following stroke. J Neuropsychiatry Clin Neurosci 1991;3:6–9.
58. Benson DF. Psychiatric aspects of aphasia. Br J Psychiatry 1973;123:555–566.
59. Signer S, Cummings JL, Benson DF. Delusions and mood disorders in patients with chronic aphasia. J Neuropsychiatry Clin Neurosci 1989;1:40–45.
60. Levine DN, Finkelstein S. Delayed psychosis after right temporoparietal stroke or trauma: relation to epilepsy. Neurology 1982;32:267–273.
61. Pakalnis A, Drake MEJ, Kellum JB. Right parieto-occipital lacunar infarction with agitation, hallucinations, and delusions. Psychosomatics 1987;28:95–96.
62. Peroutka SJ, Sohmer BH, et al. Hallucinations and delusions following a right temporoparieto-occipital infarction. Johns Hopkins Med J 1982;151:181–185.
63. Buckley P, Stack JP, et al. Magnetic resonance imaging of schizophrenia-like psychoses associated with cerebral trauma: clinicopathologic correlates. Am J Psychiatry 1993;150:146–148.
64. Nasrallah HA, Fowler RC, Jud LL. Schizophrenia-like illness following head injury. Psychosomatics 1981;22:359–361.
65. Leeks SR. A mid-brain lesion presenting as schizophrenia. NZ Med J 1967;66:311–314.
66. White JC, Cobb S. Psychological changes associated with giant pituitary neoplasms. Arch Neurol Psychiatry 1955;74:383–396.
67. Brierly JB, Cooper JE. Cerebral complications of hypotensive anesthesia in a healthy adult. J Neurol Neurosurg Psychiatry 1962;25:24–30.
68. Price TRP, Tucker GJ. Psychiatric and behavioral manifestations of normal pressure hydrocephalus. J Nerv Ment Dis 1964;164:51–55.
69. Rice E, Gendelman S. Psychiatric aspects of normal pressure hydrocephalus. JAMA 1973;223:409–412.
70. O'Flaithbheartaigh S, Williams PA, Jones GH. Schizophrenic psychosis associated with aqueduct stenosis. Br J Psychiatry 1994;164:684–686.
71. Reveley AM, Reveley MA. Aqueduct stenosis and schizophrenia. J Neurol Neurosurg Psychiatry 1983;46:18–22.
72. Vedak CS, Jampala VC, Hayashida SF. A case of aqueduct stenosis with multiple psychiatric presentations responding to surgical intervention. Neuropsychiatry Neuropsychol Behav Neurol 1988;1:147–152.
73. Bates GMJ. Lever's disease and schizophrenia. Am J Psychiatry 1964;120:1017–1019.
74. Emsley RA, Paster L. Lipoid proteinosis presenting with neuropsychiatric manifestation. J Neurol Neurosurg Psychiatry 1985;48:1290–1292.
75. Fox JTJ, Kane FTJ. Niemann-Pick's disease manifesting as schizophrenia. Dis Nerv Sys 1967;28:194.
76. Kohn R, Lilly RB, et al. Psychiatric presentations of intracranial cysts. J Neuropsychiatry Clin Neurosci 1989;1:60–66.
77. Lewis SW, Reveley MA, et al. Agenesis of the corpus callosum and schizophrenia: a case report. Psychol Med 1988;18:341–347.
78. Coren HA, Strain JJ. A case of narcolepsy with psychosis. Comp Psychiatry 1965;6:191–199.
79. Pfefferbaum A, Berger PA. Narcolepsy, paranoid psychosis, and tardive dyskinesia: a pharmacological dilemma. J Nerv Ment Dis 1977;164:293–297.
80. Burks JS, Huddlestone J, et al. A fatal encephalopathy in chronic haemodialysis patients. Lancet 1976;1:764–768.
81. Chokroverty S, Breytman ME, et al. Progressive dialytic encephalopathy. J Neurol Neurosurg Psychiatry 1976;39:411–419.
82. Read AE, Sherlock S, et al. The neuropsychiatric syndromes associated with chronic liver disease and extensive portal-systemic collateral circulation. Q J Med 1967;36:135–150.
83. Rothermich NO, von Haam E. Pancreatic encephalopathy. J Clin Endocrinol 1941;1:872–881.
84. Burnell GM, Foster TA. Psychosis with low sodium syndrome. Am J Psychiatry 1972;128:133–134.
85. Weizman A, Eldar M, et al. Hypercalcemia-induced psychopathology in malignant diseases. Br J Psychiatry 1979;135: 363–366.
86. Goldstein MG. Intensive care unit syndromes. In: Stoudemire A, Fogel BS, eds. Principles of Medical Psychiatry. New York: Grune & Stratton, 1987:403–421.
87. Gotze P, Dahme B. Psychopathological syndromes and neurological disturbances before and after open-heart surgery. In: Speidel H, Rodewald G, eds. Psychic and Neurological Dys-

functions after Open-Heart Surgery. New York: Thieme Stratton, 1980:48–67.
88. Kornfeld DS, Zimberg S, Malm J. Psychiatric complications of open-heart surgery. N Engl J Med 1965;273:287–292.
89. Roth N. The neuropsychiatric aspects of porphyria. Psychosom Med 1945;7:291–301.
90. Tishler PV, Woodward B, et al. High prevalence of intermittent acute porphyria in a psychiatric patient population. Am J Psychiatry 1985;142:1430–1436.
91. King PH, Bragdon AC. MRI reveals multiple reversible cerebral lesions in an attack of acute intermittent porphyria. Neurology 1991;41:1300–1302.
92. Alarcon RD, Franceschini JA. Hyperparathyroidism and paranoid psychosis: case report and review of the literature. Br J Psychiatry 1984;145:477–486.
93. Bursten N. Psychoses associated with thyrotoxicosis. Arch Gen Psychiatry 1981;4:262–273.
94. Hanna SM. Hypopituitarism (Sheehan's syndrome) presenting with organic psychosis. J Neurol Neurosurg Psychiatry 1970;33:192–193.
95. Sanders V. Neurologic manifestations of myxedema. N Engl J Med 1962;166:547–552.
96. Wijsenbeck H, Kriegel Y, Landau B. A paranoid state in a patient suffering from Sheehan syndrome. Am J Psychiatry 1964;120:1120–1122.
97. Berlin FS, Bergey GK, Money J. Periodic psychosis of puberty: a case report. Am J Psychiatry 1982;139:119–120.
98. Davidson J, Robertson E. A follow-up study of post-partum illness, 1946–1978. Acta Psychiatr Scand 1985;71:451–457.
99. Lindstrom LH, Nyberg F, et al. CSF and plasma beta-casomorphin-like opioid peptides in postpartum psychosis. Am J Psychiatry 1984;141:1059–1066.
100. Cutting J. The relationship between Korsakov's syndrome and "alcoholic dementia". Br J Psychiatry 1978;132:240–251.
101. Smith ADM. Megaloblastic madness. BMJ 1960;2:1840–1845.
102. Strachan RW, Henderson JG. Dementia and folate deficiency. Q J Med 1967;36:189–204.
103. Bluestein HG, Pischel KD, Woods VLJ. Immunopathogenesis of the neuropsychiatric manifestation of systemic lupus erythematosus. Springer Semin Immunopathol 1986;9:237–249.
104. Cares RM, Gordon BS, Kreuger E. Boeck's sarcoid in chronic meningo-encephalitis. J Neuropathol Exp Neurol 1957;16: 544–554.
105. Gorman DG, Cummings JL. Organic delusional syndrome. Semin Neurol 1990;10:229–238.
106. Grant HC, McMenemey WH. Giant cell encephalitis in a dement. Neuropatol Pol 1960;4:735–740.
107. MacNeill A, Grennan DM, et al. Psychiatric problems in systemic lupus erythematosus. Br J Psychiatry 1976;128:442–445.
108. Bonfa E, Golombek SJ, et al. Association between lupus psychosis and anti-ribosomal P protein antibodies. N Engl J Med 1987;317:265–271.
109. Beamish P, Kiloh LG. Psychoses due to amphetamine consumption. J Ment Sci 1960;106:337–343.
110. Cohen S, Ditman KS. Prolonged adverse reactions to lysergic acid diethylamide. Arch Gen Psychiatry 1963;8:475–480.
111. Estroff TW, Gold MS. Medication-induced and toxin-induced psychiatric disorders. In: Extein I, Gold MS, eds. Medical Mimics of Psychiatric Disorders. Washington, DC: American Psychiatric Press, 1986:163–198.
112. Hausner RS. Amantadine-associated recurrence of psychosis. Am J Psychiatry 1980;137:240–242.
113. Jefferson JW, Marshall JR. Neuropsychiatric Features of Medical Disorders. New York: Plenum Medical Book Company, 1981.
114. McConnell H, Duffy J. Neuropsychiatric aspects of medical therapies. In: Coffey CE, Cummings JL, eds. Textbook of Geriatric Neuropsychiatry. Washington, DC: American Pyschiatric Press, 1994:549–593.
115. Mehta M. PDR Guide To Drug Interactions, Side Effects, and Indications. Montvale, NJ: Medical Economics Data Production Company, 1994.
116. Paykel ES, Fleminger R, Watson JP. Psychiatric side effects of antihypertensive drugs other than reserpine. J Clin Psychopharmacol 1982;2:14–39.
117. Deveaugh-Geiss J, Pandurangi A. Confusional paranoid psychosis after withdrawal from sympathomimetic amines: two case reports. Am J Psychiatry 1982;130:1190–1191.
118. Carpenter WTJ, Strauss JS, Muleh S. Are there pathognomonic symptoms in schizophrenia? Arch Gen Psychiatry 1973;28: 847–852.
119. Ianzito BM, Cadoret RJ, Pugh DP. Thought disorder in depression. Am J Psychiatry 1974;131:703–707.
120. Cohn CK, Rosenblatt S, Faillace LA. Capras' syndrome presenting as a postpartum psychosis. South Med J 1977;70:942.
121. Enoch MD. The Capgras syndrome. Acta Psychiatr Scand 1963;39:437–462.
122. Merrin EL, Silverfarb PM. The Capgras phenomenon. Arch Gen Psychiatry 1976;33:965–968.
123. Alexander MP, Stuss DT, Benson DF. Capgras syndrome: a reduplicative phenomenon. Neurology 1979;29:334–339.
124. Ardila A, Rosseli M. Temporal lobe involvement in capgras syndrome. Int J Neurosci 1988;43:219–234.
125. Crichton P, Lewis S. Delusional misidentification, AIDS, and the right hemisphere. Br J Psychiatry 1990;157:608–610.
126. Weston MJ, Whitlock FA. The Capgras syndrome following head injury. Br J Psychiatry 1971;119:25–31.
127. Feinberg TE, Shapiro RM. Misidentification-reduplication and the right hemisphere. Neruopsychiatry Neuropsychol Behav Neurol 1989;2:39–48.
128. Quinn D. The Capgras syndrome: two case reports and a review. Can J Psychiatry 1981;26:126–129.
129. Fishbain DA, Rosomoff H. Capgras syndrome associated with metrizamide myelography. Int J Psychiatry Med 1986–1987; 16:131–136.
130. Hay GG, Jolley DJ, Jones RG. A case of the Capgras syndrome in association with pseudo-hypoparathyroidism. Acta Psychiatr Scand 1974;50:73–77.
131. Pies R. Capgras phenomenon, delirium, and transient hepatic dysfunction. Hosp Community Psychiatry 1982;33:382–383.
132. Santiago JM, Stoker DL, et al. Capgras' syndrome in a myxedema patient. Hosp Community Psychiatry 1987;38:199–201.
133. Zucker DK, Livingston RL, et al. B12 deficiency and psychiatric disorders: case report and literature review. Biol Psychiatry 1981;16:197–205.
134. De Pauw KW, Szulecka TK. Dangerous delusions: violence and the misidentification syndromes. Br J Psychiatry 1988;152:91–96.
135. De Pauw KW, Szulecka TK, Poltock TL. Fregoli syndrome after cerebral infarction. J Nerv Ment Dis 1987;175:433–437.
136. Enoch MD, Trethowan WH. Uncommon Psychiatric Syndromes. Bristol, England: John Wright and Sons, 1979.
137. Malliaras DE, Kossouvitsa YT, Christodoulou GN. Organic contributors to the intermetamorphosis syndromes. Am J Psychiatry 1978;135:985–987.

138. Damas-Mora JMR, Jenner FA, Eacott SE. On heutoscopy or the phenomenon of the double: case presentation and review of the literature. Br J Med Psychol 1980;53:75–83.
139. Drevets WC, Rubin EH. Erotomania and senile dementia of Alzheimer type. Br J Psychiatry 1987;151:400–402.
140. El Gaddal YY. De Clerambault's syndrome (erotomania) in organic delusional syndrome. Br J Psychiatry 1989;154:714–716.
141. Signer S, Cummings JL. De Clerambault's syndrome in organ affective disorder. Br J Psychiatry 1987;151:404–407.
142. Shepherd M. Morbid jealousy: some clinical and social aspects of a psychiatric symptom. J Ment Sci 1961;107:687–753.
143. Pope FM. Parasitophobia as the presenting symptom in vitamin B12 deficiency. Practitioner 1970;204:421–422.
144. Wilson JW. Delusions of parasitosis (acraphobia). Arch Dermatol 1952;66:577–585.
145. Surawicz FG, Banta R. Lycanthropy revisited. Can Psychiatr Assoc J 1975;20:537–542.
146. Berrios GE, Brook P. Visual hallucinations and sensory delusions in the elderly. Br J Psychiatry 1984;144:662–684.
147. Flynn FG, Cummings JL, Gornbein J. Delusions in dementia syndromes: investigation of behavioral and neuropsychological correlates. J Neuropsychiatry Clin Neurosci 1991;3:364–370.
148. Bernstein RL, Gaw AC. Koro: proposed classification for DMS-IV. Am J Psychiatry 1990;147:1670–1674.
149. Anderson DN. Koro: the genital retraction syndrome after stroke. Br J Psychiatry 1990;157:142–144.
150. Durst DR. Koro secondary to a tumor of the corpus callosum. Br J Psychiatry 1988;153:251–254.
151. Joseph AB, O'Leary DH. Brain atrophy and interhemispheric fissure enlargement in Cotard's syndrome. J Clin Psychiatry 1986a;47:518–520.
152. Young AW, Robertson IH, et al. Cotard delusion after brain injury. Psychol Med 1992;22:799–804.
153. Frederiks JAM. Disorders of the body schema. In: Frederiks JAM, ed. Clinical Neuropsychology. New York: Elsevier Science Publishers, 1985:373–392.
154. Weinstein EA, Kahn RL, et al. Delusional reduplication of parts of the body. Brain 1954;77:45–60.
155. Benson DF, Gardner H, Meadows JC. Reduplicative paramnesia. Neurology 1976;26:147–151.
156. Hakim H, Verma NP, Greiffenstein MF. Pathogenesis of reduplicative paramnesia. J Neurol Neurosurg Psychiatry 1988;51:839–841.
157. Cummings JL, Gorman DG, Shapira J. Physostigmine ameliorates the delusions of Alzheimer's disease. Biol Psychiatry 1993;33:536–541.
158. Perry EK, Marshall E, et al. Evidence of a monoaminergic-cholinergic imbalance related to visual hallucinations in Lewy body dementia. J Neurochem 1990;55:1454–1456.
159. Feinstein A, Ron MA. Psychosis associated with demonstrable brain disease. Psychol Med 1990;20:793–803.
160. Johnstone EC, Macmillan JF, Crow TJ. The occurrence of organic disease of possible or probable aetiological significance in a population of 268 cases of first episode schizophrenia. Psychol Med 1987;17:371–379.
161. Cornelius JR, Day NL, et al. Characterizing organic delusional syndrome. Arch Gen Psychiatry 1991;48:749–753.
162. Cutting J. Physical illness and psychosis. Br J Psychiatry 1980;136:109–119.
163. Albert E. On organically based hallucinatory-delusional psychoses. Psychopathology 1987;20:144–154.
164. Daniels EK, Shenton ME, et al. Patterns of thought disorder associated with right cortical damage, schizophrenia, and mania. Am J Psychiatry 1988;145:944–949.
165. Trimble MR. First-rank symptoms of Schneider: A new perspective? Br J Psychiatry 1990;156:195–200.
166. Joseph AB. Focal central nervous system abnormalities in patients with misidentification syndromes. In: Christodoulou GN, ed. The Delusional Misidentification Syndromes. New York: Karger: Bibliotheca Psychiatrica, 1986:68–79.
167. Buchanan RW, Kirkpatrick B, et al. Clinical correlates of the deficit syndrome of schizophrenia. Am J Psychiatry 1990;147:290–294.
168. Fenton WS, McGlashan TH. Antecedents, symptoms progression, and long-term outcome of the deficit syndrome in schizophrenia. Am J Psychiatry 1994;151:351–356.
169. Merriam AE, Kay SR, et al. Neurological signs and the positive negative dimension in shizophrenia. Biol Psychiatry 1990;28:181–192.
170. Cadet JL, Rickler KC, Weinberger DR. The clinical neurologic examination in schizophrenia. In: Nasrallah HA, Weinberger DR, eds. Handbook of Schizophrenia, Vol. 1: The Neurology of Schizophrenia. New York: Elsevier Science Publishers, 1986:1–47.
171. Mackert A, Woyth C, et al. Increased blink rate in drug-naive acute schizophrenic patients. Biol Psychiatry 1990;27:1197–1202.
172. Bartko G, Zador G, et al. Neurological soft signs in chronic schizophrenic patients: clinical correlates. Biol Psychiatry 1988;24:458–460.
173. Gupta S, Andreasen NC, et al. Neurologic soft signs in neuroleptic-naive and neuroleptic-treated schizophrenic patients and in normal comparison subjects. Am J Psychiatry 1995;152:191–196.
174. Sanders RD, Keshavan MS, Schooler NR. Neurological examination abnormalities in neuroleptic naive patients with first-break schizophrenia: preliminary results. Am J Psychiatry 1994;151:1231–1233.
175. Goldberg TE, Weinberger DR, et al. Further evidence for dementia of the prefrontal type in schizophrenia? Arch Gen Psychiatry 1987;44:1008–1014.
176. Paulsen JS, Heaton RK, et al. The nature of learning and memory impairments in schizophrenia. J Int Neuropsychol Soc 1995;1:88–99.
177. Sullivan EV, Shear PK, et al. A deficit profile of executive, memory, and motor functions in schizophrenia. Biol Psychiatry 1994;36:641–653.
178. Sheaton ME, Wible CG, McCarley RW. A review of magnetic resonance imaging studies of brain abnormalities in schizophrenia. In Brain imaging in clinical psychiatry. Krishann KR, Dokaiswamy PU (eds). Marcel Dekker, Inc., New York, 1997;297–380.
179. Shelton RC, Karson CN, et al. Cerebral structural pathology in schizophrenia: evidence for a selective prefrontal cortical defect. Am J Psychaitry 1988;145:154–163.
180. Andreasen NC, Ehrhardt JC, et al. Magnetic resonance imaging of the brain in schizophrenia: the pathophysiologic significance of structural abnormalities. Arch Gen Psychiatry 1990;47:35–44.
181. Lieberman J, Bogerts B, et al. Qualitative assessment of brain morphology in acute and chronic schizophrenia. Am J Psychiatry 1992;149:784–794.
182. Suddath RL, Christison GW, et al. Anatomical abnormalities in the brain of monozygotic twins discordant for schizophrenia. N Engl J Med 1990;322:789–794.

183. Andreasen NC, Arndt S, et al. Thalamic abnormalities in schizophrenia visualized through magnetic resonance image averaging. Science 1994;266:294–298.
184. Suddath RL, Casanova MF, et al. Temporal lobe pathology in schizophrenia: a quantitative magentic resonance imaging. Am J Psychiatry 1989;146:464–472.
185. Andreasen NC, Flashman L, et al. Regional brain abnormalities in schizophrenia measured with magnetic resonance imaging. JAMA 1994;272:1763–1769.
186. Nestor PG, Shenton ME, et al. Neuropsychological correlates of MRI temporal lobe abnormalities in schizophrenia. Am J Psychiatry 1993;150:1849–1855.
187. Shenton ME, Kikinis R, et al. Abnormalities of the left temporal lobe and thought disorder in schizophrenia: a quantitative magnetic resonance imaging study. N Engl J Med 1992;327:604–612.
188. Wilson WH, Mathew RJ. Asymmetry of RCBF in schizophrenia: relationship to AP-gradient and duration of illness. Biol Psychiatry 1993;33:806–814.
189. Gur RE, Jaggi JL, et al. Cerebral blood flow in schizophrenia: effects of memory processing on regional activation. Biol Psychiatry 1994;35:3–15.
190. Wolkin A, Angrist B, et al. Low frontal glucose utilization in chronic schizophrenia: a replication study. Am J Psychiatry 1988;145:251–253.
191. Brown R, Colter N, et al. Postmortem evidence of structural brain changes in schizophrenia: differences in brain weight, temporal horn area, and parahippocampal gyrus compared with affective disorder. Arch Gen Psychiatry 1986;43:36–42.
192. Benes FM, Davidson J, Bird ED. Quantitative cytoarchitectural studies of the cerebral cortex of schizophrenics. Arch Gen 1986;43:31–35.
193. Altshuler LL, Conrad A, et al. Hippocampal pyramidal cell orientation in schizophrenia. Arch Gen Psychiatry 1987;44:1094–1098.
194. Bogerts B, Meertz E, Schonfeldt-Bausch R. Basal ganglia and limbic system pathology in schizophrenia. Arch Gen Psychiatry 1985;42:784–791.
195. Seeman P, Guan H-C, van Tol HHM. Dopamine D4 receptors elevated in schizophrenia. Nature 1993;365:441–445.
196. Weinberger DR. The pathogenesis of schizophrenia: a neurodevelopmental theory. In: Nasrallah HA, Weinberger DR, eds. Handbook of Schizophrenia. New York: Elsevier Science Publishers, 1986:397–406.
197. Davidson M, Harvey PD, et al. Severity of symptoms in chronically institutionalized geriatric schizophrenic patients. Am J Psychiatry 1995;152:197–207.
198. Harris MJ, Jeste DV. Late-onset schizophrenia: an overview. Schizophrenia Bull 1988;14:3955.
199. Harris MJ, Cullum CM, Jeste DV. Clinical presentation of late-onset schizophrenia. J Clin Psychiatry 1988;49:356–360.
200. Kendler KS, Tsuang MT, Hays P. Age at onset in schizophrenia. Arch Gen Psychiatry 1987;44:881–890.
201. Corey-Bloom J, Jernigan T, et al. Quantitative magnetic resonance imaging of the brain in late-life schizophrenia. Am J Psychiatry 1995;152:447–449.
202. Craig TJ, Bregman Z. Late onset schizophrenia-like illness. J Am Geriatr Soc 1988;36:104–107.
203. Miller BL, Lesser IM, et al. Brain lesions and cognitive function in late-life psychosis. Br J Psychiatry 1991;157:76–82.

Chapter 13

Hallucinations

Hallucinations are sensory experiences that occur without external stimulation of the relevant sensory organ.[1] They may occur in any sensory modality—visual, auditory, tactile, olfactory, or gustatory. Hallucinations may be appreciated objectively as false events not corresponding to external reality (sometimes called *pseudohallucinations*) or they may be thought to represent actual external events. The latter are hallucinations with delusional endorsement and comprise part of a psychotic experience. Hallucinations occur in the course of a large number of pathological processes and may occasionally occur in normal individuals in the absence of any disease. The neuropsychiatric differential diagnosis and pathophysiology of hallucinations in each sensory modality are presented in this chapter.

VISUAL HALLUCINATIONS

A *visual hallucination* has been defined operationally as a symptom in which the patient claims to see something or behaves as though having seen something that the observer cannot see.[2] Visual hallucinations occur in a wide variety of ophthalmologic, neurological, toxic-metabolic, and idiopathic psychiatric disorders.[3]

Differential Diagnosis

Table 13.1 presents the differential diagnosis of visual hallucinations. *Ophthalmologic disorders* that reduce or eliminate the patient's vision frequently produce hallucinations. They have been described by patients suffering acute blindness from traumatic enucleation and in patients with poor vision secondary to cataract formation or diseases of the macula, choroid, or retina.[4–8] Brief, unformed flashes of light may also occur with ocular pathology and are noted with vitreous detachment (Moore's lightning streaks), with sudden ocular motion in individuals with no ocular pathology (flick phosphenes), or may be induced by sound in patients with a variety of ocular disorders (auditory–visual synesthesia).[9] Hallucinations must be distinguished from entoptic phenomena such as particles floating in the vitreous humor, Haidingers brushes (nerve bundles made visible by macular edema), and Scheerer's phenomenon (red blood cells circulating in the paramacular region.[8]

Hallucinations commonly occur following ocular surgery, particularly if both eyes are patched in the postoperative period. Such postsurgical hallucinations are more likely to occur if a toxic psychosis is also present.[10] The hallucinations associated with reduced vi-

TABLE 13.1. *Differential Diagnosis of Visual Hallucinations*

Opthalmologic disorders
Enucleation
Cataract formation
Macular degeneration
Choroidal/retinal disease
Vitreous traction
Central nervous system disorders
Optic nerve disease
Thalamic lesions
Brainstem lesions (peduncular hallucinosis)
Hemispheric lesions (especially medial occipital–temporal area lesions)
Degenerative diseases
Alzheimer's disease
Dementia with Lewy bodies
Creutzfeldt–Jakob disease
Parkinson's disease (following dopaminergic therapy)
Encephalitis
Epilepsy
Narcolepsy
Migraine
Medical illnesses
Delirium
Toxic disturbances
Delirium
Alcohol and drug withdrawal
Hallucinogens
Persisting perception disorder after lysergic acid diethylamide (LSD) use
Psychiatric disorders
Schizophrenia
Mania
Psychotic depression
"Normal" individuals
Dreams
Hypnagogic hallucinations
Hypnosis
Imaginary companions
Sensory deprivation
Sleep deprivation
Bereavement

sion are usually fully formed, brightly colored images of people, animals, flowers, or scenery.

Although formerly considered a condition with spontaneous visual hallucinations in the elderly, the *Charles Bonnet syndrome*, now appears to be associated with ocular pathology in most cases, and the onset of hallucinations in aged individuals should lead to a search for eye disease.[11–14] Suggested criteria for the identification of the Charles Bonnet syndrome include the occurrence of visual hallucinations that are formed, complex, persistent or repetitive, and stereotyped; fully or partially retained insight; absent delusions; and no hallucinations in other modalities.[13] Both ocular and brain disorders can produce the syndrome, although most patients will be found to suffer from ocular abnormalities. Charles Bonnet syndrome can occur in the absence of cognitive changes or emotional disturbances.[14]

Retinal ischemia in the course of amaurosis fugax associated with carotid artery disease may be manifested as blindness (typically a "descending curtain") or as visual hallucinations. The hallucinations are typically unformed and appear as scintillations, colored streaks or blobs, or flashing lights.[15,16] These visual changes closely resemble those of migraine (described below) and may lead to misdiagnosis.

Optic nerve disease, particularly *optic neuritis*, is also associated with brief unformed visual hallucinations. The phosphenes occur with movement of the eyes or may be induced by sounds.[17,18]

Focal lesions in a number of locations in the brain are associated with visual hallucinations. Lesions in the midbrain region produce the syndrome of *peduncular hallucinosis*. In this condition there are complex visual hallucinations that typically occur in the evenings, are associated with disturbances of the sleep–wake cycle, and are usually viewed as benign, entertaining phenomena. Other evidence of brainstem dysfunction is usually present. The syndrome of peduncular hallucinosis has been associated with bilateral lesions of the substantia nigra (pars reticulata), infarction of the right paramedian or posterior thalamus, and focal insults of the right midbrain tegmentum and cerebral peduncle.[19–23] Auditory–visual synesthesia has also occasionally been described by patients with lesions involving the upper brainstem. The upper region of the brainstem is perfused by the basilar artery and occlusion of the upper basilar produces the "top of the basilar syndrome."[24] This distinctive syndrome includes abnormalities of ocular movement, eyelid retraction, and skew deviation. Peduncular hallucinosis is frequently present and patients are often somnolent, report excessive dreaming, and have difficulty distinguishing dreams from reality (oneroid states). When the tempo-

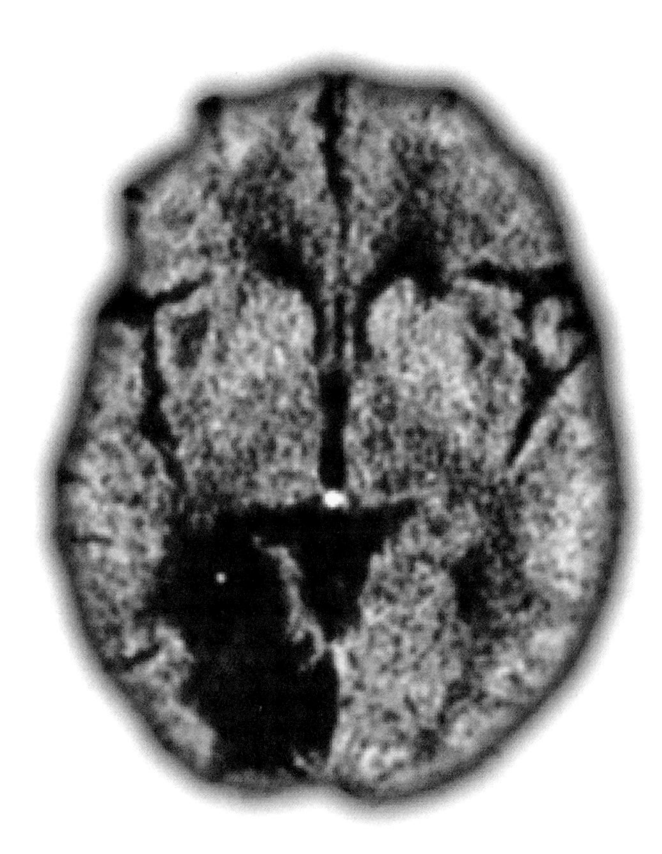

FIGURE 13.1 Computed tomogram showing a right occipital infarction associated with release hallucinations.

ral or occipital lobes are infarcted, an agitated delirium may occur.

Thalamic disorders have occasionally been associated with visual hallucinations. Focal lesions and degenerative conditions such as fatal familial insomnia have produced visual hallucinations.[19,25,26]

Hemispheric lesions can produce visual hallucinations in two clinical circumstances: as part of focal seizure activity or as release phenomena associated with visual field defects. *Release hallucinations* are generally formed images, lasting from minutes to hours. They are variable in content, may be modified by altering the visual input such as opening or closing the eyes, and tend to occur within the field defect.[27–33] The underlying pathological lesion is usually an infarction, but any focal lesion within the visual pathways in the temporal, parietal, or occipital lobes may produce release hallucinations (Fig. 13.1). Hallucinations have occasionally been reported with frontal lobe lesions, but this is rare and, when described, they have usually been associated with mass lesions capable of exerting distant effects on the temporal lobes or brainstem.

Ictal hallucinations occurring as expressions of seizure activity can usually be distinguished from release hallucinations on the basis of their clinical features (Table 13.2).[3,34] Ictal hallucinations are brief, stereotyped visual experiences.[35] There may or may not be an associated visual field defect; they rarely are lateralized to one portion of the visual field; and when they are formed, they may consist of visual recollections of past experiences. Epileptic lesions in primary and associative cortex give rise to unformed or semiformed hallucinations (flashes, lights, colors, etc.), whereas more anterior foci situated in the temporal lobe produce formed hallucinations and remembered scenes.[36] Consciousness may be altered during or after an ictal hallucination, and there may be ictal motor phenom-

TABLE 13.2. *Distinction between Ictal and Release Hallucinations Associated with Hemispheric Lesions*

Ictal Hallucinations	*Release Hallucinations*
Brief (seconds to minutes)	Longer duration (minutes to hours)
Stereotyped	Variable content
Unformed with posterior lesions; formed with temporal lesions	Usually formed regardless of lesion location
Visual field defect not necessarily present	Visual field defect usually present
Seldom lateralized	Typically lateralized to side of visual field defect
Hallucinations content (if formed) is often a visual memory	Content is usually novel
Unaltered by environmental activities	May be modified by environmental alterations (e.g., opening or closing the eyes or deviating the gaze)
Consciousness altered during or after ictal event	No associated alteration of consciousness

ena such as head or eye deviation. A variety of visual distortions, including macropsia, micropsia, and metamorphopsia, also occur as ictal visual phenomena, either without hallucinations or concomitantly with them.[37] Ictal hallucinations are more often associated with right-than left-sided lesions. These observations regarding ictal hallucinations have largely been confirmed by stimulation of the occipitotemporal cortex during surgical procedures.[38]

Diseases affecting the cerebral hemispheres can also produce visual hallucinations. Formed hallucinations are particularly common in dementia with Lewy bodies,[39,40] and they also occur, though less frequently, in Alzheimer's disease,[41,42] and Creutzfeldt-Jakob disease.[43]

Two syndromes causing visual hallucinations and unassociated with focal central nervous system (CNS) lesions are migraine and narcolepsy. Visual hallucinations occur in about half of all patients with migraine. The most common hallucination is a fortification-type zig-zag spectrum often associated with a scotoma, but fully formed complex hallucinations may also occur.[44–47]

The migrainous aura may also include macropsia and micropsia, a clinical complex called the "Alice in Wonderland syndrome" after Lewis Carroll, whose personal experiences with migraine were utilized in that famous tale.[48–50] A few patients with migraine experience persistent positive visual phenomena, including perceptual experiences such as television static, snow, lines of ants, dots, or rain lasting months to years after migrainous attacks.[51] The visions of Hildegard of Bingen, a mystic nun who lived from 1098 to 1180, provide an unusually rich account of migrainous hallucinations.[52]

Narcolepsy, is in its fully expressed form, consists of a tetrad of sleep attacks, cataplexy, sleep paralysis, and hallucinations (Chapter 23).[53,54] Hallucinations occur in 20%–50% of narcoleptics and are noted most often in the drowsy period as the patient is falling asleep (*hypnagogic hallucinations*). Hypnopompic hallucinations experienced just as the patient is awakening occur in a smaller number of cases. Hypnagogic visions occur most commonly in association with nocturnal sleep and rarely occur with daytime narcoleptic attacks. The hallucinations are primarily visual, but auditory and somesthetic hallucinations have been reported, as well as macropsia and micropsia.[55] The hallucinations are accompanied by electroencephalographic (EEG) changes characteristic of rapid eye movement (REM) or dreaming sleep and represent the intrusion of dreams into the drowsy state.[56–58] Hypnagogic hallucinations unaccompanied by other evidence of narcolepsy can occur in normal individuals and have been described in a variety of psychiatric disorders, including schizophrenia, depression, paranoid state, and puerperal psychosis.[59]

Hallucinations frequently occur in the course of acute confusional states induced by medical illnesses. They have been noted in 40%–75% of delirious patients. In delirium, the hallucinations are relatively brief, are often nocturnal, and may be regarded as real. Hallucinations in delirium are usually clear, brightly colored visions, that are experienced as three-dimensional images in nearby space.[60] The patients are often fearful and respond to the hallucinations with self-protective measures. The visions are typically formed, moving, silent images but in some cases may be accompanied by auditory or tactile hallucinations.

Among toxin-related disturbances, hallucinations occur in four major circumstances: nonspecific acute confusional states, alcohol and sedative withdrawal, hallucinogen-induced conditions, and states induced by dopaminergic agents (Table 13.3). Any drug when taken in excess may produce an acute confusional state with concomitant hallucinations. In some cases hallucinations may occur as the primary manifestation of toxicity with little evidence of delirium or confusion. Among agents that have been reported to produce hallucinations and may be particularly prone to cause this symptom are cimetidine, antidepressants, inhalants, lithium, cyclosporin, digoxin, sympathomimetics, quinidine, anticholinergics, antibiotics, hormonal agents, narcotics, antimalarials, phenacetin, disulfiram, propranolol, heavy metals, metrizamide, and bromide.[3] Patients with Parkinson's disease may experience visual hallucinations following treatment with dopaminergic agents. This is particularly common in patients with a dementia syndrome.[61–63]

Hallucinations also are common during withdrawal states, occurring in up to 75% of those manifesting an acute abstinence syndrome.[64] Withdrawal syndromes develop following abrupt cessation of intake of a variety of agents, including alcohol, barbiturates, benzodiazepines, chloral hydrate, paraldehyde, meprobamate, methaqualone, opioid compounds, and cocaine. The type of hallucination is very similar to that occurring in medical illnesses. *Zoopsia*, or hallucinations of animals, is particularly common in, although not limited to, withdrawal syndromes. Alcohol and sedative-hypnotics produce chronic suppression of REM sleep, and withdrawal is associated with REM rebound. When visual hallucinations emerge during alcohol withdrawal, REM sleep accounts for most of sleep time and begins almost immediately when the patient falls asleep. This suggests that visual hallucinations associated with alcohol withdrawal, like those occurring in narcolepsy,

TABLE 13.3. *Pharmacologic Agents Associated with Visual Hallucinations*

Hallucinogens	Hormonal agents
Lysergic acid diethylamide (LSD)	Steroids
Dimethyltryptamine (DMT)	Thyroxin
PsilocinAntibiotics	
Harmine	Antibiotics
Mescaline	Sulfonamides
MDMA (ectasy)	Penicillin
Tetrahydrocannabinol	Tetracycline
Phencyclidine (PCP)	
Ketamine	Miscellaneous agents
Abused inhalants	Antidepressants
Ether	Imipramine
Gasoline	Maprotiline
Glue	Nitrous oxide
Nitrites	Cocaine
Nitrous oxide	Amphetamines
	Lithium
Antiparkinsonian agents	Cyclosporin
Anticholinergic drugs	Aminophylline
Amantadine hydrochloride	Bismuth
Levodopa	Corticosteroids
Bromocriptine	Disulfram
Pergolide	Indomethacin
Ropinerole	Ranitidine
Pramipexole	Bromide
Entacapone	Digoxin
Tolcapone	Sympathomimetics
Selegiline	Cimetidine
	Propranolol
Drugs associated with withdrawal syndromes	Phencaetin
Glutethimide	Disulfiram
Ethyl alcohols	Narcotics
Barbiturates	Antimalarials
Benzodiazepines	Heavy metals
Chloral hydrate	Metrizamide
Paraldehyde	
Meprobamate	
Methaqualone	
Opiates	
Cocaine	

may be a product of dream phenomena intruding into the waking state.[65]

Hallucinogens (psychotomimetics, utopiates) are pharmacological agents that produce perceptual distortions and hallucinations. The latter may be associated with concomitant alterations in affect and cognition resembling those occurring in the psychoses, as well as with physiological alterations such as mydriasis, elevated heart rate and blood pressure, increased muscle tone, tachypnea, and nausea.[66] In general, hallucinogens do not produce an acute confusional state, and hallucinations are disproportionately prominent, compared to other drug-induced mental changes. Several classes of agent are included among the hallucinogens: compounds with an indole structure such as lysergic acid diethylamide (LSD), dimethyltryptamine, psilocybin, pscilocin, and harmine; cannabinols such as tetrahydrocannabinol; and a variety of other agents such as phencyclidine and ketamine (Table 13.3).[33,66] Self-reports and studies of subjects who have ingested LSD and mescaline reveal that in the preliminary stages one sees nonpatterned or geometric shapes; this progresses to more structured geometric images such as lattices, chessboards, cobwebs, funnels, and spirals; finally, fully formed images of landscapes, people, and animals may appear. Visual distortions, movement of patterns, and alterations in color intensity are common. Brief recurrences of the visual experiences ("flashbacks" or *persisting perception disorder*) may occur for several years following exposure to LSD.[67–69]

Auditory hallucinations are far more characteristic of the idiopathic psychiatric illnesses than visual hallucinations, but the latter are not uncommon. Between 24% and 46% of acute schizophrenics and up to 72% of chronic schizophrenics report having had visual hallucinations at some time during the course of their illnesses.[70–72] The visual hallucinations of schizophrenia may occur with auditory hallucinations, or they may be visual memories, bizarre, fragmented images, or even unformed flashes of light. Substance abuse in schizophrenic patients with dual diagnoses is associated with higher rates of visual hallucinations.[72] Visual hallucinations have been reported in 10%–70% of patients with affective disorders.[73] Hallucinations occurring in the context of major affective disorders usually are mood-congruent.

Finally, visual hallucinations occur in normal individuals in specific circumstances (Table 13.1). The best examples of unbidden spontaneous visual images are dreams. These unique visual hallucinations are common to the experience of all intact individuals, meet most definitions of visual hallucinations, and are pathological only when not confined to the sleep state. Children may have imaginary companions and play objects that they "see" with reality-like clarity,[73] and they also appear to be more likely to respond to emotional stress with hallucinatory syndromes.[74,75] In adults, visual hallucinations occur during periods of sleep deprivation, as a product of suggestion during hypnosis, and as hypnagogic phenomena.[59,76] In addition, hallucinations occur during sensory deprivation, a state that may share essential features with hallucinations reported with blindness of all types.[77,78] Hallucinations occur during intense emotional experiences such as in the course of grief reactions and appear to be influenced by the cultural experiences of the involved individuals.[59,79,80] Hallucinations have been described by many "visionaries" who found them to be sources of guidance and inspiration; thus, hallucinations played a role in the lives of Socrates, St. Paul, Joan of Arc, Mohammed, Luther, Moses, Pascal, Swedenborg, George Fox, Shelley, William Blake, Bunyan, Napoleon, Raphael, Schumann, Goethe, Byron, and Walter Scott.[81]

Phenomenology

There are no etiologically specific or pathognomonic types of hallucinations, but features of visual hallucinations may facilitate identification of the clinical disorders from which they originate. The characteristics that distinguish ictal from release hallucinations are described in Table 13.2. The form of epileptic hallucinations has localizing significance: posterior occipital lesions produce unformed, simple flashes; the patterns become more complex if the focus is located in visual association cortex; and more anteriorly placed lesions in the medial temporal lobe produce complex, formed images and visual memories.[35,36] Release hallucinations, by contrast, have little localizing value. They occur with lesions of the eye, optic nerve and chiasm, and cerebral hemispheres.

Lilliputian hallucinations, visions of tiny human and animal figures named for the diminutive inhabitants of the Isle of Lilliput described by Jonathan Swift in *Gulliver's Travels*,[82] are distinctive but appear to have little etiologic significance. They have been described in toxic and metabolic disorders, hypnagogic states, structural CNS disturbances, epilepsy, ocular diseases, affective disturbances, and schizophrenia.[83–85] Hallucinations of giants, or *Brobdingnagian hallucinations*, have been recorded in a small number of confusional states.[86] Lilliputian and Brobdingnagian hallucinations involve seeing small and large individuals, respectively, and must be distinguished from micropsia and macropsia, where the entire scene, along with an included figure, appears altered in size.

Autoscopy (*heutoscopy*) is another striking hallucinatory experience in which one sees one's own image. Such hallucinations occur in epilepsy, brain tumors, cerebral trauma, subarachnoid hemorrhage, cerebral syphilis, migraine, postencephalitic parkinsonism, typhus and other infectious diseases, drug intoxications, schizophrenia, and depression.[87–89] If the patient endorses the vision as a true double or believes that a double exists, even though invisible, the syndrome merges into the delusion of the double, or the doppelgänger.[90] The syndrome is put to literary use in Dostoyevsky's *The Double,* Edgar Allen Poe's *William Wilson*, Steinbeck's *Great Valley,* and Oscar Wilde's *The Portrait of Dorian Gray*.[12,87,89]

"Psychedelic" hallucinations consisting of geometric forms, spirals, funnels, and chessboards are most characteristic of the hallucinogenic drugs[91,92] but also occur with sensory deprivation[77,78] and have been described in CNS disorders, such as during recovery from acute viral encephalitis and with acute occipital lobe insults.[93]

Illusions are misperceptions or perceptual distortions of an existing external stimulus.[1] They differ from hallucinations in that the perceptual activity has its original stimulus in the external environment. Many simple and complex distortions of visual perceptions have been described among the metamorphoses. Changes in size, shape, position, motion, number, or personal relationship to the object have been described (Table 13.4).[94–98] Although illusions and hallucinations can be differentiated clinically, the distinction has little etiologic significance, since both phenomena are found in the same disorders. Illusions occur in epilepsy, migraine, narcolepsy, and infections disorders, and with hallucinogens.

Synesthesias are cross-modal experiences reported by patients who have one sensory-type experience (e.g., color) when another is stimulated (e.g., sound).[99,100]

Palinopsia is a unique form of visual hallucination that involves the persistence or recurrence of visual images after the exciting stimulus has been removed.[101–103] The image remains when the patient changes direction of gaze and may spontaneously recur for up to several hours. The phenomenon usually begins abruptly with a cerebral infarction, neoplasm, or trauma and persists only a few days, but in some cases it has endured for years.[101,102,104,105] Palinopsia can occur with lesions of either hemisphere but is most common with acute damage to the posterior aspect of the right hemisphere.[106] Originally considered an ictal manifestation, palinopsia shares many features with hallucinations originat-

TABLE 13.4. *Types of Metamorphopsia*

Type	*Manifestations*
Simple	Micropsia (objects appear small)
	Macropsia (objects appear large)
	Squeezed, tilted objects
	Wavy edges or other changes in contour
	Color enhanced, changed, or dimmed
	Apparent motion of nonmoving object
	Objects pulsating
	Objects appear far away
	Objects appear unusually close
Complex	Polyopia/entomopia (multiple simultaneous objects)
	Palinopsia (preservation of visual stimulus)
	Allesthesia (lateral transpositions)
	Upside-down reversal of vision
	Prosopo-metamorphopsia (face-specific metamorphopsia)
	Auditory-visual synesthesia
Perceptions with altered affect or memory	Deja vú (unusual familiarity)
	Jamais vú (unusual strangeness)
	Unreality (derealization)
	Undue personal significance

ing from hemispheric lesions involving the visual radiations, which suggests that it is a unique type of release hallucination.[107] Palinopsia has been reported as a side effect of trazodone and can occur during the course of mescaline and LSD intoxication.[108]

Pathophysiology

The remarkable similarity of many aspects of visual hallucinations suggests that a few basic CNS mechanisms are responsible for generation of most of the images. The most common situation in which hallucinations merge involves the reduction of visual input. This occurs in sensory isolation, enucleation, cataract formation, retinal disorder, choroidal change, macular disease, and optic nerve and tract disease, and with hemispheric lesions involving the geniculocalcarine pathways. The blindness may be partial or total, monocular or hemianoptic, and in all cases may be associated with hallucinations of similar character. West[81] proposed a "perceptual release theory," which suggests that decreased sensory input results in release of spontaneous activity of CNS structures normally mediating perceptual experience. This basic mechanism might also account for hallucinations associated with diminished arousal in narcolepsy, hypnotic and trance states, confusional states, and some idiopathic psychiatric disorders. Studies with functional magnetic resonance imaging (fMRI) demonstrate that patients with blindness due to ocular disease who experience spontaneous visual hallucinations have increased activity in the ventral occipital region.[109]

Another basic mechanism associated with several types of visual hallucination involves the appearance of dreams in the waking state. Dreams themselves are a unique variety of visual hallucinations occurring in normal individuals. The emancipation of dreams from the sleep state appears to account for hypnagogic hallucinations occurring in normal individuals and in narcoleptics and for hallucinations occurring during alcohol withdrawal, and it may play a role in the hallucinations of peduncular hallucinosis where a consistent diurnal pattern has been observed. The typical occurrence of vivid dreams before the emergence of visual hallucination in parkinsonian patients treated with dopaminergic agents suggests a relationship to sleep mechanisms.

The ability of the hallucinogens to induce vivid hallucinatory experiences is not understood. An effect on retinal function has been demonstrated but is insufficient to account for the diversity of their actions or the similarity between hallucinogen-induced visions and those of certain CNS lesions.[93,110] Lysergic acid diethylamide is also capable of inducing hallucinations in patients who have been enucleated, demonstrating the independence of the LSD effects of ocular function. The hallucinogenic effects of these agents appear to be mediated by interactions between the drugs and 5-HT_2 serotonin receptors.[66] Removal of the temporal lobes diminishes the hallucinogenic effects of LSD, suggesting that temporal lobe structures play a major role in mediating the effects of hallucinogenic agents.[111]

Cholinergic mechanisms may also have an important role in mechanisms associated with hallucinosis. Anticholinergic toxicity is accompanied by prominent hallucinations, and diseases with cholinergic deficits, including dementia with Lewy bodies, Alzheimer's disease, and Parkinson's disease with dementia, may produce visual hallucinations.[40] Treatment with cholinergic agents may reduce visual hallucinations.[112]

Antipsychotic agents reduce both hallucinations and delusions in patients with psychotic disorders. Dopaminergic and serotonergic mechanisms have been implicated. Hallucinations can thus be correlated with a few basic mechanisms—perceptual release, ictal discharges, dream intrusions, and neurochemical effects. These few mechanisms provide the foundations for understanding many aspects of these unique experiential phenomena.

Treatment

Visual hallucinations respond poorly to treatment except in specific circumstances. Those associated with epilepsy, migraine, and narcolepsy may respond to anticonvulsants, antimigrainous, and stimulant medications, respectively. Visual hallucinations occurring with degenerative disorders, such as Alzheimer's disease, dementia with Lewy bodies, and Parkinson's disease with dementia, may improve with therapy with cholinesterase inhibitors.[112] There are anecdotal reports of improvement in post-LSD persisting perceptual disorder following treatment with naltrexone.[113] Antipsychotic agents relieve hallucinations in patients with schizophrenia and other psychoses.

AUDITORY HALLUCINATIONS

Auditory hallucinations, unlike visual hallucinations, are more characteristic of idiopathic psychiatric disorders than of neuromedical or toxic disorders. An important exception to this observation is the common occurrence of auditory hallucinations in schizophrenia-like psychoses that may be associated with a variety of neurological and toxic-metabolic disorders (Chapter 12). Table 13.5 presents the disorders to be considered in the differential diagnosis of auditory hallucinations.

TABLE 13.5. *Etiologies of Auditory Hallucinations*

Peripheral lesions
Middle ear disease
Inner ear disease
Auditory nerve disease
Central nervous system disorders
Temporal lobe epilepsy
Pontine lesions
Stroke
Arteriovenous malformation
Syncope
Toxic-metabolic disturbances
Chronic alcoholic hallucinosis
Delirium
Hallucinogens
Psychiatric disorders
Schizophrenia
Mania
Psychotic depression
Multiple personality
Post-traumatic stress disorder

Deafness produced by disease of the middle ear (e.g., otosclerosis) or of the inner ear and auditory nerve can produce both unformed and formed hallucinations. The unformed hallucinations are referred to as *tinnitus* and consist of buzzing or tones of varying pitch and timbre. The formed hallucinations consist of melodies, organ music, hymns, songs, and, occasionally, voices. In some cases the content can be influenced by imagining or vocalizing a song or melody.[114,115] Deafness and auditory hallucinations appear to predispose to the development of a paranoid syndrome, particularly in the elderly.[116] The association between auditory hallucinations and deafness resembles the similar correlation of visual hallucinations with blindness and of phantom limbs with amputation.

Musical hallucinations are a unique type of auditory hallucination that often surprise the individual who begins to experience them for the first time; the patient may try to find the radio or television that they think is the origin of the sound. Musical hallucinations have a wide differential diagnosis and can been seen in any disorder producing other types of auditory hallucinations, but they are most common with deafness. Typically the individual is an elderly woman, has long-standing sensorineural deafness, and begins to experience vocal or instrumental music known from childhood.[117,118] Musical hallucinations also occur with hemispheric and brainstem lesions,[117,119,120] as well as with epilepsy and alcoholic hallucinosis.[117,121] Depression and schizophrenia have also been accompanied by musical hallucinations.[117,118,121,122] Musical hallucinations are treatment resistant; they may improve with treatment of any associated psychopathology or, in the case of hallucinations with deafness, an improvement in hearing.[118,122]

Among CNS disorders, partial seizures may give rise to auditory hallucinations. Currie and colleagues[123] studied 514 patients with temporal lobe epilepsy and found that 17% had auditory hallucinations as one component of their seizures. Crude sensations were five times more common than elaborate, formed sounds or voices. The hallucinations are brief, stereotyped sensory impressions and, if formed, may be trivial sentences, previously heard phrases, or commands.[123]

Central nervous system neoplasms also give rise to auditory hallucinations in 3%–10% of cases.[124] The hallucinations may be either formed or unformed and are associated predominantly with frontal and temporal lobe tumors. Vascular lesions have rarely been associated with auditory hallucinations. Hemorrhages and arteriovenous malformations in the region of the pontine tegmentum and lower midbrain have been associated with the acute onset of auditory hallucinations. The sounds have typically been unformed mechanical or "seashell-like" noises or music.[125,126]

Syncope may be associated with hallucinations. Sixty percent of individuals who experience transient cerebral hypoxia report visual changes such as gray haze, colored patches, or bright lights. Thirty-six had auditory hallucinations consisting or rushing or roaring noises, screaming, or voices.[127]

Auditory hallucinations may occur as part of delirious psychoses in the course of toxic and metabolic encephalopathies or during withdrawal states. Alcoholic hallucinosis is a unique auditory hallucinatory syndrome related to chronic alcoholism and alcohol withdrawal. The hallucinations are usually vocal in nature and typically consist of accusatory, threatening, and critical voices directed at the patient. The voices begin in the withdrawal period in most cases and usually cease within a few days. In a few cases, however, the auditory hallucinations become chronic, persisting for years. The patient may have no insight into the unreality of the experiences and may act on directions received from the voices or may seek protection from them with relatives or police.[128–130]

Auditory hallucinations are most characteristic of the idiopathic psychoses. They occur in 60%–90% of schizophrenic patients and up to 80% of patients with affective psychoses.[131] The hallucinations vary from "inner voices" sensed by the patient to vivid hallucinations heard as if coming from the outside.[132] Some patients may hear their own thoughts spoken aloud (*Gedankenlautwerden*). Schizophrenic patients experience the voices as objective, involuntarily present and having unique and immediate relevance to them, but a minority expect that other members of the public should be able to hear them. There is a tendency for some patients with affective disorders to lateralize their hallucinations to the right side of space.[133] Auditory hallucinations, like visual hallucinations, reflect the predominant mood state of the individual when they occur in the affective disorders.[1]

Studies of glucose metabolism in schizophrenic patients with hallucinations have demonstrated diminished metabolism in the posterior superior temporal region of the left hemisphere (auditory cortex and Wernicke's area).[134] Hallucinations were positively correlated with metabolism of anterior cingulate and striatal regions. Dichotic listening studies have also suggested left hemisphere dysfunction in hallucinating schizophrenics.[135] Evoked-response studies during auditory hallucinations in schizophrenia have shown response delays similar to those observed with externally originating sounds.[136] Thus, emerging physiologic investigations suggest that auditory hallucinations in schizophrenic patients are associated with left temporal dysfunction.

Wearing headphones has decreased auditory hallucinations in some patients and others report temporary amelioration of the hallucinations with humming or mouth opening (maneuvers that disrupt subvocalization).[70,137,138] Antipsychotic agents relieve auditory hallucinations in most psychoses.

In addition to their occurrence in the psychoses, auditory hallucinations also occur as manifestations of hysterical conversion reactions, multiple personality disorder, and post-traumatic stress disorder.[139–141] The presence of auditory hallucinations in these syndromes contributes to their frequent misdiagnosis as schizophrenia.

Auditory hallucinations must be distinguished from *palinacousis*, in which there is a persistence or late recurrence of existing auditory stimuli. Palinacousis has been associated with a variety of cerebral lesions (particularly neoplasms and infarctions), and the lesions usually involve the temporal lobe.[96,142]

OLFACTORY, GUSTATORY, AND TACTILE HALLUCINATIONS

Olfactory and gustatory (taste) hallucinations are the least common hallucinations recorded in clinical investigations. Olfactory hallucinations are well known in epilepsy; they are associated with medial temporal lobe lesions and complex partial seizures (uncinate seizures).[71,143] Such hallucinations may also occur in migraine and in both situations are usually described as unpleasant.[144] Olfactory hallucinations have also been described in multi-infarct dementia, Alzheimer's disease, and alcoholic psychosyndromes.[145] Among psychiatric disorders, olfactory hallucinations occur in schizophrenia, post-traumatic stress disorder, depression, and in at least 20%–25% of patients with Briquet's syndrome. In depression, the hallucinations are commonly mood-congruent smells of death, decay, and personal filth. Similar delusional beliefs that one is emitting a foul odor characterize the olfactory reference syndrome[146–148] In this disorder, the patient develops the monosymptomatic belief that they are the source of an offensive odor. The syndrome may be a

TABLE 13.6. *Phantom Phenomena*

Phantom Organs	*Phantom Sensations*
Limb, following amputation	Peripheral nerve injuries
Breast, following mastectomy	Spinal cord transection
Testes, following orchiectomy	Lesions of the brainstem
Nose, with mutilating diseases of the face	Lesions of the thalamus Lesions of the parietal lobes
Eyeball, after enucleation	

stable monosymptomatic delusion or a form of hypochondriasis or may progress to an affective disorder.[146]

Gustatory hallucinations occur in manic-depressive illness, schizophrenia, Briquet's syndrome, and partial seizures.

Tactile hallucinations (formication hallucinations; haptic hallucinations) are most common in withdrawal states but also can occur in association with other brain disorders.[149,150]

Somewhat related to tactile hallucinations are abnormal perceptions regarding body size and shape. These experiences are described in patients with peripheral nerve lesions, brain disorders, and psychiatric illnesses such as body dysmorphic disorder and anorexia nervosa.[151–156]

Patients who have lost limbs or other body parts may have "phantom" sensations based on hallucinations that the body part remains.[157–162] Table 13.6 lists types of phantoms reported in patients with lost appendages.

REFERENCES

1. American Psychiatric Association. Diagnostic and Statistical Manual of Mental Disorders, 4th ed. Washington, DC: American Psychiatric Press, 1994.
2. Lessell S. Higher disorders of visual function: positive phenomena. In: Glaser JS, Smith JL, eds. Neuro-ophthalmology. St. Louis: C.V. Mosby, 1975:27–44.
3. Cummings JL, Miller BL. Visual hallucinations: clinical occurrence and use in differential diagnosis. West J Med 1987; 146:46–51.
4. Bartlet JEA. A case of organized visual hallucinations in an old man with cataract and their relation to the phenomena of the phantom limb. Brain 1951;79:363–373.
5. Berrios GE, Brook P. Visual hallucinations and sensory delusions in the elderly. Br J Psychiatry 1984;144:662–684.
6. Holroyd S, Rabins PV, et al. Visual hallucinations in patients with macular degeneration. Am J Psychiatry 1992;149:1701–1706.
7. Levine AM. Visual hallucinations and cataracts. Ophthalmol Surg 1980;11:95–98.
8. Priestly BS, Foree K. Clinical significance of some entoptic phenomena. Arch Ophthalmol 1955;53:390–397.
9. Vike J, Jabbari B, Maitland CG. Auditory–visual synesthesia. Arch Neurol 1984;41:680–681.
10. Ziskind E, Jones H, et al. Observations on mental symptoms in eye patched patients: hypnagogic symptoms in sensory deprivation. Am J Psychiatry 1960;116:893–900.
11. Berrios GE, Brook P. The Charles Bonnet syndrome and the problem of visual perceptual disorders in the elderly. Age Ageing 1982;11:17–23.
12. Damas-Mora J, Skelton-Robinson M, Jenner FA. The Charles Bonnet syndrome in perspective. Psychol Med 1982;12:251–261.
13. Gold K, Rabins PV. Isolated visual hallucinations and the Charles Bonnet syndrome: a review of the literature and presentation of six cases. Comp Psychiatry 1989;30:90–98.
14. Schultz G, Melzack R. Visual hallucinations and mental state: a study of 14 Charles Bonnet syndrome hallucinators. J Nerv Ment Dis 1993;181:639–643.
15. Goodwin JA, Gorelick PB, Helgason CM. Symptoms of amaurosis fugax in atherosclerotic carotid artery disease. Neurology 1987;37:829–832.
16. Ramadan NH, Tietjen GE, et al. Scintillating scotoma associated with internal carotid artery dissection: report of three cases. Neurology 1991;41:1984–1087.
17. Davis FA, Bergen D, et al. Movement phosphenes in optic neuritis: a new clinical sign. Neurology 1976;26:1100–1104.
18. Lessell S, Cohen MM. Phosphenes induced by sound. Neurology 1979;29:1524–1527.
19. Catafau JS, Rubio F, Peres Serra J. Peduncular hallucinosis associated with posterior thalamic infarction. J Neurol 1992;239:89–90.
20. Dunn DW, Weisberg LA, Nadell J. Peduncular hallucinations caused by brainstem compression. Neurology 1983;33:1360–1361.
21. Feinberg WM, Rapcsak SZ. Peduncular hallucinosis following paramedian thalamic infarction. Neurology 1989;39:1535–1536.
22. Geller TJ, Bellur SN. Peduncular hallucinosis: magnetic resonance imaging confirmation of mesencephalic infarction during life. Ann Neurol 1987;21:602–604.
23. McKee AC, Levine DN, et al. Peduncular hallucinosis associated with isolated infarction of the substantia nigra pars reticulata. Ann Neurol 1990;27:500–504.
24. Caplan LR. Top of the basilar: syndrome. Neurology 1980;30: 72–79.
25. Noda S, Mizoguchi M, Yamamoto A. Thalamic experiential hallucinosis. J Neurol Neurosurg Psychiatry 1993;56:1224–1226.
26. Rancurel G, Garma L, et al. Familial thalamic degeneration with fatal insomnia. In: Guilleminault C, Lugaresi E, Montagna P, Gambetti P, eds. Fatal Familial Insomnia: Inherited Prion Diseases, Sleep, and the Thalamus. New York: Raven Press, 1994:15–25.
27. Benson MT, Rennie IG. Formed hallucinations in the hemianopic field. Postgrad Med J 1989;65:756–757.
28. Brust JCM, Behrens MM. "Release hallucinations" as the major symptom of posterior cerebral artery occlusion: a report of 2 cases. Ann Neurol 1977;2:432–436.
29. Cogan DG. Visual hallucinations as release phenomena. Albrecht Von Graefes Arch Klin Exp Ophthalmol 1973;188:139–150.
30. Kolmel HW. Coloured patterns in hemianopic fields. Brain 1984;107:155–167.

31. Kolmel HW. Complex visual hallucinations in the hemianopic field. J Neurol Neurosurg Psychiatry 1985;48:29–83.
32. Lance JW. Simple formed hallucinations confined to the area of a specific visual field defect. Brain 1976;99:719–734.
33. Vaphiades M, Celesia G, Brigell M. Positive spontaneous visual phenomena limited to the hemianopic field in lesions of central visual pathways. Neurology 1996;47:408–417.
34. Gloor P, Olivier A, et al. The role of the limbic system in experiential phenomena of temporal lobe epilepsy. Ann Neurol 1982;12:129–144.
35. Russell WR, Whitty CWM. Studies in traumatic epilepsy: visual fits. J Neurol Neurosurg Psychiatry 1955;18:79–96.
36. Karagulla S, Robertson EE. Psychical phenomena in temporal lobe epilepsy and in the psychoses. BMJ 1955;1:748–752.
37. Sowa MV, Pituck S. Prolonged spontaneous complex visual hallucinations and illusions as ictal phenomena. Epilepsia 1989;30: 524–526.
38. Penfield W, Perot P. The brain's record of auditory and visual experience. Brain 1963;86:595–696.
39. McKeith IG, Perry RH, et al. Operational criteria for senile dementia of Lewy body type (SDLT). Psychol Med 1992;22: 911–922.
40. Perry EK, Kerwin J, et al. Cerebral cholinergic activity is related to the incidence of visual hallucinations in senile dementia of Lewy body type. Dementia 1990;1:2–4.
41. Burns A, Jacoby R, Levy R. Psychiatric phenomena in Alzheimer's disease, II: Disorders of perception. Br J Psychiatry 1990;157:76–81.
42. Lerner AJ, Koss E, et al. Concomitants of visual hallucinations in Alzheimer's disease. Neurology 1994;44:523–527.
43. Mizutani T, Okumura A, et al. Panencephalopathic type of Creutzfeldt-Jakob disease: primary involvement of the cerebral white matter. J Neurol Neurosurg Psychiatry 1981;44:103–115.
44. Haas DC. Prolonged migraine aural status. Ann Neurol 1982;11:197–199.
45. Hachinski VC, Porchawka J, Steele JC. Visual symptoms in the migraine syndrome. Neurology 1973;23:570–579.
46. Klee A, Willanger R. Disturbances of visual perception in migraine. Acta Neurol Scand 1966;42:400–414.
47. Peatfield RC, Rose C. Migrainous visual symptoms. Arch Neurol 1981;38:466.
48. Lahat E, Eshel G, Arlazoroff A. Alice in Wonderland syndrome: a manifestation of infectious mononucleosis in children. Behav Neurol 1991;4:163–166.
49. Rolak LA. Literary neurologic syndromes: Alice in Wonderland. Arch Neurol 1991;48:649–651.
50. Todd J. The syndrome of Alice in Wonderland. Can Med Assoc J 1955;73:701–704.
51. Liu G, Schatz N, et al. Persistent positive visual phenomena in migraine. Neurology 1995;45:664–668.
52. Sacks O. Migraine: Understanding a Common Disorder. Los Angeles: University of California Press, 1985.
53. Guilleminault C. Narcolepsy syndrome. In: Kryger MH, Roth T, Dement WC, eds. Principles and Practice of Sleep Medicine, 2nd ed. Philadelphia: W.B. Saunders, 1994:549–561.
54. Roth B. Narcolepsy and Hypersomnia. New York: S. Karger, 1980.
55. Selby G, Lance JW. Observations on 500 cases of migraine and allied vascular headache. J Neurol Neurosurg Psychiatry 1960; 23:23–32.
56. Dement W, Rechschaffen A, Gulevich G. The nature of the narcoleptic sleep attack. Neurology 1966;16:18–33.
57. Sours JA. Narcolepsy and other disturbances in the sleep–waking rhythm: a study of 115 cases with review of the literature. J Nerv Ment Dis 1963;137:525–542.
58. Zarcone V. Narcolepsy. N Engl J Med 1973;288:1156–1166.
59. McDonald C. A clinical study of hypnagogic hallucinations. Br J Psychiatry 1971;118:543–547.
60. Lipowski ZJ. Delirium: Acute Confusional States. New York: Oxford University Press, 1990.
61. Sanchez-Ramos J, Ortoll R, Paulson G. Visual hallucinations associated with Parkinson disease. Arch Neurol 1996;53:1265–1268.
62. Cummings JL. Behavioral complications of drug treatment of Parkinson's disease. J Am Geriatr Soc 1991;39:708–716.
63. Goetz CG, Tanner CM, Klawans HL. Pharmacology of hallucinations induced by long-term drug therapy. Am J Psychiatry 1982;139:494–497.
64. Gross MM, Goodenough D, et al. Sleep disturbances and hallucinations in the acute alcoholic psychoses. J Nerv Ment Dis 1966;142:492–514.
65. Greenberg R, Pearlman C. Delirium tremens and dreaming. Am J Psychiatry 1967;124:133–142.
66. Underleider JT, Rechnick RN. Hallucinogens. In: Lowinson JH, Ruiz P, Millman RB, eds. Substance Abuse: A Comprehensive Textbook, 2nd ed. Baltimore: Williams and Wilkins, 1992:280–289.
67. Abraham HD. Visual phenomenology of the LSD flashback. Arch Gen Psychiatry 1983;40:884–889.
68. Horowitz MH. Flashbacks: recurrent intrusive images after the use of LSD. Am J Psychiatry 1969;126:565–569.
69. Rosenthal SH. Persistent hallucinosis following repeated administration of hallucinogenic drugs. Am J Psychiatry 1964; 121:238–244.
70. Bick PA, Kinsbourne M. Auditory hallucinations and subvocal speech in schizophrenia patients. Am J Psychiatry 1987;144: 222–225.
71. Burstein A. Olfactory hallucinations. Hosp Community Psychiatry 1987;30:72–79.
72. Sokolski KN, Cummings JL, et al. Effects of substance abuse on hallucination rates and treatment responses in chronic psychiatric patients. J Clin Psychiatry 1994;55:380–387.
73. Ali S, Denicoff K, et al. Psychosensory symptoms in bipolar disorder. Neuropsychiatry Neuropsychol Behav Neurol 1997;10: 223–231.
74. Egdell HG, Kolvin I. Childhood hallucinations. J Child Psychol Psychiatry 1972;13:279–287.
75. Weiner MF. Hallucinations in children. Arch Gen Psychiatry 1961;5:544–553.
76. West LJ, Janszen HH, et al. The psychosis of sleep deprivation. Ann Acad Sci 1962;96:66–70.
77. Flynn WR. Visual hallucinations in sensory deprivation. Psychiat Q 1962;36:55–59.
78. Heron W. The pathology of boredom. Sci Am 1957;196:52–56.
79. Matchett WF. Repeated hallucinatory experiences as part of the mourning process among Hopi Indian women. Psychiatry 1972;35:185–194.
80. Olson PR, Suddeth JA, et al. Hallucinations of widowhood. J Am Geriatr Soc 1985;33:543–547.
81. West LJ. A general theory of hallucinations and dreams. In: West LJ, ed. Hallucinations. New York: Grune & Stratton, 1962:275–291.
82. Swift J. Gulliver's Travels. New York: New American Library, 1960.
83. Alexander MC. Lilliputian hallucinations. J Ment Sci 1926;72: 187–191.
84. Fleming GWTH. A case of Lilliputian hallucinations with a sub-

sequent single macroscopic hallucination. J Ment Sci 1923;69: 86–89.
85. Goldin S. Lilliputian hallucinations. J Ment Sci 1955;101:569–576.
86. Thomas CJ, Fleming GWTH. Lilliputian and Brobdingnagian hallucinations occurring simultaneously in a senile patient. J Ment Sci 1934;80:94–102.
87. Dening RE, Berrios GE. Autoscopic phenomena. Br J Psychiatry 1994;165:808–817.
88. Devinsky O, Feldmann E, et al. Autoscopic phenomena with seizures. Arch Neurol 1989;46:1080–1088.
89. Lhermitte J. Visual hallucinations of the self. BMJ 1951;1:431–434.
90. Christodoulou GN. Syndrome of subjective doubles. Am J Psychiatry 1978;135:249–251.
91. Klüver H. Mescal and Mechanisms of Hallucinations. Chicago: University of Chicago Press, 1966.
92. Malitz S, Wilkens B, Esecover H. A comparison of drug-induced hallucinations with those seen in spontaneously occurring psychoses. In: West LJ, ed. Hallucinations. New York: Grune & Stratton, 1962:50–63.
93. Mize K. Visual hallucinations following viral encephalitis: a self report. Neuropsychologia 1980;18:193–202.
94. Cutting J. The Right Cerebral Hemisphere and Psychiatric Disorders. New York: Oxford University Press, 1990.
95. Grusser O-J, Landis T. Visual Agnosias and Other Disturbances of Visual Perception and Cognition. London: Macmillan Press, 1991.
96. Jacobs L. Visual allesthesia. Neurology 1980;30:1059–1063.
97. Lopez JR, Adornato BT, Hoyt WF. Entomopia: a remarkable case of cerebral polyopia. Neurology 1993;43:2145–2146.
98. Steiner I, Shahin R, Melamed E. Acute "upside-down" reversal of vision in transient vertebrobasilar ischemia. Neurology 1987;37:1685–1686.
99. Cytowic RE. Synesthesia and mapping of subjective sensory dimensions. Neurology 1989;39:849–850.
100. Cytowic RE. The Man Who Tasted Shapes. New York: G.P. Putnam's Sons, 1993.
101. Bender MB, Feldman M, Sobin AJ. Palinopsia. Brain 1968;91: 321–338.
102. Critchley M. Types of visual perseveration: "paliopsia" and "illusory visual spread". Brain 1951;74:267–299.
103. Stagno SJ, Gates TJ. Palinopsia: a review of the literature. Behav Neurol 1991;4:67–74.
104. Meadows JC, Munro SSF. Palinopsia. J Neurol Neurosurg Psychiatry 1977;40:5–8.
105. Young WB, Heros DO, et al. Metamorphopsia and palinopsia. Arch Neurol 1989;46:820–822.
106. Michel EM, Troost BT. Palinopsia: cerebral localization with computed tomography. Neurology 1980;30:887–889.
107. Cummings JL, Syndulko K, et al. Palinopsia reconsidered. Neurology 1982;32:444–447.
108. Geleberg AJ. Trazodone and palinopsia. Biol Ther Psychiatry 1990;13:37–38.
109. Ffytche D, Howard RJ, et al. The anatomy of conscious vision: an fMRI study of visual hallucinations. Nat Neurosci 1998;1:738–742.
110. Watson SJ. Hallucinogens and other psychotomimetics: biological mechanisms. In: Barchas JD, Berger PA, eds. Psychopharmacology. New York: Oxford University Press, 1977: 341–354.
111. Serafetinides EA. The significance of the temporal lobes and of hemispheric dominance in the production of the LSD-25 symptomatology in man: a study of epileptic patients before and after temporal lobectomy. Neuropsychologia 1965;3:69–79.
112. Cummings J, Back C. The cholinergic hypothesis of neuropsychiatric symptoms in Alzheimer's disease. Am J Geriatr Psychiatry 1998;6:S64–S78.
113. Lerner A, Oyffe I, et al. Naltrexone treatment of hallucinogen persisting perception disorder. Am J Psychiatry 1997;153: 437.
114. Hammeke TA, McQuillen MP, Cohen BA. Musical hallucinations associated with acquired deafness. J Neurol Neurosurg Psychiatry 1983;4:570–572.
115. Nayani T, David A. The auditory hallucination: a phenomenological survey. Psychol Med 1996;26:177–189.
116. Cooper AF, Curry AR. The pathology of deafness in the paranoid and affective psychoses of later life. J Psychosomat Res 1976;20:97–105.
117. Keshavan MS, David AS, et al. Musical hallucinations: a review and synthesis. Neuropsychiatry Neuropsychol Behav Neurol 1992;5:211–223.
118. Wengel SP, Burke WJ, Holeman D. Musical hallucinations. The sounds of silence? J Am Geriatr Soc 1989;37:163–166.
119. Keshavan MS, Kahn EM, Brar JS. Musical hallucinations following removal of a right frontal meningioma. J Neurol Neurosurg Psychiatry 1988;51:1235–1241.
120. Murata S, Naritomi H, Sawada T. Musical auditory hallucinations caused by brainstem lesion. Neurology 1994;44:156–158.
121. Duncan R, Mitchell JD, Critchley EMR. Hallucinations and music. Behav Neurol 1989;2:115–124.
122. Fenton GW, McRae DA. Musical hallucinations in a deaf elderly woman. Br J Psychiatry 1989;155:401–403.
123. Currie S, Healthfield KWG, et al. Clinical course and prognosis of temporal lobe epilepsy. Brain 1971;94:173–190.
124. Tarachow S. The clinical value of hallucinations in localizing brain tumors. Am J Psychiatry 1941;97:1434–1442.
125. Cascino GD, Adams RD. Brainstem auditory hallucinosis. Neurology 1986;36:1042–1047.
126. Lanska DJ, Lanska MJ, Mendez MF. Brainstem auditory hallucinosis. Neurology 1987;37:1685.
127. Lempert T, Bauer M, Schmidt D. Syncope: a videometric analysis of 56 episodes of transient cerebral hypoxia. Ann Neurol 1994;36:233–237.
128. Deiker T, Chambers HE. Structure and content of hallucinations in alcohol withdrawal and functional psychosis. J Stud Alcohol 1978;39:1831–1840.
129. Surawicz FG. Alcoholic hallucinosis: a missed diagnosis. Can J Psychiatry 1980;25:57–63.
130. Victor M, Hope JM. The phenomenon of auditory hallucinations in chronic alcoholism. J Nerv Ment Dis 1958;126:451–481.
131. Junginger J, Frame CL. Self-report of the frequency and phenomenology of verbal hallucinations. J Nerv Ment Dis 1985; 173:149–155.
132. Sedman G. "Inner voices." Phenomenological and clinical aspects. Br J Psychiatry 1966;112:485–290.
133. Gruber LN, Mangat BS, Abou-Taleb H. Laterality of auditory hallucinations in psychiatric patients. Am J Psychiatry 1984; 141:586–588.
134. Cleghorn JM, Franco S, et al. Toward a brain map of auditory hallucinations. Am J Psychiatry 1992;149:1062–1069.
135. Green MF, Hugdahl K, Mitchell S. Dichotic listening during auditory hallucinations in patients with schizophrenia. Am J Psychiatry 1994;151:357–362.

136. Tiihonen J, Hari R, et al. Modified activity of the human auditory cortex during auditory hallucinations. Am J Psychiatry 1992;149:255–257.
137. Feder R. Auditory hallucinations treated by radio headphones. Am J Psychiatry 1982;139:1188–1190.
138. Green MF, Kinsbourne M. Auditory hallucinations in schizophrenia: does humming help? Biol Psychiatry 1989;25:633–635.
139. Levinson H. Auditory hallucinations in a case of hysteria. Br J Psychiatry 1966;112:19–26.
140. McKegney FP. Auditory hallucination as a conversion symptom. Comp Psychiatry 1967;8:80–89.
141. Mueser KT, Butler RW. Auditory hallucinations in combat-related chronic posttraumatic stress disorder. Am J Psychiatry 1987;144:299–302.
142. Malone GL, Leiman HI. Differential diagnosis of palinacousis in a psychiatric patient. Am J Psychiatry 1983;140:1067–1068.
143. Bromberg W, Schilder P. Olfactory imagination and olfactory hallucinations. Arch Neurol Psychiatry 1934;32:467–492.
144. Wolberg FL, Ziegler DK. Olfactory hallucinations in migraine. Arch Neurol 1982;39:382.
145. Rupert SL, Hollender MH, Mehrhof EG. Olfactory hallucinations. Arch Gen Psychiatry 1961;5:313–318.
146. Davidson M, Mukherjee S. Progression of olfactory reference syndrome to mania: a case report. Am J Psychiatry 1982;139:1623–1624.
147. Pryse-Phillips W. An olfactory reference syndrome. Acta Psychiatr Scand 1971;47:484–509.
148. Videbach TH. Chronic olfactory paranoid syndromes. Acta Psychiatr Scand 1967;42:187–213.
149. Berrios GE. Tactile hallucinations: conceptual and historical aspects. J Neurol Neurosurg Psychiatry 1982;45:285–293.
150. Reilly TM. Delusional infestation. Br J Psychiatry 1988;153:44–46.
151. Birtchnell SA. Dysmorphobia—a centenary discussion. Br J Psychiatry 1988;153:41–43.
152. Cooper PH, Taylor MJ. Body image disturbance in bulimia nervosa. Br J Psychiatry 1988;153:32–36.
153. Lukianowicz N. "Body image" disturbances in psychiatric disorders. Br J Psychiatry 1967;113:31–47.
154. Lunn V. On body hallucinations. Acta Psychiatr Scand 1965;41:387–399.
155. Stacy CB. Complex haptic hallucinations and palinaptia. Cortex 1987;23:337–340.
156. Whitehouse AM, Freeman CPL, Annandale A. Body size estimation in anorexia nervosa. Br J Psychiatry 1988;153:23–26.
157. Carlen PL, Wall PD, et al. Phantom limbs and related phenomena in recent traumatic amputations. Neurology 1978;28:211–217.
158. Frederiks JAM. Phantom limb and phantom limb pain. In: Frederiks JAM, ed. Clinical Neuropsychology. New York: Elsevier Science Publishing, 1985:395–404.
159. Henderson WR, Smyth GE. Phantom limbs. J Neurol Neurosurg Psychiatry 1948;11:88–112.
160. Mayeux R, Benson DF. Phantom limb and multiple sclerosis. Neurology 1979;29:724–726.
161. Weinstein EA, Kahn RL, et al. Delusional reduplication of parts of the body. Brain 1954;77:45–60.
162. Weinstein S, Sersen EA, Vetter RJ. Phantoms following orchiectomy. Neuropsychologia 1968;6:61–63.

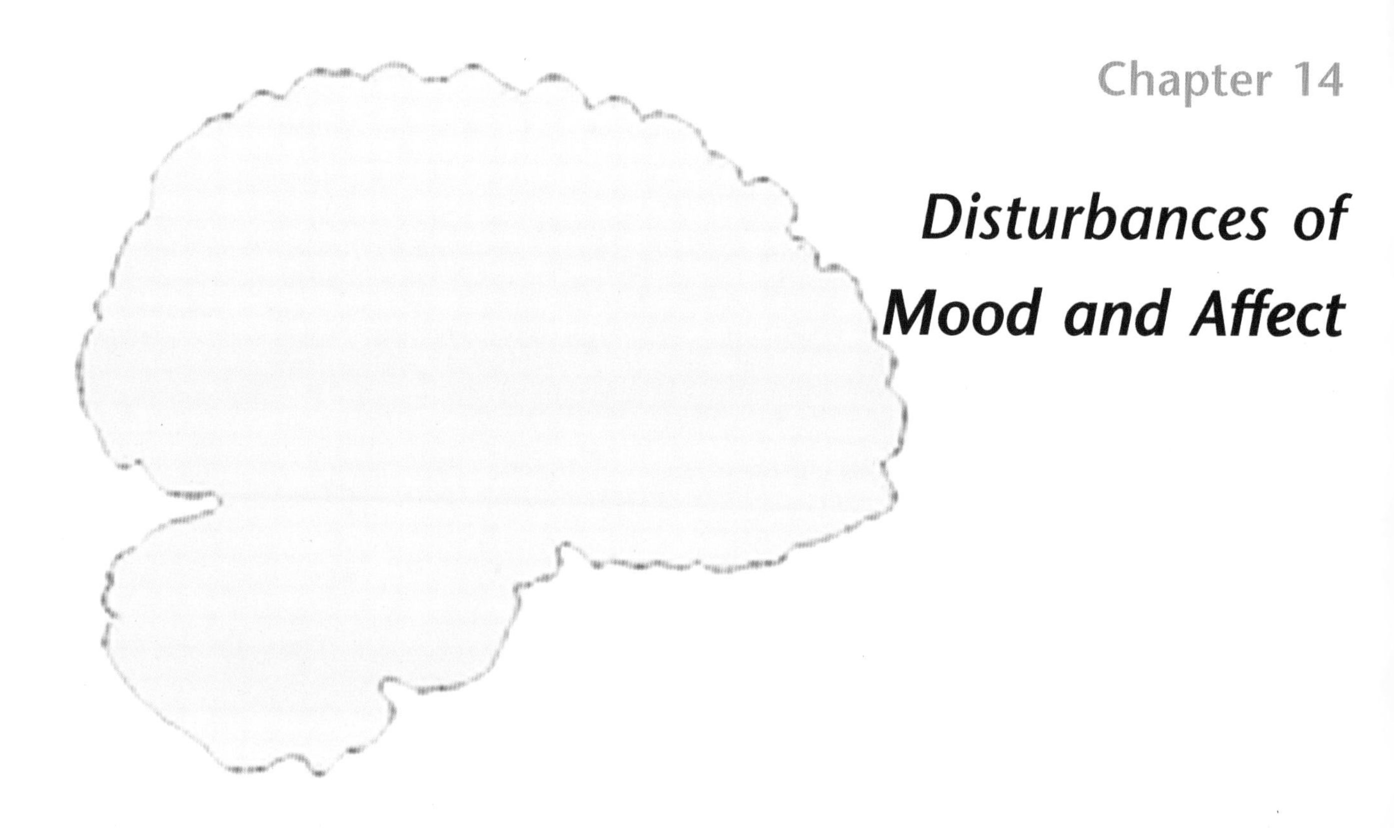

Chapter 14

Disturbances of Mood and Affect

Mood refers to the internal emotional experience of an individual, whereas *affect* refers to the external expression of emotion. Mood is a pervasive and sustained emotion that influences the perception of the world. Changes of mood include dysphoric mood (an unpleasant mood including sadness), elevated mood (an exaggerated feeling of well-being), expansive mood (lack of restraint in expressing one's feelings frequently with an overevaluation of one's significance or importance), and irritable mood (easily annoyed and provoked to anger).[1] Affective changes include laughter and weeping.

In most cases mood and affect are congruent with the external affect accurately reflecting the individual's internal mood. In some cases of neurological disease, however, mood and affect may be dissociated, providing a special challenge in neuropsychiatry. In pseudobulbar palsy and certain types of epileptic seizures, for example, patients may cry or laugh when the corresponding mood is absent.

Abnormalities of mood and affect are ubiquitous in patients with neurological disease and are among the most common neuropsychiatric disturbances observed and requiring care. In addition, there is growing evidence that "idiopathic" mood disorders (both depression and mania) have neurological bases, and in many cases of late-onset mood disorders, specific structural alterations in the brain have been identified. There is a striking convergence of evidence regarding the distribution of central nervous system (CNS) lesions producing depression or mania and the anatomical localization of changes on neuroimaging observed in the idiopathic forms of these disorders. Together these observations are beginning to provide a more comprehensive understanding of the pathophysiology of disturbances of mood and affect.

In this chapter, depression and its characteristics, causes, and treatment are discussed first. Next, mania and secondary mania and their management are presented. Finally, abnormalities of affect are described. Other types of mood changes such as catastrophic reactions and rage are discussed briefly. Information relevant to depression can be found in the chapter on Alzheimer's disease and other dementias (Chapter 10), movement disorders (Chapter 18), epilepsy and temporal lobe syndromes (Chapter 21) and focal and infectious central nervous system disorders (Chapter 26).

DEPRESSION

Clinical Manifestations of Depression

Depression is a broad term that encompasses changes in mood as well as a complex clinical syndrome (major depressive episode). Depression as a mood change typically includes sadness and anhedonia or an impaired ability to experience pleasure. In severe cases there are frequent thoughts of suicide or death; depression is frequently accompanied by feelings of guilt, helplessness, hopelessness, and worthlessness.

A major depressive episode is multidimensional (Table 14.1) and includes disturbances of mood, abnormalities of verbal expression, affective changes, motoric manifestations, cognitive alterations, motivational changes, neurovegetative disturbances (sleep, appetite, and sexual behavior), and neuroendrocrinologic changes. The mood alterations of a major depressive episode include sadness and anhedonia. Abnormalities of verbal expression include a delay in responding verbally, short verbal responses, a trailing off in verbal volume in the course of a sentence, slowness of speech, and a reduced initiation of conversation. In addition there is frequently a lack of emotional inflection of the voice (*dysprosody*). The patient frequently returns to themes of guilt, hopelessness, and worthlessness in the course of the conversation.

Affective alterations of depression include reduced facial mobility; eyebrows tend to be furrowed and drawn together, deepening the vertical furrow between them. Eye contact with the examiner is avoided and the patient may weep. Motoric manifestations of depression include retardation (or periods of agitation), catatonia, postural slumping, body immobility, and slowed movements, including walking.

Cognitive impairment is also present in depression (described in more detail below) and includes a poverty of associations, a tendency toward depressive thoughts, executive dysfunction, and disproportionate impairment of visuospatial skills. Motivational disturbances include reduced interest and difficulty in initiating new activities.

Neurovegetative abnormalities present in depression include alterations in appetite (usually reduced although occasionally increased) and sleep abnormalities; the latter include difficulty falling asleep, multiple awakenings, early morning awakening, and diminished rapid eye movement (REM) latency. Occasionally, patients sleep excessively. Patients often evidence diurnal variability in mood manifested by more severe sadness in the morning that remits partially by the afternoon and evening. Loss of libido is characteristic of depression.

Neuroendocrine disturbances notable in depression include failure to suppress endogenous cortisol secretion with administration of exogenous dexamethasone in the dexamethasone suppression test (DST) and abnormalities on the thyrotropin-releasing hormone (TRH) test.

Neuroimaging changes associated with depression include increased periventricular white matter abnormalities on computerized tomography (CT) or magnetic resonance imaging (MRI) and decreased cerebral blood flow or cerebral metabolism involving primarily the frontal regions (discussed in greater detail below) observed on single photon emission computerized tomography (SPECT) or positron emission tomography (PET).[1–3]

The presence of CNS disease may make the recognition of a depression syndrome more difficult and frequently modifies the manifestations of the disorder. Aphasic patients, for example, cannot voice the usual propositional expression of their internal emotional state; patients with right hemisphere injury or basal ganglia disease have dysprosodia and frequently lack the ability to inflect their voice. They lack the emotional vocal changes that allow the clinician to gauge the severity of the patient's mood changes. Patients with parkinsonism may have slowed movements, vocal changes, and facial hypomimia similar to depression without any corresponding mood abnormality. Likewise, patients with Alzheimer's disease and other dementias may experience apathy, cognitive disturbances, retardation or agitation, changes in appetite, and changes in sleep pattern in the absence of a depressed mood. Patients with pseudobulbar palsy may sob or tear, expressing a sad affect that is dissociated from any underlying mood changes. Apathy is common in many neurological diseases in the absence of depression[4] (Chapter 15). Assessment of the core psychological features of depression, including sadness, feelings of guilt, worthlessness, hopelessness, and helplessness, is critical to recognizing the depression syndrome in patients with neuropsychiatric disorders. Rating scales that include clinical features confounding the manifestation of depression and neurological disease must be used with caution in this population.

Classification of Depressive Disorders

The mood disorders currently recognized include major depressive disorders (characterized by one or more major depressive episodes), dysthymic disorder, bipolar I disorder (characterized by one or more manic or mixed episodes usually associated with periods of major depressive episodes), bipolar II disorder (characterized by one or more depressive episodes accompanied by at least one hypomanic episode), cyclothymic disorder (featuring periods with hypomanic symptoms and periods with depressive symptoms that do not meet criteria for a major depressive episode or bipolar disorder), mood disorders due to a general medical con-

TABLE 14.1. *Characteristics of a Depression Syndrome (Not All Manifestations Are Present in All Patients)*

General Characteristic	*Manifestations*
Mood alterations	Sadness
	Anhedonia
	Feelings of guilt
	Feelings of helplessness
	Feelings of hopelessness
	Feelings of worthlessness
	Thoughts of death and suicidal ideation
Changes in verbal expression	Delay in responding verbally
	Shortened length of verbal responses
	Slow rate of speech
	Reduced initiation of conversation
	Hypophonia (increases toward the end of sentences)
	Lack of emotional inflection (dysprosody)
	Stereotyped speech themes involving sad content
Disturbances of affect	Sad expression
	Crying
	Avoidance of eye contact
	Furrowed brow
	Facial immobility and hypomimia
Motor alterations	Bending of the head forward
	Slumped shoulders
	Reduced gestures in movement
	Slowness of gait and movement (occasional agitation)
	Catatonia
Cognitive impairments	Reduced associations
	Tendency toward recurrent depressive thoughts
	Impaired executive function
	Reduced visuospatial function
Motivational changes	Reduced interest
	Reduced initiation of new activities
Neurovegetative disturbances	Appetite changes (usually reduced, occasionally increased)
	Sleep disturbances
	Early morning awakening
	Difficulty falling asleep
	Multiple awakenings
	Reduced REM sleep latency
	Diurnal variation in mood
	Loss of libido
Neuroendocrinologic changes	Abnormal dexamethasone suppression test (DST)
	Abnormal thyrotropin releasing hormone (TRH) test
Associated changes in neuroimaging	Increased periventricular white matter changes
	Reduced frontal lobe blood flow or metabolism

REM, rapid eye movement.

dition including neurological disorders, and substance-induced mood disorders.[1] The features required for a diagnosis of mood disorder due to a general medical condition are not specified in the *Diagnostic and Statistical Manual of Mental Disorders* (DSM),[1] but most researchers have used the criteria of major depressive episodes or minor depression, with the latter defined according to the same criteria as dysthymic disorder but beginning coincidentally or after the onset of the CNS disorder.

Interactions between Depression and Neurological Disorders

The interaction between depression and neurological disease is complex. When depression precedes the onset of neurological disease, it is often unclear whether the depression is the first manifestation of the illness or coincidentally preceded the onset of an ensuing brain disease; in some cases depression and neurological signs or dementia may coexist as independent products of a CNS condition (this appears to be the case in Parkinson's disease, for example); depression may produce abnormalities of cognition, including a dementia syndrome of depression (described below); neurological disorders contain many elements of a depressive disorder, even in the absence of a mood change, and may imitate a depressive disorder (pseudodepression); depression may reduce motivation and reduce the patient's interest in answering questions or engaging in examinations, thus imitating a dementia syndrome (depressive pseudodementia); patients may experience sadness and grief when they recognize declining cognitive or functional abilities as a consequence of a neurological disease; and neurological disease may precipitate depression in predisposed individuals (Table 14.2).

Depression has not been systematically assessed in many neurological disorders and very few studies have been based on community samples. Available information is based on assessment of depressive symptoms in convenience samples of patients attending clinics and reflect the referral biases of the reporting sites. In addition, the methodology for assessing depression has varied widely, producing discrepant results in the literature. Increasing awareness of these difficulties and advances in assessment methodologies used in recent studies have allowed more insight into mood changes in patients with neurological disorders.

Observations that have emerged consistently across neurological populations include the following: depression is underecognized in neurological conditions, even when recognized; depression tends to be undertreated in patients with neurological disorders; depres-

TABLE 14.2. *Relationships between Depression and Neurologic Disorders*

Depression is underrecognized and undertreated in neurological disorders (e.g., those with a personal or family history of depression).
Neurologic disorders may precipitate depression in predisposed individuals.
Neurologic disorders may produce depression as a sentinel event preceding other manifestations of brain disease.
Depression may accompany the onset of a neurologic disorder as a neurobiological component of the condition.
Depression may emerge in the course of a neurological disorder.
Depression may occur as psychological response to disability.
Depression can produce cognitive changes (dementia of depression).
Depression can reduce motivation and decrease willingness to answer questions, imitating a dementia syndrome (depressive pseudodementia).
Neurological disorders can produce signs—slowness, hypomima, apathy—that imitate depression (pseudodepression).
Depression commonly co-occurs with other types of psychopathology (psychosis, agitation) in patients with neurologic disorders.
Neurological disorders can produce affective changes of depression without a corresponding mood change (sham emotions; pseudobulbar palsy).
Neurological disorders can obscure recognition of a depressive disorder (aphasia, dysprosodia).
Depression exacerbates functional and cognitive impairment in neurologic disorders.
Depression hinders rehabilitation after brain injury.
Depression is most common in patients with conditions affecting the basal ganglia, frontal cortex, or temporal cortex.
Depression symptom profiles may vary among neurologic disorders (depression is associated with paranoia in epilepsy and psychomotor retardation in post-stroke syndromes; depression has less guilt and self-blame in Parkinson's disease compared to that in idiopathic depressive disorders).
Neurologic disorders produce depression in the caregivers of affected individuals.
Drugs used to treat neurological disorders can cause depression (phenobarbital, interferon).
Drugs used to treat neurological disorders can improve or prevent the emergence of depression (pramipexole).
Depression in neurologic disease often responds poorly to pharmacotherapy and ECT; treatment-related delirium is common.

ECT, electroconvulsive therapy.

sion hinders rehabilitation of patients with acute neurological conditions such as stroke or traumatic brain injury; depression exacerbates functional abnormalities and disturbances of activities of daily living; depression increases cognitive disturbances; depression and apathy are frequently inadequately distinguished. Patients with diseases affecting the basal ganglia and frontal and temporal lobes appear to be at particularly high risk for the development of depression, but depressive symptoms are common in nearly all types of brain disorders that have been assessed. There have been very few randomized double-blind, placebo-controlled trials of antidepressants in patients with neurological disease and depression, and antidepressant treatment is based primarily on extrapolation from treatment approaches developed for idiopathic depressive disorders. These extrapolations must be accepted with caution since the presence of a brain disorder may modify the response to antidepressant therapies.

Depression in Neurological Disorders

Table 14.3 lists neurological disorders in which depression has been described as a prominent manifestation.

• Parkinson's Disease and Parkinsonian Syndromes Depression has been studied intensively in basal ganglia disorders, especially Parkinson's disease (Chapter 18). The frequency of depression is increased in Parkinson's disease: major depression has been observed in 5%–25% of cases and minor depression in 25%–50% of cases. In all, approximately 50% of patients with Parkinson's disease will manifest depressive symptoms of varying severity.[5,6] Female gender and a past history of depression increase the likelihood of developing depression during the course of the illness.[5] Depression in Parkinson's disease is associated with more impaired cognitive functioning, the presence of psychotic features, and greater physical disability.[7–11] Several studies have found that depression is more common in patients with left hemisphere involvement and right hemi-Parkinsonism.[12,13] The profile of depressive symptoms in Parkinson's disease includes dysphoria, pessimism, and prominent somatic symptoms with less evidence of guilt and self-blame.[5,7] Patients with major depression show more rapid cognitive decline, deterioration of activities of daily living, and more severe disability than patients with no mood changes or minor depression.[14] In patients with the "on–off" phenomenon, depressive symptoms are more common when the patient is "off" and mood normalizes or may even become elevated during "on" periods.[15,16]

TABLE 14.3. *Principal Neurological and Systemic Disorders Producing Depression*

Neurological Disorders	*Systemic Disorders*
Cortical degenerations	Infections
Alzheimer's disease	Viral
Frontotemporal dementias	Bacterial
Dementia with Lewy bodies	
	Endocrine disorders
Extrapyramidal diseases	Hyperthyroidism
Parkinson's disease	Hypothyroidism
Huntington's disease	Hyperparathyroidism
Progressive supranuclear palsy	Hypoparathyroidism
Corticobasal degeneration	Cushing's syndrome
Wilson's disease	Addison's disease
Idiopathic basal ganglia calcification (Fahr's disease)	Hyperaldosteronism
	Premenstrual depression
	Diabetes mellitus
Cerebrovascular disease	Acromegaly
Stroke	Prolactinonia
White matter ischemia	Hypopituatarism
Arteriovenous malformations	Postpartum affective disorders
Cerebral neoplasms	
	Inflammatory disorders
Cerebral trauma	Systemic lupus erythematosus
	Temporal arteritis
CNS infections	Sjogren's syndrome
Lyme encephalitis	
Creutzfeldt-Jakob disease	Vitamin deficiencies
Viral encephalitis	Folate
	Vitamin B_{12}
Multiple sclerosis	Niacin
	Vitamin C
Epilepsy	
	Miscellaneous systemic disorders
Narcolepsy	
	Cardiopulmonary disease
	Renal disease and uremia
Hydrocephalus	
	Systemic neoplasms
	Porphyria
Inherited metabolic disorders	
	Klinefelter's syndrome

Neurobiological studies provide insight into the pathophysiological basis of depression in Parkinson's disease. Levels of 5-hydroxyindoleacetic acid (5-HIAA) are significantly lower in depressed than in nondepressed patients.[17] At autopsy, patients with Parkinson's disease and depression have been found to have significantly greater involvement of the locus coeruleus, the nuclear origin of norepinephrine, than nondepressed patients.[18] Studies of regional cerebral glucose metabolism using [^{18}F]-fluorodeoxyglucose (FDG) PET reveal diminished metabolism in the caudate nucleus and the inferior frontal regions[19] or the medial frontal and anterior cingulate regions.[20]

Depression is reported with variable frequency among other parkinsonian syndromes. Depressive symptoms were detected in approximately 20% of patients with progressive supranuclear palsy in a study using a prospective interview evaluation technique.[21] Through a similar approach, depression was found in approximately 75% of patients with corticobasal degeneration.[22] Depression has been reported in approximately 50% of patients with dementia with Lewy bodies.[23–25]

In Wilson's disease, 10%–25% of patients manifest psychiatric disturbances before the onset of neurological abnormalities, and in 10%–15% depression is the initial symptom.[26,27] Up to 40 or 50% of patients eventually manifest a psychiatric illness; 20%–40% show depressive symptoms.[28,29] Psychiatric symptoms are more common in patients with the neurological form of the disease than in those with the hepatic form. There is little relationship between motor, psychiatric, and cognitive symptoms, suggesting that the psychiatric manifestations are independent manifestations of the underlying neuropathological changes.[28,30]

• Huntington's Disease Approximately 40% of patients with Huntington's disease have mood disorders: 20%–30% meet criteria for major depressive episodes, 10% have depression with hypomanic episodes, and 5% have dysthymic disorder.[31–33] The association between mood disorder and Huntington's disease appears to be more common in specific families in whom the two disorders occur together on a recurrent basis.[33] Huntington's disease is associated with a marked increase in suicide rate; suicide was eight times more common among Huntington's disease patients aged 50–69 years than in a control group of unaffected individuals. Suicide is most common early in the course of the illness.[34] No correlation has been identified between the dementia syndrome, severity of motor abnormalities, or CAG repeat length in the causative mutation of Huntington's disease on chromosome 4.[35] The independence of psychiatric symptoms from cognitive and motor symptoms in Huntington's disease suggests that the mood changes are not a reaction to the illness but are a neuropsychiatric manifestation of the underlying pathophysiological alterations. Studies with FDG PET show reduced metabolism in the orbitofrontal–inferior prefrontal cortex in depressed patients with Huntington's disease compared to that in nondepressed Huntington's disease patients.[36] Studies of cerebrospinal fluid (CSF) have indicated no differences in concentrations of 5-HIAA between patients with Huntington's disease and normal volunteers. However, corticotropin-releasing factor (CRF) is elevated in Huntington's disease, and there is a positive correlation between CRF levels and the severity of major depression.[37]

• Alzheimer's Disease and Frontotemporal Dementias Patients with cortical dementias such as Alzheimer's disease and the frontotemporal dementias tend to exhibit less severe depression than those with subcortical disorders such as Parkinson's disease and Huntington's disease (Chapter 10). Depressive symptoms, however, are common in these disorders, occurring in approximately 40% of patients.[38,39]

Relatively few patients with Alzheimer's disease meet all criteria for a major depressive episode, but depressive symptoms such as tearfulness, sad affect, and expression of feelings of hopelessness and worthlessness are common. Ten to twenty percent of patients meet criteria for major depression.[39–41] Depression is a common harbinger of Alzheimer's disease, occurring up to 5 years prior to the onset of the illness.[42,43] Depression is associated with physical aggression and impaired instrumental activities of daily living and executive dysfunction.[44–47] Some studies have found depression to be more common in women with Alzheimer's disease and those with earlier onset of the dementia.[48,49] Patients with a family history of mood disorder are more likely to experience depression following the onset of Alzheimer's disease.[50,51] Patients with Alzheimer's disease tend to underestimate their own depressive symptoms and a surrogate reporter should be interviewed to assess mood changes.[52–54] Depression in Alzheimer's disease must be distinguished from apathy, which is a common symptom in this disorder.[4] Deep white matter lesions visualized on MRI have been found to correlate with ideational aspects of depression (low self-esteem, suicidal thoughts),[55] and functional imaging has shown correlations between depression and reduced metabolism and cerebral blood flow (CBF) in either the parietal or frontal cortical regions.[56–58] At autopsy, patients with Alzheimer's disease and depression often have greater involvement of the locus ceruleus than those without depression.[59,60] Depressed patients with Alzheimer's disease also have greater reduction in cortical norepinephrine levels, with relative preservation of cholinergic markers in subcortical regions, than those without depression.[61]

Cerebrovascular Disease Depression following stroke has been investigated and relationships between time of assessment, lesion location, lesion size, and a variety of clinical features have emerged. In the immediate post-stroke period (7–10 days) there is a higher frequency of depression with left hemispheric lesions than with right hemispheric lesions. This relationship is observed for both major depression (approximately 15% of patients) and minor depression (approximately 25% of patients).[62,63] After 3 to 6 months, this hemispheric asymmetry is no longer evident: patients with both right and left hemispheric lesions evidence depressive disorders and there is an inverse correlation between the distance of the lesion from the frontal pole of either hemisphere and the severity of depression (Fig. 14.1).[62] After 1 year, patients with right hemisphere lesions are more likely to be depressed than patients with left hemisphere strokes.[62]

When anxiety complicates depression, the lesion is more likely to involve cortical than subcortical structures.[9] Among patients with subcortical lesions, depression is more common following injury to the left caudate nucleus than injury to the right basal ganglia or thalamic nuclei.[63,64] Several investigators have documented a relationship between lesion size and severity of depressive disorders.[65,66] Patients with depression and left hemisphere stroke have greater cognitive impairment than patients with depression and right hemisphere stroke or patients with stroke without depressive disorders.[67–69]

Post-stroke depression is more common in patients who have enlarged ventricles, which suggests that atrophy may predispose to the occurrence of post-stroke mood changes.[64] Women are more likely to exhibit post-stroke depression than men.[70] The profile of depressive symptoms is essentially identical in post-stroke depression and idiopathic late-onset depression, although psychomotor retardation is more prominent in post-stroke disorders.[71] Depression is correlated with impaired activities of daily living, both acutely and at long-term follow-up.[72,73]

Patients with post-stroke depression have significantly lower concentrations of CSF 5-HIAA than nondepressed post-stroke patients and have higher serotonin receptor (S_2) binding in uninjured regions of right parietal and temporal cortex than do nondepressed subjects.[74] Post-stroke depression may be mediated in part through serotonergic mechanisms, with reduced serotonin availability and failure to up-regulate serotonin receptors in the injured hemisphere.

Vascular depression refers to the accruing evidence that many cases of late-onset depression are related to overt stroke, silent cerebral infarction, or subcortical

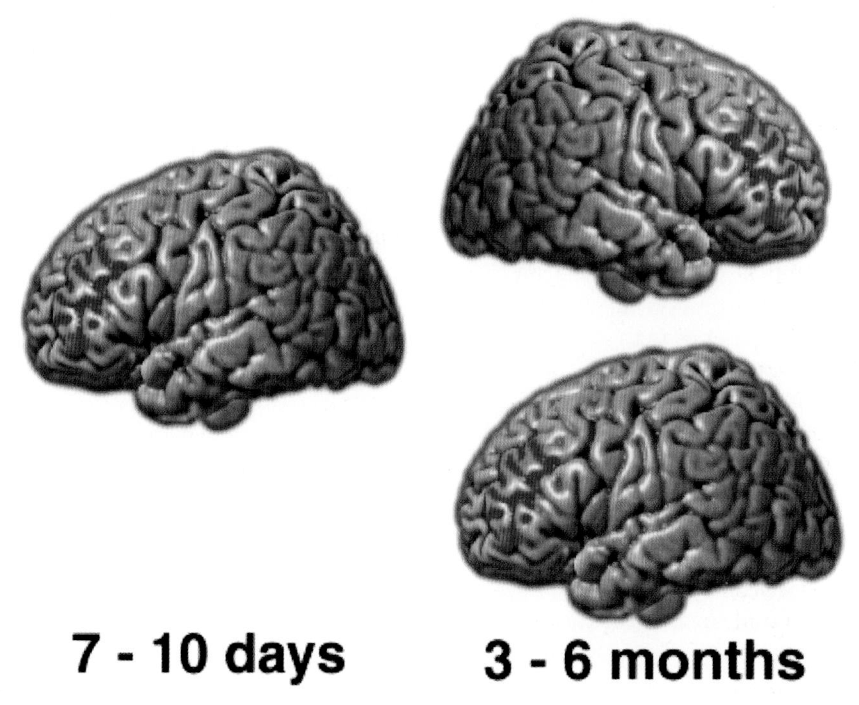

FIGURE 14.1 Lesion location and time course in post-stroke depression.[62] Regions in red were more commonly associated with depression for each time period.

TABLE 14.4. *Features of Vascular Depression**

Cardinal features

- Clinical and/or neuroimaging evidence of cerebrovascular disease
  Clinical evidence includes a history of transient ischemia attack or focal signs, gait disturbance, or incontinence
  Neuroimaging evidence includes structural or functional evidence of focal ischemia, hemorrhagic brain injury, or subcortical ishcemia injury.
- Onset of depression after age 65 or change in course of early-onset depressive disorder

Secondary features

- Vascular risk factors (hypertension, history of myocardial infarction, atrial fibrillation, carotid stenosis, hyperlipidemia, hypercholesterolemia, obesity, smoking habit)
- Cognitive impairment with executive dysfunction
- Psychomotor retardation
- Mood abnormalities with less guilt
- Poor insight
- Impaired activities of daily living
- Absence of a family history of mood disorder

Prognosis

- Mood changes respond to antidepressant pharmacotherapy but the response is less predicatble than when these agents are used in idiopathic depression. Confusion following initiation of treatment is not uncommon.
- Mood changes may respond to ECT; prolonged post-ECT confusion is more common than when applying ECT to idiopathic depression.

ECT, electroconvulsive therapy.
*Identified from Alexopoulos et al. (1997).[78]

white matter ischemic injury.[75,76] Table 14.4 presents a summary of the characteristics of vascular depression. Onset is typically after the age of 65 or there is a change in the course of an early-onset depressive disorder. There is clinical or imaging evidence of cerebrovascular disease, and vascular risk factors (especially hypertension) are commonly present. Neuropsychological dysfunction, particularly executive abnormalities, is evident. The profile of mood symptoms includes less prominent guilt than is typical in most cases of idiopathic depression, poor insight, psychomotor retardation, and an absence of family history of a mood disorder.[77,78] From a treatment perspective, vascular depression may respond to either pharmacotherapy or treatment with electroconvulsive therapy (ECT); the response is less robust than in patients with idiopathic depression and complications such as confusion are more common.[79–82] Increased white matter hyperintensities consistent with ischemic injury and vascular damage to the basal ganglia are particularly common substrates for vascular depression.[83–85]

• Epilepsy The neuropsychiatry of epilepsy is discussed more comprehensively in Chapter 21. Depression is common in this disorder; it may occur as a prodromal emotional change prior to a seizure, comprise part of an aura at the beginning of an ictal event, be present as an ictal manifestation during the course of the seizure, follow a seizure as part of a postictal state or may be an interictal manifestation. The interictal depression associated with epilepsy is common in those with complex partial seizures, particularly those with left-sided foci.[86,87] Interictal depression is the most common type of psychopathology observed among epilepsy patients. It is characterized by mood changes, psychotic traits, paranoia, chronic dysthymia, anxiety, and hostility.[86,88] Although suicide and suicide attempts are common among epileptic patients, they are more often associated with personality disorders and psychosis than with depression.[89] Studies with FDG-PET reveal reduced metabolism in the inferior frontal cortex bilaterally[90] or unilaterally in the left temporal lobe.[87]

In treating depression and epilepsy, clinicians must take into account the interactions of the seizures and the therapeutic interventions. Amoxapine and maprotiline lower seizure threshold and may exacerbate seizures. Monoamine oxidase inhibitors (MAO-I) are the agents least likely to increase seizure frequency.[91] Interactions between anticonvulsants and antidepressants also must be considered. Tricyclic antidepressants may increase carbamazepine and phenytoin levels, while phenobarbital and phenytoin may decrease tricyclic antidepressant levels. Selective serotonin reuptake inhibitors (SSRIs) may increase levels of carbamazepine, phenobarbital, phenytoin, and valproate; and carbamazepine, phenobarbital, and phenytoin may all decrease paroxetine levels.[91] Patients with epilepsy requiring ECT to relieve their depression should remain on their anticonvulsant regimen during the course of ECT treatment. Their anticonvulsant drug should not be administered on the morning prior to the treatment. The seizure threshold should be determined at the time of the first treatment and a stimulant dosage at least moderately above seizure threshold should be administered during the subsequent treatments.[91]

• Multiple Sclerosis The neuropsychiatry of multiple sclerosis is described in Chapter 26. Mood abnormalities are common manifestations of multiple sclerosis and a management challenge for the clinician providing care to these patients. Up to 80% of patients with multiple sclerosis have depressive symptoms, and approximately 20% have symptom severity sufficient to warrant treatment with antidepressants.[92] The profile of mood abnormalities in multiple sclerosis includes sadness, self-reproach, and somatic features. Neurovegetative and somatic abnormalities are common in multiple sclerosis without mood abnormalities and may mislead clinicians into overestimating the prevalence of depression.[93,94] Depression is unrelated to physical disability, but mood changes are more likely to occur during exacerbation or progression.[93,95] Rates of mood disorders are not elevated among first-degree relatives of patients with multiple sclerosis, suggesting that the high rate of mood changes in these patients is not genetically determined.[96] Depression is one of the principal determinants of quality of life of patients with multiple sclerosis.[97] Patients with cerebral lesions are much more likely to suffer depressive episodes than patients with spinal lesions who are equally disabled.[98] Patients with mood abnormalities have higher plaque counts in frontal and temporal regions than patients with multiple sclerosis and no associated mood changes.[99–101]

A variety of treatment issues arise with regard to multiple sclerosis and depression. Treatment with interferon β-1b (IFN-β-1b) has been reported to induce new-onset depression or worsen depressive symptoms.[102] Treatment with antidepressants or psychotherapy improves adherence to IFN-β-1b therapy.[103] Depression in multiple sclerosis has been reported to respond to both tricyclic antidepressants and SSRIs.[104,105] Psychotherapy has an important role in aiding patients with this complex relapsing and remitting disorder to adjust to the challenges presented by the disease.[106]

• **Traumatic Brain Injury** The neuropsychiatry of traumatic brain injury is discussed in more detail in Chapter 26. Depression is a common sequelae of traumatic brain injury. Depression is present in approximately 25% of patients in the acute post-traumatic period and additional patients have delayed-onset depression. The frequency of depression continues to rise for several years following the brain injury.[107–109] Anxious depression is common following brain injury.[108,110] The severity of brain injury is a significant predictor of post-traumatic psychopathology; pre-injury family function and a past history of psychiatric disorder also increase the risk of a posttraumatic neuropsychiatric syndrome.[111,112] Most studies have shown a predominance of frontal lesions in patients with post-traumatic mood disorders. In some cases the lesions involved the dorsolateral prefrontal cortex,[111] in others, the lesions involved the orbitofrontal cortex.[113] Depression associated with post-traumatic brain injury may improve following treatment with SSRIs and there is often a concominant improvement in anger, aggression, and post-concussive symptoms.[114]

• **Other Neurological Disorders** There have been few studies of the relative frequencies of neurologic causes of depressive disorders. Patients on a neurological inpatient service found to evidence depression included those with epilepsy, stroke, multiple sclerosis, Parkinson's disease, aqueductal stenosis, and brain tumors.[115] Neurologically ill patients manifesting depression and seen by a psychiatric consultation liaison service included patients with stroke, human immunodefiency virus (HIV) infection, cerebritis due to systemic lupus, erythematosus Parkinson's disease, hypothyroidism, multiple sclerosis, brain trauma, brain tumors, arteriovenous malformations, and epilepsy.[116] Depression is most likely to accompany cerebral tumors when either heteromodal frontal or parietal association cortex is involved in combination with paralimbic regions.[117] Depression has also been associated with frontal arachnoid cysts, third ventricular colloid cysts, Lyme encephalopathy, and hereditary storage diseases including GM2, gangliosidosis, and Gaucher's disease.[118–122]

Treatment of depression in patients with neurological illnesses is approached with the usual pharmacological agents. Electroconvulsive therapy is an important therapeutic option but should be avoided in those with increased intercranial pressure, headache, or focal neurological deficits.[123]

Toxic and Metabolic Disorders with Depression

• **Endocrine and Systemic Disorders** A wide variety of endocrinopathies and systemic disorders have been associated with depression (Table 14.3). Medical disorders producing depression include cardiovascular syndromes, such as hypoxia and mitral valve prolapse. Abnormalities secondary to renal dysfunction including uremia, hyponatremia and hypoglycemia, nutritional deficits, and collagen vascular diseases.[124] Systemic neoplasms have also been associated with the presence of depressive symptoms.[124] Many endocrine disorders produce depression, including hypo- and hyperthyroidism, hypo- and hyperparathyroidism, adrenal cortical insufficiency (Addison's disease), adrenal cortical excess (Cushing's syndrome), diabetes, and pituitary diseases such as agromegaly, prolactinoma, and hypopituitarism.[125–127]

● **Drugs Associated with Depression** Many drugs have been related to the induction of depression (Table 14.5). Classes of agents most likely to induce depression include antihypertensive drugs, analgesics, anticonvulsants, sedative-hypnotics, antimicrobial drugs, antineoplastic compounds, and endocrine agents.

Neurobiology of Idiopathic Depression

The clinical factors of depressive disorders are described above. The phenomenological similarities of the clinical syndrome of idiopathic depression and depressive disorders occurring in conjunction with neurologic and medical illnesses suggest a common underlying pathophysiology (Chapter 5). Differences in the anatomical distribution of lesions in different neurological disorders producing depression may account for the degree of phenomenologic hetereogeneity observed in these depressive disorders. The existence of shared mechanisms in idiopathic and acquired depression syndromes is supported by the convergence of anatomic and physiologic abnormalities in the two groups of disorders.

Structural imaging with CT or MRI demonstrate diminished size of the frontal lobe and of the caudate nuclei in depressed patients, compared to that in nondepressed patients.[128–130] In patients with late-onset depression, deep white matter and periventricular hyperintensities are typically present. These lesions are significantly less common among patients with early-onset depression.[130,131] Studies of CBF using SPECT or PET reveal diminished blood flow in the inferior frontal, dorsolateral prefrontal, and anterior cingulate regions.[132–134] Studies with FDG-PET show diminished activity of the left dorsal anterior lateral prefrontal cortex and the caudate nuclei.[135] Positron emission tomography studies using ligands to identify serotonin receptors (5-HT_2) indicate reduced receptor density in the posterolateral orbital frontal cortex and the anterior insular cortex of the right hemisphere. While not completely consistent, these studies demonstrate abnormalities of the frontal lobes and basal ganglia across different metabolic, perfusion, and structural studies.

Neuropsychological testing also suggests frontal and frontal–subcortical dysfunction in patients with major depression. Executive abnormalities are consistently found and skills mediated predominantly by the right hemisphere tend to be more impaired than those mediated by the left. Nonverbal memory abnormalities and a discrepancy in the verbal/performance I.Q., with more pronounced deficit in the performance I.Q., have been described.[136,137] Verbal memory abnormalities have been identified in some studies.[138,139] Executive abnormalities become more apparent in more severe depressions.[136] These neuropsychological deficits correlate with reduced blood flow in the medial prefrontal cortex.[140,141] In addition, depressed patients fail to activate cingulate cortex and striatum in the course of a test demanding executive function.[142]

There have been relatively few postmortem studies of patients with major depression. Histopathologic studies have revealed diminished cortical thickness and decreased neural and glial densities in the rostral orbitofrontal cortex, reductions in glial densities in the caudal orbitofrontal cortex, and reductions in the density and size of neurons and glial cells in the dorsolateral prefrontal cortex.[143] Findings from postmortem studies of receptor binding are not completely consistent but have usually shown increased serotonergic binding involving the 5-HT_{2A} and 5-HT_{1A} receptors.[144] Several studies have identified a reduced number of neurons in the locus coeruleus of suicide victims.[144]

Studies in idiopathic and acquired depressive disorders suggest dysfunction of complex neurological circuits mediating mood and responsible for depression (Fig. 14.2). Orbitofrontal and anterior temporal paralimbic cortex are commonly involved in idiopathic and secondary depressions. Similarly, the caudate nucleus has often been observed to be abnormal in depression syndromes. Frontal–subcortical circuits linking the orbital frontal cortex and caudate are involved in many conditions with depression. Orbitofrontal cortex is joined via white matter tracts to adjacent anterior temporal paralimbic cortex. These cortical regions are also connected to subcortical and brainstem structures involved in Parkinson's disease and parkinsonian syndromes. Mayberg et al.[145] proposed that together these structures comprise a network that produces symptoms of depression when it is functionally compromised. This network includes paralimbic structures responsible for the emotional/experiential aspect of depression, limbic–hypothalamic influence mediating neurovegetative manifestations, limbic–basal ganglia connections responsible for affective and motor features, and limbic–dorsolateral prefrontal interactions

TABLE 14.5. *Drugs and Other Compounds Associated with Depression*

Category	*Agents*	*Category*	*Agents*
Antihypertensive drugs	Clonidine	Withdrawal syndromes	Barbiturates
	Oxprenolol		Alcohol
	Propranolol		Benzodiazepines
	Reserpine		Amphetamines
	Methyldopa		Corticosteroids
	Guanethidine	Hallucinogens	Indole hallucinogens
	Hydralazine	Antineoplastic drugs	Azathioprine
	Bethanidine		C-Asparaginase
	Nifedipine		Plicamycin (mithramycin)
	Prazosin		Vincristine
Antiarrhythmics	Cardiac glycosides		6-Azauridine
	Procainamide		Bleomycin
	Lidocaine		Trimethoprim
Antiparkinsonian drugs	Amantadine		Interferon
	Bromocriptine	Endocrine agents	Corticosteroids
	Levodopa		Oral contraceptives
Analgesics	Non-steroidal anti-inflammatory agents		Androgens
			Triamcinolone
	Narcotics		Norethisterone
	Phenacetin		Danazol
	Indomethacin	Central nervous system stimulants	Amphetamines
Anticonvulsants	Phenytoin		Fenfluramine
	Barbiturates		Diethylpropion
	Ethosuximide		Phenmetrazine
	Lamotrigine	Miscellaneous drugs	Acetazolamide
	Felbamate		Anticholinesterase
Sedative-hypnotics	Benzodiazepines		Baclofen
	Chloral hydrate		Choline
	Ethanol		Cyproheptadine
	Clomethiazole		Diltiazem
	Clorazepate		Diphenoxylate
Neuroleptics	Butyrophenones		Disulfiram
	Phenothiazines		Goserelin
Antibiotics	Penicillins		Halothane
	Sulfamethoxazole		Interferon α-2a
	Clotrimazole		Interferon β-2
	Cycloserine		Mebeverine
	Dapsone		Meclizine
	Ethionamide		Methoserpidine
	Tetracycline		Methysergide
	Griseofulvin		Metoclopramide
	Metronidazole		Metrizamide
	Streptomycin		Tacrolimus
	Nitrofurantoin		Naltrexone
	Nalidixic acid		Phenindione
	Sulfonamides		Pizotifen
	Procaine penicillin		Procaine
	Thiocarbanilide		Salbutamol
	Acyclovir		Tetrabenazine
	Isoniazid		Veratrum

Modified from Blank and Duffy (2000).[189]

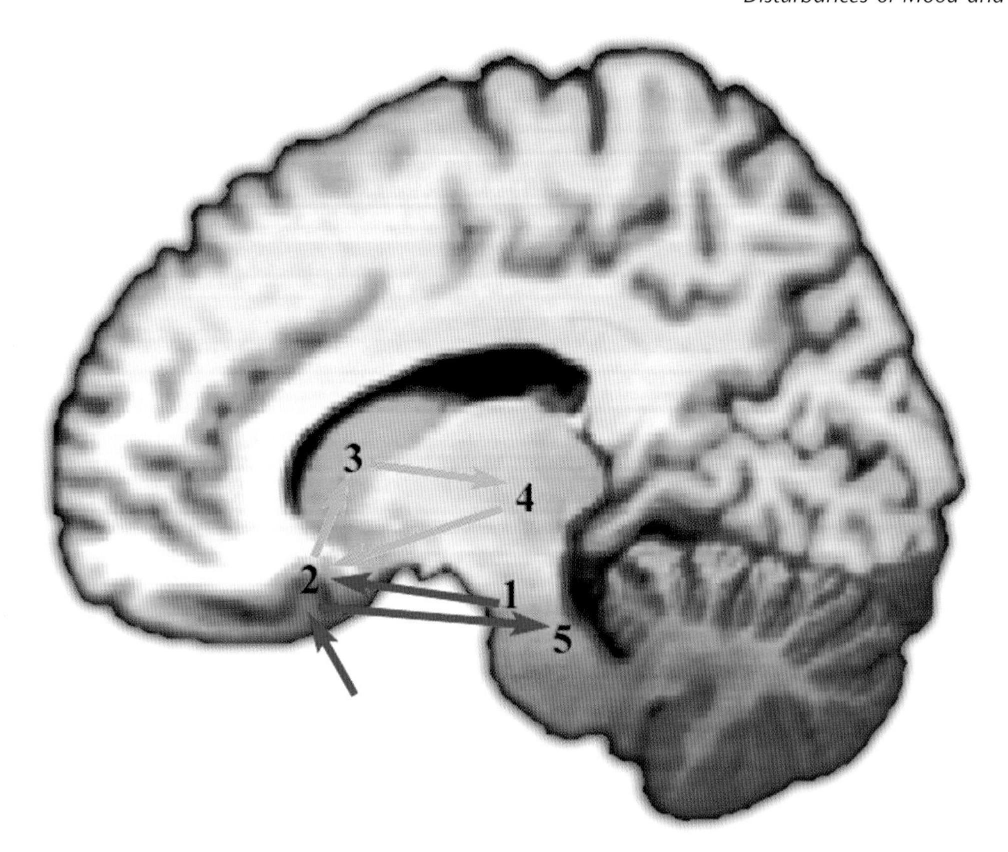

FIGURE 14.2 Depression model. Possible mechanisms for common paralimbic cortex hypometabolism in primary and secondary depressions include *1*. Connections of mesencephalic monamine neurons (vta, dr, lc) and their cortical projections; *2*. projections to basotemporal limbic regions and amygdala; *3* and *4*, cortical–basal ganglia circuits; and *5*. serotonergic neuron projections from orbital cortex. BG, basal ganglia; Cd, caudate; Cg, anterior cingulate; dr, dorsal raphe; iPF, inferior prefrontal cortex; lc, locus cerulerus; OF, orbital frontal cortex; T, temporal cortex; Th, thalamus; vta, ventral tegmental area. Modified from Mayberg et al.[145]

contributing to the cognitive characteristics of the depression syndrome.

Treatment of Depression

Table 14.6 outlines the typical starting dose and dosage range of a variety of classes of antidepressant agents. 1The choice of antidepressant is dictated largely by side effects. Clinicians should familiarize themselves with the use of a few agents from each class and become expert in the anticipation of adverse events and in administration regimens. There have been few studies of antidepressants in secondary depression associated with neurological disorders, and the efficacy of these agents may vary from that observed in clinical trials of patients with idiopathic mood disorders. Electroconvulsive therapy represents an alternative treatment intervention for patients requiring rapid resolution of their mood disorder, those who are intolerant of pharmacotherapy, or those who are treatment resistant.[146]

MANIA

Mania is a mood disorder featuring an elevated or expansive mood, increased physical activity, accelerated thought and speech, and neurovegetative changes. Idiopathic mania appears to be a genetically determined disorder and secondary mania is produced by a wide variety of neurological and endocrine disorders. Many types of drugs have also been associated with induction of secondary mania.

Currently recognized types of mania and cyclical mood disorders include bipolar I disorder, characterized by recurrent manic episodes or mixed manic and depressive episodes; bipolar II disorder, characterized by the occurrence of one or more major depressive episodes accompanied by at least one hypomanic episode; cyclothymic disorder, evidenced by numerous periods with hypomanic symptoms and numerous periods with depressive symptoms; mood disorder due to a neurological or medical condition and featuring an elevated expansive or irritable mood; and a substance-

TABLE 14.6. *Agents Used to Treat Depression*

Class *Generic Agent (Brand Name)*	*Initial Dose*	*Usual Dosage (mg/day)*
Tricyclic antidepressants and related compounds		
Amitriptyline (Elavil, Enovil)	25–75 mg qhs	100–300
Amoxapine (Asendin)	50 mg bid	100–400
Desipramine (Norpramin)	25–75 mg qhs	100–300
Doxepin (Adapin, Sinequan)	25–75 mg qhs	100–300
Imipramine (Janimine, Tofranil)	25–75 mg qhs	100–300
Maprotiline (Ludiomil)	25–75 mg qhs	100–225
Nortriptyline (Aventyl, Pamelor)	25–50 mg qhs	50–150
Protriptyline (Vivactil)	15 mg qAM	15–60
Trimipramine (Surmontil)	25–75 mg qhs	100–300
Selective serotonin reuptake inhibitors		
Citalopram (Celexa)	20 mg qAM	20–60
Fluoxetine (Prozac)	10–20 mg qAM	20–80
Fluvoxamine (Luvox)	50 mg qhs	100–300
Paroxetine (Paxil)	10–20 qAM	20–50
Sertraline (Zoloft)	25–50 mg qAM	50–150
Dopamine-reuptake blocking compounds		
Bupropion (Wellbutrin,	100 mg tid IR	300–450*
Wellbutrin SR, Zyban)	150 mg for 3–7 days, then 150 mg bid SR	
Serotonin/norepinephrine reuptake inhibitors		
Venlafaxine (Effexor, Effexor-XR)	25 mg bid-tid IR	75–375
	37.5 mg qd XR	
Nefazodone (Serzone)	100 mg bid	300–600
Trazodone (Desyrel)	50 mg tid	150–600
Mixed receptor effect		
Mirtazapine (Remeron)	15 mg qhs	15–45
Monoamine oxidase inhibitors		
Phenelzine (Nardil)	15 mg tid	15–90
Tranylcypromine (Parnate)	10 mg bid	10–60

*Not to exceed 150 mg/dose to minimize seizure risk for IR and 200 mg/dose for SR.

induced mood disorder characterized by elevated, expansive, or irritable mood.[1] Manic episodes may have psychotic features or catatonic features and may sometimes begin in the postpartum period. Rapid-cycling bipolar disorders feature the occurrence of four or more mood episodes during the previous 12-month period.[1]

Clinical Features of Mania and Hypomania

Mania is characterized by an elevated or euphoric mood, irritability (often with aggressiveness), hyperactive motor activity, pressured speech, flight of ideas and racing thoughts, distractibility, grandiosity, decreased need for sleep, excessive energy, poor judgment, disordered thinking, and poor insight. In psychotic cases there may be mood-congruent delusions and hallucinations.[1,147] Objectively patients feel more cheerful, more self-confident, and more energetic. They have a desire for increased social activity at work, at home, or in school and often have increased libido.[148] Poor judgment may result in engagement in high-risk activities. Hypomanic episodes are more brief, less intense, and do not have psychotic features; the symptoms are not severe enough to cause marked impairment in social or occupational functioning or to necessitate hospitalization.[1] In mixed episodes both manic and depressive features are present over the course of a period lasting at least 1 week. The thought disorders associated with mania may include racing thoughts, complex associations, clanging associations, rhyming puns, and flight of ideas. Speech may be pressured with short response latencies, extended soliloquies, and excessive volume. Expansive facial features and gestures may be noted.[149] Delusions of grandeur or beliefs in special powers or an annointed mission are typical mood-consistent delusions of mania. Table 14.7 presents a list of behaviors characteristic of mania.

TABLE 14.7. *Characteristic Features of Mania*

Aspect	*Features*
Mood	Elation
	Euphoria
	Irritability
	Impatience
Self-perception	Enhanced self-esteem
	Denial of illness
	Decreased somatic complaints
	Certainty of success
Social interactions	Unable to empathize
	Intrusive
	Provocative
Judgement	Impaired
	Disregard for potential adverse consequences of behavior
	Rash confidence
	Reckless
Thought processes	Racing thoughts
	Flight of ideas
	Increased associations
	Clang associations
	Rhyming puns
Speech	Short latencies
	Pressured
	Loud
Motor behaviors	Smiling, laughing
	Expansive facial expressions
	Expansive gestures
	Assertive posture
	Stereotypy and catatonic features (some cases)
Neurovegetative features	Decreased need for sleep
	High energy
	Increased libido
	Increased appetite
Psychosis (not present in all cases)	Mood-congruent grandiose delusions
	Mood-congruent hallucinations

Secondary Mania in Neurological Disorders

Krauthammer and Klerman[150] offered one of the first criterion-driven descriptions of secondary mania. They suggested that secondary mania is characterized by (*1*) episode duration of at least 1 week; (*2*) elated and/or irritable mood; and (*3*) at least two of the following behaviors: (*a*) hyperactivity, (*b*) pressured speech, (*c*) flight of ideas, (*d*) grandiosity, (*e*) decreased sleep, (*f*) distractibility, (*g*) lack of judgement. Exclusionary criteria included (*1*) a clear previous history of manic-depressive or other mood disorder or (*2*) the coexistence of the manic syndrome with symptomatology of an acute confusional state or delirium.

Table 14.8 lists neurological disorders reported to produce secondary mania. Several generalizations emerge from review of the focal and degenerative disorders that have been associated with secondary mania. First, right-sided lesions are far more common than left-sided lesions. Second, within the right hemisphere, involvement of the frontal lobe, basal ganglia (particularly the inferior caudate nucleus), thalamus, and inferior temporal regions are the most common locations for lesions associated with secondary mania. Parietal and occipital lesions have rarely produced this condition.[151] Third, many patients exhibiting secondary mania in association with a focal neurological disorders have a family history of psychiatric disturbance or mood disorder. This suggests that the combination of a strategically placed lesion with a genetic vulnerability may be necessary to produce the syndrome of secondary mania. Fourth, the presence of cerebral atrophy as evidenced by enlargement of the anterior lateral

TABLE 14.8. *Neurologic and Systemic Etiologies of Secondary Mania*

Neurologic Disorders	*Neurologic Disorders (Cont'd)*	*Systemic Disorders*
Extrapyramidal diseases	Infections	Uremia and hemodialysis
Idiopathic basal ganglia calcification	Neurosyphilis	Dialysis dementia
Parkinson's disease	Herpes simplex virus	Hyperthyroidism
After dopaminergic treatment	St. Louis encephalitis	Pellagra
After thalamotomy	Q fever	Carcinoid syndrome
After pallidotomy	Influenza	Vitamin B_{12} deficiency
With deep brain stimulation	Cryptococcosis	Postpartum mania
Huntington's disease	Mononucleosis	Cerebral anoxia
Postencephalitic Parkinson's disease	HIV encephalopathy	
Wilson's disease	Following streptococcal infections with Sydenham's chorea	
Idiopathic dystonia		
	Congenital disorders	
Cerebrovascular disorders	Fragile X syndrome	
Pontine infarction	Velo-cardio-facial syndrome (mutation of 22q11)	
Premedian thalamus (right)		
Arteriovenous malfunctions	Miscellaneous conditions	
Vascular dementia (right temporal, left temporal)	Cerebral trauma	
Basal ganglia (right)	Multiple sclerosis	
	Temporal lobe epilepsy	
Cerebral neoplasms	Frontotemporal dementia	
Craniopharyngioma	Hemispherectomy (right)	
Temporal glioma (right)	Cerebellar degenerations	
Parasagittal meningioma (compressing right frontal cortex)	Kleine-Levin syndrome	
Intraventricular meningioma (right)	Klinefelter's syndrome	
Diencephalic glioma (right)	Cerebral sarcoidosis	
Multiple metastases		

ventricles or third ventricle also appears to be a predisposing factor to secondary mania, indicating that reduced cerebral reserve may have a role in the syndrome. Finally, patients with seizures in association with their neurological disease appear to be at particularly high risk for the development of secondary mania.

Extrapyramidal disorders are among the most common syndromes associated with secondary mania, although only a minority of patients develop the disorder. Hypomania, occasionally progressing to mania, occurs in approximately 10% of patients with Huntington's disease.[32] In patients with Parkinson's disease, mania may occur following treatment with dopaminergic therapy, after pallidotomy or thalamotomy, or with deep brain stimulation. Mania also has been observed in patients with idiopathic basal ganglia calcification, postencephalitic Parkinson's disease, Wilson's disease, and idiopathic dystonia.[152,153]

Cerebrovascular disease is another common cause of secondary mania. The most common location for a lesion producing secondary mania is in the right perithalamic area.[154–157] Hemiballismus, hemidystonia, postural tremor, or left-sided chorea frequently accompany the thalamic and subthalamic lesions.[156,158] In some cases a combination of a strategically placed lesion and a precipitating medication such as a SSRI was simultaneously present.[158] Pontine infarctions and temporal arteriovenous malformations have been associated with secondary mania.[159–161] Vascular dementia is a rare cause of secondary mania.[162]

A variety of cerebral neoplasms have produced secondary mania, including diencephalic tremors, medial frontal meningiomas, subtemporal bone tumors, multiple metastases, hypothalamic teratomas, craniopharyngiomas, acoustic neurinomas, and brainstem tumors.[155,163–165]

Traumatic brain injury is another potential cause of secondary mania. Most lesions have involved the midline orbitofrontal regions or thalamic or parathalamic structures including the temporal lobe, thalamus, or paralimbic striatum.[9,166–169] The mania may appear immediately following the brain injury or may be delayed for up to 3 years. Most cases appear within the first few months following the brain injury.[169] Irritable mood is more common than euphoria in post-traumatic bipolar disorder and psychosis occurs in approximately 15% of patients.[170] Post-traumatic mania is not associated with overall severity of brain injury, degree of physical impairment, degree of cognitive impairment, or level of social function.[171]

Euphoria or eutonia (an unwarranted sense of well-being) are second only to depression as mood manifestations in patients with multiple sclerosis. Frank mania is more unusual, and has been recorded in approximately 10% of patients with multiple sclerosis. Determination of the prevalence of mania in multiple sclerosis is complicated by the common use of steroids to treat this disorder and these agents are known to precipitate mania in some patients.[104,172]

A variety of infections have also been associated with secondary mania, including neurosyphilis (the general paretic form), herpes simplex encephalitis, St. Louis encephalitis, Q fever, influenza, cryptococcosis, mononucleosis, HIV encephalopathy, and following streptococcal infection with Sydenham's chorea.[173–178]

Other disorders that have been associated with secondary mania include cerebellar degenerations, cerebral sarcoidosis, cardiac arrest, temporal lobe epilepsy, hemispherectomy, Kleine-Levin syndrome and Klinefelter's syndrome, and frontotemporal dementia.[179–185] Certain chromosomal abnormalities, including fragile X syndrome and velo-cardio-facial syndrome resulting from a microdeletion on chromosome 22q11, are associated with bipolar mood disorders.[186,187] These mutation associations may provide insight into the genetic basis of mood disorders.

Metabolic and Systemic Disorders with Secondary Mania

A variety of systemic disorders have been reported to produce secondary mania (Table 14.8). Uremia and hemodialysis, dialysis-related dementia, hyperthyroidism, pellagra, carcinoid syndrome, vitamin B_{12} deficiency, and postpartum conditions have all been associated with secondary mania.[179,182]

Drug-Induced Mania

A large number of drugs have been reported to produce secondary mania (Table 14.9). Some classes of agents are particularly likely to produce manic syndromes, including antihypertensive and cardiovascular compounds, antiparkinsonian compounds (especially dopaminergic agents), anticonvulsants, antidepressants, atypical antipsychotics, CNS stimulants, some types of antimicrobial agents, and hallucinogens. Withdrawal syndromes may also include secondary ma-

TABLE 14.9. *Drugs and Toxins Reported to Produce Mania*

Category	*Agents*
Antihypertensive drugs	Clonidine
	Propranolol
	Methyldopa
	Hydralazine
	Captopril
Cardiac glycosides, antiarrhythmics, cardiovascular drugs	Digitalis
	Procainamide
	Lidocaine
	Propafenone
	Disopyramide
Antiparkinsonian drugs	Amantadine
	Bromocriptine
	Levodopa
	Lisuride
	Piribedil
	Procyclidine
	Selegeline
Analgesics	Non-steroidal anti-inflammatory (NSAIDs)
	Tramadol
	Indomethacin
Anticonvulsants	Carbamazepine
	Phenytoin
	Barbiturates
	Ethosuximide
	Clonazepam
	Phenacemide

(*continued*)

TABLE 14.9. *Drugs and Toxins Reported to Produce Mania (Continued)*

Category	*Agents*
Sedative-hypnotics	Alprazolam
	Triazolam
	Buspirone
	Meprobamate
Antidepressants	Bupropion
	Fluoxetine
	Fluvoxamine
	Monoamine oxidase inhibitors (MAOIs)
	Mirtazapine
	Nefazodone
	Paroxetine
	Sertraline
	Tricyclic agents
	Trazodone
Antipsychotics	Olanzapine
	Risperidone
Antimicrobial agents	Cycloserine
	Dapsone
	Penicillins
	Podophyllin
	Zidovudine (AZT)
	Antimalarials
	Isoniazid
	Iproniazid
	Cephalosporins
H_2 blockers	Cimetidine
Withdrawal syndromes	Alcohol
	Amphetamines
	Baclofen
	Diethylpropion
	Monoamine oxydase inhibitors (MAOIs)
	Nicotine
	Opiates
	Propranolol
	Tricyclic agents
Hallucinogens	Cannabinols
	Indole hallucinogens
	Phencyclidine
Antineoplastic drugs	Procarbazine
Endocrine agents	Corticotropin (ACTH)
	Corticosteroids
	Dihydroepiandrosterone
	Lupron
	Thyroid hormones
Central nervous system stimulants	Amphetamine
	Cocaine
	Diethylpropion
	Ephedrine
	Fenfluramine
	Isoetharine
	Methylphenidate
	Pemoline
	Phenylephrine
	Phenylpropanolamine
	Pseudoephedrine
	Sympathomimetics
Miscellaneous drugs	Albuterol
	Aspartame
	Baclofen
	Bromide
	Calcium
	Cyclobenzaprine
	Cyclosporin A
	Cyproheptadine
	Dithyl-*m*-toluamide
	Diltiazem
	Flutamide
	Interferon α
	L-Glutamine
	Mepacrine
	Metrizamide
	Oxandrolone
	Oxymetholone
	Procarbazine
	Procyclidine
	Propafenone
	Theophylline
	Tryptophan
	Yohimbine
	Zidovudine (AZT)

nia.[188–193] In patients with an underlying bipolar illness, treatment with antidepressants may precipitate a manic episode and change the pattern of cycling to more accelerated cyclic variations or change the severity of the manic behavior. L-dopa has been noted to have a similar property in some patients.[194]

In a review of 128 case reports of drug-induced mania, Sultzer and Cummings[194a] noted that steroids, levodopa and other dopaminergic agents, proniazid, sympathomimetic amines, triazolobenzodiazepines, and hallucinogens were the agents most likely to cause secondary mania. Common characteristics of drug-induced manic episodes included hyperactivity, rapid speech, elevated mood, and insomnia. A majority of the patients had a personal history of mood disorder, a family history of a psychiatric disturbance, or symptoms of a mood abnormality at the time the inciting agent precipitated the manic episode.

Idiopathic Bipolar Illness

The lifetime prevalence of idiopathic bipolar disorder is 1.2%. It represents 20% of all cases of major mood disorder.[195] Mean age of onset is 18 years in men and 20 years in women with substantial variability in the periodicity and pattern of manic episodes, but nearly all cases eventually recur. As the number of episodes increases, the cycle length tends to decrease.

There is a marked familial preponderance of bipolar illness. Monozygotic twins have a concordance rate of 65% while dizygotic twins exhibit a 20% concordance rate. No specific mutation has been identified but linkage has been demonstrated and replicated to chromosomes 18 and 21.[196] While onset of bipolar illness in adolescence and mid-life is typically an inherited condition, late-onset mania (over 65) is typically associated with an underlying neurological condition.[44,197–199]

Neuroimaging has revealed abnormalities in patients with bipolar illness. Those with bipolar I disorder more commonly have periventricular hyperintensities, unlike bipolar II patients or normal controls.[200] Magnetic resonance imaging studies also reveal enlarged amygdalae in bipolar patients.[201] Studies with SPECT show marked right–left asymmetries of temporal perfusion during mania.[202,203]

Treatment of Mania

Mania usually impairs social and occupational function and requires treatment. Antimanic agents are outlined in Table 14.10. In the acute phase of management of a manic patient, an antipsychotic or sedating agent may be necessary. For control of mania and reduction of recurrence of manic episodes, lithium, carbamazapine, or valproate are typically used. Gabapentin and lamotrigine are newer anticonvulsants that may have mood-stabilizing properties.[204,205] Clonazepam may be useful as an adjunct to mood stabilizer therapy.[204,206] Electroconvulsive therapy produces remission of mania in 80% of patients who have responded poorly to pharmacotherapy.[170]

OTHER TYPES OF MOOD DISORDERS

A variety of other types of mood disorders have been described in specific neuropsychiatric conditions. This section briefly describes these disorders. Mood changes occurring with epilepsy are also described in Chapter

TABLE 14.10. *Agents Used to Treat Mania*

Generic Name (Brand Name)	*Usual Dosage*	*Therapeutic Serum Level*
Lithium (Eskalith; Lithane; Lithobid; Lithonate; Lithotabs)	600–1800 mg/day	0.5–105 mEq/L
Carbamazepine (Epitrol; Tegretol; Tegretol–XR)	600–1800 mg/day	N/A
Valproate (Depakene; Depakote)	1–3 g/day	50–125 μg/mL
Gabapentin (Neurontin)	500–3600 mg/day	N/A
Lamotrigine (Lamictal)	100–400 mg/day	N/A

N/A, not available.

21 and mood disorders associated with violence and aggression are presented in Chapter 24.

Ictal Fear

In his classic paper on emotions occurring in the course of epileptic seizures, Dennis Williams[207] called attention to the common occurrence of fear as an ictal experience. Of patients who experienced an ictal emotional disorder, 61% experienced fear as their primary emotion. Fear was most common when the epileptic lesion was in the anterior temporal region. Fear has been observed in patients with foci in either temporal lobe, but it appears to be more common in patients whose seizures originate in the right temporal cortex.[208] Fear is occasionally the only manifestation of temporal lobe seizures or status epilepticus and must be distinguished from panic attacks (Chapter 17).[209]

Catastrophic Reactions

The frequency of catastrophic reactions in patients with brain disorders is controversial. Agitation and aggression outbursts are common in dementia syndromes (Chapter 10) but they are less common in patients with focal brain injuries. Gainotti[210] described anxiety reactions as complex emotional syndromes including anxiety, tears, aggressive behavior, swearing, displacement activities, and refusal. He found such reactions to be more common in patients with left hemisphere lesions. More recent studies fail to confirm a lateral predominance of lesions but note that catastrophic reactions are more common in patients with depression, individuals with a personal or family history of psychiatric disorder, and patients with lesions in the basal ganglia.[211]

Rage Reactions with Hypothalamic Lesions

Patients with lesions of the ventromedial hypothalamus may manifest a complex syndrome evidenced by a triad of neurobehavioral findings: rage, hyperphagia with obesity, and memory impairment. Rage reactions have the features of an exaggerated frontal disinhibition syndrome with impulsive property destruction and interpersonal violence that is unprovoked. The syndrome has been seen with hypothalamic hamartomas, craniopharyngiomas, astrocytomas, and gangliocytomas.[212–215] In some cases, rage and intermittent explosive behavior are the primary manifestation of the hypothalamic lesion.[216]

Mood Changes in the Kleine-Levin Syndrome

Kleine-Levin syndrome occurs in adolescent boys and is characterized by two cardinal symptoms of episodic hypersomnia and inordinate hunger (megaphagia) while awake. In addition to these two key symptoms, patients also exhibit cognitive impairment, perplexity, mood disorders, and disinhibited behavior. In most cases, the syndrome is self-limited and resolves spontaneously without a definite etiologic diagnosis being determined. Symptomatic cases resulting from head trauma and encephalitis have been reported.[217–219] The primary mood changes observed in the syndrome include irritability and aggression. The few autopsied cases of patients who have succumbed to illnesses causing the Kleine-Levin syndrome suggest involvement of the hypothalamus.

Emotional Placidity in the Klüver-Bucy Syndrome

The Klüver-Bucy syndrome is characterized by visual agnosia or "psychic blindness," strong oral tendencies, hypermetamorphosis or excessive tendency to attend and react to visual stimuli, altered sexual activity, changes in dietary habits, and emotional placidity characterized by the loss of both aggressive and fear responses.[220,221] The syndrome typically follows damage to the amygdalae and their cortical connections.[222] Emotionally, patients show a diminution in the amplitude of emotional responses. They fail to exhibit anger in provocative circumstances, are indifferent toward family and friends, and lack emotion-related interests. Facial amimia and monotony of vocal expression are common accompaniments of the absence of emotion.[223] The characteristics of the syndrome overlap with apathy (Chapter 16) and the lack of engagement observed in patients with Alzheimer's disease (Chapter 10) and some frontal lobe disorders (Chapter 9).

Eutonia in Multiple Sclerosis

The concept of eutonia was originally introduced into the literature by Cottrell and Wilson[224] who found that 84% of their cases had a mood syndrome characterized by a sense of well-being and unconcern over physical disability. Eutonia can be distinguished from euphoria by the latter's elevated mood, increased energy, and exaggerated flow of ideas and speech.[104] The eutonic state bears some resemblance to anosognosia for illness (Chapter 8).

Miscellaneous Mood Syndromes

In addition to the mood syndromes described above, patients with neuropsychiatric syndromes may exhibit a variety of other mood disorders or unusual emotional states. *Alexithymia* refers to a syndrome in which individuals have difficulty verbalizing symbols, lack the ability to express feelings verbally, have an impoverished fantasy life, and show an over-conformity in their interpersonal relations.[225] This syndrome has been observed in patients following commissurotomy with division of the corpus callosum.

Seizure foci provide a veritable laboratory of human emotions and provide tentative localizing information (Chapter 21). Among mood-related symptoms that have been described in the course of epileptic seizures are embarrassment[226] and preoccupation with death.[227]

DISTURBANCES OF AFFECT

Disturbances of mood—the emotion experienced by the patient—must be distinguished from abnormalities of affect—the emotion expressed by the patient but not necessarily corresponding to an underlying mood state. In normal circumstances, mood and affect are congruent, but in many neuropsychiatric conditions the two may be dissociated. Three principal situations in which brain lesions result in alterations of affect are (*1*) pseudobulbar palsy; (*2*) ictal effective changes; and (*3*) motor disturbances such as dysprosody, effecting vocal inflection, and parkinsonism, effecting facial emotional display.

Pseudobulbar Palsy

Exaggerated emotional expression and unintended laughing or unmotivated weeping occur in pseudobulbar palsy. The emotion expressed may be completely unrelated to the mood of the patient or may reflect the appropriate emotion but is out of proportion to the intensity of the feeling experienced.[228,229] Patients may have episodes of either laughing or crying or both, and in some cases one affect may gradually change into the other, making it difficult to judge from the facial contortion which emotion is being displayed. Once initiated, the emotional expression is difficult or impossible to arrest, although with redirection of the conversation the patient's affect may be changed from weeping to laughter or vice versa. In general, older patients with pseudobulbar palsy tend to exhibit weeping as the primary affected manifestation, whereas younger patients tend to have exaggerated mirthless laughter.

Pseudobulbar palsy is typically produced by bilateral lesions located in the corticonuclear tracks between the cerebral cortex and the brainstem. Table 14.11 presents the phenomenology of pseudobulbar palsy, demonstrating that the syndrome results from a combination of impaired supranuclear control and preserved or exaggerated function of subcortical and brainstem mechanisms. Volitional facial paresis occurs with lesions of the descending pyramidal tract, and pseudobulbar palsy results when lesions release motor programs of the limbic system and subcortical structures. The converse, the preservation of volitional facial movements with paresis of emotionally related facial expression, occurs with lesions of the limbic system including the frontal white matter, anterior thalamus, and insula.[230]

In addition to the pseudobulbar affect, patients with pseudobulbar palsies have several other clinical features (Table 14.12). The three cardinal features of pseudobulbar palsy are disturbed emotional expression, dysarthria, and dysphagia.[229] Dysarthria is often severe and is characterized by imprecise articulation, slow speech rate, low pitch, and a strained-strangled sound of effortful phonation. Occasionally the dysarthria is so profound that the patient is completely unintelligible or mute. Lieberman and Benson[231] reported a patient with amyotrophic lateral sclerosis and severe pseudobulbar palsy who could not speak and eventually communicated only through eye movement Morse code.

Examination of the face of pseudobulbar patients reveals a variety of signs of upper motor neuron dysfunction. The face is masked, showing little spontaneous emotional expression in the resting state and partial paralysis of the facial muscles is common. Weakness of forced lip and eye closure is present and the patient has difficulty grimacing and voluntarily contracting the platysma. Tongue protrusion is weak. The jaw-jerk is exaggerated and the muscle stretch reflexes of the face, including the orbicularis oris and orbicularis oculi, are increased. Palatal movement is diminished when the patient is asked to phonate (say "aah") but the gag reflex response to posterior pharyngeal stimulation is hyperactive. Many patients also have supranuclear ophthalmoplegia, limb weakness and spasticity, exaggerated muscle stretch reflex of the limb, and extensor plantar responses.

Martin[232] reported a 25-year-old man who had an attack of uncontrollable laughter while attending his mother's funeral and a 23-year-old woman who began laughing uncontrollably when a boxer was knocked

TABLE 14.11. *Phenomenology of the Pseudobulbar Palsy Syndrome**

Level/Function	*Impaired Supranuclear Control*	*Exaggerated/Preserved Subcortical and Brainstem Mechanisms*
CN III, IV, VI	Impaired volitional saccades	Intact oculocephalic reflexes
CN V	Reduced masseter strength	Brisk jaw jerk
CN VII	Impaired facial muscle strength	Brisk facial muscle scretch reflexes
CN IX, X	Reduced volitional palate control (e.g., impaired raising of uvula when saying "ahh")	Exaggerated gag reflex
CN XII	Tongue weakness	—
Swallowing	Dysphagia	Intact reflex swallowing with pharyngeal stimulation
Speech articulation	Dysarthria	Intact reflex movements such as yawning
Motor programs for facial emotional expression	Hypomimia	Laughter and weeping dissociated from mood (pseudobulbar affect)
Pyramidal motor function	Limb weakness, spasticity	Exaggerated muscle stretch reflexes; Babinski signs

*The clinical features reflect loss of supranuclear control and preservation or exaggeration of subcortical and brainstem mechanisms. CN, cranial nerve.

from the ring and fell almost at her feet. In both cases, pseudobulbar palsy was associated with a brainstem aneurysm and acute subarachnoid hemorrhage. A few such patients have literally died laughing when acute pseudobulbar palsy with sustained laughter was initiated by a fatal event. Martin[232] commented that "this is the greatest mockery of all when the patient should be forced to laugh as a portent of his own doom."

TABLE 14.12. *Clinical Characteristics of the Pseudobulbar Syndrome*

Unmotivated laughing or crying
Dysarthria
Dysphagia
Drooling
Weakness of volitional facial movements
Facial amimia/hypomimia
Brisk jaw jerk
Exaggerated facial muscle stretch reflexes
Increased gag reflex
Associated neurological findings
Supranuclear ophthalmoplegia
Limb weakness
Increased limb muscle stretch reflexes
Limb rigidity with spasticity
Extensor plantar responses

Pseudobulbar palsy has been produced by multiple sclerosis, traumatic brain injury, amyotropic lateral sclerosis, progressive supranuclear palsy, meningiomas of the clivus, vascular dementia, anoxic brain injury, encephalitis, basal artery aneurysms, chordomas of the clivus, brainstem tumors, Creutzfeldt-Jakob disease, and neurosyphilis.[233,234] Bilateral lacunar infarctions in the internal capsule disconnecting the descending supranuclear pathways from brainstem mechanisms are a particularly common cause of pseudobulbar palsy.[235,236] Rarely, apparently unilateral lesions of the limbic system may result in the acute onset of sustained laughter, which gradually subsides over a period of days or weeks.[237]

A variety of types of medications may be successful in reducing the affective dyscontrol associated with pseudobulbar palsy. Levodopa, tricyclic antidepressants including amitriptyline and nortriptyline, and SSRIs have all been successful in ameliorating pseudobulbar palsy.[238–241]

Ictal Affect

Alterations of emotional expression in epilepsy may be secondary to mood changes caused by the focal seizure or may be sham emotional responses induced by the focal epileptic activity. Patients are frequently amnestic for the period of the seizure and distinction between these two causes of altered emotional expression may be impossible in some cases. The two principal affective alterations described in epilepsy involve either laughing or crying. Ictal seizures with pronounced changes in affect during the ictus include complex partial seizures, infantile spasms, and seizures associated with hypothalamic lesions. In adults, ictal affect is typically a manifestation of limbic epilepsy and complex partial seizures. Seizures manifesting laughter as an ictal automatism have been called "gelastic" seizures, and those with ictal crying have been called quiritirian (meaning *to scream* or *cry*) or "dacrystic" (meaning *to tear*).[242–244] Ictal laughter is more common in epileptic patients than ictal crying. The laughter is inappropriate, stereotyped, and occurs without a precipitating stimulus. There is frequently an associated alteration of consciousness, and other seizure manifestations may be evident. Postictal confusion is typically present and the patient is usually amnestic for the event. Psychiatric and psychomotor phenomena accompanying the laughter may include running (cursive epilepsy), macropsia and metamorphopsia, deja-vu feelings, olfactory hallucinations, feelings of sexual orgasm, or limb or verbal automatisms. When consciousness persists during the attacks, the laughter is usually recalled as being disagreeable and incongruent, not the pleasurable experience the affect would suggest.[243] Lesions producing gelastic seizures have most commonly been in the right hemisphere and involve the mesiofrontal–anterior cingulate or the anteromesial temporal regions.[245,246]

Treatment of gelastic and dacrystic seizures is with the same medications indicated for the treatment of complex partial seizures (Chapter 21).

Infantile spasms (Chapter 21) is a form of primary generalized seizure disorder occurring in infants and young children. It results from a wide variety of CNS disturbances and laughter may be one manifestation of the seizures. The laughing varies from prolonged violent attacks lasting for as long as several minutes or short periods of giggling or grinning. The epileptic nature of the laughter is indicated by the lack of external precipitants, the odd nature of the laughter (mothers recognize it as uncharacteristic of their children), the accompanying manifestations of epilepsy, motoric evidence of seizure activity, and the response to anticonvulsants.

The third clinical syndrome in which ictal laughter occurs is in association with hamartomas involving the hypothalamus. The characteristic syndrome in which hypothalamic hamartomas and ictal laughter are associated is characterized by onset of laughing seizures in infancy, normal early childhood psychomotor development, increasing length of seizures and appearance of other seizure types between ages 4 and 10, and progressive cognitive deterioration and severe behavioral problems in late childhood and early adolescence. Seizure control with anticonvulsants or cortical resection is poor.[247] The hamartomas can typically be visualized by MRI and SPECT performed during the epileptic attack, and such studies show hyperperfusion in the hypothalamic region and adjacent thalamus without cortical or cerebellar changes.[248] The laughter associated with the epileptic syndrome of hypothalamic hamartomas is described as pleasant laughter or giggling lasting a few seconds and occurring two to six times per day. The seizures continue daily throughout early childhood and may persist into later childhood and adolescence.[247] Treatment with anticonvulsants is usually partially effective at best; corticectomies have produced little improvement, and anterior corpus callosotomy may reduce some types of associated seizures.[249]

Angelman Syndrome

Angelman syndrome, disparagingly referred to as the "happy puppet syndrome,"[250] is an unusual disorder characterized by mental deficiency, abnormal puppet-like gait, a characteristic facies (described below), and frequent paroxysms of laughter. Also characteristic are microcephaly, maxillary hypoplasia, a large mouth with tongue protrusion, and widely spaced teeth. Prognathia is present and there is decreased pigmentation of the choroid and iris, the latter resulting in pale blue eyes. Seizures are common.

Dysprosody and Facial Hypomimia

Prosody refers to the affective and inflectional aspects of speech based on syllable and word stress, speech rhythm and cadence, and syllable and word pitch shifts (Chapter 6). These aspects of language expression are responsible for investing speech with its emotional content, and any modification of prosodic ability distorts patients' communication of their emotional states.

Prosody is altered by lesions in a variety of locations, including right anterior hemispheric regions and the basal ganglia.[251,252] Speech is typically monotonic and boring without inflection. Dysprosody reduces patients' abilities to elaborate their feelings and makes it difficult for their families and clinicians to appreciate their emotional state.

Hypomimia refers to the lack of emotional expression of the face. Normally, the face communicates substantial information about the internal emotional state of the individual. Surprise, happiness, sadness, anger, boredom, apprehension, and irritability are all revealed on the face of the individual. Facial hypomimia reduces the patients' ability to display their internal emotional states and makes it difficult for family members and others to deduce the emotions experienced by the patient. Lesions of the limbic cortical structures, including frontal white matter, anterior thalamus, insula, and medial temporal region, result in emotional facial paresis and hypomimia.[230]

REFERENCES

1. American Psychiatric Association. Diagnostic and Statistical Manual of Mental Disorders, 4th ed. Washington, DC: American Psychiatric Press, 1994.
2. Leff JP, Isaacs AD. Psychiatric Examination in Clinical Practice, 3rd ed. London: Blackwell Scientific Publications, 1990.
3. Parker G, Hadzi-Pavlovic D. Melancholia: A Disorder of Movement and Mood; A Phenomenological and Neurobiological Review. Cambridge, UK: Cambridge University Press, 1996.
4. Levy ML, Cummings JL, et al. Apathy is not depression. J Neuropsychiatry Clin Neurosci 1998;10:314–319.
5. Cummings JL. Depression and Parkinson's disease: a review. Am J Psychiatry 1992;149:443–454.
6. Tandberg E, Larsen JP, et al. The occurrence of depression in Parkinson's disease: a community-based study. Arch Neurol 1996;53:175–179.
7. Brown RG, MacCarthy B, et al. Depression and disability in Parkinson's disease: a follow-up of 132 cases. Psychol Med 1988;18:49–55.
8. Menza MA, Mark MH. Parkinson's disease and depression: the relationship to disability and personality. J Neuropsychiatry Clin Neurosci 1994;6:165–169.
9. Starkstein SE, Cohen BS, et al. Relationship between anxiety disorders and depressive disorders in patients with cerebrovascular injury. Arch Gen Psychiatry 1990;47:246–251.
10. Tandberg E, Larsen JP, et al. Risk factors for depression in Parkinson disease. Arch Neurol 1997;54:625–630.
11. Troster AI, Stalp LD, et al. Neuropsychological impairment in Parkinson's disease with and without depression. Arch Neurol 1995;52:1164–1169.
12. Fleminger S. Left-sided Parkinson's disease is associated with greater anxiety and depression. Psychol Med 1991;21:629–638.
13. Starkstein SE, Preziosi TJ, et al. Depression in Parkinson's disease. J Nerv Ment Dis 1990;178:27–31.
14. Starkstein SE, Mayberg HS, et al. A prospective longitudinal study of depression, cognitive decline, and physical impairments in patients with Parkinson's disease. J Neurol Neurosurg Psychiatry 1992;55:377–382.
15. Friedenberg DL, Cummings JL. Parkinson's disease, depression, and the on–off phenomenon. Psychosomatics 1989;30:94–99.
16. Nissenbaum H, Quinn NP, et al. Mood swings associated with the 'on–off' phenomenon in Parkinson's disease. Psychol Med 1987;17:899–904.
17. Mayeux R, Stern Y, et al. Altered serotonin metabolism in depressed patients with Parkinson's disease. Neurology 1984;34: 642–646.
18. Chan-Palay V, Asan E. Alterations in catecholamine neurons of the locus coeruleus in senile dementia of the Alzheimer type and in Parkinson's disease with and without dementia and depression. J Comp Neurol 1989;287:373–392.
19. Mayberg HS, Starkstein SE, et al. Selective hypometabolism in the inferior frontal lobe in depressed patients with Parkinson's disease. Ann Neurol 1990;28:57–64.
20. Ring HA, Bench CJ, et al. Depression in Parkinson's disease: a positron emission study. Br J Psychiatry 1994;165:333–339.
21. Litvan I, Mega MS, et al. Neuropsychiatric aspects of progressive supranuclear palsy. Neurology 1996;47:1184–1189.
22. Litvan I, Cummings JL, et al. Neuropsychiatric features of corticobasal degeneration. J Neurol Neurosurg Psychiatry 1998; 65:717–721.
23. Ballard C, Holmes C, et al. Psychiatric morbidity in Dementia with Lewy bodies: a prospective clinical and neuropathological comparative study with Alzheimer's disease. Am J Psychiatry 1999;156:1039–1045.
24. Klatka LA, Louis ED, et al. Psychiatric features in diffuse Lewy body disease: a clinicopathologic study using Alzheimer's disease and Parkinson's disease comparison groups. Neurology 1996;47:1148–1152.
25. Weiner MF, Risser RC, et al. Alzheimer's disease and its Lewy body variant: a clinical analysis of postmortem verified cases. Am J Psychiatry 1996;153:1269–1273.
26. Scheinberg IH, Sternlieb I. Wilson's Disease. Philadelphia: W.B. Saunders, 1984.
27. Starosta-Rubinstein S, Young AB, et al. Clinical assessment of 31 patients with Wilson's disease. Arch Neurol 1987;44:365–370.
28. Dening DC, Berrios GE. Wilson's disease: psychiatric symptoms in 195 cases. Arch Gen Psychiatry 1989;46:1126–1134.
29. Medalia A, Scheinberg IH. Psychopathology in patients with Wilson's disease. Am J Psychiatry 1989;146:662–664.
30. Medalia A, Galynker I, et al. The interaction of motor, memory, and emotional dysfunction in Wilson's disease. Biol Psychiatry 1992;31:823–826.
31. Caine ED, Shoulson I. Psychiatric syndromes in Huntington's disease. Am J Psychiatry 1983;140:728–733.
32. Folstein SE. Huntington's Disease: A Disorder of Families. Baltimore: The Johns Hopkins University Press, 1989.
33. Folstein SE, Abbott MH, et al. The association of affective disorder with Huntington's disease in a case series and in families. Psychol Med 1983;13:537–542.
34. Schoenfeld M, Myers RH, et al. Increased rate of suicide among patients with Huntington's disease. J Neurol Neurosurg Psychiatry 1984;47:1283–1287.
35. Zappacosta B, Monza D, et al. Psychiatric symptoms do not correlate with cognitive decline, motor symptoms, or CAG repeat length in Huntington's disease. Arch Neurol 1996;53: 493–497.
36. Mayberg HS, Starkstein SE, et al. Paralimbic frontal lobe hypometabolism in depression associated with Huntington's disease. Neurology 1992;42:1791–1797.
37. Kurlan R, Caine E, et al. Cerebrospinal fluid correlates of depression in Huntington's disease. Arch Neurol 1988;45:881–883.
38. Levy ML, Miller BL, et al. Alzheimer disease and frontotemporal dementias: behavioral distinctions. Arch Neurol 1996;53: 687–690.
39. Mega M, Cummings JL, et al. The spectrum of behavioral changes in Alzheimer's disease. Neurology 1996;46:130–135.
40. Lyketsos CG, Steele C, et al. Major and minor depression in Alzheimer's disease: prevalence and impact. J Neuropsychiatry Clin Neurosci 1997;9:556–561.

41. Mendez MF, Martin RJ, et al. Psychiatric symptoms associated with Alzheimer's disease. J Neuropsychiatry Clin Neurosci 1990;2:28–33.
42. Devanand DP, Sano M, et al. Depressed mood and the incidence of Alzheimer's disease in the elderly living in the community. Arch Gen Psychiatry 1996;53:175–182.
43. Jost BC, Grossberg GT. The evolution of psychiatric symptoms in Alzheimer's disease: a natural history study. J Am Geriatr Soc 1996;44:1078–1081.
44. Chen ST, Altshuler LL, et al. Bipolar disorder late in life: a review. J Geriatr Psychiatry Neurol 1998;11:29–35.
45. Fitz AG, Teri L. Depression, cognition, and functional ability in patients with Alzheimer's disease. J Am Geriatr Soc 1994;42: 186–191.
46. Forsell Y, Winblad B. Major depression in a population of demented and nondemented older people: prevalence and correlates. J Am Geriatr Soc 1998;46:27–30.
47. Lyketsos CG, Steele C, et al. Physical aggression in dementia patients and its relationship to depression. Am J Psychiatry 1999;156:66–71.
48. Lawlor BA, Ryan TM, et al. Clinical symptoms associated with age at onset in Alzheimer's disease. Am J Psychiatry 1994;151: 1646–1649.
49. Migliorelli R, Teson A, et al. Prevalence and correlates of dysthymia and major depression among patients with Alzheimer's disease. Am J Psychiatry 1995;152:37–44.
50. Pearlson GD, Roos CA, et al. Association between family history of affective disorder and the depressive syndrome of Alzheimer's disease. Am J Psychiatry 1990;147:452–456.
51. Strauss ME, Ogrocki PK. Confirmation of an association between family history of affective disorder and the depressive syndrome in Alzheimer's disease. Am J Psychiatry 1996;153: 1340–1342.
52. Cummings JL, Ross W, et al. Depressive symptoms in Alzheimer disease: assessment and determinants. Alzheimer Dis Assoc Disord 1995;9:87–93.
53. Burns A, Jacoby R, et al. Psychiatric phenomena in Alzheimer's disease, I: disorders of thought content; II: disorders of perception. Br J Psychiatry 1990;157:72–81.
54. Ott BR, Fogel BS. Measurement of depression in dementia: self vs clinician rating. Int J Geriatr Psychiatry 1992;7:889–904.
55. Lopez OL, Becker JT, et al. Psychiatric correlates of MR deep white matter lesions in probable Alzheimer's disease. J Neuropsychiatry Clin Neurosci 1997;9:246–250.
56. Hirono N, Mori E, et al. Frontal lobe hypometabolism and depression in Alzheimer's disease. Neurology 1998;50:380–383.
57. Starkstein SE, Vazquez S, et al. A SPECT study of depression in Alzheimer's disease. Neuropsychiatry Neuropsychol Behav Neurol 1995;8:38–43.
58. Sultzer DL, Mahler ME, et al. The relationship between psychiatric symptoms and regional cortical metabolism in Alzheimer's disease. J Neuropsychiatry Clin Neurosci 1995;7:476–484.
59. Forstl H, Burns A, et al. Clinical and neuropathological correlates of depression in Alzheimer's disease. Psychol Med 1992; 22:877–884.
60. Zubenko GS, Moossy J. Major depression in primary dementia. Arch Neurol 1988;45:1182–1186.
61. Zubenko GS, Moossy J, et al. Neurochemical correlates of major depression in primary dementia. Arch Neurol 1990;47:209–214.
62. Shimoda K, Robinson RG. The relationship between poststroke depression and lesion location in long-term follow-up. Biol Psychiatry 1999;45:187–192.
63. Morris PLP, Robinson RG, et al. Lesion location and poststroke depression. J Neuropsychiatry Clin Neurosci 1996;8:399–403.
64. Starkstein SE, Robinson RG, et al. Differential mood changes following basal ganglia vs thalamic lesions. Arch Neurol 1988;45:725–730.
65. Schwartz JA, Speed NM, et al. ^{99}mTc-hexamethylpropyleneamine oxime single photon emission CT in poststroke depression. Am J Psychiatry 1990;147:242–244.
66. Sharpe M, Hawton K, et al. Mood disorders in long-term survivors of stroke: associations with brain lesion location and volume. Psychol Med 1990;20:815–828.
67. Bolla-Wilson K, Robinson RG, et al. Lateralization of dementia of depression in stroke patients. Am J Psychiatry 1989;146: 627–634.
68. Robinson RG, Bolla-Wilson K, et al. Depression influences intellectual impairment in stroke patients. Br J Psychiatry 1986; 148:541–547.
69. Starkstein SE, Robinson RG, et al. Comparison of patients with and without poststroke major depression matched for size and location of lesion. Arch Gen Psychiatry 1988;45:247–252.
70. Paradiso S, Robinson RG. Gender differences in poststroke depression. J Neuropsychiatry Clin Neurosci 1998;10:41–47.
71. Lipsey JR, Spencer WC, et al. Phenomenological comparison of poststroke depression and functional depression. Am J Psychiatry 1986;143:527–529.
72. Parikh RM, Robinson RG, et al. The impact of poststroke depression on recovery in activities of daily living over a 2-year follow-up. Arch Neurol 1990;47:785–789.
73. Sinyor D, Amato P, et al. Post-stroke depression: relationships to functional impairment, coping strategies, and rehabilitation outcome. Stroke 1986;17:1102–1107.
74. Mayberg HS, Robinson RG, et al. PET imaging of cortical S2 serotonin receptors after stroke: lateralized changes and relationship to depression. Am J Psychiatry 1988;145:937–943.
75. Fujikawa T, Yamawaki S, et al. Incidence of silent cerebral infarction in patients with major depression. Stroke 1993;24: 1631–1634.
76. Krishnan KRR, Hays JC, et al. MRI-defined vascular depression. Am J Psychiatry 1997;154:497–501.
77. Alexopoulos GS, Meyers BS, et al. Vascular depression hypothesis. Arch Gen Psychiatry 1997;54:915–922.
78. Alexopoulos GS, Meyers BS, et al. Clinically defined vascular depression. Am J Psychiatry 1997;154:562–565.
79. Simpson S, Baldwin RC, et al. Is subcortical disease associated with a poor response to antidepressants? Neurological, neuropsychological and neuroradiological findings in late-life depression. Psychol Med 1998;28:1015–1026.
80. Figiel GS, Coffey CE, et al. Brain magnetic resonance imaging findings in ECT-induced delirium. J Neuropsychiatry Clin Neurosci 1990;2:53–58.
81. Figiel GS, Krishnan KRR, et al. Radiologic correlates of antidepressant-induced delirium: the possible significance of basal-ganglia lesions. J Neuropsychiatry Clin Neurosci 1989;1:188–190.
82. Hickie I, Scott E, et al. Subcortical hyperintensities on magnetic resonance imaging: clinical correlates and prognostic significance in patients with severe depression. Biol Psychiatry 1995;37:151–160.
83. Coffey CE, Figiel GS, et al. White matter hyperintensity on magnetic resonance imaging: clinical and neuroanatomic correlates in the depressed elderly. J Neuropsychiatry Clin Neurosci 1989;1:135–144.
84. Greenwald BS, Kramer-Ginsberg E, et al. MRI signal hyperin-

tensities in geriatric depression. Am J Psychiatry 1996;153: 1212–1215.
85. Lesser IM, Boone KB, et al. Cognition and white matter hyperintensities in older depressed patients. Am J Psychiatry 1996;153:1280–1287.
86. Mendez MF, Cummings JL, et al. Depression in epilepsy: significance and phenomenology. Arch Neurol 1986;43:766–770.
87. Victoroff JI, Benson DF, et al. Depression in complex partial seizures. Arch Neurol 1994;51:155–163.
88. Robertson MM, Trimble MR, et al. Phenomenology of depression in epilepsy. Epilepsia 1987;4:364–372.
89. Mendez MF, Lanska DJ, et al. Causative factors for suicide attempts by overdose in epileptics. Arch Neurol 1989;46:1065–1068.
90. Bromfield EB, Altshuler L, et al. Cerebral metabolism and depression in patients with complex partial seizures. Arch Neurol 1992;49:617–623.
91. McConnell H, Duncan D. Treatment of psychiatric comorbidity in epilepsy. In: McConnell HW, Snyder PJ, eds. Psychiatric Comorbidity in Epilepsy: Basic Mechanisms, Diagnosis and Treatment. Washington DC: American Psychiatric Press, 1998: 245–361.
92. Diaz-Olavarrieta C, Cummings JL, et al. Neuropyschiatric manifestations of multiple sclerosis. J Neuropsychiatry Clin Neurosci 1999;11:51–57.
93. Huber SJ, Rammohan KW, et al. Depressive symptoms are not influenced by severity of multiple sclerosis. Neuropsychiatry Neuropsychol Behav Neurol 1993;6:177–180.
94. Nyenhuis DL, Rao SM, et al. Mood disturbance versus other symptoms of depression in multiple sclerosis. J Int Neuropsychol Soc 1995;1:291–296.
95. Dalos NP, Rabins PV, et al. Disease activity and emotional state in multiple sclerosis. Ann Neurol 1983;13:573–577.
96. Sadovnick AD, Remick RA, et al. Depression and multiple sclerosis. Neurology 1996;46:628–632.
97. Vickrey BG, Hays RD, et al. A health-related quality of life measure for multiple sclerosis. Qual Life Res 1995;4:187–206.
98. Schiffer RB, Caine ED, et al. Depressive episodes in patients with multiple sclerosis. Am J Psychiatry 1983;140:1498–1500.
99. Honer WG, Hurwitz T, et al. Temporal lobe involvement in multiple sclerosis patients with psychiatric disorders. Arch Neurol 1987;44:187–190.
100. Pujol J, Bello J, et al. Lesions in the left arcuate fasciculus region and depressive symptoms in multiple sclerosis. Neurology 1997;49:1105–1110.
101. Reischies FM, Baum K, et al. Psychopathological symptoms and magnetic resonance imaging findings in multiple sclerosis. Biol Psychiatry 1993;33:676–678.
102. Neilley LK, Goodin DS, et al. Side effect profile of interferon β-1b in MS: results of an open label trial. Neurology 1996;46: 552–554.
103. Mohr DC, Goodkin DE, et al. Treatment of depression improves adherence to interferon β-1b therapy for multiple sclerosis. Arch Neurol 1997;54:531–533.
104. Feinstein A. The Clinical Neuropsychiatry of Multiple Sclerosis. Toronto: Cambridge University Press; 1999.
105. Schiffer RB, Wineman NM. Antidepressant pharmacotherapy of depression associated with multiple sclerosis. Am J Psychiatry 1990;147:1493–1497.
106. Minden SL. Psychotherapy for people with multiple sclerosis. J Neuropsychiatry Clin Neurosci 1992;4:198–213.
107. Deb S, Lyons I, et al. Rate of psychiatric illness 1 year after traumatic brain injury. Am J Psychiatry 1999;156:374–378.
108. Jorge RE, Robinson RG, et al. Depression and anxiety following traumatic brain injury. J Neuropsychiatry Clin Neurosci 1993;5:369–374.
109. Goldstein FC, Levin HS, et al. Cognitive and behavioral sequelae of closed head injury in older adults according to their significant others. J Neuropsychiatry Clin Neurosci 1999;11:38–44.
110. Fann JR, Katon WJ, et al. Psychiatric disorders and functional disability in outpatients with traumatic brain injuries. Am J Psychiatry 1995;152:1493–1499.
111. Jorge RE, Robinson RG, et al. Comparison between acute- and delayed-onset depression following traumatic brain injury. J Neuropsychiatry Clin Neurosci 1993;5:43–49.
112. Max JE, Robin DA, et al. Traumatic brain injury in children and adolescents: psychiatric disorders at one year. 1998;10: 290–297.
113. Grafman J, Vance SC, et al. The effects of lateralized frontal lesions on mood regulation. Brain 1986;109:1127–1148.
114. Fann JR, Uomoto JM, et al. Sertraline in the treatment of major depression following mild traumatic brain injury. J Neuropsychiatry Clin Neurosci 2000;12:226–232.
115. Berrios GE, Samuel C. Affective disorder in the neurological patient. J Nerv Ment Dis 1987;175:173–176.
116. Rundell JR, Wise MG. Causes of organic mood disorder. J Neuropsychiatry Clin Neurosci 1989;1:398–400.
117. Irle E, Peper M, et al. Mood changes after surgery for tumors of the cerebral cortex. Arch Neurol 1994;51:164–174.
118. Clemons J, Jasser MZ, et al. Gaucher's disease initially diagnosed as depression. Am J Psychiatry 1997;154:290.
119. Jane J, Doraiswamy M, et al. Frontal arachnoid cyst associated with depression: case report and literature review. Depression 1993;1:275–277.
120. Kaplan RF, Meadows M-E, et al. Memory impairment and depression in patients with Lyme encephalopathy. Neurology 1992;42:1263–1267.
121. Renshaw PF, Stern TA, et al. Electroconvulsive therapy treatment of depression in a patient with adult GM2 gangliosidosis. Ann Neurol 1992;31:342–344.
122. Upadhyaya AK. Psychiatric presentation of third ventricular colloid cyst: a case report. Br J Psychiatry 1988;152:567–569.
123. Mattingly G, Figiel GS, et al. Prospective uses of ECT in the presence of intracranial tumors. J Neuropsychiatry Clin Neurosci 1991;3:459–463.
124. Pies RW. Medical "mimics" of depression. Psychiatry Ann 1994;24:519–520.
125. Gadde KM, Krishnan KRR. Endocrine factors in depression. Psychiatry Ann 1994;24:521–524.
126. Hutto B. The symptoms of depression in endocrine disorders. CNS Spectrums 1999;4:51–61.
127. Krystal A, Krishnan KRR, et al. Differential diagnosis and pathophysiology of Cushing's syndrome and primary affective disorder. J Neuropsychiatry Clin Neurosci 1990;2:34–43.
128. Krishnan KRR, McDonald WM, et al. Magnetic resonance imaging of the caudate nuclei in depression. Arch Gen Psychiatry 1992;49:553–557.
129. Soares JC, Mann JJ. The anatomy of mood disorders—review of structural neuroimaging studies. Biol Psychiatry 1997;41:86–106.
130. Coffey CE, Wilkinson WE, et al. Quantitative cerebral anatomy in depression. Arch Gen Psychiatry 1993;50:7–16.
131. Figiel GS, Krishnan KRR, et al. Subcortical hyperintensities on brain magnetic resonance imaging: a comparison between late age onset and early onset elderly depressed subjects. Neurobiol Aging 1991;12:245–247.

132. Bench CJ, Friston KJ, et al. The anatomy of melancholia—focal abnormalities of cerebral blood flow in major depression. Psychol Med 1992;22:607–615.
133. Lesser IM, Mena I, et al. Reduction of cerebral blood flow in older depressed patients. Arch Gen Psychiatry 1994;51:677–686.
134. Mayberg HS. Frontal lobe dysfunction in secondary depression. J Neuropsychiatry Clin Neurosci 1994;6:428–442.
135. Baxter LR, Phelps ME, et al. Cerebral metabolic rates for glucose in mood disorders. Arch Gen Psychiatry 1985;42:441–447.
136. Boone KB, Lesser IM, et al. Cognitive functioning in older depressed outpatients: relationship of presence and severity of depression to neuropsychological test scores. Neuropsychology 1995;9:390–398.
137. Sackeim HA, Freeman J, et al. Effects of major depression on estimates of intelligence. J Clin Exp Neuropsychol 1992;14: 268–288.
138. Bornstein RA, Baker GB, et al. Depression and memory in major depressive disorder. J Neuropsychiatry Clin Neurosci 1991;3:78–80.
139. Lichtenberg PA, Ross T, et al. The relationship between depression and cognition in older adults: a cross-validation study. J Gerontol 1995;50B:25–32.
140. Bench CJ, Friston KJ, et al. Regional cerebral blood flow in depression measured by positron emission tomography: the relationship with clinical dimensions. Psychol Med 1993;23:579–590.
141. Dolan RJ, Bench CJ, et al. Neuropsychological dysfunction in depression: the relationship to regional cerebral blood flow. Psychol Med 1994;24:849–857.
142. Elliott R, Baker SC, et al. Prefrontal dysfunction in depressed patients performing a complex planning task: a study using positron emission tomography. Psychol Med 1997;27:931–942.
143. Rajkowska G, Miguel-Hildago JJ, et al. Morphometric evidence for neuronal and glial prefrontal cell pathology in major depression. Biol Psychiatry 1999;45:1085–1098.
144. Mann JJ, Arango V. Abnormalities of brain structure and function in mood disorders. In: Charney DS, Nestler EJ, Bunney BS, eds. Neurobiology of Mental Illness. New York: Oxford University Press, 1999: 385–393.
145. Mayberg HS, Lewis PJ, et al. Paralimbic hypoperfusion in unipolar depression. J Nucl Med 1994;35:929–934.
146. Nobler MS, Sackeim HA, et al. Electroconvulsive therapy: current practice and future directions. In: Halbreich U, Montgomery SA, eds. Pharmacotherapy for Mood, Anxiety, and Cognitive Disorders. Washington, DC: American Psychiatric Press, 2000: 167–187.
147. Altman EG, Hedeker DR, et al. The clinician-administered rating scale for mania (CARS-M): development, reliability, and validity. Biol Psychiatry 1994;36:124–134.
148. Altman EG, Hedeker D, et al. The Altman self-rating mania scale. Biol Psychiatry 1997;42:948–955.
149. Carroll BJ. Psychopathology and neurobiology of manic-depressive disorders. In: Carroll BJ, Barrett JE, eds. Psychopathology and the Brain. New York: Raven Press, 1991: 265–282.
150. Krauthammer C, Klerman GL. Secondary mania: manic syndromes associated with antecedent physical illness or drugs. Arch Gen Psychiatry 1978;35:1333–1339.
151. Carroll BT, Goforth HW, et al. Mania due to general medical conditions: frequency, treatment, and cost. Int J Psychiatry Med 1996;26:5–13.
152. Lauterbach EC, Spears TE, et al. Bipolar disorder in idiopathic dystonia: clinical features and possible neurobiology. J Neuropsychiatry Clin Neurosci 1992;4:435–439.
153. Trautner RJ, Cummings JL, et al. Idiopathic basal ganglia calcification and organic mood disorder. Am J Psychiatry 1988;145:350–353.
154. Bogousslavsky J, Ferrazzini M, et al. Manic delirium and frontal-like syndrome with paramedian infarction of the right thalamus. J Neurol Neurosurg Psychiatry 1988;51:116–119.
155. Cummings JL, Mendez MF. Secondary mania with focal cerebrovascular lesions. Am J Psychiatry 1984;141:1084–1087.
156. Kulisevsky J, Berthier ML, et al. Hemiballismus and secondary mania following a right thalamic infarction. Neurology 1993;43:1422–1424.
157. Vuilleumier P, Ghika-Schmid F, et al. Persistent recurrence of hypomania and prosopoaffective agnosia in a patient with right thalamic infarct. Neuropsychiatry Neuropsychol Behav Neurol 1998;11:40–44.
158. Berthier ML, Kulisevsky J, et al. Poststroke bipolar affective disorder: clinical subtypes, concurrent movement disorders, and anatomical correlates. J Neuropsychiatry Clin Neurosci 1996;8: 160–167.
159. Benjamin S, Kirsch D, et al. Hypomania from left frontal AVM resection. Neurology 2000;54:1389–1390.
160. Drake ME, Pakalnis A, et al. Secondary mania after ventral pontine infarction. J Neuropsychiatry Clin Neurosci 1990;2: 322–325.
161. Starkstein SE, Berthier ML, et al. Emotional behavior after a Wada test in a patient with secondary mania. J Neuropsychiatry Clin Neurosci 1989;1:408–412.
162. Smeraski PJ. Clonazepam treatment of multi-infarct dementia. J Geriatr Psychiatry Neurol 1988;1:47–48.
163. Binder RL. Neurologically silent brain tumors in psychiatric hospital admissions: three cases and a review. J Clin Psychiatry 1983;44:94–97.
164. Greenberg DB, Brown GL. Mania resulting from brain stem tumor. J Nerv Ment Dis 1985;173:434–436.
165. Mazure CM, Leibowitz K, et al. Drug-responsive mania in a man with a brain tumor. J Neuropsychiatry Clin Neurosci 1999;11:114–115.
166. Bakchine S, Lacomblez L, et al. Manic-like state after bilateral orbitofrontal and right temporoparietal injury: efficacy of clonidine. Neurology 1989;39:777–781.
167. Robinson RG, Boston JD, et al. Comparison of mania and depression after brain injury: causal factors. Am J Psychiatry 1988;145:172–178.
168. Starkstein SE, Pearlson GD, et al. Mania after brain injury: a controlled study of causative factors. Arch Neurol 1987;44: 1069–1073.
169. Wright MT, Cummings JL, et al. Bipolar syndromes following brain trauma. Neurocase 1997;3:111–118.
170. Shukla S, Cook BL, et al. Mania following head trauma. Am J Psychiatry 1987;144:93–96.
171. Jorge RE, Robinson RG, et al. Secondary mania following traumatic brain injury. Am J Psychiatry 1993;150:916–921.
172. Minden SL, Schiffer RB. Depression and affective disorders in multiple sclerosis. In: Halbreich U, ed. Multiple Sclerosis: A Neuropsychiatric Disorder. Washington, DC: American Psychiatric Press, 1993: 33–54.
173. Black KJ, Perlmutter JS. Septuagenarian Sydenham's with secondary hypomania. Neuropsychiatry Neuropsychol Behav Neurol 1997;10:147–150.

174. Fisher CM. Hypomanic symptoms caused by herpes simplex encephalitis. Neurology 1996;47:1374–1378.
175. Goldney RD, Temme PB. Case report: manic depressive psychosis following infectious mononucleosis. J Clin Psychiatry 1980;41:322–323.
176. Lyketsos CG, Schwartz J, et al. AIDS mania. J Neuropsychiatry Clin Neurosci 1997;9:277–279.
177. Thienhaus OJ, Khosla N. Meningeal cryptococcosis misdiagnosed as a manic episode. Am J Psychiatry 1984;141:1459–1460.
178. Ross RL, Smith GR, et al. Neurosyphilis and organic mood syndrome: a forgotten diagnosis. Psychosomatics 1990;31:448–450.
179. Cummings JL. Organic psychoses: delusional disorders and secondary mania. Psychiatr Clin N Am (Neuropsychiatry) 1986;9: 378–382.
180. Forrest DV. Bipolar illness after right hemispherectomy. Arch Gen Psychiatry 1982;39:817–819.
181. Kumar A, Agarwal M. Secondary affective disorder in survivor of cardiac arrest: a case report. Br J Psychiatry 1988;153:836–839.
182. Larson EW, Richelson E. Organic causes of mania. Mayo Clin Proc 1988;63:906–912.
183. Lyketsos CG, Stoline AM, et al. Mania in temporal lobe epilepsy. Neuropsychiatry Neuropsychol Behav Neurol 1993;6: 19–25.
184. Walbridge DG. Rapid-cycling disorder in association with cerebral sarcoidosis. Br J Psychiatry 1990;157:611–613.
185. Yadalam KG, Jain AK, et al. Mania in two sisters with similar cerebellar disturbance. Am J Psychiatry 1985;142:1067–1069.
186. Jeffries FM, Reiss AL, et al. Bipolar spectrum disorder and fragile X syndrome: a family study. Biol Psychiatry 1993;33:213–216.
187. Papolos DF, Faedda GI, et al. Bipolar spectrum disorders in patients diagnosed with velo-cardio-facial syndrome: does a hemizygous deletion of chromosome 22q11 result in bipolar affective disorder? Am J Psychiatry 1996;153:1541–1547.
188. Aubry J-M, Simon AE, et al. Possible induction of mania and hypomania by olanzapine or risperidone: a critical review of reported cases. J Clin Psychiatry 2000;61:649–655.
189. Blank K, Duffy JD. Medical therapies. In: Coffey CE, Cummings JL, Lovell MR, Pearlson GD, eds. Textbook of Geriatric Neuropsychiatry, 2nd ed. Washington, DC: American Psychiatric Press, 2000: 699–728.
190. Kurlan R, Dimitsopulos T. Selegiline and manic behavior in Parkinson's disease. Arch Neurol 1992;49:1231.
191. Markowitz JS, Carson WH, et al. Possible dihydroepiandrosterone-induced mania. Biol Psychiatry 1999;45:241–242.
192. Rachman M, Garfield DAS, et al. Lupron-induced mania. Biol Psychiatry 1999;45:243–244.
193. Strite D, Valentine AD, et al. Manic episodes in two patients treated with interferon alpha. J Neuropsychiatry Clin Neurosci 1997;9:273–276.
194. Murphy DL, Brodie HKH, et al. Regular induction of hypomania by L-dopa in "bipolar" manic-depressive patients. Nature 1971;229:135–136.
194a. Sultzer D, Cummings J. Drug-induced mania; causative agents, clinical characteristics and management of retrospective analysis of the literature. Med Toxicol 1988;4:127–143.
195. Jefferson JW, Greist JH. Mood disorders. In: Hales RE, Yudofsky SC, Talbott JA, eds. The American Psychaitric Press Textbook of Psychiatry, 2nd ed. Washington, DC: American Psychiatric Press, 1994:465–494.
196. McMahon FJ, DePaulo JR. Affective disorders. In: Jameson JL, ed. Principles of Molecular Medicine. Totowa, NJ. Humana Press, 1998:995–1003.
197. Shulman KI, Tohen M, et al. Mania compared with unipolar depression in old age. Am J Psychiatry 1992;149:341–345.
198. Tohen M, Shulman KI, et al. First-episode mania in late life. Am J Psychiatry 1994;151:130–132.
199. Van Gerpen MW, Johnson JE, et al. Mania in the geriatric patient population. Am J Geriatr Psychiatry 1999;7:188–202.
200. Altshuler LL, Curran JG, et al. T2 hyperintensities in bipolar disorder: magnetic resonance imaging comparison and literature meta-analysis. Am J Psychiatriy 1995;152:1139–1144.
201. Strakowski SM, DelBello MP, et al. Brain magnetic resonance imaging of structural abnormalities in bipolar disorder. Arch Gen Psychiatry 1999;56:254–260.
202. Gyulai L, Alavi A, et al. I-123 iofetamine single-photon computed emission tomography in rapid cycling bipolar disorder: a clinical study. Biol Psychiatry 1997;41:152–161.
203. Migliorelli R, Starkstein SE, et al. SPECT findings with patients with primary mania. J Neuropsychiatry Clin Neurosci 1993;5:379–383.
204. Dunn RT, Frye MS, et al. The efficacy and use of anticonvulsants in mood disorders. Clin Neuropharmacol 1998;21:215–235.
205. Fuller MA, Sajatovic M. Drug Information Handbook for Psychiatry, 2nd ed. Cleveland: Lexi-Comp, 2000.
206. Keck PE, McElroy SL, et al. Anticonvulsants in the treatment of bipolar disorder. J Neuropsychiatry Clin Neurosci 1992;4: 395–405.
207. Williams D. The structure of emotions reflected in epileptic experiences. Brain 1956;79:29–67.
208. Hermann BP, Wyler AR, et al. Ictal fear: lateralizing significance and implications for understanding the neurobiology of pathological fear states. Neuropsychiatry Neuropsychol Behav Neurol 1992;5:205–210.
209. McLachlan RS, Blume WT. Isolated fear in complex partial status epilepticus. Ann Neurol 1980;8:639–641.
210. Gainotti G. Emotional behavior and hemi-spheric side of the brain. Cortex 1972;8:41–55.
211. Starkstein SE, Federoff P, et al. Catastrophic reaction after cerebrovascular lesions: frequency, correlates, and validation of a scale. J Neuropsychiatry Clin Neurosci 1993;5:189–194.
212. Beal MF, Kleinman GM, et al. Gangliocytoma of third ventricle: hyperphagia, somnolence, and dementia. Neurology 1981;31:1224–1228.
213. Flynn FG, Cummings JL, et al. Altered behavior associated with damage to the ventromedial hypothalamus: a distinctive syndrome. Behav Neurol 1988;1:49–58.
214. Haugh RM, Markesbery WR. Hypothalamic astrocytoma. Arch Neurol 1983;40:560–563.
215. Reeves AG, Plum F. Hyperphagia, rage, and dementia accompanying a ventromedial hypothalamic neoplasm. Arch Neurol 1969;20:616–624.
216. Tonkonogy JM, Geller JL. Hypothalamic lesions and intermittent explosive disorder. J Neuropsychiatry Clin Neurosci 1992;4:45–50.
217. Critchley M. Periodic hypersomnia and megaphagia in adolescent males. Brain 1962;85:627–657.
218. Carpenter S, Yassa R, et al. A pathologic basis for Kleine-Levin syndrome. Arch Neurol 1982;39:25–28.
219. Salter MS, White PD. A variant of the Kleine-Levin syndrome precipitated by both Epstein-Barr and Varicella-Zoster virus infections. Biol Psychiatry 1993;33:388–390.

220. Klüver H, Bucy PC. Preliminary analysis of functions of the temporal lobes in monkeys. Arch Neurol Psychiatry 1939;42: 979–1000.
221. Lilly R, Cummings JL, et al. The human Klüver-Bucy syndrome. Neurology 1983;33:1141–1145.
222. Hayman LA, Rexer JL, et al. Kluver-Bucy syndrome after bilateral selective damage of amygdala and its cotical connections. J Neuropsychiatry Clin Neurosci 1998;10:354–358.
223. Terzian H, Ore GD. Syndrome of Kluver and Bucy reproduced in man by bilateral removal of the temporal lobes. Neurology 1955;5:373–380.
224. Cottrell SS, Wilson SAK. The affective symptomatology of disseminated sclerosis. J Neurol Psychopathol 1926;7:1–30.
225. Tenhauten WD, Hoppe KD, et al. Alexithymia: an experimental study of cerebral commissurotomy patients and normal control subjects. Am J Psychiatry 1986;143:312–316.
226. Devinsky O, Hafler DA, et al. Embarrassment as the aura of a complex partial seizure. Neurology 1982;32:1284–1285.
227. Greenberg DB, Hochberg FH, et al. The theme of death in complex partial seizures. Am J Psychiatry 1984;141:1587–1589.
228. Ironside R. Disorders of laughter due to brain lesions. Brain 1956;79:589–609.
229. Langworthy OR, Hesser FH. Syndrome of pseudobulbar palsy. Arch Intern Med 1940;65:106–121.
230. Hopf H, Muller-Forrell W, et al. Localization of emotional and volitional facial paresis. Neurology 1992;42:1918–1923.
231. Lieberman A, Benson DF. Control of emotional expression in pseudobulbar palsy. Arch Neurol 1977;34:717–719.
232. Martin JP. Fits of laughter (show mirth) in organic cerebral disease. Brain 1950;73:453–464.
233. Mendez MF, Nakawatase TV, et al. Involuntary laughter and inappropriate hilarity. J Neuropsychiatry Clin Neurosci 1999; 11:253–258.
234. Shaibani AT, Sabbagh MN, et al. Laughter and crying in neurologic disorders. Neuropsychiatry Neuropsychol Behav Neurol 1994;7:243–250.
235. Besson G, Bogousslavsky J, et al. Acute pseudobulbar palsy and suprabulbar palsy. Arch Neurol 1991;48:501–507.
236. Kim JS, Choi-Kwon S. Poststroke depression and emotional incontinence: correlation with lesion location. Neurology 2000;54:1805–1810.
237. Swash M. Released involuntary laughter after temporal lobe infarction. J Neurol Neurosurg Psychiatry 1972;35:108–113.
238. Nahas Z, Arlinghaus KA, et al. Rapid response of emotional incontinence to selective serotonin reuptake inhibitors. J Neuropsychiatry Clin Neurosci 1998;10:453–455.
239. Robinson RG, Parikh RM, et al. Pathological laughing and crying following stroke: validation of a measurement scale and a double-blind treatment study. Am J Psychiatry 1993;150:286–293.
240. Schiffer RB, Herndon RM, et al. Treatment of pathologic laughing and weeping with amitriptyline. N Engl J Med 1985;312: 1480–1482.
241. Udaka F, Yamao S, et al. Pathologic laughing and crying treated with levodopa. Arch Neurol 1984;42:1095–1096.
242. Daly DD, Mulder DW. Gelastic epilepsy. Neurology 1957;7:189–192.
243. Sethi PK, Rao S. Gelastic, quiritarian, and cursive epilepsy. J Neurol Neurosurg Psychiatry 1976;39:823–828.
244. Offen ML, Davidoff DW, et al. Dacrystic epilepsy. J Neurol Neurosurg Psychiatry 1976;39:829–834.
245. Arroyo S, Lesser RP, et al. Mirth, laughter and gelastic seizures. Brain 1993;116:757–780.
246. Luciano D, Devinsky O, et al. Crying seizures. Neurology 1993;43:2113–2117.
247. Berkovic SF, Andermann F, et al. Hypothalamic hamartomas and ictal laughter: evolution of a characteristic epileptic syndrome and diagnostic value of magnetic resonance imaging. Ann Neurol 1988;23:429–439.
248. Kuzniecky R, Guthrie B, et al. Intrinsic epileptogenesis of hypothalamic hamartomas in gelastic epilepsy. Ann Neurol 1997;42:60–67.
249. Cascino GD, Andermann F, et al. Gelastic seizures and hypothalamic hamartomas: evaluation of patients undergoing chronic intracranial EEG monitoring and outcome of surgical treatment. Neurology 1993;43:747–750.
250. Jones KL. Smith's Recognizable Patterns of Human Malformation, 4th ed. Philadelphia: W.B. Saunders, 1988.
251. Ross ED. The aprosodias. Arch Neurol 1981;38:561–569.
252. Ross ED, Mesulam M-M. Dominant language functions of the right hemisphere? Prosody and emotional gesturing. Arch Neurol 1979;36:144–148.

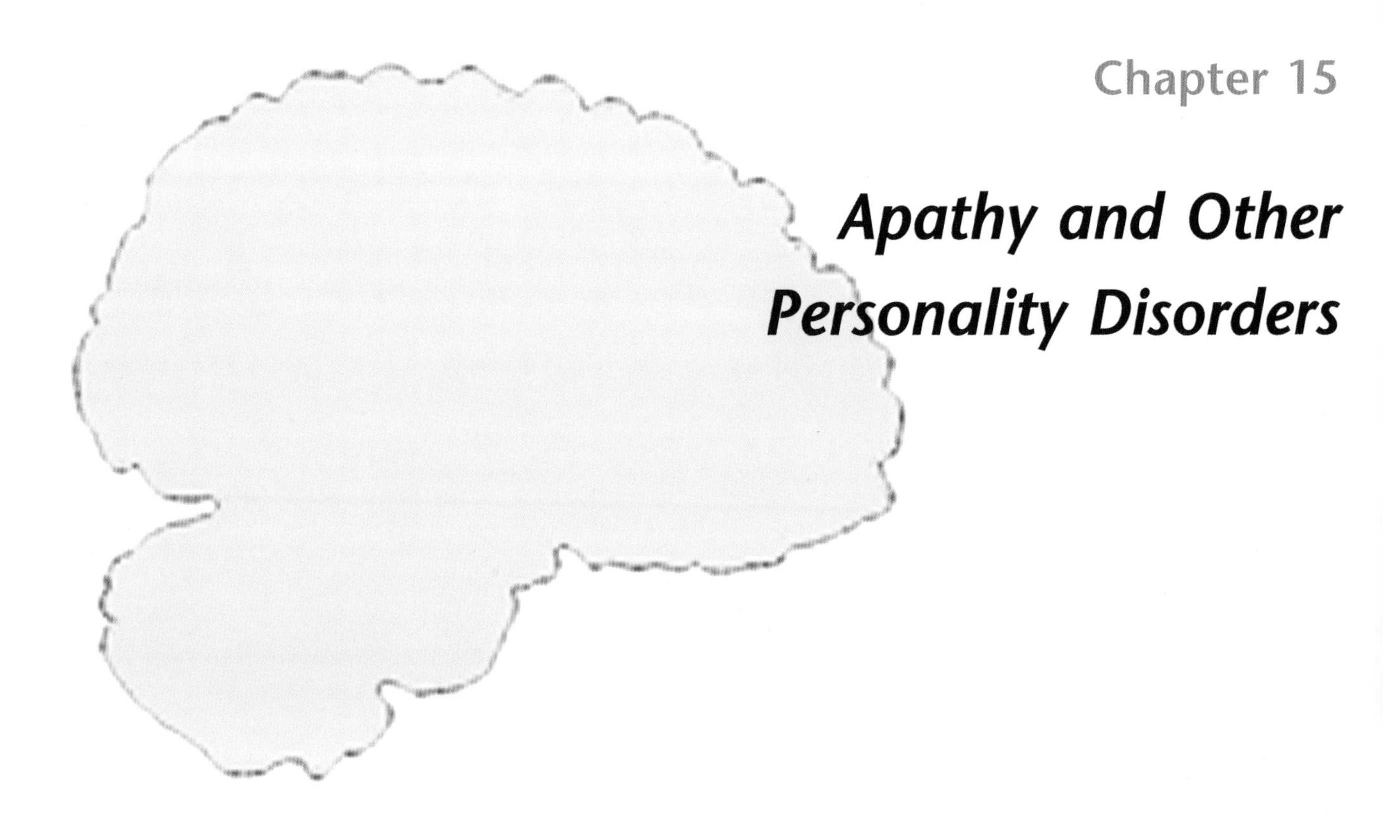

Chapter 15

Apathy and Other Personality Disorders

Personality is subserved by a widely distributed network of neural connections enabling the unique behavioral repertoire of the interpersonal interactions, subjective reactions, aversions, and goal direction that define personality. The subtle intonations in voice, characteristic mannerisms, style of comportment, and priority of desires all combine to make each individual unique.[1] All personality styles are influenced by genetic and environmental factors that guide the development of brain structure and function and ultimately contribute to personality function. When neurological diseases affect brain function it is often the widely distributed networks supporting these subtle behaviors that are the first to herald the diseases'presence. Often before the more restricted "modular" networks supporting motor, sensory, or even language and memory functions are affected, personality evidences the burden of a neurological disease. This chapter surveys our understanding of the neuroanatomy of apathy and common personality alterations. The major personality alterations, obsessive-compulsive disorder, hysteria and anxiety, are addressed in Chapters 16 and 17.

APATHY

Apathy or decreased motivation can affect performance on nearly all aspects of cognitive, instrumental, or interpersonal functions. The loss of motivation to engage in previously enjoyable behaviors is sometimes associated with depression; however, apathy can be present in the absence of sadness.[2,3] Apathy is perhaps most severe in akinetic mutism resulting from anterior cingulate lesions or from damage to this region's frontal-subcortical circuit.[4] In patients suffering from cerebrovascular disease, apathy is greatest in those with subcortical and right frontal lesions involving the anterior cingulate and related structures.[4] Apathy is the most prominent neuropsychiatric symptom in degenerative dementias including Alzheimer's disease (AD),[5] frontotemporal degeneration (FTD),[6] progressive supranuclear palsy (PSP)[7] and the dementia of Parkinson's disease (PD).[8] In AD and in individuals with mild cognitive impairment (MCI), functional defects in the anterior cingulate appear to subserve the manifestation of apathy, thus supporting a major role for the anterior cingulate in general motivation across disease etiologies (see Fig. 15.1).[9,10] Placidity is a personality alteration that has frequently been observed in association with bilateral limbic system lesions; it has been reported in patients with the Klüver-Bucy syndrome, the Wernicke-Korsakoff syndrome, and following frontal lobotomy.

Apathy may respond to psychostimulants[11] and apathy in AD often improves following treatment with cholinesterase inhibitors (Chapter 10).

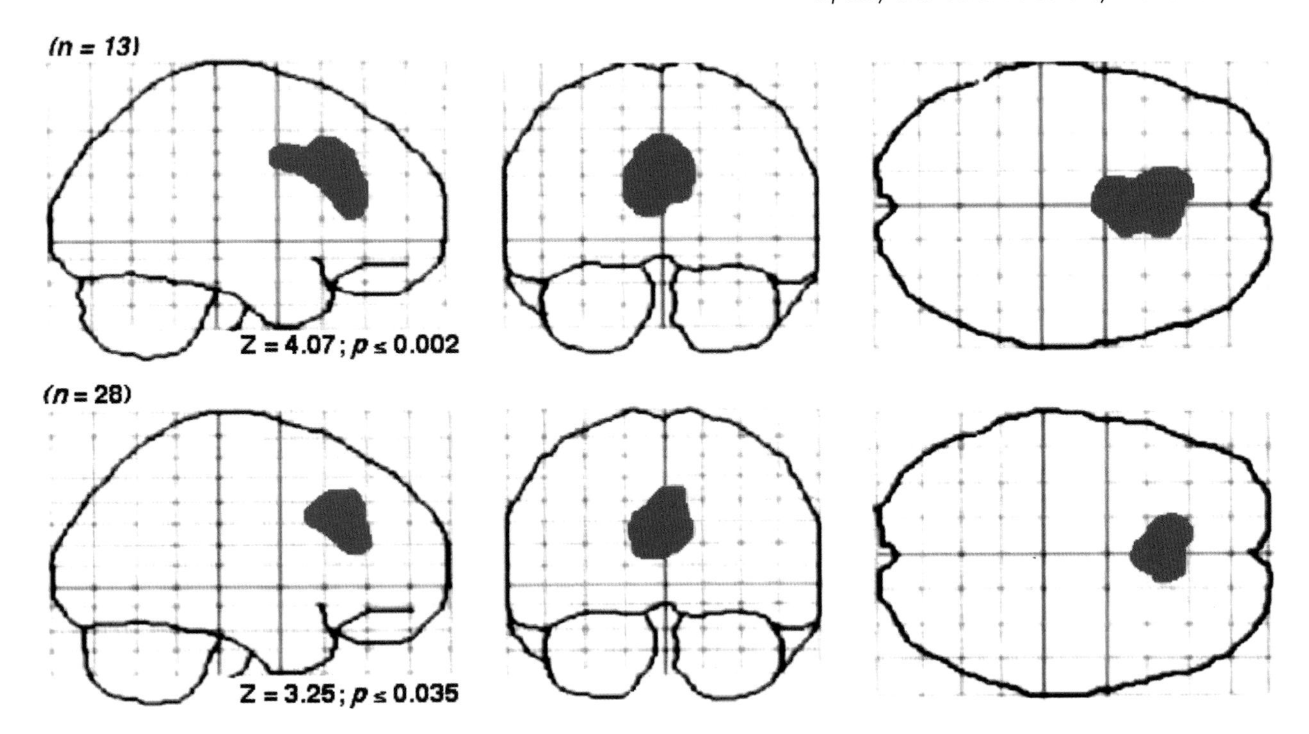

FIGURE 15.1 Statistical parametric map derived from technetium-99m D,L hexamethylpro-peleneamine oxime (^{99m}Tc-HMPAO) single photon emission computed tomography (SPECT) of patients with mild cognitive impairment (*top row*) and patients with Alzheimer's disease (*bottom row*). The map shows the location of anterior cingulate hypoperfusion in those with high apathy scores compared to those with low apathy scores as measured by the Neuropsychiatric Inventory. Adapted from Migneco et al. (2001).[10]

OTHER PERSONALITY ALTERATIONS

Personality disorders reflect behavioral patterns sufficiently inflexible and maladaptive to impair efficient functioning or produce significant subjective distress.[1] The identification of personality disorders is difficult and controversial, and combinations of different personality characteristics are common. Table 15.1 presents one approach to the classification of personality disorders and lists the neurological factors that may contribute to their occurrence. Of patients with antisocial personalities, 40%–60% exhibit electroencephalographic (EEG) abnormalities—usually temporal lobe dysrhythmia or slowing of posterior background rhythms (Chapter 21).[12] Minor abnormalities on neurological examination, the occurrence of epilepsy-like experiences, and deficits on neuropsychological tests are also common among individuals with antisocial personalities.[12] Children with hyperactivity and learning disorders are predisposed to develop antisocial personalities in adulthood; this predisposition supports a relationship between brain dysfunction and antisocial behavior.[12] Perhaps the most severe example of antisocial personality disorder is that of the predatory murderer. Metabolic increases in the right subcortical region of predatory murderers, compared to controls, contrasts with the orbitofrontal defect in affective-impulsive murderers (see Chapter 24).[13]

Borderline personality disorders have also been associated with childhood hyperactivity and with a variety of other neurological disorders, including epilepsy, cerebral trauma, and encephalitis. Impulsivity, unstable and intense interpersonal relationships, intense anger and poor control of temper, identity disturbances, affective instability, physically self-damaging activities, and chronic feelings of emptiness or boredom characterize the borderline syndrome.[1] Patients with borderline personality associated with brain dysfunction are more likely to be male and develop aberrant personality patterns at an earlier age of onset than patients with idiopathic borderline syndromes. Early computerized tomography (CT) evaluation of patients with borderline personality disorder has failed to reveal any distinct abnormalities; but when assessment is conducted with magnetic resonance imaging (MRI), significant atrophy of the frontal lobes has been found.[14] Patients suffering from borderline personality disorder are more likely to have suffered repeated traumatic experiences in childhood, and repeated stress produces hippocampal atrophy; an MRI evaluation of medial

TABLE 15.1. *Personality Disorders*[1] *and Their Neurological Correlates*

Personality Disorder	*Associated Neurological Condition*
Dependent personality	Genetic neurobiologic contributions
Histrionic personality	Genetic neurobiologic contributions
Narcissistic personality	Genetic neurobiologic contributions
Compulsive personality	Genetic neurobiologic contributions
Passive-aggressive personality	Genetic neurobiologic contributions
Schizoid personality	Genetic neurobiologic contributions
Avoidant personality	Genetic neurobiologic contributions
Schizotypal personality	Genetic neurobiologic contributions
Antisocial personality	EEG abnormalities common; hyperactivity in childhood
Borderline personality	Brain damage; hyperactivity in childhood; epilepsy
Paranoid personality	Brain damage
Organic personality	Orbitofrontal lobe damage

EEG, electroencephalogram.

temporal volume conducted in female patients with borderline personality disorder showed significant hippocampal and amygdalar atrophy in patients compared to controls.[15] These findings may represent the effect of stress on the medial temporal lobe and not the relationship of these brain regions to the borderline personality, since hippocampal atrophy was significantly related to the extent and duration of early traumatic experiences.

Functional defects implicate medial frontal, orbitofrontal, superior temporal, insular, thalamic, and striatal metabolic dysfunction[16,17] in patients with borderline personality disorder on positron emission tomography (PET). Patients using fenfluramine have a more blunted metabolic response to serotonergic challenge than controls in bilateral anterior cingulate/orbitofrontal, left superior and middle temporal, and left parietal cortex as well as the left body of the caudate, this response implicates a serotonergic defect in patients.[17] Support for a serotonergic defect in borderline personality disorder has been found with α-[^{11}C]methyl-L-tryptophan (α^{11}C-MTrp) PET, reflecting serotonergic synthetic capability. In one study, patients demonstrated significantly decreased α^{11}C-MTrp-PET in medial frontal, superior temporal, and striatal regions (Fig. 15.2).[18] The overlap of these functional imaging studies supports a medial and orbitofrontal abnormality that may subserve the impulsive aggression demonstrated by patients with the borderline personality disorder.

The personality change syndrome described in the *Diagnostic and Statistical Manual of Mental Disorders*, 4th ed. (1994)[1] is characterized by emotional lability and impaired impulse control. It most closely resembles the personality alteration that follows damage to the orbitofrontal aspects of the brain, as discussed in the following paragraphs and in Chapter 9.

The principal clue that an acquired neurological disorder is responsible for a particular personality pattern is the occurrence of a sudden change in the patient's behavioral style. Regardless of the type of change manifested, the patient who suddenly exhibits an altered personality is often harboring a neurological illness. The types of personality change induced by neurological disease may be variable, and most current personality nosologies are inadequate to account for the diversity of personality changes that have been noted in patients with neurological disturbances. Table 15.2 lists some of the personality changes reported in patients with specific neurological conditions.

Damage to the frontal lobe may produce different behavioral alterations, depending on the site of the lesion (Chapter 26). Patients with orbitofrontal injuries lack tact and restraint. They are often irritable, facetious, and euphoric and tend to be uncritical, disinhibited, and impulsive. Patients with lesions of the medial frontal lobes, by contrast, are apathetic and indifferent, and they lack initiative. Combinations of frontal lobe symptoms are common.[19] Frontal lobe syndromes occur with frontal neoplasms, demyelinating

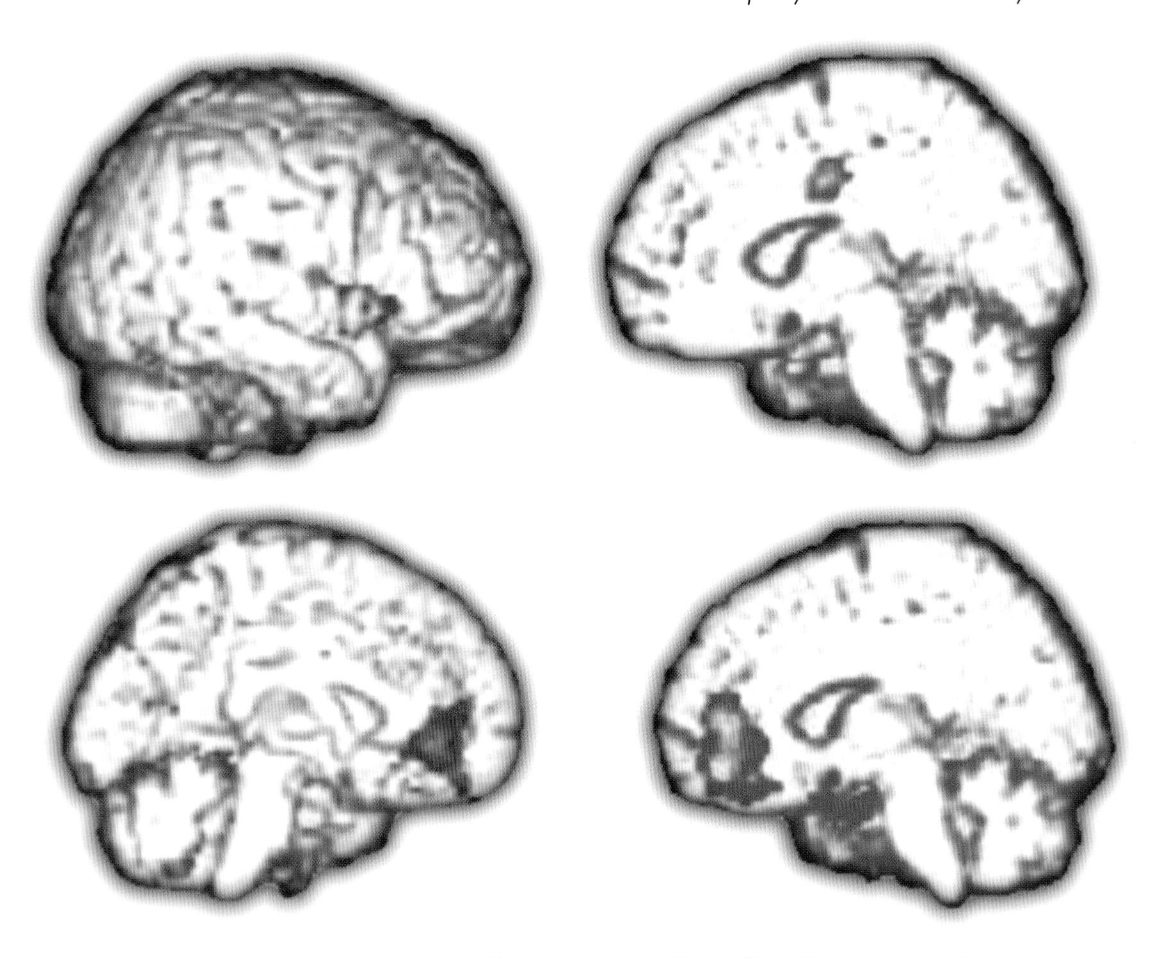

FIGURE 15.2 Statistical maps of alpha-[^{11}C]methyl-L-tryptophan (α^{11}C-MTrp) positron emission tomography (PET) reflecting significantly decreased serotonergic synthetic capability in patients with borderline personality disorder compared to controls. The top maps reveal significantly decreased regions of α^{11}C-MTrp-PET signal in eight female patients, while the bottom maps show regions of significant decrease in five males. Adapted from Leyton et al. (2001).[18]

disorders, degenerative diseases, and infectious and inflammatory illnesses and are particularly common following closed head injury.

Correlation of personality alterations with the lateralization of brain injuries has received little study, but a few observations are available. Patients with left hemisphere lesions may become depressed or paranoid.[20]

Patients who sustain right hemisphere damage as adults manifest denial of illness and tend to be abnormally euthymic.[21] Injury to the right hemisphere occurring early in life may lead to a personality pattern characterized by shyness, depression, isolation, and schizoid behavior.[19] The schizotypal personality disorder has been associated with atrophy of the prefrontal region,[22] increased sulcal space[23] and lateral ventricles[24,25] with predominant enlargement of the left temporal and frontal ventricles;[26] decreased temporal lobe volume[27] with possible focal gray matter loss in the left superior temporal gyrus,[28] and decreased mediodorsal thalamic and pulvinar volumes[29] compared to controls.

The existence of a personality pattern specifically associated with epilepsy is controversial, but considerable evidence suggests that patients with seizure disorders, particularly those with complex partial seizures, are subject to several types of personality alteration.[30] One characteristic personality change consists of interpersonal viscosity, circumstantial speech, religiosity and increased attention to nascent philosophical concerns, hyposexuality, and hypergraphia (Chapter 21).[31] The personality syndrome is not pathognomonic of epilepsy and occurs in other psychiatric disorders, but its emergence after the onset of a limbic system injury implies the existence of a relationship between the lesion site and the development of the behavioral characteristics. Pond and Bidwell[32] found this type of personality alteration in approximately 5% of all epileptics followed

TABLE 15.2 *Neurological Disorders with Associated Personality Alterations*

Neurological Condition	*Personality Alteration*
Frontal lobe damage	
Orbitofrontal	Tactless; facetious; euphoric; impulsive and disinhibited
Mediofrontal	Apathetic, indifferent; lack initiative
Lateralized hemispheric injury	
Left	Paranoia, depression
Right	Adult onset: denial, neglect, paranoia
	Childhood onset: shyness, depression, social isolation
Epilepsy	Epileptic personality: viscosity, circumstantiality, religiosity, hyposexuality, hypergraphia
	Borderline personality: paranoia
Bilateral limbic system lesions	Placidity
	Hyperactivity, learning disability
	Antisocial personality
	Borderline personality

in general medical practices in Great Britain. Another personality pattern found in epileptics is the borderline personality disorder described above.

Personality alterations occur with many neurological disorders in addition to those described here, but this aspect of neuropsychiatry has received little systematic study. Alterations in personality also overlap with the chronic changes in mood, described in Chapter 14.

REFERENCES

1. American Psychiatric Association. Diagnostic and Statistical Manual of Mental Disorders, 4th ed. Washington, DC: American Psychiatric Press, 1994.
2. Andersson S, Krogstad JM, Finset A. Apathy and depressed mood in acquired brain damage: relationship to lesion localization and psychophysiological reactivity. Psychol Med 1999;29:447–456.
3. Levy ML, Cummings JL, Fairbanks LA, et al. Apathy is not depression. J Neuropsychiatry Clin Neurosci 1998;10:341–319.
4. Mega MS, Cohenour RC. Akinetic mutism: a disconnection of frontal–subcortical circuits. Neurol Neuropsychol Behav Neurol 1997;10:254–259.
5. Mega MS, Cummings JL, et al. The spectrum of behavioral changes in Alzheimer's disease. Neurology 1996;46:130–135.
6. Levy ML, Miller BL, et al. Alzheimer disease and frontotemporal dementias. Behavioral distinctions. Arch Neurol 1996;53:687–690.
7. Litvan I, Mega MS, et al. Neuropsychiatric aspects of progressive supranuclear palsy. Neurology 1996;47:1184–1189.
8. Aarsland D, Cummings JL, Larsen JP. Neuropsychiatric differences between Parkinson's disease with dementia and Alzheimer's disease. Int J Geriatr Psychiatry 2001;16:184–191.
9. Benoit M, Dygai I, et al. Behavioral and psychological symptoms in Alzheimer's disease. Relation between apathy and regional cerebral perfusion. Dement Geriatr Cogn Disord 1999;10:511–517.
10. Migneco O, Benoit M, et al. Perfusion brain SPECT and statistical parametric mapping analysis indicate that apathy is a cingulate syndrome: a study in Alzheimer's disease and nondemented patients. Neuroimage 2001;13:896–902.
11. Watanabe MD, Martin EM, et al. Successful methylphenidate treatment of apathy after subcortical infarcts. J Neuropsychiatry Clin Neurosci 1995;7:502–504.
12. Raine A. The Psychopathology of Crime. San Diego, CA: Academic Press, 1993.
13. Raine A, Meloy RJ, et al. Reduced prefrontal and increased subcortical brain functioning assessed using positron emission tomography in predatory and affective murderers. Behav Sci Law 1998;16:319–332.
14. Lyoo IK, Han MH, Cho DY. A brain MRI study in subjects with borderline personality disorder. J Affect Disord 1998;50:235–243.
15. Driessen M, Herrmann J, et al. Magnetic resonance imaging volumes of the hippocampus and the amygdala in women with borderline personality disorder and early traumatization. Arch Gen Psychiatry 2000;57:1115–1122.
16. De La Fuente JM, Goldman S, et al. Brain glucose metabolism

in borderline personality disorder. J Psychiatr Res 1997;31:531–541.
17. Soloff PH, Meltzer CC, et al. A fenfluramine-activated FDG-PET study of borderline personality disorder. Biol Psychiatry 2000; 47:540–7.
18. Leyton M, Okazawa H, et al. Brain regional α-[11C]methyl-L-tryptophan trapping in impulsive subjects with borderline personality disorder. Am J Psychiatry 2001;158:775–782.
19. Eslinger PJ, Geder L. Behavioral and emotional changes after focal frontal lobe damage. In: Behavior and Mood Disorders in Focal Brain Lesions. Bogousslavsky J, Cummings J (eds). Cambridge, England Univesity Press, 2000:217–260.
20. Benson DF. Psychiatric aspects of aphasia. Br J Psychiatry 1973;123:555–566.
21. Filley CM. Neurobehavioral Anatomy. Denver, Colorado: University of Colorado Press, 1995.
22. Raine A, Sheard C, et al. Prefrontal structural and functional deficits associated with individual differences in schizotypal personality. Schizophr Res 1992;7:237–247.
23. Dickey CC, Shenton ME, et al. Large CSF volume not attributable to ventricular volume in schizotypal personality disorder. Am J Psychiatry 2000;157:48–54.
24. Siever LJ, Rotter M, et al. Lateral ventricular enlargement in schizotypal personality disorder. Psychiatry Res 1995;57:109–118.
25. Kurokawa K, Nakamura K, et al. Ventricular enlargement in schizophrenia spectrum patients with prodromal symptoms of obsessive-compulsive disorder. Psychiatry Res 2000;99:83–91.
26. Buchsbaum MS, Yang S, et al. Ventricular volume and asymmetry in schizotypal personality disorder and schizophrenia assessed with magnetic resonance imaging. Schizophr Res 1997;27: 45–53.
27. Downhill JE Jr, Buchsbaum MS, et al. Temporal lobe volume determined by magnetic resonance imaging in schizotypal personality disorder and schizophrenia. Schizophr Res 2001;48: 187–199.
28. Dickey CC, McCarley RW, et al. Schizotypal personality disorder and MRI abnormalities of temporal lobe gray matter. Biol Psychiatry 1999;45:1393–1402.
29. Byne W, Buchsbaum MS, et al. Magnetic resonance imaging of the thalamic mediodorsal nucleus and pulvinar in schizophrenia and schizotypal personality disorder. Arch Gen Psychiatry 2001;58:133–140.
30. Trimble MR. Personality disturbances in epilepsy. Neurology 1983;33:1332–1334.
31. Bear DM, Fedio P. Quantitative analysis of interictal behavior in temporal lobe epilepsy. Arch Neurol 1977;34:454–467.
32. Pond DA, Bidwell BH. A survey of epilepsy in fourteen general practices. II. Social and psychological aspects. Epilepsia 1960;1: 285–299.

Obsessive-Compulsive Disorder and Syndromes with Repetitive Behaviors

Obsessive-compulsive disorder (OCD) is a condition of great neuropsychiatric interest, characterized by intrusive subjective obsessions, involuntary compulsions, and ritualized behavior. This disorder can be inherited or acquired and many cases respond well to treatment. The pathophysiology of the disorder is increasingly well understood and linked to dysfunction of frontal–subcortical circuits. This chapter presents the clinical features, neurodiagnostic assessment, pathophysiology, and management of OCD. In addition, OCD-spectrum disorders are considered and neurobiobehavioral and neuropsychiatric syndromes with prominent repetitive behaviors are presented.

OBSESSIVE-COMPULSIVE DISORDER

Clinical Features

Obsessions are recurring impulses, thoughts, or impulses that are experienced as intrusive and inappropriate and cause marked distress. *Compulsions* are repetitive behaviors (e.g., handwashing, ordering, checking) or mental acts (e.g., praying, counting, repeating words silently) that the person feels driven to perform in response to an obsession or according to rules that must be applied rigidly.[1] Obsessive-compulsive disorder is defined by the presence of obsessions or compulsions that are excessive or unrealistic, cause marked distress, and are not limited to the specific content of a co-occurring disorder (e.g., preoccupation with food in an eating disorder or hair pulling in trichotillomania).[1] The incidence of OCD is approximately 7.5 new cases per 1000 persons; 1-year prevalence rates of approximately 1.65% have been identified and lifetime prevalence rates in North American populations have been found to be between 2.6% and 3.2%.[2] The mean age of onset of OCD symptoms is 21 to 25 years, although 20% of cases begin in childhood and 30% begin in adolescents. Long-term follow-up studies indicate that the course is chronic and fluctuating, with most patients experiencing some degree of improvement after several decades of symptoms. Twenty percent of patients exhibit complete recovery and an additional 30% have only residual subclinical symptoms.[3]

Table 16.1 provides a list of obsessions and compulsions observed in OCD patients that is derived from the Yale-Brown Obsessive Compulsive Scale.[4,5] Among the most common obsessions occurring in patients with OCD are concerns regarding contamination, pathologic doubt (e.g., concerning having hurt or contami-

TABLE 16.1. *Symptoms of Obsessive-Compulsive Disorder*[4,5]

Obsessions	*Compulsions*
Aggressive obsessions	Cleaning and washing compulsions
Fear of harming self	Excessive hand washing
Fear of harming others	Excessive showering, bathing, or grooming
Violent or horrific images	Excessive cleaning of household items or other inanimate objects
Fear of blurting out obscenities or insults	Excessive measures to prevent or remove contact with contaminants
Fear of performing embarrassing acts	Checking compulsions
Fear that one will perform unwanted acts	Checking locks, stove, appliances, etc.
Fear that one will steal things	Checking for harm of others
Contamination obsessions	Checking for mistakes
Concerns with bodily waste or secretions	Repeating rituals
Concerns with dirt or germs	Rereading or rewriting
Concerns with environmental contaminants	Need to repeat routine activities (e.g., going in and out of a door, or getting up and down from a chair)
Concerns regarding animals (e.g., insects)	Counting compulsions
Concerns about getting ill	Ordering and arranging compulsions
Sexual obsessions	Miscellaneous compulsions
Forbidden sexual thoughts, images, or impulses	Mental rituals
Aggressive sexual thoughts toward others	Excessive list making
Hoarding and saving obsessions	Need to tell, ask, or confess
Religious obsessions	Need to touch, tap, or rub
Concern with sacrilege and blasphemy	Ritualized eating behavior
Excessive concern with right and wrong, morality	Superstitious behaviors
Obsession with need for symmetry or exactness	Trichotillomania
Somatic obsessions	Other self-damaging or self-mutilating behaviors
Concern with illness or disease	
Excessive concern with body part or appearance	
Miscellaneous obsessions	
Need to know or remember	
Fear of saying certain things	
Fear of not saying exactly the right thing	
Fear of losing things	
Intrusive (non-violent) images, sounds, words, or music	
Lucky or unlucky numbers	
Colors with special significance	
Superstitious fears	

nated someone or something), excessive somatic concerns (e.g., regarding illness, disease, or bodily appearance), need for symmetry, aggressive obsessions, and sexual obsessions. Among the most common compulsions are checking, washing, counting, experiencing the need to ask or confess, excessive need for symmetry and precision, and hoarding.[6] Factor analysis of the symptoms collected using symptom rating scales suggests that four general areas of concern and behavior account for most symptoms of OCD: checking, symmetry and ordering, cleanliness and washing, and hoarding.[7] The obsessional and compulsive symptoms of OCD vary in severity from mild to extremely disabling. Patients with severe OCD may take many hours to perform toileting rituals, cleaning, washing, and arranging. The symptoms of childhood and adolescent onset of OCD are essentially identical to those of adult OCD.[8–10]

Patients with Gilles de la Tourette's syndrome frequently suffer from OCD (Chapter 19 and discussed below). Patients with OCD and Gilles de la Tourette syndrome have significantly more violent, sexual, and symmetry obsessions, and more touching, blinking, counting, and self-damaging compulsions. Patients with OCD without a concurrent tic disorder have more obsessions concerning dirt or germs and more cleaning compulsions.[11]

Neuropsychological deficits have been observed consistently in patients with OCD. Executive abilities, verbal memory, attention, and intelligence are normal when compared to unaffected controlled populations. However, spatial working memory, spatial recall, and speed of motor initiation and execution (on executive tasks, such as the Tower of London planning task) are impaired.[12–15] These neuropsychological abnormalities distinguish patients with OCD from those with social phobia, unipolar depression, and panic disorder.[13,14] The deficits suggest an abnormality of frontal–striatal processing of nonverbal information.

Patients with OCD frequently have coexisting major mental illnesses. Among OCD patients the lifetime risk of major depressive disorder is approximately 70%; simple phobia, 20%; social phobia, 20%; eating disorder, 20%; alcohol dependence, 14%; panic disorder, 12%; separation anxiety disorder, 2%; and Gilles de la Tourette syndrome, 10%.[6] Comorbid psychiatric disorder must be distinguished from OCD-spectrum disorders and from illnesses that have repetitive behaviors as major manifestations (discussed below).

The heredity of OCD has been difficult to define with precision and no single mutation has been found to be responsible for the disorder. Twin studies and family studies support a model of polygenic inheritance with at least one major gene effect.[16,17] Familial studies of patients with Gilles de la Tourette syndrome and chronic multiple tics show an increased prevalence of OCD among first-degree relatives, suggesting that some forms of OCD may be alternate expressions of a pathophysiology shared with tic syndromes.[18]

Neuroimaging Abnormalities in Obsessive-Compulsive Disorder

Caudate nuclear volume is subtly but significantly smaller in patients with OCD than in healthy controls when measured with either X-ray computed tomography (CT) or magnetic resonance imaging (MRI).[19,20] Functional imaging using fluorodeoxyglucose (FDG) positron emission tomography (PET) consistently reveals hypermetabolic activity of the brain in OCD compared to the activity in normal controls. Brain regions most consistently observed to be hypermetabolic include the orbitofrontal cortex, anterior cingulate cortex, and heads of the caudate nuclei.[8–10,21–23] Similarly, studies with technetium-hexamethyl-propyleneamine-oxime (Tc-HMPAO) single-photon emission computed tomography (SPECT) indicate increased cerebral blood flow in the orbitofrontal cortex and medial frontal cortex as well as in dorsal parietal regions.[24,25]

Regional activation of cortical structures has been investigated in patients with OCD using PET or functional MRI to measure changes in regional cerebral blood flow (rCBF).[26,27] Exposure to provocative stimuli (pictures of a person about whom the patient had sexual and violent obsessions or objects considered to be contaminated and offensive to patients with contamination obsessions) resulted in increased blood flow in anterior cingulate orbitofrontal, anterior temporal, insular, and caudate nuclear regions.

Significant reductions in regional brain metabolism and blood flow have been documented following pharmacotherapy or behavioral therapy in patients who exhibited a clinical response to the treatment intervention.[28–31]

Pathophysiology

There is no established pathology of OCD and the putative pathophysiology of OCD is inferred from abnormalities observed on structural and functional imaging studies, as well as from the anatomical distribution of lesions associated with OCD in patients with acquired OCD syndromes (described below). Substantial evidence impugns frontal–striatal circuitry in the mediation of OCD. Five frontal–striatal thalamic circuits have been described[32] connecting supplementary motor cortex, frontal eye fields, dorsolateral prefrontal cortex, anterior cingulate cortex, and orbital frontal cortex to striatal and thalamic structures. In each circuit a direct excitatory link is in dynamic balance with an indirect inhibitory link.[32] Orbitofrontal cortex is involved in modulating socially appropriate behaviors and when impaired, results in tactless, impulsive, disinhibited behavior (Chapter 9). Medial frontal cortex mediates motivation, persistence, and environmental engagement.[33,34] Hyperactivity of these regions may result in abnormal capture of cognitive themes relating to violence, hygiene, order, and sex, the subjects of obsessions and compulsions in OCD.[35] This abnormality may result from an imbalance between inhibitory and excitatory pathways within frontal–striatal circuits and could reflect abnormalities at frontal, striatal, pallidal, or thalamic nuclear levels.[35–37] The response of these symptoms to serotonergic therapy (described below) suggests that serotonergic input in these circuits is critical to normal function and is disturbed in patients with OCD.[38] Figure 16.1 demonstrates the frontal–striatal pathways implicated in mediation of OCD.

Differential Diagnosis of Obsessive-Compulsive Disorder

The differential diagnosis of OCD includes two different classes of disorder. First, there are OCD-spectrum disorders in which there is accumulating evidence that the underlying pathophysiology of the clinical manifestations is identical or similar to that of OCD (Table 16.2). Second, there are conditions in which OCD occurs as a manifestation of another underlying pathophysiology, such as Huntington's disease or frontotemporal dementia (Table 16.3).

The OCD-spectrum disorders share with OCD an obsession-like intrusive preoccupation and repetitive, time-consuming, and often self-destructive, compulsion-like behaviors. They differ from OCD in being focused nearly exclusively on a single dimension of behavior. Table 16.2 lists and defines some of these types of behaviors, including body dysmorphic disorder,

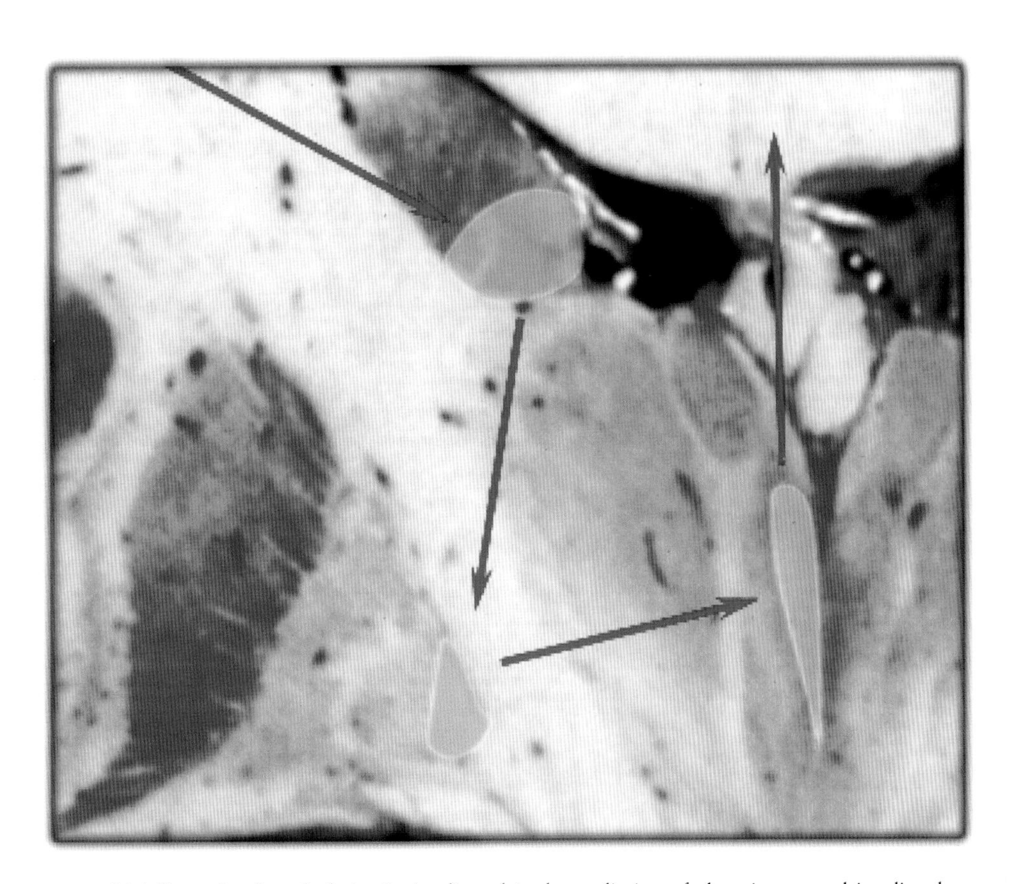

FIGURE 16.1 Frontal–subcortical circuits implicated in the mediation of obsessive-compulsive disorder (orbitofrontal cortex, globus pallidus, medial thalamus).

TABLE 16.2. *Obsessive-Compulsive Disorder–Spectrum Disorders*

Disorder	*Obsessional Symptoms*
Body dysmorphic disorder	A preoccupation with some imagined defect in appearance in a normal-appearing person
Anorexia nervosa	Intense fear of gaining weight or becoming fat even though underweight
Compulsive drinking behavior	Compulsive polydypsia with excessive water drinking
Hypochondriasis	Obsessions about being ill and compulsions to check with others for either diagnosis and treatment or reassurance that one is not ill
Olfactory reference syndrome	Persistent preoccupation with personal odor undetectable by others
Trichotillomania	Compulsive hair-pulling in which the patient is unable to resist impulses to pluck hair from scalp, eyebrows, eyelashes, or other body regions
Sexual obsessions and compulsions	Ego-dystonic intrusive thoughts about sexual matters or compulsive sexual hyperactivity such as promiscuity or compulsive masturbation
Pathologic gambling	Preoccupation with gambling and urges to gamble resulting in irresistible gambling impulses
Alcohol and substance abuse	Uncontrollable and unpleasant urge to engage in alcohol and substance abuse
Compulsive skin picking	Compulsions to remove small irregularities on the skin (e.g. blemishes, dry skin), usually associated with extensive cleaning rituals
Pseudologia fantastica	Compulsive telling of untruths that are easily discovered
Impulse control disorders	
Intermittent explosive disorder	Episodic loss of control of aggressive impulses
Pyromania	Impulsive, deliberate fire-setting
Kleptomania	Impulsive theft

anorexia nervosa, hypochondriasis, trichotillomania, sexual obsessions and compulsions, pathological gambling, self-mutilation, alcohol and substance abuse, compulsive skin-picking, compulsive drinking, and pathological lying.[39–48] The relationship of these disorders to OCD is a subject of current research; recognizing these disorders to be within the OCD spectrum may facilitate treatment choices for some patients.

Impulse control disorders may also bear a relationship to OCD and OCD-spectrum disorders. They share with OCD the inability to resist specific behaviors although their phenomenology differs in many other respects. Impulse control disorders include intermittent explosive disorder (episodic loss of control of aggressive impulses), pyromania (impulsive, repetitive, deliberate fire-setting), and kleptomania (impulsive stealing).[49]

A variety of neurological and psychiatric disorders have been associated with obsessive-compulsive behaviors (Table 16.3). The most common clinical circumstances in which OCD is encountered include mental retardation syndromes, particularly those with autistic features; post-encephalitic Parkinson's disease; Sydenham's chorea; Gilles de La Tourette syndrome; and frontotemporal degenerations.[8–11,50–62] In addition, OCD has been induced by cocaine, psychostimulants, and manganese intoxication.[63,64] Obsessions and compulsions may occasionally be seen in patients with schizophrenia or major depression.[65,66]

Management of Obsessive-Compulsive Disorder

Patients with OCD may respond to behavioral therapy, cognitive therapy, psychotherapy, pharmacotherapy, electroconvulsive therapy (ECT), or neurosurgical intervention. Pharmacotherapy is the principal means of treatment of OCD. Selective serotonin reuptake inhibitors (fluvoxamine, fluoxetine, sertraline, paroxetine, citalopram) or the tricyclic clomipramine (which has pronounced effects on serotonin reuptake) are the principal agents used in the treatment of OCD. Higher doses of these drugs are typically required for the treatment of OCD than for treatment of mood disorders.[46,67–70] The same agents are used when OCD occurs in conjunction with a neurological disorder such as Huntington's disease or frontotemporal dementia. Addition of haloperidol or an other neuroleptic agent may further improve the response in treatment-refrac-

TABLE 16.3. *Neurological Conditions in Which Obsessive-Compulsive Disorder Has Been Reported*

Condition/Disorder	*Subtype(s)*
Degenerative disorders	Frontotemporal dementia
	Huntington's disease
	Parkinson's disease
	Progressive supranuclear palsy
	Neuroacanthocytosis
Brain tumors	Frontal
	Striatal
Cerebral infarction	Striatal
Gilles de la Tourette syndrome	
Fragile X syndrome	
Epilepsy	Temporal lobe
	Cingulate
Sydenham's chorea	
Pediatric autoimmune neuropsychiatric disorder associated with strep (PANDAS)	
Idiopathic striatal calcification	
Meige's syndrome (idiopathic blepharospasm)	
Miscellaneous structural lesions	Carbon monoxide poisoning with bilateral caudate lesions
	Neonatal hypoxia with caudate lesions
	Anoxic injury to striatum
	Wasp sting with injury to globus pallidus
	Carbon monoxide intoxication with pallidal injury
	Anoxia with pallidal damage
	Head injury
	Multiple sclerosis
	Hypoglycemia
Intoxications	Amphetamine
	Cocaine
	Manganese
Inherited childhood syndromes	Lesch-Nyhan syndrome
	Prader-Willi syndrome
Mental retardation	
Autism	
Asperger's syndrome	

tory patients, particularly those with a comorbid chronic tic disorder.[71]

OCD-spectrum disorders may respond to the same treatments used for OCD. In some cases, self-injurious behavior (discussed below) has responded to treatment with opioid antagonists such as naloxone. In patients refractory to pharmacotherapy, ECT may provide an effective alternative.[72] Patients with treatment-refractory and very disabling OCD may be considered for psychosurgery. Stereotactic anterior capsulotomy, cingulotomy, and ventromedial leukotomy may produce beneficial effects.[73]

SYNDROMES WITH REPETITIVE BEHAVIORS

Self-Injurious Behavior

Self-injurious behavior (SIB) is a distressing set of behaviors both for the patient involved and for family members and caregivers working with the patient. Self-injurious behaviors include trichotillomania, repetitive skin picking and excoriations, self-induced ocular injuries including enucleation, lip biting and finger chewing, head banging, self-burning (usually with cigarettes), autocastration, and autosurgeries.[74] Self-injurious behavior can be classified into major self-mutilation, stereotypic self-mutilation, and superficial or moderate self-mutilation[75] (Table 16.4). Major self-mutilation includes sexual self-castration, auto-enucleation, and other major forms of self-injury that threaten life. This extreme sort of behaviors is associated with psychotic states, acute intoxication, transsexualism, and mental retardation. Stereotypic self-mutilation involves those cases of SIB associated with OCD and OCD-spectrum disorders. Head banging and other forms of SIB observed in mentally retarded and autistic individuals are included in this category. Superficial or moderate self-mutilation is common and occurs in extraordinary emotional circumstances including post-traumatic stress disorder, combat, and personality disorders (borderline, histrionic, and antisocial). Sexually motivated sadomasochistic behaviors with SIB are also included here.

Neurological disorders in which self-mutilation may occur include Gilles de la Tourette syndrome, idiopathic and inherited mental retardation syndromes, and frontotemporal dementia.[51,52,76–79] Head banging (jactatio capatis) is a specific type of stereotypic movements observed most commonly in mental retardation but may occur in any of the SIB disorders.[77] Self-injury syndromes have been treated with neuroleptic agents, opioid antagonists, selective serotonin reuptake inhibitors, proponalol, carbamazepine, and lithium.[80]

Obsessive-compulsive behavior must occasionally be distinguished from other types of repetitive motor acts (Table 16.5). These include echolalia, palolalia, coprolalia, echopraxia, copropraxia, perseveration, stereotypies, mannerisms, chorea, dyskinesia, stereotypic hand waving, stereotypic catatonia, and tics. Diogenes' syndrome (also known as *senile squalor*) refers to a disorder observed in the elderly of hoarding or abnormal accumulation of trash and useless objects.[81] Kluver-Bucy syndrome and the temporal lobe epilepsy personality are two other neurological conditions in which obsessive and compulsive-like behaviors are observed.

TABLE 16.4. *Neurologic and Psychiatric Syndromes Associated with Self-Injurious Behavior*

Major self-mutilation
Psychotic states
Schizophrenia, depression, mania
Acute intoxications
Transexualism
Mental retardation
Gilles de la Tourette syndrome
Stereotypic self-mutilation
Mental retardation
Autism
Lesch-Nyhan syndrome
Gilles de la Tourette syndrome
Rett syndrome
Aicardi syndrome
Goubert syndrome
Brachmann De-Lange syndrome
Prader-Willi syndrome
Smith Maganis syndrome
Oculocerebrorenal syndrome of lowe
Fragile X syndrome
Asperger's syndrome
Cornelia de Lange syndrome
Neuroacanthocytosis
Frontotemporal dementia
Superficial or moderate self-mutilation
Personality disorders
Histrionic personality disorder
Borderline personality disorder
Antisocial personality disorder
Sadomasochistic disorder
Flagellation
Piercing
Other forms of pain infliction
Post-traumatic stress disorder
Combat disorder

Kluver-Bucy patients exhibit hypermetamorphosis, a syndrome characterized by compulsive exploration of high-stimulus items in the environment.[82] Patients with temporal lobe epilepsy personality exhibit circumstantiality, hypergraphia, and overinclusive involvement with religious and philosophical topics (Chapter 21).

TABLE 16.5. *Repetitive Movements to Be Distinguished from Obsessive-Compulsive Behavior*

Movement	*Manifestation*
Echolalia	Repeating the words of others
Palolalia	Repeating one's own words
Coprolalia	Involuntary cursing
Myoclonus	Asymmetric, random repetitive jerks
Tics	Stereotyped repetitive jerks involving the same movement each time
Chorea	Irregular but stereotyped movement disorder
Dyskinesia	Irregular, stereotyped form of chorea, usually drug-induced
Mannerisms	Purposeless repetitive movements
Perseveration	Involuntary repetition of a motor act (usually elicited in writing and copying)
Catatonia	Prolonged maintenance of postures
Hoarding	Hoarding and accumulation of trash or useless material; also known as *Diogenes syndrome*
Hypermetamorphosis	Component of the Kluver-Bucy syndrome with compulsive exploration of high-stimulus items in the environment
Temporal lobe epilepsy personality	Personality alteration associated with focal temporal dysfunction, characterized by over detailed explanations, writing, and concern with philosophical and religious topics

In some cases, these phenomena may respond to treatment used to manage OCD.

REFERENCES

1. American Psychiatric Association. Diagnostic and Statistical Manual of Mental Disorders, 4th ed. Washington, DC: American Psychiatric Press, 1994.
2. Angst J. The epidemiology of obsessive-compulsive disorder. In: Hollander E, Zohar J, Marazzati D, Olivier B, eds. Current Insights in Obsessive-Compulsive Disorder. New York: John Wiley & Sons, 1994:93–104.
3. Skoog G, Skoog I. A 40-year follow-up of patients with obsessive-compulsive disorder. Arch Gen Psychiatry 1999;56:121–127.
4. Goodman WK, Price LH, et al. The Yale-Brown Obsessive-Compulsive Scale I: development, use, and reliability. Arch Gen Psychiatry 1989;46:1006–1011.
5. Goodman WK, Lawrence HP, et al. The Yale-Brown Obsessive Compulsive Scale II: validity. Arch Gen Psychiatry 1989;46: 1012–1016.
6. Rasmussen SA, Eisen JL. The epidemiology and clinical features of obsessive-compulsive disorder. In: Obsessive-Compulsive Disorders, 3rd ed. Chicago: Mosby, 1998:12–43.
7. Leckman JF, Grice DE, et al. Symptoms of obsessive-compulsive disorder. Am J Psychiatry 1997;154:911–917.
8. Swedo SE, Rapoport JL, et al. High prevalence of obsessive-compulsive symptoms in patients with Sydenham's chorea. Am J Psychiatry 1989;146:246–249.
9. Swedo SE, Schapiro MB, et al. Cerebral glucose metabolism in childhood-onset obsessive-compulsive disorder. Arch Gen Psychiatry 1989;46:518–523.
10. Swedo SE, Rapoport JL, et al. Obsessive-compulsive disorder in children and adolescents. Arch Gen Psychiatry 1989;46:335–341.
11. George MS, Trimble MR, et al. Obsessions in obsessive-compulsive disorder with and without Gilles de la Tourette's syndrome. Am J Psychiatry 1993;150:93–97.
12. Christensen KJ, Kim SW, et al. Neuropsychological performance in obsessive-compulsive disorder. Biol Psychiatry 1992;31:4–18.
13. Cohen LJ, Hollander E, et al. Specificity of neuropsychological impairment in obsessive-compulsive disorder: a comparison with social phobic and normal control subjects. J Neuropsychiatry Clin Neurosci 1996;8:82–85.
14. Purcell R, Maruff P, et al. Neuropsychological deficits in obsessive-compulsive disorder. Arch Gen Psychiatry 1998;55:415–423.
15. Savage CR, Keuthen NJ, et al. Recall and recognition memory in obsessive-compulsive disorder. J Neuropsychiatry Clin Neurosci 1996;8:99–103.
16. Pauls DL. Phenotypic variability in obsessive-compulsive disorder and its relationship to familial risk. CNS Spectrums 1999;4: 57–61.
17. Rasmussen SA. Genetic studies of obsessive-compulsive disorder. In: Hollander E, Zohar J, Marazzati D, Olivier B, eds. Current Insights in Obsessive-Compulsive Disorder. New York: John Wiley & Sons, 1994:105–114.
18. Pauls DL, Towbin KE, et al. Gilles de la Tourette's syndrome and obsessive-compulsive disorder. Arch Gen Psychiatry 1986; 43:1180–1182.
19. Luxenberg JS, Swedo SE, et al. Neuroanatomical abnormalities in obsessive-compulsive disorder detected with quantitative X-ray computed tomography. Am J Psychaitry 1988;145: 1089–1093.
20. Robinson D, Houwei W, et al. Reduced caudate nucleus volume in obsessive-compulsive disorder. Arch Gen Psychiatry 1995;52: 393–398.

21. Baxter LR, Phelps ME, et al. Local cerebral glucose metabolic rates in obsessive-compulsive disorder: a comparison with rates in unipolar depression and normal controls. Arch Gen Psychiatry 1987;44:211–218.
22. Baxter LR, Schwartz JM, et al. Cerebral glucose metabolic rates in nondepressed patients with obsessive-compulsive disorder. Am J Psychiatry 1988;145:1560–1563.
23. Perani D, Colombo C, et al. [^{18}F] FDG PET study in obsessive-compulsive disorder: a clinical/metabolic correlation study after treatment. Br J Psychiatry 1995;166:244–250.
24. Machlin SR, Harris GJ, et al. Elevated medial-frontal cerebral blood flow in obsessive-compulsive patients: a SPECT study. Am J Psychiatry 1991;148:1240–1242.
25. Rubin RT, Villaneuva-Meyer J, et al. Regional xenon 133 cerebral blood flow and cerebral technetium 99m HMPAO uptake in unmedicated patients with obsessive-compulsive disorder and matched normal control subjects. Arch Gen Psychiatry 1992;49: 695–702.
26. Breiter JC, Rauch SL, et al. Functional magnetic resonance imaging of symptom provocation in obsessive-compulsive disorder. Arch Gen Psychiatry 1996;53:595–606.
27. Rauch Sl, Jenike MA, et al. Regional cerebral blood flow measured during symptom provocation in obsessive-compulsive disorder using oxygen ^{15}-labeled carbon dioxide and positron emission tomography. Arch Gen Psychiatry 1994;51:62–70.
28. Baxter LR, Schwartz JM, et al. Caudate glucose metabolic rate changes with both drug and behavior therapy for obsessive-compulsive disorder. Arch Gen Psychiatry 1992;49:681–689.
29. Hoehn-Saric R, Pearlson GD, et al. Effects of fluoxetine on regional cerebral blood flow in obsessive-compulsive patients. Am J Psychiatry 1991;148:1243–1245.
30. Schwartz JM, Stoessel PW, et al. Systematic changes in cerebral glucose metabolic rate after successful behavior modification treatment of obsessive-compulsive disorder. Arch Gen Psychiatry 1996;53:109–113.
31. Swedo SE, Pietrini P, et al. Cerebral glucose metabolism in childhood-onset obsessive compulsive disorder. Arch Gen Psychiatry 1992;49:690–694.
32. Alexander GE, DeLong MR, Strick PL. Parallel organization of functionally segregated circuits linking basal ganglia and cortex. Annu Rev Neurosci 1986;9:357–381.
33. Cummings JL. Frontal–subcortical circuits and human behavior. Arch Neurol 1993;50:873–880.
34. Mega MS, Cummings JL. Frontal–subcortical circuits and neuropsychiatric disorders. J Neuropsychiatry Clin Neurosci 1994; 6:358–370.
35. Saxena S, Brody AL, et al. Neuroimaging and frontal–subcortical circuitry in obsessive-compulsive disorder. Br J Psychiatry 1998;173 (Suppl 35):26–37.
36. Cummings JL, Frankel M. Gilles de la Tourette syndrome and the neurological basis of obsessions and compulsions. Biol Psychiatry 1985;20:1117–1126.
37. Rosenberg DR, Matcheri SK. Toward a neurodevelopmental model of obsessive-compulsive disorder. Biol Psychiatry 1998; 43:623–640.
38. Blier P, de Montigny C. Possible serotonergic mechanisms underlying the antidepressant and anti-obsessive-compulsive disorder responses. Biol Psychiatry 1998;44:313–323.
39. Hollander E, Phillips KA. Body image and experience disorders. In: Yaryura-Tobias JA, Neziroglu FA, eds. Obsessive-Compulsive Disorder Spectrum: Pathogenesis, Diagnosis, and Treatment. Washington, DC: American Psychiatric Press, 1993:17–48.
40. Kaye WH, Weltzin T, Hsu LKG. Anorexia nervosa. In: Yaryura-Tobias JA, Neziroglu FA, eds. Obsessive-Compulsive Disorder Spectrum: Pathogenesis, Diagnosis, and Treatment. Washington, DC: American Psychiatric Press, 1993:49–70.
41. Fallon BA, Rasmussen SA, Liebowitz MR. Hypochondriasis. In: Yaryura-Tobias JA, Neziroglu FA, eds. Obsessive-Compulsive Disorder Spectrum: Pathogenesis, Diagnosis, and Treatment. Washington, DC: American Psychiatric Press, 1993:71–92.
42. Swedo SE. Trichotillomania. In: Yaryura-Tobias JA, Neziroglu FA, eds. Obsessive-Compulsive Disorder Spectrum: Pathogenesis, Diagnosis, and Treatment. Washington, DC: American Psychiatric Press, 1993:93–111.
43. Anthony DT, Hollander E. Sexual compulsions. In: Yaryura-Tobias JA, Neziroglu FA, eds. Obsessive-Compulsive Disorder Spectrum: Pathogenesis, Diagnosis, and Treatment. Washington, DC: American Psychiatric Press, 1993:139–150.
44. DeCaria CM, Hollander E. Pathological gambling. In: Yaryura-Tobias JA, Neziroglu FA, eds. Obsessive-Compulsive Disorder Spectrum: Pathogenesis, Diagnosis, and Treatment. Washington, DC: American Psychiatric Press, 1993:151–177.
45. Neziroglu F, Anemone R, Yaryura-Tobias JA. Onset of obsessive-compulsive disorder in pregnancy. Am J Psychiatry 1992;149: 947–950.
46. Jenicke MA, Wilhelm S. Illnesses related to obsessive-compulsive disorder: introduction. In: Jenicke MA, Baer L, Minichiello WE, eds. Obsessive-Compulsive Disorders: Practical Management, 3rd ed. Chicago: Mosby, 1998:121–142.
47. Goldman MB, Janecek HM. Is compulsive drinking a compulsive behavior? A pilot study. Biol Psychiatry 1991;29:503–505.
48. Modell JG, Mountz JM, Ford CV. Pathological lying associated with thalamic dysfunction demonstrated by [^{99m}Tc] HMPAO SPECT. J Neuropsychiatry Clin Neurosci 1992;4:442–446.
49. Kavoussi RJ, Coccaro EF. Impulsive personality disorders and disorders of impulse control. In: Hollander E, ed. Obsessive-Compulsive Related Disorders. Washington, DC: American Psychiatric Press, 1993:179–202.
50. Ames D, Cummings JL, et al. Repetitive and compulsive behavior in frontal lobe degenerations. J Neuropsychiatry Clin Neurosci 1994;6:100–113.
51. Berthier ML, Campos VM, Kulisevsky J. Echopraxia and self-injurious behavior in Tourette's syndrome: a case report. Neuropsychiatry Neuropsychol Behav Neurol 1996;9:280–283.
52. Berthier ML, Kulisevsky J, et al. Obsessive-compulsive disorder associated with brain lesions: clinical phenomenology, cognitive function, and anatomic correlates. Neurology 1996;47:353–361.
53. Cummings JL, Cunningham K. Obsessive-compulsive disorder in Huntington's disease. Biol Psychiatry 1992;31:263–270.
54. Frankel M, Cummings JL, et al. Obsessions and compulsions in Gilles de la Tourette's Syndrome. Neurology 1986;36:378–382.
55. George MS, Melvin JA, Kellner CH. Obsessive-compulsive symptoms in neurologic disease: a review. Behav Neurol 1992;5:3–10.
56. Grimshaw L. Obsessional disorder and neurological illness. J Neurol Neurosurg Psychiatry 1964;27:229–231.
57. Holzer JC, Goodman WK, et al. Obsessive-compulsive disorder with and without a chronic tic disorder. Br J Psychiatry 1994; 164:469–473.
58. Laplane D, Levasseur M, et al. Obsessive-compulsive and other behavioural changes with bilateral basal ganglia lesions. Brain 1989;112:699–725.
59. McDougle CJ, Kresch LE, et al. A case-controlled study of repetitive thoughts and behavior in adults with autistic disorder and obsessive-compulsive disorder. Am J Psychiatry 1995;152:772–777.
60. Miguel EC, Baer L, et al. Phenomenological differences appear-

ing with repetitive behaviours in obsessive-compulsive disorder and Gilles de la Tourette's syndrome. Br J Psychiatry 1997;170: 140–145.

61. Pitman RK, Green RC, et al. Clinical comparison of Tourette's disorder and obsessive-compulsive disorder. Am J Psychiatry 1987;144:1166–1171.
62. Zohar AH, Pauls DL, et al. Obsessive-compulsive disorder with and without tics in an epidemiological sample of adolescents. Am J Psychiatry 1997;154:274–276.
63. Frye PE, Arnold LE. Persistent amphetamine-induced compulsive rituals: response to pyridoxine (B6). Biol Psychiatry 1981;16: 583–587.
64. Rosse RB, Fay-McCarthy M, et al. Transient compulsive foraging behavior associated with crack cocaine use. Am J Psychiatry 1993;150:155–156.
65. Kindler S, Kaplan Z, Zohar J. Obsessive-Compulsive symptoms in schizophrenia. In: Hollander E, ed. Obsessive-Compulsive Related Disorders. Washington, DC: American Psychiatric Press, 1993:203–214.
66. Stein DJ, Le Roux L, et al. Is olfactory reference syndrome an obsessive-compulsive spectrum disorder? Two cases and a discussion. J Neuropsychiatry Clin Neurosci 1998;10:96–99.
67. Goodman WK, Murphy T. Obsessive-compulsive disorder and Tourette's syndrome. In: Enna SJ, Coyle JT, eds. Pharmacological Management of Neurological and Psychiatric Disorders. New York: McGraw-Hill, 1998:177–211.
68. Jenicke MA, Baer L, Minichiello WE. Obsessive-Compulsive Disorders: Practical Management. Chicago: Mosby, 1998.
69. Jenicke MA. Drug treatment of obsessive-compulsive disorders. In: Jenike MA, Baer L, Minichiello WE, eds. Obsessive-Compulsive Disorders: Practical Management, 3rd ed. St. Louis: Mosby, 1998.
70. Tollefson GD, Rampey AH, et al. A multicenter investigation of fixed-dose fluoxetine in the treatment of obsessive-compulsive disorder. Arch Gen Psychiatry 1994;51:559–567.
71. McDougle CJ, Goodman WK, et al. Haloperidol addition in fluvoxamine-refractory obsessive-compulsive disorder. Arch Gen Psychiatry 1994;51:302–308.
72. Mellman LA, Gorman JM. Successful treatment of obsessive-compulsive disorder with ECT. Am J Psychiatry 1984;141: 596–597.
73. Baer L, Rauch SL, et al. Cingulotomy for intractable obsessive-compulsive disorder. Arch Gen Psychiatry 1995;52:384–392.
74. Pies RW, Popli AP. Self-injurious behavior: pathophysiology and implications for treatment. Clin Psychiatry 1995;56:580–588.
75. Favazza AR, Rosenthal RJ. Diagnostic issues in self-mutilation. Hosp Community Psychiatry 1993;44:134–140.
76. Deb S. Self-injurious behaviour as part of genetic syndromes. Br J Psychiatry 1998;172:385–388.
77. Mendez MF, Mirea A. Adult head-banging and stereotypic movement disorders. Mov Disord 1998;13:825–828.
78. Mendez MF, Bagert BA, Edwards-Lee T. Self-injurious behavior in frontotemporal dementia. Neurocase 1997;3:231–236.
79. Robertson MM, Trimble MR, Lees AJ. Self-injurious behaviour and the Gilles de la Tourette syndrome: a clinical study and review of the literature. Psychol Med 1989;19:611–625.
80. Clarke DJ. Psychopharmacology of severe self-injury associated with learning disabilities. Br J Psychiatry 1998;172:389–394.
81. Reyes-Ortiz CA, Mulligan T. Letters to the editor. J Geriatr Soc 1996;44:1486–1488.
82. Lilly R, Cummings JL, et al. Clinical features of the human Kluver-Bucy syndrome. Neurology l983;33:1141–1145.

Chapter 17

Anxiety Disorders

Anxiety disorders can be among the most disabling of neuropsychiatric conditions. They have an emotional urgency and intrusiveness that makes normal function impossible, disables the patient, and fills the patient with feelings of fear, dread, failure, or death. When occurring as discrete attacks, anxiety can occur in relation to specific environmental precipitants or can occur unpredictably without identifiable triggers. Anxiety disorders present important connections between neurobiology, learning theory, and dynamic psychiatry. Anxiety can be a learned reaction and can be conditioned and deconditioned with learning strategies. Anxiety also has a recognizable neurobiology as a fear response that has become disconnected from life-threatening exigencies. Pharmacologic approaches to the management of anxiety have provided both insight into the neurobiological basis of the disorder and relief to anxiety disorder patients.

This chapter presents the neuropsychiatry of anxiety disorders, beginning with a classification of the recognized anxiety states. Neurologic and medical disorders associated with anxiety are described, the neurobiological basis of anxiety is summarized, and current approaches to the pharmacologic management of anxiety are discussed. Related discussions can be found in chapters devoted to disturbances of mood and affect (Chapter 14), obsessive-compulsive disorder and repetitive behaviors (Chapter 16), epilepsy and temporal lobe syndromes (Chapter 21), and dissociative states and hysteria (Chapter 22). Anxiety occurring in individual neurological disorders is described in the chapters on Alzheimer's disease and other dementias (Chapter 10), movement disorders (Chapter 18), and focal brain lesions (Chapter 26).

CLASSIFICATION OF ANXIETY DISORDERS

Currently recognized anxiety disorders include panic attack, agoraphobia, panic disorder, social phobia, specific phobia, obsessive-compulsive disorder, posttraumatic stress disorder, acute stress disorder, generalized anxiety disorder, substance-induced anxiety disorder, and anxiety due to general medical conditions.[1] Obsessive-compulsive disorder (OCD) appears to have relatively few relationships with other types of anxiety disorders, has a distinct neurobiology, and typically responds to distinct pharmacotherapeutic interventions. For these reasons, OCD is treated separately in this volume and is discussed with other repetitive behaviors in Chapter 16.

Panic attacks are discrete periods of intense fear or discomfort accompanied by a variety of somatic or cognitive symptoms that begin suddenly and build to a peak rapidly (usually 10 minutes or less). The subject has a sense of imminent danger or impending doom and has a desire to escape the situation. Panic attacks occur in several different anxiety disorders including panic disorder, social phobia, specific phobia, post-traumatic stress disorder, and acute stress disorder. *Panic disorder* is characterized by recurrent unexpected panic attacks.

Agoraphobia is anxiety about being in places or situations from which escape might be difficult or embarrassing and in which help may not be available in the event of a panic attack or panic symptoms. Agoraphobia leads to avoidance of these situations, such as being outside the home alone, being in a crowd of people, traveling in an automobile, bus, or airplane, or being on a bridge or in an elevator.

Social phobia features marked and persistent fear of social or performance situations in which embarrassment may occur. Social phobia often occurs when anxious anticipation is normal but the individual's response in this situation is excessive and disabling.

Specific phobia (also called *simple phobia*) is characterized by marked and persistent fear of clearly discernable circumscribed objects or situations. Individuals may fear travel in airplanes because of concern about crashing, may fear dogs because of concerns about being bitten, or may fear driving because of concerns about being hit by other vehicles on the road. They may be afraid of blood and injury, or fear being in closed-in situations.

Post-traumatic stress disorder includes persistent re-experiencing of a traumatic event, persistent avoidance of stimuli associated with the trauma, numbing of general responsiveness, and persistent symptoms of increased arousal. The syndrome follows exposure to an extreme traumatic stressor that entails direct personal experience of an event in which actual or threatened death or serious injury occurred or the witnessing of an event that involves death, injury, or threat to the physical integrity of another person.

Acute stress disorder comprises anxiety, dissociative, and other symptoms that occur within 1 month after exposure to an extreme traumatic stressor. In *generalized anxiety disorder*, patients experience excessive anxiety and worry, on most days for a period of at least 6 months, about numerous events or activities. *Substance-induced anxiety disorder* is characterized by prominent anxiety symptoms judged to be due to the direct physiological effects of a substance (discussed below). *Anxiety disorder due to a general medical condition* is clinically significant anxiety judged to be due to the direct physiological effects of a general medical condition (described below).[1] Neurological, medical, and drug-induced anxiety are emphasized in this chapter and would usually be classified as anxiety disorder due to a general medical condition or substance-induced anxiety disorder. These conditions provide insight into the neurobiology of anxiety by implicating structural and biochemical changes in brain regions involved in mediating anxiety symptoms.

PREVALENCE OF ANXIETY DISORDERS

The lifetime prevalence estimate for anxiety disorders (all types combined) is 14.6% and the median age at onset is 15 years.[2] The lifetime prevalence of panic disorder is 1.6% and occurs significantly more commonly in women than in men. Lifetime prevalence rates for generalized anxiety disorder range from 4.1% to 6.6% when concurrent or major depression are excluded.[3] Between 15% and 23% of all children meet criteria for some type of anxiety disorder: 4%–5% have separation anxiety disorder; 1%–2%, panic disorder; 1%–2.5%, agoraphobia; 10%–12%, specific phobia; 3%–20%, social phobia; 5%–14%, post-traumatic stress disorder; and 3% have generalized anxiety disorder.[4] Anxiety disorders tend to decline in old age: 5.5%–10.2% have some type of anxiety disorder, 3.1%–4.8% meet criteria for phobic disorder, and 0.1%–0.3% have panic disorder.[5] Anxiety has been identified in 20% of patients following stroke, in 20% of a sample of community-dwelling patients with Parkinson's disease, and in 37% of patients with multiple sclerosis attending a multiple sclerosis clinic.[6–8]

SIGNS AND SYMPTOMS OF ANXIETY

The signs and symptoms of anxiety can be divided into cognitive, behavioral, and physiological dimensions[9] (Table 17.1). The cognitive dimension includes feelings of apprehension and dread, racing thoughts, fearfulness, and distractibility. Depersonalization and derealization may occur during the course of a panic attack. The behavioral manifestations of anxiety include excessive movements, repetitive motor acts, pressured speech, excessive startle response, widened palpebral fissures, and a facial expression consistent with fear or apprehension. The physiological signs and symptoms of anxiety include tachycardia, palpitations, chest tightness, dry mouth, hyperventilation, paresthesias, lightheadedness, sweating, urge to urinate, and tremor.

TABLE 17.1. *Signs and Symptoms of Anxiety*[9]

Cognitive Symptoms	*Behavioral Signs*	*Physiological Signs and Symptoms*
Nervousness	Hyperkinesis	Tachycardia
Apprehension	Repetitive motor acts/habits/ mannerisms	Palpitations
Racing thoughts	Phobias and avoidance	Chest tightness
Worry	Pressured speech	Dry mouth
Fearfulness	Startle response	Hyperventilation
Irritability	Widened palpebral fissures	Paresthesias
Distractibility	Facial expression of apprehension	Light-headedness
Dread	Sweating	Urge to urinate
Desire to escape		Tremor
Depersonalization/ derealization		

Anxiety tremors are typically high-frequency, low-amplitude tremors involving the hands and occasionally the lower extremities (Chapter 18).

NEUROLOGIC, MEDICAL, AND TOXIC CAUSES OF ANXIETY

Anxiety accompanies a large number of neurological disorders, has been observed in many medical conditions, and may be induced by a wide variety of substances (Table 17.2). Much of the available information derives from case reports and convenience samples and there have been few epidemiologically based studies of anxiety in neurological conditions.

Individual case reports of patients with anxiety disorders following local brain regions suggest anatomical regions that may be critical to mediating anxiety syndromes. Post-stroke anxiety frequently accompanies depression and is more likely to occur with left anterior cortical lesions than with left frontal subcortical lesions.[10] In a study of patients hospitalized for stroke, 27% experienced early-onset generalized anxiety disorder and an additional 23% experienced the onset of anxiety symptoms 3 or more months after the stroke. Depressive symptoms were present in 75% of the patients. Anxiety symptoms commonly lasted 1 and $^1/_2$ to 3 months and were not related to social, cognitive, or physical disability.[11]

Anxiety was recorded in 20% of a population-based study of Parkinson's disease[6] and has been observed in 30%–40% of clinically based samples.[12,13] In Parkinson's disease, anxiety commonly co-occurs with depression and when the on–off phenomenon develops, it is more severe during the off period.[14,15] Among patients with focal lesions, a relationship has emerged between anxiety or panic and temporal lobe lesions, predominantly those affecting right-sided structures.[16–20] These focal brain lesions have included post-traumatic encephalomalacia, compressive meningiomas, metastatic brain tumors, and primary intraparenchymal brain tumors. Anxiety has been observed in patients with Wilson' disease, central nervous system syphilis, and

TABLE 17.2. *Neurologic Disorders Associated with Anxiety Syndromes*

Stroke
Parkinson's disease
Multiple sclerosis
Huntington's disease
Wilson's disease
Post-traumatic encephalopathy
Encephalitis
Syphilis
Meningiomas
Metastatic brain tumors
Primary brain tumors
Epilepsy

encephalitis.[21] Anxiety has been recorded in 10%–50% of patients with Huntington's disease.[22]

Typical anxiety symptoms can occur in the course of partial complex seizures originating in the temporal lobes.[23,24] Ictal attacks with panic symptoms are usually more brief and more stereotyped than classical panic attacks, and in some cases they progress to partial complex or generalized seizures. Patients are amnestic for the attacks, in contrast to panic disorders occurring in the absence of epilepsy, which can be recalled. Patients with atypical panic attacks should be investigated for possible partial complex seizures.[25] Electroencephalographic (EEG) abnormalities are uncommon in classical anxiety disorders.[26] In some cases, panic symptoms in patients with EEG abnormalities in the temporal regions may respond to anticonvulsant therapy even in the absence of a diagnosis of epilepsy.[27,28]

Anxiety has been reported in a large number of medical illnesses (Table 17.3). Cardiopulmonary disease and hypoxia are commonly associated with anxiety, and anxiety has been associated with angina, cardiac arrhythmias, congestive heart failure, asthma, acute pulmonary emboli, mitral valve prolapse, and pneumothorax. Endocrine disorders associated with anxiety include hyperthyroidism, hypothyroidism, parathyroid dysfunction, Cushing's disease (hyperadrenalism) and other adrenal abnormalities, pheochromocytoma, and virilization disorders of females. Gastrointestinal disorders associated with anxiety include peptic ulcer, gastrointestinal reflux, ulcerative colitis, and irritable bowel syndrome. Inflammatory conditions that have induced anxiety include lupus erythematosus, rheumatoid arthritis, polyarteritis nodosa, and temporal arteritis. Deficiency states, including deficiencies of vitamin B_{12}, vitamin B_1, vitamin B_6, niacin, and folic acid, have also been related to anxiety states. Other conditions that have been related to anxiety include hypoglycemia, carcinoid syndrome, systemic malignancies, premenstrual syndrome, febrile illnesses, porphyria, infectious mononucleosis, post-hepatic syndrome, uremia, hepatic failure, tuberculosis, brucellosis, cerebral malaria, and mastocytosis.[21,29–36]

Many types of drugs have been associated with the emergence of anxiety disorders (Table 17.4). In some cases it has been difficult to distinguish anxiety from the threats to life and survival that are associated with being ill, anxiety induced by the medical illness itself, and anxiety induced by treatment. In the conditions described here, anxiety was reported to emerge with introduction of therapy and to subside with its withdrawal. Classes of drugs associated with monoaminer-

TABLE 17.3. *Medical and Systemic Disorders Associated with Anxiety Syndromes*

Disorder Type	*Specific Disorders*
Cardiopulmonary diseases	Angina
	Cardiac arrythmias
	Congestive heart failure
	Mitral valve prolapse
	Acute asthmatic attacks
	Pulmonary embolis
	Pneumothorax
Endocrinopathies	Hyperthyroidism
	Hypoglycemia
	Pheochromocytoma
	Hypoparathyroidism
	Carcinoid syndrome
	Cushings's syndrome with \hyperadrenalism
	Multiple endocrine neoplasia
	Virilization disorder of females
Gastrointestinal disorders	Peptic ulcer
	Gastrointestinal reflux
	Ulcerative colitis
	Irritable bowel syndrome
Deficiency states	Vitamin B_{12}
	Vitamin B_6
	Vitamin B_1
	Niacin
	Folic acid
Collagen vascular disorders	Systemic lupus erythematosis
	Rheumatoid arthritis
	Polyarteritis nodosa
	Temporal arteritis
Infections and post-infectious states	Mononucleosis
	Chronic fatigue syndrome
	Tuberculosis
	Brucellosis
	Cerebral malaria
	Febrile illnesses
Miscellaneous conditions	Premenstrual syndrome
	Prophyria
	Nephritis
	Uremia
	Hepatic failure
	Hypo- and hypercalcemia

TABLE 17.4. *Medications, Illicit Substances, Withdrawal Syndromes, and Toxic Compounds Associated with Anxiety Syndromes*

Medications	*Illicit Substances*	*Withdrawal Syndromes*	*Toxic Substances*
Oxprenolol	Cannabinols	Barbiturates	Benzene
Captopril	Alcohol	Alcohol	Carbon disulfide
Amantadine	Ecstasy	Benzodiazepines	Heavy metals
Bromocriptine		Amphetamines	Mercury
Levodopa		Corticosteroids	Organophosphates
Salicylates		Propanolol	Organic solvents
Narcotics		Reserpine	
Barbiturates		Monoamine oxidase inhibitors (MAOIs)	
Ethosuximide		Tricyclic antidepressants	
Tricyclic antidepressants		Caffeine	
Buproprion		Meprobamate	
Fluoxetine		Nicotine	
Penicillin		Opiates	
Cycloserine		Anticholinergics	
Dapsone		Aspartame	
Sulfonamides			
Procaine penicillin			
Isoniazid			
Interferon			
Lidocaine			
Monoamine oxidase inhibitors (MAOIs)			
Digitalis			
Corticosteroids			
Thyroid hormones			
Centrol nervous system stimulants			
Cocaine			
Sympathomimetics			
Amphetamine			
Fenfluramine			
Diethylpropion			
Caffeine			
Methylphenidate			
Pemoline			
Phenylpropanolamine			
Isoproterenol			
Theophylline			
Yohimbine			
Nicotinic acid			

gic metabolism induce anxiety more frequently than other types of compounds, which suggests an important role for monoaminergic substances in the mediation of anxiety symptoms and signs. Agents associated with anxiety include oxprenolol, captopril, amantadine, bromocriptine, levodopa, salicylates, narcotics, barbiturates, ethosuximide, tricyclic antidepressants, buproprion, fluoxetine, penicillin, cycloserine, dapsone, sulfonamides, procaine penicillin, isoniazid, interferon, lidocaine, monoamine oxidase inhibitors, digitalis, corticosteroids, thyroid hormones, all classes of central nervous system (CNS) stimulants, theophylline, yohimbine, and nicotinic acid.[21,33,34,37] Intramuscular injection of procaine G penicillin has been associated with Hoigne's syndrome, which is characterized by doom, anxiety, psychosis, and seizures.[38]

Illicit substances associated with anxiety include cannabinols, cocaine, amphetamines, alcohol, and ecstasy.[37,39,40] Withdrawal syndromes are frequently associated with autonomic hyperactivity and anxiety. Withdrawal from the following substances has been observed to produce signs and symptoms of anxiety: barbiturates, alcohol, benzodiazepines, amphetamines, chlorohydrate, corticosteroids, propanolol, reserpine, monoamine oxidase inhibitors, tricyclic antidepressants, caffeine, meprobamate, nicotine, opiates, and anticholinergics.[34,37]

Central nervous system stimulants associated with anxiety include cocaine, sympathomimetics, amphetamine, fenfluramine, diethylpropion, caffeine, methylphenidate, pemoline, phenylpropanolamine, and isoproterenol.[37]

Exposure to toxic substances in the course of a variety of occupations has been related to the occurrence of anxiety. Compounds thought to induce anxiety signs and symptoms include aspartame, benzene, carbon disulfide, heavy metals, mercury, organophosphates, phosphorus, and organic solvents.[37,41]

NEUROLOGICAL STUDIES IN IDIOPATHIC ANXIETY DISORDERS

Studies of patients with idiopathic panic disorder have revealed a high frequency (40%) of focal temporal lobe abnormalities including asymmetric atrophy and areas of abnormal signal intensity.[42] Positron emission tomographic (PET) measurements of cerebral glucose metabolism reveal lower metabolic rates in basal ganglia and white matter of patients with generalized anxiety disorder.[43] Results from PET studies in patients with panic disorder showed increased glucose metabolism in the left hippocampus and perihippocampal area.[44] Patients with social phobia were found to have markedly decreased striatal dopamine reuptake site density compared to that in normals.[45]

Patients vulnerable to lactate-induced panic have abnormal cerebral blood flow compared to control subjects, and they have abnormally high perfusion in the inferior frontal cortex and left occipital cortex and significantly lower perfusion in the hippocampal regions bilaterally.[46] Lactate infusion results in greater increases of brain lactate in patients with lactate-induced panic than in control subjects. Elevations are also more prolonged in the brain and are decoupled from falling blood lactate levels.[47] Hyperventilation also causes a disproportionate increase in brain lactate levels in subjects with panic disorder.[48]

NEUROLOGICAL BASIS OF ANXIETY

Consensus has not been reached on the neurobiological basis of anxiety disorders. In addition, the different anxiety syndromes likely have differing underlying pathogenic components contributing to a common syndromic outcome. Genetic, biochemical, and structural elements may all contribute to the emergence of anxiety syndromes. Thirty percent of the monozygotic co-twins of panic disorder probands have panic disorder, while dizygotic co-twins are no more likely than unrelated individuals to manifest panic disorder. This observation supports a genetic contribution to this form of anxiety.[49] There is no increased concordance rate among monozygotic twin pairs for generalized anxiety disorder.

A variety of transmitter substances are capable of inducing panic, including yohimbine and withdrawal from opiates, antidepressant drugs, and clonidine.[49,50] Yohimbine, isoproterenol, and psychostimulants that have been widely associated with anxiety states stimulate noradrenergic function. Elevation of dopamine levels in the mesolimbic dopamine system by psychostimulants and other anxiogenic substances also appears to play a role in anxiety syndromes.[50] Peptides implicated in anxiety and fear include corticotropin-releasing factor (CRF) and cholecystokinin (CCK). Ventilatory dysregulation may lead to hyperventilation, panic, and the urge to flee in patients with panic disorder and lactate vulnerability.[50,51]

The success of benzodiazepines in ameliorating anxiety implies a role for benzodiazepine receptors in modifying brain mechanisms mediating anxiety symptoms. Benzodiazepine receptors are concentrated in cortical gray matter, subcortical areas, and hippocampus. Neuropeptide Y also has anxiolytic effects.[50]

TABLE 17.5. *Anxiolytic Agents*

Class *Drug (Brand Name)*	*Usual Daily Dose (mg)*
Benzodiazapines	
Alprazolam (Xanax)	1–4
Chlordiazepoxide (Librium)	15–40
Clonazepam (Klonopin)	1–6
Clorazepate (Tranxene)	15–60
Diazepam (Valium)	5–40
Halazepam (Paxipam)	60–160
Lorazepam (Ativan)	1–6
Oxazepam (Serax)	45–120
Serotonin reuptake inhibitors	
Fluoxetine (Prozac)	20–60
Fluvoxamine (Luvox)	100–200
Paroxetine (Paxil)	20–50
Sertraline (Zoloft)	50–200
Citalopram (Celexa)	10–30
Tricyclic antidepressants	
Desipramine (Norpramin)	150–300
Amitriptyline (Elavil)	150–300
Clomipramine (Anafranil)	100–250
Imipramine (Tofranil)	150–300
Nortriptyline (Pamelor)	50–150
Azapirones	
Buspirone (BuSpar)	15–60
Monoamine oxidase inhibitors	
Phenelzine (Nardil)	45–90
Tranylcypromine (Parnate)	30–50
Other	
Trazodone (Desyrel)	150–300
Nefazodone (Serzone)	300–500
Venlafaxine (Effexor)	75–375
Propanolol (Inderal)	60–160
Clonidine (Catapres)	0.2–0.6
Hydroxyzine (Atarax)	300–400
Amobarbital (Amytal)	60–150

The high frequency of abnormalities in the temporal regions discovered by neuroimaging and the relationship of focal temporal lobe disorders to anxiety suggest that the temporal brain regions may mediate aspects of anxiety symptoms.

Cumulative observations suggest that anxiety is a complex behavioral response programmed into the nervous system to facilitate the individual's survival. The fear response involves complex neural networks involving perception, assessment of threat potential, and planning a response. The anatomical network includes limbic, neocortical, and subcortical structures. In addition, neural activation must coordinate a complex peripheral response consisting of a sympathetic activation of cardiopulmonary mechanisms and neuroendocrine responses involving the pituitary–adrenal axis. An abnormally low threshold for activating this response or failure to extinguish the response following exposure to a traumatic event may underlie clinical anxiety disorders.[52]

TREATMENT OF ANXIETY SYNDROMES

Anxiety may respond to nonpharmacologic, pharmacologic, and surgical intervention. Psychotherapies that have been useful in treating anxiety include relaxation training, rebreathing training, systematic desensitization, education and counseling, and cognitive psychotherapy.[53]

Several classes of drugs have anxiolytic properties (Table 17.5). Benzodiazepines and serotonin reuptake inhibitors are the drugs most commonly used for the treatment of generalized anxiety disorder. Buspirone is an azapirone anxiolytic that may have advantages over benzodiazepines in patients with neurological disorders and anxiety symptoms. Buspirone is less likely to induce sedation, memory or psychomotor interaction, or disinhibition phenomena. However, it has a slower onset of anxiolytic action (1–2 weeks) than benzodiazepines, and requires dosing three times per day.[54]

Selective serotonin reuptake inhibitors, tricyclic antidepressants, and monoamine oxidase inhibitors have an anxiolytic potential.[55] Serotonin reuptake inhibitors are the agents of choice for long-term management of anxiety disorders. β-Adrenergic receptor blockers may also benefit some patients, particularly those who have primarily somatic manifestations. α_2 Adrenergic receptor agonists such as clonidine may also be helpful in anxiety, particularly anxiety symptoms associated with withdrawal syndromes. Valproic acid, trazodone, venlafaxine, and nefazodone have also been observed in preliminary studies to have anxiolytic effects.[55,56]

Psychosurgery has been used to ameliorate chronic, severe, medically intractable anxiety. Limbic leukotomy and subcaudate tractotomy alleviate anxiety in 50%–60% of patients who have failed pharmacotherapy.[53]

REFERENCES

1. American Psychiatric Association. Diagnostic and Statistical Manual of Mental Disorders, 4th ed. Washington, DC: American Psychiatric Press, 1994.
2. Young JG, Brasic JR, et al. The developing brain and mind: advances in research techniques. In: Lewis M, ed. Child and Adolescent Psychiatry: A Comprehensive Textbook. Philadelphia: Williams & Wilkins, 1996:1209–1234.
3. Hollander E, Simeon D, Gorman JM. Anxiety disorders. In: Hales RE, Yudofsky SC, Talbott JA, eds. The American Psychiatric Press Textbook of Psychiatry, 2nd ed. Washington, DC: American Psychiatric Press, 1994:495–563.
4. Perry BD. Anxiety disorders. In: Coffey CE, Brumback RA, eds. Textbook of Pediatric Neuropsychiatry. Washington, DC: American Psychiatric Press, 1998:579–594.
5. Flint AJ. Epidemiology and comorbidity of anxiety disorders in the elderly. Am J Psychiatry 1994;151:640–649.
6. Aarsland D, Larsen JP, et al. Range of neuropsychiatric disturbances in patients with Parkinson's disease. J Neurol Neurosurg Psychiatry 1999;67:492–496.
7. Birkett DP. The Psychiatry of Stroke. Washington, DC: American Psychiatric Press, 1996.
8. Diaz-Olavarrieta C, Cummings JL, et al. Neuropsychiatric manifestations of multiple sclerosis. J Neuropsychiatry Clin Neurosci 1999;11:51–57.
9. Sheikh JI. Anxiety disorders. In: Coffey CE, Cummings JL, Lovell MR, Pearlson GD, eds. Textbook of Geriatric Neuropsychiatry, 2nd ed. Washington, DC: American Psychiatric Press, 2000:347–366.
10. Starkstein SE, Robinson RG. Depression in Neurologic Disease. Baltimore: The Johns Hopkins University Press, 1993.
11. Castillo CS, Schultz SK, Robinson RG. Clinical correlates of early-onset and late-onset poststroke generalized anxiety. Am J Psychiatry 1995;152:1174–1179.
12. Menza MA, Robertson-Hoffman DE, Bonapace AS. Parkinson's disease and anxiety: comorbidity with depression. Biol Psychiatry 1993;34:465–470.
13. Richard IH, Schiffer RB, Kurlan R. Anxiety and Parkinson's disease. Mov Disord 1996;8:383–392.
14. Siemers ER, Shekhar A, et al. Anxiety and motor performance in Parkinson's disease. Mov Disord 1993;8:501–506.
15. Starkstein SE, Robinson RG, et al. Anxiety and depression in Parkinson's disease. Behav Neurol 1993;6:151–154.
16. Capwell RR, Carter R. Organic anxiety syndrome secondary to metastatic brain tumor. Psychosomatics 1991;32:231–233.
17. Drubach DA, Kelly MP. Panic disorder associated with a right paralimbic lesion. Neuropsychiatry Neuropsychol Behav Neurol 1989;2:282–289.
18. George MS, McLeod-Bryant S, et al. Panic attacks and agoraphobia associated with a giant right cerebral arteriovenous malformation. Neuropsychiatry Neuropsychol Behav Neurol 1990; 3:206–212.
19. Ghadirian AM, Gauthier S, Bertrand S. Anxiety attacks in a patient with a right temporal lobe meningioma. J Clin Psychiatry 1986;47:270–271.
20. Ontiveros A, Fontaine R, et al. Correlation of severity of panic disorder and neuroanatomical changes on magnetic resonance imaging. J Neuropsychiatry Clin Neurosci 1989;1:404–408.
21. Othmer E, Othmer SC. The Clincial Interview Using DSM-IV, Vol. I: Fundamentals. Washington DC: American Psychiatric Press, 1994.
22. Cummings JL. Behavioral and psychiatric symptoms associated with Huntington's disease. In: Advances in Neurology. New York: Raven Press, 1995:179–186.
23. Nickell PV. Panic attacks, complex partial seizures, and multiple meningiomas. Anxiety 1994;1:40–42.
24. Young GB, Chandarana PC, et al. Mesial temporal lobe seizures presenting as anxiety disorders. J Neuropsychiatry Clin Neurosci 1995;7:352–357.
25. Weilburg JB, Schachter S, et al. Focal paroxysmal EEG changes during atypical panic attacks. J Neuropsychiatry Clin Neurosci 1993;5:50–55.
26. Stein MB, Uhde TW. Depersonalization disorder: effects of caffeine and response to pharmacotherapy. Biol Psychiatry 1989;26: 315–320.
27. Matthews K, Bell JS, Fowlie DG. Panic symptoms and cerebral electrical disturbance: restoration of function with carbamazepine. Behav Neurology 1996;9:37–40.
28. McNamara ME, Fogel BS. Anticonvulsant-responsive panic attacks with temporal lobe EEG abnormalities. J Neuropsychiatry Clin Neurosci 1990;2:193–196.
29. Moore DP. Textbook of clinical neuropsychiatry. Arnold, London, 2001.
30. Karajgi B, Rifkin A, et al. The prevalence of anxiety disorders in patients with chronic obstructive pulmonary disease. Am J Psychiatry 1990;147:200–201.
31. Liberthson R, Sheehan DV, et al. The prevalence of mitral valve prolapse in patients with panic disorders. Am J Psychiatry 1986; 143:511–515.
32. Matuzas W, Al-Sadir J, et al. Mitral valve prolapse and thyroid abnormalities in patients with panic attacks. Am J Psychiatry 1987;144:493–496.
33. Moore DP, Jefferson JW. Handbook of Medical Psychiatry: Mosby, 1996.
34. Pies R. Medical differential diagnosis of anxiety disorders. Psychiatric Times 1997:60–68.
35. Pine DS, Weese-Mayer DE, et al. Anxiety and congenital central hypoventilation syndrome. Am J Psychiatry 1994;151:864–870.
36. Wilcox JA. Pituitary microadenoma presenting as panic attacks. Br J Psychiatry 1991;158:426–427.
37. Blank K, Duffy JD. Medical therapies. In: Coffey CE, Cummings JL, Lovell MR, Pearlson GD, eds. Textbook of Geriatric Neuropsychiatry, 2nd ed. Washington DC: American Psychiatric Press, 2000:699–728.
38. Silber T, D'Angelo L. Doom anxiety and Hoigne's syndrome. Am J Psychiatry 1987;144:1365.
39. Louie AK, Lannon RA, et al. Clinical features of cocaine-induced panic. Biol Psychiatry 1996;40:938–940.
40. Pallanti S, Mazzi D. MDMA (ecstasy) precipitation of panic disorder. Biol Psychiatry 1992;32:91–95.
41. Dager SR, Holland JP, et al. Panic disorder precipitated by exposure to organic solvents in the work place. Am J Psychiatry 1987;144:1056–1058.
42. Fontaine R, Breton G, et al. Temporal lobe abnormalities in panic disorder: an MRI study. Biol Psychiatry 1990;27:304–310.
43. Wu JC, Buchsbaum MS, et al. PET in generalized anxiety disorder. Biol Psychiatry 1991;29:1181–1199.

44. Bisaga A, Katz JL, et al. Cerebral glucose metabolism in women with panic disorder. Am J Psychiatry 1998;155:1178–1183.
45. Tiihonen J, Kuikka J, et al. Dopamine reuptake site densities in patients with social phobia. Am J Psychiatry 1997;154:239–242.
46. De Cristofaro MT, Sessarego A, et al. Brain perfusion abnormalities in drug-naive, lactate-sensitive panic patients: a SPECT study. Biol Psychiatry 1993;33:505–512.
47. Dager SR, Marro KI, et al. Preliminary application of magnetic resonance spectroscopy to investigate lactate-induced panic. Am J Psychiatry 1994;151:57–63.
48. Dager SR, Strauss WL, et al. Proton magnetic resonance spectroscopy investigation of hyperventilation in subjects with panic disorder and comparison subjects. Am J Psychiatry 1995;152: 666–672.
49. Keshavan MS, Yeragani VK. Neurobiology of anxiety. In: Prasad AJ, ed. Biological Basis and Therapy of Neuroses. Boca Raton, FL: CRC Press, 1988:1–21.
50. Charney DS, Nagy LM, et al. Neurobiological mechanisms of human anxiety. In: Fogel BS, Schiffer RB, Rao SM, eds. Neuropsychiatry. Philadelphia: Williams & Wilkins, 1996:257–286.
51. Coplan JD, Lydiard RB. Brain circuits in panic disorder. Biol Psychiatry 1998;44:1264–1276.
52. Charney DS, Bremner JD. The neurobiology of anxiety disorders. In: Charney DS, Nestler EJ, Bunney BS, eds. Neurobiology of Mental Illness. New York: Oxford University Press, 1999: 494–517.
53. Taylor MA. The Fundamentals of Clinical Neuropsychiatry. New York: Oxford University Press, 1999.
54. Janicak PG, Davis JM, et al. Principles and Practice of Psychopharmacotherapy. Philadelphia: Williams & Wilkins, 1997.
55. Pine DS, Grun J, Gorman JM. Anxiety disorders. In: Enna SJ, Coyle JT, eds. Pharmacological Management of Neurological and Psychiatric Disorders. New York: McGraw-Hill, 1998:53–94.
56. Schatzberg AF, Cole JO, DeBattista C. Manual of Clinical Psychopharmacology. Washington DC: American Psychiatric Press, 1997.

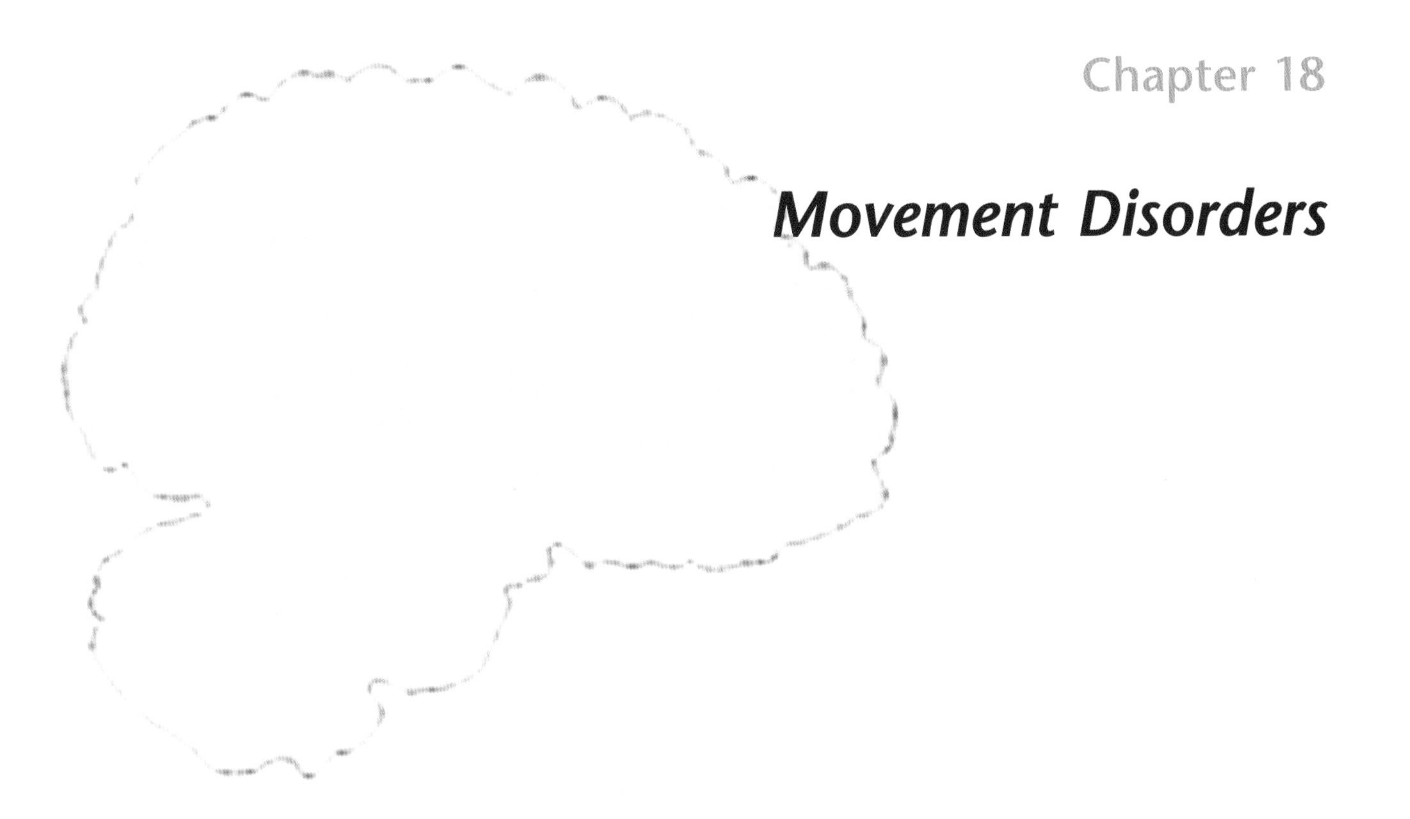

Chapter 18

Movement Disorders

Movement disorders are commonly encountered in neuropsychiatry. Basal ganglia diseases are commonly accompanied by neuropsychiatric disturbances; primary psychiatric illnesses such as schizophrenia and mania may be manifested by disorders of movement such as stereotypies, mannerisms, and catatonia; treatment of behavioral disturbances may produce abnormalities of movement such as parkinsonism, dystonia, and tremor. Basal ganglia diseases have been associated with a wide range of neuropsychiatric symptoms including depression, anxiety, delusions, apathy, irritability, and disinhibition. In addition, cognitive impairment, such as executive dysfunction and dementia, is present in many basal ganglia diseases. The frontal subcortical circuits linking regions of the frontal cortex to basal ganglia and thalamus and mediating motor function, ocular motor function, executive abilities, motivation, and social–emotional functions provide a rational anatomical and biochemical framework for understanding the relationship of extrapyramidal disorders to cognitive and emotional dysfunction. The frontal subcortical circuits are discussed later in this chapter.

Extrapyramidal disorders and their neuropsychiatric manifestations are described in this chapter. Information relevant to extrapyramidal diseases is included in chapters in this volume devoted to psychosis (Chapter 12), hallucinations (Chapter 13), disturbances of mood and affect (Chapter 14), apathy and personality alterations (Chapter 15), obsessive-compulsive disorder and repetitive behavior (Chapter 16), and anxiety disorders (Chapter 17). Chapter 19 presents the tic syndromes, startle disorders, and myoclonus. Chapter 20 discusses catatonia. Childhood movement disorders are described briefly in Chapter 25. Movement disorders occurring with focal and infectious brain disorders are described in Chapter 26.

CLASSIFICATION AND DEFINITION OF MOVEMENT DISORDERS

Abnormalities of movement can occur with disorders at every level of the neuraxis, from the cortex through basal ganglia, cerebellum, spinal cord, peripheral nerves, myoneural junction, and muscles (Fig. 18.1). Weakness and spasticity are hallmarks of the interaction of the pyramidal motor system while weakness and diminished muscle tone are characteristic of nerve and muscle disorders. Extrapyramidal or basal ganglia diseases do not produce weakness but cause abnormalities in the initiation of movement (hypokinetic parkin-

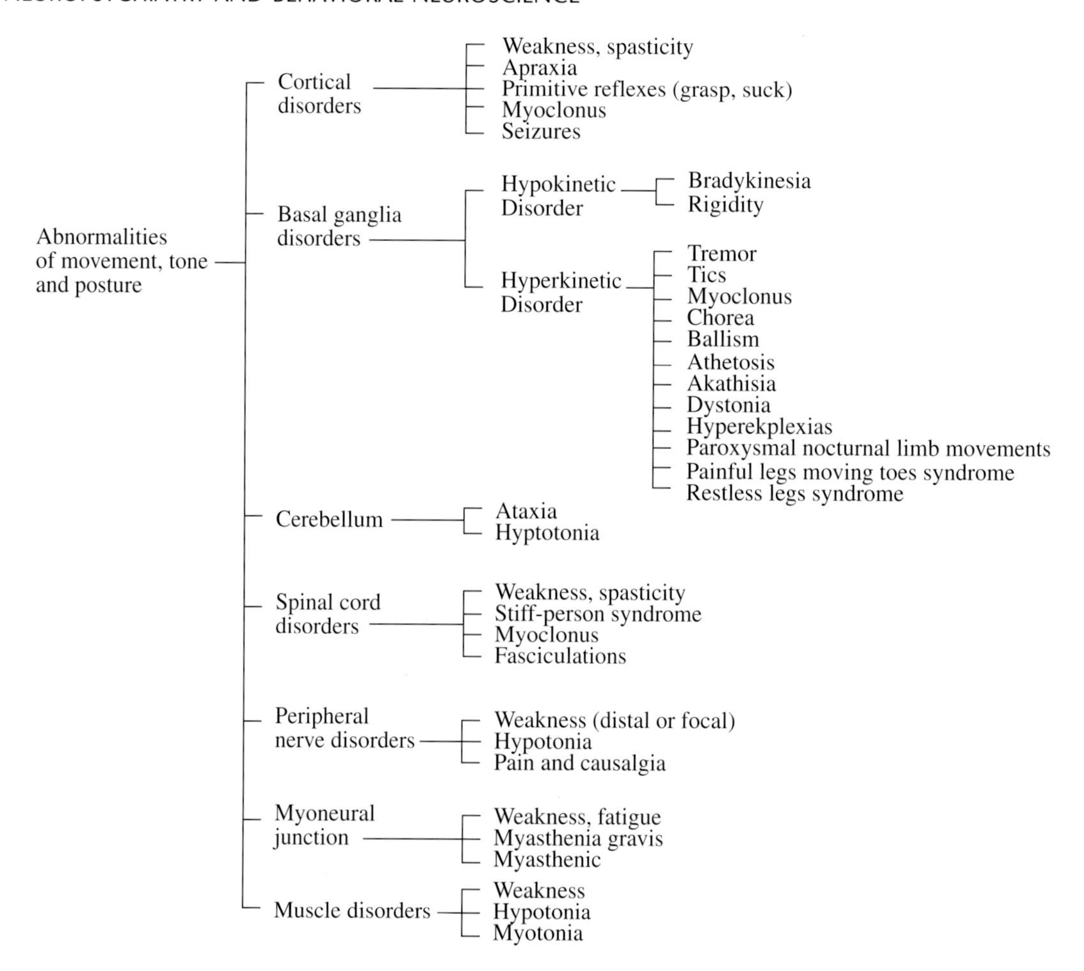

FIGURE 18.1 Classification of disorders of the motor system.

sonian disorders) or exhibit abnormal activation of muscle programs and plans leading to hyperkinetic disorders.

Figure 18.2 provides an approach to the classification of extrapyramidal movement disorders. Akinetic rigid syndromes are parkinsonian disorders characterized by reduced initiation of movement (akinesia), a slowed execution of movement (bradykinesia), and plastic or lead pipe rigidity. Cog wheel or a ratchet type of rigidity is present if the disorder also has a tremor component. The typical tremor of Parkinson's disease has a rest tremor as well as akinesia and rigidity. Other parkinsonian syndromes may or may not be accompanied by tremor. *Dystonia* refers to a movement disorder caused by sustained muscle contractions causing twisting and repetitive movement or abnormal postures.[1] Dystonia may be focal (e.g., blepharospasm or torticollis); segmental, affecting two or more contiguous body parts; or multifocal; it may affect one-half of the body (hemidystonia) or be generalized. Tremor results from involuntary oscillations of a body part produced by alternating or synchronous contractions of reciprocally enervated muscles.[1,2] Several types of tremor are recognized, including essential tremor, intentional or cerebellar tremor, rest tremor, often associated with parkinsonism, physiologic tremor, and tremors associated with specific conditions (rubral with midbrain disease, dystonic in association with dystonia syndromes, wing beating with Wilson's disease, and orthostatic). *Myoclonus* refers to sudden shock-like muscle contractions that may be focal, multifocal, or generalized. It is typically random and irregular.[1] *Chorea* consists of arrhythmic, rapid, and often jerky movements that may be simple or complex and involve one body part or another in a continuous random sequence.[1,2] Choreic movements are purposeless but may be incorporated into deliberate movements to make them less obvious. Fidgety movements, facial grimacing, and abnormal respiration may all be manifestations of chorea. *Tardive dyskinesia* is a special case of a choreaform disorder following exposure to dopamine blocking agents. *Athetosis* refers to slow, sinuous, writhing movements of the distal parts of limbs. *Ballism* or *ballismus* describes wild flinging or throwing

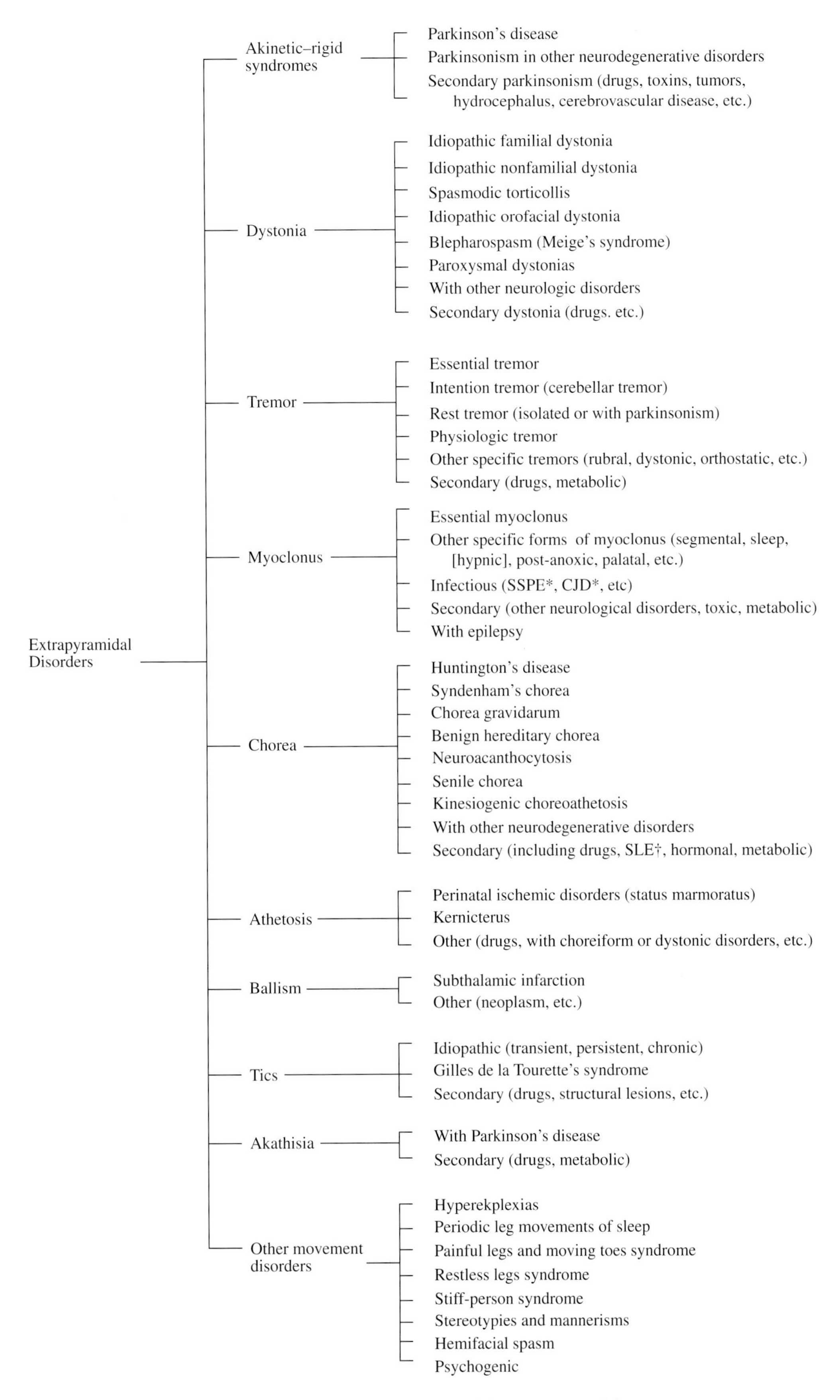

FIGURE 18.2 Classification of disorders of the extrapyramidal system.

movements usually involving the proximal musculature of one side of the body.[2] *Tics* are repetitive, irregular, stereotyped movements or vocalizations that are subject to partial voluntary control.[2] *Akathisia* is a subjective sensation of restlessness associated with an inability to remain still. Patients move to relieve involuntary sensations and the syndrome is expressed as changes in body positions, standing, or pacing.[1,2] Other movement disorders include hyperaplexias or startle syndromes (Chapter 19), periodic leg movements of sleep, the painful legs and moving pose syndrome, restless leg syndrome, stiff persons syndrome, stereotypies and mannerisms, hemifacial spasm, and psychogenic movement disorders.[1]

BASAL GANGLIA DISEASES

Parkinson's Disease

Parkinson's disease is the most common movement disorder encountered in the elderly. Parkinson's disease has a prevalence of approximately 160 per 100,000 persons (range 100–250 per 100,000 in different surveys).[3] It may be somewhat less common among African blacks and Asians than among Caucasians. This disease is more common in men than women. Before the introduction of contemporary therapies for Parkinson's disease, the average duration of the illness was 8 to 10 years; following routine use of levodopa

TABLE 18.1. *Diagnostic Criteria for Parkinson's Disease*[6]

Group A: features characteristic of Parkinson's disease
Resting tremor
Bradykinesia
Rigidity
Asymmetric signs
Group B: features suggestive of alternative diagnoses
Features unusual early in the clinical course
Prominent postural instability in the first 3 years after symptom onset
Freezing phenomena in the first 3 years
Hallucinations unrelated to medications in the first 3 years
Dementia preceding motor symptoms or in the first year
Supranuclear gaze palsy (other than restriction of upward gaze or slowing of vertical saccades
Severe, symptomatic dysautonomia unrelated to medications
Documentation of a condition known to produce parkinsonism and plausible connected to the patient's symptoms (such as suitably located focal brain lesions or neuroleptic use within the past 6 months)
Criteria for definite diagnosis of Parkinson's disease
All criteria for possible Parkinson's disease (below) are met *and*
Histopathologic confirmation of the diagnosis is obtained at autopsy
Criteria for probable diagnosis of Parkinson's disease
At least three of the four features in group A are present *and*
None of the features in group B is present (note: symptom duration of at least 3 years is necessary to meet this requirement) *and*
Substantial and sustained response to levodopa or a dopamine agonist has been documented
Criteria for possible diagnosis of Parkinson's disease
At least two of the four features in group A are present: at least one of these is tremor or bradykinesia, *and*
None of the features in group B is present (or symptoms have been present for less than 3 years, and none of the features in group B is present to date) *and*
Substantial and sustained response to levodopa or a dopamine agonist has been documented (or patient has not had an adequate trial of levodopa or dopamine agonist)

and related therapies the survival increased to approximately 15 years following diagnosis.[3]

Clinical Features The classical clinical features of Parkinson's disease include rest tremor, rigidity, and bradykinesia. Abnormalities of postural righting reflexes and symmetry of onset are also characteristic features. Diagnosis is supported by a substantial and sustained response to levodopa therapy.[4,5] Table 18.1 provides an approach for the identification of definite, probable, and possible Parkinson's disease.[6] In addition to the core clinical features, patients with Parkinson's disease also manifest a hypophonic and monotonic verbal output, micrographia with progressive diminution in the size of written figures, reduced armswing, flexed posture and enblock turns, shuffling gait (reduced stride length and reduced step height) festination (a tendency to accelerate; the phenomenon may occur in speech or gait), a failure to suppress blinking with repeated tapping of the forehead between the eyes (Glabellar tap reflex or Myerson's sign), and a marked increase in tone in the ipsilateral limb when the contralateral limb is active (Froment's sign). Pursuit eye movements are broken into multiple saccadic steps and up-gaze and convergence may be moderately limited. There is no impairment of vertical gaze, an observation that helps distinguish Parkinson's disease from progressive supranuclear palsy (discussed below).[7]

Neuropathology Pathologically, Parkinson's disease is characterized by nerve cell loss and the formation of Lewy bodies in remaining neurons in a variety of brainstem nuclei and, to a lesser extent, cortical structures.[8] The structures with the greatest involvement include the substantia nigra, ventral tegmental area, locus coeruleus, and nucleus basalis. In the cerebral cortex, Lewy bodies tend to be distributed in the temporal and frontal cortex, entorhinal areas, anterior cingulate cortex, and insular cortex. Cell loss in the substantia nigra correlates well with akinesia and rigidity, whereas cognitive deficits and neuropsychiatric features of Parkinson's disease are more likely related to the cholinergic deficit associated with nucleus basalis lesions, cortical dopaminergic deficits related to cell loss in the ventral tegmental area, serotonergic deficits secondary to involvement of the raphe nuclei, and Lewy body formation in the cerebral cortex. From a neurochemical perspective, the principal abnormality is the deficiency of dopamine. This central deficit is accompanied by abnormalities in serotonin, norepinephrine, and acetylcholine.

The differential diagnosis of Parkinson's disease includes a large number of parkinsonian syndromes (Table 18.2). The most common disorders considered in the differential diagnosis include dementia with Lewy bodies (DLB), vascular parkinsonism, progressive supranuclear palsy (PSP), cortical basal degeneration, and drug-induced Parkinsonism. Dementia with Lewy bodies typically has a less pronounced parkinsonian

TABLE 18.2. *Differential Diagnosis of Parkinsonism*

Idiopathic (Lewy body) Parkinson's disease
Familial Parkinson's disease
Post-encephalitic parkinsonism
Encephalitis lethargica
Other encephalitides, including syphilis
Parkinsonism in other "degenerative" diseases
Multiple system atrophies
Olivopontocerebellar degeneration
Shy-Drager syndrome
Striatonigral degenerations
Pallidal degenerations
Progressive supranuclear palsy
Wilson's disease
Acquired hepatocerebral degeneration
Machado-Joseph-Azorean disease
Alzheimer's disease
Dementia with Lewy bodies
Creutzfeldt-Jakob disease
Cortical–basal degeneration
Frontotemporal dementia
Parkinson–dementia–ALS complex of Guam
Rigid variant of Huntington's disease
Hallervorden-Spatz disease
Calcification of the basal ganglia (idiopathic and symptomatic)
Ceroid lipofuscinosis (Kuf's disease)
Neuroacanthocytosis
GM1 gangliosidosis
Gaucher's disease
Mitochondrial encephalopathies
Secondary parkinsonism
Drug- or toxin-induced parkinsonism
Postanoxic, carbon monoxide intoxication, cyanide poisoning, carbon disulfide, methanol, ethanol
"Punch drunk" syndrome (dementia puglistica)
Hydrocephalus (normal pressure and high pressure)
Space-occupying lesions: tumors, etc.
Multiple cerebral infarcts (atherosclerotic parkinsonism, including Binswanger's disease, amyloid angiopathy, lacunar state)

ALS, amyotrophic lateral sclerosis.

syndrome, often without tremor. Onset of the dementia is within 1 year of the recognition of the parkinsonian syndrome. Patients with DLB tend to have a limited or no response to levodopa therapy. Vascular parkinsonism can be distinguished from Parkinson's disease by a history of stroke, transient ischemic attacks, vascular risk factors such as hypertension, and combined pyramidal and extrapyramidal signs on physical examination. Patients with PSP have abnormalities of vertical gaze, more pronounced axial rigidity, and more marked bradykinesia, apathy, and dysarthria. Corticobasal degeneration is distinguished by the presence of apraxia and more severe cognitive disturbances.

● **Treatment** A variety of agents are available to relieve the symptoms of Parkinson's disease (Table 18.3). Selegiline, a monoamine oxidase B inhibitor, may have a neuroprotective effect in Parkinson's disease, but its efficacy in this regard is controversial. The goal of the use of all other agents is to relieve the symptoms of Parkinson's disease. Anticholinergic agents decrease the tremor and have less marked effects on bradykinesia and rigidity. The common occurrence of adverse side effects, particularly in elderly and cognitively compromised patients, makes these agents less useful. Dopaminergic agents available for the treatment of Parkinson's disease include amantadine, an agent that facilitates dopamine release and inhibits its reuptake into the synaptic terminal, levodopa and combined levodopa/carbidopa, dopamine receptor agonists (bromocriptine, pergolide, ropinerole, pramipexole), and Catechol-O-methyltransferase inhibitors (tolcapone, entacapone). Treatment is typically initiated with a dopamine receptor agonist and levodopa/carbidopa is added when increased efficacy is required. Addition of a catechol-O-methyltransferase inhibitor may further enhance the effect of levodopa/carbidopa.[9] Patients who develop severe "on–off" symptoms with prolonged dopaminergic therapy may benefit from surgical therapy with pallidotomy or deep brain stimulation. Patients with tremor may have their motor disorder ameliorated by thalmotomy.

TABLE 18.3. *Drugs Used to Treat Parkinson's Disease*

Class	*Drug (Brand Name)*	*Usual Final Dosage*	*Initial Dose*	*Half-Life*
Dopamine release facilitation and reuptake blockade	Amantadine (Symmetrel)	100 mg bid or tid	100 mg once daily	2–4 hr
Dopamine precursor	Levodopa; given with peripheral dopa decarboxylase inhibitor (carbidopa [in 4:1 and 10:1 ratios]) (Sinemet)	Varies Late in disease, patients may require multiple doses/day (sometimes >2 g levodopa/day)	25/100 mg tid	1.5 hr
	Controlled-release (CR) formulations (with carbidopa [4:1]) (Sinemet CR)		50/200 mg bid (CR formulation)	8–12 hr
Dopamine agonist, ergot-derived	Bromocriptine (Parlodel)	30–40 mg/day	1.25 mg bid	3–8 hr
	Pergolide (Permax)	3–5 mg/day	0.05 mg once daily	27 hr
Dopamine agonist, non–ergot-derived	Ropinerole (Requip)	Up to 24 mg/day in 3 divided doses	0.25 mg tid	6 hr
	Pramipexole (Mirapex)	Up to 4.5 mg/day in 3 divided doses	0.125 mg tid	8–12 hr
Monoamine oxidase B inhibitor	Selegiline (Deprenyl, Eldepryl)	5 mg bid	5 mg bid (in A.M.)	8–10 hr
Catechol O-methyl-transferase inhibitor	Tolcapone (Tasmar)	100 or 200 mg tid (at 6-hr intervals)	100 mg tid	2–3 hr
	Entacapone (Comtan)	200 mg 4–10 times/day (given with levodopa)	200 mg administered with each levodopa dose	2.4 hr

Cholinergic disturbances are present in most patients with Parkinson's disease and dementia. This suggests a role for cholinesterase inhibitors in ameliorating the cognitive symptoms. Preliminary reports suggest that some cholinesterase inhibitors may be useful.[10]

Neuropsychiatric Aspects Patients with Parkinson's disease exhibit a complex repertoire of neuropsychiatric symptoms occurring as a product of the disease state or following treatment with dopaminergic agents (Table 18.4). Cognitive deficits are ubiquitous in Parkinson's disease and are less common among patients who have tremor at onset or who have tremor-predominant syndrome.[11] The most common form of neuropsychological deficit observed in patients with Parkinson's disease is an executive function disorder or mild subcortical dementia characterized by difficulty with word list generation, a retrieval deficit disorder type of memory impairment, abnormalities of organizational skills when copying complex figures, and difficulty with set switching on tests such as the Wisconsin Card Sort Test or Trails B.[12–15] Some studies have suggested a specific deficit in visuospatial dysfunction in patients with Parkinson's disease, but further study suggests that these visuospatial abnormalities may be attributable to executive dysfunction.[16]

Overt dementia meeting criteria of the *Diagnostic and Statistical Manual of Mental Disorders*[17] is present in 30%–40% of patients with Parkinson's disease.[18–20] The features of this dementia include more marked memory impairment, subtle to marked language abnormalities, variable visuospatial deficits, and variable degrees of executive dysfunction. The heterogeneity of the clinical syndrome associated with dementia in Parkinson's disease reflects the heterogeneous neuropathological underpinnings of this disorder (Table 18.4). Executive deficits are associated with dopaminergic abnormalities.[21] Many patients with Parkinson's disease have atrophy of the nucleus basalis and a cortical cholinergic deficiency complicating the dopaminergic deficit. In addition, patients with Parkinson's disease may have cortical Lewy bodies, Alzheimer-type pathology in the cerebral cortex, or a combination of Alzheimer-type pathology and cortical Lewy

TABLE 18.4. *Neuropsychiatric Aspects of Parkinson's Disease*

Cognitive impairment
Executive dysfunction
Dementia
Subcortical type
Subcortical and cortical features
Cholinergic deficit (atrophy of nucleus basalis)
Cortical Lewy bodies
Alzheimers-type pathology
Alzheimer's pathology and Lewy bodies
Depression
Anxiety
Apathy
Sleep attacks
REM behavior disorder
Personality featuring reduced novelty-seeking
Drug-related behavioral disturbances
Hedonistic homeostatic dysregulation
Sleep disorders (nightmares)
Hallucinations
Euphoria hypomania or mania
Hypersexuality, paraphilia
Obsessive-compulsive behaviors
Delirium

REM, rapid eye movement.

bodies. Clinical findings that reliably distinguish these syndromes have not been identified.[15,22]

Depression is the most common psychiatric disturbance identified in patients with Parkinson's disease, occurring in approximately 40% of patients.[23,24] Major depression, observed in 7%–10% of patients with Parkinson's disease, is more uncommon, and occurs more frequently in patients who are cognitively impaired and those who exhibit the akinetic, rigid variant of Parkinson's disease.[24,25] Correlates of depression in Parkinson's disease include the presence of psychosis, greater impairment in activities of daily living, the presence of motor fluctuations, and early onset of Parkinson's disease.[26,27] Depression correlates poorly with disability in patients with Parkinson's disease and appears to be an independent neurobiological manifestation of the disorder rather than a reaction to physical impairment.

Depressed patients with Parkinson's disease exhibit diminished metabolism and cerebral blood flow in the medial frontal, orbital frontal, and anterior cingulate cortices as well as in the head of the caudate nuclei.[28,29] Depressed patients also exhibit reductions in cerebrospinal fluid (CSF), 5-hydroxyindoleacetic acid (5-HIAA), the principal metabolite of serotonin, in the CSF compared to patients without depression.[30] At autopsy, depressed patients exhibit more marked cell loss in the locus coeruleus.[31]

Selective serotonin reuptake inhibitors (SSRIs) are the most commonly used agents for the treatment of depression in Parkinson's disease.[32–34] Tricyclic antidepressant agents and buproprion have also been utilized.[23] Levodopa and anticholinergic agents appear to have little effect on depression; bromocriptine, a dopamine receptor agonist, has been reported to have antidepressant activity.[35] The D_3 selective dopamine agonists such as pramipexole and ropinirole may reduce the emergence of depression in patients with Parkinson's disease and reduce existing symptoms.[36,37] Patients who experience intolerable side effects from antidepressant medications or who are treatment unresponsive may have their depression ameliorated with electroconvulsive therapy (ECT),[38,39] which produces transient benefit in motor function as well as more sustained improvement in mood. Anxiety is a common symptom in Parkinson's disease, occurring in 20%–40% of patients,[18,40–42] and is more common among patients with depression than in those without mood changes. Apathy has also been described in patients with Parkinson's disease both with and without accompanying mood abnormalities.[43] Depression and irritability occur in approximately 10% of patients with Parkinson's disease.[18] From a personality perspective, Parkinson's disease patients are often described as rigid, stoic, slow tempered, frugal, and orderly, characteristics that may relate to damage to the mesal limbic dopaminergic system.[41,44] Rapid eye movement (REM) sleep behavior disorder (RBD) may herald Parkinson's disease and is eventually present in up to 15% of patients.[45]

A variety of neuropsychiatric symptoms have been associated with dopaminergic therapy of Parkinson's disease. The earliest manifestations of levodopa excess include nightmares, nocturnal vocalization, and myoclonus.[46] Patients with more severe responses experience visual hallucinations and those with the most severe adverse events manifest delusions. Hallucinations occur in approximately 30% of patients treated with dopaminergic agents and delusions emerge in approximately 10%.[47] Cognitive impairment, age, duration of disease, apolipoprotein E, ϵ-4 allele, history of depression, and history of a sleep disorder are associated with the emergence of visual hallucinations.[48–50] Hallucinations tend to be well formed and relatively stereotyped, are images of animals or humans, are typically silent, and are more likely to be experienced in the night or twilight hours than during the day.[51] Psychotic symptoms are also associated with age, stage of Parkinson's disease, severity of depression, and presence of cognitive impairment.[52] It is notable that in several studies the dose of dopaminergic drugs was similar in patients with and without hallucinations and delusions, suggesting that while dopaminergic therapy is critical for the emergence of these phenomena, the group that experienced these symptoms is determined by host factors (presence of dementia, age, etc.).[52] Clozapine is the most efficacious treatment for dopamine-associated psychosis in Parkinson's disease;[53] olanzapine, quetiapine, and risperidone are other atypical antipsychotics potentially useful in the treatment of Parkinson's disease.[54–56] Patients must be closely monitored for the worsening of parkinsonism.[57] Cholinesterase inhibitors have been reported to reduce delusions and hallucinations in some patients with Parkinson's disease, dementia, and psychosis.[10]

Mood swings are common, in concert with the "on–off" phenomenon in Parkinson's disease. Many patients experience more sadness and depression while *off* and up to 10% experience euphoria during *on* periods.[47,58] Rarely, mania occurs during the on period.[59] Hypersexuality and paraphillic sexual behavior (particularly sadomasochism) have been reported to occur with the introduction or elevation of levodopa dosage.[60–62] Obsessive and compulsive behaviors have also been described in conjunction with levodopa therapy. Some patients exhibit an addiction syndrome labeled "hedonistic homeostatic dysregulation" in association with dopamine therapy. This syndrome is most likely to occur in men with early-onset Parkinson's disease who take increasing quantities of dopaminergic agents despite increasingly severe drug-related dyskinesias, social impairment, and disturbed occupational functioning. They may exhibit cyclical mood changes with hypomania or mania.[63]

Dementia with Lewy Bodies

Dementia with Lewy bodies is characterized by progressive dementia syndrome and at least two of the following three features: (*1*) fluctuating cognition with pronounced variation in attention and alertness, (*2*) recurrent visual hallucinations that are typically well formed and detailed, and (*3*) parkinsonism[64] (Table 18.5). Dementia with Lewy bodies is just one of the dementia syndromes in which a neuropsychiatric assessment and the identification of prominent neuropsychiatric symptoms is a necessary part of the diagnostic process. Features supportive of the diagnosis DLB include repeated falls, syncope, transient loss of consciousness, systematized delusions, nonvisual hallucinations, REM behavior disorder, depression, and neuroleptic sensitivity. The latter is particularly important from a treatment perspective since DLB patients have prominent delusions and hallucinations, often prompting therapy with antipsychotic medications. Use of traditional antipsychotics in this setting may result in marked parkinsonism.[65] The sensitivity of the diagnostic criteria for DLB (Table 18.5) has ranged from 35% to 75%, whereas the specificity has ranged from 80% to 95%.[66–68] Thus when the criteria are present, they accurately predict the pathological diagnosis in most cases. However, a larger number of patients will have more cortical Lewy bodies at autopsy than are identified by the criteria.

The critical pathological feature of DLB is the presence of Lewy bodies. In many cases the Lewy bodies are accompanied by Alzheimer-type pathology, particularly senile plaques, regional neuronal loss affecting the substantia nigra, locus coeruleus, and nucleus basalis, microvacuolation, synapse loss, and neurotransmitter deficits.[64,69] Nearly all patients with cortical Lewy bodies have brainstem Lewy bodies involving the substantia nigra. The cell loss is less severe in the substantia nigra and other brainstem nuclei in DLB than in classical Parkinson's disease. It tends to be more severe than is typical of Alzheimer's disease without Lewy body pathology.[69] At the level of the cortex, Lewy bodies are preferentially distributed in the limbic and paralimbic regions including the anterior cingulate cortex, amygdaloid complex, insula, and entorhinal and transentorhinal cortex. The hippocampal formation is spared. Neocortical involvement is usually most severe in the temporal lobe and least severe in the occipital lobe, with parietal and frontal cortex showing intermediate levels of involvement.[69,70] Many patients with DLB have prominent Alzheimer-type neuritic plaques in the cortex and modest presence of neurofibrillary tangles. This observation has given rise to some

TABLE 18.5. *Consensus Criteria for Clinical Diagnosis of Probable and Possible Dementia with Lewy Bodies*[64]

1. Progressive cognitive decline of sufficient magnitude to interfere with normal social or occupational function. Prominent or persistent memory impairment may not necessarily occur in the early stages but is usually evident with progression. Deficits on tests of attention and of frontal–subcortical skills and visuospatial ability may be especially prominent.
2. Two of the following core features are essential for a diagnosis of probable DLB, and one is essential for possible DLB:
   a. Fluctuating cognition with pronounced variations in attention and alertness
   b. Recurrent visual hallucinations that are typically well formed and detailed
   c. Spontaneous motor features of parkinsonism
3. The following features are supportive of the diagnosis:
   a. Repeated falls
   b. Syncope
   c. Transient loss of consciousness
   d. Neuroleptic sensitivity
   e. Systematized delusions
   f. Hallucinations in other modalities
   g. REM behavior disorder[259]
   h. Depression[259]

DLB, dementia with Lewy bodies; REM, rapid eye movement.

TABLE 18.6. *Disorders with Lewy Bodies*[69,70]

Disorder	*Location of Lewy Bodies*
Parkinson's disease	Brainstem and rare cortical
Parkinson's disease with dementia	Brainstem and cortex
Dementia with Lewy bodies	Brainstem and cortex
Pure Lewy body disease	Brainstem and cortex (without co-occurring Alzheimer-type pathology)
Lewy body dysphagia	Dorsal vagal nuclei
Primary autonomic failure	Sympathetic neurons in spinal cord
Conditions with incidental Lewy bodies	
Ataxia telangiectsia	
Corticobasal degeneration	
Down's syndrome	
Familial early-onset Alzheimer's disease	
Hallervorden-Spatz disease	
Motor neuron disease	
Multiple system atrophy	
Neuroaxonal dystrophy	
Progressive supranuclear palsy	
Subacute sclerosing panencephalitis	

uncertainty as to whether DLB is best regarded as a variant of Alzheimer's disease[71] or a distinct clinical pathological entity.[72] At autopsy patients with DLB have marked deficits in cholinergic markers as well as reductions in basal ganglia dopamine comparable to those in patients with Parkinson's disease.[73,74] Lewy bodies have been observed in a variety of disorders in addition to Parkinson's disease and DLB (Table 18.6).[69,75]

The dementia of DLB is characterized by memory impairment with disproportionately severe visuospatial abnormalities and disturbances of executive function.[76,77] Visual hallucinations occur in 60%–70% of patients with DLB, auditory hallucinations in approximately 50%, depression in 50%–70%, and delusions in 50%–70%.[78,79] Delusional misidentification such as Capgras symptoms is particularly common in DLB.[78,80] The REM behavior disorder is a common feature of DLB, particularly among men with this disorder.[81,82] Parkinsonism of DLB is characterized by rigidity, bradykinesia, and dystonia. Myoclonus is more common than in typical Parkinson's disease and rest tremor is less frequent.[83] Patients with DLB are less likely to respond to levodopa therapy than patients with classical Parkinson's disease, but a substantial number exhibit at least moderate responses to therapy.[83] Imaging studies aid in distinguishing DLB from Alzheimer's disease. Hippocampal atrophy is less severe in DLB than in patients with Alzheimer's disease.[84,85] Studies of cerebral blood flow and brain glucose metabolism reveal greater involvement of the occipital lobes in DLB than in Alzheimer's disease. Other cortical regions have similar levels of involvement.[86] Studies of the dopaminergic system assessing either uptake of labeled fluoradopa or measurement of postsynaptic dopamine receptors have the greatest specificity for distinguishing DLB from Alzheimer's disease. Patients with DLB have abnormalities of subcortical dopaminergic systems whereas patients with Alzheimer's disease do not.[87,88] Patients with DLB experiencing visual hallucinations have greater preservation of cortical serotonin and more severe depletion of cortical cholinergic markers than patients without hallucinations.[89,90]

Treatment of DLB involves use of cholinesterase inhibitors and judicious administration of atypical antipsychotics. Cholinesterase inhibitors have been reported to improve cognition and reduce psychotic symptoms in DLB.[91,92] Conventional neuroleptics must be avoided in DLB, given the risk of exaggerated toxicity; atypical antipsychotics may reduce delusions and hallucinations without exacerbating parkinsonism.[93,94] All patients treated with antipsychotic agents must be carefully monitored for the emergence or exacerbation of parkinsonism. Patients with disabling degrees of spontaneous parkinsonism should be treated with dopaminergic agents. The response is typically less robust than that observed in Parkinson's disease but may be sufficient to warrant continuation of therapy.

Progressive Supranuclear Palsy

Progressive supranuclear palsy (also known as *Steele-Richardson-Olszewski syndrome*) is characterized by the tetrad of dementia supranuclear gaze palsy, pseudobulbar palsy, and axial rigidity. Classical opthalmoplegia of PSP is characterized by paralysis of voluntary vertical gaze, beginning with reduced volitional down-gaze followed by a reduction in voluntary up-gaze capacity. Pursuit movements and lateral volitional and pursuit movements are involved as the disease progresses. Oculocephalic reflexes remain intact. Microsquare wave jerks are evident in many cases. Progressive supranuclear palsy features a masked expressionless face and a prominent dysarthria with hypophonia, a strained-strangled quality, and low intelligibility, eventually leading to anarthria.[95] The rigidity of PSP is more pronounced in the truncal and neck muscles than in the limbs. Neck extension contrast with the neck flexion is characteristic of Parkinson's disease. Tremor is typically absent. The dementia of PSP has a classical subcortical pattern with prominent executive deficits and relative preservation of memory.[96–99] Table 18.7 presents criteria for the diagnosis of definite, probable, and possible PSP.[100]

TABLE 18.7. *National Institute of Neurological Diseases and Stroke–Society for Progressive Supranuclear Palsy (NINDS-SPSP) Criteria for Diagnosis of Progressive Supranuclear Palsy*[100]

Diagnostic Class	*Criteria*
Diagnostic, definite	Clinically probable or possible PSP and histopathologic evidence of typical PSP
Diagnostic, probable	Gradually progressive disorder
	Onset at age 40 or later
	Vertical (upward or downward gaze) supranuclear palsy and prominent postural instability with falls in the first year of disease onset
	No evidence of other diseases that could explain the foregoing features, as indicated by mandatory exclusion criteria
Diagnostic, possible	Gradually progressive disorder
	Onset at age 40 or later
	Either vertical (upward or downward gaze) supranuclear palsy *or* both slowing of vertical saccades and prominent postural instability with falls in the first year of disease onset
Supportive criteria	Symmetric akinesia or rigidity, proximal more than distal
	Abnormal neck posture, especially retrocollis
	Poor or absent response of parkinsonism to levodopa therapy
	Early dysphagia and dysarthria
	Early onset of cognitive impairment, including at least two of the following: apathy, impairment in abstract thought, decreased verbal fluency, utilization or imitation behavior, or frontal release signs
Exclusionary criteria	Recent history of encephalitis
	Alien limb syndrome, cortical sensory deficits, focal frontal or temporoparietal atrophy
	Hallucinations or delusions unrelated to dopaminergic therapy
	Cortical dementia of Alzheimer's type (severe amnesia and aphasia or agnosia, according to NINCDS-ADRDA criteria)
	Prominent, early cerebellar symptoms or prominent, early-unexplained dysautonomia (marked hypotension and urinary disturbances)
	Severe, asymmetric parkinsonian signs (i.e., bradykinesia)
	Neuroradiologic evidence of relevant structural abnormality (i.e., basal ganglia or brainstem infarcts, lobar atrophy)
	Whipple's disease, confirmed by polymerase chain reaction, if indicated

NINCDS-ADRDA, National Institute of Neurologic as Communication; Disorder as Stroke-Alzheimer's Disease and Rotated Disorders Association; PSP, progressive supranuclear palsy.

TABLE 18.8. *Principal Features Distinguishing Progressive Supranuclear Palsy from Other Basal Ganglia Disorders*

Disorder	*Characteristic Clinical Features*
Progressive supranuclear palsy	Vertical supranuclear gaze palsy
	Pseudobulbar palsy with movement dysarthria
	Axial rigidity (greater than limb rigidity)
	Dementia
Corticobasal ganglionic degeneration	Alien limb syndrome
	Severe limb apraxia (inability to use correctly mimed objects, or perform symbolic gestures on command)
	Cortical sensory deficits
	Markedly asymmetric onset of bradykinesia
	Focal frontal or temporoparietal atrophy
Parkinson's disease	Asymmetric onset of bradykinesia symptoms
	Tremor-dominant disease
	Marked and prolonged levodopa benefit
	If dementia is present, cognitive changes follow motor abnormalities by several years
Dementia with Lewy bodies	Hallucinations or delusions unrelated to dopaminergic therapy
	Fluctuating cognition/consciousness
	Dementia onset within 1 year of other features
Multiple system atrophy	Prominent cerebellar symptomatology or unexplained early and prominent incontinence, impotence, or marked postural hypotension
Vascular parkinsonism	Focal neurologic signs
	Multiple strokes, one of which involves the brainstem and basal ganglia
	Lacunar infarcts and white matter ischemic injury on neuroimaging
Whipple's disease	Ocular-masticatory myorhythmia, laboratory confirmation (e.g., polymerase chain reaction), if indicated
Post-encephalitic parkinsonism	History of encephalitis, oculogyric crisis
Creutzfeldt-Jakob disease	Disease course of <1 year, myoclonus, EEG abnormalities

Neuropsychiatric assessment of patients with PSP reveals marked apathy in a majority of patients; depression, anxiety, and disinhibition are relatively common and rare patients exhibit psychosis.[41,100,101]

Computerized tomography (CT) and magnetic resonance imaging (MRI) in PSP patients indicate atrophy of the midbrain and quadrigeminal plate with prominent CSF spaces around the brainstem and dilatation of the aqueduct and posterior third ventricle.[102] Assessment of fructose metabolism with positron emission tomography (PET) reveals reduced metabolic activity in the caudate nucleus, putamen, thalamus, pons, and cerebral cortex, particularly the cortical, motor, and premotor regions.[103] Frontal association cortex and paralimbic regions are only moderately affected. Studies of the dopaminergic system reveal reduced fluoradopa uptake in the caudate nucleus and anterior and posterior putamen in PSP; this contrasts with Parkinson's disease, in which the posterior putamen is disproportionately severely affected.[104] A similar pattern of involvement of the dopaminergic system is found with studies assessing the distribution of the dopamine transporter.[105]

Neuropathologically, PSP is characterized by the presence of neurofibrillary tangles or neurophil threads and brainstem, diencephalic, and basal ganglia structures including the pallidum, subthalamic nucleus, substantia nigra, pons, striatum, oculomotor complex, medulla, and ventate nucleus of the cerebellum.[106,107] The tangles of PSP are composed of straight filaments, contrasting with the paired helical filaments characteristic of Alzheimer's disease. Neuronal loss also occurs in the areas of tangle-bearing neurons. Cholinergic marker activity is reduced throughout the brain.[108]

Progressive supranuclear palsy must be distinguished from cortical basal degeneration (discussed below), Parkinson's disease, DLB, Alzheimer's disease, multiple system atrophy, parkinsonism associated with multi-infarct states, Whipple's disease, post-encephalitic parkinsonism, Creutzfeldt-Jakob disease, frontotemporal degenerations, pallidal systems degenerations, progressive subcortical glyosis, and hydrocephalus.[23,106,109,110] Table 18.8 presents a summary of features that assist in distinguishing PSP from Parkinsonian syndromes with overlapping features.

There is no consistently efficacious treatment for PSP. Some patients have limited responses to dopaminergic therapy for variable periods of time and a few patients have responded to treatment with methysergide or tricyclic antidepressants.[111–113] Cholinesterase inhibitors have not proven to be beneficial.[114] Intramuscular botulinum toxin may relieve painful neck and limb spasms.[115]

Corticobasal Degeneration

Corticobasal degeneration is a parkinsonian syndrome characterized by asymmetric rigidity, dystonia, and myoclonus associated with apraxia of the affected limb and cortical sensory loss (Table 18.9).[116] Tremor, hyperreflexia, Babinski signs, oculomotor paralysis, dysarthria, and dysphagia may also occur. Myoclonus of corticobasal degeneration is primarily distal, occurring in the limb affected by apraxia and parkinsonism. Action and reflex myoclonus are typical; attempts to move the limb voluntarily are interrupted by repetitive myoclonic jerks.[117] Apraxia of corticobasal degeneration is an ideiomotor type (Chapter 6) characterized by spatial, temporal, and sequencing errors that result in unrecognizable action products when the patient is asked to perform specific transitive and intransitive actions.[118] Patients may exhibit a frank alien hand phenomenon in which they can exert little control over the actions of a limb and experience the limb activity as involuntary. This type of alien limb phenomenon must be distinguished from a marked grasp reflex, which also lacks voluntary control and may result in unusual groping and grasping movements of the affected limb. This type of alien hand may also occur in corticobasal degeneration.

TABLE 18.9. *Diagnostic Criteria for Corticobasal Degeneration*[116]

Inclusion criteria
Rigidity plus one cortical sign (apraxia, cortical sensory loss, or alien limb phenomenon); or
Asymmetric rigidity, dystonia, and focal reflex myoclonus
Qualification of inclusion factors
Rigidity: easily detectable without reinforcement
Apraxia: more than simple use of limb as object; clear absence of cognitive or motor deficit sufficient to explain disturbance
Cortical sensory loss: preserved primary sensation; asymmetric
Alien limb phenomenon: more than simple levitation
Dystonia: focal in limb; present at rest at onset
Myoclonus: reflex myoclonus spreads beyond stimulated digits
Exclusion criteria
Early dementia (this will exclude some patients who have corticobasal degeneration, but whose illness cannot clinically be distinguished from other primary dementing diseases
Early vertical gaze palsy
Rest tremor
Severe autonomic disturbances
Sustained responsiveness to levodopa
Lesions on imaging studies indicating another responsible pathologic process

Neuropsychologically, patients with corticobasal degeneration exhibit modest decline in most cortical functions, a disproportionately severe dysexecutive syndrome, a retrieval deficit disorder, and marked visuospatial disturbances (particularly if the left limb and right hemisphere are predominantly involved).[119,120] Patients with corticobasal degeneration are subject to depression and may exhibit apathy, irritability, and agitation.[121] Anxiety, disinhibition, delusions, and hallucinations are uncommon. The specificity of the clinical syndrome described here is high for the diagnosis of corticobasal degeneration but the sensitivity is low, with many patients with this diagnosis going undetected.[122]

Studies of cerebral metabolism and cerebral blood flow with PET and single photon emission computerized tomography (SPECT) reveal a marked asymmetry of cerebral activity, reduced opposite the most affected limb; dorsolateral frontal, medial frontal, inferior parietal, sensory motor, and lateral temporal cortices exhibit reduced metabolism. At the subcortical level, the striatum and thalamus are also affected.[123–125]

Neuropathological examination reveals swollen achromatic neurons and cellular loss in the affected regions. At the cortical level, the pre- and postcentral gyri are most affected. Balloon cells may also be found in the anterior cingulate gyrus and insular cortex. The hippocampus and perihippocampal regions are typically unaffected. Cell loss and tau-immunoreactive lesions are evident in the caudate nucleus, putamen, substantia nigra, locus coeruleus, raphe nuclei, and tegmental gray areas. The red nucleus, subthalamic nucleus, and thalamic nuclei are more mildly affected.[126,127]

No treatment has been found to be efficacious in ameliorating the symptoms of corticobasal degeneration.

Idiopathic Basal Ganglia Calcification

Idiopathic basal ganglia calcification (also known as *Fahr's disease*) is a familial disorder with at least two identified phenotypes. The first is characterized by an early-onset psychotic disorder with minimal or no parkinsonism and limited cognitive abnormalities. The second syndrome begins in the fifth or sixth decade of life and is characterized by subcortical dementia, parkinsonism, and depression or mania.[128–130]

Pathologically, idiopathic basal ganglia calcification is characterized by the occurrence of ferrocalcific deposits primarily in the blood vessels of the basal ganglia (Fig. 18.3). Idiopathic basal ganglia calcification must be distinguished from a variety of other disorders that may produce intercranial calcification (Table 18.10). A few patients with basal ganglia calcification exhibit obsessive-compulsive behaviors and neuropsychological assessment reveals diminished motor speed, executive dysfunction, and visuospatial and memory changes.[131]

Other Parkinsonian Syndromes

Neuropsychiatric and neuropsychological features of other parkinsonian syndromes have been subject to lim-

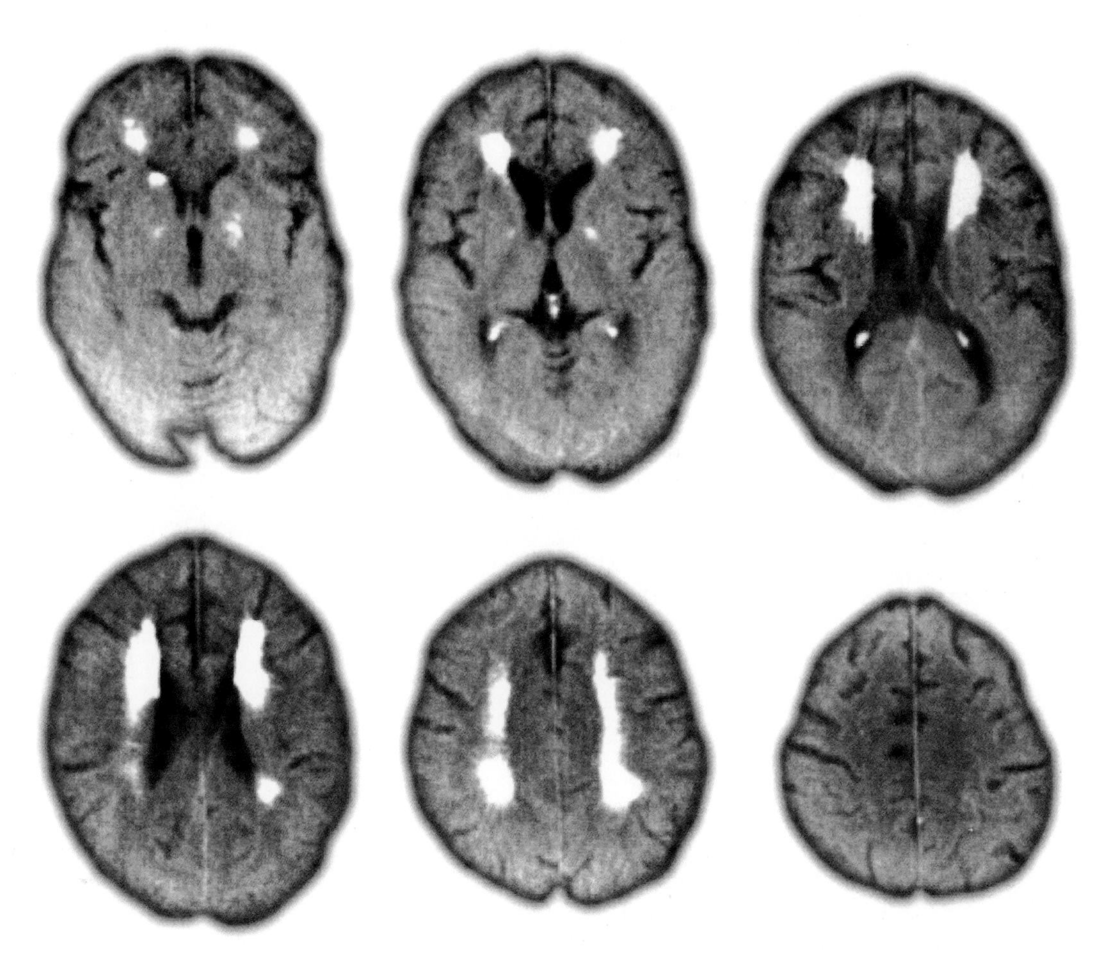

FIGURE 18.3 Basal ganglia calcifications with idiopathic basal ganglia calcification syndrome (Fahr's disease).

TABLE 18.10. *Conditions Associated with Intracranial Calcification*

Type of Condition	*Disorders*
Idiopathic	Idiopathic basal ganglia calcification (Fahr's disease)
	Hallervorden-Spatz disease
	Autosomal dominant dystonia
	Paroxysmal dystonic choreoathetosis
	Lipomembranous polycystic osteodysplasia
	Adrenoleukodystrophy
Endocrine	Hypoparathyroidism (primary and secondary)
	Pseudohypoparathyroidism
	Pseudo-pseudohypoparathyroidism
	Hyperparathyroidism
	Hypothyroidism
Infections	AIDS
	Post-encephalitic parkinsonism
	Cytomegalic inclusion disease
	Encephalitic (measles, chicken pox, rubella, pertussis, coxsackie B)
	Toxoplasmosis
	Cysticercosis
	Trichinosis
	Nocardia
	Tuberculosis
	Echinococcosis
	Coccidiodomycosis
	Epstein-Barr virus infection
	Malaria
	Cerebral syphilis
Toxic and physical injury	Anoxia
	Carbon monoxide poisoning
	Lead poisoning
	Radiation
	Methotrexate exposure
Congenital disorders	Familial apoceruloplasmin deficiency
	Biotinidase deficiency
	Carbonic anhydrase II deficiency
	Down's syndrome
	Kearn's-Sayre syndrome
	Cockeyne's syndrome
	Tuberous sclerosis
	Dihydropteridine reductase deficiency
	Sturge-Weber syndrome
	Lissencephaly
	Choroidocerebral calcification syndrome
	Marinesco-Sjögren syndrome
	Neurofibromatosis
	Leigh's disease

(*continued*)

TABLE 18.10. *Conditions Associated with Intracranial Calcification (Continued)*

Type of Condition	*Disorders*
Vascular disorders	Infarctions
	Aneurysms
	Arteriovenous malformations
Brain tumors	Meningiomas
	Astrocytomas
	Craniopharyngiomas
	Choroid plexus papilloma
	Metastases
	Medulloblastoma
	Teratomas
	Ependymomas
	Oligodendroglioma
Miscellaneous	Subdural hematomas
	Hamartomas/lipomas
	Lipoid proteinosis
Physiological	Choroid plexus
	Pineal
	Dural (especially falcine)
	Globus pallidus interna (limited extent)

ited study. *Hallervorden-Spatz disease* (also known as *neurodegeneration with brain iron accumulation type I*) is an autosomal recessive disease characterized by the accumulation of iron-containing pigments in the globus pallidus and substantia nigra. In adults, onset form of the disease with parkinsonism and a subcortical dementia pattern has been described.[132,133] α-Synuclein containing Lewy bodies have been observed in this syndrome.[134] The basal ganglia deposits are readily seen on CT or MRI.[135,136]

Vascular parkinsonism is a common cause of the parkinsonian syndrome in the elderly.[137,138] Gait abnormalities in vascular parkinsonism are often identical to those of Parkinson's disease, with short stride length and reduced step height producing a shuffling gait. Patients with vascular parkinsonism are less likely to manifest tremor, do not have levodopa-responsive syndromes, and often have a combination of pyramidal and extrapyramidal findings on neurological examination. Neuropsychologically, patients with subcortical lacunar infarctions and Vingschwanger's disease exhibit executive dysfunction with deficits in shifting mental set, response inhibition and generative cognition.[139] Depression and apathy are common neuropsychiatric manifestations observed in patients with subcortical vascular disease.[140–142] Magnetic resonance imaging readily reveals subcortical ischemic injury (Fig. 18.4).[143]

Multiple system atrophy (MSA) includes three syndromes: olivopontocerebellar atrophy, striatonigral degeneration, and Shy-Drager syndrome. Clinical and pathological features of these three disorders overlap. When parkinsonian features predominate the term *striatonigral degeneration* is generally used; if cerebellar features predominate, *sporadic olivopontocerebellar atrophy* is commonly used. When autonomic failure predominates, the term *Shy-Drager syndrome* is invoked. Table 18.11 presents the diagnostic features of MSA.[144] Parkinsonism occurs in 87% of patients, autonomic dysfunction in 74%, and cerebellar signs in 54%. Pyramidal signs are almost common, occurring in 49% of patients, and dysarthria and dystonia occur in a minority. Patients appear to be vulnerable to levodopa-induced kinesias. The diagnostic criteria have moderate sensitivity (approximately 70%) and good specificity (97%).[145] Approximately 25% of patients exhibit overt dementia and a larger number have abnormal performance on test of executive function.[120,145,146] Depression occurs at about the same rate in MSA as in idiopathic Parkinson's disease, where it is a common feature.[147] Magnetic resonance imaging reveals putamenal abnormalities with dorsolateral hy-

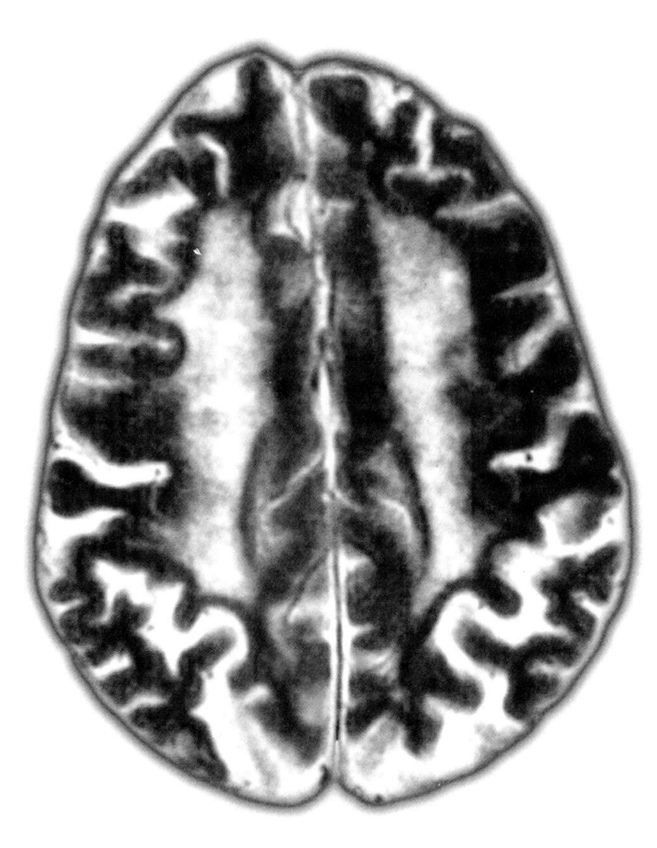

FIGURE 18.4 Magnetic resonance image of patient with Binswanger's disease showing extensive ischemic injury of the cerebral white matter.

pointensity and a lateral hyperintense rim.[148] Bilateral hypometabolism is evident in the putamen and caudate when MSA patients are studied with [^{18}F]-fluorodeoxyglucose (FDG) PET.[149] Pathologically, cell loss and glyosis are evident in the putamen, caudate, nucleus, external pallidum, substantia nigra, locus coeruleus, inferior olives, pontine nuclei, cerebellar purkinje cells, and intermediolateral cell columns of the spinal cord.[150] Glial and neuronal cells contain cytoplasmic inclusions that stain positively for α-synuclein.[151] α-Adrenergic agonists such as midodrine may be useful in ameliorating the postural hypotension observed in MSA.[152]

The Chamorro population of Guam suffers from an endemic parkinsonism dementia complex (known locally as "bodig" and a disorder resembling amyotrophic lateral sclerosis (ALS) (lytico). Similar pathology is found in the two syndromes consisting of neurofibrillary tangles in the neocortex, basal ganglia, and thalamus.[153] A similar syndrome is occasionally seen on a sporadic basis outside of Guam. Parkinsonism and motor neuron disease can also be seen in frontotemporal dementia (Chapter 10).

Post-encephalitic parkinsonism was a common seguelae of the epidemic of von Economo's encephalitis occurring between 1917 and 1927. These patients exhibited parkinsonism, frequently complicated by catatonia, tic disorders, and chorea. A plethora of neuropsychiatric symptoms accompanied the syndrome, including mood disorders, obsessions and compulsions, and a subcortical dementia syndrome in adults; children manifested hyperactivity, conduct disorders, and paraphillic sexual behavior.[154]

Wilson's Disease

Wilson's disease is an autosomal recessive disorder produced by a mutation on chromosome 13q14.3. The gene encodes a transport protein and the mutations associated with Wilson's disease result in abnormal copper deposition in the liver, the basal ganglia, and the cornea of the eyes.[155] Wilson's disease typically begins in childhood or adolescence, with a modal age of 19 years for those with neurological symptoms. In some cases, the disease may have its onset delayed until as late as the fifth or sixth decade.[156] Roughly one-third

TABLE 18.11. *Diagnostic Features of Multiple System Atrophy**

Multiple system atrophy is a sporadic, progressive, adult-onset disorder characterized by autonomic dysfunction, parkinsonism, and ataxia in any combination. The features of this disorder include the following:

A. Parkinsonism (bradykinesia with rigidity or tremor or both), usually with a poor or unsustained motor response to chronic levodopa therapy

B. Cerebellar or corticospinal signs

C. Orthostatic hypotension, impotence, urinary incontinence, or retention, usually preceding or within 2 years after the onset of the motor symptoms

Exclusionary criterion: These features cannot be explained by medications or other disorders.

*When parkinsonian features predominate, the term *striatonigral degeneration* is often used; when cerebellar features predominate, *sporadic olivopontocerebellar atrophy* is often used; when autonomic failure predominates, the term *Shy-Drager syndrome* is used.[144]

of patients with Wilson's disease present first with psychiatric symptoms, one-third present with neurological symptoms, and one-third with evidence of hepatic disease. Dysarthria, dystonia, abnormalities of rapid, alternating movements, parkinsonian-type rigidity, disturbances of gait and posture, and a fixed facial expression are common manifestations of the disease. About one-third of patients have a rhythmic proximal tremor (wing-beating tremor), abnormal eye movements, and hypereflexia. Drooling, bradykinesia, frontal release signs, and athetosis are less common manifestations.[157] Examination of the eye reveals a golden brown or greenish ring of pigmentation at the margin of the cornea near the limbus. It can often be detected with the naked eye, but certain identification requires a slit-lamp examination. While highly characteristic of Wilson's disease, the Kayser-Fleischer ring is not completely specific for the disease and is also found in non-Wilsonian liver disorders. A sunflower cataract is present in approximately 20% of patients with Wilson's disease, and rarely there is progressive loss of vision associated with retinal dysfunction.[156]

From a neuropsychiatric perspective, four symptom clusters have been identified in patients with Wilson's disease: mood and affect, behavior/personality, schizophrenia-like symptom, and cognitive impairment.[158] Personality and mood changes are the most common behavioral manifestations. Patients may manifest depression, suicidal behavior, incongruous behavior suggestive of a frontal lobe syndrome, aggression, and irritability.[159,160] At least half of the patients have neuropsychiatric symptoms early in the disease course. Personality alterations, frontal lobe symptoms, irritability, and aggression are highly correlated with specific neurological symptoms including dysarthria, dysphasia, drooling, and rigidity. Neuropsychologically, patients with Wilson's disease exhibit retrieval deficit–type memory abnormalities, impairment on trail-making tests, and poor word list generation, consistent with disturbance of the frontal–subcortical systems.[161,162]

Computerized tomography reveals characteristic hypodense areas in the basal ganglia in approximately half of Wilson's disease patients.[163] In nearly all patients with Wilson's disease MRI shows abnormalities. High–signal intensity lesions on key_2-weighted images are seen in the thalamic nuclei, brainstem, and lenticular nuclei.[164] Investigations with glucose PET indicate a diffusely reduced brain metabolism that is most marked in the lenticular nuclei.[165] Investigation of the dopaminergic system using fluorodopa PET demonstrates that most patients have abnormally low dopamine uptake into basal ganglionic regions.[166]

Pathologically, there is cell loss and marked glial proliferation, particularly in the putamen, with less marked changes in the globus pallidus and caudate nucleus. The deeper layers of the cerebral cortex and adjacent white matter are also affected. Opalski cells are large, round cells characteristic of Wilson's disease that are most prominent in the thalamus, pallidum, and substantia nigra.[167]

Treatment of Wilson's disease is aimed at reducing copper availability by decreasing copper intake, limiting copper absorption, increasing copper elimination, or liver transplantation.[168] Ammonium tetrathiomolybdate or zinc is commoly used to reduce gastrointestinal absorption of copper and penicillamine or trientine is used as copper chelators to enhance copper elimination. A penicillamine may precipitate neurological worsening.[169] Psychiatric symptoms may improve following chelation therapy, but many patients require management with psychotropic agents.

Huntington's Disease

Huntington's disease is an autosomal dominant disorder resulting from a mutation on chromosome 4 at a novel 4p16.3 and consisting of an increased number of CAG trinucleotide repeats.[170] Normal chromosomes possess 6–34 CAG repeats at this location, whereas patients with Huntington's disease exhibit from 39 to 86 repeating units. The average age of onset is between 35 and 40; the earliest cases manifest symptoms at 2 or 3 years of age and patients with onset as late as the seventh or eighth decade have been recorded. Patients with longer trinucleotide repeat lengths have an earlier age of onset and more rapid progression than those with fewer repeats.[171] Clinically, Huntington's disease is manifested by the triad of choreaform movements, dementia, and neuropsychiatric disturbances. A more uncommon rigid (Westfall) variant of Huntington's disease also occurs and is more likely to occur in children. Choreiform movements of Huntington's disease tend to be more proximal than distal and first appear in the form of fidgety or restless movements. Patients often incorporate the movements into semi-intentional gestures creating unusual mannerisms. Both the upper and lower face are involved with intermittent wrinkling of the forehead as well as perioral and tongue movements. Truncal and proximal leg chorea may impose a lurching and high stepping (peacock) quality to the gait. Hyperkinetic movements of the abdominal, chest, and diaphragmatic musculature may produce irregularities of speech.[172] Movement abnormalities include slow and hypometric saccades, increased saccadic latencies, difficulty maintaining fixation, or performing antisaccades.[173,174] Huntington's disease must be distinguished from a variety of other causes of chorea (Table 18.12).

TABLE 18.12 *Causes of Chorea*

Neurodegenerative disorders	**Other systemic disorders**
Huntington's disease	Sydenham's chorea
Dentatorubropallidoluysian atrophy	Lupus erythematosus
Cerebellar system degenerations	Antiphospholipid antibody syndrome
Neuroacanthocytosis	Chorea gravidarum
Hallervorden-Spatz disease	Polycythemia vera
Pallidal degenerations	Periarteritis nodosa
	Behcet's disease
Lesions of the basal ganglia	Henoch-Schonlein syndrome
Vascular (stroke)	Acquired hepatocerebral degeneration
Neoplastic	Renal failure
Infectious (encephalitis)	Paraneoplastic syndrome
Inflammatory	
	Essential chorea syndromes
Drugs	Benign familial chorea
Tardive dyskinesia (dopamine antagonists)	Senile chorea
Antiparkinsonian agents	
Stimulants	**Paroxysmal chorea**
Opiates	Paroxysmal kinesigenic choreoathetosis
Antiepileptic agents	Paroxysmal dystonic choreoathetosis
Exogenous hormones (estrogens)	
Lithium	**Inherited metabolic disorders**
	Wilson's disease
Metabolic conditions	Neuronal ceroid lipofuscinosis
Hyperthyroidism	Lesch-Nyhan syndrome
Hyperglycemia	Galactosemia
Hypoglycemia	GM1 and GM2 gangliosidosis
Hypernatremia	Glutaric acidemia
Hyponatremia	
Hypoparathyroidism	**Intoxications**
Hypocalcemia	Carbon monoxide
Hypomagnesemia	Alcohol
Thiamine deficiency	Manganese
Niacin deficiency	Toluene
	Organophosphates
	Miscellaneous
	Subdural hematoma
	Multiple sclerosis
	Postvaccinial
	Cerebral palsy
	Kernicterus
	Creutzfeldt-Jakob disease

Mood disorders and personality alterations are the most common behavioral changes exhibited by patients with Huntington's disease.[175,176] Approximately 40% of patients exhibit major depressive disorders or meet criteria for dysthymia. The relationship of mood disorder to Huntington's disease is consistent across families, with some kindreds manifesting this combination of symptoms and others evidencing chorea and dementia without concomitant mood changes.[177] Irritability, anger, and intermittent explosive disorder are the most common types of personality alterations exhibited.[175,178,179] Antisocial behavior and conduct disturbances are also common.[175,176] Approximately 10% of patients exhibit hypomania and a few may have manic episodes.[175] Apathy, irritability, and disinhibition are other behaviors reported in patients with Huntington's disease.[180] Psychosis is relatively unusual in Huntington's disease, but a schizophrenia-like disorder with delusions, auditory hallucinations, and thought disorder has been described.[175,176] Obsessive-compulsive behavior is a reported but uncommon manifestation of Huntington's disease.[181] Sexual misconduct is more common, occurring in up to 20% of Huntington's disease patients.[182,183] The rate of suicide is increased in Huntington's disease. It is at least twice that in the general population and is increased at least fourfold in those after the diagnosis is suspected or established.[184] Psychiatric symptoms do not correlate with the CAG repeat length[185] but the occurrence of anger and hostility are among the earliest clinical features to predict the emergence of Huntington's disease in mutation carriers.[186] Huntington's disease patients with depression exhibit relative hypometabolism of the orbital and inferior prefrontal cortex compared to nondepressed patients.[187] Psychotic patients with Huntington's disease exhibit diminished frontal lobe metabolism compared to those without psychosis, and patients with minor behavioral changes have reduced metabolic activity in the ventrobasal regions of the striatum.[188]

Dementia eventually supervenes in all patients with Huntington's disease. The pattern of cognitive impairment is consistent with a subcortical dementia and involvement of frontal–subcortical circuits.[189] Memory is affected early in the disease course but changes are less profound than those observed in dementias such as Alzheimer's disease. Patients generally exhibit a retrieval deficit–type disorder with relatively intact memory storage.[190,191] Learning abnormalities may occur in Huntington's disease and reflect deficiencies in perceptual analysis rather than an aphasic type of anomia.[192] Executive functioning is compromised early in the disease course.[193,194] Psychomotor speed and attentional abilities are highly correlated with deficits and activities of daily living in Huntington's disease.[195] There is a relationship between severity of cognitive decline and number of CAG repeats.[196]

Structural imaging in patients with Huntington's disease reveal reduced size of the caudate nuclei in a majority of patients, with low signal intensity in the caudate and lentiform nuclei (Fig. 18.5).[197] Thalamus and medial temporal structures are also reduced in size and there is evidence of white matter degeneration.[197] Glucose PET demonstrates diminished metabolic activity in the caudate nuclei prior to the occurrence of structural atrophy.[198,199] Subcortical as well as cortical hypometabolism and hypoperfusions are detectable as the disease progresses.[200,201]

At autopsy, Huntington's disease is characterized by striking atrophy of the caudate nucleus and less dramatic changes in the putamen and globus pallidus.[202] Changes begin in the medial paraventricular portions of the caudate nucleus, the tail of the caudate, and the dorsal part of the putamen.[203] Medium-sized spiny neurons of the striatum are most vulnerable. Striatal levels of γ-aminobutyric acid (GABA) and its synthetic enzyme glutamate decarboxylase are decreased. Choline acetyltransferase and several striatal neuropeptides are also decreased. Dopamine and glutamate levels are largely preserved.[202] Striatal projections to the globus pallidus are more affected in the akinetic rigid form of Huntington's disease than in the more common choreic form.[204] The CAG repeat comprising the mutation of Huntington's disease appears to lead to an alteration in the huntingtin protein and its accumulation in neuronal intranuclear inclusions in the cortex and the striatum.[205]

No therapy that halts or reduces the progression of Huntington's disease has been discovered. D_2 blocking agents and benzodiazepines may reduce the severity of chorea. Clozapine has been reported to be useful in reducing the psychosis of Huntington's disease,[206] and SSRIs and buspirone have been reported to reduce aggressive behavior.[207,208] Conventional antidepressants are used to address the depression of Huntington's disease and ECT may be useful in treatment-resistant patients.[209,210] Luprolide may be useful in reducing sexual aggression in male patients with Huntington's disease. Mania and hypomania can be treated with mood-stabilizing agents such as carbamazapine and valproic acid.

Neuroacanthocytosis

Neuroacanthocytosis is an autosomal recessive disorder characterized by involuntary movements, cognitive and behavioral changes, peripheral neuropathy, and the presence of acanthocytes (red cells with multiple spiny

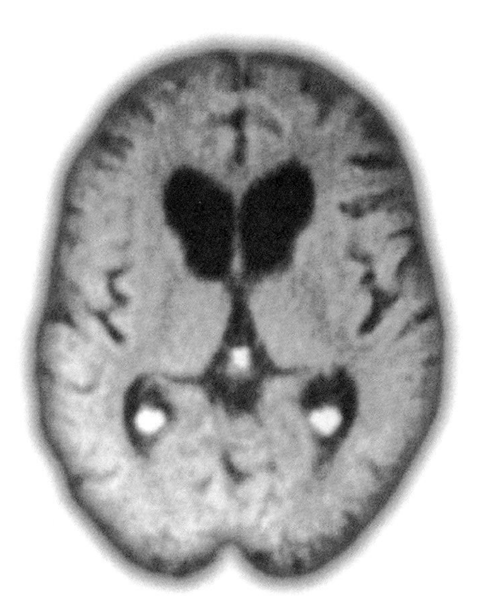

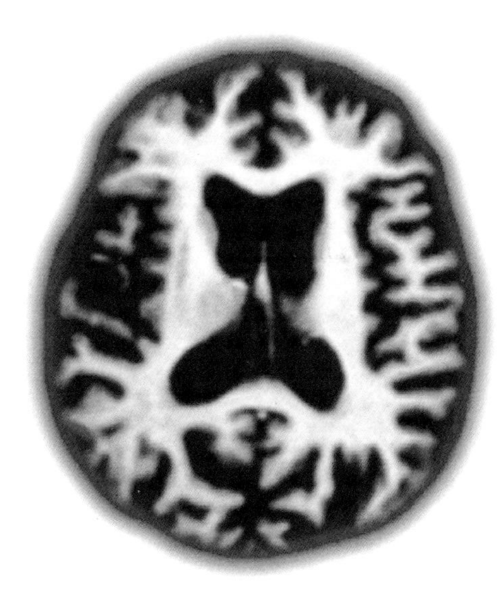

FIGURE 18.5 Computerized tomogram (*A*) showing reduced caudate volume and enlargement of the anterior horns of the lateral ventricles. Magnetic resonance imaging (*B*) reveals caudate and cortical atrophy in a patient with more advanced Huntington's disease.

projections).[211–213] The movement disorder is typically choreiform in nature, but dystonia, tics, involuntary vocalization, and parkinsonian features also occur. In some cases, lip biting has been prominent. Dementia is a prominent feature of most cases as the disease progresses. Personality changes consistent with a frontal lobe syndrome and depression have been described in patients with neuroacanthocytosis.[212,214,215] A retrieval deficit–type memory abnormality is present and executive dysfunction may be prominent.[212,216,217] At autopsy, patients have extensive neuronal loss and glyosis involving the stiratum, pallidum, and substantia nigra. Dopamine and substance P are depleted in the affected areas. Neuroacanthocytosis can be distinguished from Huntington's disease by its autosomal recessive inheritance, presence of acanthocytes on peripheral blood smears, and peripheral neuropathy.[211,218]

Sydenham's Chorea

Sydenham's chorea has its onset between the ages of 5 and 15 years of age and follows infection with group A b-hemolytic streptococcus. It is more common in girls than boys and evidence of the preceding infection can be elicited in approximately one-third of cases. The streptococcal infection, usually a pharyngitis, precedes movement disorder by 1 to 6 months. The chorea is primarily distal, affecting the hands and feet, and produces a characteristic milking movement of the examiner's hands when the patient grips the examiner (milkmaid's grip). Infection-related carditis and arthritis may accompany the chorea. Hypometric saccades, an occasional oculogyric crises, have also been reported.[219,220] Cognitive impairment and personality alterations, particularly irritability and obsessive-compulsive behavior, frequently accompany the choreiform syndrome. It is now recognized that obsessive-compulsive disorder and Gilles de la Tourette's syndrome may represent late complications of streptococcal infections. This complex set of disorders have been labeled *pediatric autoimmune neuropsychiatric disorders associated with streptococcal infections* (PANDAS).[221] The basal ganglia, including putamen, caudate, and globus pallidus, are enlarged in subjects manifesting Sydenham's chorea.[222] Glucose PET may show increased metabolism in the basal ganglia at the time of the chorea.[223] There is also an increased occurrence of psychosis among patients with Sydenham's chorea.[224,225]

Dystonia

There is a wide variety of primary and secondary dystonias (Table 18.13). Substantial progress has been made in defining the genetic basis of the early-onset dystonias (onset under age 26); a mutation in the *DYT1* gene on 9q34is responsible for most cases.[226,227] The genes responsible for a few late-onset dystonic syndromes have been defined, but most familial cases are currently produced by unknown mutations. In addition, sporadic cases of primary dystonia are not infrequent. Late-onset dystonias, including blepharospasm, oral mandibular dystonia, spasmodic dystonia (laryn-

TABLE 18.13. *Classification of Dystonia*

Primary dystonia
- Sporadic
- Inherited
  - Classic autosomal dominant (Oppenheim's) dystonia (*DYT1*, 9q34)
  - Adult-onset cranial-cervical dystonia (*DYT6*, 8p21)
  - Adult-onset cervical dystonia (18p)
  - Familial dystonia with unknown mutations

Secondary dystonia
- Dystonia-plus syndromes
  - Sporadic
  - Inherited
  - Myoclonic dystonia (18)
  - Dopa-responsive dystonia
    - GTP cyclohydrolase I deficiency (DYT5);
    - Tyrosine hydroxylase deficiency;
    - Other biopterin deficient diseases
  - Rapid-onset dystonia-parkinsonism

Associated with neurodegenerative disorders
- Sporadic
  - Parkinson's disease
  - Progressive supranuclear palsy
  - Multiple system atrophy
  - Corticobasal degeneration
- Inherited
  - Wilson's disease
  - Huntington's disease
  - Juvenile parkinsonism-dystonia
  - Progressive pallidal degeneration
  - Hallervorden-Spatz disease
  - Hypoprebetalipoproteinemia, acanthyocytosis, retinitis pigmentosa, and pallidal degeneration (HARP syndrome)
  - Machado-Joseph-Azorean disease
  - Ataxia telangiectasia
  - Neuroacanthocytosis
  - Rett's syndrome
  - Intraneuronal inclusion disease
  - Infantile bilateral striatal necrosis
  - Familial basal ganglia calcifications
  - Spinocerebellar degeneration
  - Olivopontocerebellar atrophy
  - Hereditary spastic paraplegia with dystonia
  - X-linked dystonia parkinsonism or Lubag (*DYT3*, xq13)
  - Deletion of 18q

Associated with metabolic disorders
- Amino acid disorders
  - Glutaric acidemia
  - Methylmalonic acidemia
  - Homocystinuria
  - Hartnup's disease
  - Tyrosinosis
- Lipid disorders
  - Metachromatic leukodystrophy
  - Ceroid lipofuscinosis
  - Dystonic lipidosis ("sea blue" histiocytosis)
  - Gangliosidoses—GM1, GM2 variants
  - Hexosaminidase A and B deficiency
- Miscellaneous metabolic disorders
  - Mitochondrial encephalopathies
  - Leigh's disease
  - Lesch-Nyhan syndrome
  - Leber's disease
  - Triosephosphate isomerase deficiency
  - Vitamin E deficiency
  - Biopterin deficiency

Secondary to a known specific cause
- Perinatal cerebral injury and kernicterus: athetoid cerebral palsy, delayed-onset dystonia
- Infection: viral encephalitis, encephalitis lethargica, Reye's syndrome, subacute sclerosing panencephalitis, Creutzfeldt-Jakob disease, HIV
- Other: tuberculosis, syphilis, acute infectious torticollis
- Paraneoplastic brainstem encephalitis
- Cerebrovascular or ischemic injury
- Brain tumor
- Arteriovenous malformation
- Head trauma and brain surgery
- Cervical injury
- Peripheral trauma
- Brainstem lesion
- Primary antiphospholipid syndrome
- Hypoxia
- Multiple sclerosis
- Central pontine myelinolysis
- Toxicants: manganese, carbon monoxide, carbon disulfide, cyanide, methanol, disulfiram, 3-nitroproprionic acid
- Metabolic: hypoparathyroidism
- Drugs: levodopa, bromocriptine, antipsychotics (acute and tardive dystonia), metoclopramide, fenfluramine, flecainide, anticonvulsants, certain calcium channel blockers, ergots, anxiolytics, serotonergic antidepressants

(*continued*)

TABLE 18.13. *Classification of Dystonia (Continued)*

Other hyperkinetic syndromes associated with dystonia
Tic disorders with dystonic tics
Paroxysmal dyskinesias
Paroxysmal kinesigenic dyskinesia
Paroxysmal nonkinesigenic dyskinesia
Paroxysmal exertion-induced dyskinesia
Paroxysmal hypnogenic dyskinesia
Hypnogenic dystonia (probably a seizure disorder)
Psychogenic
Pseudodystonia
Atlanto-axial subluxation
Syringomyelia
Arnold-Chiari malformation
Trochlear nerve palsy
Congenital nystagmus
Head thrust with oculomotor apraxia
Vestibular torticollis
Posterior fossa mass
Soft tissue neck mass
Congenital postural torticollis
Isaac's syndrome
Sandifer's syndrome
Satoyoshi syndrome
Stiff-person syndrome
Spinal deformities
Platybasia and basilar impression
Spasticity
Seizure producing twisting postures

gial dystonia), pharyngeal dystonia, spasmodic torticollis, truncal dystonia, and writer's cramp, are more likely to be focal in nature.[2] Many secondary dystonias occur, including dystonias associated with movement disorders such as Parkinson's disease and corticobasal dystonia and dystonic syndromes induced by long-term use of dopamine-blocking agents (tardive dystonia). The pathophysiology of dystonia is not certain. No pathological or biochemical abnormalities have been consistently identified in the brains of patients suffering from primary dystonic disorders. Positron emission tomographic studies suggest abnormal brain networks involving the frontal lobe and basal ganglia. There is a dissociation between activity of the lentiform nucleus and the thalamic nuclei as well as hyperactivity of supplementary motor and prefrontal areas.[228,229]

The pathways involved in the production of dystonia appear to spare circuitry involved in cognition and emotion. There are few consistent emotional, mood, behavioral, personality, or psychiatric syndromes observed in patients with dystonia.[230–232] Formal personality assessments have shown elevated subscale scores on mesaures of somatization, interpersonal sensitivity, and depression. The changes were modest in severity and likely reflected the presence of the disabling movement disorder and the secondary psychosocial consequences.[233] A few cases of bipolar disorder have been associated with idiopathic dystonia.[234] Dystonia is often misdiagnosed as a conversion disorder. Its unusual nature misleads many clinicians into believing that the disorder has a psychogenic basis. Psychogenic causes of movement disorders are rare, and criteria for identifying them are presented in Chapter 22.

Ataxias

Ataxia is a manifestation of cerebellar or sensory dysfunction leading to disequilibrium. Cerebellar ataxias are accompanied by additional signs such as ataxia on the finger-to-nose and heel-to-shin test, dysdiadokokinesia on tests of rapid alternating movements, past pointing when asked to touch a point in space, abnormalities of rebound check when the pressure against which a patient is pushing is suddenly released, and hypotonia and a variety of eye signs, reflecting loss of coordination of ocular movements. A wide range of disorders can cause ataxic syndromes (Table 18.14). These include autosomal recessive disorders; autosomal dominant ataxias; mitochondrial diseases with ataxia; prion diseases leading to spongiform encephalopathies; congenital malformations of the cerebellum; early-onset cerebellar diseases; inborn errors of metabolism; acquired disorders including paraneoplastic cerebellar syndromes; vitamin deficiencies; hypothyroidism; infectious illnesses; vascular disorders; multiple sclerosis; traumatic brain injury; heatstroke; and a variety of toxins. Substantial progress has been made in defining the genetic basis of the autosomal dominant cerebellar ataxias. At least nine syndromes have been identified with specific mutations (Table 18.15). There is phenotypic variety within each family carrying one of the mutations, but there are also shared neurologic and neuropsychiatric features. Five to ten percent of patients with spinocerebellar ataxia (SCA)-1 have a dementia syndrome and other members of the kindreds have had frontal lobe syndromes with euphoria and emotional lability.[235–237] Five to thirty percent of patients with

TABLE 18.14. *Ataxias*

Hereditary ataxias

- Autosomal recessive
  - Friedreich's ataxia
  - Ataxia telangiectasia
  - A-b lipoproteinemia
  - Ataxia with vitamin E deficiency
  - Retsum's disease
  - Cerebrotendinous xanthomatosis
  - Adrenoleukodystrophy
  - Metachromatic leukodystrophy
  - Globoid cell leukodystrophy
  - Ceroid lipofuscinosis
  - GM2 gangliosidosis
  - Niemann-Pick disease type C
  - Autosomal recessive spastic ataxia (Charlevoir-Saguenay syndrome)
- Autosomal dominant cerebellar ataxias (ADCA)
  - ADCA I
    - SCA-1
    - SCA-2
    - SCA-3/Machado-Joseph-Azorean disease
    - SCA-4
  - ADCA II
    - SCA-7
  - ADCA III
    - SCA-5
    - SCA-6
  - Dentatorubral-pallidoluysian atrophy
  - Periodic ataxia
    - Type I
    - Type 2
  - Mitochondrial disorders with ataxia
    - Mitochondrial encephalomyopathy, lactic acidosis, and stroke-like episodes (MELAs)
    - Leigh's disease
    - Myoclonus epilepsy with ragged red fibers (MERRF)
    - Kearns-Sayre syndrome
- Spongiform encephalopathies (prion disorders)
  - Creutzfeldt-Jakob disease
  - Kuru
  - Gerstmann-Straussler-Scheinker disease

Non-hereditary ataxias

- Multiple system atrophy (olivopontocerebellar atrophy type)
- Paraneoplastic cerebellar degeneration
- Metabolic conditions
  - Vitamin B_1 deficiency (Wernicke's encephalopathy)
  - Vitamin E deficiency
  - Vitamin B_{12} deficiency
  - Hypothyroidism
  - Hypoparathyroidism
  - Gluten ataxia
- Cerebellar encephalitis and post-infectious ataxia
  - Ataxia
  - Viral (mumps, Epstein-Barr virus, others)
  - Bacterial (pertussis, mycoplasma, others)
  - Miller Fisher syndrome (form of Guillain-Barré syndrome with ophthalmoplegia, ataxia, and areflexia)
- Focal cerebellar disorders
  - Vascular (ischemia, hemorrhage, arteriovenous malformation, including von Hippel-Lindau syndrome)
  - Neoplasms
  - Multiple sclerosis
  - Traumatic injury
- Ataxia due to toxins and adverse physical circumstances
  - Alcohol
  - Heat stroke
  - Neuroleptic malignant syndrome
  - Phenytoin
  - Phenobarbital
  - Thallium
  - Lead
  - Mercury
- Congenital ataxic disorders with cerebellar malformations
  - Cogan's syndrome (oculomotor apraxia, motor retardation, and ataxia)
  - Joubert's syndrome (neonatal hyperpnea, abnormal ocular movements, mental retardation, ataxia)
  - COACH syndrome (cerebellar vermis agenesis, oligophrenia, ataxia, coloboma, hepatic fibrosis)
  - Vermian hypoplasia with coloboma and hepatic fibrosis (oligophrenia, cerebellar ataxia)
  - Gillespie's syndrome (oligophrenia, aniridia, mental retardation, ataxia)
  - Angelmann syndrome (retardation, ataxia, spasticity, seizure, microencephaly)
- Early-onset ataxic disorders (before age 20)
  - Cerebellar ataxia, Holmes type (hypogonadism, mental retardation, deafness, retinopathy, corticospinal signs, neuropathy)
  - Unverricht-Lundborg disease (myoclonus, seizures)

(*continued*)

TABLE 18.14. *Ataxias (Continued)*

Louis Bar syndrome (telangiectases, tremor, choreoathetosis, oculomotor apraxia, immune deficiency, increased a-fetoprotein level)
Xeroderma pigmentosum (skin photosensitivity, mental retardation, dwarfism, spasticity, choreoathetosis, deafness, hypogonadism)
Cockayne's syndrome (skin photosensitivity, mental retardation, retinopathy, extrapyramidal features, deafness, neuropathy)
Ataxia with parkinsonism (hypomimia, rigidity, parkinsonian tremor)
Behr's syndrome (optic atrophy, mental retardation, spasticity)
Marinesco-Sjögren syndrome (mental retardation, cataracts, short stature, skeletal abnormalities)
Ataxia and pigmentary retinopathy (pigmentary retinopathy, hearing loss, mental retardation)
Lichtenstein-Knorr disease (hearing loss)

SCA, spinocerebellar ataxia.

SCA-2 manifest dementia and in some cases a frontal dysexecutive syndrome has been prominent.[238,239] Most patients with SCA-3, also known as *Machado-Joseph disease*, have normal intellectual function but a few cases with cognitive impairment have been recorded.[240] Dementia has been reported in approximately 15% of patients with SCA-6 and a few patients with SCA-7 have exhibited dementia and psychosis.[241,242] Dementia is common in dentatorubropallidoluysian atrophy.[243,244] There is pathologic as well as clinical heterogeneity within the families with the various autosomal dominant cerebellar atrophy mutations. In general, those mutations associated with more widespread pathology, such as a mutation producing dentatotrubropallidoluysian atrophy, have more widespread pathology involving the dentate nucleus invariably; the globus pallidus, thalamic nucleus and red nucleus typically; and other areas, including the thalamus and striatum, variably.[243] All of the autosomal dominant cerebellar atrophy syndromes described in this section are due to CAG repeat expansions on specific chromosomes (Table 18.15) and the extent of pathology varies with the length of the CAG repeat.[241]

Friedreich's ataxia is the most common hereditary ataxia, occurring with a prevalence of 1 per 50,000. The disease typically has its onset between 5 and 15 years of age and features progressive ataxia of all limbs, loss of deep tendon reflexes and a vibration sense in the lower extremities, cerebellar-type dysarthria and pyramidal signs, muscle weakness, and Babinski signs. A majority of patients with Friedreich's ataxia are ho-

TABLE 18.15. *Neuropsychiatric Features of the Autosomal Dominant Cerebellar Ataxias*

Gene	*Chromosome*	*Usual Neuropathology*	*Neuropsychiatric Features*
SCA1	6	Olivopontocerebellar atrophy	Dementia late; mental status changes early in 10%; mood disorders and personality changes
SCA2	12	Olivopontocerebellar atrophy	Dementia in 5%–30% of cases; one family had prominent dementia, another had a consistent dysexecutive syndrome
SCA3/Machado-Joseph disease	14	Spinopontine atrophy	Intellect typically preserved; dementia in 5%–20%
SCA4	16	Unknown	None described
SCA5	11	Cortical cerebellar atrophy	None described
SCA6	19	Cortical cerebellar atrophy	Dementia in 25%
SCA7	3	Olivopontocerebellar atrophy	Dementia in 10%; psychosis in some
SCA10	22	Unknown	None described
Dentatorubro-pallido-luysian atrophy	12	Multisystem degeneration	Dementia common; psychosis in some

TABLE 18.16. *Classification of Principal Types of Tremors*

Tremor	*Action Characteristic*	*Amplitude*	*Frequency*	*Etiology*
Physiological	Action	Small	High (8–12 Hz)	Exaggerated by anxiety, fatigue, toxic-metabolic disorders
Essential	Action	Small	High (8–12 Hz)	Senile, hereditary, sporadic
Parkinsonian	Resting	Large	Low (4–6 Hz)	Parkinsonian syndromes
Cerebellar	Action	Crescendo	Low	Cerebellar or midbrain disease

mozygous for a GAA repeat expansion of the x25 gene.[245] Pathologically, there is degeneration of the posterior columns of the spinal cord resulting from loss of the primary sensory neurons of the dorsal ganglia. The spinocerebellar tracks are atrophic and the descending corticospinal tracks are also atrophied, although the motor neurons in the ventral horns are well preserved. The deep cerebellar nuclei are severely affected and there may be mild loss of Purkinje cells in the cerebellar cortex. Other cerebral structures are typically unaffected.[130] Given the limited involvement of neuronal structures outside of the spinal cord and cerebellum, few neurobehavioral changes would be expected among patients with Friedreich's ataxia. Slowing of information-processing speed and deficits on some tests of visuospatial cognition have been suggested in some studies.[246,247]

Tremor

Tremor is a common product of many disturbances affecting the central nervous system. Like dystonia, tremor appears to be generated by pathways that lie outside of circuits critical to cognition and emotion and tremor has few links to specific neuropsychiatric disorders. However, it is common for tension, anxiety, and distress to exaggerate any existing tremor. Moreover, many drugs used to treat psychiatric illnesses induce tremor syndromes. Thus, knowledge of tremor is critical to appropriate patient assessment and treatment in neuropsychiatry.

There are four major varieties of tremors (Table 18.16). *Parkinsonian tremors* are large-amplitude, low-frequency resting tremors that accompany Parkinson's disease and other parkinsonian syndromes. *Physiological tremors* are small-amplitude, high-frequency action tremors that occur during activity and disappear at rest. Physiological tremors may be exaggerated by anxiety, fatigue, and a variety of drugs, toxins, and metabolic disorders. *Essential tremors* share the same amplitude and frequency characteristics of physiological tremors but occur in sporadic, familial, or "senile" situations. *Cerebellar tremors* are also action (intention) tremors. They are of low frequency and increase in amplitude as the target is approached (crescendo pattern). Cerebellar tremors are produced by degenerative, demyelinating, toxic, and traumatic disorders of the cerebellum and midbrain.[248,249] Two other circumstances in which tremors occur include task-specific tremors, particularly orthostatic tremors, and primary writing tremors that appear only when the patient is in specific provocative circumstances (standing or writing). In addition, tremor has been associated with peripheral neuropathies, and disruption of peripheral feedback mechanisms may be responsible for some tremor disorders. Table 18.17 lists the principle etiologies of tremors. Beta-blockers such as propranolol and metoprolol, primidone, or benzodiazepines are the principle treatments for central tremor. Dopaminergic or anticholinergic agents may provide relief to parkinsonian-type rest tremors.[248]

PATHOPHYSIOLOGY OF NEUROPSYCHIATRIC SYMPTOMS

Basal ganglia diseases are accompanied by a wide array of neuropsychiatric disturbances including depression, anxiety, irritability, and disinhibition. When patients with Parkinson's disease are treated with dopaminergic agents, new symptoms emerge, including nightmares, hallucinations, delusions, mood elevation, and, occasionally, sexual behavior disturbances. This marked overlap of movement disorders, extraocular movement control disturbances, neuropsychiatric symptoms, and cognitive changes in patients with basal ganglia disorders reflects the disruption of function of a complex set of frontal subcortical circuits that link

TABLE 18.17. *Etiologies of Tremor*

Physiological tremor (normal and accentuated)
Psychotropic agents: lithium, tricyclic antidepressants, phenothiazines, butyrophenones
Nonpsychotropic drugs: epinephrine, isoproterenol, metaproterenol, terbutaline, xanthine (coffee, tea) theophylline, levodopa, amphetamines, thyroid hormone, hypoglycemic agents, adrenocorticosteriods, valproate sodium
Endocrine: thyrotoxicosis, hypoglycemia, pheochromocytoma
Stress-induced: anxiety, fright, fatigue, cold-shivering
Metabolic disorders: uremia, hepatic encephalopathy, anoxia
Miscellaneous agents and conditions: alcohol or sedative withdrawal, mercury, lead, arsenic, bismuth, carbon monoxide, methyl bromide, monosodium glutamate
Essential tremor
Autosomal dominant, senile, sporadic
With other movement disorders: parkinsonism, torsion dystonia, torticollis, writer's cramp, hereditary nonprogressive chorea
Cerebellar tremor
Cerebellar degenerations
Multiple sclerosis
Wilson's disease
Drugs and toxins: lithium, phenytoin, barbiturates, alcohol, mercury, 5-fluorouracil
Midbrain lesions ("rubral" tremor)
Parkinsonian tremor
Parkinson's disease
Parkinsonian syndrome
Dopamine-blocking agents
Task-specific tremors
Orthostatic tremor
Primary writing tremor
Tremor associated with peripheral neuropathies
With Charcot-Marie-Tooth disease (Roussy-Lévy syndrome)
With hereditary sensory neuropathy (Déjerine-Sottas disease)
With acquired neuropathies

the cortex (predominantly frontal areas) to regions of the striatum, globus pallidus/substantia nigra, and thalamus. Dysfunction in basal ganglia disorders at the level either of the striatum or the globus pallidus can produce a complex array of cognitive, behavioral, motoric, and oculomotor symptoms, depending on which of the frontal subcortical circuits are disrupted. The similarity between frontal cortical dysfunction and disorders of the basal ganglia can also be linked plausibly to this frontal–subcortical architecture.

There are five basic frontal–subcortical circuits that link regions of the frontal cortex to areas within the striatum, subregions within the globus pallidus and substantia nigra, and nuclei within the thalamus.[250–252] Each of the circuits includes a direct pathway that connects the striatum to the globus pallidus/substantia nigra and thalamus before projecting via thalamal–cortical projections back to the frontal cortex and an indirect pathway that connects the striatum to the globus pallidus externa, which projects to the subthalamic nucleus and then to the thalamus. A dynamic balance between the direct circuit and the indirect circuit on the thalamus produces the final common thalamal cortical output processed at the level of the basal ganglia. Each of these five major circuits contains multiple subchannels that connect more restricted regions of anatomy within each of the member structures of the frontal–subcortical circuits.[253] Four subchannels (medial, lateral, dorsal, and ventral) have been identified within the dorsolateral prefrontal projection; five subchannels have been identified within the orbitofrontal projection; and two subchannels have been identified within the anterior cingulate–subcortical circuit.

Each of the major circuits has a similar anatomy (Figs. 18.6 and 18.7). The motor circuit begins in the primary motor areas as well as the supplementary motor regions and projects to the putamen, which in turn projects to globus pallidus interna and lateral portions of the substantia nigra. These structures project to the ventral anterior and centromedian nuclei of the thalamus, which in turn projects to premotor cortex and supplementary motor area, closing the frontal–subcortical loop. The indirect pathway within the motor circuit projects to globus pallidus externa and then to the subthalamic nucleus, which in turn projects to the ventral anterior regions of the thalamus.[250,254] The oculomotor circuit originates in the frontal eye fields, which project to the body of the caudate nucleus then to the dorsal medial sector of globus pallidus interna and ventral lateral substantia nigra. These project to the ventral anterior and medial dorsal nuclei of the thalamus, which project in turn to the frontal eye fields. An indirect pathway connects the body of the caudate with globus pallidus externa and subthalamic nucleus. Dysfunction of the motor circuit results in hypokinetic or hyperkinetic disorders, whereas dysfunction of the ocular motor circuit produces supranuclear ocular motor disorders.

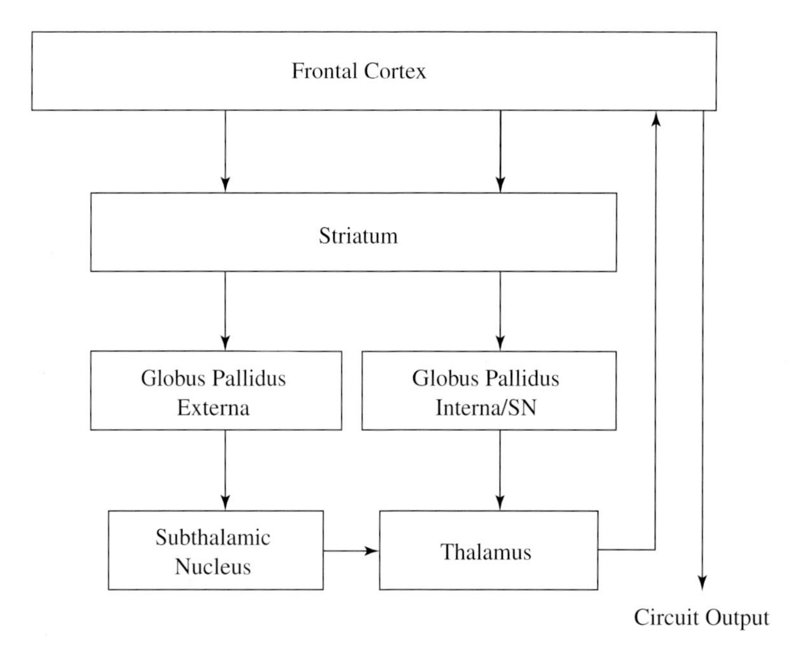

FIGURE 18.6 Principal structures included in the five principal subcortical circuits.

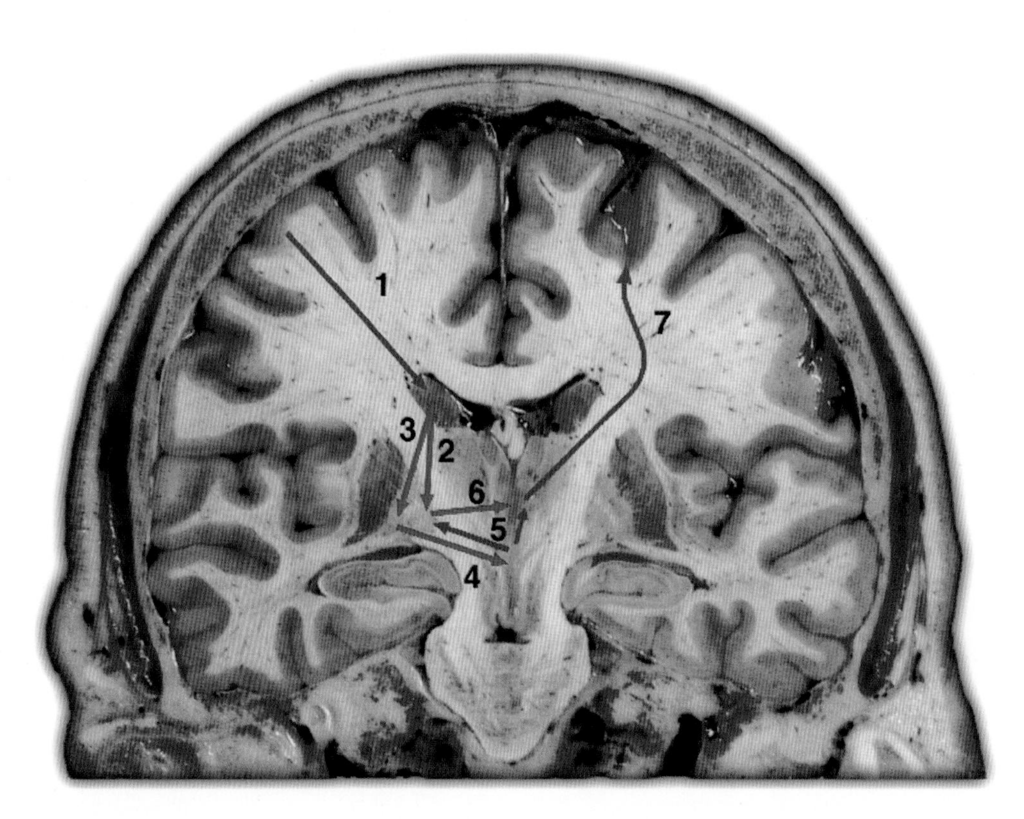

FIGURE 18.7 Anatomical structures involved in the frontal–subcortical circuits. 1) cortico-striato, 2) striato-globus pallidum interna, 3) striato-globus pallidus esterna, 4) globus pallidus externa-subthalamic nucleus, 5) subthalamic nuclear-globus plallidus interna, 6) globus pallidus interna-thalamus, 7) the femur-cortical projections.

The dorsolateral prefrontal circuit, orbitofrontal–subcortical circuit, and anterior cingulate–subcortical circuit mediate important aspects of cognition and emotion (Fig. 18.8). The dorsolateral prefrontal circuit mediates executive function, and executive control disorders result from interruption of the circuit or dysfunction of any member structure of the dorsolateral prefrontal–subcortical circuitry. The orbitofrontal circuit mediates aspects of civil behaviors and impulse control; dysfunction of this region results in disinhibition, tactlessness, impulsiveness, and disrupted social interaction. The anterior cingulate–subcortical circuit mediates motivation, and dysfunction of this circuitry results in apathy, disinterest, affective flattening, loss of affection, and reduced emotional valence in environmental interactions.[255,256] The dorsolateral prefrontal circuit begins in dorsolateral prefrontal cortical regions projecting to the dorsolateral head of the caudate nucleus and throughout the body and tail of the caudate. The caudate then projects to dorsomedial globus pallidus and rostral substantia nigra. The internal globus pallidus sends projections to ventral anterior thalamic nuclei and the substantia nigra projects to the medial dorsal thalamus. The thalamic nuclei project to the dorsolateral prefrontal cortex. An indirect circuit connects the dorsolateral caudate to globus pallidus interna, subthalamic nucleus, and medial dorsal thalamus. A lateral orbitofrontal–subcortical circuit projects to the ventral medial segment of the caudate nucleus and from there to the globus pallidus interna and substantia nigra before projecting to the ventral anterior and medial dorsal nuclei of the thalamus, which in turn project back to the lateral orbitofrontal cortex. This prefrontal–subcortical circuit may mediate the ability to alternate between behavioral sets.[251]

"Limbic" circuits include projections from both the anterior cingulate region and the medial orbitofrontal region. The anterior cingulate projects to the nucleus

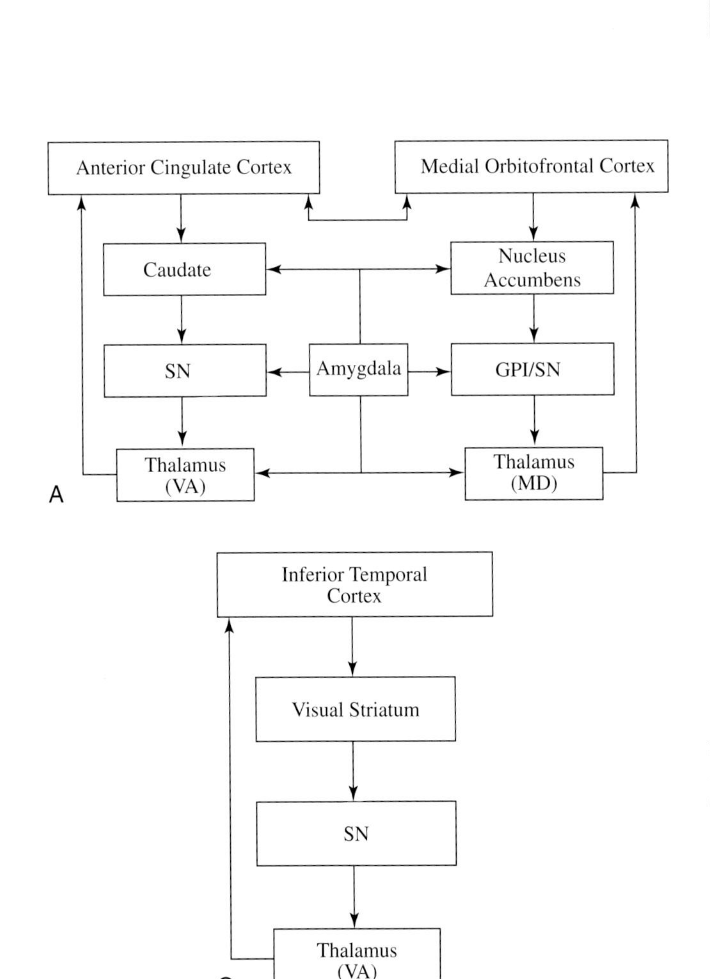

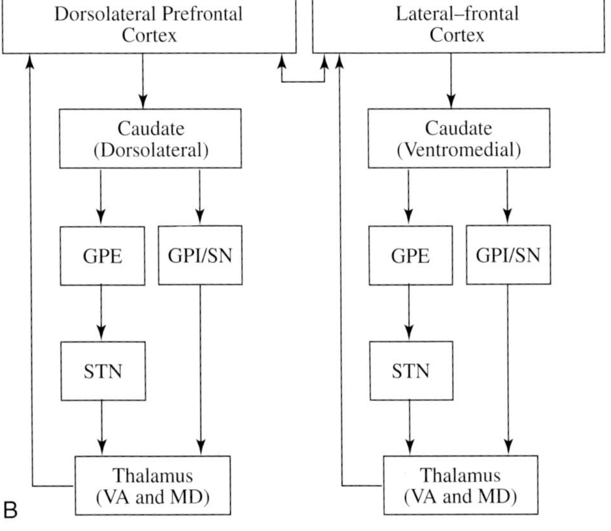

FIGURE 18.8 Illustration of the three principal behaviorally relevant frontal–subcortical circuits. (*A*) anterior cingulate-medial orbitofrontal cortex; (*B*) dorsolateral prefrontal and lateral orbitofrontal cortex, (*C*) interior temporal center. The anterior cingulate–medial orbitofrontal circuit (*A*) may not have direct and indirect pathways present in other frontal–subcortical circuits. Short association fibers connect the circuits at the level of the prefrontal cortex. GPE, globus pallidus external; GPI, globus pallidus interna; MD, medial dorsal nucleus of the thalamus; SN, substantia nigra; STN, subthalamic nucleus; VA, ventral interior nucleus of the thalamus.

accumbens and the ventral or limbic striatum. This ventral striatal region in turn projects to the ventral pallidum, which connects to the medial dorsal thalamus projecting back to frontal regions. The ventral pallidum is not clearly differentiated into internal and external segments and the existence of a direct and indirect pathway in this circuit is less certain.[251] The limbic circuitry receives input from the amygdala at several points including the ventral striatum, medial dorsal nucleus of the thalamus, and anterior cingulate and medial orbital frontal cortex. The limbic circuit also has extensive efferent connections from the ventral pallidum to limbic subcortical structures, including the hypothalamus and ventral tegmental area.[251]

The frontal–subcortical circuits represent parallel processing loops and remain largely discrete at subcortical levels. The frontal portions of the frontal–subcortical circuits are connected via short association fibers, a feature that emphasizes the integrative activity of the frontal cortex, where executive and limbic function can be integrated with motivational activation to produce volitional motor and oculomotor activity.

The frontal–subcortical circuits receive noncircuit information from related cortical regions. For example, posterior association cortex projects to the anterior dorsolateral prefrontal association cortex, where it can be integrated into the activity of the dorsolateral prefrontal–subcortical circuit. Similarly, limbic structures of the anterior and medial temporal cortex project widely into the limbic–frontal–subcortical circuit. Thus, the frontal–subcortical circuits serve not only to link frontal and subcortical structures in integrated loops but also to integrate information from posterior hemispheric structures, projecting to frontal regions via long, intrahemispheric tracks.

Frontal–subcortical circuits may intersect with other complex cortical, subcortical circuits relevant to behavior. For example, connections from visual association regions of the temporal cortex project to visual striatum, substantia nigra, ventral anterior thalamus, and back to visual association cortex.[257] This circuit may provide the neurobiological basis for complex visual phenomena, including hallucinations observed with basal ganglia disorders and their treatment.

The principal transmitters employed in the frontal–subcortical circuits have been defined (Fig. 18.9).[258] The excitatory transmitter glutamate projects to the striatum from the cortex and to the thalamus from the cortex and to the cortex from the thalamus. Projections from the striatum to the globus pallidus (both internal and external segments) utilize the inhibitory transmitter GABA. Similarly, projections from globus pallidus externa to subthalamic nucleus and from globus pallidus interna to thalamus utilize GABA for inhibitory transmission. The subthalamic nucleus employs the excitatory transmitter glutamate and its connection to thalamus. Thus, the direct pathway has two consecutive inhibitory neurons, whereas the indirect pathway has two consecutive inhibitor neurons and one excitatory neuron. The influence on the thalamus via the direct pathway will be excitatory (inhibition of inhibition) whereas the thalamic influence via the indirect pathway will be inhibitory (via inhibition of the subthalamic excitatory connection). This arrangement emphasizes the dynamic balance between inhibitory and excitatory pathways in the frontal–subcortical circuit and the critical role of the thalamus as the subcortical structure that integrates the inhibitory and excitatory influences prior to its cortical projection, representing the final integrated balance. Dopaminergic projections from substantia nigra inhibit the indirect pathway and facilitate the direct subcortical pathway (Fig. 18.7). Serotonergic fibers from the raphe nuclei also project widely to regions within the frontal–subcortical circuits. Enkephalin is a peptide transmitter coexisting in the striatal globus pallidus externa projection. Substance P is a peptide transmitter coexisting in the GABAergic projection from striatum to globus pallidus interna.[254] This chemoarchitecture of the frontal–subcortical circuits provides a basis for understanding the complex, motoric, cognitive, and emotional effects of dopamine depletion in Parkinson's disease and therapeutic intervention with dopaminergic and serotonergic compounds in basal ganglia disorders.

Disturbances of the balance between direct (excitatory) and indirect (inhibitory) pathways in the frontal–subcortical circuits may provide a means of understanding some behavioral disturbances commonly observed in patients with basal ganglia disorders. Overactivity of thalamal–cortical projections may lead to agitation, irritability, euphoria, and anxiety common in hyperkinetic disorders and similar to the excitatory thalamal–cortical stimulation producing chorea. In contrast, apathy may be a product of reduced thalamal–cortical activation and parallel the reduced thalamocortical stimulation occurring with parkinsonism.[121] The imbalance occurring between direct and indirect pathways in the motor circuit led to the introduction of pallidotomy to reduce the hyperkinetic disorders occurring with the "on–off" phenomenon in patients chronically treated with levodopa. It also led to the development of deep brain stimulation to equalize excitatory and inhibitory input within these frontal–subcortical structures. Similar interventions may eventually allow treatment of neuropsychiatric and cognitive disturbances associated with frontal–subcortical circuit dysfunction.

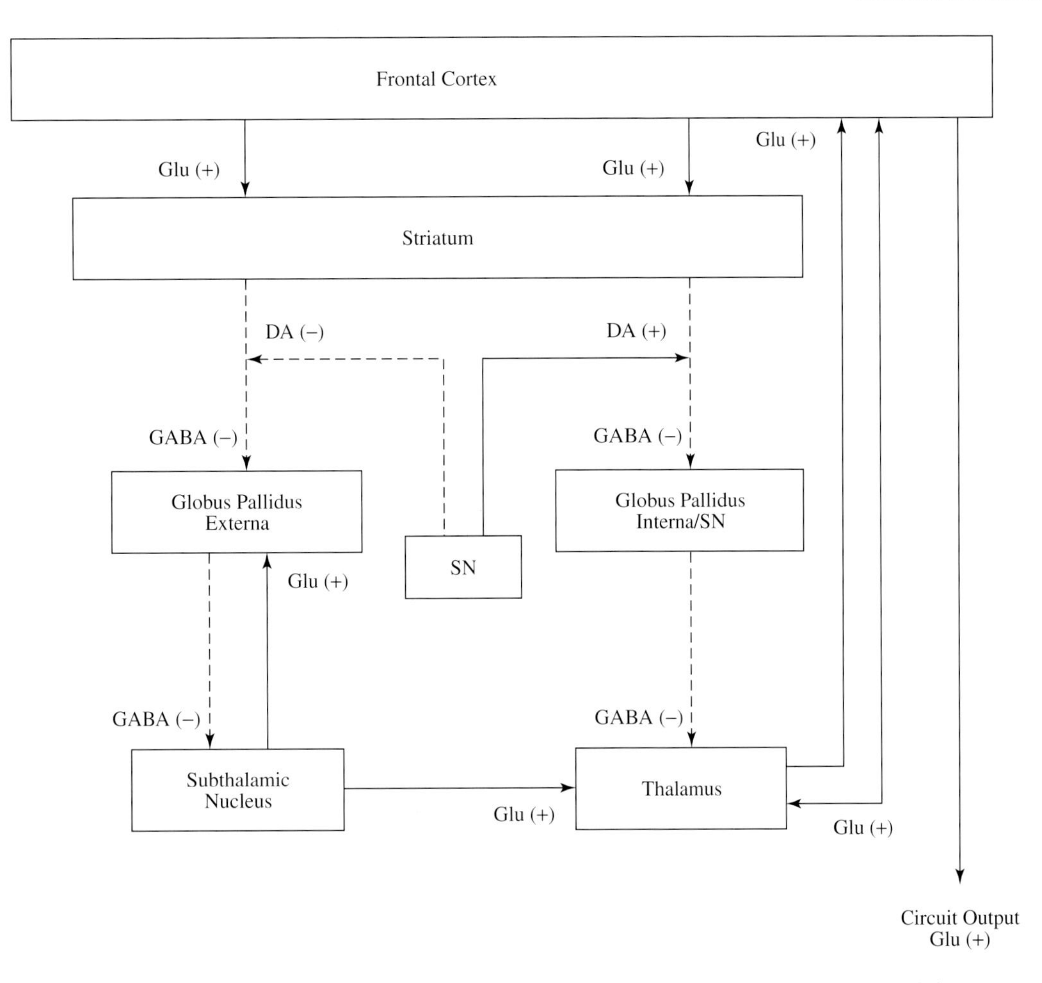

FIGURE 18.9 Multiple transmitters including glutamate (Glu), γ-aminobutyric acid (GABA), and dopamine (DA) play major roles in the frontal–subcortical circuits. Excitatory connections are shown with solid arrows and inhibitory connections are shown with striped arrows. SN, substantia nigra.

REFERENCES

1. Kishore A, Calne DB. Approach to the patient with a movement disorder and overview of movement disorders. In: Watts RL, Koller WC, eds. Movement Disorders: Neurologic Principles and Practice. New York: McGraw-Hill, 1997:3–14.
2. Weiner WJ, Lang AE. Movement Disorders: A Comprehensive Survey. Futura Publishing Company, Mount Kisco, New York, 1989.
3. Marttila RJ. Epidemiology. In: Koller WC, ed. Handbook of Parkinson's Disease, 2nd ed. New York: Marcel Dekker, 1992: 33–57.
4. Langston JW, Widner H, et al. Core assessment program for intracerebral transplantations (CAPIT). Mov Disord 1992;7:2–13.
5. Hughes AJ, Daniel SE, et al. Accuracy of clinical diagnosis of idiopathic Parkinson's disease: a clinico-pathological study of 100 cases. J Neurol Neurosurg Psychiatry 1992;55:181–184.
6. Gelb DJ, Oliver E, Gilman S. Diagnostic criteria for Parkinson disease. Arch Neurol 1999;56:33–39.
7. Vidailhet M, Rivaud S, et al. Eye movements in parkinsonian syndromes. Ann Neurol 1994;35:420–426.
8. Fearnley J, Lees A. Pathology of Parkinson's disease. In: Calne DB, ed. Neurodegenerative Diseases. Philadelphia: W.B. Saunders, 1994:545–554.
9. Lang AE, Lozano AM. Parkinson's disease. N Engl J Med 1998;339:1130–1143, 1044–1053.
10. Hutchinson M, Fazzini E. Cholinesterase inhibition in Parkinson's disease. J Neurol Neurosurg Psychiatry 1996;61:324–325.
11. Hershey LA, Feldman BJ, et al. Tremor at onset: predictor of cognitive and motor outcome in Parkinson's disease. Arch Neurol 1991;48:1049–1051.
12. Gabrieli JDE, Singh J, et al. Reduced working memory span in Parkinson's disease: evidence for the role of a frontostriatal system in working and strategic memory. Neuropsychology 1996; 10:322–332.
13. Huber SJ, Shuttleworth EC, et al. Cortical vs subcortical dementia: neuropsychological differences. Arch Neurol 1986;43: 392–394.
14. Kitagawa M, Fukushima J, Tashiro K. Relationship between antisaccades and the clinical symptoms in Parkinson's disease. Neurology 1994;44:2285–2289.
15. Ross GW, Mahler ME, Cummings JL. The dementia syndromes of Parkinson's disease: cortical and subcortical features. In: Huber SJ, Cummings JL, eds. Parkinson's Disease: Neurobehav-

ioral Aspects. New York: Oxford University Press, 1992: 132–148.

16. Ogden JA, Growdon JH, Corkin S. Deficits on visuospatial tests involving forward planning in high-functioning parkinsonians. Neuropsychiatry Neuropsychol Behav Neurol 1990;3:125–139.
17. American Psychiatric Association. Diagnostic and Statistical Manual of Mental Disorders, 4th ed. Washington, DC: American Psychiatric Press, 1994.
18. Aarsland D, Larsen JP, et al. Range of neuropsychiatric disturbances in patients with Parkinson's disease. J Neurol Neurosurg Psychiatry 1999a;67:492–496.
19. Mayeux R, Denaro J, et al. A population-based investigation of Parkinson's disease with and without dementia: relationship to age and gender. Arch Neurol 1992;49:492–497.
20. Tison F, Dartigues JF, et al. Dementia in Parkinson's disease: a population-based study in ambulatory and institutionalized individuals. Neurology 1995;45:705–708.
21. Gotham AM, Brown RG, Marsden CD. 'Frontal' cognitive function in patients with Parkinson's disease 'on' and 'off' levodopa. Brain 1988;111:299–321.
22. Haroutunian V, Serby M, et al. Contribution of Lewy body inclusions to dementia in patients with and without Alzheimer disease neuropathological conditions. Arch Neurol 2000;57: 1145–1150.
23. Cummings JL, Benson DF. Dementia: A Clinical Approach. Boston: Butterworth-Heinemann, 1992.
24. Tandberg E, Larsen JP, et al. The occurrence of depression in Parkinson's disease: a community-based study. Arch Neurol 1996;53:175–179.
25. Starkstein SE, Petracca G, et al. Depression in classic versus akinetic-rigid Parkinson's disease. Mov Disord 1998;13:29–33.
26. Starkstein SE, Berthier ML, et al. Depression in patients with early versus late onset of Parkinson's disease. Neurology 1989; 39:1441–1445.
27. Tandberg E, Larsen JP, et al. Risk factors for depression in Parkinson disease. Arch Neurol 1997;54:625–630.
28. Mayberg HS, Starkstein SE, et al. Selective hypometabolism in the inferior frontal lobe in depressed patients with Parkinson's disease. Ann Neurol 1990;28:57–64.
29. Ring JA, Bench CJ, et al. Depression in Parkinson's disease: a positron emission study. Br J Psychiatry 1994;165:333–339.
30. Mayeux R, Stern Y, et al. Altered serotonin metabolism in depressed patients with Parkinson's disease. Neurology 1984;34: 642–646.
31. Chan-Palay V, Asan E. Alterations in catecholamine neurons of the locus coeruleus in senile dementia of the Alzheimer type and in Parkinson's disease with and without dementia and depression. J Comp Neurol 1989;287:373–392.
32. Ceravolo R, Nuti A, et al. Paroxetine in Parkinson's disease: effects on motor and depressive symptoms. Neurology 2000;55: 1216–1218.
33. Hauser R, Zesiewicz TA. Sertraline for the treatment of depression in Parkinson's disease. Mov Disord 1997;12:756–759.
34. Richard IH, Kurlan R, Parkinson Study Group. A survey of antidepressant drug use in Parkinson's disease. Neurology 1997; 49:1168–1170.
35. Jouvent R, Abensour P, et al. Antiparkinsonian and antidepressant effects of high doses of bromocriptine: an independent comparison. J Affect Disord 1983;5:141–145.
36. Cummings JL. D-3 receptor agonists: combined action neurologic and neuropsychiatric agents. J Neurol Sci 1999;163:2–3.
37. Pogarell O, Kunig G, Oertel WH. A non-ergot dopamine agonist, pramipexole, in the therapy of advanced Parkinson's disease: improvement of parkinsonian symptoms and treatment-associated complications: a review of three studies. Clin Neuropharmacol 1997;20:S28–S35.
38. Douyon R, Serby M, et al. ECT and Parkinson's diseaes revisited: a "naturalistic" study. Am J Psychiatry 1989;146:1451–1455.
39. Rasmussen KG, Abrams R. The role of electroconvulsive therapy in Parkinson's disease. In: Huber SJ, Cummings JL, eds. Parkinson's Disease: Neurobehavioral Aspects. New York: Oxford University Press, 1992:255–270.
40. Henderson R, Kurlan R, et al. Preliminary examination of the comorbidity of anxiety and depression in Parkinson's disease. J Neuropsychiatry Clin Neurosci 1992;4:257–264.
41. Menza MA, Cocchiola J, Golbe LI. Psychiatric symptoms in progressive supranuclear palsy. Psychosomatics 1995;36:550–554.
42. Starkstein SE, Robinson RG, et al. Anxiety and depression in Parkinson's disease. Behav Neurol 1993;6:151–154.
43. Starkstein SE, Mayberg HS, et al. Reliability, validity, and clinical correlates of apathy in Parkinson's disease. J Neuropsychiatry Clin Neurosci 1992;4:134–139.
44. Glosser G, Clark C, et al. A controlled investigation of current and premorbid personality: characteristics of Parkinson's disease patients. Mov Disord 1995;10:201–206.
45. Comella C, Nardine T, et al. Sleep-related violence, injury, and REM sleep behavior disorder in Parkinson's disease. Neurology 1998;51:526–529.
46. Nausieda PA, Weiner WJ, et al. Sleep disruption in the course of chronic levodopa therapy: an early feature of the levodopa psychosis. Clin Neuropharmacol 1982;5:183–194.
47. Cummings JL. Behavioral complications of drug treatment of Parkinson's disease. J Am Geriatr Soc 1991;39:708–716.
48. de la Fuente-Fernandez R, Nunez MA, Lopez E. The apolipoprotein E-4 allele increases the risk of drug-induced hallucinations in Parkinson's disease. Clin Neuropharmacol 1999; 22:226–230.
49. Naimark D, Jackson E, et al. Psychotic symptoms in Parkinson's disease patients with dementia. J Am Geriatr Soc 1996;44: 296–299.
50. Sanchez-Ramos JR, Ortoll R, Paulson GW. Visual hallucinations associated with Parkinson disease. Arch Neurol 1996;53: 1265–1268.
51. Goetz CG, Tanner CM, Klawans HL. Pharmacology of hallucinations induced by long-term drug therapy. Am J Psychiatry 1982;139:4.
52. Aarsland D, Larsen JP, et al. Prevalence and clinical correlates of psychotic symptoms in Parkinson's disease: a community-based study. Arch Neurol 1999b;56:595–601.
53. The Parkinson Study Group. Low-dose clozapine for the treatment of drug-induced psychosis in Parkinson's disease: the Parkinson Study Group. N Engl J Med 1999;340:757–763.
54. Ellis T, Cudkowicz ME, et al. Clozapine and risperidone treatment of psychosis in Parkinson's disease. J Neuropsychiatry Clin Neurosci 2000;12:364–369.
55. Parsa MA, Bastani B. Quetiapine (Seroquel) in the treatment of psychosis in patients with Parkinson's disease. J Neuropsychiatry Clin Neurosci 1998;10:216–219.
56. Wolters EC, Jansen ENH, et al. Olanzapine in the treatment of dopeminomimetic psychosis in patients with Parkinson's disease. Neurology 1996;47:1085–1087.
57. Goetz CG, Blasucci LM, et al. Olanzapine and clozapine: comparative effects on motor function in hallucinating PD patients. Neurology 2000;55:789.

58. Nissenbaum H, Quinn MP, et al. Mood swings associated with the 'on–off' phenomenon in Parkinson's disease. Psychol Med 1987;17:899–904.
59. Keshavan MS, David AS, et al. 'On–off" phenomena and manic-depresive mood shifts: a case report. J Clin Psychiatry 1986;47:93–94.
60. Harvey NS. Serial cognitive profiles in levodopa-induced hypersexuality. Br J Psychiatry 1988;153:833–836.
61. Quinn NP, Toone B, et al. Dopa dose-dependent sexual deviation. Br J Psychiatry 1983;142:296–298.
62. Uitti RJ, Tanner CM, et al. Hypersexuality with antiparkinsonian therapy. Clin Neuropharmacol 1989;12:375–383.
63. Giovannoni G, O'Sullivan JD, et al. Hedonistic homeostatic dysregulation in patients with Parkinson's disease on dopamine replacement therapies. J Neurol Neurosurg Psychiatry 2000;68: 423–438.
64. McKeith IG, Galaski D, et al. Consensus guidelines for the clinical and pathologic diagnosis of dementia with Lewy bodies (DLB): report of the consortium on DLB international workshop. Neurology 1996;47:1113–1124.
65. McKeith I, Fairbairn A, et al. Neuroleptic sensitivity in patients with senile dementia of Lewy body type. BMJ 1992;305:673–678.
66. Lopez OL, Litvan I, et al. Accuracy of four clinical diagnostic criteria for the diagnosis of neurodegenerative dementias. Neurology 1999;53:1292–1299.
67. McKeith IG, Fairbairn AF, et al. An evaluation of the predictive validity and inter-rater reliability of clinical diagnostic criteria for senile dementia of Lewy body type. Neurology 1994; 44:872–877.
68. Mega MS, Masterman DL, et al. Dementia with Lewy bodies: reliability and validity of clinical and pathologic criteria. Neurology 1996;47:1403–1409.
69. Ince PG, Perry EK, Morris CM. Dementia with Lewy bodies. A distinct non-Alzheimer dementia syndrome. Brain Pathol 1998;8:299–324.
70. Gomez-Tortosa E, Newell K, et al. Clinical and quantitative pathologic correlates of dementia with Lewy bodies. Neurology 1999;53:1284–1291.
71. Hansen L, Salmon D, et al. The Lewy body variant of Alzheimer's disease: a clinical and pathologic entity. Neurology 1990; 40:1–8.
72. Perry RH, Irving D, et al. Senile dementia of Lewy body type: a clinically and neuropathologically distinct form of Lewy body dementia in the elderly. J Neurol Sci 1990a;95:119–139.
73. Langlais PJ, Thal L, et al. Neurotransmitters in basal ganglia and cortex of Alzheimer's disease with and without Lewy bodies. Neurology 1993;43:1927–1934.
74. Tiraboschi P, Hansen LA, et al. Cholinergic dysfunction in diseases with Lewy bodies. Neurology 2000;54:407–411.
75. Gomez-Tortosa E, Ingraham AO, et al. Dementia with Lewy bodies. J Am Geriatr Soc 1998;46:1449–1458.
76. Downes JJ, Priestley NM, et al. Intellectual, mnemonic, and frontal functions in dementia with Lewy bodies: a comparison with early and advanced Parkinson's disease. Behav Neurol 1998/1999;11:173–183.
77. Shimomura T, Mori E, et al. Cognitive loss in dementia with Lewy bodies and Alzheimer disease. Arch Neurol 1998;55: 1547–1552.
78. Ballard C, Holmes C, et al. Psychiatric morbidity in dementia with Lewy bodies: a prospective clinical and neuropathological comparative study with Alzheimer's disease. Am J Psychiatry 1999;156:1039–1045.
79. Weiner MF, Risser RC, et al. Alzheimer's disease and its Lewy body variant: a clinical analysis of postmortem verified cases. Am J Psychiatry 1996;153:1269–1273.
80. Hirono N, Mori E, et al. Distinctive neurobehavioral features among neurodegenerative dementias. J Neuropsychiatry Clin Neurosci 1999;11:498–503.
81. Boeve BF, Silber MH, et al. REM sleep behavior disorder and degenerative dementia: an association likely reflecting Lewy body disease. Neurology 1998;51:363–370.
82. Ferman TJ, Boeve BF, et al. REM sleep behavior disorder and dementia: cognitive differences when compared with AD. Neurology 1999;52:951–957.
83. Louis ED, Klatka LS, et al. Comparison of extrapyramidal features in 31 pathologically confirmed cases of diffuse Lewy body disease and 34 pathologically confirmed cases of Parkinson's disease. Neurology 1997;48:376–380.
84. Barber R, Gholkar A, et al. Medial temporal lobe atrophy in MRI in dementia with Lewy bodies. Neurology 1999;52: 1153–1158.
85. Hashimoto M, Kitagaki H, et al. Medial temporal and whole-brain atrophy in dementia with Lewy bodies: a volumetric MRI study. Neurology 1998;51:357–362.
86. Albin RL, Minoshima S, et al. Fluoro-deoxyglucose positron emission tomography in diffuse Lewy body disease. Neurology 1996;47:462–466.
87. Hu XS, Okamura N, et al. ^{18}F-fluorodopa PET study of striatal dopamine uptake in the diagnosis of dementia with Lewy bodies. Neurology 2000;55:1575–1576.
88. Walker Z, Costa DC, et al. Dementia with lewy bodies: a study of post-synaptic dopaminergic receptors with iodine-123 iodobenzamide single-photon emission tomography. Eur J Nucl Med 1997;24:609–614.
89. Cheng AVT, Ferrier IN, et al. Cortical serotonin-S_2 receptor binding in Lewy body dementia, Alzheimer's and Parkinson's diseases. J Neurol Sci 1991;106:50–55.
90. Perry EK, Marshall E, et al. Evidence of a monoaminergic-cholinergic imbalance related to visual hallucinations in Lewy body dementia. J Neurochem 1990b;55:1454–1456.
91. Kaufer DI, Catt KE, et al. Dementia with Lewy bodies: response of delirium-like features to donepezil. Neurology 1998;51: 1512.
92. McKeith I, Del Ser T, et al. Efficacy of rivastigmine in dementia with Lewy bodies: a randomised, double-blind, placebo-controlled international study. Lancet 2000;356:2031–2036.
93. Chacko RC, Hurley RA, Jankovic J. Clozapine use in diffuse Lewy body disease. J Neuropsychiatry Clin Neurosci 1993;5: 206–208.
94. Walker Z, Grace J, et al. Olanzapine in dementia with lewy bodies: a clinical study. Int J Geriatr Psychiatry 1999;14:459–466.
95. Steele JC. Progressive supranuclear palsy. Brain 1972;95:693–704.
96. Albert ML, Feldman RG, Willis AL. The 'subcoritcal dementia' of progressive supranuclear palsy. J Neurol Neurosurg Psychiatry 1974;37:121–130.
97. Grafman J, Litvan I, et al. Frontal lobe function in progressive supranuclear palsy. Arch Neurol 1990;47:553–558.
98. Milberg W, Alber M. Cognitive differences between patients with progressive supranuclear palsy and Alzheimer's disease. J Clin Exp Neuropsychol 1989;11:605–614.
99. Pillon B, Dubois B, et al. Heterogeneity of cognitive impairment in progressive supranuclear palsy, Parkinson's disease, and Alzheimer's disease. Neurology 1986;36:1179–1185.

100. Litvan I, Agid Y, et al. Clinical research criteria for the diagnosis of progressive supranuclear palsy (Steele-Richardson-Olszewski syndrome): report of the NINDS-SPSP international workshop. Neurology 1996;47:1–9.
101. Ovsiew F, Schneider J. Schizophrenia and atypical motor features in a case of progressive supranculear palsy (the Steele-Richardson-Olszewski syndrome). Behav Neurol 1993;6:243–247.
102. Ambrosetto P. CT in progressive supranuclear palsy. Am J Neuroradiol 1987;8:849–851.
103. Foster NL, Gilman S, et al. Cerebral hypometabolism in progressive supranuclear palsy studied with positron emission tomography. Ann Neurol 1988;24:399–406.
104. Brooks DJ, Ibanez V, et al. Differing patterns of striatal ^{18}f-dopa uptake in Parkinson's disease, multiple system atrophy, and progressive supranuclear palsy. Ann Neurol 1990;28:547–555.
105. Ilgin N, Zubieta J, et al. PET imaging of the dopamine transporter in progressive supranuclear palsy and Parkinson's disease. Neurology 1999;52:1221–1226.
106. Hauw JJ, Daniel SE, et al. Preliminary NINDS neuropathologic criteria for Steele-Richardson-Olszewski syndrome (progressive supranuclear palsy). Neurology 1994;44:2015–2019.
107. Steele JC, Richardson JC, Olszewski J. Progressive supranuclear palsy: a heterogenous degeneration involving the brain stem, basal ganglia, and cerebellum with vertical gaze and pseudobulbar palsy, muchal dystonia and dementia. Arch Neurol 1964;10:333–359.
108. Ruberg M, Javoy-Agid F, et al. Dopaminergic and cholinergic lesions in progressive supranuclear palsy. Ann Neurol 1985;18: 523–529.
109. Dubinsky RM, Jankovic J. Progressive supranuclear palsy and a multi-infarct state. Neurology 1987;37:570–576.
110. Litvan I, Mega MS, et al. Neuropsychiatric aspects of progressive supranuclear palsy. Neurology 1996;47:1184–1189.
111. Engel PA. Treatment of progressive supranuclear palsy with amitriptyline: therapeutic and toxic effects. J Am Geriatr Soc 1996;44:1072–1074.
112. Kompoliti K, Goetz CG, et al. Pharmacological therapy in progressive supranuclear palsy. Arch Neurol 1998;55:1099–1102.
113. Rafal RD, Grimm RJ. Progressive supranuclear palsy: functional analysis of the response to methysergide and antiparkinsonian agents. Neurology 1981;31:1507–1518.
114. Litvan I, Blesa R, et al. Pharmacological evaluation of the cholinergic system in progressive supranuclear palsy. Ann Neurol 1994;36:55–61.
115. Polo KB, Jabbari B. Botulinum toxin-A improves the rigidity of progressive supranuclear palsy. Ann Neurol 1994;35:237–239.
116. Riley DE, Lang AE. Clinical diagnostic criteria. In: Litvan I, Goetz CG, Lang AE, eds. Corticobasal Degeneration and Related Disorders. Philadelphia: Lippincott & Williams, 2000:29–34.
117. Thompson PD, Shibasaki H. Myoclonus in corticobasal degeneration and other neurodegenerations. In: Litvan I, Goetz CG, Lang AE, eds. Corticobasal Degeneration: Advances in Neurology. Philadelphia: Lippincott & Williams, 2000:69–81.
118. Leiguarda R, Merello M, Balej J. Apraxia. In: Litvan I, Goetz CG, Lang AE, eds. Corticobasal Degeneration and Related Disorders. Philadelphia: Lippincott Williams & Wilkins, 2000: 103–121.
119. Mimura M, White RF, Albert MLM. Corticobasal degeneration: neuropsychological and clinical correlates. J Neuropsychiatry Clin Neurosci 1997;9:94–98.
120. Pillon B, Blin J, et al. The neuropsychological pattern of corticobasal degeneration: comparison with progressive supranuclear palsy and Alzheimer's disease. Neurology 1995;45:1477–1483.
121. Litvan I, Cummings JL, Mega M. Neuropsychiatric features of corticobasal degeneration. J Neurol Neurosurg Psychiatry 1998a;65:717–721.
122. Litvan I, Agid Y, et al. Accuracy of the clinical diagnosis of corticobasal degeneration: a clinicopathologic study. Neurology 1997;48:119–125.
123. Eidelberg D, Dhawan V, et al. The metabolic landscape of cortico-basal ganglionic degeneration: regional asymmetries studied with positron emission tomography. J Neurol Neurosurg Psychiatry 1991;54:856–862.
124. Markus HS, Lees AJ, et al. Patterns of regional cerebral blood flow in corticobasal degeneration studied using HMPAO SPECT: comparison with Parkinson's disease and normal controls. Mov Disord 1995;10:179–187.
125. Nagahama Y, Fukuyama J, et al. Cerebral glucose metabolism in corticobasal degeneration: comparison with progressive supranuclear palsy and normal controls. Mov Disord 1997;12: 691–696.
126. Dickson DW, Liu WK, et al. Neuropathologic and modern consideration. In: Litvan I, Goetz CG, Lang AE, eds. Corticobasal Degeneration. Advances in Neurology. Philadelphia: Lippincott & Williams, 2000:9–27.
127. Schneider JA, Watts RL, et al. Corticobasal degeneration: neuropathological and clinical heterogeneity. Neurology 1997;48: 959–969.
128. Cummings JL, Gosenfeld LF, et al. Neuropsychiatric disturbances associated with idiopathic calcification of the basal ganglia. Biol Psychiatry 1983;18:591–601.
129. Trautner RJ, Cummings JL, et al. Idiopathic basal ganglia calcification and organic mood disorder. Am J Psychiatry 1988; 145:350–353.
130. Koenig M, Durr A. Friedreich's ataxia. In: Klockgether T, ed. Handbook of Ataxia Disorders. New York: Marcel Dekker, 2000:151–161.
131. Lopez-Villegas D, Kulisevsky J, et al. Neuropsychological alterations in patients with computed tomography–detected basal ganglia calcification. Arch Neurol 1996;53:251–256.
132. Eidelberg D, Sotrel A, et al. Adult onset Hallervorden-Spatz disease with neurofibrillary pathology: a discrete clinicopathological entity. Brain 1987;110:993–1013.
133. Sethi KD, Adams RJ, et al. Hallervorden-Spatz syndrome: clinical magnetic resonance imaging correlations. Ann Neurol 1988;24:692–694.
134. Arawaka S, Saito Y, et al. Lewy body in neurodegeneration with brain iron accumulation type 1 is immunoreactive for a-synuclein. Neurology 1998;51:887–889.
135. Dooling EC, Richardson EP, Davis KR. Computed tomography in Hallervorden-Spatz disease. Neurology 1980;30:1128–1130.
136. Littrup PJ, Gebarski SS. MR imaging of Hallervorden-Spatz disease. J Comput Assist Tomogr 1985;9:491–493.
137. Murrow RW, Schweiger GD, et al. Parkinsonism due to a basal ganglia lacunar state: clinicopathologic correlation. Neurology 1990;40:897–900.
138. Winikates J, Jankovic J. Clinical correlates of vascular parkinsonism. Arch Neurol 1999;56:98–102.
139. Wolfe N, Linn R, et al. Frontal systems impairment following multiple lacunar infarcts. Arch Neurol 1990;47:129–132.
140. Dian L, Cummings JL, et al. Personality alterations in multi-infarct dementia. Psychosomatics 1990;31:415–419.

141. Alexopoulos GS, Meyers BS, et al. 'Vascular depression' hypothesis. Arch Gen Psychiatry 1997;54:915–922.
142. Sultzer DL, Levin HS, et al. A comparison of psychiatric symptoms in vascular dementia and Alzheimer's disease. Am J Psychiatry 1993;150:1806–1812.
143. Zijlmans JCM, Thijssen HOM, et al. MRI in patients with suspected vascular parkinsonism. Neurology 1995;45:2183–2188.
144. Concensus Committee of the American Autonomic Society and the American Academy of Neurology. Consensus statement on the definition of orthostatic hypotension, pure autonomic failure, and multiple system atrophy. Neurology 1996;46: 1470.
145. Litvan I, Goetz CG, et al. What is the accuracy of the clinical diagnosis of multiple system atrophy? A clinicopathologic study. Arch Neurol 1997b;54:937–944.
146. Robbins TW, James M, et al. Cognitive performance in multiple system atrophy. Brain 1992;115:271–291.
147. Pilo L, Ring H, et al. Depression in multiple system atrophy and in idiopathic Parkinson's disease: a pilot comparative study. Biol Psychiatry 1996;39:803–807.
148. Kraft E, Schwarz J, et al. The combination of hypointense and hyperintense signal changes on T_2-weighted magnetic resonance imaging sequences. Arch Neurol 1999;56:225–228.
149. De Volder AG, Francart J, et al. Decreased glucose utilization in the striatum and frontal lobe in probable stratonigral degeneration. Ann Neurol 1989;26:239–247.
150. Wenning GK, Tison F, et al. Multiple system atrophy: a review of 203 pathologically proven cases. Mov Disord 1997;12:133–147.
151. Giasson BI, Duda JE, et al. Oxidative damage linked to neurodegeneration by selective a-synuclein nitration in synucleinopathy lesions. Science 2000;290:985–989.
152. Wright RA, Kaufmann HC, et al. A double-blind, dose-response study of midodrine in neurogenic orthostatic hypotension. Neurology 1998;51:120–124.
153. McGeer PL, Schwab C, et al. Familial nature and continuing morbidity of the amyotrophic lateral sclerosis-parkinsonism dementia complex of Guam. Neurology 1997;49:400–409.
154. Cheyette S, Cummings JL. Encephalitis lethargica: lessons for contemporary neuropsychiatry. J Neuropsychiatry Clin Neurosci 1995;7:125–134.
155. Rosenberg RN, Prusiner SB, et al. Clinical Companion to the Molecular and Genetic Basis of Neurological Disease. Boston: Butterworth Heinemann, 1998.
156. Hoogenraad TU. Wilson's Disease. Philadelphia: W.B. Saunders, 1996.
157. Starosta-Rubinstein S, Young AB, et al. Clinical assessment of 31 patients with Wilson's disease. Arch Neurol 1987;44:365–370.
158. Dening TR. Psychiatric aspects of Wilson's disease. Br J Psychiatry 1985;147:677–682.
159. Dening TR, Berrios GE. Wilson's disease. Arch Gen Psychiatry 1989;46:1126–1134.
160. Akil M, Schwartz JA, et al. The psychiatric presentations of Wilson's disease. J Neuropsychiatry Clin Neurosci 1991;3:377–382.
161. Isaacs-Glaberman K, Medalia A, Scheinberg IH. Verbal recall and recognition abilities in patients with Wilson's disease. Cortex 1989;25:353–361.
162. Medalia A, Isaacs-Glaberman K, Scheinberg IH. Neuropsychological impairment in Wilson's disease. Arch Neurol 1988;45:502–504.
163. Williams JB, Walshe JM. Wilson's disease: an analysis of the cranial computerized tomographic appearances found in 60 patients and the changes in reponse to treatment with chelating agents. Brain 1981;104:735–752.
164. Roh JK, Lee TG, et al. Initial and follow-up brain MRI findings and correlation with the clinical course in Wilson's disease. Neurology 1994;44:1064–1068.
165. Hawkins RA, Mazziotta JC, Phelps ME. Wilson's disease studied with FDG and positron emission tomography. Neurology 1987;37:1707–1711.
166. Snow BJ, Bhatt M, et al. The nigrostriatal dopaminergic pathway in Wilson's disease studied with positron emission tomography. J Neurol Neurosurg Psychiatry 1991;54:12–17.
167. LeWitt PA, Brewer GJ. Wilson's disease (progressive hepatolenticular degeneration). In: Calne DB, ed. Neurodegenerative Diseases. Philadelphia: W.B. Saunders, 1994:667–683.
168. Gwinn-Hardy KA. Wilson's disease. In: Adler CH, Ahlskog JE, eds. Parkinson's Disease and Movement Disorders: Diagnosis and Treatment Guidelines for the Practicing Physician. Totowa, NJ: Humana Press, 2000:397–410.
169. Brewer GJ, Turkay A, Yuzbasiyan-Gurkan V. Development of neurologic symptoms in a patient with asymptomatic Wilson's disease treated with penicillamine. Arch Neurol 1994;51:304–305.
170. Myers RH, Marans KS, MacDonald ME. Huntington's disease. In: Wells RD, Warren ST, eds. Genetic Instabilities and Hereditary Neurological Diseases. San Diego: Academic Press, 1998: 301–323.
171. Brandt J, Bylsma FW, et al. Trinucleotide repeat length and clinical progression in Huntington's disease. Neurology 1996; 46:527–531.
172. Caviness JN. Huntington's disease and other choreas. In: Adler CH, Ahlskog JE, eds. Parkinson's Disease and Movement Disorders. Totowa, NJ: Humana Press, 2000:321–330.
173. Currie J, McArthur C, et al. High-resolution eye movement recording in the early diagnosis of Huntington's disease. Neuropsychiatry Neuropsychol Behav Neurol 1992;5:46–52.
174. Lasker AG, Zee DS, et al. Saccades in Huntington's disease: slowing and dysmetria. Neurology 1988;38:427–431.
175. Folstein SE. Huntington's Disease: A Disorder of Families. Baltimore: Johns Hopkins University Press, 1989.
176. Caine ED, Shoulson I. Psychiatric syndromes in Huntington's disease. Am J Psychiatry 1983;140:728–733.
177. Folstein SE, Abbott MG, et al. The association of affective disorder with Huntington's disease in a case series and in families. Psychol Med 1983;13:537–542.
178. Burns A, Folstein S, et al. Clinical assessment of irritability, aggression, and apathy in Huntington and Alzheimer disease. J Nerv Ment Dis 1990;178:20–26.
179. Shiwach RS, Patel V. Aggressive behaviour in Huntington's disease: a cross-sectional study in a nursing home population. Behav Neurol 1993;6:43–47.
180. Litvan I, Paulsen JS, et al. Neuropsychiatric assessment of patients with hyperkinetic and hypokinetic movement disorders. Arch Neurol 1998b;55:1313–1319.
181. Cummings JL, Cunningham K. Obsessive-compulsive disorder in Huntington's disease. Biol Psychiatry 1992b;31:263–270.
182. Nance MA, Sanders G. Characteristics of individuals with Huntington disease in long-term care. Mov Disord 1996;11: 542–548.
183. Federoff JP, Peyser C, et al. Sexual disorders in Huntington's disease. J Neuropsychiatry Clin Neurosci 1994;6:147–153.
184. Schoenfeld M, Myers RH, et al. Increased rate of suicide among patients with Huntington's disease. J Neurol Neurosurg Psychiatry 1984;47:1283–1287.
185. Zappacosta B, Monza D, et al. Psychiatric symptoms do not

correlate with cognitive decline, motor symptoms, or CAG repeat length in Huntington's disease. Arch Neurol 1996;53:493–497.

186. Baxter LR, Mazziotta JC, et al. Psychiatric, genetic, and positron emission tomographic evaluation of persons at risk for Huntington's disease. Arch Gen Psychiatry 1992;49:148–154.
187. Mayberg HS, Starkstein SE, et al. Paralimbic frontal lobe hypometabolism in depression associated with Huntington's disease. Neurology 1992;42:1791–1797.
188. Kuwert T, Lange HW, et al. Cerebral glucose consumption measured by PET in patients with and without psychiatric symptoms of Huntington's disease. Psychiatry Res 1989;29: 361–362.
189. Morris M, Scourfield J. Psychiatric aspects of Huntington's disease. In: Harper PS, ed. Huntington's Disease, 2nd ed. Philadelphia: W.B. Saunders, 1996:73–121.
190. Hodges JR, Salmon DP, Butters N. Differential impairment of semantic and episodic memory in Alzheimer's and Huntington's diseases: a controlled prospective study. J Neurol Neurosurg Psychiatry 1990;53:1089–1095.
191. Lange KW, Sahakian BJ, et al. Comparison of executive and visuospatial memory function in Huntington's disease and dementia of Alzheimer's type matched for degree of dementia. J Neurol Neurosurg Psychiatry 1995;58:598–606.
192. Hodges JR, Salmon DP, Butters N. The nature of the naming deficit in Alzheimer's and Huntington's disease. Brain 1991; 114:1547–1558.
193. Lawrence AD, Sahakian BJ, et al. Executive and mnemonic functions in early Huntington's disease. Brain 1996;119:1633–1645.
194. Rich JB, Bylsma FW, Brandt J. Self-ordered pointing performance in Huntington's disease patients. Neuropsychiatry Neuropsychol Behav Neurol 1996;9:99–106.
195. Rothlind JC, Bylsma FW, et al. Cognitive and motor correlates of everyday functioning in early Huntington's disease. J Nerv Ment Dis 1993;181:194–199.
196. Jason GW, Suchowersky O, et al. Cognitive manifestations of Huntington disease in relation to genetic structure and clinical onset. Arch Neurol 1997;54:1081–1088.
197. Jernigan T, Salmon DP, et al. Cerebral structure on MRI, Part II: specific changes in Alzheimer's and Huntington's diseases. Biol Psychiatry 1991;29:68–81.
198. Hayden MR, Martin WRW, et al. Positron emission tomography in the early diagnosis of Huntington's disease. Neurology 1986;36:888–894.
199. Young AB, Penney JB, et al. PET scan investigations of Huntington's disease: cerebral metabolic correlates of neurological features and functional decline. Ann Neurol 1986;20:296–303.
200. Kuwert T, Lange HW, et al. Cortical and subcortical glucose consumption measured by PET in patients with Huntington's disease. Brain 1990;113:1405–1423.
201. Sax DS, Powsner R, et al. Evidence of cortical metabolic dysfunction in early Huntington's disease by single-photon-emission computed tomography. Mov Disord 1996;11:671–677.
202. Greenamyre JT, Shoulson I. Huntington's disease. In: Calne DB, ed. Neurodegenerative Diseases. Philadelphia: W.B. Saunders, 1994:685–704.
203. Vonsattel JP, Myers RH, et al. Neuropathological classification of Huntington's disease. J Neuropathol Exp Neurology 1985; 44:559–577.
204. Albin RL, Reiner A, et al. Striatal and nigral neuron subpopulations in rigid Huntington's disease: implications for the functional anatomy of chorea and rigidity-akinesia. Ann Neurol 1990;27:357–365.
205. DiFiglia M, Sapp E, et al. Aggregation of huntingtin in neuronal intranuclear inclusions and dystrophic neurites in brain. Science 1997;277:1990–1993.
206. Sajatovic M, Verbanac P, et al. Clozapine treatment of psychiatric symptoms resistant to neuroleptic treatment in patients with Huntington's chorea. Neurology 1991;41:156.
207. Byrne A, Martin W, Hnatko G. Beneficial effects of buspirone therapy in Huntington's disease. Am J Psychiatry 1994;151: 1097.
208. Ranen NG, Lipsey JR, et al. Sertraline in the treatment of severe aggressiveness in Huntington's disease. J Neuropsychiatry Clin Neurosci 1996;8:338–340.
209. Lewis CF, DeQuardo JR, Tandon R. ECT in genetically confirmed Huntington's disease. J Neuropsychiatry Clin Neurosci 1996;8:209–210.
210. Ranen NG, Peyser CE, Folstein SE. ECT as a treatment for depression in Huntington's disease. J Neuropsychiatry Clin Neurosci 1994;6:154–159.
211. Bird TD, Cederbaum S, et al. Familial degeneration of the basal ganglia with acanthocytosis: a clinical, neuropathological, and neurochemical study. Ann Neurol 1978;3:253–258.
212. Hardie RJ, Pullon HWH, et al. Neuroacanthocytosis: a clinical, haematological and pathological study of 19 cases. Brain 1991;114:13–49.
213. Spitz MC, Jankovic J, Killian JM. Familial tic disorder, parkinsonism, motor neuron disease, and acanthocytosis: a new syndrome. Neurology 1985;35:366–370.
214. Kartsounis LD, Hardie RJ. The pattern of cognitive impairments in neuroacanthocytosis. Arch Neurol 1996;53:77–80.
215. Wyszynski B, Merriam A, et al. Choreoacanthocytosis: report of a case with psychiatric features. Neuropsychiatry Neuropsychol Behav Neurol 1989;2:137–144.
216. Delecluse F, Deleval J, et al. Frontal impairment and hypoperfusion in neuroacanthosytosis. Arch Neurol 1991;48:232–234.
217. Medalia A, Merriam A, Sandberg M. Neuropsychological deficits in choreoacanthocytosis. Arch Neurol 1989;46:573–575.
218. Feinberg TE, Cianci CD, et al. Diagnostic tests for choreoacanthocytosis. Neurology 1991;41:1000–1006.
219. Cardoso F, Eduardo C, et al. Chorea in fifty consecutive patients with rheumatic fever. Mov Disord 1997;12:701–702.
220. Trinidad KS, Kurlan R. Chorea, athetosis, dystonia, tremor and parkinsonism. In: Robertson MM, Eapen V, eds. Movement and Allied Disorders. New York: John Wiley & Sons, 1995: 105–147.
221. Swedo SE, Leonard HL, et al. Identification of children with pediatric autoimmune neuropsychiatric disorders associated with streptococcal infections by a marker associated with rheumatic fever. Am J Psychiatry 1997;154:110–112.
222. Giedd JN, Rapoport JL, et al. Sydenham's chorea: magnetic resonance imaging of the basal ganglia. Neurology 1995;45: 2199–2202.
223. Goldman S, Amrom D, et al. Reversible striatal hypermetabolism in a case of Sydenham's chorea. Mov Disord 1993;8:355–358.
224. Wilcox JA, Nasrallah HA. Sydenham's chorea and psychosis. Neuropsychobiology 1986;15:13–14.
225. Wilcox JA, Nasrallah H. Sydenham's chorea and psychopathology. Neuropsychobiology 1988;19:6–8.
226. Bressman SB, de Leon D, et al. Clinical–genetic spectrum of

primary dystonia. In: Fahn S, Marsden CD, DeLong M, eds. Dystonia 3: Advances in Neurology. Philadelphia: Lippincott-Raven, 1998:79–91.
227. Bressman SB, Sabatti C, et al. The DYT1 phenotype and guidelines for diagnostic testing. Neurology 2000;54:1746–1752.
228. Eidelberg D. Abnormal brain networks in DYT1 dystonia. In: Fahn S, Marsden CD, DeLong M, eds. Dystonia 3: Advances in Neurology. Philadelphia: Lippincott-Raven, 1998:127–133.
229. Ceballos-Baumann AO, Brooks DJ. Activation positron emission tomography scanning in dystonia. In: Fahn S, Marsden CD, DeLong M, eds. Dystonia 3: Advances in Neurology. Philadelphia: Lippincott-Raven, 1998:135–152.
230. Grafman J, Cohen LG, Hallett M. Is focal hand dystonia associated with psychopathology? Mov Disord 1991;6:29–35.
231. Jahanshahi M, Marsden CD. Personality in torticollis: a controlled study. Psychol Med 1988;18:375–387.
232. Harrington RC, Wieck A, et al. Writer's cramp: not associated with anxiety. Mov Disord 1988;3:195–200.
233. Scheidt CE, Heinen F, et al. Spasmodic torticollis—a multicentre study on behavioural aspects, IV: psychopathology. Behav Neurol 1996;9:97–103.
234. Lauterbach EC, Spears TE, Price ST. Bipolar disorder in idiopathic dystonia: clinical features and possible neurobiology. J Neuropsychiatry Clin Neurosci 1992;4:435–439.
235. Dubourg O, Durr A, et al. Analysis of the SCA1 CAG repeat in a large number of families with dominant ataxia: clinical and molecular correlations. Ann Neurol 1995;37:176–180.
236. Genis D, Matilla T, et al. Clinical, neuropathologic, and genetic studies of a large spinocerebellar ataxia type 1 (SCA1) kindred: $(CAG)_n$ expansion and early premonitory signs and symptoms. Neurology 1995;45:24–30.
237. Khati C, Stevanin G, et al. Genetic heterogeneity of autosomal dominant cerebellar ataxia type 1: clinical and genetic analysis of 10 French families. Neurology 1993;43:1131–1137.
238. Schols L, Amioridis G, et al. Autosomal dominant cerebellar ataxia: phenotypic differences in genetically defined subtypes? Ann Neurol 1997a;42:924–932.
239. Storey E, Forrest SM, et al. Spinocerebellar ataxia type 2. Arch Neurol 1999;56:43–50.
240. Schols L, Gispert S, et al. Spinocerebellar ataxia type 2. Arch Neurol 1997;54:1073–1080.
241. Ikeuchi T, Takano H, et al. Spinocerebellar ataxia type 6: CAG repeat expansion in a_{1A} voltage-dependent calcium channel gene and clinical variations in Japanese population. Ann Neurol 1997;42:879–884.
242. Benton CS, de Silva R, et al. Molecular and clinical studies in SCA-7 define a broad clinical spectrum and the infantile phenotype. Neurology 1998;51:1081–1086.
243. Becher MW, Rubinsztein DC, et al. Dentatorubral and pallidoluysian atrophy (DRPLA). Mov Disord 1997;12:519–530.
244. Warner TT, Williams LD, et al. A clinical and molecular genetic study of dentatorubropallidoluysian atrophy in four European families. Ann Neurol 1995;37:452–259.
245. Geschwind DH, Perlman S, et al. Friedreich's ataxia GAA repeat expansion in patients with recessive or sporadic ataxia. Neurology 1997;49:1004–1009.
246. Fehrenbach RA, Wallesch CW, Claus D. Neuropsychologic findings in Friedreich's ataxia. Arch Neurol 1984;41:306–308.
247. Hart RP, Kwentus JA, et al. Information processing speed in Friedreich's ataxia. Ann Neurol 1985;17:612–614.
248. Koller WC. Treatment of tremor disorders. In: Kurlan R, ed. Treatment of Movement Disorders. Philadelphia: J.B. Lippincott, 1995:407–427.
249. Matsumoto JY. Tremor disorders: overview. In: Adler CH, Ahlskog JE, eds. Parkinson's Disease and Movement Disorders: Diagnosis and Treatment Guidelines for the Practicing Physician. Totowa, NJ: Humana Press, 2000:273–281.
250. Alexander GE, DeLong MR, Strick PL. Parallel organization of functionally segregated circuits linking basal ganglia and cortex. Ann Rev Neurosci 1986;9:357–381.
251. Alexander GE, Crutcher MD, DeLong MR. Basal ganglia-thalamocortical circuits: parallel substrates for motor, oculomotor, "prefrontal" and "limbic" functions. Progress in Brain Research. 1990;85:119–146.
252. Parent A, Hazrati LN. Functional anatomy of the basal ganglia, I: The cortico-basal ganglia-thalamo-cortical loop. Brain Res 1995;20:91–127.
253. Middleton FA, Strick PL. A revised neuroanatomy of frontal-subcortical circuits. In: Lichter DG, Cummings JL, eds. Frontal-Subcortical Circuits in Psychiatric and Neurological Disorders. New York: The Guilford Press, 2000:44–58.
254. Alexander GE, Crutcher MD. Functional architecture of basal ganglia circuits: neural substrates of parallel processing. Trends Neurosci 1990;13:260–276.
255. Cummings JL. Frontal–subcortical circuits and human behavior. Arch Neurol 1993;50:873–880.
256. Mega MS, Cummings JL. Frontal–subcortical circuits and neuropsychiatric disorders. J Neuropsychiatry Clin Neurosci 1994;6:358–370.
257. Middleton FA, Strick PL. The temporal lobe is a target of output from the basal ganglia. Proc Natl Acad Sci USA 1996;93:8683–8687.
258. Graybiel AM. Neurotransmitters and neuromodulators in the basal gangia. Trends Neurosci 1990;13:244–254.
259. McKeith IG, Perry EK, Perry RG. Report of the second dementia with Lewy body international workshop: diagnosis and treatment: Consortium on Dementia with Lewy Bodies. Neurology 1999;53:902–905.

Tics, Startle Syndromes, and Myoclonus

Tics, startle syndromes, and myoclonus share the common phenomenological feature of being brief, rapid, hyperkinetic disorders. Tic syndromes are among the most complex of neuropsychiatric disorders and are considered in greatest detail in this chapter. Startle syndromes are novel and uncommon conditions. Myoclonic disorders are relatively nonspecific indicators of brain dysfunction. An overall approach to movement disorders and a discussion of other conditions with hyperkinetic manifestations can be found in Chapter 18. Obsessive-compulsive disorder (OCD) seen with Gilles de la Tourette syndrome (GTS) and self-injurious behaviors also sometimes seen in GTS are discussed in the chapter devoted to OCD (Chapter 16).

TICS AND GILLES DE LA TOURETTE SYNDROME

Tics are brief, involuntary movements, vocalizations, or sensations that are recurrent, repetitive, and stereotyped. They are not rhythmic and tend to occur at irregular intervals. They are purposeless, inappropriate, and usually experienced as irresistible, although they can be suppressed volitionally for varying periods of time.[1] Tics may be single or multiple and may be simple or complex. They tend to change over time in their severity, location, and complexity. Tics are typically decreased by distraction and increased by calling attention to them.

Tic disorders can be primary or secondary to a variety of other conditions (Table 19.1). The primary tic disorders to be considered include transient tic disorder, chronic motor or vocal tic disorder, adult-onset tic disorder, and GTS.

Transient tic disorder consists of single or multiple motor or vocal tics that occur many times per day, nearly every day, for at least 4 weeks, but persist for no longer than 12 consecutive months. The disorder begins before the age of 18 and typically occurs in early childhood. It affects approximately 12% of children.[2,3] *Chronic motor or vocal tic disorder* is characterized by single or multiple motor or vocal tics but not both, occurring several times a day, nearly every day or intermittently throughout a period of more than 1 year. This disorder has its onset before the age of 18. *Adult tic disorder* is similar to chronic motor or vocal tic disorder but begins later in life, usually after age 30.[2,3] *Gilles de la Tourette syndrome* is characterized by multiple motor plus one or more vocal tics that occur many times per day, nearly every day for more than a year. The onset of the disorder is before the age of 18 and

TABLE 19.1. *Etiologic Classification of Tics*

Primary tic disorders
Transient tic disorder
Chronic motor or vocal tic disorder
Adult-onset tic disorder
Gilles de la Tourette syndrome
Secondary tic disorders
Chromosomal abnormalities
Down's syndrome
Fragile X syndrome
18q22 translocations
XYY male
XXX + 9p mosaicism
Lesch-Nyhan disease
Developmental
Autistic syndrome
Rett syndrome
Pervasive developmental delay
Static encephalopathies
Degenerative
Neuroacanthocytosis
Huntington's disease
Wilson's disease
Infectious/post-infectious
PANDAS
Sydenham's chorea
Encephalitis lethargica and post-encephalitic disorders
Encephalitis
Rubella syndrome
Drug-induced or drug-precipitated
Neuroleptics (tardive Tourette's syndrome)
Stimulants (pemoline, dextroamphetamine, methylphenidate, cocaine)
Levodopa and related dopaminergic agents
Anticonvulsants
Other
Carbon monoxide poisoning
Traumatic brain injury
Stroke

PANDAS, pediatric autoimmune neuropsychiatric disorders after streptococcal infection.

the mean age of onset is approximately age 7 years (Table 19.2).[2]

Secondary tics occur in a variety of neurological disorders (Table 19.1). Hereditary and chromosomal abnormalities can produce tic syndromes, and developmental disorders such as the autistic disorders (autism, Asperger's syndrome), Rett's syndrome, pervasive developmental delay, and static encephalopathies have all been associated with tic disorders. Neuroacanthocytosis is a neurodegenerative disease affecting primarily the caudate nuclei and producing tics, chorea, and a dementia syndrome. The disorder typically begins in early adulthood and is associated with a peripheral neuropathy. Huntington's disease and Wilson's disease have also been associated with tics. Among infectious and post-infectious syndromes associated with tics are Sydenham's chorea, encephalitis lethargica, viral encephalitis, and rubella syndromes. A variety of medications have been noted to induce tics including chronic treatment with neuroleptics (tardive Tourette's syndrome), levodopa and related dopaminergic agents, and anticonvulsants. Stimulants such as pemoline, dextroamphetamine, methylphenidate, and cocaine may either induce tics or precipitate them in predisposed individuals. Given the high rate of attention-deficit hyperactivity disorder in GTS (discussed below), it is not unusual that the tics are first observed in children treated with stimulants for their attention deficit disorder. Other conditions that have been associated with tic syndromes include carbon monoxide poisoning, traumatic brain injury, and stroke.[4]

TABLE 19.2. *Diagnostic Criteria for Gilles de la Tourette's Syndrome*

Multiple motor and vocal tics
Tics occur many times per day nearly every day for more than 1 year
Tics cause marked distress or significant impairment in social or occupational function
Onset before age 18 years
Tics are not due to direct effects of a substance (e.g., stimulants) or a general medical condition (e.g., Huntington's disease)

Source: Adapted from the *Diagnostic and Statistical Manual of Mental Disorders, 4th ed.*[2]

GILLES DE LA TOURETTE SYNDROME

Gilles de la Tourette Syndrome is a lifelong tic disorder characterized by multiple motor and vocal tics that wax and wane over time (Table 19.2). The disorder has been recognized since 1885, when Georges Albert Edouard Brutus Gilles de la Tourette (1857–1904) published a two-part article describing nine French men and women with the disorder.[5] The tics of GTS usually begin around the age of 7 and vocalizations are most likely to begin at approximately 11 years of age.[1] Eye blinking is the most common symptom heralding the onset of the disorder, but rolling of the eyes, opening eyes widely, facial grimacing, nose twitching, or licking or biting of the lips are not uncommon. Presenting vocalizations include sniffing, throat clearing, coughing, panting, or spitting.[6] The tics tend to become more frequent and more complex through adolescence and then may partially abate.[7] Although temporary remissions may occur, permanent disappearance of the tics is unusual.

Simple motor tics consist of eye blinking, shoulder shrugs, nose flare, or arm jerking while complex motor tics include tossing of the head, touching, rubbing, jumping, hopping, hitting oneself, squatting, sniffing one's hands, sniffing objects, licking, and echoing the movements of others and copxopraxia (Table 19.3). Simple local tics include throat clearing, sniffing, grunting, clicking, squeaking, coughing, and snorting while complex local tics include whistling, belching, humming, sucking, coprolalia, echolalia, and pallilalia (Table 19.3). Dystonic tics (repetitive, slow, head turning, or repeated dystonic moving of limbs) are common in GTS, and a few patients have coexistent persistent dystonia.[8]

Gilles de la Tourette syndrome is not a purely motor disorder. Sensory tics and premonitory urges frequently precede the movements. The regions where premonitory urges are most common include the palms, shoulders, midline, abdomen, and throat. Performance of motor acts following the premonitory urge is accompanied by a sense of relief.[9,10] There is no associated dementia syndrome and most patients complete high school and find employment despite substantial challenges posed by the tics and related disorders.

TABLE 19.3. *Phenomenology of Tics*

Types of Tics	*Common Examples*
Simple motor tics	Eyeblink
	Shoulder shrug
	Nose flare
	Arm jerking
Complex motor tics	Tossing the head
	Touching
	Rubbing
	Spitting
	Copropraxia
	Jumping
	Hopping
	Hitting self
	Squatting
	Sniffing hands
	Sniffing objects
	Licking
	Echokinesis
Simple vocal tics	Throat clearing
	Sniffing
	Grunting
	Clicking
	Squeaking
	Coughing
	Snorting
Complex vocal tics	Whistling
	Belching
	Humming
	Sucking
	Coprolalia
	Echolalia
	Palilalia
Sensory tics	Pulling sensation
	Popping sensation
	Urge to tic
	Feeling that a tic is imminent
	Feeling that a sensation will be relieved by a tic
	Itch
	Tightness
	Feeling that something is touching the body

Neuropsychiatric Features

Gilles de la Tourette syndrome is characterized by a plethora of neuropsychiatric disorders (Table 19.4). Approximately 50% of children with GTS have comorbid attention deficit or attention deficit-hyperactivity disorder.[11]

Fifty to sixty percent of patients with GTS manifest OCD.[12] Obsessions and compulsions in GTS overlap phenomenologically with complex tics and include blurting obscenities, imitating the movements of others, counting compulsions, and compulsions to hurt oneself. When compared with OCD patients without tics, GTS patients with OCD have more violent, sex-

TABLE 19.4. *Neuropsychiatric Manifestations of Gilles de la Tourette's Syndrome*

Attention deficit-hyperactivity disorder
Obsessive-compulsive disorder
Conduct disorder
Oppositional defiant behavior
Coprolalia, copropraxia, mental coprolalia
Non-obscene, complex, inappropriate behaviors
Commenting negatively on another's weight, intelligence, appearance, dress, etc.
Uttering insults
Self-injurious behavior
Head banging
Body or head pinching/slapping
Striking body or head with an object
Poking sharp objects into the body
Scratching body
Striking the eyes

ual, and symmetrical obsessions and more touching, blinking, counting, and self-damaging compulsions. They also have more sensory symptoms and more autonomic symptoms. Patients with OCD have more obsessions concerning dirt or germs and more cleaning compulsions.[13–15] "Just right" perceptions are also common in GTS, characterized by the need to perform acts until they are felt to be exactly right.[16]

Behavioral disorders including conduct disorder and oppositional defiant disorder also can complicate the course of childhood GTS. These conditions are present in approximately one-third of children first presenting for assessment of GTS.[16]

Coprolalia and copropraxia are particularly problematic and dramatic symptoms in GTS. Coprolalia has been reported in 21% to 37% of series of patients with GTS.[1] The coprolalia tends to involve common visceral-type swear words; religious and deity-derived swearing is uncommon (Table 19.5). Copropraxia occurs in approximately 10%–15% patients and includes touching one's own genitals, obscene finger gestures, masturbatory movements, attempting to touch other's genitals, and staring or looking at the crotch. Mental coprolalia has been reported in a variable number of patients (5%–35%) and involves obsessions with repetitive, intrusive consideration of the same words involved in coprolalia. Repetitive and compulsive exhibitionism has also been reported as a manifestation of GTS.[17] In addition to obscene words and gestures, non-obscene complex, socially inappropriate behavior can be a manifestation of GTS.[18] Insulting others, adversely commenting on the characteristics of others, or the need to suppress the urge to insult others occurs in up to 40% of patients in clinical samples.

Self-injurious behavior can complicate the course of GTS. Thirty-three percent of patients in one clinical investigation manifested such behaviors. Typical behaviors included head banging, body or head punching and slapping, striking the body or head with hard objects, piercing the body or poking sharp objects into it, scratching body parts, and putting hands through windows or other hard or dangerous surfaces.

Neuroimaging and Laboratory Studies

The diagnosis of GTS depends exclusively on clinical criteria (Table 19.2) and there are no diagnostic neuroimaging or laboratory findings. Nevertheless, abnormalities have been found with specific investigations of central nervous system structure and function. Smaller volumes of basal ganglia (putamen or caudate) have been identified in children with GTS than in normal controls.[19] Metabolic imaging using [^{18}F]-fluorodeoxyglucose positron emission tomography (FDG-PET) has shown increased metabolism in the lateral premotor and supplementary motor regions and decreased metabolism in the caudate and thalamic nu-

TABLE 19.5. *Coprolalia, Copropraxia, and Related Disorders Occurring in Gilles de la Tourette's Syndrome*

Disorder	*Manifestations*
Coprolalia (most common utterances)	Fuck
	Shit
	Mother-fucker
	Cunt
	Prick
	Cocksucker
	Cockey
Copropraxia	Touching one's own genitals
	Obscene finger gesture
	Masturbating movements
	Touching others' genitals
	Looking at one's or another's crotch
Mental coprolalia	Same words as listed above
Other	Exhibitionism

clei.[20] Cerebral perfusion studies using single photon emission computerized tomography (SPECT) similarly revealed diminished blood flow in the caudate and anterior cingulate of patients with GTS.[21] Studies of dopamine metabolism also showed abnormalities in GTS: binding to D_2 dopamine receptors in the caudate nucleus was greater in children with GTS than in their discordant twin,[22] and dopamine transporter activity (as studied with ligand-based SPECT) was increased in patients with GTS.[23]

Electroencephalographic studies show that voluntary jerks that mimic the tics of GTS are prefaced by premovement, negative potentials, whereas no such premovement potential is evident prior to spontaneous tics.[24] These observations are consistent with the involuntary nature of tics.

Consistent with the elevated dopamine receptor activity observed in the basal ganglia, cerebrospinal fluid levels of homovanillic acid (HVA) are reduced. This suggests that the increased receptor and transporter activity results in reduced dopamine turnover.[25,26]

Inheritance

Genetic investigations suggest that OCD, chronic motor tics, and GTS are alternate expressions of an autosomal dominant trait. The trait is highly penetrant when all three syndromes are considered. Penetrance is gender influenced, with males more likely to manifest the syndrome than females.[27,28] Maternal transmission of GTS seems to be associated with greater motor tic complexity and more frequent rituals, whereas paternal transmission is associated with increased vocal tic frequency, earlier onset of vocal tics relative to motor tics, and more permanent attention deficit-hyperactivity disorder.

Pathology

There have been relatively few postmortem examinations of patients with GTS. Haber and colleagues[29] examined one patient at autopsy and found a striking absence of dynorphin-like fibers in the dorsal part of the external segment of the globus pallidus. The ventral pallidum also showed only faint staining for this compound. Singer and colleagues[30] examined three patients with GTS pathologically and found that dopamine uptake carrier sites were significantly increased in the caudate and putamen.

Pathophysiology

The available clinical neuroimaging and neuropathological data allow construction of a possible model for the pathophysiology of GTS. Autopsy and neuroimaging studies suggest a hyperactivity of dopaminergic systems in the basal ganglia, including the caudate and putamen. Genetic studies indicate that this hyperactivity is inherited as an autosomal dominant condition that may become manifest as a chronic motor tic syndrome, OCD, GTS, or GTS plus OCD. Hyperactivity within the putamen would mediate tic symptoms, as this structure is involved primarily in the motor circuitry of the frontal–subcortical circuits.[31,32] Obsession, whose content is highly emotional in nature, would be mediated through limbic portions of the frontal–subcortical circuitry, particularly the ventral striatum.[31,32]

Coprolalia and copropraxia are aspects of GTS that have been particularly difficult to explain. These emotional vocalizations may be related phylogenetically to limbic vocalizations in animals and designed to communicate fear or anger.[33,34] Emotional communications differ from propositional language that has as its primary intention the provision of information in spoken or written form. Hyperactivity within limbic areas of the striatum may result in abnormal, involuntary activation of mechanisms mediating limbic vocalization with ejaculation of the usual human means of expressing fear and anger (cursing).

Differential Diagnosis

Secondary etiologies of GTS and tic syndromes were discussed above and they are listed in Table 19.1. The relationship between Syndenham's chorea or group A hemolytic streptococcal infection without chorea has emerged as being particularly important. Children experiencing this infection may develop pediatric autoimmune neuropsychiatric disorders associated with streptococcal infections (PANDAS).[35,36] Diagnostic criteria for PANDAS include (*1*) presence of OCD or a tic disorder; (*2*) onset between age 3 years and the beginning of puberty; (*3*) abrupt onset of symptoms or a course characterized by dramatic exacerbations of symptoms; (*4*) onset or exacerbation of symptoms temporally related to infection with group A β-hemolytic streptococcus; and (*5*) abnormal neurological examination during exacerbation (hyperactivity, choreiform movements, tics).[37] Evidence of a group A hemolytic streptoccocal infection may be garnered through throat culture, ASO titer, or antistreptococcal deoxyribonuclease-B titer.[38] Many children with this syndrome also have a trait marker for rheumatic fever susceptibility (D8/17).[39] A substantial number of cases currently diagnosed as idiopathic OCD, GTS, or tic syndrome likely can be explained by the PANDAS syndrome.

TABLE 19.6. *Pharmacotherapy of Tics and Gilles de la Tourette's Syndrome (GTS)*

Disorder(s) Treated	*Agent*	*Daily Dosage*
Tics	Pimozide	2–18 mg/day
	Haloperidol	2–1 mg/day
	Clonidine	0.05–0.8 mg/day
	Risperidone	6–10 mg/day
Obsessive-compulsive disorder in GTS	Fluoxetine	10–40 mg/day
	Fluvoxamine	100–200 mg/day
Attention-deficit hyperactivity disorder in GTS	Methylphenidate	5–30 mg/day
	Dextroamphetamine	5–20 mg/day

Examples are provided; similar agents in each class may also be effective.

Treatment

Pharmacotherapy of GTS is summarized in Table 19.6 and examples of medications frequently used are provided (alternative agents in the same classes may have equal efficacy). Dopamine-blocking agents are the mainstay of the treatment of the tic disorder of GTS. Pimozide is the most commonly used compound, although haloperidal is a frequent alternative.[40] Side effects of pimozide include sedation, weight gain, depression, parkinsonism, and akathisia.[41] Clonidine has been used to treat GTS and may be most useful in those patients with associated behavioral disturbances.[42] Some patients, however, do not respond to this agent,[43] and one double-blind, placebo-controlled study failed to demonstrate any effect of this agent in GTS.[44] Obsessional and compulsive symptoms in GTS have responded to treatment with selective serotonin reuptake inhibitors (SSRIs), including fluoxetine and fluvoxamine.[45,46] Treatment of the attention-deficit hyperactivity disorder that may accompany GTS depends on use of psychostimulants, including methylphenidate or dextroamphetamine. Twenty to twenty-five percent of patients will have an increase of tics with treatment; the majority of patients, however, will tolerate these agents well with improvement in their attention-deficit disorder. Tricyclic antidepressants, clonidine, and fluoxetine have benefitted attention deficit or hyperactivity in GTS in selected patients.

DISORDERS WITH EXCESSIVE STARTLE RESPONSES

Disorders with excessive startle responses are a rare and unusual group of disorders that must be distinguished from both tics and myoclonus. Table 19.7 provides a classification of disorders with excessive startle responses. These responses are normal in several circumstances; hypnic jerks occur just as one is falling asleep and consist of a single excessive myoclonic jerk, usually accompanied by an internal sense of startle as if one were falling. A surprise will normally elicit a startle response, particularly if one is anxious or fearful when the surprise occurs.

TABLE 19.7. *Disorders with Excessive Startle Responses*

Type of Disorder	*Disorders*
Physiologic	Hypnic jerks
	Surprise
Pathologic	Hyperekplexia
	Startle syndromes
	Jumping Frenchman of Maine
	Latah
	Miryachit
	Startle epilepsy
	Gilles de la Tourette syndrome
	Startle myoclonus (with Creutzfeldt-Jakob disease)
	Sedative-hypnotic withdrawal
	Anxiety disorders
	Generalized anxiety
	Post-traumatic stress disorder
Related disorders	Cataplexy with the narcolepsy syndrome
	Paroxysmal kinesigenic choreoathetosis

Hyperekplexias

Primary hyperekplexias are autosomal dominant disorders with two phenotypic expressions. The major form includes (*1*) hyperekplexia and hypertonia in infants manifested by hesitant, wide-based gait when they begin walking; (*2*) prominent hypnagogic and sometimes generalized myoclonic jerks; (*3*) an enhanced startle response to any stimulus; (*4*) a typical autosomal dominant familial occurrence; and (*5*) amelioration with clonazepam in most cases.[47] The startle responses in this major form of the disease are characterized by a generalized stiffening, sometimes accompanied by a glottic sound and frequently leading to a fall without loss of consciousness. In some cases, the disorder does not become apparent until early adulthood.[48] In the minor form of the disease, excessive startle is triggered by acute febrile illness in childhood or stress in adult life; this is the only feature of the condition.[49] The disorder has been linked to chromosome 5Q.[50]

Startle Syndrome

Of substantial neuropsychiatric interest are the three classical startle syndromes: the Jumping Frenchman of Maine, latah, and miryachit. These three syndromes all share the four common features of echopraxia, echolalia, automatic obedience, and increased startle, and in many cases coprolalia is present.[49] Whether these disorders are neurological diseases or represent culture-bound syndromes has been the subject of debate. These conditions first came to scientific attention when George Beard observed "jumpers" in the Moosehead Lake region of Maine in 1880. He emphasized two features. First, he remarked on the temporariness and momentariness of the phenomenon, which was over in a second after the startle response had occurred. Second, he commented on the persistence of the syndrome, which was lifelong and usually occurred in several members of a family. Latah is a similar disorder, first described in the Malay Archipelago and miryachit was first observed in Siberia.

Startle epilepsy occurs in patients with severe brain damage such as prenatal anoxic injury or Tay-Sach's disease.[47] Loud sounds precipitate a violent jerk with limb and trunk flexion. There is coincident desynchronization of the electroencephalographic pattern.

Startle myoclonus occurs in Creutzfeldt-Jakob disease. It may occur early in the illness; patients have a startle response when the telephone rings or some other unexpected sound occurs. The response may persist until late in the disease when touching the patient, tapping their reflexes, or bumping the bed may result in an excessive startle response.

Startle reactions have been seen in GTS when sudden surprise precipitates an excessive response, with or without associated exacerbation of tics.

Delirious patients, particularly those experiencing withdrawal from sedative-hypnotics or depressants such as alcohol, have autonomic hyperarousal and may exhibit excessive startle responses.

Excessive Startle with Psychiatric Illnesses

Excessive startle also appears in anxiety disorders, particularly generalized anxiety disorder and post-traumatic stress disorder.[51] Post-traumatic stress disorder is characterized by exposure to a traumatic event with re-experiencing of that event, avoidance of stimuli associated with the trauma, and persistence of symptoms of increased arousal, including an exaggerated startle response.[2]

Related Disorders

Two related conditions that must be distinguished from startle reactions are catalepsy, which occurs when patients with narcolepsy are suddenly surprised, resulting in a loss of limb tone and a subsequent fall, and paroxysmal kinesiogenic choreoathetosis, in which patients have a brief choreoatheototic episode in response to movement. Surprise may precipitate the choreoathetotic events.[4]

MYOCLONUS

Myoclonus consists of brief, asymmetric, and usually asynchronous jerks of sufficient intensity to move a limb or body part. Table 19.8 provides an etiologic classification of myoclonus.[4,52] Most of the causes of myoclonus are discussed in other sections of this book and will not be re-addressed here. The neuropsychiatric syndromes associated with myoclonus depend on the underlying etiology of the myoclonic disorder. The epilepsies, dementias, basal ganglia degenerations, and central nervous system infections associated with myoclonus are described in the Chapters 21, 10, 18, and 26; respectively.

Palatal myoclonus differs from other forms of myoclonus in that it is very regular, beating rhythmically at rates of 60 to 200 per minute. The myoclonic movement involves the palate and in some cases also affects the larynx, pharynx, eyeballs, corner and floor of the mouth, and diaphragm.[53,54] The lesions associated with palatal myoclonus are located in the brainstem and frequently involve the central segmental tract. At autopsy there is enlargement of the inferior olivary nucleus ip-

TABLE 19.8. *Etiologic Classification of Myoclonus*[4,54]

Physiologic myoclonus
- Hypnic (sleep) jerks
- Hiccup (singultus)

Essential myoclonus
- Hereditary
- Sporadic

Epileptic myoclonus
- Isolated myoclonic epileptic jerks
- Idiopathic stimulus-sensitive epilepsy
- Epilepsia partialis continua
- Photosensitive myoclonus
- Myoclonic absences in petit mal epilepsy
- Infantile spasms
- Myoclonic astatic epilepsy (Lennox-Gastaut syndrome)
- Cryptogenic myoclonus epilepsy (Aicardi syndrome)
- Juvenile myoclonic epilepsy (of Janz)
- Benign familial myoclonic epilepsy (of Rabot)
- Progressive myoclonic epilepsy (Unverricht-Lundberg disease)

Symptomatic myoclonus
- Inherited metabolic encephalopathies
  - Lafora body disease
  - Tay-Sachs disease (GM2 gangliosidosis)
  - Gaucher's disease
  - Neuronal ceroid lipofuscinosis
  - Sialidosis
  - Myoclonus epilepsy and ragged red fibers (MERRF)
- Ataxic syndromes
  - Dyssynergia cerebellaris myoclonica (Ramsey-Hunt syndrome)
  - Friedreich's ataxia
  - Ataxia-telangiectasia
  - Other spinocerebellar degenerations
- Basal ganglia degenerations
  - Corticobasal degeneration
  - Dentatorubral-pallidoluysian atrophy (DRPLA)
  - Torsion dystonia
  - Wilson's disease
  - Hallervorden-Spatz disease
  - Huntington's disease
  - Multiple system atrophy
  - Progressive supranuclear palsy
  - Parkinson's disease

Symptomatic myoclonus (cont'd)
- Cortical dementias
  - Creutzfeldt-Jakob disease
  - Alzheimer's disease
- Viral encephalopathies
  - Subacute sclerotic panencephalitis (SSPE)
  - Encephalitis lethargica
  - Arbovirus encephalitis
  - Herpes encephalitis
  - Human immunodeficiency virus (HIV) encephalopathy
  - Post-infectious encephalitis
- Metabolic disorders
  - Hepatic failure
  - Renal failure
  - Dialysis dysequilibrium syndrome
  - Hyponatremia
  - Hypoglycemia
  - Non-ketotic hyperglycemia
  - Biotin deficiency
  - Post-anoxic myoclonus (Lance-Adams syndrome)
- Toxic encephalopathies
  - Drug-induced
    - Lithium
    - Tricyclic antidepressants
    - Selective serotonin reuptake inhibitors
    - Clozapine
    - Tardive myoclonus
    - Sedative-hypnotic withdrawal
  - Other intoxications
    - Bismuth
    - Aluminum
    - Mercury
    - Lead
    - Bromide
    - Strychnine
- Miscellaneous myoclonic syndromes
  - Palatal myoclonus
  - Segmental (spinal) myoclonus
  - Myoclonus/opsoclonus with neuroblastoma
  - Exaggerated startle reactions
  - Periodic movements of sleep
  - Asterixis (negative myoclonus)
  - Myoclonus with stiff man syndrome
  - Psychogenic myoclonus

TABLE 19.9. *Differentiation of Tics and Myoclonus*

Factor	*Tics*	*Myoclonic Jerks**
Duration	Brief	Brief
Repetitive	Yes	No
Stereotyped in location	Yes	No
Premonitory urge	Yes	No
Complex behaviors or vocalizations	Sometimes	No
Distraction	Decreases	No effect
Voluntarily suppressible	Yes	No
Movement-related	No	Often
Treatment	Dopamine-blocking agent (e.g., pimozide)	Benzodiazepines (e.g., clonazepam)†

*These features do not apply to the special case of palatal myoclonus.
†Postanoxic myoclonus often responds to treatment with 5-hydroxytryptophan.

silateral to the damage of the tract and contralateral to the most affected side of the palate. Herrmann and Brown[56] and Yakovlev[57] noted that the muscles involved in palatal myoclonus are those derived from the branchial arches and suggested that the movements represent a release of primitive gill motions programmed into the central nervous system and released from phylogenetic suppression by lesions of the central segmental tract.

Occasionally, it will be necessary to distinguish tics from myoclonus, particularly in the adult who has the new onset of a tic disorder. Table 19.9 presents the distinguishing features of these two disorders, both of which are characterized by brief, hyperkinetic movements. Tics tend to be repetitive and stereotyped compared to myoclonic jerks; they are preceded by a premonitory urge or other sensation that is lacking in myoclonus. Distraction tends to decrease tic frequency and has no effect on myoclonic jerks; tics are often voluntarily suppressible, at least for brief periods of time, whereas myoclonic jerks are not subject to voluntary control. Myoclonic jerks are often precipitated by movement, whereas tics are independent of other motor activities. Tics may be accompanied by other complex motor activities, whereas myoclonic jerks are always simple motor acts. Dopamine-blocking agents are the treatment of choice for tic disorders, whereas benzodiazapines such as clonazepam provide relief in some myoclonic disorders.

Rarely, myoclonic movement disorders have a psychogenic basis; clinical features and a thorough evaluation may help identify these conditions. The character of the movement is often inconsistent in contrast to the brief, repetitive jerks characteristic of neurologically based myoclonus. There may be associated myoclonic symptoms such as nonanatomic sensory loss, nonepileptic seizures, or transient blindness or paralysis. Psychogenic myoclonus is often decreased by distraction, may be reduced by placebo treatment, and can be enhanced by suggestion. Psychogenic myoclonus often has an acute onset, sudden resolution, or spontaneous remission. Finally, there must be evidence of underlying psychopathology and no evidence of an identifiable cause for the myoclonus (Table 19.10).[55] Patients with psychogenic myoclonus should be assessed and treated for their psychiatric conditions.

TABLE 19.10. *Features Indicative of Psychogenic Etiology of Myoclonus*[59]

Inconsistent character of the movements
Associated psychogenic symptoms (sensations, seizures, paralysis, blindness, etc.)
Reduction of myoclonus with distraction
Reduction of myoclonus with placebo treatment
Enhancement of myoclonus with suggestion
Spontaneous remission
Acute onset
Sudden resolution
Evidence of relevant psychopathology
No other cause of myoclonus identifiable

REFERENCES

1. Shapiro AK, Shapiro ES, et al. Gilles de la Tourette Syndrome. New York: Raven Press, 1988.
2. American Psychiatric Association. Diagnostic and Statistical Manual of Mental Disorders, 4th ed. Washington, DC: American Psychiatric Press, 1994.
3. Lohr JB, Wisniewski AA. Movement Disorders. A Neuropsychiatric Approach. New York: The Guilford Press, 1987.
4. Weiner WJ, Lang AE. Movement Disorders. A Comprehensive Survey. Mount Kisco, NY: Futura Publishing Company, 1989.
5. Lajonchere C, Nortz M, Finger S. Gilles de la Tourette and the discovery of Tourette syndrome. Arch Neurol 1996;53:567–574.
6. Brun A. Frontal lobe degeneration of the non-Alzheimer type, I: neuropathology. Arch Gerontol Geriatr 1988;6:193–208.
7. Goetz CG, Tanner CM, et al. Adult tics in Gilles de la Tourette's syndrome: description and risk factors. Neurology 1992;42:784–788.
8. Stone LA, Jankovic J. The coexistence of tics and dystonia. Arch Neurol 1991;48:862–865.
9. Kurlan R, Lichter D, Hewitt D. Sensory tics in Tourette's syndrome. Neurology 1989;39:731–734.
10. Leckman JF, Walker DE, Cohen DJ. Premonitory urges in Tourette's syndrome. Am J Psychiatry 1993;150:98–102.
11. Towbin KE, Riddle MA. Obsessive-compulsive disorder. In: Lewis M, ed. Child and Adolescent Psychiatry: A Comprehensive Textbook. Baltimore: Williams & Wilkins, 1991:685–697.
12. Frankel M, Cummings JL, et al. Obsessions and compulsions in Gilles de la Tourette's syndrome. Neurology 1986;36:378–382.
13. George MS, Trimble MR, et al. Obsessions in obsessive-compulsive disorder with and without Gilles de la Tourette's syndrome. Am J Psychiatry 1993;150:93–97.
14. Holzer JC, Goodman WK, et al. Obsessive-compulsive disorder with and without a chronic tic disorder. Br J Psychiatry 1994;164:469–473.
15. Miguel EC, Rauch SL, Jenike MA. Obsessive-compulsive disorder. Psychiatr Clin North Am 1997;20:863–883.
16. Leckman JF, Walker DE, et al. "Just right" perceptions associated with compulsive behavior in Tourette's syndrome. Am J Psychiatry 1994;151:675–680.
17. Comings DE, Comings BG. A case of familial exhibitionism in Tourette's syndrome successfully treated with haloperidol. Am J Psychiatry 1982;139:913–915.
18. Kurlan R, Daragjati C, et al. Non-obscene complex socially inappropriate behavior in Tourette's syndrome. J Neuropsychiatry Clin Neurosci 1996;8:311–317.
19. Singer HS, Reiss AL, et al. Volumetric MRI changes in basal ganglia of children with Tourette's syndrome. Neurology 1993;43:950–956.
20. Eidelberg D, Moeller JR, et al. The metabolic anatomy of Tourette's syndrome. Neurology 1997;48:927–934.
21. Moriarty J, Costa DC, et al. Brain perfusion abnormalities in Gilles de la Tourette's syndrome. Br J Psychiatry 1995;167:249–254.
22. Wolf SS, Jones DW, et al. Tourette syndrome: prediction of phenotypic variation in monozygotic twins by caudate nucleus D_2 receptor binding. Science 1996;273:1225–1227.
23. Malison RT, McDougle CJ, et al. [^{123}I]β-CIT SPECT imaging of striatal dopamine transporter binding in Tourette's disorder. Am J Psychiatry 1995;152:1359–1361.
24. Obeso JA, Rothwell JC, Marsden CD. Simple tics in Gilles de la Tourette's syndrome are not prefaced by a normal premovement EEG potential. J Neurol Neurosurg Psychiatry 1981;44:735–738.
25. Butler IJ, Koslow SH, et al. Biogenic amine metabolism in Tourette syndrome. Ann Neurol 1979;6:37–39.
26. Singer HS, Butler IJ, et al. Dopaminergic dysfunction in Tourette syndrome. Ann Neurol 1982;12:361–366.
27. Pauls DL, Leckman JF. The inheritance of Gilles de la Tourette's syndrome and associated behaviors. Evidence for autosomal dominant transmission. N Engl J Med 1986;315:993–997.
28. Pauls DL, Towbin KE, et al. Gilles de la Tourette's syndrome and obsessive-compulsive disorder. Evidence supporting a genetic relationship. Arch Gen Psychiatry 1986;43:1180–1182.
29. Haber SN, Kowall NW, et al. Gilles de la Tourette's syndrome: a postmortem neuropathological and immunohistochemical study. J Neurol Sci 1986;75:225–241.
30. Singer HS, Hahn IH, Moran TH. Abnormal dopamine uptake sites in postmortem striatum from patients with Tourette's syndrome. Ann Neurol 1991;30:558–562.
31. Alexander MP, Baker E, Kaplan E. Dimensions of performance in patients with ideomotor apraxia. Neurology 1986;36:345.
32. Cummings JL. Frontal–subcortical circuits and human behavior. Arch Neurol 1993;50:873–880.
33. Devinsky O, Bear D, et al. Perception of emotion in patients with Tourette's syndrome. Neuropsychiatry Neuropsychol Behav Neurol 1993;6:166–169.
34. MacLean PD. The Triune Brain in Evolution: Role in Paleocerebral Functions. New York: Plenum Press, 1990.
35. Swedo SE, Leonard HL, et al. Pediatric autoimmune neuropsychiatric disorders associated with streptococcal infections: clinical description of the first 50 cases. Am J Psychiatry 1998;155:264–271.
36. Asbahr FR, Negrao AB, et al. Obsessive-compulsive and related symptoms in children and adolescents with rheumatic fever with and without chorea: a prospective 6-month study. Am J Psychiatry 1998;155:1122–1124.
37. Kurlan R. Tourette's syndrome and 'PANDAS': will the relation bear out? Pediatric autoimmune neuropsychiatric disorders associated with streptococcal infection. Neurology 1998;50:1530–1534.
38. Perlmutter SJ, Garvey MA, et al. A case of pediatric autoimmune neuropsychiatric disorders associated with streptococcal infections. Am J Psychiatry 1998;155:1592–1598.
39. Swedo SE, Leonard HL, et al. Identification of children with pediatric autoimmune neuropsychiatric disorders associated with streptococcal infections by a marker associated with rheumatic fever. Am J Psychiatry 1997;154:110–112.
40. Mesulam MM, Petersen RC. Treatment of Gilles de la Tourette's syndrome: eight-year, practice-based experience in a predominantly adult population. Neurology 1987;37:1828–1833.
41. Regeur L, Pakkenberg B, et al. Clinical features and long-term treatment with pimozide in 65 patients with Gilles de la Tourette's syndrome. J Neurol Neurosurg Psychiatry 1986;49:791–795.
42. Cohen DJ, Detlor J, et al. Clonidine ameliorates Gilles de la Tourette syndrome. Arch Gen Psychiatry 1980;37:1350–1357.
43. Leckman JF, Detlor J, et al. Short- and long-term treatment of Tourette's syndrome with clonidine: a clinical perspective. Neurology 1985;35:343–351.
44. Goetz CG, Tanner CM, et al. Clonidine and Gilles de la Tourette's syndrome: double-blind study using objective rating methods. Ann Neurol 1987;21:307–310.
45. McDougle CJ, Goodman WK, et al. The efficacy of fluvoxamine in obsessive-compulsive disorder: effects of comorbid chronic tic disorder. J Clin Psychopharmacol 1993;13:354–358.
46. Como PG, Kurlan R. An open-label trial of fluoxetine for ob-

sessive-compulsive disorder in Gilles de la Tourette's syndrome. Neurology 1991;52:872–874.
47. Wilkins DE, Hallet M, Wess MM. Audiogenic startle reflex of man and its relationship to startle syndromes. A review. Brain 1986;109:561–573.
48. Saenz-Lope E, Herranz FJ, Masdeu JC. Startle epilepsy: a clinical study. Ann Neurol 1984;16:78–81.
49. Lees AJ. Tics and Related Disorders. London: Churchill Livingstone, 1985.
50. Ryan SG, Sherman SL, et al. Startle disease, or hyperekplexia: response to clonazepam and assignment of the gene (*STHE*) to chromosome 5q by linkage analysis. Ann Neurol 1992;31: 663–668.
51. Howard R, Ford R. From the jumping Frenchmen of Maine to post-traumatic stress disorder: the startle response in neuropsychiatry. Psychol Med 1992;22:695–707.
52. Caviness JN. Myoclonus. Mayo Clin Proc 1996;71:679–688.
53. Soso MJ, Nielsen VK, Jannetta PJ. Palatal myoclonus. Reflex activation of contractions. Arch Neurol 1984;41:866–869.
54. Tahmoush AJ, Brooks JE, Keltner JL. Palatal myoclonus associated with abnormal ocular and extremity movements. A polygraphic study. Arch Neurol 1972;27:431–440.
55. Monday K, Jankovic J. Psychogenic myoclonus. Neurology 1993;43:349–352.
56. Hermann C Jr, Brown JW. Palatal myoclonus: A reappraisal. J Neurol Sci 1967;5:473–492.
57. Yakovlev P. Discussion. Trans Am Neurol Assoc 1957;82:87–89.

Catatonia, Motoric Manifestations of Psychiatric Illnesses, and Drug-Induced Motor System Disturbances

Movement of body, limbs, mouth, and tongue provide the only evidence for human mental life. Without speech and movement the completely paralyzed but alert individual cannot be distinguished from one in a coma. Movement provides the means by which human cognition is made manifest in the world through engineering, art, and science. Movement allows persons to meet their basic needs for food, drink, and pleasure. The complex interaction between the limbic system and the motoric system provides the basis for relating one's emotional state to speech and movement (hence the term *e-motion*) and is of particular relevance in neuropsychiatry.

Structures critical to normal movement include the dorsolateral prefrontal cortex (mediating willed movement), anterior cingulate (mediating motivational aspects of motoric activities), pyramidal motor system (responsible for cortically mediated movements), and extrapyramidal motor system, including the basal ganglia (responsible for mediating nonpyramidal influences on motor activity). Frontal–subcortical circuits mediating motor function are arranged in parallel with circuits involved in volitional eye movements, executive function, civil and socially integrated behavior, and motivation.[1,2] The representations of motor, executive, and emotional function at the level of the basal ganglia account for the common occurrence of motor disturbances in patients with psychiatric illnesses and the frequent emergence of neuropsychiatric abnormalities in patients with basal ganglia diseases (Chapter 18). Psychotropic medications useful in controlling psychiatric symptoms commonly exert their effects on basal ganglionic structures of frontal–subcortical circuits and hence motor abnormalities (such as parkinsonism and tardive dyskinesia) are common side effects of these agents.

This chapter addresses the principal motor abnormalities occurring in psychiatrically ill patients, including catatonia, syndrome-specific motor disturbances, and movement abnormalities induced by psychotropic substances.

CATATONIA

Diagnosis and Clinical Features

Catatonia refers to a cluster of striking motor signs occurring with idiopathic psychosis (mood disorders, schizophrenia), intrinsic brain disease, metabolic disorders affecting brain function, and drug-induced syndromes. The motor symptoms include abnormalities of posture, tone, volitional motor activity, and speech and there may be periods of extreme hyperactivity or hypoactivity.[3,4] Typical features of the catatonic syndrome include mutism, stupor, stereotypy, posturing, catalepsy, automatic obedience, negativism, echolalia, or echopraxia. Table 20.1 provides a classification of motor signs observed in catatonic patients.

Signs of catatonia involve abnormalities of spontaneous behavior and disturbances affecting compliance elicited in the course of the examination. The principal abnormalities of spontaneous movement include stereotypies and mannerisms. *Stereotypies* are non–goal-directed movements that are carried out in a uniform way. *Mannerisms* are goal-directed movements executed in an abnormal stilted manner, although the purpose is usually apparent to the examiner.[5] *Mitgehen* is an abnormality of induced movement in which the patient moves in response to the slightest pressure by the examiner. This may occur in spite of verbal instructions to resist. *Mitmachen* is a motor disorder in which the patient acquiesces to every passive movement of the body made by the examiner, but as soon as the examiner releases the body part, the patient returns it to the resting position. Dramatic abnormalities of posture include *waxy flexibility*, in which the patient allows the posturing of the limbs with prolonged maintenance of that posture. *Catalepsy* is the temporary maintenance of spontaneous postures that are often abnormal and bizarre,[5,6] with gradual resumption of normal resting positions.

Catatonic stupor refers to a temporary reduction or obliteration of both reactive and spontaneous relational movements (action and speech), but with evidence of retained consciousness.[7] Behavioral features such as eye pursuit, sudden switching from stupor to activity, "last minute" responses, and retention of normal reflexes (pupillary, oculocephalic, corneal) suggest the absence of obtundation or coma. *Negativism*, including gegenhalten (active resistance to movement), may accompany the stuporous state. Stupor appears to be the extreme form of the retarded type of catatonia, characterized in its more mild form by negativism, mutism, rigidity, catalepsy, and staring. Catatonic excitement (furor) is characterized by the abrupt or rapid onset of impulsive, combative behavior.[8]

Catatonia is sometimes described as a rare phenomenon that is not often seen in contemporary neuropsychiatric practice. However, adequate examination for catatonic signs is rarely performed, and when specific observations for spontaneous and solicited catatonic phenomena are made, the syndrome is not uncommon in psychiatric, neurologic, and toxic-metabolic disorders. In the course of the examination it may also be observed that catatonic patients perform externally

TABLE 20.1. *Motor Features of the Catatonic Syndrome*[3,5,80–82]

Abnormalities of posture
Catalepsy (tendency of maintenance of postures for long periods)
Psychological pillow (patients retain their head in an elevated position as if lying on a pillow)
Persistent abnormal posture (flexed, lordotic, twisted, tilted, awkward)
Abnormal spontaneous movements
Limb stereotypies (movement that is not goal directed and is carried out in a uniform way)
Handling (patient touches and handles everything within reach)
Intertwining (patient continually intertwines fingers or grasps clothes or kneads cloth)
Abnormal trunk movements (such as rocking)
Grimacing and facial movements (including Schnauzkrampf, consisting of marked wrinkling of the nose with pouting of the lips)
Abnormalities of tone and motor compliance
Waxy flexibility (patient allows posturing similar to bending a warmed candle)
Gegenhalten (opposition movement in which patient resists passive movements of the body to precisely the same degree as the pressure exerted by examiner)
Automatic obedience (patient carries out every command given in an automatic manner)
Mitgehen (patient moves body in response to light pressure by examiner)
Mitmachen (patient allows passive movement of the body made by examiner but returns limb to resting position when the examiner releases the patient)
Ambitendency (intermittent cooperation and withdrawal of cooperation)
Echopraxia (tendency to echo movements of the examiner)
Negativism (active lack of cooperation and defiance of attempts to influence behavior)
Abnormalities of speech
Mutism (absence or severe reduction of speech)
Echolalia (repeating what the examiner says)
Palalalia (repeating what the patient says)
Perseveration (repetition of single words or phrases)
Verbigeration (senseless incomprehensible sounds that are frequently repeated)
Speech-prompt catatonia (patient answers with intelligible words that are the first thing that comes to mind)
Abnormalities of arousal
Hyperactivity (catatonic excitement)
Hypoactivity (catatonic stupor)

guided tasks (catching a ball) better than internally guided movements (throwing, kicking).[9]

Etiologies of Catatonia

The catatonic syndrome may be a manifestation of idiopathic psychiatric disorders, neurological diseases, metabolic encephalopathies, or toxic conditions (Table 20.2). Catatonia is more common in mania than any other psychiatric illness and has been reported in approximately 30% of manic patients systematically examined for catatonic signs.[10,11] Subtle types of catatonia (clumsiness, awkwardness, or postural disturbance) are common among schizophrenics, whereas more severe catatonic signs such as stereotypies, mannerisms, ambitendency, catalepsy, automatic obedience, excitement, and stupor are rarely observed in contemporary psychiatric practice.[12] When present in schizophrenia, catatonic symptoms commonly co-occur with features of formal thought disorder, affective blunting, and neu-

TABLE 20.2 *Differential Diagnosis of Catatonia*

Idiopathic psychiatric disorders	Systemic and metabolic disturbances
Schizophrenia	Diabetic ketoacidosis
Mania	Hypercalcemia (hyperparathyroidism)
Depression	Pellagra
	Porphyria
Neurological disorders	Homocystinuria
Basal ganglia disturbances	Glomerulonephritis
Post-encephalitic parkinsonism	Hepatic encephalopathy
Globus pallidus lesions (bilateral)	Thrombotic thrombocytopenic purpura
Basal ganglia calcification	Systemic lupus erythematosus
Primary pallado-nigro-subthalamic atrophy	Hypernatremia
Limbic system disorders	Typhoid fever
Herpes encephalitis	Mononucleosis
Temporal lobe infarction	
Neoplasms	Toxic agents and drug reactions
Subacute sclerosing panencephalitis	Mescaline
Lyme disease with encephalitis	Ethyl alcohol
Diencephalic lesions	Amphetamine
Thalamotomy for parkinsonism	Phencyclidine
Hemorrhage	Glutethimide withdrawal
Wernicke's encephalopathy	Morphine
Neoplasm	Disulfiram
Frontal lobe disorders	Aspirin intoxication
Anterior cerebral artery aneurysm	Dopamine-depleting agents (reserpine)
Traumatic contusion	Cortisone
Arteriovenous malformation	Dopamine withdrawal (levodopa)
General paresis (syphilis)	Neuroleptics
Neoplasms	Hallucinogens
Cortical venous thrombosis	Neuroleptic malignant syndrome
Epilepsy and postictal states	Benzodiazepine withdrawal
Mental retardation	
Pinealoma	
Brainstem lesions	
Paraneoplastic encephalopathy	
Right-sided stroke	

rological soft signs. Among depressed patients, catatonia occurs in individuals who are older, more cognitively impaired, more disabled in activities of daily living, and more severely depressed.[13] Patients with psychiatric illnesses and catatonia are more likely to have had a history of previous brain injury or to have a physical illness at the onset of the catatonic period than patients not exhibiting catatonic phenomena.[14]

Neurological disorders exhibiting catatonia include basal ganglia disorders such as post-encephalitic parkinsonism, bilateral lesions of the globus pallidi, primary pallido-nigro-subthalamic atrophy, right hemisphere stroke, akinetic mutism, hydrocephalus, bacterial meningitis, acquired immunodeficiency syndrome (AIDS), and idiopathic basal ganglia calcification. Disorders of the limbic system producing catatonia include herpes encephalitis, temporal lobe infarction, temporal lobe neoplasms, subacute sclerosis panencephalitis, and lyme disease with encephalitis. Diencephalic lesions can also produce catatonic behaviors including thalamotomy for Parkinson's disease, thalamic hemorrhage, Wernicke's encephalopathy, and thalamic brain tumors. Frontal lobe disorders also present with or produce catatonic syndromes, including anterior cerebral artery aneurysms, traumatic contusions, arteriovenous malformations, general paresis of the insane, frontal lobe neoplasms, and cortical venous thrombosis with frontal infarction.[15–25]

Miscellaneous neurological conditions reported to produce catatonia include epilepsy and postictal states, brainstem lesions, multiple sclerosis, paraneoplastic encephalopathy, pinealomas, tuberous sclerosis, and brain disorders associated with mental retardation. Systemic and metabolic disturbances described as producing catatonic symptoms include diabetic ketoacidosis, hypercalcemia (hyperparathyroidism), pellagra, porphyria, homocystinuria, postpartum psychosis, Addison's disease, Cushing's disease, hyperthyroidism, glomerulonephritis with renal failure, hepatic encephalopathy, hyponatremia, typhoid fever, mononucleosis, thrombotic, thrombocytopenic purpura, and systemic lupus erythematosus. Toxic agents and drug reactions capable of inducing catatonia include mescaline, ethyl alcohol, amphetamines, phencyclidine, diazepam, aspirin, steroids, neuroleptics, hallucinogens, dopamine-depleting agents (e.g., tetrabenazine), dopamine withdrawal (e.g., levodopa), morphine, glutethimide withdrawal, and benzodiazepine withdrawal.[15–25] Table 20.2 presents a comprehensive differential diagnosis of disorders reported to produce catatonic phenomena.

Two syndromes require special discussion in the context of catatonia: neuroleptic malignant syndrome and lethal or malignant catatonia. *Neuroleptic malignant syndrome* (described in detail below) is a disorder characterized by severe muscle rigidity and fever in association with the use of neuroleptic medication and including at least two of the following signs: diaphoresis, dysphasia, tremor, incontinence, changes in level of consciousness ranging from confusion to coma, mutism, tachycardia, elevated or labile blood pressure, leukocytosis, or laboratory evidence of muscle injury (e.g., elevated creatine phosphokinase).[26] Neuroleptic malignant syndrome can have catatonic features and it occurs primarily in patients with psychiatric illnesses receiving treatment with neuroleptic medications. The presence of a psychiatric disorder may lead to misdiagnosis as a catatonic syndrome secondary to psychotic episode. Onset of the disorder with extrapyramidal-type muscle rigidity progressing to obtundation and coma in the presence of substantial autonomic instability usually comprises an accurate diagnosis.

Lethal catatonia associated with an idiopathic psychiatric disorder typically begins with extreme psychotic excitement leading to fever, exhaustion, and death mediated by dehydration and cardiac arrest.[21,27,28] Neurologic, metabolic, and toxic causes of catatonia can also be lethal and herpes encephalitis, intracranial neoplasms, systemic organ failure, and drug intoxication or withdrawal must be considered in all cases presenting with catatonia and serious medical comorbidity.

Management of Catatonia

There are three aspects to the management of patients exhibiting catatonic syndromes. First, patients must undergo comprehensive evaluation for neurologic, metabolic, and toxic causes of catatonic phenomenon. This may include neuroimaging, lumbar puncture, electroencephalography (EEG), serum tests, and drug screens, depending on the clinical circumstances. A high index of suspicion of comorbid disease must be maintained in patients presenting with a catatonic syndrome.

The second aspect of management of a patient manifesting catatonic phenomena addresses the potential medical consequences of catatonia. These include aspiration, dehydration, pulmonary emboli, thrombophlebitis, urinary tract infection, and, in extreme cases, acute renal failure and cardiopulmonary arrest.[21]

Third, in patients whose catatonia does not resolve in response to treatment of an identified underlying disorder or in whom no specific neuromedical etiologic process is identified, therapy of the catatonia itself may be warranted. Benzodiazepines (diazepam and lorazepam) and sodium amobarbital produce rapid and dramatic relief of catatonia in 50%–75% of patients receiving the agents intravenously. There may be immediate restoration of the ability to talk and move normally. The effect tends to resolve as the action of the agent resides; some patients benefit from long-term benzodiazepine therapy.[29–33] Catatonic patients who are refractory to treatment with benzodiazepines or lorazepam usually respond well to treatment with electroconvulsive therapy (ECT).[34–37] Occasional catatonic patients have responded to treatment with lithium, anticholinergic agents, atypical antipsychotics, or carbamazepine.

MOTOR DISTURBANCES IN PSYCHIATRIC ILLNESSES

Schizophrenia

Schizophrenic patients may manifest a variety of motor abnormalities. In many cases, these are induced by chronic treatment with neuroleptic medications but spontaneous, hyperactive movements were well recorded prior to the introduction of these therapies.[38] Table 20.3 lists the abnormal hyperkinetic mannerisms and stereotypies that may be observed in schizophrenia. Features that distinguish these movements from those of drug-induced tardive dyskinesia are described below.

In addition to catatonic type movements, schizophrenic patients may evidence a variety of ocular motor abnormalities. Many patients exhibit abnormalities in smooth pursuit eye tracking of targets and blink rates are also increased. Blink rate is normalized by successful treatment with antipsychotic agents.

Schizophrenic patients also evidence a variety of neurological "soft" signs on examination (soft signs are described in Chapter 3). Thirty to fifty percent of patients exhibit abnormalities of motor sequencing and complex motor activity and sensory integration.[39–41] Schizophrenic patients exhibiting violent behavior evidence more neurological soft signs and neuropsychological deficits than schizophrenic patients without aggressive behavior.[42]

Mania

Manic patients also exhibit observable motoric behaviors that are diagnostically useful (Chapter 14). Their speech is often rapid and circumstantial, revealing flight of ideas characterized by clang associations, rhyming, punning, joking, and grandiosity.[6] These patients are hyperactive, restless, and agitated. They may be aggressive, threatening, or menacing when upset. In some cases, they are sexually provocative or seductive and may engage in nudity or sexual exposure. Their dress may be bizarre, outlandish, or peculiar and they may exhibit catatonic features.[6,43] Neurological soft signs may be elicited on examination more commonly in patients with mania than in normal controls.[44,45]

Depression

Depressed patients also display a rich repertoire of motor system abnormalities relevant to clinical diagnosis

TABLE 20.3. *Abnormal Movements Observed in Schizophrenia*

Area Affected	*Movements*
Eyes	Opening wide, squeezing shut, abnormal blinking, rapid lateral glances, staring, gaze deviation away from examiner
Nose	Wrinkling, sniffing, flaring nostrils
Mouth and jaw	Pouting (Schnauzkrampf), lip smacking, grinning, grimacing, biting, chewing
Tongue	Protrusion, licking, clicking
Face	Wrinkling forehead
Head and neck	Torsion movements, hyperextension, shaking, nodding
Extremities	Picking, pulling, handling, twisting, kneading, grasping, tapping, rubbing, intertwining fingers, wringing hands, folding hands, spreading fingers, flinging arms
Trunk and whole body	Shoulder shrugging, contortionist movements, back-arching, rocking, shuffling, hopping, turning, skipping, running, excessive leg lifting, marionette-like movements

and an understanding of their psychopathology (Chapter 14). Diagnostic schemes such as those of the fourth edition of the *Diagnostic and Statistical Manual of Mental Disorders*[26] include only superficial reference to agitation or retardation and do not draw attention to the importance and utility of careful observation and examination of motor behaviors in depressed patients. Depressed patients may exhibit abnormalities of gait, with slowing or diminished stride length. In conversation they frequently do not initiate interaction or react normally to the examiner. Postural slumping, body immobility, slowed body movements, delay in initiation of motor activity, and slowness of movements are evident in patients with retarded depression. Facial immobility, staring, and loss of emotional expression are also characteristic. The verbal output of depressed patients features delay of responses, shortened responses, and a reduction in the variety of themes developed in conversation or produced in response to queries by the examiner.[46,47] Patients with agitated depression evidence motor and facial agitation and often exhibit speech perseveration. Retarded depressions must be distinguished from apathetic syndromes (Chapter 14).

Anxiety

Anxious patients also exhibit motility disturbances. The eyebrows are raised and there is deepening of the furrows of the forehead and widening of the palpebral fissures. The mouth is often held slightly open, the body is rigidly upright, and when the patient is seated the knees are typically pressed together. Respiratory movements are fast and shallow, perspiration is increased, and the pupils may be dilated[48] (Chapter 17).

Obsessive-Compulsive Disorder

Patients with obsessive-compulsive disorder may exhibit their compulsions and rituals during the course of the examination (Chapter 16). Patients may be unable to resist cleaning, touching, or avoiding certain objects, repeating specific acts, checking, counting, or pursuing other involuntary activities. They may act with obsessional slowness, devoting long periods of time to repetitive acts or trying to achieve the "just right" sensation. Gilles de la Tourette syndrome and frontotemporal dementia are often accompanied by obsessive-compulsive disorders (Chapter 16).

DRUG-INDUCED MOVEMENT DISORDERS

Psychotropic medications exert their beneficial effects on structures mediating emotional function and portions of these same structures are often members of circuits involved in motor activity. Thus side effects associated with psychotropic medications frequently involve motor system disturbances. Conventional antipsychotic medications, antidepressants, and lithium all have movement disorders as common side effects. Table 20.4 presents an overall classification of drug-induced movement disorders. These must be considered in any patient presenting with behavioral disturbances who is currently being treated or has been treated in the past with psychotropic agents.

TABLE 20.4. *Drug-Induced Movement Disorders*

Acute dystonic reactions
Akathisia
Parkinsonism
Akinetic-rigid syndrome
Rest tremor
Rabbit syndrome
Tremor
Myoclonus
Choreoathetosis
Tardive dyskinesia
Classic tardive dyskinesia
Tardive dystonia
Toricollis
Blepharospasm
Tardive Meige's syndrome
Spasmodic dysphonia
Segmental/generalized dystonia
Tardive akathisia
Tardive tics
Invariant tics
Tardive Gilles de la Tourette syndrome
Tardive myoclonus
Tardive complex
Withdrawal-emergent dyskinesia
Neuroleptic malignant syndrome

Acute Dystonic Reaction

Acute drug-induced dystonia is most commonly a side effect of treatment with conventional neuroleptic medications such as phenothiazines or butyrophenones. Dystonia has occasionally been reported as a consequence of other medications including carbamazepine, buspirone, phenytoin, fluoxetine, metoclopramide, and ondansetron.[49] Dystonic reactions most commonly involve the facial and neck musculature, producing jaw clenching, tongue protrusion, oculogyric deviation, difficulty in swallowing, torticollis, or retrocollis. In some cases, flexion, extension, or torsion of the trunk occurs, and the complete axial extension with opisthotonus may be observed.[50] Over 95% of acute dystonic episodes occur within the first 4 days of therapy and many occur within the first few hours. They are more commonly associated with higher doses of medications and are more frequent in younger patients than older patients.[50,51] No definite relationship to the underlying psychiatric diagnosis has been demonstrated.[52] Treatment with parenteral anticholinergic medications (benzotropine) or antihistamines (e.g., diphenhydramine) usually provides relief within 15 minutes.[50] Benzodiazepines are also effective in ameliorating akathisia.

Akathisia

Akathisia is a syndrome consisting of a subjective sense of restlessness and a need to move and objective restless movements, including repeated leg crossing, swinging of the legs, lateral knee movements, pacing, and rocking from foot to foot while standing.[53–56] Akathisia is one of the most distressing drug-induced conditions experienced by patients and promotes both noncompliance and agitation. Akathisia typically develops within the first 6 weeks of initiating therapy with a neuroleptic agent and occurs in approximately 20% of exposed individuals.[53] Akathisia is more common with higher doses of medication.

Subacute drug-induced akathisia must be distinguished from pseudoakathisia, which features motor restlessness without the usual corresponding sense of a subjective need to move, and tardive akathisia, which begins after long-term exposure to neuroleptics or upon withdrawal from neuroleptic therapy.[57]

Akathisia has been observed following treatment with a variety of non-neuroleptic psychotropic agents, including buspirone, tricyclic antidepressants, selective serotonin reuptake inhibitors (SSRIs), monoamine oxidase inhibitors (MAOIs), trazodone, benzodiazepines, carbamazepine, levodopa, lithium, nefazodone, flunarizine, metoclopramide, prochlorperazine, and clozapine.[49,58–61] In addition, akathisia may occur in idiopathic Parkinson's disease[62] and with focal brain lesions involving the basal ganglia and frontal lobes.[63]

Beta-blockers are the agents most likely to ameliorate akathisia. In addition, treatment with clonidine, anticholinergic agents, amantadine, and benzodiazepines may be successful.[53,64]

Parkinsonism

Drug-induced parkinsonism is one of the most common side effects of neuroleptic therapy. Akinesia and rigidity are the two most common manifestations, but rest tremor of the extremities or rest tremor of the mouth and perioral structures (rabbit syndrome) may also occur. The principal differential diagnostic challenge concerns idiopathic Parkinson's disease (Chapter 18). Rigidity may either be of the cog-wheel or plastic type and is best elicited by performing flexion–extension movements of the wrist and elbow with the patient at rest. Incipient rigidity may be elicited by asking the patient to perform movements with the contralateral limb while the ipsalateral limb is being tested for rigidity. Rigidity and tremor may be asymmetrical in patients with drug-induced parkinsonism.[65]

Akinesia is manifested by a paucity of spontaneous movements and a loss of automatic movements such as blinking and swallowing. *Bradykinesia*, a slowness of movement, commonly accompanies this syndrome.[66] Loss of associated movements such as swinging the arms while walking and "en block" turns are common subtle manifestations of drug-induced parkinsonism. Tremor is less frequent in drug-induced than idiopathic parkinsonism, but a classical type of resting tremor may be observed in some cases.

Parkinsonism begins within 2 weeks of therapy with neuroleptic agents and is detectable in 50% to 80% of patients, at least in mild form. The elderly are at particularly high risk for the development of drug-induced parkinsonism.[66] Higher-potency neuroleptics such as perphenazine and haloperidol are more likely to induce parkinsonism than lower-potency agents and those with concomitant anticholinergic effects. Atypical antipsychotics are also less likely to induce drug-induced parkinsonism.

Rabbit syndrome refers to a peculiar-appearing disorder in which the perioral tremor results in rapid movements of the lips resembling those observed around a rabbit's mouth. The tongue is typically not involved, a characteristic that helps distinguish rabbit syndrome from tardive dyskinesia.

Drug-induced parkinsonism is managed by reducing the dose of the inciting agent or treatment with amantadine or an anticholinergic drug. Levodopa and related dopaminergic agents may be required in severe cases. Parkinsonism typically resolves within 2–4 weeks of intervention but may persist for 12–18 months.

Action Tremor

Many psychotropic agents are capable of inducing an action tremor. These tremors are high frequency (12

TABLE 20.5. *Classification of Tremors*

Tremor	*Action Characteristic*	*Amplitude*	*Frequency*	*Etiology*
Exaggerated physiological	Action	Small	High (8–12 Hz)	Anxiety, fatigue, toxic- metabolic disorders
Essential	Action	Small	High (8–12 Hz)	Aging, hereditary, sporadic
Parkinsonian	Resting	Large	Low (4–6 Hz)	Parkinsonian syndromes
Cerebellar	Action	Crescendo	Low	Cerebellar or midbrain disease

cycles per second) and low amplitude and represent exaggerated physiologic tremors. Tricyclic antidepressants, lithium, anticonvulsants, and neuroleptic agents have all been implicated in producing action tremors.

Management typically involves stopping the offending drug. Beta-blocking drugs, primidone, or benzodiazepines such as clonazepam may be useful if the tremor-inducing agent is critical to treatment of the underlying psychiatric disorder.[49,67–70]

Drug-induced action tremors must be distinguished from rest tremors associated with parkinsonism, essential tremors, and cerebellar intention tremors (Table 20.5). The causes of exaggerated physiological tremors are listed in Table 20.6.

Myoclonus

Myoclonus refers to brief, irregular, and usually asymmetric jerks involving a sufficient number of motor units to produce movement of a limb or body part (Chapter 19). Myoclonus has been induced by a variety of drugs, including the anticonvulsants (carba-

TABLE 20.6. *Etiologies of Tremors*

Exaggerated physiological tremor
- Psychotropic agents: lithium, tricyclic antidepressants, phenothiazines, butyrophenones
- Nonpsychotropic agents: epinephrine, isoproterenol, metaproterenol, terbutaline, xanthines (coffee, tea), theophylline, levodopa, amphetamines, thyroid hormone, hypoglycemic agents, adrenocorticosteroids, valproate sodium, lamotrigine, fenfluramine, flunarizine, cimetidine
- Endocrine: thyrotoxicosis, hypoglycemia, pheochromocytoma
- Stress-induced: anxiety, fright, fatigue, cold-shivering
- Metabolic disorders: uremia, hepatic encephalopathy, anoxia
- Miscellaneous agents and conditions: alcohol or sedative withdrawal, mercury, lead, arsenic, carbon monoxide, methyl bromide, monosodium glutamate

Essential tremor
- Autosomal dominant, age-related, sporadic
- With Charcot-Marie-Tooth disease (Roussy-Lévy syndrome)
- With other movement disorders: parkinsonism, torsion dystonia, toricollis, writer's cramp, hereditary nonprogressive chorea

Cerebellar tremor
- Cerebellar degenerations
- Multiple sclerosis
- Wilson's disease
- With hereditary sensory neuropathy (DÇjerine-Sottas disease)
- Drugs and toxins: lithium, phenytoin, barbiturates, alcohol, mercury, 5-fluorouracil
- Midbrain lesions ("rubral" tremor)

Parkinsonian tremor
- Parkinson's disease
  - Idiopathic
  - Post-encephalitic
- Other degenerative extrapyramidal syndromes
- Toxic-metabolic parkinsonian syndromes

Other tremor syndromes
- Primary writing tremor
- Tremor with dystonia
- Tremor with peripheral neuropathy

mazepine and phenytoin), tricyclic antidepressants, SSRIs, metoclopramide, neuroleptics, levodopa, opiates, MAOIs, and cocaine.

Choreoathetosis

Choreoathetotic movements are relatively fast (slower than tics or myoclonus and faster than athetosis or dystonia), asymmetric, writhing movements that are irregular but relatively stereotyped. Tardive dyskinesia (discussed below) is a particular type of chorea but in some cases choreoathetosis begins relatively soon after initiation of therapy, unlike the deferred onset characteristic of the tardive syndromes. Agents reported to induce chorea include neuroleptic agents, aminophylline, amphetamines, amoxapine, anabolic steroids, anticholinergic drugs, carbamazepine, ethosuximide, phenobarbital, phenytoin, valproic acid, baclofen, cimetidine, cocaine, cyclosporin, levodopa, lithium, methadone, oral contraceptives, theophylline, and tricyclic antidepressants.[49]

Tardive Dyskinesia

Classic tardive dyskinesia is a relatively rapid choreic movement disorder that follows chronic exposure to neuroleptic medications. Patients over age 55 are particularly vulnerable to the development of tardive dyskinesia; 25% exhibit the syndrome after 1 year of neuroleptic therapy and 50% to 60% have dyskinetic movements after 3 years of cumulative antipsychotic treatment.[69,70] Once present, tardive dyskinesia may be permanent despite cessation of antipsychotic therapy. After onset, the movements tend to be relatively stable, although they may fluctuate from examination to examination.[71,72] Female gender, mood disorder as the underlying psychiatric illness, history of neuroleptic-induced parkinsonism, and the presence of a brain disorder all increase the likelihood of developing tardive dyskinesia.[73] Likewise, higher doses of neuroleptics and increasing duration of neuroleptic exposure increase the prevalence of tardive dyskinesia.

Tardive dyskinesia produces a distinctive syndrome with stereotyped, smacking-type movements of the lips, tongue, and cheeks. In addition, piano-playing movements of the fingers are often evident, and toes may "dance" or show irregular flexion–extension and fanning movements. Truncal and respiratory musculature is involved in some cases.[73]

In many cases, the diagnosis of tardive dyskinesia is obvious whereas in others it must be distinguished from catatonia (discussed above), neurological disorders with chorea such as Huntington's disease drug-induced disorders, or idiopathic rest tremor. Table 20.7 presents features that distinguish tardive dyskinesia from other movement disorders in patients with psychiatric illnesses. The tendency of tardive dyskinesia to involve the lower face and distal limbs and to be decreased when the limb is called into action and increased when attention is distracted from the movements helps differentiate tardive dyskinesia from other movement disorders. The ability to volitionally suppress the dyskinetic movements and the minimal subjective awareness of the movements are also helpful distinguishing features. Tardive dyskinesia must be distinguished from other types of choreiform movement disorders (Chapter 18). As noted above, a variety of non-neuroleptic medications are also capable of inducing choreiform

TABLE 20.7. *Differential Diagnosis of Tardive Dyskinesia, Catatonia, and Huntington's Disease*

Features	*Tardive Dyskinesia*	*Catatonia*	*Huntington's Disease*	*Rest Tremor*
Facial distribution	Lower	Upper and lower	Upper and lower	Lower
Limb distribution	Distal	Distal	Proximal > distal	Distal
Respiratory movements affected	Frequently	Rarely	Frequently	No
Effect of action or movement	Decreases	Variable effects	No effect	Decreases
Effect of conscious attention to movement	Decreases	No effect	No effect	Increases
Volitionally suppressible	Yes	Yes	No	No
Effect of distraction on movements	Increases	No effect	No effect	Increases
Subjective awarenessof movements	Minimal	Minimal	Variable	Yes
Other catatonic phenomena present	Usually not	Yes	Usually not	No
Course over time	Stable or worsen	Wax and wane	Steadily progressive	Stable

TABLE 20.8. *Disorders with Abnormal Facial Movements*

Category	*Disorders*
Dystonia	Idiopathic cranial–cervical dystonia (Meige's syndrome)
	Blepharospasm
	Oromandibular dystonia
	Cervical dystonia
	Spasmodic dysphonia
	Adult-onset idiopathic generalized dystonia
	Inherited dystonia musculorum deformans
	Cranial dystonia caused by brainstem-diencephalic-cerebellar lesions
	Acute drug-induced dystonia
	Tongue, jaws, neck
	Oculogyric crises
	Tardive dystonia
Chorea	Tardive dyskinesia
	Spontaneous orofacial dyskinesia of elderly
	Edentulous orofacial dyskinesia
	Other choreatic disorders: Huntington's disease, Sydenham's chorea, chorea gravidarum, Lesch-Nyhan syndrome, hyperthyroidism, hypernatremia, and drug reactions (dopaminergic, anticonvulsants, oral contraceptives, and antihistamines)
Tremor	Essential tremor (head, chin, voice)
	Exaggerated physiologic tremors
	Parkinsonian tremor (jaw, chin, lips, tongue, "rabbit syndrome")
	Parkinson's disease
	Drug-induced parkinsonism
	Dystonic tremor
	Cerebellar tremor and titubation
	Head flopping
	Head nodding (congenital nystagmus, spasmus natuns)
	"Bobble-headed doll" syndrome
Tics	Gilles de la Tourette's syndrome
	Transient or chronic simple tics
	Secondary (acquired) tourettism
	Neuroacanthocytosis
Miscellaneous	Stereotypic movements with psychosis
	Mannerisms
	Habits
	Focal seizures of facial musculature
	Facial myoclonus
	Hemifacial spasm
	Synkinetic movements following Bell's palsy
	Oculomasticatory myorrhythmia in Whipple's disease
	Myokymia with multiple sclerosis, brainstem lesions
	Oculogyric crises
	Post-encephalitus parkinsonism
	Drug-induced
	Palatal myoclonus

movements, and these must be considered in the differential diagnosis.

In addition to the classic dyskinetic movements, a variety of alternate manifestations of tardive dyskinesia have been identified, including tardive dystonia, tardive akathisia, tardive tic syndrome, tardive myoclonus, and a tardive complex consisting of three or more types of tardive movement disorders occurring simultaneously.[74,75]

The most common manifestation of tardive dystonia is dystonic posturing of the neck such as toricollis or anterocollis. Blepharospasm with tonic or clonic eye blinking may occur and oral mandibular dystonia is not uncommon. The combination of oral mandibular dystonia and blepharospasm is known as *tardive Meige's syndrome*. Spasmodic dysphonia with dystonic type of speech can also be a manifestation of a tardive movement disorder. Segmental dystonia (involving two adjacent body segments such as the neck and shoulder) or generalized dystonia are rare manifestations of tardive dystonic disorder.

Tardive akathisia is a pseudoakathisia syndrome with marked restlessness and forced movement with a limited subjective sense of restlessness. Tardive tics may consist of either invariant tics or a full-blown tic and vocalization disorder similar to Gilles de la Tourette's syndrome (Chapter 19). Finally, tardive myoclonus with intermittent, asymmetric myoclonic jerking may also occur on a tardive basis (Table 20.4).

Tardive dyskinesia must be distinguished from a variety of hyperactive facial movements (Table 20.8).

The pathophysiology of tardive dyskinesia is unclear. Fluorodeoxyglucose positron emission tomography reveals marked hyperactivity in the globus pallidus and primary motor cortex of patients with the syndrome,[76] which suggests that chronic dopamine blockade has resulted in a hyperactivity of basal ganglia motor circuits.

Tardive dyskinesia is a treatment-resistant syndrome. It can be temporarily suppressed through the introduction or increased dosing of neuroleptic medications; however, this may eventually lead to exacerbation of the dyskinesia. Atypical neuroleptics such as clozapine, risperidone, olanzapine, ziprasidone, and quetiapine tend not to produce tardive dyskinesia and may allow the resolution of dyskinetic movements when administered chronically to dyskinetic patients. Dopamine-depleting agents such as reserpine may reduce the movements. b-Adrenergic-blocking drugs such as propranolol are useful in some patients. Tardive dystonia syndromes may respond to administration of high doses of anticholinergic drugs, but these agents may worsen classical dyskinesia. Baclofen, clonazapam, or diazepam have been useful in individual patients.[73] Prevention of the emergence of tardive dyskinesia through minimizing use of conventional neuroleptics and employing atypical antipsychotics when chronic therapy is necessary is a more successful strategy than suppressing or eliminating tardive dyskinesia once it has appeared.

Neuroleptic Malignant Syndrome

Neuroleptic malignant syndrome is an acute, life-threatening disorder characterized by hyperthermia, muscle rigidity, autonomic dysfunction, and mental status changes.[77] Autonomic abnormalities include tachypnea, dyspnea, cardiac arrhythymias, labile blood pressure, diaphoresis, pallor, flushing, and genitourinary dysfunction. Typical mental status changes include agitation, lethargy, delirium, and, eventually, stupor and coma. In addition to muscular rigidity there may be tremor, dystonia, chorea, or myoclonus. Occasional patients have seizures, nystagmus, and extensor plantar responses.[77] Many patients (45%–90%) have elevated serum creatine kinase levels.

Patients exhibiting this syndrome require support for vital functions, or death will ensue rapidly. Management in an intensive care unit with ventilatory support and cardiac mentoring is usually necessary. In addition to support of vital signs, treatment with dantrolene sodium or bromocriptine or a combination of the two may help bring the syndrome to a more rapid termination.[77]

Treatment with a conventional neuroleptic is the most common cause of neuroleptic malignant syndrome. However, it also has been observed following treatment with metaclopramide, in patients abusing cocaine, and in patients with Parkinson's disease who are suddenly withdrawn from dopaminergic medication.[77–79]

REFERENCES

1. Cummings JL. Frontal–subcortical circuits and human behavior. Arch Neurol 1993;50:873–880.
2. Mega MS, Cummings JL. Frontal–subcortical circuits and neuropsychiatric disorders. J Neuropsychiatry Clin Neurosci 1994;6: 358–370.
3. Taylor MA. Catatonia: a review of a behavioral neurologic syndrome. Neuropsychiatry Neurosurg Behav Neurol 1990;3:48–72.
4. Scheepers B, Rogers D, et al. Catatonia: a neuropsychiatric disorder. Behav Neurol 1995;8:157–161.
5. Hamilton M. Fish's Schizophrenia. Boston: Wright PSG, 1984.
6. Taylor MA. The Neuropsychiatric Guide to Modern Everyday Psychiatry. New York: The Free Press, 1993.
7. Berrios GE. Stupor revisited. Compr Psychiatry 1981;22:466–478.

8. Morrison JR. Retarded and excited types. Arch Gen Psychiatry 1973;28:39–41.
9. Northoff G, Wenke J, et al. Ball experiments in 32 acute akinetic catatonic patients: deficits of internal initiation and generation of movements. Mov Disord 1995;10:589–595.
10. Abrams R, Taylor MA. Catatonia. Arch Gen Psychiatry 1976; 33:579–581.
11. Taylor MA, Abrams R. Catatonia. Arch Gen Psychiatry 1977; 34:1223–1225.
12. Manschreck TC, Maher BA, et al. Disturbed voluntary motor activity in schizophrenic disorder. Psychol Med 1982;12:73–84.
13. Starkstein SE, Petracca G, et al. Catatonia in depression: prevalence, clinical correlates, and validation of a scale. J Neurol Neurosurg Psychiatry 1996;60:326–332.
14. Wilcox JA, Nasrallah HA. Organic factors in catatonia. Br J Psychiatry 1986;149:782–784.
15. Carroll BT, Anfinson TJ, et al. Catatonic disorder due to general medical conditions. J Neuropsychiatry Clin Neurosci 1994; 6:122–133.
16. Fisher CM. Catatonia due to disulfiram toxicity. Arch Neurol 1989;46:798–804.
17. Gelenberg AJ. The catatonic syndrome. Lancet 1976;1:1339–1341.
18. Herman M, Harpham D, et al. Nonschizophrenic catatonic states. NY State J Med 1942;42:624–627.
19. Lohr JB, Wisniewski AA. Movement Disorders New York: The Guilford Press, 1987.
20. Pfister HW, Preac-Mursic V, et al. Catatonic syndrome in acute severe encephalitis due to borrelia burgdorferi infection. Neurology 1993;43:433–435.
21. Philbrick KL, Rummans TA. Malignant catatonia. J Neuropsychiatry Clin Neurosci 1994;6:1–13.
22. Rogers D, Karki C, et al. The motor disorders of mental handicap: an overlap with the motor disorders of severe psychiatric illness. Br J Psychiatry 1991;158:97–102.
23. Saver JL, Greenstein P, et al. Asymmetric catalepsy after right hemisphere stroke. Mov Disord 1993;8:69–73.
24. Sternbach H, Yager J. Catatonia in the presence of mid-brain and brainstem abnormalities. J Clin Psychiatry 1981;42:352–353.
25. Tandon R, Walden M, Falcon S. Catatonia as a manifestation of paraneoplastic encephalopathy. J Clin Psychaitry 1988;49: 121–122.
26. American Psychiatric Association. Diagnostic and Statistical Manual of Mental Disorders, 4th ed. Washington, DC: American Psychiatric Press, 1994.
27. Castillo E, Rubin RT, Holsboer-Trachsler E. Clinical differentiation between lethal catatonia and neuroleptic malignant syndrome. Am J Psychiatry 1989;146:324–328.
28. Mann SC, Caroff SN, et al. Lethal catatonia. Am J Psychiatry 1986;143:1374–1381.
29. Greenfeld D, Conrad C, et al. Treatment of catatonia with low-dose lorazepam. Am J Psychiatry 1987;144:1224–1225.
30. McCall WV, Shelp FE, McDonald WM. Controlled investigation of the amobarbital interview for catatonic mutism. Am J Psychiatry 1992;149:202–206.
31. McEvoy JP, Lohr JB. Diazepam for catatonia. Am J Psychiatry 1984;141:284–285.
32. Menza MA, Harris D. Benzodiazepines and catatonia: an overview. Biol Psychiatry 1989;26:842–846.
33. Salam SA, Pillai AK, Beresford TP. Lorazepam for psychogenic catatonia. Am J Psychiatry 1987;144:1082–1083.
34. Bush G, Fink M, et al. Catatonia II: treatment with lorazepam and electroconvulsive therapy. Acta Psychiatr Scand 1996;93: 137–143.
35. Petrides G, Divadeenam KM, et al. Synergism of lorazepam and electroconvulsive therapy in the treatment of catatonia. Biol Psychiatry 1997;42:375–381.
36. Rohland BM, Carroll BT, Jacoby RG. ECT in the treatment of the catatonic syndrome. J Affect Disord 1993;29:255–261.
37. Yeung PP, Milstein RM, et al. ECT for lorazepam-refractory catatonia. Convuls Ther 1996;12:31–35.
38. Rogers D. The motor disorders of severe psychiatric illness: a conflict of paradigms. Br J Psychiatry 1985;147:221–232.
39. Gupta S, Andreasen NC, et al. Neurologic soft signs in neuroleptic-naive and neuroleptic-treated schizophrenic patients and in normal comparison subjects. Am J Psychiatry 1995;152: 191–196.
40. Heinrichs DW, Buchanan RW. Significance and meaning of neurological signs in schizophrenia. Am J Psychiatry 1988;145: 11–18.
41. Sanders RD, Keshavan MS, Schooler NR. Neurological examination abnormalities in neuroleptic-naive patients with first-break schizophrenia: preliminary results. Am J Psychiatry 1994;151:1231–1233.
42. Krakowski MI, Convit A, et al. Neurological impairment in violent schizophrenic inpatients. Am J Psychiatry 1989;146:849–853.
43. Altman EG, Hedeker DR, et al. The clinician-administered rating scale for mania (CARS-M): development, reliability, and validity. Biol Psychiatry 1994;36:124–134.
44. Mukherjee S, Shukla S, Rosen A. Neurological abnormalities in patients with bipolar disorder. Biol Psychiatry 1984;19:337–345.
45. Nasrallah HA, Tippin J, McCalley-Whitters M. Neurological soft signs in manic patients: a comparison with schizophrenic and control groups. J Affect Disord 1983;5:45–50.
46. Widlocker D. Retardation: a basic emotional response? In: Davis JM, ed. The Affective Disorders. Washington DC: American Psychiatric Press, 1983:165–181.
47. Parker G, Hadzi-Pavlovic D. Development and structure of the CORE system. In: Parker G, ed. Melancholia: A Disorder of Movement and Mood. Cambridge, UK: Cambridge University Press, 1996:82–129.
48. Leff JP, Isaacs AD. Psychiatric Examination in Clinical Practice. London: Blackwell Scientific Publications, 1990.
49. Jain KK. Drug-Induced Neurological Disorders. Seattle: Hogrefe & Huber Publishers, 1996.
50. Casey DE. Neuroleptic-induced acute dystonia. In: Lang AE, ed. Drug-Induced Movement Disorders. Mount Kisco, NY: Futura Publishing Co., 1992:21–41.
51. Aguilar EJ, Keshavan MS, et al. Predictors of acute dystonia in first episode psychotic patients. Am J Psychiatry 1994;151: 1819–1921.
52. Khanna R, Das A, Damodaran SS. Prospective study of neuroleptic-induced dystonia in mania and schizophrenia. Am J Psychiatry 1992;149:511–513.
53. Adler LA, Angrist B. Acute neuroleptic-induced akathisia. In: Lang AE, ed. Drug-Induced Movement Disorders. Mount Kisco, NY: Futura Publishing Co., 1992:85–119.
54. Barnes TRE. A rating scale for drug-induced akathisia. Br J Psychiatry 1989;154:672–676.
55. Gibb WRG, Lees AJ. The clinical phenomenon of akathisia. J Neurol Neurosurg Psychiatry 1986;49:861–866.
56. Sachdev P. A rating scale for acute drug-induced akathisia: development, reliability, and validity. Biol Psychiatry 1994;35: 263–271.

57. Lang AE. Withdrawal akathisia: case reports and a proposed classification of chronic akathisia. Mov Disord 1994;9:188–192.
58. Cohen BM, Keck PE, et al. Prevalence and severity of akathisia in patients on clozapine. Biol Psychiatry 1991;29:1215–1219.
59. Eberstein S, Adler LA, et al. Nefazodone and akathisia. Biol Psychiatry 1996;40:798–799.
60. Fleishman SB, Lavin MR, et al. Antiemetic-induced akathisia in cancer patients receiving chemotherapy. Am J Psychiatry 1994; 151:763–765.
61. Micheli F, Pardal MF, et al. Flunarizine- and cinnarizine-induced extrapyramidal reactions. Neurology 1987;37:881–884.
62. Fudge J, Wilner PJ. Akathisia in Parkinson's disease. Neuropsychiatry Neurosurg Behav Neurol 1996;9:248–253.
63. Daigneault S, Braun CMJ, et al. Pseudoakathisia from an acquired lesion. Neuropsychiatry Neuropsychol Behav Neurol 1998;11:164–170.
64. Fleischhacker WW, Roth SD, Kane JM. The pharmacologic treatment of neuroleptic-induced akathisia. J Clin Psychopharmacol 1990;10:12–21.
65. Sethi KD, Zamrini EY. Asymmetry in clinical features of drug-induced parkinsonism. J Neuropsychiatry Clin Neurosci 1990;2: 64–66.
66. Friedman JH. Drug-induced parkinsonism. In: Lang AE, ed. Drug-Induced Movement Disorders. Mount Kisco, NY: Futura Publishing Co., 1992:41–83.
67. Riley DE. Antidepressant therapy and movement disorders. In: Lang AE, ed. Drug-Induced Movement Disorders. Mount Kisco, NY: Futura Publishing Co., 1992:231–255.
68. Weiner WJ, Sanchez-Ramos J. Movement disorders and dopaminomimetic stimulant drugs. In: Lang AE, ed. Drug-Induced Movement Disorders. Mount Kisco, NY: Futura Publishing Co., 1992:315–337.
69. Jeste DV, Caligiuri MP, et al. Risk of tardive dyskinesia in older patients: a prospective longitudinal study of 266 outpatients. Arch Gen Psychiatry 1995;52:756–765.
70. Woerner MG, Alvir JMJ, et al. Prospective study of tardive dyskinesia in the elderly: rates and risk factors. Am J Psychiatry 1998;155:1521–1528.
71. Bergen JA, Eyland EA, et al. The course of tardive dyskinesia in patients on long-term neuroleptics. Br J Psychiatry 1989;154: 523–538.
72. Gardos G, Casey DE, et al. Ten-year outcome of tardive dyskinesia. Am J Psychiatry 1994;151:836–841.
73. Khot V, Egan MF, et al. Neuroleptics and classic tardive dyskinesia. In: Lang AE, Weiner WJ, eds. Drug-Induced Movement Disorders. Mount Kisco, NY: Futura Publishing Co., 1992:121–166.
74. Gardos G, Cole JO, et al. Clinical forms of severe tardive dyskinesia. Am J Psychiatry 1987;144:895–902.
75. Wirshing WC, Cummings JL. Tardive movement disorders. Neuropsychiatry Neurosurg Behav Neurology 1990;3:23–25.
76. Pahl JJ, Mazziotta JC, et al. Positron-emission tomography in tardive dyskinesia. J Neuropsychaitry Clin Neurosci 1995;7: 457–465.
77. Factor SA, Singer C. Neuroleptic malignant syndrome. In: Lang AE, ed. Drug-Induced Movement Disorders. Mount Kisco, NY: Futura Publishing Co., 1991:199–230.
78. Addonizio G, Susman VL, Roth SD. Neuroleptic malignant syndrome: review and analysis of 115 cases. Biol Psychiatry 1987; 22:1004–1020.
79. Keck PE, Pope HG Jr, et al. Risk factors for neuroleptic malignant syndrome: a case–control study. Arch Gen Psychiatry 1989;46:914–918.
80. Bush G, Fink M, et al. Catatonia I: rating scale and standardized examination. Acta Psychiatr Scand 1996;93:129–136.
81. Lund CE, Mortimer AM, et al. Motor, volitional and behavioural disorders in schizophrenia, I: assessment using the Modified Rogers scale. Br J Psychiatry 1991;158:323–327.
82. McKenna PJ, Lund CE, et al. Motor, volitional and behavioural disorders in schizophrenia II: The 'conflict of paradigms' hypothesis. Br J Psychiatry 1991;158:328–336.

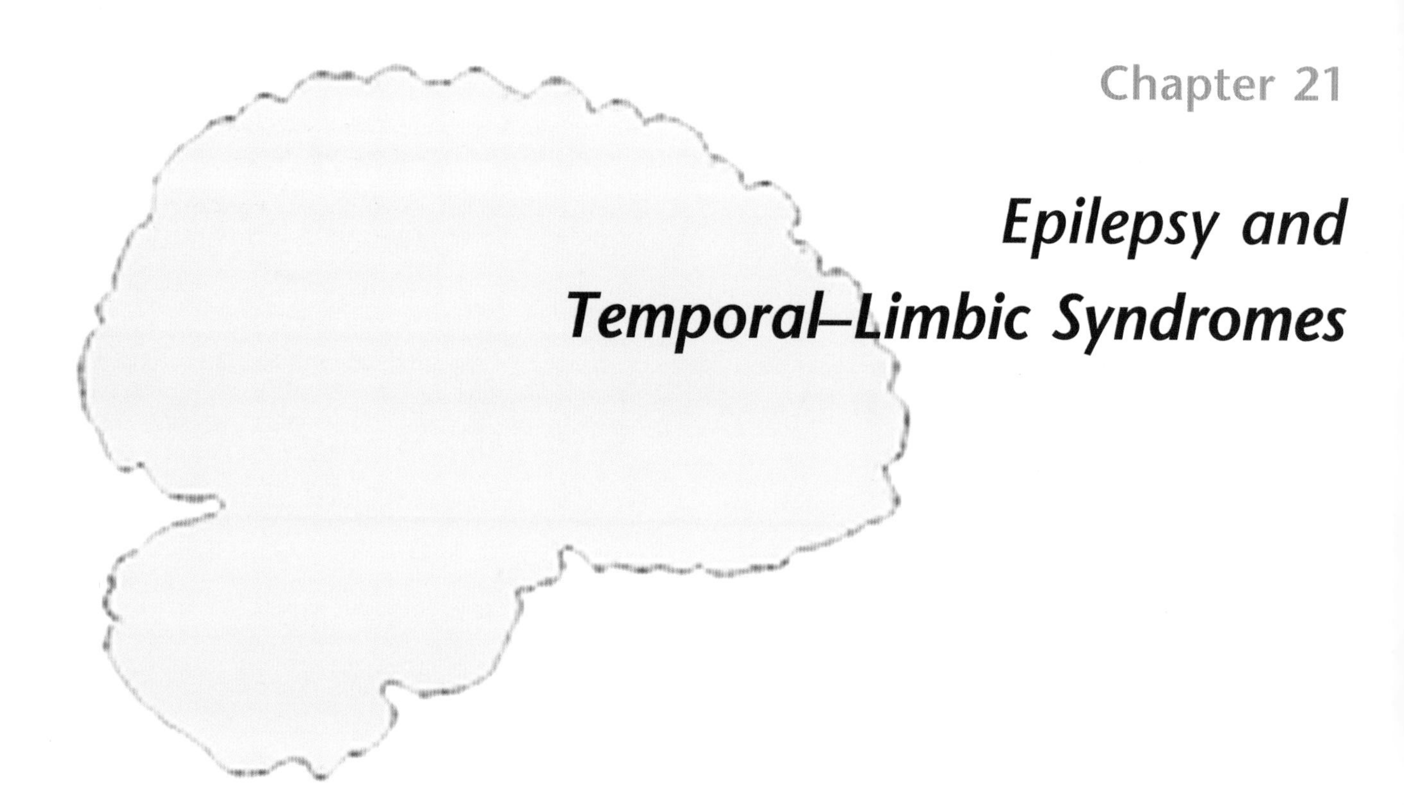

Chapter 21

Epilepsy and Temporal–Limbic Syndromes

The limbic system is the anatomical mediator of many, if not most, neuropsychiatric syndromes. The limbic system provides the anatomical basis for emotional function while the neocortex provides the anatomical basis for instrumental cognitive functions such as language, praxis, and visual recognition (Chapters 2 and 5). Dysfunction of the amygdala may provide the most rich and diverse array of neuropsychiatric syndromes of any single structure within the brain. Thus, the neuropsychiatry of temporal lobe syndromes ranks as one of the most important areas for neuropsychiatry. The limbic system embraces the orbitofrontal cortex and anterior cingulate cortex (Chapter 9), portions of the basal ganglia (Chapter 18), as well as the thalamus and hypothalamus. This chapter will present the neuropsychiatry of temporal and limbic syndromes.

Temporal and limbic syndromes include a wide array of neuropsychiatric disturbances (Table 21.1). Among the epilepsies, temporal lobe epilepsy, frontal lobe epilepsy, and gelastic seizures associated with hypothalamic hamartomas all have foci within the limbic system. Memory disorders are a common product of limbic system dysfunction. Amnesia occurs with lesions of the hippocampal–medial hemispheric limbic system and retrieval deficit syndromes occur with dysfunction of the frontal–subcortical circuits (Chapter 7). Confabulation and paramnesia may accompany amnestic disorders. Personality alterations are common with limbic system injury; apathy, disinhibition, irritability, and the Gastaut-Geschwind syndrome have been described with limbic lesions (Chapter 15). Psychotic disorders are associated with limbic system dysfunction including schizophrenia, the schizophrenia-like psychoses of epilepsy, and psychoses associated with temporal lobe tumors, infarction, neoplasms, and infections (Chapter 12). Mood disorders including depression, mania/hypomania, euphoria, anger/rage, and ecstasy or religious feelings have also been described in patients with limbic disorders (Chapter 14). The Klüver-Bucy syndrome and isolated fragments of the disorder occur with bilateral medial temporal, thalamic, or hypothalamic damage. Changes in sexual behavior including hypersexuality, hyposexuality, and paraphilias may also be manifestations of limbic dysfunction (Chapter 23). Orbitorontal abnormalities involving the limbic system produce utilization and imitation behaviors. Addiction and related disorders appear to be mediated by the nucleus accumbens and a related system of limbic structures. Limbic hypothalamic disorders include abnor-

TABLE 21.1. *Temporal Lobe and Limbic Syndromes*

Syndrome Category	*Syndromes*
Epilepsy	Temporal lobe epilepsy
	Frontal lobe epilepsy
	Hypothalamic hamartomas with gelastic seizures
Memory disorders	Amnesia
	Retrieval deficit syndrome
	Confabulation
	Paramnesia
Personality changes	Apathy
	Disinhibition
	Irritability
	Gastaut-Geschwind syndrome
Psychoses	Schizophrenia
	Schizophrenia-like psychoses of epilepsy
	Psychoses with temporal lobe trauma, infarction, neoplasm, infection
Mood disorders	Depression
	Mania/hypomania
	Euphoria
	Anger/rage
	Ecstacy/religious feelings
Klüver-Bucy syndrome and related disorders	Klüver-Bucy syndrome
	Hyperorality
Sexual behavior changes	Hypersexuality
	Hyposexuality
	Paraphilia
Frontal–limbic behavior disorders	Utilization behavior
	Imitation behavior
Addiction and related disorders	
Limbic–hypothalamic syndromes	Disorders of eating and appetite
	Disorders of sexual function
	Precocious puberty
	Disorders of thirst and drinking
	Kleine-Levin syndrome
Miscellaneous	Autonomic dysfunction
	Anxiety
	Aggression
	Agitation
	Dissociative disorder
	Gourmand syndrome

malities of eating and appetite, disorders of sexual function, precocious puberty, disorders of thirst and drinking, and the Kleine-Levin syndrome (Chapter 23). In addition, autonomic dysfunction, anxiety, agitation, aggression, dissociative disorders, and the Gourmand syndrome (preoccupation with food) also reflect limbic dysfunction.

Seizures and epileptic disorders will be presented first in this chapter, with an emphasis on temporal lobe epilepsy and related conditions, and then non-epileptic temporal–limbic and related disorders not discussed in other chapters, particularly the Klüver-Bucy syndrome, will be described.

EPILEPSY

The term *epilepsy* derives from the Greek *epilepsia*, meaning "to take hold or to seize."[1] Epilepsy is a common disorder affecting approximately 1% of the population and may involve individuals of any age. It is associated with a plethora of behavioral changes ranging from minor alterations of consciousness during a brief seizure to chronic schizophrenia-like psychoses and affective disorders persisting throughout the interictal period. Some of the behaviors observed in epileptic patients are a direct manifestation of the epileptic cerebral discharge, whereas others may be related to the existence of a lesion giving rise to both the epilepsy and the behavioral changes. The effects of chronic anticonvulsant therapy and the psychological consequences of suffering from an unpredictable, socially disabling disease also contribute to the behavioral alterations of epileptic patients.

Classification and Differential Diagnosis of Seizures and of Epilepsies

Seizures and epilepsies must be distinguished. *Seizures* are convulsions that may be produced by a wide variety of events, including alcohol and drug withdrawal syndromes, hypoglycemia, transient cerebral anoxia, and epileptic syndromes (Fig. 21.1). *Epilepsies* are characterized by recurrent seizures and their classification is based on a variety of types of information in addition to seizure type such as age of onset, intellectual development, findings on neurological examination, and results of neuroimaging studies.[2] No seizure type is pathognomonic of any epileptic syndromes and most epileptic syndrome involve more than one type of seizure. When evaluating a patient presenting with a seizure, the first consideration is whether the seizure is one manifestation of an epileptic syndrome or secondary to some other condition.

Seizures also must be distinguished from a variety of nonseizure causes of loss of consciousness or transient behavioral disturbances (Table 21.2). These include pseudoepileptic seizures, syncope, hypoglycemia, delirium, transient ischemic attacks, sleep disorders including narcolepsy and rapid eye movement (REM) be-

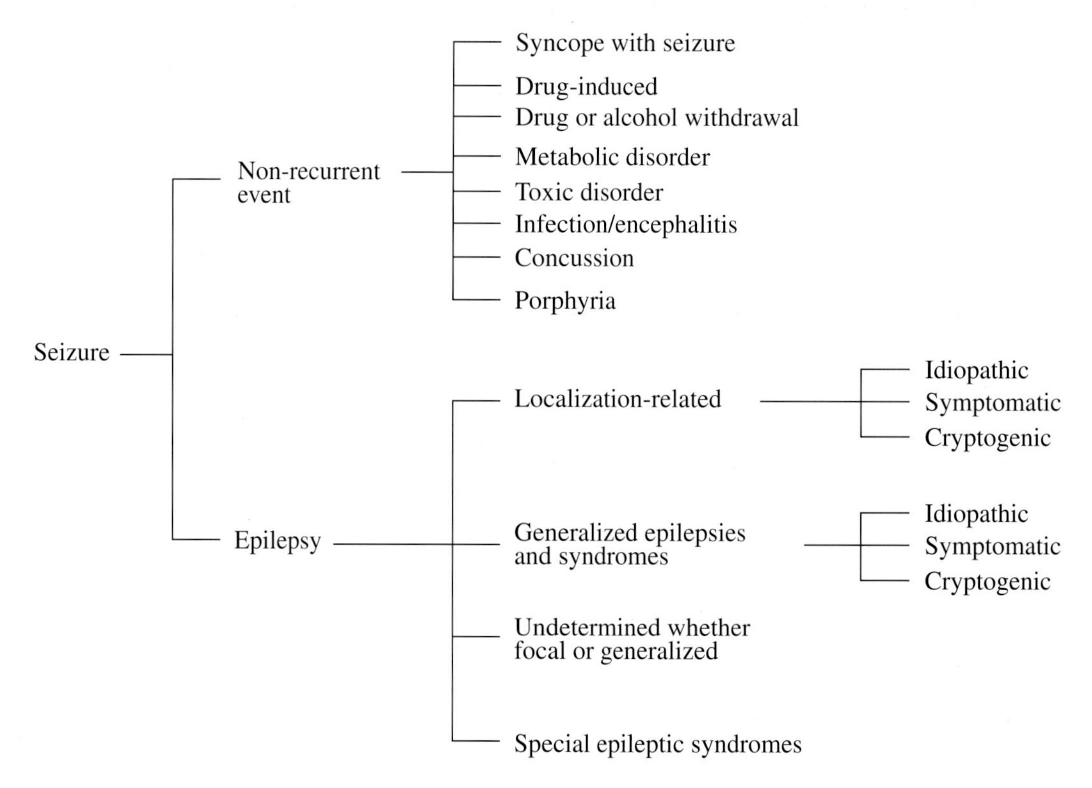

FIGURE 21.1 Differential diagnostic approach to seizures. (Table 22.3 provides additional information on seizures and Table 22.4 provides information on epilepsies.)

TABLE 21.2. *Causes of Loss of Consciousness or Transient Behavioral Changes to Be Distinguished from Epileptic Seizures*

- Systemic disorders
  - Syncope
    - Respiratory syncope
    - Hypotensive syncope
    - Cardiac syncope
      - Bradyarhythmias (sick sinus syndrome)
      - Heart block
      - Tachyarhythmias (atrial, ventricular)
    - Circulatory syncope
      - Hypovolemia
      - Pulmonary embolism
      - Autonomic nervous system dysfunction
      - Drug-related syncope
        - Nitrates
        - Calcium channel blockers
        - Vasodilators
        - Antihypertensive
    - Reflex syncope
      - Vasodepressor (vasovagal) syncope
      - Carotid sinus syncope
      - Other reflex triggers (glossopharyngeal, post-micturition, post-tussive, ocular, splanchnic, cerebral, esophageal)
  - Endocrine disorders
    - Hypoglycemia
    - Hyperglycemia
    - Pheochromocytoma
    - Carcinoid syndrome
  - Toxic and metabolic disorders
    - Waxing and waning delirium
    - Alcohol or drug-related intoxication and withdrawal
- Neurologic disorders
  - Dizziness and vertigo
  - Cerebrovascular events
    - Transient ischemic attacks
    - Transient global amnesia
    - Drop attacks
  - Migraine
  - Intermittent movement disorders
    - Acute dystonic reactions
- Neurologic disorders (cont'd)
    - Paroxysmal ataxia
    - Tics
    - Paroxysmal dystonic choreoathetosis
    - Paroxysmal kinesigenic choreoathetosis
    - Restless legs (periodic movements of sleep)
    - Myoclonus
    - Asterixis
    - Startle disease (hyperekplexia)
- Sleep disorders
  - Narcolepsy
  - Idiopathic hypersomnolence
  - Isolated cataplexy
  - Sleep apnea
  - Somnambulism and somniloquy
  - Night terrors
  - Enuresis
  - Bruxism
  - Periodic movements of sleep
  - REM behavior disorder
- Psychiatric disorders
  - Psychogenic seizures
  - Episodic dyscontrol
  - Malingering
  - Dissociative states
    - Psychogenic fugue
    - Multiple personality disorders
    - Depersonalization
  - Self-mutilatory behavior
  - Anxiety
    - Panic attacks
    - Hyperventilation
- Childhood disorders
  - Temper tantrums
  - Breath-holding spells
  - Pallid infantile syncope
  - Hiatal hernia (Sandifer's syndrome)

REM, rapid eye movement.

TABLE 21.3. *Classification of Seizures*

I. Partial seizures
  A. Simple partial seizures (consciousness not impaired)
    1. With motor signs
      a. Focal motor without march
      b. Focal motor with march (jacksonian)
      c. Versive
      d. Postural
      e. Phonatory (vocalization or arrest of speech)
    2. With somatosensory or special-sensory symptoms (simple hallucinations, e.g., tingling, light flashes, buzzing)
      a. Somatosensory
      b. Visual
      c. Auditory
      d. Olfactory
      e. Gustatory
      f. Vertiginous
    3. With autonomic symptoms or signs (including epigastric sensation, pallor, sweating, flushing, piloerection, and pupillary dilation)
    4. With psychic symptoms (disturbance of higher cerebral function); these symptoms rarely occur without impairment of consciousness and are much more commonly experienced as complex partial seizures
      a. Dysphasic
      b. Dynamic (e.g., déjà-vu)
      c. Cognitive (e.g., dreamy states, distortions of time sense)
      d. Affective (fear, anger, etc.)
      e. Illusions (e.g., macropsia)
      f. Structured hallucinations (e.g., music, scenes)
  B. Complex partial seizures (with impairment of consciousness; may sometimes begin with simple symptomatology)
    1. Simple partial onset followed by impairment of consciousness
      a. With simple-partial features (I.A.1–I.A.4) followed by impaired consciousness
      b. With automatisms
    2. With impairment of consciousness at onset
      a. With impairment of consciousness only
      b. With automatisms
  C. Partial seizures evolving to secondarily generalized seizures (may be generalized tonic-clonic, tonic, or clonic)
    1. Simple-partial seizures (I.A) evolving to generalized seizures
    2. Complex-partial seizures (I.B) evolving to generalized seizures
    3. Simple-partial seizures evolving to complex-partial seizures evolving to generalized seizures
II. Generalized Seizures
  A. Absence seizures (typical)
    1. Impairment of consciousness only
    2. With mild clonic components
    3. With atonic components
    4. With tonic components
    5. With automatism
    6. With autonomic components (2 through 6 may be used alone or in combination).
  B. Atypical absence
    1. Changes in tone that are more pronounced than in II.A.1
    2. Onset and/or cessation that is not abrupt
  C. Myoclonic seizures; myoclonic jerks (single or multiple)
  D. Clonic seizures
  E. Tonic seizures
  F. Tonic-clonic seizures
  G. Atonic seizures (astatic)
  H. Combinations of the above may occur, e.g., C and G, C and E

havior disorder, intermittent movement disorders, psychogenic seizures, rage attacks, dissociative states, and anxiety-related syndromes.[3,4] In children, seizures must be distinguished from migraine, night terrors, hyperventilation attacks, temper tantrums, Sandifer's syndrome (episodic tonic axial extension caused by hiatus hernia), breath-holding spells, pallid infantile syncope, cardiac disorders, cataplexy, and movement disorders.[5]

Seizures associated with epileptic syndromes may be either partial or generalized (Table 21.3). *Simple-partial seizures* have a focal onset and produce no impairment of consciousness. There may focal be motor manifestations (limb movement, head turning, postural changes, vocalization), somatosensory symptoms (visual hallucinations, auditory hallucinations, etc.), autonomic signs, or "psychic" symptoms such as apha-

TABLE 21.4. *Classification of Epilepsies and Epileptic Syndromes*

I. Localization-related (focal, local, partial) epilepsies and syndromes
   1. Idiopathic (with age-related onset)
      - Benign childhood epilepsy with centrotemporal spikes
      - Childhood epilepsy with occipital paroxysms
      - Primary reading epilepsy
   2. Symptomatic
      - Chronic progressive epilepsia partialis continua of childhood (Kojewnikow's syndrome)
      - Syndromes characterized by seizures with specific modes of precipitation (for example, reflex epilepsy)
      - Temporal lobe epilepsies (amygdalohippocampal, lateral)
      - Frontal lobe epilepsies (supplementary motor, cingulate, anterior frontopolar, orbitofrontal, dorsolateral, opercular, motor cortex)
      - Parietal lobe epilepsies
      - Occipital lobe epilepsies
   3. Cryptogenic

II. Generalized epilepsies and syndromes
   1. Idiopathic (with age-related onset)
      - Benign neonatal familial convulsions
      - Benign neonatal convulsions
      - Childhood absence epilepsy (pyknolepsy)
      - Juvenile absence epilepsy
      - Juvenile myoclonic epilepsy
      - Epilepsy with grand mal seizures (generalized tonic-clonic seizures) on awakening
      - Other generalized idiopathic epilepsies not defined above
      - Epilepsies with seizures precipitated by specific modes of activation
   2. Cryptogenic or symptomatic
      - West syndrome (infantile spasms, Blitz-Nick-Salaam Krämpfe)
      - Lennox-Gastaut syndrome
      - Epilepsy with myoclonic-astatic seizures
      - Epilepsy with myoclonic absences
   3. Symptomatic
      a. Nonspecific cause
         - Early myoclonic encephalopathy
         - Early infantile epileptic encephalopathy with suppression-burst
         - Other symptomatic generalized epilepsies not defined above
      b. Specific syndromes
         - Epileptic seizures complicating disease states

III. Epilepsies and syndromes undetermined, whether focal or generalized
   1. With both generalized and focal seizures
      - Neonatal seizures
      - Severe myoclonic epilepsy in infancy
      - Epilepsy with continuous spike-wave activity during slow-wave sleep
      - Acquired epileptic aphasia (Landau-Kleffner syndrome)
      - Other undetermined epilepsies not defined above
   2. Without unequivocal generalized or focal seizures

IV. Special syndromes
   1. Situation-related seizures
      - Febrile convulsions
   2. Isolated seizures or isolated status epilepticus

sia. *Complex-partial seizures* involve impairment of consciousness. Impaired consciousness may be the only manifestation of the seizure or there may be associated automatisms. Seizures with psychic symptoms commonly evolve to complex-partial seizures with impaired consciousness. Partial seizures may spread to adjacent structures, producing secondary generalized seizures of tonic-clonic, tonic, or clonic type. Generalized seizures include absence seizures, atypical absence seizures, myoclonic seizures, clonic seizures, tonic seizures, tonic-clonic seizures, and atonic (astatic) seizures or combinations of any of these.

Epilepsies consist of recurrent seizures. Table 21.4 provides the international classification of epilepsies and epileptic syndromes.[6] The primary distinction among the epileptic syndromes are those that are localization related and those that are generalized. *Localization-related epilepsies* include idiopathic syndromes such as benign childhood epilepsy with centrotemporal spikes, childhood epilepsy with occipital paroxysms, and primary reading epilepsy. Symptomatic localization-related epilepsies are those most important in neuropsychiatry and include the temporal lobe epilepsies and frontal lobe epilepsies. Cryptogenic localization-related epilepsies are those of undetermined etiology. Among the *generalized epilepsies* are a series of idiopathic syndromes, disorders that may be either cryptogenic or symptomatic, and conditions that are symptomatic and of known cause. *Idiopathic generalized epilepsies* include benign neonatal convulsions, childhood absence epilepsy, juvenile absence epilepsy, juvenile myoclonic epilepsy, and other generalized idiopathic epilepsies. Disorders that may be *either cryptogenic or symptomatic* include infantile spasms, West syndrome, Lennox-Gastaut syndrome, and epilepsy with myoclonic seizures or absences. *Symptomatic generalized epilepsies* include myoclonic encephalopathies and epileptic seizures complicating disease states such as the phakomatoses (e.g., tuberous sclerosis). Syndromes whose focal or generalized origin is unknown include acquired epileptic aphasia (Landau-Kleffner syndrome) and severe myoclonic epilepsy in infancy. Special syndromes whose origins are uncertain include febrile convulsions and isolated seizures (Table 21.4).

Among the most challenging differential diagnoses confronting the clinician caring for patients with epilepsy is identification of pseudoseizures or *pseudoepileptic seizures*. These events may closely imitate partial-complex or generalized seizures although they typically have a more gradual onset, less stereotyped body movements, less tongue biting, less micturition and defecation, less self-injury, limited impairment of consciousness, brief or no postictal confusion and a longer duration of seizures activity (Table 21.5)[4,7] (see also Chapter 22). Interictal electroencephalograms (EEGs) may be of little value in distinguishing the two conditions. The EEGs obtained during the event may capture the epileptic activity in those with epilepsy-related seizures, and telemetry with video monitoring may help to distinguish the two disorders. Prolactin is elevated in approximately 80% of patients with partial-complex and generalized seizures and is not elevated in those with nonepileptic seizures. The elevation is maximal approximately 20 minutes after the episode and is elevated from 3 to 10 times above baseline levels.[8] Reduction or elimination of anticonvulsants will exacerbate epileptic seizures and typically has little effect on pseudo-epileptic seizures. Pseudoepileptic seizures occur in up to 20% of epileptics and are a manifestation of significant concomitant psychopathology. Among the types of psychopathology occurring in patients with pseudoepileptic seizures, personality disorders such as borderline personality and antisocial personality are most common. Depression is a frequent

TABLE 21.5. *Characteristics Distinguishing Epileptic Seizures and Pseudoepileptic Seizure*

Characteristic	*Pseudoepileptic*	*Epileptic*
Clinical features		
Onset	Often gradual	Abrupt
Body movements	Struggling, asynchronous	Tonic, then tonic-clonic
Biting	Lips, arms, other areas	Tongue
Micturition	Rare	Common
Defecation	Rare	Occasional
Self-injury	Rare	Common
Postictal confusion	Absent	Present
Consciousness	Complete or partial retention	Lost
Duration	Several to many minutes	30 seconds to a few minutes
Initiation or termination by suggestion	Yes	No
EEG		
Interictal	± Abnormal	±Abnormal
Ictal	No change on artifact	Abnormal
Neuroendocrinology		
Prolactin	No change	Increase
Relation to anticonvulsant withdrawal	No increase in seizures	Increase in seizures

finding. Bipolar illness may also be observed and some patients satisfy criteria for Briquet's syndrome (somatization disorder).[9]

Behavioral Disturbances in Epileptic Patients

Behavioral alterations occurring in epileptic patients may occur in the parictal (peri-ictal) period; as a direct manifestation of epileptic discharge within the cerebral hemispheres during the ictal period; following the seizure in the postictal period; or between seizures as part of an interictal behavioral syndrome. In addition, some behavioral disturbances appear to be associated specifically with effective control of behaviors, the syndrome of "forced normalization," or the alternate psychoses (Table 21.6). Behavior disturbances are most common among patients with foci in temporal–limbic structures giving rise to temporal lobe epilepsy (TLE) and partial-complex seizures.

• **Parictal Disorder** The least intensively studied of the epilepsy-associated behavioral disorders are the parictal disturbances. These anticipate the occurrence of seizures by days or weeks, are most typical of the severe epilepsies, and are characterized by vague changes in mood, irritability, or, occasionally, psychosis.[10]

• **Ictal Behavioral Phenomena** A variety of behaviors occur in the course of simple-partial and complex-partial seizures. *Simple-partial seizures* are manifested by elementary motor phenomena; simple psychosensory or special sensory symptoms involving the somatosensory, visual, auditory, olfactory, or gustatory systems; autonomic signs or symptoms; or psychic symptoms without impairment of consciousness such as aphasia, memory distortions, cognitive alterations, mood changes, illusions, or hallucinations. If consciousness is impaired during or following these phenomena, the seizure is classified as a complex-partial seizure, whereas when consciousness is unimpaired the seizure is classified as a simple-partial seizure. Partial seizures may progress to secondary generalization. Memory distortions reported by patients in the course of the partial seizures include déjà vu, jamais vu, déjà entendu, jamais entendu, and déjà pensée. Cognitive alterations include dreamy states, depersonalization, forced thinking, and thought blocking. Mood symptoms include fear, depression, pleasant and unpleasant emotions, anxiety, embarrassment, and anger. Psychosensory symptoms include illusions (macropsia, micropsia, metamorphopsia) and complex hallucinations such as re-experiencing past events, seeing people, animals, or complex scenes, and hearing voices and complex sounds.

Partial-complex seizures may give rise to psychomotor automatisms that are usually simple, perseverative, poorly executed, purposeless motor behaviors (Table 21.7). Typical examples are pushing, groping, chewing, swallowing, spitting, lip smacking, rubbing, and plucking. Automatisms preceded by an initial motionless stare typically originate in the temporal lobe whereas automatisms without an initial stare usually originate from nontemporal lobe areas.[11] Speech automatisms may include shouting, screaming, or verbal reiteration such as repeating a brief statement over and over. Ictal changes of affect include epileptic laughter (gelastic seizures) or crying (dacrystic or quiritarian seizures). More complex automatisms include running (cursive seizures), drinking, removal of clothing, and masturbation. Directed aggression in the course of seizures is extremely rare, although injury may occur from poorly directed scratching, pushing, or agitated behavior occurring during the ictal or postictal confusional period. In some cases, status epilepticus with prolonged psychomotor automatisms may occur. Such patients have confusional behavior with automatisms persisting for hours to days.

Any combination of these phenomena may occur in TLE and in some cases a temporal lobe "march" can be discerned, with one symptom followed by another as the seizure activity spreads within temporal–limbic structures. Partial-complex seizures comprised of a

TABLE 21.6. *Classification of Behavioral Alterations Occurring in Epileptic Patients*

Parictal behavioral changes
Ictal behavioral events
Postictal behavioral changes
Interictal neuropsychiatric alterations
Personality changes
Schizophrenia-like psychoses
Mood disturbances
Dissociative disturbances
Depersonalization
Fugue states, poriomania
Multiple personality
Aggression
Altered sexual behavior
Forced normalization with behavioral alterations
Anticonvulsant-related psychopathology

TABLE 21.7. *Behavioral Alterations Occurring with Partial Seizures*

- Intellectual symptomatology
  - Aphasia
  - Memory distortion
    - Déjà vu, jamais vu
    - Déjà entendu, jamais entendu
    - Déjà penseé
  - Cognitive alterations
    - Dreamy state
    - Depersonalization
    - Forced thinking
    - Thought blocking
- Mood symptoms (ictal emotions)
  - Fear
  - Depression
  - Miscellaneous
  - Pleasant experience, serenity, ecstasy
  - Unpleasant experience
  - Anxiety
  - Embarrassment
  - Anger, irritability
- Psychosensory symptomatology
  - Illusions
  - Hallucinations
    - Auditory
    - Visual
    - Gustatory
    - Olfactory
    - Formication (tachle)
- Experential symptomatology
  - Feeling a presence
  - Feeling possessed
  - Feeling dead
  - Feeling of impending doom
- Psychomotor symptomatology
  - Simple psychomotor automatisms
    - Incoordination
    - Negativism
    - Staring
    - Pushing
    - Groping
    - Searching
    - Chewing
    - Swallowing
    - Spitting
    - Lip smacking
    - Rubbing
    - Plucking
  - Speech automatisms
    - Shouting
    - Screaming
    - Verbal reiteration
- Affective automatisms
  - Gelastic epilepsy (laughing)
  - Dacrystic or quiritarian epilepsy (crying)
- Complex automatisms
  - Cursive epilepsy (running)
  - Drinking
  - Undressing
  - Masturbation
  - Prolonged twilight states with automatisms
- Compound forms
  - Temporal lobe "march"
  - Ictal psychosis

combination of hallucinations, thought and memory disturbances, and psychomotor automatisms may produce an ictal psychosis.

Automatisms may occur during petit mal seizures as well as during complex-partial seizures and the two must be distinguished to allow appropriate therapeutic decisions. Table 21.8 contrasts the typical motor characteristics of these two types of seizures.

Postictal Phenomena Complex automatisms can also occur during the postictal confusional state immediately following a seizure. The postictal confusional period may last from minutes to hours and in some cases the initiating seizure may have been so brief as to go unnoticed. During the postictal period, the patient may talk or walk around, but is usually disoriented and amnestic for both the ictal and postictal

TABLE 21.8. *Characteristics that Distinguish Complex Partial Seizures Manifested Solely by Impaired Consciousness and Petit Mal Seizures*

Characteristic	*Complex-Partial Seizures*	*Petit Mal Seizures*
Age of onset	Any age, rare in childhood	Childhood
Aura	Yes	No
Duration	Minutes	Seconds
Frequency	Few per day or less	May be many per day
Precipitants	No	Hyperventilation
Postictal confusion	Yes	No
EEG	Focal (usually temporal) spikes or slowing	Generalized 3-second spike and wave pattern
Etiology	Acquired (trauma, neoplasm, arteriovenous malformation, etc.)	Genetic
Treatment	Phenytoin Carbamazepine Valproate	Ethosuximide Valproate

episode. It may be impossible to distinguish ictal and postictal behavioral activity unless the patient has ongoing EEG monitoring.

● **Interictal Behavioral Alterations** Interictal alterations of behavior noted in patients with epilepsy are among the most controversial and poorly understood areas of contemporary neuropsychiatry and epileptology. Some investigators relate these changes to the underlying limbic dysfunction while others ascribe them to the psychological distress associated with having an epileptic disorder. No rigorous community-based survey of the presence of these disorders among epileptics has been accomplished and their frequency remains uncertain. Published reports are based largely on convenience samples of patients attending university-related epilepsy centers. These studies suggest that the frequency of psychopathology among epileptics is substantially greater than that in the non-epileptic general population.

There are several risk factors for interictal psychopathology. Patients with TLE have disproportionate psychopathology compared to patients with other types of epilepsy. The presence of cognitive impairment and mental retardation is significantly associated with increased psychopathology among patients with epilepsy. Likewise, intractable seizures are more likely to be associated with psychopathology than are well-controlled seizure disorders.[12–14] Among patients with partial seizures and auras preceding their epileptic events, those with cognitive auras (derealization, depersonalization, dreamy states, forced thought, altered time perception) are more likely to evidence psychopathology than those with other types of auras.[15]

Psychopathology in Epilepsy

● **Personality Alterations** The existence of an "epileptic personality" is controversial. Several of the personality alterations formally ascribed to epileptics can now be recognized to be the consequences of recurrent uncontrolled seizures with repeated anoxic insults or head injuries, the result of treatment with potentially toxic agents, or the effects of chronic institutionalization and social ostracism. Many of these factors have now been modified through use of improved anticonvulsant medications.

Some types of personality traits have repeatedly been observed among patients with epilepsy, particularly those with complex-partial seizures and temporal lobe foci. Aggression, altered sexual interest, circumstantiality, decreased emotionality, dependence, passivity, lability, guilt, humorlessness, sobriety, hypergraphia, hypermoralism, irritability, obsessionalism, paranoia, philosophical interests, religiosity, sadness, sense of personal destiny, and viscosity are among the features commonly described.[16] The Gastaut-Geschwind syndrome includes many of these features and is characterized by hypergraphia, hyposexuality, hyperreligiosity, exaggerated philosophical concern, interpersonal "stickiness," and circumstantiality.[17–19] In some cases, the obsessionalism manifested in interpersonal stickiness, circumstantiality, and hypergraphia may be evidenced in other ways such as excessive painting or collecting. These characteristics differentiate patients with TLE from normal controls and are among the most distinctive personality alterations that occur in patients with epileptic disorders. The symptom complex, however, can also be seen in patients suffering from other types of psychiatric illness and is not pathognomonic of epilepsy.

The major instrument used for personality assessment in the clinical setting is the Minnesota Multiphasic Personality Inventory (MMPI). Application of this tool to patients with TLE has yielded varying results and the tool was not designed to identify psychopathology in patients with neurological disorders. In most cases, epileptic patients have been found to have elevated paranoia and schizophrenia scale scores.

An increased incidence of dissociative experiences, fugue states, poriomania (wondering epilepsy), borderline personality disorder, and multiple personality have been associated with epilepsy[20,21] (Chapter 22).

● Psychoses Among the most well-documented neuropsychiatric complications of epilepsy is a schizophrenia-like psychosis that occurs in patients with TLE. Psychosis occurs with increased frequency in all types of epilepsy, whereas the schizophrenia-like disorder is associated primarily with TLE.[22] Hill[23] was among the first to observe that some epileptic patients developed a chronic paranoid hallucinatory psychosis that resembles idiopathic schizophrenia. Slater and Beard[24] studied the occurrence of psychoses among epileptics and confirmed Hill's observation that the two disorders occur together too frequently to be ascribed to chance. In their series, the mean age of onset of psychosis was 30 years and followed the onset of epilepsy by approximately 14 years. Of the patients who had paranoid delusions and auditory hallucinations, approximately half had a formal thought disorder. There was no consistent relationship between the severity and course of the psychoses and the frequency of seizures. In several cases, the seizures had greatly diminished in frequency or ceased at the time the psychosis emerged. Slater and Beard[24] noted that the psychosis occurred specifically in patients with TLE. They observed that the patients tended to have less flattening of affect than patients with idiopathic schizophrenia, and there was an increase in obsessional traits in their personalities (pedantry, circumstantiality). The disorder was further distinguished from idiopathic schizophrenia by an absence of psychosis among family members. Most recent studies have refined, extended, and largely confirmed these original observations and have established the frequency of psychosis to be between 5% and 10% among patients with epilepsy, with higher rates found among those with complex-partial seizures and TLE.[10]

Interictal psychoses may resemble nuclear schizophrenia, paranoid psychosis, or psychosis with mood symptoms.[10,22] Risk factors for psychosis with epilepsy include complex-partial–type seizures, onset of seizures in early adolescence, female gender, left-sided seizure foci within the temporal lobe (particularly in the mediobasal region), left-handedness, cognitive impairment or mental retardation, and the presence of psychic auras.[10,22,25–27] Compared to patients with epilepsy and no psychosis, psychotic epileptic patients at autopsy have larger cerebral ventricles, more periventricular glyosis, and more focal cerebral damage.[26]

Several types of psychoses have been associated with epilepsy and must be distinguished from the interictal psychosis syndrome (Table 21.9). *Parictal psychoses* are those that occur in the parictal period in both localization-related and generalized epilepsies. They develop gradually over days to weeks, occur primarily in severe epilepsies, and are more likely to occur during periods of increased seizure activity.[10]

As noted above, ictal psychoses can occur as manifestations of psychomotor epilepsy, psychomotor status epilepticus, absence status, or spike-wave stupor. There is impairment of consciousness with delusions, hallucinations, and paranoid ideation. Typically, these

TABLE 21.9. *Causes of Psychosis in Epileptic Patients*

Parictal psychoses
Ictal psychoses
Complex-partial status epilepticus
Petit mal status epilepticus
Postictal psychosis
Interictal psychosis
Schizophrenia-like psychoses
Mood disorders with psychosis
Paranoid psychosis
Forced normalization/alternate psychoses
Anticonvulsant-related psychoses
Anticonvulsant toxicity
Anticonvulsant withdrawal
Psychosis with underlying brain disorder

ictal psychoses last hours to days, are most common with severe epilepsies, and occur with withdrawal of antiepileptic medication.[10]

Postictal psychoses occur with complex-partial seizures or primary or secondary generalized tonic-clonic seizures. There is typically a lucid interval between the seizures and the onset of the psychosis. The psychotic period commonly lasts several days. Postictal psychoses are commonly associated with withdrawal of antiepileptic drugs, a series of recurrent seizures, or status epilepticus.[10] Postictal psychosis is also associated with later age of onset, better intellectual function, and more evidence of bilateral epileptiform activity than chronic interictal psychosis.[29] Postictal psychoses may be precipitated by withdrawal of antiepileptic medication, which should be monitored as a possible hazard of discontinuing medication for video EEG monitoring.[30] Unilateral hippocampal sclerosis is frequently evident on magnetic resonance images (MRIs) of patients with postictal psychoses.[31]

Forced normalization refers to the emergence of psychopathology when the EEG becomes normal or nearly normal compared to previous and subsequent EEG findings.[22] Psychosis is one of the forms of psychopathology that may occur under these circumstances. The syndrome is also called "alternate psychosis." The phenomenon has been observed more often with primary generalized than localization-related epilepsies. Typically, there is a prodromal phase with insomnia, anxiety and social withdrawal, followed by either a delusional hallucinatory psychosis or a nonpsychotic syndrome with depressive features. Treatment of this syndrome depends on the resumption of at least occasional seizure activity.[10] Psychosis may be associated with anticonvulsant toxicity or anticonvulsant withdrawal or with an underlying brain disorder that may produce psychotic disturbances (traumatic brain injury, degenerative brain disorders, etc.).

Treatment of epilepsy-associated psychosis depends on the relationship of the psychosis to the seizures. Parictal, ictal, and postictal psychoses are all ameliorated by effective treatment of seizures. Psychosis associated with forced normalization may be treated by allowing occasional ictal events. Interictal psychoses usually require treatment with antipsychotic agents. Of the conventional neuroleptics, chlorpromazine and loxapine should be avoided because they lower the seizure threshold and may precipitate or exacerbate seizures in patients with epilepsy. Of the atypical antipsychotic agents, clozapine is most likely to precipitate seizures.[32] Table 21.9 summarizes the etiologies of psychosis that occur in patients with epilepsy.

Mood Disorders Mood disorders are the most common type of psychopathology encountered in patients with epilepsy. Among patients receiving care in specialized settings, approximately 30% exhibit depressive symptoms. Depression is the most common reason for psychiatric hospitalization of epileptic patients.[33,34] Among patients with intractable disorders, 60% have lifetime histories of depressive syndromes.[14] Similar to psychoses, mood disorders have complex relationships to ictal events. Parictal, ictal, postictal, and interictal depressive disorders have been described. Depressive twilight states associated with status epilepticus have also been noted. Forced normalization with alternate behavior disorders is frequently manifested by mood abnormalities.[22] Finally, depression may be related to the use of anticonvulsants or to an underlying brain disorder (Table 21.10). Depression is more common in patients with TLE than in those with other forms of epileptic syndromes. Most, but not all, studies of patients with complex-partial seizures and depression have found that depression is more common with left-sided than with right-sided lesions.[33,35,36] Studies that did not identify more depression among patients with left temporal foci found evidence of bilateral frontal dysfunction.[37,38] The phenomenology of the interictal depression associated with epilepsy features anxiety, hostility, and paranoia.[33,39] The characteristics are those of an endogenous depression with chronic mood changes and occasional severe depressive episodes. Suicide and suicidal attempts are common among hospitalized epileptics. Compared to non-epileptic patients attempting suicide, those with

TABLE 21.10. *Etiologies of Depression in Epilepsy*

Psychological reaction to diagnosis and associated social and occupational limitations
Parictal mood changes
Ictal depression
Depressive twilight state
Postictal depression
Interictal mood disorder
Depression coexisting with interictal psychosis
Forced normalization/alternate behavior disturbance with mood abnormalities
Anticonvulsant-induced mood alterations
Anticonvulsant withdrawal with mood symptoms
Mood disorder with underlying brain disease

epilepsy and suicidality are more likely to exhibit borderline personality disorders and impulsivity.[40]

Anticonvulsants often have beneficial effects on mood. Carbamazapine, lamotrigine, and valproate have all been shown to ameliorate mood disorders in patients with epilepsy.[34] Conversely, phenobarbital and vigabatrin have been associated with depressive episodes.[34]

Antidepressants may lower the seizure threshold and increase seizures, particularly in epileptic patients with marginal seizure control. Amoxapine, maprotiline, mianserin, and clomipramine have relatively high epileptogenic potentials.[32]

Interictal elation, mania, and hypomania are much less common than depression. When they occur, they are associated with right-sided brain lesions.[34]

Anxiety Anxiety may occur as a psychological reaction to the diagnosis of epilepsy and the associated psychosocial challenges; as a parictal psychological change in the hours to days preceding a seizure; as an aura at the onset of an ictal event, as part of a partial seizure in association with epileptic psychosis or epileptic mood disorders; in association with underlying brain injury; or in association with anticonvulsant-induced mood alterations.[41] Anxiety and panic attacks can be particularly difficult to distinguish from seizures with fear as an important subjective manifestation. In general, patients with epilepsy-related fear have more brief, stereotyped attacks with associated epileptic phenomena. Electroencephalographic tracings may be useful in distinguishing the two disorders and a response to anticonvulsant agents may also be helpful.

Altered Sexual Behavior A variety of types of altered sexual behavior may occur in either the ictal or interictal periods of patients with epilepsy. Ictal sexual behaviors and sensations include genital sensations, orgasm, coital movements with pelvic thrusting, automatic disrobing, masturbation, sexual thoughts, and sexual arousal. There may be postictal erection or undressing during the confusional state. In the interictal period, epileptic patients most commonly exhibit hyposexuality, but in some cases, hypersexuality or altered sexuality with paraphilic behavior have been reported.[16] Hyposexuality may reflect depression or intermittent prolactin elevation associated with seizures, or it may be a consequence of antiepileptic drug therapy. Hypersexuality may occur with reduced abnormal brain electrical activity in the postictal period, following temporal lobectomy, or accompanying reduced seizures secondary to optimal seizure management. Among the paraphilias that have been described in epileptic patients are fetishism, transvestitism, transsexualism, voyeurism, exhibitionism, sadism, masochism, pedophilia, frotteurism, and genital mutilation (Table 21.11).

TABLE 21.11. *Altered Sexual Behavior in Epilepsy*

Ictal
Sensory seizures
Genital sensations
Orgasmic sensations
Cognitive seizures
Sexual thoughts
Arousal
Motor seizures
Coital movements
Exhibitionism (automatic disrobing)
Masturbation
Erection
Postictal
Erection
Undressing
Interictal
Hyposexuality
Hypersexuality
Altered sexuality (occasional association)
Fetishism
Transvestitism
Transsexualism
Voyeurism
Exhibitionism
Sadism
Masochism
Pedophilia
Frotteurism
Genital mutilation

Cognitive Function and Epilepsy Epilepsy is compatible with normal cognition and many epileptic patients function well intellectually. However, neuropsychological test batteries consistently show impairments in patients with epilepsy, particularly those with early-onset seizures, repeated generalized tonic-clonic seizures, or status epilepticus.[42,43] Patients with intractable, generalized, or left focal seizures often have

TABLE 21.12. *Neuropathological Diagnosis in Patients with Temporal Lobe Epilepsy Treated with Lobectomy*[47]

Diagnostic Group	*Cases* n *(%)*		*First Seizures (years)*	*Operation (years)*
Ammon's horn sclerosis	107	(43)	5.19	22.64
Inflammatory	8	(3)	8.62	24.75
Double pathology	18	(7)	9.53	22.33
Indefinite	25	(10)	14.72	31.24
No apparent lesion	41	(16)	15.37	28.56
"Alien tissue" lesion*	38	(15)	16.93	26.77
Trauma	7	(3)	17.31	33.14
Developmental lesion	5	(2)	22.80	38.20

*Neoplasm, hamartoma, cortical dysplasia.

memory defects and those with left temporal lobe abnormalities may exhibit anomia or difficulties with verbal fluency. Executive abnormalities are often present in patients with complex-partial seizures of frontal origin. Memory performance has been correlated with both MRI, measurements of hippocampus, and postsurgical volumetric cell densities in regions CA1 and CA2 of hippocampus.[44,45]

Anticonvulsants may further impair cognition, particularly when high blood levels must be obtained for seizure control or when multiple drug regimens are required. Phenobarbital has disproportionate adverse neuropsychological effects compared to other anticonvulsants.[46]

Etiologies of Seizures

Identification of seizures in an epilepsy syndrome should be followed by a careful search for the cause of the recurrent epileptic attacks. Table 21.12 presents the causes of complex-partial seizures in patients subjected to temporal lobe resection.[47] Mesial temporal sclerosis was the most common cause followed by "alien tissue" lesions (neoplasms, hemartomas, cortical dysplasia), indefinite pathology and no apparent lesions, double pathology, inflammatory lesions, trauma, and developmental lesions. Ammon's horn sclerosis consists of loss of nerve cells in the hippocampus with accompanying fibrous glyosis and a variable degree of shrinkage and atrophy. In most cases, the nerve cell loss is most severe in the Sommer sector (H1) of the hippocampus. Ammon's horn sclerosis has been associated with febrile convulsions, birth injury, and episodes of status epilepticus.[47] Vascular malformations, congenital cerebral malformations, and a variety of miscellaneous conditions including encephalitis, cerebral infarctions, abscesses, meningitis, and aneurysms account for the remaining cases of TLE. Neoplasms, (meningiomas, astrocytomas, glioblastomas or metastases), trauma, vascular malformations, infections, and infarctions can also produce simple-partial seizures. Most primary generalized epilepsies are genetically determined disorders.

Pathophysiology of Neuropsychiatric Syndromes in Patients with Epilepsy

The pathophysiology of neuropsychiatric disorders in epilepsy is no more understood than that of the idiopathic psychiatric disorders. However, information regarding the pathophysiologic basis of these conditions can be inferred from clinical circumstances that comprise risk factors for neuropsychiatric symptoms, the anatomical focus of the lesions most likely to be associated with neuropsychiatric conditions, and the effects of drugs that ameliorate both epilepsy and behavioral disturbances. Risk factors for psychopathology in epilepsy include onset of seizures in early adolescence, female gender, and the presence of cognitive impairment. These observations suggest that the state of maturation of the nervous system, hormonal effects, and interaction with cognition all influence the emergence of psychopathology. The higher prevalence of psychopathology among patients with foci in the limbic system, particularly the medial temporal regions and in those with left-sided foci, indicates that limbic system dysfunction and lateralization within the language-dominant hemisphere are important predisposing factors. The tendency for behavioral changes to be deferred for several years after the onset of seizures indicates that dynamic changes induced by the aberrant functional activity within the brain may be necessary in many patients for the occurrence of neuropsychiatric disorders. A limbic system "hyperconnection" syndrome has been proposed as a contributing factor. It is posited that the abnormal limbic activation results in the suffusion of experience with abnormal emotional valence.[48] Finally, it appears increasingly likely that many epilepsies are the result of ion channel dysfunction involving sodium and potassium channels and their regulation by the inhibitory transmitter γ-aminobutyric acid (GABA). Several anticonvulsants that also exhibit mood-stabilizing properties exert their effects through influencing channel function.[49] This suggests channel dysfunction as part of the molecular basis of mood-related psychopathology in epilepsy and perhaps of other types of epilepsy-related psychopathological symptoms as well.

Treatment of Epilepsy

Many agents are currently available for the treatment of epilepsy. The mechanism of action through which they exert their anticonvulsant effects is not completely understood, but most of these agents affect ion channels and appear to decrease neuronal excitation or increase neuronal inhibition. First-line agents for partial seizures and generalized tonic-clonic seizures include carbamazapine, phenytoin, and valproate (Table 21.13).

TABLE 21.13. *Anticonvulsant Agents*

Agent	*Indication*	*Starting Dose*	*Maintenance Dose*
Phenytoin (Dilantin)	First-line or add-on for partial seizures and GTCS	100–200 mg/day	200–400 mg/day
Carbamazepine (Tegretol)	First-line or add-on for partial seizures and GTCS	100 mg/day	400–1600 mg/day
Valproate (Depakote)	First-line or add-on for partial seizures, GTCS, and Lennox-Gastaut syndrome	400–800 mg/day	500–2500 mg/day
Phenobarbital (Bellatal)	Alternate therapy for partial seizures, GTCS, Lennox-Gastaut syndrome, childhood epilepsy syndrome	30 mg/day	30–180 mg/day
Clonazepam (Klonopin)	Adjunctive therapy for partial seizure and GTCS	0.25 mg/day	0.5–4 mg/day
Ethosuximide (Zarontin)	First-line or adjunctive therapy for generalized absence seizures	250 mg/day	750–2000 mg/day
Felbamate (Felbatol)	Adjunctive therapy for refractory partial and secondarily generalized seizure	1200 mg/day	1200–3600 mg/day
Gabapentin (Neurontin)	Adjunctive therapy in adults with partial or secondarily generalized seizures	300 mg/day	900–3600 mg/day
Lamotrigine (Lamictal)	Adjunctive or monotherapy for partial and generalized epilepsy	12.5–25 mg/day	100–200 mg/day
Levetiracetam (Keppra)	Adjunctive therpay for partial seizures	1000 mg/day	1000–3000 mg/day
Oxcarbazepine (Trileptal)	Adjunctive or monotherapy for partial and secondarily generalized seizures	600 mg/day	900–2400 mg/day
Tiagabine (Gabatril)	Adjunctive therapy for partial and secondary generalized seizure	15 mg/day	30–45 mg/day
Topiramate (Topomax)	Adjunctive therapy for partial, secondarily, generalized seizures, Lennox-Gastaut syndrome, and primary generalized GTCS	25–50 mg/day	200–600 mg/day
Vigabatrin (Sabril)	Adjunctive therapy for partial and secondarily generalized seizures, Lennox-Gastaut syndrome, and infantile spasms	1000 mg/day	1000–3000 mg/day
Zonisamide (Exegran)	Adjunctive therapy for partial and secondarily generalized seizures	100–200 mg/day	400–600 mg/day

GTCS, generalized, tonic-clonic seizures.

First-line therapy for generalized absence seizures is ethosuximide. Adjunctive therapies that have become available include felbamate, gabapentin, lamotrigine, levetiracetam, oxcarbazepine, tiagabine, topiramate, vigabatrin, and zonisamide (Table 21.13).[50,51]

Surgical treatments for patients with epilepsy are used when seizures prove to be medically refractory. Limbic resections include anterior temporal lobectomies and amygdalohippocampectomies. Neocortical resections include extratemporal resections of lesions occurring in other cortical regions and lesionectomies aimed at specific neocortical targets (such as neoplasms or arteriovenous malformations). Other less commonly used interventions include hemispherectomy and multilobar resections, corpus callosotomy, subpial transections (used for treatment of epileptic foci in eloquent cortex), and vagus nerve stimulation.[32] There is typically substantial improvement in seizure control following surgery, even in these medically intractable cases, and there may be concomitant improvement in depression, psychosis, and aggression. Not all patients evidence behavioral improvement and in some cases depression and psychosis emerge following surgery.[22,52] Vagus nerve stimulation has an increasing role in seizure treatment and appears to ameliorate depression as well as epileptic activity.

NON-EPILEPTIC LIMBIC DISORDERS

Table 21.1 provides an overview of temporal and limbic syndromes. Most of these have been addressed in other chapters of this volume. In this section, the Klüver-Bucy syndrome and oral behaviors are described.

Klüver-Bucy Syndrome

Klüver and Bucy[53] first reported the syndrome that bears their names in 1939. They described the behavioral effects of removal of both temporal lobes in Macaque monkeys. The monkeys exhibited (*1*) "psychic blindness" or visual agnosia; (*2*) strong oral tendencies in examining available objects (licking, gently biting, chewing, touching with lips, smelling); (*3*) a marked tendency to attend and react to visual stimuli (hypermetamorphosis); (*4*) a marked change in emotional behavior or absence of emotional reactions in the sense that motor and vocal reactions generally associated with anger and fear were not exhibited; and (*5*) an increase in sexual activities. Later reports of animals observed for longer periods of time after the surgery included descriptions of dietary changes and ingestion of food items of a type not eaten prior to the lobectomy. Further investigations revealed that the syndrome is more obvious in animals kept in captivity than in animals subjected to the same surgery and observed in more natural social circumstances. Anatomical studies of the syndrome indicate that production of the disorder depends critically on bilateral dysfunction of the amygdaloid nuclei. Surgery targeted on the amygdala has reproduced the syndrome, and surgeries of surrounding cortical regions but leaving the amygdala and its limbic connections intact failed to cause the syndrome. Fragments of the condition, or partial Klüver-Bucy syndromes, have been observed with thalamic and hypothalamic lesions.

Any condition with bilateral medial temporal–amygdaloid nuclear injury can produce the Klüver-Bucy syndrome. The most common causes include herpes encephalitis, frontotemporal dementias, traumatic brain injury, late-stage Alzheimer's disease, paraneoplastic limbic encephalitis, bilateral temporal lobe infarction, and seizure-related disorders (as a manifestation of psychomotor status or occurring in the postictal state).[54–60]

The Klüver-Bucy syndrome is difficult to modify but the behavior of some patients has improved following treatment with carbamazepine.[61,62] The sexual aggression sometimes observed in patients with the Klüver-Bucy syndrome may respond to treatment with leuprolide, a gonadotropin-releasing hormone agonist that decreases testosterone levels.[63]

Hyperoral Behaviors

Hyperoral behaviors are not confined to Klüver-Bucy syndrome and occur in a variety of central nervous system conditions (Table 21.14). Alterations in diet and eating are seen in frontotemporal dementia, the Gourmand syndrome, and eating disorders such as anorexia and bulimia. Hyperphagia occurs in acquired and congenital syndromes including ventromedial hypothalamic injury, the Kleine-Levin syndrome, Prader-Willi syndrome, and the Laurence-Moon-Biedle syndrome. Lip biting and self-mutilation occur in Gilles de la Tourette syndrome, Lesch-Nyhan syndrome, neuroacanthocytosis, and obsessive-compulsive disorder. Oral movements are present in a variety of epileptic conditions and basal ganglia syndromes. Rett's syndrome includes hand and nail biting as manifestations. Pica, or the ingestion of non-food items, occurs in autism, mental retardation, and certain dietary deficiency states. Polydipsia occurs in major psychiatric illnesses, and belching, spitting, and rumination also occur in varied neuropsychiatric circumstances.[64–67]

TABLE 21.14. *Conditions with Altered Oral Behaviors*

Klüver-Bucy syndrome (complete or partial) Hyperorality Altered dietary preferences Frontotemporal dementia Dietary/eating compulsions Carbohydrate craving Advanced dementia with stuffing of excessive food into mouth Gourmand syndrome (preoccupation with food following right anterior brain injury) Eating disorders Anorexia Bulimia Hyperphagia Kleine-Levin syndrome Prader-Willi syndrome Laurence-Moon-Biedle syndrome Ventromedial hypothalamic injury Hunger with temporal lobe injury Lip biting and self-mutilation Gilles de la Tourette syndrome Lesch-Nyhan syndrome Neuroacanthocytosis Obsessive-compulsive disorder	Oral movements Ictal oral automatisms (in temporal lobe epilepsy) Tardive dyskinesia Rabbit syndrome (parkinsonian tremor of lips) Oculomasticatory myorhythmia (in Whipple's disease) Facial myokymia (in multiple sclerosis, brainstem glioma) Hemifacial spasm Meige's syndrome Facial chorea in choreiform syndromes Facial tics Aberrant innervation of orbicular oris muscles following VII nerve injury or Bell's palsy Miscellaneous conditions Rett syndrome—hand and nail biting Pica (ingestion of non-food substances) Autism Mental retardation Dietary deficiency states Polydipsia—excessive drinking Schizophrenia and other psychoses Belching/eructations Spitting Agitation syndrome Obsessive-compulsive disorder Mental retardation Rumination (repetitive regurgitation of small amounts of food) Infants Mental retardation Gastroesophageal disorders Suck reflex Infants Frontal lobe disorders Dementia

Table construction aided by Dean Foti and Mario Mendez.

REFERENCES

1. Glaser GH. Historical perspectives and future directions. In: Wyllie E, ed. The Treatment of Epilepsy: Principles and Practice, 2nd ed. Baltimore: Williams & Wilkins, 1996:3–8.
2. Wyllie E, Luders H. Classification of seizures. In: Wyllie E, ed. The Treatment of Epilepsy: Principles and Practice, 2nd ed. Baltimore: Williams and Wilkins, 1996:355–357.
3. Barry E, Fisher RS. Introduction. In: Fisher RS, ed. Imitators of Epilepsy. New York: Demos Publications, 1994:1–10.
4. Porter RJ. Diagnosis of psychogenic and other nonepileptic seizures in adults. In: Devinsky O, Theodore WH, eds. Epilepsy and Behavior. New York: Wiley-Liss, 1991:237–249.
5. Sassower K, Duchowny M. Psychogenic seizures and nonepileptic phenomena in childhood. In: Devinsky O, Theodore WH, eds. Epilepsy and Behavior. New York: Wiley-Liss, 1991:223–235.
6. Mosewich RK, So EL. A clinical approach to the classification of seizures and epileptic syndromes. Mayo Clin Proc 1996;71: 405–414.
7. Rowan AJ. Diagnosis of non-epileptic seizures. In: Gates JR, Rowan AJ, eds. Non-epileptic Seizures, 2nd ed. Boston: Butterworth Heinemann, 2000:15–30.
8. Schachter SC. Neuroendocrine aspects of epilepsy. In: Devinsky

O, Theodore WH, eds. Epilepsy and Behavior. New York: Wiley-Liss, 1991:303–333.
9. Stewart RS, Lovitt R, Stewart M. Psychopathology associated with hysterical seizures. In: Gross M, ed. Pseudoepilepsy: The Clinical Aspects of False Seizures. Lexington, MA: Lexington Books, 1982:97–108.
10. Schmitz B. Psychosis and epilepsy: the link to the temporal lobe. In: Trimble MR, Bolwig TG, eds. The Temporal Lobes and the Limbic System. Petersheld, Hampshire: Wrightson Biomedical Publishing, 1992:149–167.
11. Escueta AVD, Bacsal FE, Treiman DM. Complex partial seizures on closed-circuit television and EEG: a study of 691 attacks in 79 patients. Ann Neurol 1982;11:292–300.
12. Edeh J, Toone BK, Corney RH. Epilepsy, psychiatric morbidity, and social dysfunction in general practice: comparison between hospital clinic patients and clinic nonattenders. Neuropsychiatry Neuropsychol Behav Neurol 1990;3:180–192.
13. Fiordelli E, Beghi E, et al. Epilepsy and psychiatric disturbance: a cross sectional study. Br J Psychiatry 1993;163:446–450.
14. Victoroff J. DSM-III-R psychiatric diagnoses in candidates for epilepsy surgery: lifetime prevalence. Neuropsychiatry Neuropsychol Behav Neurol 1994;7:87–97.
15. Mendez MF, Engebrit B, et al. The relationship of epileptic auras and psychological attributes. J Neuropsychiatry Clin Neurosci 1996;8:287–292.
16. Devinsky O. Interictal behavioral changes in epilepsy. In: Devinsky O, Theodore WH, eds. Epilepsy and Behavior. New York: Wiley and Liss, 1991:1–21.
17. Trimble MR, Mendez MF, Cummings JL. Neuropsychiatric symptoms from the temporolimbic lobes. In: Salloway S, Malloy P, Cummings JL, eds. The Neuropsychiatry of Limbic and Subcortical Disorders. Washington, DC: American Psychiatric Press, 1997:123–132.
18. Waxman SG, Geschwind N. The interictal behavior syndrome of temporal lobe epilepsy. Arch Gen Psychiatry 1975;32: 1580–1586.
19. Bear DM, Fedio P. Quantitative analysis of interictal behavior in temporal lobe epilepsy. Arch Neurol 1977;34:454–467.
20. Ahern GL, Herring AM, et al. The association of multiple personality and temporolimbic epilepsy: intracarotid amobarbital test observations. Arch Neurol 1993;50:1020–1025.
21. Benson DF, Miller BL, Signer SF. Dual personality associated with epilepsy. Arch Neurol 1986;43:471–474.
22. Trimble MR. The Psychoses of Epilepsy. New York: Raven Press, 1991.
23. Hill D. Psychiatric disorders of epilepsy. Med Press 1953;229: 473–475.
24. Slater E, Beard AW. The schizophrenic-like psychoses of epilepsy. Psychiatric aspecs. Br J Psychiatry 1963;109:95–150.
25. Adachi N, Matsuura M, et al. Predictive variables of interictal psychosis in epilepsy. Neurology 2000;55:1310–1314.
26. Mendez MF, Grau R, et al. Schizophrenia in epilepsy: seizure and psychosis variables. Neurology 1993;43:1073–1077.
27. Umbricht D, Degreef G, et al. Postictal and chronic psychoses in patients with temporal lobe epilepsy. Am J Psychiatry 1995;152:224–231.
28. Bruton CJ, Stevens JR, Frith CD. Epilepsy, psychosis and schizophrenia: clinical and neruopathologic correlations. Neurology 1994;44:34–42.
29. Szabo CA, Lancman M, Stagno S. Postictal psychosis: a review. Neuropsychiatry Neuropsychol Behav Neurol 1996;9:258–264.
30. Kanner AM, Stagno S, et al. Postictal psychiatric events during prolonged video-electroencephalographic monitoring studies. Arch Neurol 1996;53:258–263.
31. Kanemoto K, Takeuchi J, et al. Charactristics of temporal lobe epilepsy with mesial temporal sclerosis, with special reference to psychotic episodes. Neurology 1996;47:1199–1203.
32. McConnell H, Duncan D. Treatment of psychiatric comorbidity in epilepsy. In: McConnell HW, Snyder PJ, eds. Psychiatric Comorbidity in Epilepsy: Basic Mechanisms, Diagnosis and Treatment. Washington, DC: American Psychiatric Press, Inc., 1998:245–361.
33. Mendez MF, Cummings JL, Benson DF. Depression in epilepsy: significance and phenomenology. Arch Neurol 1986;43:766–770.
34. Robertson M. Mood disorders associated with epilepsy. In: McConnell HW, Snyder PJ, eds. Psychaitric Comorbidity in Epilepsy. Washington, DC: American Psychiatric Press, 1998: 133–167.
35. Altshuler LL, Devinsky O, et al. Depression, anxiety, and temporal lobe epilepsy: laterality of focus and symptoms. Arch Neurol 1990;47:284–288.
36. Victoroff JI, Benson DF, et al. Depression in complex partial seizures. Arch Neurol 1994;51:155–163.
37. Bromfield EB, Altshuler L, et al. Cerebral metabolism and depression in patients with complex partial seizures. Arch Neurol 1992;49:617–623.
38. Hermann BP, Seidenberg M, et al. Mood state in unilateral temporal lobe epilepsy. Biol Psychiatry 1991;30:1205–1218.
39. Robertson MM, Trimble MR, Townsend HRA. Phenomenology of depression in epilepsy. Epilepsia 1987;28:364–372.
40. Mendez MF, Lanska DJ, et al. Causative factors for suicide attempts by overdose in epileptics. Arch Neurol 1989;46:1065–1068.
41. Betts TA. Depression, anxiety and epilepsy. In: Reynolds EH, Trimble MR, eds. Epilepsy and Psychiatry. New York: Churchill Livingstone, 1981:60–71.
42. Perrine K, Gershengorn J, Brown ER. Interictal neuropsychological function in epilepsy. In: Devinsky O, Theodore WH, eds. Epilepsy and Behavior. New York: Wiley-Liss, 1991:181–193.
43. Hermann BP, Seidenberg M, et al. Neuropsychological characteristics of the syndrome of mesial temporal lobe epilepsy. Arch Neurol 1997;54:369–376.
44. Lencz T, McCarthy G, et al. Quantitative magnetic resonance imaging in temporal lobe epilepsy: relationship to neuropathology and neuropsychological function. Ann Neurol 1992;31:629–637.
45. Sass KJ, Buchanan CP, et al. Verbal memory impairment resulting from hippocampal neuron loss among epileptic patients with structural lesions. Neurology 1995;45:2154–2158.
46. Meador KJ, Loring DW. Cognitive effects of antiepileptic drugs. In: Devinsky O, Theodore WH, eds. Epilepsy and Behavior. New York: Wiley-Liss, 1991:151–170.
47. Bruton CJ. The Neuropathology of Temporal Lobe Epilepsy. Oxford: Oxford University Press, 1988.
48. Bear DM. Temporal lobe epilepsy—a syndrome of sensory-limbic hyperconnection. Cortex 1979;15:357–384.
49. Dunn RT, Frye MS, et al. The efficacy and use of anticonvulsants in mood disorders. Clin Neuropharmacol 1998;21:215–235.
50. Dichter MA, Brodie MJ. New antiepileptic drugs. N Engl J Med 1996;334:1583–1590.
51. Shorvon SD. Handbook of Epilepsy Treatment. London: Blackwell Science, 2000.
52. Strauss E, Wada J. Psychiatric and psychosocial changes associated with anterior temporal lobectomy. In: Devinsky O, Theodore WH, eds. Epilepsy and Behavior. New York: Wiley Liss, 1991:135–149.

53. Klüver H, Bucy PC. Preliminary analysis of function of the temporal lobes in monkeys. Arch Neurol Psychiatr 1939;42:979–1000.
54. Burns A, Jacoby R, Levy R. Psychiatric phenomena in Alzheimer's disease, IV: disorders of behaviour. Br J Psychiatry 1990; 157:86–94.
55. Cummings JL, Duchen LW. Klüver-Bucy syndrome in Pick disease: clinical and pathologic correlations. Neurology 1981;31: 1415–1422.
56. Hayman LA, Rexer JL, et al. Klüver-Bucy syndrome after bilateral selective damage of amygdala and its cortical connections. J Neuropsychiatry Clin Neurosci 1998;10:354–358.
57. Lilly R, Cummings JL, et al. The human Klüver-Bucy syndrome. Neurology 1983;33:1141–1145.
58. Lopez OL, Becker JT, et al. The nature of behavioral disorders in human Klüver-Bucy syndrome. Neuropsychiatry Neuropsychol Behav Neurol 1995;8:215–221.
59. Marlowe WB, Mancall EL, Thomas JJ. Complete Klüver-Bucy syndrome. Cortex 1975;11:53–59.
60. Müller A, Baumgartner RW, et al. Persistent Klüver-Bucy syndrome after bilateral thalamic infarction. Neuropsychiatry Neuropsychol Behav Neurol 1999;12:136–139.
61. Hooshmand H, Sepdham T, Vries JK. Klüver-Bucy syndrome: successful treatment with carbamazepine. JAMA 1974;229: 1782.
62. Stewart JT. Carbamazepine treatment of a patient with Klüver-Bucy syndrome. J Clin Psychiatry 1985;46:496–497.
63. Ott BR. Leuprolide treatment of sexual aggression in a patient with dementia and the Klüver-Bucy syndrome. Clin Neuropharmacol 1995;18:443–447.
64. Malcolm A, Thumshirn MB, et al. Rumination syndrome. Mayo Clin Proc 1997;72:646–652.
65. Regard M, Landis T. "Gourmand syndrome": eating passion associated with right anterior lesions. Neurology 1997;48:1185–1190.
66. Smith G, Vigen V, et al. Patterns and associates of hyperphagia in patients with dementia. Neuropsychiatry Neuropsychol Behav Neurol 1998;11:97–102.
67. Fisher CM. Hunger and the temporal lobe. Neurology 1994;44: 1577–1579.

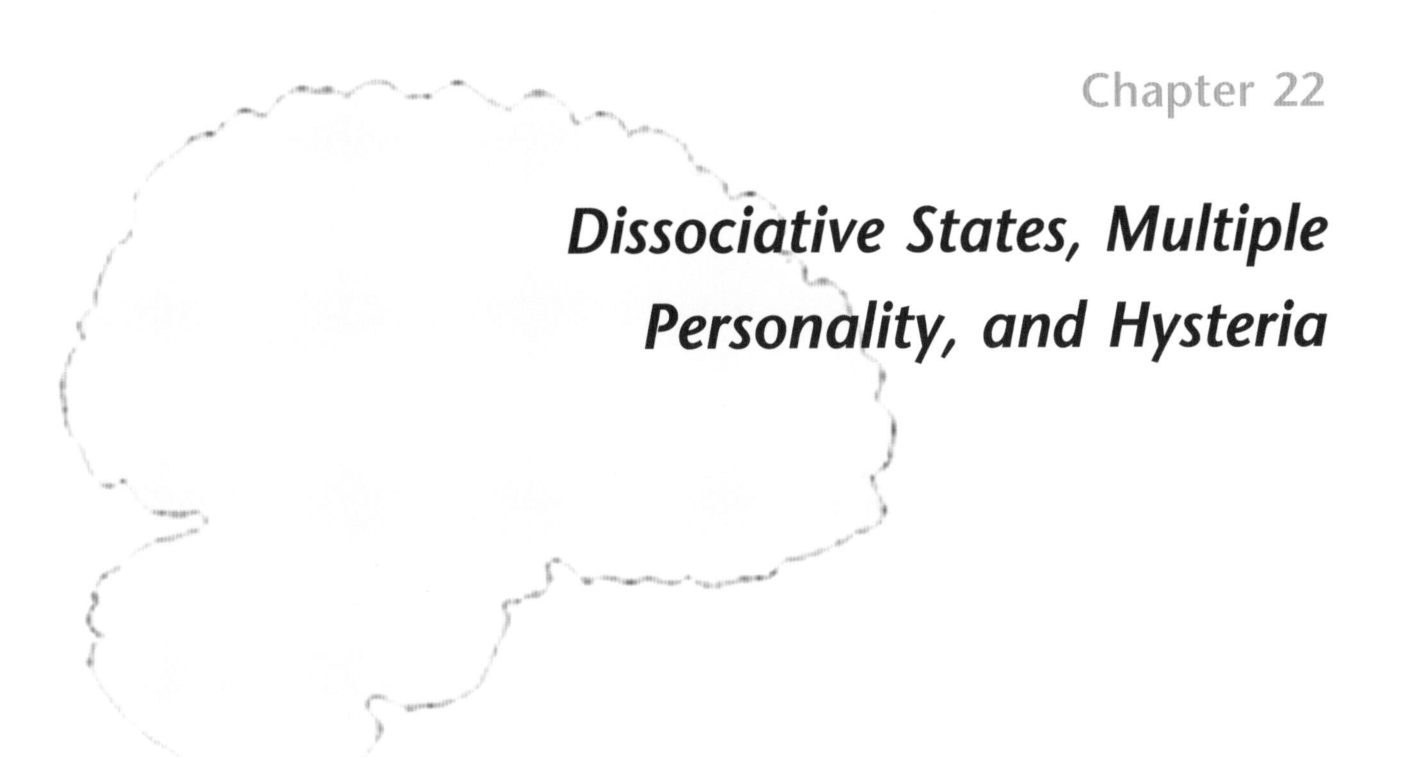

Chapter 22

Dissociative States, Multiple Personality, and Hysteria

This chapter presents a fascinating and complex area of neuropsychiatry in which there is a complicated blending of neurologic and psychiatric issues. In each case, a differential diagnosis includes both classical neurological disorders, such as epilepsy, and idiopathic psychiatric conditions, primarily personality disorders. The set of disorders described in this chapter is challenging for the clinician. The examiner may feel that the patient is practicing deliberate deception, attempting to mislead the observer, seeking to gain attention or sympathy, and diverting attention from patients with more life-threatening neuropsychiatric disorders. It is critical for the clinician to avoid this interpersonal stance in dealing with patients suffering from dissociative states, conversion reactions, or multiple personality. These disorders are among the most disabling of neuropsychiatric conditions and present a great challenge to the physician's diagnostic acumen and therapeutic commitment.

In this chapter, the related conditions of dissociative states and multiple personality are considered first. Then conversion disorders and hysteria are discussed. The differential diagnosis of conditions producing temporary memory impairment is described and syndromes such as malingering, Munchausen syndrome, sleep disorders, pseudoepileptic seizures, and Ganser's syndrome are briefly noted. Chapters relevant to this information include memory disorders (Chapter 7), dementias (Chapter 10), delirium (Chapter 11), anxiety disorders (Chapter 17), epilepsy (Chapter 21), and sleep disturbances (Chapter 23).

DISSOCIATIVE DISORDERS AND MULTIPLE PERSONALITY

Dissociative disorders currently recognized include dissociative amnesia, dissociative fugue, dissociative identity disorder (multiple personality disorder), depersonalization disorder, and a variety of more limited dissociative phenomena such as derealization, dissociation occurring in individuals undergoing torture or coercion, trance states, and possession syndromes (Table 22.1).[1]

Dissociative experiences are common in the general population and many normal individuals report minor dissociative or inattentive episodes such as being unable to recall part of a conversation or feeling abnormally absorbed by television programs or movies. Other experiences, however, are more unique and usually indicative of a dissociative disorder. These include being approached by people one doesn't know who call

TABLE 22.1. *Classification of Dissociative Disorders*[1]

Dissociative amnesia
Dissociative fugue
Dissociative identity disorder (multiple personality)
Depersonalization disorder
Other dissociative conditions
Derealization
Depersonalization during torture and coercion
Trance states
Hypnosis
Possession states
Ganser syndrome

one by a different name; feeling as though one is two different people; driving a car and realizing one doesn't remember part of the trip; not remembering important events in one's life; being in a familiar place but finding it unfamiliar; finding drawings that one must have done but doesn't remember doing; seeing oneself as if looking at another person; feeling as though other people and objects are not real; finding unfamiliar things among one's belongings; feeling as though one's body is not one's own; finding oneself in a place without being aware of how one got there; or finding oneself dressed in clothes one doesn't remember buying or putting on.[2]

In one large Canadian study the overall prevalence of dissociative disorder was 12.2%: 6% had dissociative amnesia, 2.8% had depersonalization disorder, and 3% had multiple personality disorder.[2]

Dissociative Amnesia

Dissociative (psychogenic) amnesia refers to the inability, too extensive to be explained by normal forgetfulness, to recall important personal information.[1] Several patterns of memory impairment have been observed. In *localized amnesia* the individual fails to recall events that occurred during a circumscribed period of time. In *selective amnesia* the person can recall some but not all events during a circumscribed period of time, and the amnesia usually involves information of a highly personal nature. *Generalized amnesia* is characterized by a failure to recall all of a person's past life. These are the classical "amnesiacs" brought to police stations unable to state their personal identity. *Continuous amnesia* is a condition featuring an inability to recall events subsequent to a specific time up to and including the present. *Systematized amnesia* is characterized by a loss of memory for a certain category of information such as material relating to one's family or a particular person.[1] Careful testing of individuals with localized, selective, or systematized amnesia often reveals dissociations between explicit and implicit recall involving the same periods of time. For example, a patient may be unable to recall information regarding their own past life whereas they can remember details of a family member or friend's life occurring during the same period.[2a] The amnesia usually begins and ends abruptly and its onset is typically related to emotionally stressful events. Psychogenic amnesia differs from amnestic disorders associated with identifiable neurological disorders by the absence of typical anterograde and retrograde aspects of memory loss (Chapter 7).

Psychogenic amnesia is a conversion syndrome occurring at times of stress in patients predisposed by an underlying psychiatric or neurologic condition. Depression is the most common predisposing illness but psychogenic amnesia also occurs in patients with mania, schizophrenia, and personality disorders.[1] Malingering accounts for a portion of cases.

Dissociative Fugue

Dissociative (psychogenic) fugue is characterized by sudden unexpected travel away from home or customary place of daily activities with an inability to recall some or all of one's past. There may be confusion about personal identity or the assumption of a new identity but patients do not regularly alternate between distinct identities as in dissociative identity (multiple personality) disorder.[1] The behavior is more purposeful and integrated than that occurring in patients in psychogenic amnesia, but the new identity is less complete than in multiple personality disorder. As in the case of dissociative amnesia, depression is the most common condition predisposing to dissociative fugue. Fugues are also more likely to occur in certain personality disorders, including histrionic, compulsive, schizoid, avoidant, and borderline personalities.[3]

Dissociative Identity Disorder

Dissociative identity disorder (multiple personality) features the presence of two or more distinct identities or personalities that recurrently take control of behavior. There is an inability to recall important personal information, and one identity may deny knowledge of another. Thus, the personality being interviewed is amnestic for periods during which other personalities were in control.[1] All of the personalities are aware of discontinuities in experience with periods of unrecallable behavior. The individual personalities are usually discrepant, displaying behavior unlike that of the primary or host personality. Subpersonalities may identify themselves as a child, a helper, an individual of the opposite sex, or a person of different sociocultural background. Dissociative identity disorder is most frequently associated with personality disorders (60%) but may also occur with mood disorders, psychoses, anxiety disorders, eating disorders, somatization disorders, substance abuse, adjustment disorders, and a variety of neurological conditions (discussed below).[4] Severe childhood abuse or neglect is an historical feature shared by many patients with multiple personality disorder.[5–7]

In addition to behavioral changes exhibited by alternate personalities, there may also be changes in pain sensitivity, galvanic skin responses, electroencephalographic (EEG) patterns, evoked response patterns, and handedness.[8–10]

Depersonalization Disorder

Depersonalization disorder is characterized by persistent or recurrent episodes of feeling detachment or estrangement from oneself. The individual may feel like an automaton or as if he or she is living in a dream or movie. There may be a sensation of being an outside observer of one's own body.[1] Some studies using functional neuroimaging have shown dysfunction of the left frontotemporal structures.[11]

Possession States

Possession states are culturally syntonic dissociative phenomena. They include ritualized trance states such as religious possession, mediumship and channeling, and phenomena such as voodoo, witchcraft, and faith healing.[12] Some cultures have given specific names to these phenomena: Amok (Indonesia), Bebainin (Indonesia), Latah (Malaysia), Pibloktoq (Arctic), Ataquedenerbious (Latin America), and possession (India).[1] Demonic possession may be a special example of a ritualized trance state.

DIFFERENTIAL DIAGNOSIS OF DISSOCIATIVE STATES

Table 22.2 summarizes the principal differential diagnostic considerations involved in the production of dissociative phenomena or conditions that must be distinguished from dissociative states.

TABLE 22.2. *Differential Diagnosis of Dissociative States*

Idiopathic psychiatric disorders and unusual psychosocial circumstances
Childhood abuse or neglect
Acute stress disorder
Post-traumatic stress disorder
Substance abuse
Borderline personality disorder
Histrionic personality disorder
Compulsive personality disorder
Schizoid personality disorder
Avoidant personality disorder
Briquet's syndrome (somatization disorder)
Eating disorders
Depression
Schizophrenia
Neurologic disorders
Epilepsy
Déjà vú, Jaimais vú
Poriomania
Petit mal status
Partial-complex status
Seizure-related identity shifts in dissociative identity disorder
Migraine
Transient global amnesia
Post-concussion syndrome
Anosognosia with hemidepersonalization
Toxic-metabolic disorders
Alcoholic blackouts
Drug-induced amnestic episodes
Delirium
Sleep disorders
REM behavior disorder
Narcolepsy
Somnambulism
Malingering

REM, rapid eye movement.

Idiopathic Psychiatric Disorders

As noted above, many patients with dissociative states have suffered from extreme childhood abuse or neglect, often including sexual abuse. Psychiatric disorders to be considered in a differential diagnosis of dissociative states include acute stress disorders, post-traumatic stress disorders, substance abuse, borderline personality disorder, histrionic personality disorder, obsessive-compulsive personality disorder, schizoid personality disorder, avoidant personality disorder, eating disorders, depression, and schizophrenia.

Neurological Disorders

Neurological disorders are among the most common causes of episodic alterations in behavior that cannot be recalled, and the evaluation of any patient with periodic memory lapses must include a careful search for central nervous system disease. Epilepsy, migraine, transient global amnesia, and post-concussion syndrome all produce temporary impairment of memory. Epilepsy and migraine may be associated with dissociative phenomena and right parietal lesions with anosognosia may produce a hemidepersonalization.

Epilepsy

The diagnosis of epilepsy is not difficult if the patient manifests a typical convulsion (Chapter 21). In some cases, however, the patient may be amnestic for any aura preceding the ictus and may manifest psychomotor automatisms during the ictal or postictal period without a generalized convulsion. In such cases, diagnostic confusion may arise. There are several clues to the ictal nature of these events. Seizures begin abruptly in a stereotyped manner and end suddenly, although there is usually a variable period of postictal confusion. Seizures are typically short-lived, although more prolonged confusional periods may occur in epileptic twilight states and status epilepticus. The patient is often fatigued and postictal headache is common. Behavior during the seizure or postictal state is typically simple and repetitive, lacking complexity and purpose. The patient is either partially responsive or unresponsive to external stimuli. Most patients have a history of previously diagnosed seizures or a predisposing brain injury. Patients in status epilepticus may have petit mal seizures or partial complex seizures as the underlying epileptic disorder.[13–16]

Patients may have prolonged confusional states following seizures, which must be distinguished from dissociative episodes.[17] Nonconvulsive status epilepticus from seizures originating in the frontal lobe may be particularly difficult to identify, since the motor automatisms are often bizarre. Features that may help identify frontal lobe complex-partial seizures include occurrence in clusters of many per day, brief episodes lasting less than 1 minute, sudden onset and termination with little postictal confusion, prominent complex motor automatisms, complex vocalizations, nonspecific warnings preceding the attacks, bizarre behavior during the attack that is stereotyped for individual patients, a history of similar episodes, and onset following a brain injury.[18,19]

Epilepsy also can be a cause of depersonalization and dissociative identity disorders. Dissociative phenomena are not unusual among patients with epilepsy, particularly those with temporal lobe epilepsy.[10,20] In addition, patients with multiple personality and possession states have an unusual frequency of abnormal EEGs,[13] and switching from one personality to another has been associated with spontaneous seizures[21] or with lateralized hemispheric inactivation occurring in the course of the intracarotid amobarbital test.[22] Drake and colleagues[23] studied 15 patients with clinical histories of epilepsy and multiple personality and suggested the following general principles relating the two conditions: (*1*) temporary personality disintegration may occur as part of postictal confusion or psychosis; (*2*) elaboration of multiple independent personalities is more often a manifestation of personality disorder than epilepsy; (*3*) epilepsy and multiple personality disorder may in some cases be related to right hemispheric dysfunction; and (*4*) multiple or disordered personalities may be precipitated or exacerbated by anticonvulsant medications.

A rare dissociated state associated with epilepsy is poriomania,[24] in which the patient experiences prolonged episodes of aimless wandering followed by retrograde amnesia for the experience.

Temporal lobe epilepsy can produce depersonalization. Experiences common in complex-partial seizures include dreamy states, micropsia, macropsia, déjà vú, jamais vú, metamorphopsia, and anxiety. These phenomena may also occur in idiopathic depersonalization syndromes, complicating diagnostic efforts. The two syndromes can usually be distinguished by the occurrence in epileptic patients of brief and stereotyped depersonalization experiences, a history of generalized seizures, epileptiform EEG abnormalities, and a predisposing brain disease. Compared with partial-complex seizures, anxiety-depersonalization syndromes are more likely to have emotional precipitants, occur more frequently (at least daily), resolve more slowly, lack postictal confusion, have an earlier age at onset, and are more likely to involve patients from families with psychiatric illness.[2,25] Patients with anxiety-depersonalization also experience more depression, anxiety, irrational fears, phobias, and hypochondriacal symptoms.

Migraine

Migraine may produce recurrent episodes of automatic behavior, depersonalization, or even loss of consciousness. These patients may be amnestic for the episode. The attacks are usually preceded by visual distortions, hallucinations, vertigo, numbness, or other migrainous phenomena and typically are followed by a prominent, unilateral throbbing headache.

Transient Global Amnesia

Transient global amnesia (Chapter 7) is a true amnestic disorder manifested by disturbances of new learning in the absence of significant alterations of remote memory, general intellectual function, or personality. The amnesia begins abruptly and is characterized by difficulty in learning new information or remembering what is said. The same questions are asked repeatedly. The amnesic episode usually lasts less than 24 hours and the retrograde portion shrinks to within a few minutes of onset of the amnestic period when the attack terminates. Most cases of transient amnesia represent ischemic attacks with hypoperfusion of temporal structures. Other causes include migraine, intoxication with sedatives, cerebral neoplasms, and seizures. Precipitating factors have included sexual intercourse, sudden alterations in body temperature, highly emotional circumstances, and pain. Transient global amnesia can usually be distinguished from psychogenic amnesia by the retention of personal identity, the character of the retrograde amnesia, and the patient's emotional upset regarding the amnesia.[26]

Post-Concussion Amnesia

Amnesia is a frequent consequence of trauma and is associated with medial temporal lobe injury (Chapters 7 and 26). These periods of amnesia typically last a few minutes to a few hours and follow cerebral concussion. During the amnestic period, the patient may accomplish complex activities and appear to behave normally. The onset of the amnesia is associated with a blow to the head, but loss of consciousness may not necessarily occur. The most dramatic cases of post-concussion amnesia have been reported in boxers.[27,28] Winterstein[29] described several such cases including one boxer who was amnestic for a period of several hours between the fourth round of a match and his return home that evening. During the unrecalled period he completed and won the fight, washed and dressed, collected his money, bought train tickets for himself and three friends, and traveled home. Gene Tunney, a former heavyweight champion of the world, had a similar episode in association with a training bout while preparing for his second fight with Jack Dempsey, and the experience significantly influenced his desire to retire from the ring.[30] The amnesia includes both a retrograde period preceding the concussion and a longer anterograde period following the concussion-producing blow. The victim does not recall being struck and, therefore, unless external trauma is apparent, the etiology of the amnestic episode may not be obvious.

Other Neurological Disorders

Temporary periods of memory impairment or depersonalization have also been reported with brain tumors, stroke, encephalitis, degenerative dementias, and extrapyramidal disturbances. In addition, toxic and metabolic conditions associated with depersonalization include hypoglycemia, hypoparathyroidism, hypothyroidism, hyperventilation, botulism, carbon monoxide intoxication, mescaline ingestion, and following use of indomethacin or marijuana.[31,32]

Hemidepersonalization refers to the unique experience of some patients with focal brain disease in which they view the affected side as unreal or distorted in size or shape, or as belonging to someone else. The depersonalization syndrome is one manifestation of anosognosia and occurs primarily in patients with right-sided parietal lobe lesions.[33,34]

Left-sided brain injuries are more likely to be associated with a general sense of depersonalization and derealization.[35] Depersonalization must be distinguished from delusions in which one believes that either oneself or the world has changed and from hallucinations in which one has an actual visual perception of one's own body (autoscopy) (Chapter 13). These experiences may simulate the delusional syndrome of subjective doubles and autoscopic hallucinations, but the depersonalized patient is usually aware that it is the experience of the self or external reality that is altered and not the self or external reality per se.

Sleep Disorders

Unrecallable behavior may occur with sleep–wakefulness disorders. Automatic unrecalled behavior may occur during the period of sleep in the case of rapid eye movement (REM) sleep disorder or somnambulism or during apparently wakeful periods in the case of narcolepsy. The REM behavior disorder is a parasomnia characterized by the occurrence of complex motor behaviors during REM sleep. Punching, kicking, and leaping from the bed have been described. The syndrome occurs predominantly in men (90%) in their 60's and 70's. An association between REM behavior disorder and Parkinsons's disease and dementia with Lewy bodies (Chapter 10) has been described; the disorder also has been observed following stroke and as an idiopathic condition.[36]

Somnambulism (sleepwalking) occurs in 1% to 6% of the population, is more common in children than adults, and affects males more than females. In 25% of cases there is a family history of sleepwalking. Somnambulism may last for only a few seconds, with the patient simply sitting up in bed, or it may last up to an hour and include modestly complex behavior such as walking around objects, dressing, and opening doors. The patient's eyes are open and efforts to communicate may elicit slurred or mumbled responses. There is complete amnesia for the episode. Kleitman[37] described a man who walked along a window ledge 12 stories above the ground and returned to bed without awakening. Somnambulism usually occurs within the first 3 hours after going to bed when the patient is in stage III or stage IV sleep. In some cases, called "night terrors," the patient sits up in bed crying or screaming uncontrollably with a fast heart rate and rapid breathing.[36] Unrecalled automatic behavior may also occur in narcolepsy and other hypersomnias. Patients may perform complex behavior such as walking or driving without mishap. These patients are completely amnestic for the episodes and are surprised by their behavior when normal consciousness resumes. The episodes have been attributed to multiple micro-episodes of disturbed consciousness.[38–40]

Ganser Syndrome

Ganser syndrome has been labeled an hysterical pseudodementia with features of depersonalization and dissociation.[41,42] The syndrome includes hallucinations, prominent sensory changes of an hysterical type, alterations of consciousness with amnesia for the episodes, and verbal responses that are either illogical or "near misses."[42] Whittlock[42] noted that "a Ganser reaction is a hysterical pseudo-stupidity which occurs almost exclusively in jails and in old-fashioned German textbooks. It is now known to be almost always due more to conscious malingering than to unconscious stupefaction." However, the Ganser syndrome has occurred in association with toxic confusional states, head injuries, alcohol-related dementias, general paresis, postpartum psychoses, and a variety of psychiatric conditions, including schizophrenia and depression.[42]

Also known as the syndrome of approximate answers, Ganser patients tend to reply to simple questions with answers that are incorrect but whose nature and consistency suggest that there is knowledge of the correct answer. For example, when asked how many legs a dog has, the patient may respond "3". When asked how many inches in a foot, they may respond "13"; or asked how many days in the week, they may reply "8".[43] Recovery from the syndrome is usually abrupt. The patient is amnestic for the period encompassing the Ganser behavior.

Treatment of Dissociative Phenomena

There has been little exploration of the pharmacologic treatment of depersonalization. Improvement with clonazepam, carbamazepine, and fluoxetine has been described.[8,11,44,45] Long-term psychotherapy is usually necessary to intervene in more severe dissociative disorders such as dissociative identity disorder.[2]

HYSTERIA AND SOMATIFORM DISORDERS

A variety of somatiform disorders are currently grouped together and include somatization disorder (hysteria or Briquet's syndrome), undifferentiated somatiform disorder, conversion disorder, pain disorder, hypochondriasis, body dysmorphic disorder, and unusual somatiform disorders such as pseudocyesis (a false belief of being pregnant) and other unexplained physical complaints.[1] *Somatization disorder,* or *Briquet's syndrome,* is a polysymptomatic disorder that begins before age 30 and is characterized by a combination of pain, gastrointestinal, sexual, and pseudoneurological symptoms. *Undifferentiated somatiform disorder* is characterized by a variety of physical complaints lasting at least 6 months but does not meet the threshold for a diagnosis of somatization disorder. *Conversion disorder* (discussed below) involves unexplained symptoms or deficits affecting voluntary motor or sensory function judged to be associated with psychological factors. *Pain disorder* is diagnosed when pain is the predominant focus of clinical attention and psychological factors are judged to have an important role in its onset, severity, exacerbation, or maintenance.

Hypochondriasis refers to preoccupation with having a serious disease, based on one's interpretation of bodily symptoms or functions. *Body dysmorphic disorder* features preoccupation with an imagined or exaggerated defect in physical appearance generally involving the hair, nose, and skin.[1,46]

The somatiform disorders must be distinguished from factitious illness consisting of the intentional production or feigning of physical or psychological signs or symptoms to assume the sick role.[1] The best known of the factitious illnesses is *Munchausen's syndrome*, whose central features are pathological lying, peregrination (traveling or wandering), and recurrent, feigned, or simulated illness. Common accompanying features include borderline or antisocial personality traits, a history of childhood deprivation, equanimity for diagnostic procedures, equanimity for treatments or operations, evidence of self-induced physical signs, knowledge of or experience in a medical field, history of multiple hospitalizations, multiple scars, a record of encounters with the police, and unusual or dramatic presentations.[47]

Conversions Disorders

Of the somatiform disorders, it is *conversion disorder* that is most likely to come to neuropsychiatric attention. The older term *hysteria* has been used to describe a plethora of clinical phenomena. It is applied to a particular personality disorder (*histrionic personality*) characterized by self-dramatization, excessive attention seeking, overreaction to minor events, irrational emotional outbursts, and a tendency toward manipulative suicide threats, gestures, or attempts. *Epidemic* or *mass hysteria* refers to the occurrence of similar physical symptoms or unusual behaviors that affect a group of individuals and have no identifiable neurological or medical cause. In the lay literature, *hysterical* has been applied to excessive emotional displays that may occur in individuals who receive unexpected favorable or unfavorable news. *Briquet's syndrome* is characterized by many physical complaints beginning before age 50, persisting for a period of several years, and including at least four pain symptoms, gastrointestinal symptoms, one sexual symptom, and one pseudoneurological symptom.[1] *Conversion disorders* (or *hysterical conversion disorders*) feature the occurrence of neurological symptoms in the absence of confirmatory signs of neurological disease or in excess of any disability attributable to an existing neurological condition.

Conversion symptoms reported in the literature include mutism, paralysis, amnesia, blindness, ataxia, and seizures.[47] Neuro-ophthalmic signs consistent with hysteria include spurious blindness, nonphysiologic visual field defects such as cylindrical tunnel-shaped fields, and convergence spasm.[48] A conversion disorder is suspected when the motor or sensory disturbance fails to conform to anatomic and physiologic patterns, results change on multiple examination, or responses change with suggestion.[48]

Fahn[49] has provided criteria for documented, clinically established, probable, and possible psychogenic movement disorders (Table 22.3). Features that suggest a psychogenic movement disorder include a sudden onset; inconsistent movements (changing characteristics over time); incongruous movements and postures (not fitting within recognized patterns or normal physiological profiles); bizarre features of the movement disorder; control of the movement disorder with suggestion; exhaustion and fatigue of the movements; spontaneous remissions; disappearance of the movements with distraction; response to placebo, psychotherapy, or physiotherapy; and history of a psychiatric disorder. The presence of psychogenic weakness, sensory complaints, multiple somatization, self-inflicted injuries, obvious psychiatric illness, the presence of secondary gain, or involvement in litigation or compensation are also suggestive of circumstances facilitating the emergence of psychogenic movement disorders.

Tremors and gait disorders are the most common psychogenic movement disorders observed in neuropsychiatric practice. Studies of psychogenic tremors reveal unusual clinical and temporal profiles, absence of other neurological signs, inconsistent and incongruous phenomenology, selective disability with preserved performance of some functions despite severe tremors, lessening or abolition of the tremor with distraction, unusual handwriting and drawing, presence of multiple somatizations, unresponsiveness to treatment, absence of evidence of neurological disease, presence of a psychiatric condition, spontaneous remission, or recovery with psychotherapy.

Pseudoepileptic seizures are another relatively common manifestation of conversion disorders and can be difficult to distinguish from epileptic seizures (Chapter 21). Table 22.4 summarizes features that help distinguish pseudoepileptic from epileptic convulsions.[50–53] Pseudoepileptic seizures tend to last longer and exhibit more variability, occur in response to emotional precipitants, are rarely associated with incontinence or self-injury, and often terminate without evidence of postictal confusion. Out-of-phase clonic movements, forward pelvic thrusting, and side-to-side head move-

TABLE 22.3. *Criteria for Identifying Psychogenic Movement Disorders*[49]

Documented psychogenic movement disorder
Movements are completely relieved by psychotherapy, psychological suggestion (including physiotherapy), or placebos. Remission is often dramatic, with sudden improvement occurring within a few hours or days of intervention.
Clinically established psychogenic movement disorder
The disorder is inconsistent over time (the features are different when the patient is serially examined) or is incongruent with a classical movement disorder, and at least one of the following is present:
Other neurological signs are present that are definitely psychogenic in nature (false weakness, false sensory findings).
Multiple somatizations are present.
An obvious psychiatric disturbance is present.
The movement disorder disappears with distraction.
Excessive (appearing deliberate) slowing of movement is present.
Probable psychogenic movement disorder
The movements are inconsistent or incongruent but there are no other associated features.
The movements are consistent and congruent with a known neurological disorder but the movements can be made to disappear with distraction.
The movements are consistent and congruent with a known neurological disorder but other signs are present that are definitely psychogenic in origin.
The movements are consistent and congruent with a known neurological disorder but multiple somatizations are also present.
Possible psychogenic disorder
Movements are consistent and congruent with a neurological disorder and an obvious emotional disturbance is present.

ments are substantially more common in pseudoepileptic seizures than in epileptic seizures.[54]

The clinician must be wary of the diagnosis of hysteria despite the apparent nonphysiologic nature of symptoms presented by the patient. Classical features of conversion disorders, including "la belle indifference," nonanatomical sensory loss, splitting of the midline to pain or vibratory stimulation, changing sensory loss, and "give way" weakness, have all been documented in patients with stroke or brain tumors.[55–57]

Follow-up studies confirm the need for caution in the diagnosis of conversion disorder and attributing neurological symptoms to somatiform conditions. In his original study, Slater[58] found that 60% of patients diagnosed with conversion hysteria developed diagnosable neurological illness in the ensuing decade. More recently, Mace and Trimble[59] found that 15% of patients with a diagnosis of hysteria and investigated for neurological symptoms had an established neurological diagnosis 10 years later. Only 3% of the patients had relief from their original symptoms, indicating that the prognosis for recovery is poor.

The significance of conversion symptoms is controversial. Psychoanalytic schools of thought conceived of the symptoms as symbolically significant, and believed that physical symptoms substituted for repressed, instinctual, and unacceptable impulses. An alternative psychodynamic explanation suggests that the symptoms represent a form of nonverbal communication between patient and physician by which the patient can covertly transmit personal needs or distress. Some believe that the patient simply exaggerates existing symptoms to focus the physician's attention, whereas others suggest that the symptoms arise directly from existing physical disturbances too subtle to be detected by the clinician.

In view of the difficulties of defining, understanding, and diagnosing conversion symptoms, as well as the high rate of chronicity and emergence of medical, neurological, and psychiatric disturbances among patients

TABLE 22.4. *Differential Diagnostics of Pseudoepileptic and Epileptic Seizures*[50–53]

Clinical Features	*Pseudoepileptic Seizures*	*Epileptic Seizures*
Age	Adolescence, early adulthood	Common in childhood following brain injury
Gender	1 M:4 F	1 M:1 F
Psychiatric history	Common	Variable
Place	Usually at home or in school	Anywhere
Presence of others	Yes (almost never alone)	Sometimes when patient is alone
During sleep	Rare	Frequently
Frequency	Frequent, especially under stress	Infrequent, except in petit mal
Pattern of seizure	Changes from seizure to seizure	Stereotyped
Prodromal signs (aura)	Variable with anxiety and overbreathing at times	Stereotyped
Onset	Often gradual	Usually sudden
Duration	Many minutes or longer	Brief (except status epilepticus)
Incontinence	Rare	Micturition common
Biting	Lips	Usually tongue
Scream	During seizure	At onset mainly
Motor activity	Variable and bizarre (struggle or sexual movements)	Stereotyped tonic-clonic
Consciousness	Often retained	Lost with generalized seizure
Injury	Infrequent	Frequent
Pupillary reflexes	Normal and reactive to light	Sluggish or nonreactive
Babinski's sign	Not seen	May be seen
Memory of the seizure	Present at times	Amnesia (except in simple-partial seizures)
Orientation after seizure	Oriented	Disoriented temporarily or confused
Effect of suggestion	Precipitates or terminates an attack	No effect
Postictal stupor or sleep	Rare	Frequent
Postictal headache/pain	Rare	Common

presenting with these symptoms, the most defensible clinical approach is to regard conversion phenomena as important harbingers of an underlying condition that must be identified and treated.

Perhaps the best understanding of the neural mechanisms of hysteria come from functional imaging's confirmation of a classical theory. Charcot (1889)[60] theorized that hysterical paralysis reflected "*dynamic* or *functional* lesions . . . of which no trace is found after death."[61] General support of Charcot's position has been provided from regional cerebral blood flow (rCBF) mapping using PET in a patient with hysterical left hemiparesis of 2 years duration.[62] Functional maps revealed similar activations of the appropriate premotor and cerebellar regions when the patient attempted to move either her paretic or good leg, suggesting good effort, but showed a localized failure to increase rCBF in right motor cortex with attempted movements of the left paretic limb. When she tried to move the left paretic limb, significant activations, compared to those during the contralateral condition, were found in the right anterior cingulate and right orbitofrontal cortex; these limbic regions appear to have actively inhibited movement of the left leg despite premotor and downstream cerebellar activation (see Fig. 22.1). These findings suggest that the orbitofrontal cortex may be the source of unconscious inhibition while the anterior cingulate functions to decouple premotor from primary motor cortex in hysterical paralysis. These same structures are activated in hypnotic paralysis,[63] lending further support to the subconscious limbic suppression of consciously directed motor systems.

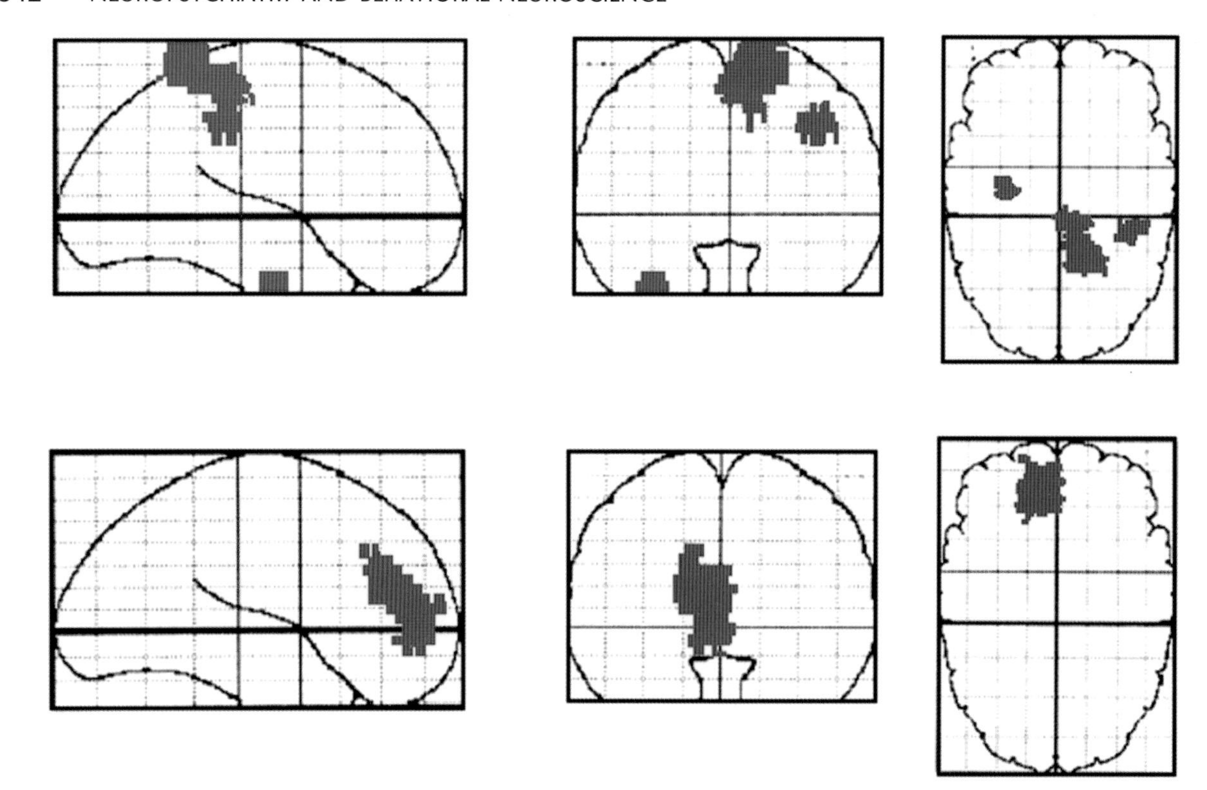

FIGURE 22.1 Statistical maps demonstrating relative regional blood flow increases ($p < 0.001$) as measured by $H_2{}^{15}O$ positron emission tomography (PET) resulting from movement of the right (good) leg in a patient with hysterical paralysis of the left leg. When the patient prepared to move the bad left leg, normal activation was seen in the left lateral premotor cortex and the cerebellar hemispheres bilaterally (relative to baseline), indicating the patient's "readiness" to move the paralyzed leg and that the hemiparalysis was not feigned. The top maps show activated regions when the normal (right) leg is moved that do not occur when attempts to move the bad (left) leg are made. The bottom maps show activated regions during attempts to move the bad (left) leg that did not occur when the good (right) leg was moved. Adapted from Marshall et al. (1997).[62]

REFERENCES

1. American Psychiatric Association. Diagnostic and Statistical Manual of Mental Disorders, 4th ed. Washington, DC: American Psychiatric Press, 1994.
2. Ross CA. Dissociative Identity Disorder: Diagnosis, Clinical Features, and Treatment of Multiple Personality. New York: John Wiley & Sons, 1997.

2a. Cambodonico JR, Rediess S. Dissociation of implicit and explicit knowledge in a case of psychogenic retrograde amnesia. J Internat Neuropsychol Soc 1996;2:146–158.

3. Millon T. Disorders of Personality. DSM III: Axis III. New York: John Wiley & Sons, 1981.
4. Boon S, Draijer N. Multiple personality disorder in The Netherlands: a clinical investigation of 71 patients. Am J Psychiatry 1993;150:489–494.
5. Chu JA, Dill DL. Dissociative symptoms in relation to childhood physical and sexual abuse. Am J Psychiatry 1990;147:887–892.
6. Rose CA, Miller SD, et al. Structured interview data on 102 cases of multiple personality disorder from four centers. Am J Psychiatry 1990;147:596–601.
7. Sanders B, Giolas MH. Dissociation and childhood trauma in psychologically disturbed adolescents. Am J Psychiatry 1991; 148:50–54.
8. Coons PM, Multstein V, Marley C. EEG studies of two multiple personalities and its control. Arch Gen Psychiatry 1982;39: 823–825.
9. Matthew RJ, Jack RA, West WS. Regional cerebral blood flow in a patient with multiple personality. Am J Psychiatry 1985;142: 504–505.
10. Schenk L, Bear D. Multiple personality and related dissociative phenomena in patients with temporal lobe epilepsy. Am J Psychiatry 1981;138:1311–1316.
11. Hollander E, Carrasco JL, et al. Left hemisphere activation in depersonalization disorder: a case report. Biol Psychiatry 1992; 31:1157–1162.
12. McCormick S, Golf DC. Possession states: approaches to clinical evaluation and classification. Behav Neurol 1992;5:161–167.
13. Mesulam M-M. Dissociative states with abnormal temporal lobe effects: multiple personality and the illusion of possession. Arch Neurol 1981;38:176–181.
14. Escueta AV, Boxley J, et al. Prolonged twilight state and automatisms: a case report. Neurology 1974;24:331–334.

15. Goldensohn ES, Gold AP. Prolonged behavioral disturbances as ictal phenomena. Neurology 1960;10:1–9.
16. Lugaresi E, Pazzaglia P, Tassinari CA. Differentiation of "absence status" and "temporal lobe status". Epilepsia 1971;12:77–87.
17. Biton V, Gates JR, Depauda Sussman S. Prolonged postical encephalopathy. Neurology 1990;40:963–966.
18. Thomas P, Zifkin B, et al. Nonconvulsive status epilepticus of frontal origin. Neurology 1999;52:1174–1183.
19. Williamson PD. Psychogenic non-epileptic seizures and frontal seizures: diagnostic considerations. In: Rowan JA, Gates JR, eds. Non-Epileptic Seizures. Boston: Butterworth-Heinemann, 1993: 55–72.
20. Devinsky O, Putman F, et al. Dissociative states and epilepsy. Neurology 1989;39:835–840.
21. Benson DF, Miller BL, Signer SF. Dual personality associated with epilepsy. Arch Neurol 1986;43:471–474.
22. Ahern GL, Herring AM, et al. The association of multiple personality and temporolimbic epilepsy: intracarotid amobarbital test observations. Arch Neurol 1993;50:1010–1025.
23. Drake ME, Pakalnis A, Denio LC. Differential diagnosis of epilepsy and multiple personality: clinical and EEG findings in 15 cases. Neuropsychiatry Neuropsychol Behav Neurol 1988;1: 131–140.
24. Mayeux R, Alexander MP, et al. Poriomania. Neurology 1979; 29:1616–1619.
25. Harper M, Roth M. Temporal lobe epilepsy and the phobic anxiety–depersonalization syndrome, Part I: a comparative study. Compr Psychiatry 1962;3:129–151.
26. Evans JH. Transient loss of memory, an organic mental symptom. Brain 1966;89:539–548.
27. Critchley M. Medical aspects of boxing, particularly from a neurological standpoint. BMJ 1957;1:357–362.
28. Spillane JD. Five boxers. BMJ 1962;2:1205–1210.
29. Winterstein CE. Head injuries attributable to boxing. Lancet 1937;2:719–720.
30. Martland HS. Punch drunk. JAMA 1928;91:1103–1107.
31. Kenna JC, Sedman G. Depersonalization in temporal lobe epilepsy and the organic psychoses. Br J Psychiatry 1965;111:293–299.
32. Matthew RJ, Wilson WH, et al. Depersonalization after marijuana smoking. Biol Psychiatry 1993;33:431–441.
33. Critchley M. The Parietal Lobes. New York: Hafner Press, 1953.
34. Cutting J. Study of anosognosia. J Neurol Neurosurg Psychiatry 1978;41:548–555.
35. Cutting J. The Right Cerebral Hemisphere and Psychiatric Disorders. New York: Oxford University Press, 1990.
36. Reite M, Ruddy J, Nagel K. Concise Guide to Evaluation and Management of Sleep Disorders. London: American Psychiatric Press, 1997.
37. Kleitman N. Sleep and Wakefulness. Chicago: University of Chicago Press, 1963.
38. Guilleminault C, Billiard M, et al. Altered states of consciousness in disorders of daytime sleepiness. J Neurol Sci 1975;26: 377–393.
39. Guilleminault C, Phillips R, Dement WC. A syndrome of hypersomnia with automatic behavior. Electroencephalogr Clin Neurophysiol 1975;26:377–393.
40. Ross ED. Sensory-specific and fractional disorders of recent memory in man, I: isolated loss of visual recent memory. Arch Neurol 1980;37:193–200.
41. Cocores JA, Santa WG, Patel MD. The Ganser syndrome: evidence suggesting its classification as a dissociative disorder. Int J Psychiatry Med 1984;14:47–56.
42. Whitlock FA. The Ganser syndrome. Br J Psychiatry 1967;113: 19–29.
43. Goldin S, MacDonald JE. The Ganser state. J Ment Sci 1955;101: 267–285.
44. Fichtner CG, Kuhlma DT, et al. Decreased episodic violence and increased control of dissociation in a carbamazepine-tested case of multiple personality. Biol Psychiatry 1990;27:1045–1052.
45. Stein MB, Uhde TW. Depersonalization disorder: effects of caffeine and response to pharmacotherapy. Biol Psychiatry 1989;26: 315–320.
46. Phillips KA, McElroy SL, et al. Body dysmorphic disorder: 30 cases of imagined ugliness. Am J Psychiatry 1993;150:302–308.
47. Folks DG, Freeman AM. Munchausen's syndromes and other factitious illnesses. Psychiatr Clin North Am 1985;8:263–278.
48. Keane JR. Neuro-opthalamic signs and symptoms of hysteria. Neurology 1982;32:757–762.
49. Fahn S. Psychogenic movement disorders. In: Marsden CD, ed. movement disorders 3. Boston: Butterworth Heinnemann, 1994: 359–372.
50. Bergey KB. Psychogenic seizures. In: Fisher RS, ed. Imitators of Epilepsy. New York: Demos Publications, 1994:283–306.
51. Gates JR, Luciano D, Devinsky O. The classification and treatment of nonepileptic events. In: Devinsky O, Theodore WH, eds. Epilepsy and Behavior. New York: Wiley-Liss, 1991:251–263.
52. Gross M. The clinical diagnosis of psychogenic seizures. In: Gross M, ed. Psuedoepilepsy: The Clinical Aspects of False Seizures. Toronto: Lexington Books, 1983:79–96.
53. Lesser RP, Krauss G. Psychogenic non-epileptic seizures resembling complex partial seizures. In: Rowan AJ, Gates JR, eds. Non-Epileptic Seizures. Boston: Butterworth-Heinemann, 1993: 39–45.
54. King DW, Gallagher BB, et al. Convulsive non-epileptic seizures. In: Rowan AJ, Gates JR, eds. Non-Epileptic Seizures. Boston: Butterworth-Heinemann, 1993:31–37.
55. Chandarana P, Young GB, et al. Unusual neuropsychiatric symptoms as a manifestation of a frontotemporal tumor. Neuropsychiatry Neuropsychol Behav Neurol 1992;5:53–55.
56. Gould R, Miller BL, et al. The validity of hysterical signs and symptoms. J Nerv Ment Dis 1986;174:593–597.
57. Jones JB, Barklage NE. Conversion disorder: camouflage for brain lesions in two cases. Arch Intern Med 1990;150:1343–1345.
58. Slater E. Diagnosis of 'hysteria'. BMJ 1965;1:1395–1399.
59. Mace CJ, Trimble MR. Ten-year prognosis of conversion disorder. Br J Psychiatry 1996;169:282–288.
60. Charcot JM. Clinical Lectures on Diseases of the Nervous System. London: New Sydenham Society, 1889.
61. Freud S. Quelques considérations pour une étude comparative des paralysies motrices organiques et hystériques. Arch Neurol 1893;26:29–43.
62. Marshall JC, Halligan PW, et al. The functional anatomy of a hysterical paralysis. Cognition 1997;64:B1–B8.
63. Halligan PW, Athwal BS, et al. Imaging hypnotic paralysis: implications for conversion hysteria. Lancet 2000;355:986–987.

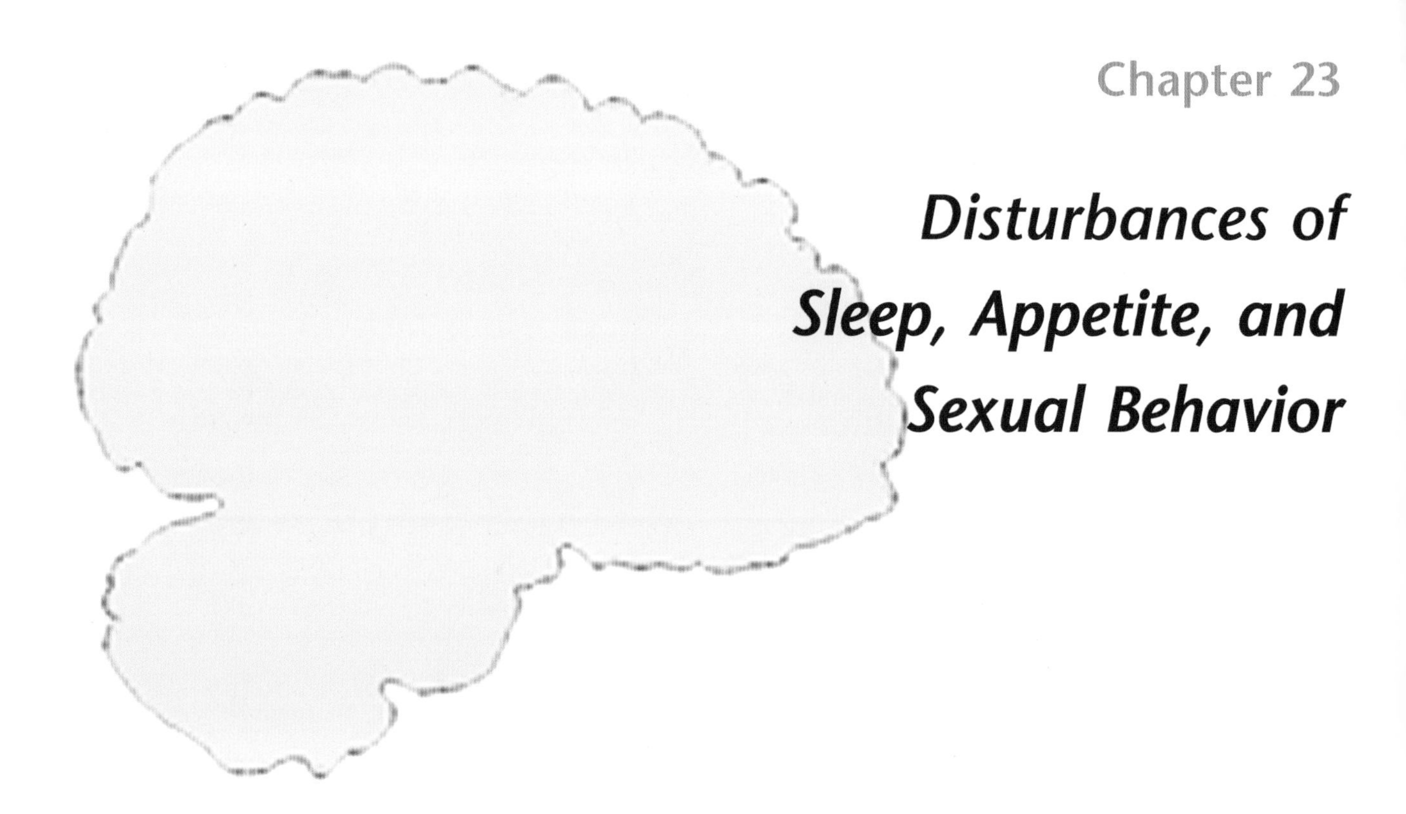

Chapter 23

Disturbances of Sleep, Appetite, and Sexual Behavior

Neurological, medical, and idiopathic psychiatric disorders often affect the basic life functions of sleep, appetite, and sexuality. These basic drives are called "vegetative" functions, to contrast them with more volitional aspects of behavior, but they are not passively automatic. Limbic influences, especially arising from the medial orbitofrontal cortex and amygdala, imbue objects and persons in our environment with appetitive valence—whether as a potential meal or mate. Sleep is subserved by brainstem and limbic regions[1,2] that control consciousness and may be required to consolidate memories and recently learned associations of external stimuli and internal drives. This coupling of internal milieu with external objects, or persons, strengthens our behavioral patterns and is often rehearsed or reexperienced in dreams. Rapid eye movement (REM) sleep may be necessary for memory consolidation through the re-engagement of networks initially stimulated during wakeful learning (see Fig. 23.1).[3]

SLEEP DISORDERS

Sleep disturbances can be divided into disorders of excessive sleepiness (hypersomnias), disorders in which the patient is unable to sleep adequately (insomnias), and a variety of sleep-related conditions that do not alter the total amount of sleep but are nocturnal in occurrence (parasomnias) (Table 23. 1).

Hypersomnias

● Narcolepsy *Narcolepsy* is a disorder of excessive somnolence characterized by sudden irresistible attacks of sleep. The sleep attacks often occur in combination with other features of a clinical tetrad including cataplexy, sleep paralysis, and hypnagogic hallucinations.[4] Narcoleptic sleep attacks most commonly begin between the ages of 15 and 25 and persist throughout life. Onset of this disorder rarely occurs before the age of 10 or after the age of 50. Most patients have between one and six attacks per day. The disorder has a prevalence in the population of 0.3% to 1% and is more common in men than women. Most narcolepsy cases are sporadic, associated with a loss of the hypothalamic neuropeptide orexin; rarely, focal lesions in the pons[5] or diencephalon[6,7] can cause narcolepsy. Familial occurrence of narcolepsy is also common, as 10%–30% of patients demonstrate autosomal dominant inheritance of the disorder, with genetic defects localized to the *HLA-DR2, DRW15,* and *DW6* genes. Cerebrospinal fluid level of hypocroten are decreased.[8]

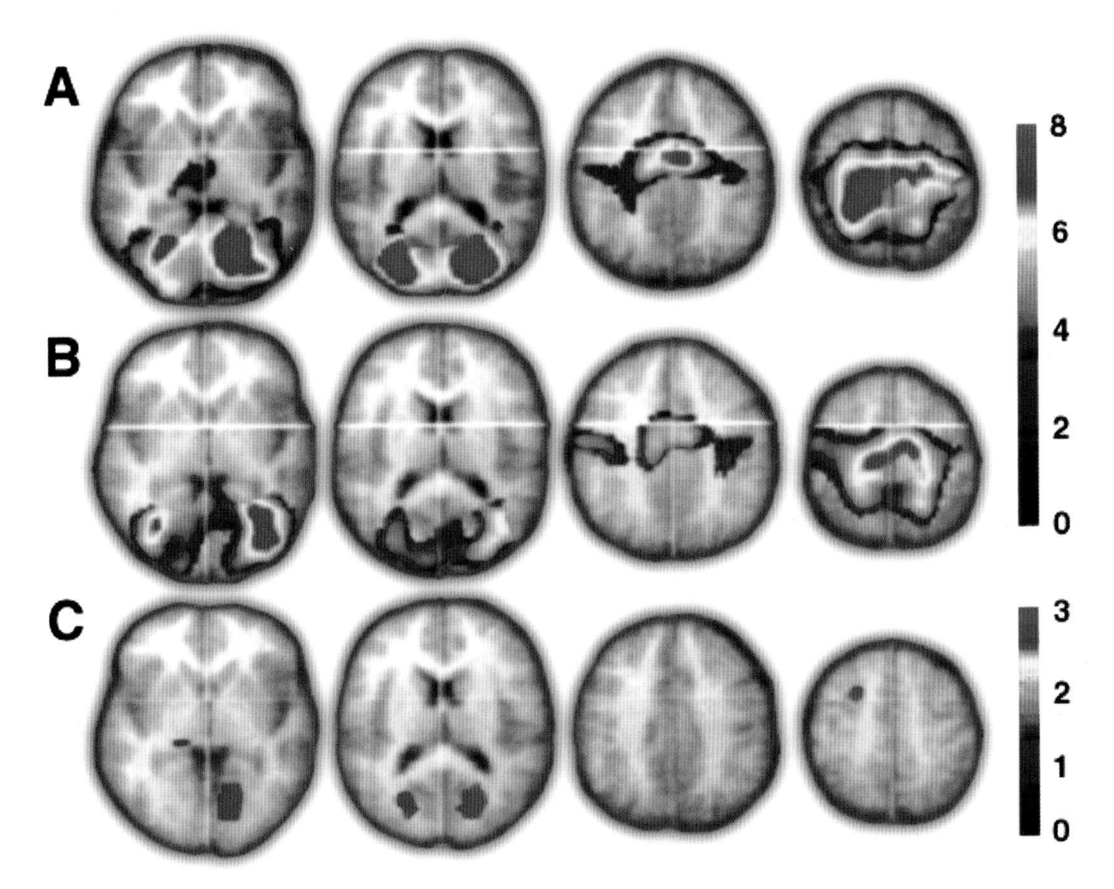

FIGURE 23.1 Statistical parametric maps of regional cerebral blood flow increases as measured by $H_2{}^{15}O$ positron emission tomography (PET) resulting from (*A*) rapid eye movement (REM) sleep compared to wakefulness in six subjects who were trained on a serial reaction time task; (*B*) REM sleep compared to wakefulness in five different subjects who were not trained on the serial reaction time task; (*C*) regions activated more during REM sleep versus wakefulness in trained subjects compared to untrained subjects. Color scale reflects *t* values; subject's left is on the left. Note that during REM sleep in trained subjects greater blood flow occurs in the left premotor cortex, left thalamus, and bilateral cuneus compared to untrained subjects. These regions are also activated by learning the task that requires visual attention and right hand output, supporting the theory that REM sleep strengthens neural circuits entrained by learning during wakefulness. Adapted from Maquet et al. (2000).[3]

Sleep attacks are the first manifestation of the disorder in 90% of cases, but most patients eventually develop other elements of the syndrome, with the narcolepsy-cataplexy combination being most common.[4] *Cataplexy* refers to the sudden loss of muscular tone and occurs in 60% of narcoleptic patients; *sleep paralysis* is the inability to move during the transition period from sleep and wakefulness and occurs in 20%–40% of patients; and *hypnagogic* and *hypnopopic hallucinations* are visual or auditory hallucinations occurring upon falling asleep or awakening, respectively, and are experienced by 15%–50% of narcolepsy patients.

Pathophysiologically, narcolepsy results from an aberrant intrusion of rapid eye movement (REM) sleep into the waking state. The sleep attack itself represents the sudden onset of REM sleep, the sleep paralysis and cataplexy result from the loss of muscle tone that accompanies REM sleep, and the hallucinations reflect the dreams that occur during REM periods.[4] Electroencephalographic (EEG) studies also reveal the abrupt onset of REM sleep is coincident with the occurrence of the narcoleptic sleep attack. In addition to the classic clinical and EEG characteristics, narcoleptic patients also have disturbed nocturnal sleep and episodes of unrecalled automatic behavior. Narcoleptic patients are at increased risk for personality disorders, depression, and, occasionally, psychoses.[9]

In addition to idiopathic and hereditary narcolepsy, symptomatic forms of hypersomnia secondary to structural neurological insults have also been reported. The majority of central nervous system (CNS) disorders associated with hypersomnia involve the brainstem or thalamus, where they can influence neurological mechanisms mediating sleep processes. Narcolepsy has been

TABLE 23.1. *Classification of Sleep Disorders*

Hypersomnias	*Insomnias*	*Parasomnias*
Narcolepsy	Psychiatric disorders with insomnia	Disorders of arousal
Primary	Mood disturbances	Somnambulism
Secondary	Other psychiatric disorders	Somniloquy
Sleep apnea	Transient and situational disturbances	Night terrors
Occlusive type	Toxic-metabolic disorders with insomnia	Enuresis
Central type	Sleep apnea	Nightmares and other dream alterations
Mixed forms	Miscellaneous insomnias	REM sleep behavior disorders
Toxic-metabolic hypersomnias	With nocturnal myoclonus	Miscellaneous conditions
Medical illnesses	With restless legs	Nocturnal seizures
Toxic conditions	With CNS disorders	Cluster headache
Psychiatric disorders with hypersomnia	Chronic primary insomnia	Asthma
Depression		Cardiovascular symptoms
Other psychiatric disturbances		Gastrointestinal disturbances
Periodic hypersomnia		Painful erections
Kleine-Levin syndrome		Head banging
Menstruation-associated syndrome		Bruxism
Miscellaneous hypersomnias		
Sleep drunkenness		
With nocturnal myoclonus		
With restless legs		
With CNS lesions		
Idiopathic hypersomnias		

REM, rapid eye movement.

noted to occur with von Economo encephalitis lethargica, with malarial brainstem infections, and with a variety of other forms of viral encephalitis. Brainstem trauma, cerebrovascular disease, neoplasms, shunt failure without hydrocephalus, developmental disorders, and multiple sclerosis have been associated with hypersomnolence in rare cases.[10–12] Idiopathic hypersomnolence has also been described that is resistant to treatment.

Narcolepsy usually requires treatment with amphetamines or modafinil methylphenidate, whereas cataplexy, sleep paralysis, and hypnagogic hallucinations respond best to imipramine hydrochloride.

● Sleep Apnea *Sleep apnea* is a potentially life-threatening illness characterized by multiple episodes of nocturnal apnea-excessive snoring and daytime sleepiness. The apnea is a product of insufficient air exchange and may be of central origin, reflecting inadequate stimulation of the respiratory muscles, a product of peripheral or occlusive conditions with obstruction of the oropharynx, or the result of mixed central and peripheral components. The patients may have frequent nocturnal awakenings and complain of nocturnal insomnia with daytime fatigue and somnolence. Systemic hypertension, cardiac arrhythmias, and morning headaches are common consequences of sleep apnea. Although extreme obesity may produce occlusive sleep apnea (Pickwickian syndrome), few sleep apnea patients are obese, and the diagnosis should be considered in any individual complaining of excessive daytime sleepiness. The diagnosis of sleep apnea is based on observation of a minimum of 30 episodes of apnea lasting for at least 10 seconds during a 7-hour period of sleep. The number of abnormal respiratory events per hour of sleep is recorded by polysomnography and reported as the apnea–hypopnea index.[13] Obstructive episodes are more common during REM than non-REM sleep stages, probably because of muscle atonia that occurs during REM stage sleep, allowing the upper airway collapse.

For relief of sleep apnea, a variety of treatment modalities are used. Aggravating agents such as hypnotics and propranolol should be discontinued, sleeping upright may reduce symptoms, and loss of excessive weight should be encouraged. Protriptyline enhances ventilatory drive and may help patients with moderately severe apnea syndromes. Advanced sleep apnea may require tracheostomy to ensure adequate nocturnal ventilation. Like narcolepsy, the central form of sleep apnea may be idiopathic or may result from a variety of CNS disorders, including syringomyelia, posterior fossa neoplasms, bulbar poliomyelitis, brainstem infarction, Shy-Drager syndrome (a parkinsonian syndrome with prominent autonomic dysfunction), and olivopontocerebellar degeneration.[14,15]

• Toxic-Metabolic Hypersomnias Tolerance of CNS stimulants (amphetamines, methylphenidate, and caffeine) or their withdrawal may result in a paradoxical increase in daytime somnolence. Likewise, sustained use of depressants such as opiates, barbiturates, alcohol, antihistamines, and anxiolytics may result in excessive daytime sleep. A wide variety of medical illnesses, including systemic infections, hormonal disorders, and environmental toxins, can also produce hypersomnia either directly or by disturbing nocturnal sleep.

• Psychiatric Disorders with Hypersomnia Insomnia is more characteristic of depression than hypersomnia, but in some cases the patient experiences a pathological increase in sleepiness, particularly early in the course of the depression. Hypersomnia in the elderly may predate the manifestation of depression, with an associated relative risk of 3.46 being reported in a large prospective population study.[16] Sleep studies indicate a decreased latency to REM onset, as in depression with insomnia. Patients with personality disorders, dissociative disorders, hypochondriasis, and schizophrenia may complain of excessive sleepiness, but sleep studies reveal that the total daily amount of sleep seldom is increased.

• Periodic Hypersomnias *Periodic hypersomnias* are disorders of excessive sleep that recur at prolonged intervals. Kleine-Levin syndrome is a periodic hypersomnia that involves primarily adolescent males. These patients have irregular episodes lasting for a few days or weeks and characterized by increased somnolence, increased hunger, exaggerated sexual activity, and a confusional state with hallucinations, delusions, and poor attention and memory. The episodes recur at approximately 5-month intervals and eventually spontaneously disappear[17,18] or may be due to focal lesions of the pituitary–hypothalamic axis.[19] A syndrome sharing many of the features of the Kleine-Levin syndrome has been described in women, in which the periodic sleep disturbance is temporally linked to the menstrual cycle.[20]

• Miscellaneous Hypersomnias In addition to the hypersomnias previously discussed, a variety of other disorders with excessive somnolence have also been described. *Sleep drunkenness* refers to a syndrome characterized by extended sleep with otherwise normal sleep architecture. Patients have difficulty awakening completely and are confused, uncoordinated, and slow for the first few hours after arising.[21] Nocturnal myoclonus and the restless legs syndrome usually cause insomnia but in some cases may present as a compensatory daytime somnolence. Excessive sleep may also occur with CNS lesions, particularly those involving the brainstem and diencephalon. Increased daytime sleepiness may follow trauma or may occur with tumors or infarctions affecting the brainstem, hypothalamus, or thalamus.[10–12] A few patients demonstrate an idiopathic hypersomnia unlike any of the primary or symptomatic hypersomnias described. Also called *slow-wave narcolepsy* or *hypersomnia with normal sleep*, the syndrome is characterized solely by increased daytime sleepiness. The sleep is of the slow-wave type and, unlike the sleep of narcolepsy, does not refresh the patient.[22]

Insomnias

• Psychiatric Disorders Depression is the most prevalent cause of insomnia, accounting for approximately 20% of patients referred to sleep disorder clinics with a chief complaint of inability to sleep. Sleep during a major depressive episode is characterized by a decreased latency period between sleep onset and the beginning of the first REM period and by an increased density of REMs during REM periods.[23] Patients also have less total REM sleep and less deep sleep and complain of difficulty in falling and staying asleep as well as early morning awakening. Insomnia is also common in mania but rarely is the dominant clinical feature.

Insomnia is a frequent complaint of patients with personality disorders and may occur in individuals with somatoform disorders, obsessive-compulsive disorders, or schizophrenia. Situational disturbances with anxiety also produce insomnia, and their greatest effect is on the time interval between retiring and sleep onset.

• Toxic-Metabolic Disorders The most common toxic conditions with insomnia involve the use of CNS stimulants and the withdrawal of CNS depressants. Amphetamines, methylphenidate, and caffeine all produce increased arousal and diminished sleep. Withdrawal of alcohol or of sedative-hypnotic agents produces a withdrawal insomnia that may persist for up to 6 weeks following cessation of drug use. Even mildly sedating agents such as benzodiazepines, anticonvulsants, steroids, antipsychotic agents, antidepressant drugs, opiates, marijuana, and propranolol can be associated with insomnia when they are withdrawn.[23]

A wide variety of medical conditions can produce disturbances of nocturnal sleep. Pain syndromes can be particularly disruptive and produce insomnia in patients with arthritis, headache disorders, and other chronic pain problems. Hyperthyroidism, pregnancy, gastrointestinal diseases, eating disorders, and cardiovascular disorders may also cause repeated awakening.

• Sleep Apnea Sleep apnea, as noted earlier, usually presents with excessive daytime sleepiness, but some patients, particularly those with central-type apneas, have multiple nocturnal arousals and may have insomnia as their chief complaint.

• Miscellaneous Disorders *Sleep-related periodic myoclonus* is a syndrome characterized by difficulty in falling asleep and sustained nocturnal awakenings associated with repetitive myoclonic jerking of the legs. The jerking occurs every 20–40 seconds for periods of a few minutes to as long as 2 hours. The disorder occurs between ages 30 and 60 and is a chronic, persistent problem.[24] Periodic nocturnal myoclonus must be distinguished from the common myoclonic jerks that occur just as one is falling asleep, myoclonic jerks occurring in the toxic-metabolic disturbances, epileptic myoclonic jerks, and flexor spasms associated with cervical spondylosis and other spinal cord diseases.[24]

The *restless legs syndrome* is an idiopathic disturbance characterized by creeping-crawling sensations in the distal lower extremities that are most disagreeable when the patient is at rest and are relieved by walking. The sensations are most marked in the evening and night and may prevent the patient from sleeping.[24] The restless legs syndrome and periodic nocturnal myoclonus may present with excessive daytime sleepiness if nocturnal insomnia is severe. Brainstem lesions such as neoplasms and basilar artery strokes can produce almost complete loss of REM sleep, slow-wave sleep, or all phases of sleep.[25–27] Similarly, degenerative brainstem disorders such as Parkinson's disease and progressive supranuclear palsy produce diminished REM sleep and decreased total sleep time. Epileptic patients may have decreased sleep even in the absence of overt nocturnal seizures.

Chronic primary insomnia is an idiopathic sleep disturbance characterized by prolonged sleep latencies, diminished total sleep time, and decreased slow-wave sleep. The syndrome occurs in patients without identifiable psychiatric, neurological, or toxic-metabolic disturbances.

• Treatment of Insomnia The treatment of insomnia is fraught with difficulty. Many sedative-hypnotics, although effective in the first few weeks of administration, soon lose efficacy and may even exaggerate sleep problems through tolerance and withdrawal effects. Benzodiazepine sedative agents have fewer side effects than barbiturate or antihistaminic drugs but are not without adverse consequences and may lead to dependency and intellectual impairment and exaggerate the effects of alcohol or other CNS depressants. To the fullest extent possible, the clinician should direct efforts at eliminating the underlying cause of the insomnia (depression, sleep apnea, medical illness, stimulant use), encouraging the patient to regularize sleeping habits and avoiding the chronic use of sedating medications.

Parasomnias

Parasomnias are a diverse group of disorders that occur during sleep or are exacerbated by sleep but do not necessarily result in either hypersomnia or insomnia (Table 23.1).

• Disorders of Arousal Sleepwalking (somnambulism), sleep-talking (somniloquy), enuresis (bedwetting), and night terrors (pavor nocturnus, incubus) are all disorders of nocturnal arousal.[28] Each of these nighttime events is initiated during stage IV slow-wave sleep and represents automatic behavior with incomplete arousal. The patient is in a confusional state while the behavioral automatons are executed. Although sleepwalking and sleep-talking were originally suspected to be dream related, they do not occur during REM sleep when dreams are most prevalent, and patients rarely report dream memories when awakened from a somnambulistic episode. Night terrors differ from nightmares in that the patient suddenly cries out and exhibits signs of acute anxiety, such as tachypnea, tachycardia, diaphoresis, and dilated pupils. Frequently, the patient does not awaken and has no memory of the episode the following morning. If awakened at the time of the attack, the patient may give a vague description of an apprehensive feeling but lacks the detailed dream recall of patients awakened from REM sleep. Stage IV sleep and the associated disorders of arousal can be suppressed by administration of benzodiazepines.

• Nightmares and Other Dream Alterations

Nightmares are unpleasant dreams that, like most other dreams, occur in periods of REM sleep. The unpleasant quality usually correlates with the presence of anxiety or some situational disturbance. Occasionally, complex partial seizures may give rise to terrifying dreams as part of a complex psychosensory seizure.[29] Epileptic sleep terrors respond to anticonvulsant therapy. Dreaming and dream recall are affected by focal lesions of the nervous system. Amnesic patients with the Wernicke-Korsakoff syndrome have normal to low amounts of REM time and have little dream recall when awakened.[30] Parieto-occipital lesions produce diminished REMs, loss of electroencephalographic (EEG) sleep spindles ipsilateral to the lesion, and impaired dreaming or dream recall. Dreams and dream recall may be exaggerated by some brainstem lesions and, as noted earlier, abolished by others.

REM sleep behavior disorder is characterized by acting-out dream sequences and occurs in Parkinson's disease and dementia with Lewy bodies (Chapter 18).

• Miscellaneous Conditions Finally, there are a group of disorders whose occurrence is exaggerated during periods of nocturnal sleep. These include nocturnal seizures, cluster headache, asthma attacks, some cardiovascular and gastrointestinal symptoms, head banging (jactatio capitis nocturnus), and bruxism (tooth grinding). Penile erections occur during nocturnal REM periods and may occasionally be sustained and painful.

APPETITE DISTURBANCES

Profound loss of appetite (anorexia) or increased appetite (hyperphagia) may be produced by a number of neurological, medical, and psychiatric disorders (Table 23.2). Although hyperphagia frequently leads to obesity, the latter also has complex genetic, dietary, psychosocial, and activity-level determinants and is not considered separately here.

Anorexia

Loss of appetite is more commonly a product of idiopathic psychiatric conditions than of neurological or medical illness, but a few neurological diseases can produce anorexia and must be considered in the differential diagnosis. Hypothalamic lesions are more likely to cause hyperphagia than anorexia, but when the lateral hypothalamic region is involved, there may be a marked loss of appetite. Tumors are the usual cause of hypothalamic injury, but appropriately placed vascular in-

TABLE 23.2. *Disorders Producing Alterations in Appetite*

Diminished Appetite (Anorexia)	*Increased Appetite (Hyperphagia)*
Neurological disorders	Neurological disorders
Hypothalamic lesions	Hypothalamic lesions
Advanced degenerative brain diseases	Kleine-Levin syndrome
Systemic medical illnesses	Bilateral temporal lobe injury (Klüver-Bucy syndrome)
Psychiatric disorders	Psychiatric disorders
Depression	Mania
Anorexia nervosa	Depression
	Bulimia

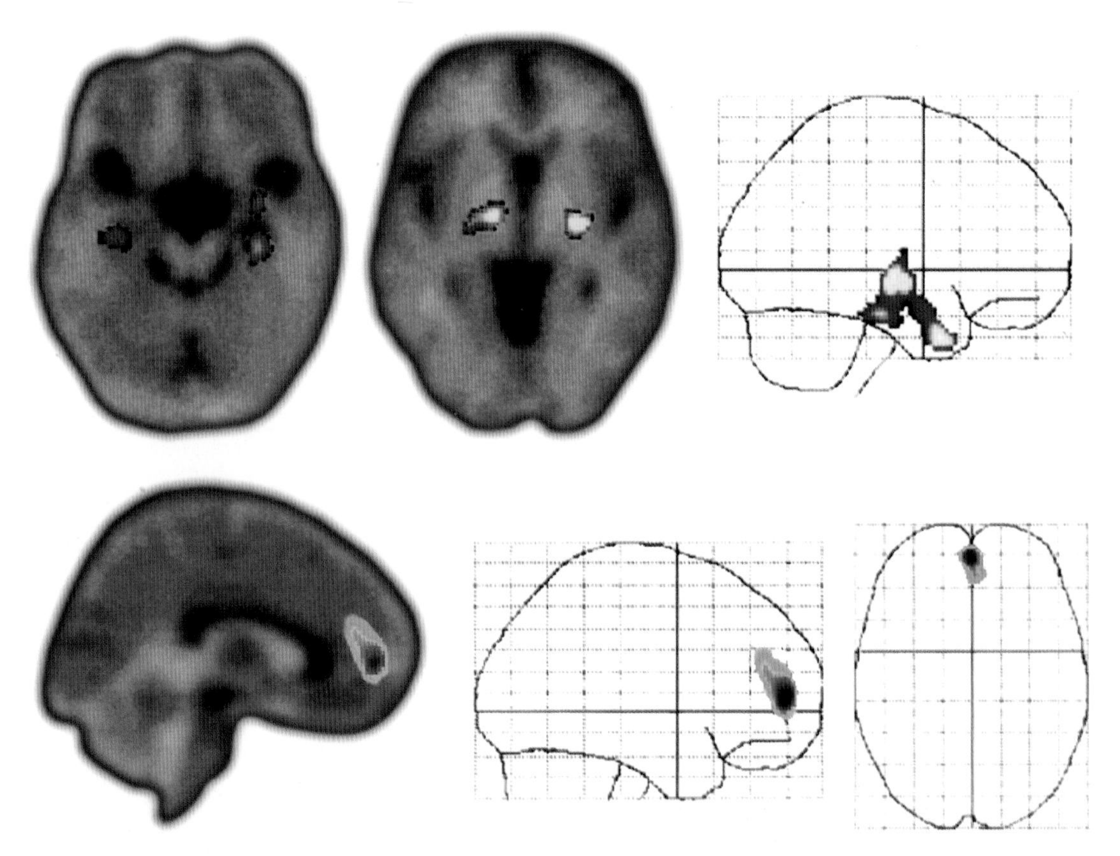

FIGURE 23.2 Statistical parametric maps of resting regional cerebral blood flow increases (top row) and decreases (bottom row) as measured by {^{123}I}-iodoamphetamine (I^{123}IMP) single photon emission computed tomography (SPECT) in 14 females with anorexia nervosa compared to 8 normal control females. Color maps reflect uncorrected significance values with a threshold of $p < 0.05$. Note that anterior cingulate hypoperfusion occurs with hyperperfusion of medial temporal and thalamic regions, suggesting general limbic dysfunction in anorexia nervosa compared to controls. Adapted from Takano et al. (2001).[34]

farctions and infectious lesions may also produce anorexia. When such conditions occur in adolescents, they may be misdiagnosed as anorexia nervosa.[31] A related disorder is the diencephalic inanition syndrome that occurs in infants as a consequence of anterior hypothalamic neoplasms.[32] Patients typically have a clinical triad consisting of marasmus (severe weight loss), euphoria, and nystagmus. The illness ends in death by the age of 2 years. Anorexia is also seen in many degenerative brain diseases and may be particularly severe in the advanced stages of Alzheimer's disease and Huntington's disease.

Medical illnesses, including cardiopulmonary diseases, liver and kidney failure, endocrine disturbances, and infections, may produce anorexia leading to severe loss of weight. Systemic cancer may have particularly profound effects on appetite, and neoplasms of prostate, pancreas, lung, or gastrointestinal tract may present with weight loss as the first indication of their presence. Depression is the most common cause of anorexia. The loss of appetite may accompany acute grief reactions and is a principal feature of major depressive episodes. The anorexia is accompanied by sleep disturbances, loss of libido, and neuroendocrinologic alterations consistent with limbic–hypothalamic–pituitary dysfunction.

Anorexia nervosa is the most dramatic of the disorders of eating and weight control. Although actual loss of appetite is uncommon until the late phases of the disorders, the patients take extreme measures to lose weight, including avoidance of high-calorie foods, self-induced vomiting, use of diuretics and laxatives, and excessive exercising. Despite these efforts, the patients manifest an intense fear of obesity and continue to feel as though overweight even when emaciated. The disorder commonly begins in adolescence, although onset may occur in the third decade and rarely even later. It rarely involves males. In most cases there is a single episode that resolves with full recovery; a few patients have a relapsing and remitting course with recurrent episodes; and a few have an unremitting course ending in death by starvation. There is a familial predisposition to the disorder, and patients with urogenital abnormalities and Turner's syndrome appear to be particularly vulnerable to the development of anorexia nervosa. Functional imaging in patients has implicated limbic dysfunction with hypoperfusion in the caudate and anterior cingulate, and hyperperfusion in the medial temporal and thalamic regions, compared to controls (see Fig. 23.2).[33,34] There is also a relationship between anorexia nervosa and affective disorder, and follow-up studies show an increased incidence of mood disturbances among patients with previous episodes of anorexia nervosa.

A wide variety of clinical and metabolic alterations accompany anorexia nervosa (Table 23.3). Most of the changes appear to be secondary to the severe weight loss and occur with starvation of any etiology. The clinically evident abnormalities include hypothermia, dependent edema, bradycardia, hypotension, constipation, and the development of lanugo.[35] Many endocrinologic alterations have also been described in patients with anorexia nervosa. There are diminished levels of thyroid-stimulating hormone (thyrotropin) (TSH), thyroxin (T4), luteinizing hormone (luteotropin) (LH), follicle-stimulating hormone (FSH), and gonadal steroids. There is a prepubertal LH secretory pattern, diminished responses to luteinizing hormone–releasing hormone (LHRH), impaired dexamethasone–induced suppression of cortisol secretion, and decreased response to insulin-induced hypoglycemia. Plasma cortisol and growth hormone levels are elevated. Amenorrhea occurs in all females and is the feature least likely to normalize after weight has been restored. Cerebrospinal fluid (CSF) abnormalities include decreased homovanillic acid (HVA) and 5-hydroxyindoleacetic acid (5-HIAA) levels during the anorectic episode. Studies of sleep architecture reveal decreased REM latency similar to but less marked than the shortened REM latency found in depression. Treatment of anorexia nervosa depends on a combination of behavioral therapy, psychotherapy, and psychopharmacological treatment. In some cases, crisis-oriented intervention and forced feedings may be required to prevent death from starvation.

TABLE 23.3. *Characteristics of Anorexia Nervosa*

Characteristics
Clinical features
Onset between adolescence and age 30 years
Female predominance (95%)
Mortality rate of 15%–20%
Familial predisposition
Sleep alterations (decreased REM latency)
Increased occurrence of urogenital abnormalities and Turner syndrome
Intense fear of becoming overweight
Disturbed body images, "feel fat" even when emaciated
Eating behavior
Anorexia uncommon until late in clinical course
Decreased intake of high-calorie food, self-induced vomiting, use of laxatives and diuretics, excessive exercising, use of stimulants and appetite suppressants
Physical signs
Hypothermia
Dependent edema
Cardiovascular changes
Bradycardia
Hypotension
Lanugo (neonatal-like hair)
Constipation
Endocrinologic alterations
Amenorrhea
Decreased TSH and T4
Diminished levels of LH, FSH, gonadal steroids
Prepubertal LH secretory pattern
Diminished response to LHRH
Increased plasma cortisol levels
Impaired responsiveness to insulin-induced hypoglycemia
Impaired dexamethasone-induced suppression of cortisol secretion
Elevated growth hormone levels
Cerebrospinal fluid abnormalities
Decreased homovanillic acid
Decreased 5-hydroxyindoleacetic acid

FSH, follicle-stimulating hormone; LH, luteinizing hormone; LHRH, luteinizing hormone–releasing hormone; REM, rapid eye movement; TSH, thyroid-stimulating hormone.

Hyperphagia

Hyperphagia, the pathological increase in appetite, is a relatively rare symptom that may complicate neurological or psychiatric disorders. Lesions of the ventromedial hypothalamus commonly produce hyperphagia that may be associated with diabetes insipidus, rage, somnolence, hypogonadism, and/or memory loss.[36] The Kleine-Levin syndrome (discussed earlier) is a product of presumed hypothalamic dysfunction characterized by periodic episodes of hyperphagia, somnolence, and altered sexual behavior in adolescent males.[17,18] Hyperphagia is also seen in association with noncommunicating hydrocephalus and may be relieved by ventriculoperitoneal shunting.[37]

The Klüver-Bucy syndrome is a unique behavioral syndrome that was first described in monkeys subjected to bilateral anterior temporal lobectomy. The syndrome complex includes changes in dietary habits with bulimia, emotional placidity, psychic blindness, hypermetamorphosis (compulsive exploration of objects in the environment), hypersexuality, and hyperorality (Table 23.4).[38] In humans, the core symptoms are frequently complicated by the co-occurrence of amnesia, aphasia, dementia, or seizures.[39] The Klüver-Bucy syndrome may occur with any etiologic process producing bilateral temporal dysfunction and has been reported in herpes encephalitis, trauma, bitemporal surgery, paraneoplastic disorders, adrenoleukodystrophy, bilateral temporal infarction, Pick's disease, Alzheimer's disease, hypoglycemia, temporal lobe seizures, and toxoplasmosis (Table 23.5). Hyperphagia may also occur in the course of affective disorders. Anorexia is the most common appetite alteration in depression, but a few patients have a paradoxical increase in appetite. Exaggerated hunger is frequent in mania.

Bulimia is an idiopathic disorder manifest by episodic binge eating.[35] Patients are aware that the eating pattern is abnormal and attempt to eat inconspicuously. They are fearful of not being able to stop eating voluntarily, and depressed mood and self-deprecatory thoughts are common following the binges. The binging is interspersed with programs for weight loss, including restrictive diets, self-induced vomiting, and use of cathartics, stimulants, and diuretics. The bulimia may coexist with anorexia nervosa (bulimorexia) or may occur as an independent disease entity. The disorder begins in adolescence or early adult life, occurs primarily in females, and has a chronic intermittent course. Most bulimics are of normal weight, but a few are obese and a few are slightly underweight. A few patients have improved with treatment with anticonvulsants, and monoamine oxidase inhibitors (MAOIs) have been successful in those with prominent depression and anxiety.

TABLE 23.4. *Human Klüver-Bucy Syndrome*

Core Features	*Additional Features Common in Humans*
Emotional placidity	Aphasia
Hyperorality	Amnesia
Hypermetamorphosis	Dementia
Dietary changes	Seizures
Altered sexual activity	
Psychic blindness (sensory agnosia)	

TABLE 23.5. *Differential Diagnosis of the Human Klüver-Bucy Syndrome*

Herpes encephalitis
Amygdalotomy (bilateral)
Temporal lobectomy (bilateral)
Post-traumatic encephalopathy
Paraneoplastic limbic encephalitis
Adrenoleukodystrophy
Bilateral temporal lobe infarction
Pick's disease
Alzheimer's disease
Hypoglycemia
Toxoplasmosis
Bilateral epileptic foci

ALTERED SEXUAL BEHAVIOR

Sexual urges are among the strongest drives for all life forms. The association of external stimuli with these urges is integrated, along with hypothalamic function, in the paleocortex of the posterior medial orbitofrontal cortex. A functional imaging study comparing male marmosets presented with female scents from ovulating and ovariectomized animals supports these paleocortical regions' participation in sexual arousal (see Fig. 23.3).[40] Three categories of altered sexual behavior may occur when brain systems are disrupted: decreased sexual activity, increased sexual activity, and sexual deviations.

Decreased Sexual Activity

Decreased sexual drive may accompany both neurological and psychiatric disorders (Table 23.6). It is common among epileptics, in whom it appears to be a product of several converging influences (Table 23.7).

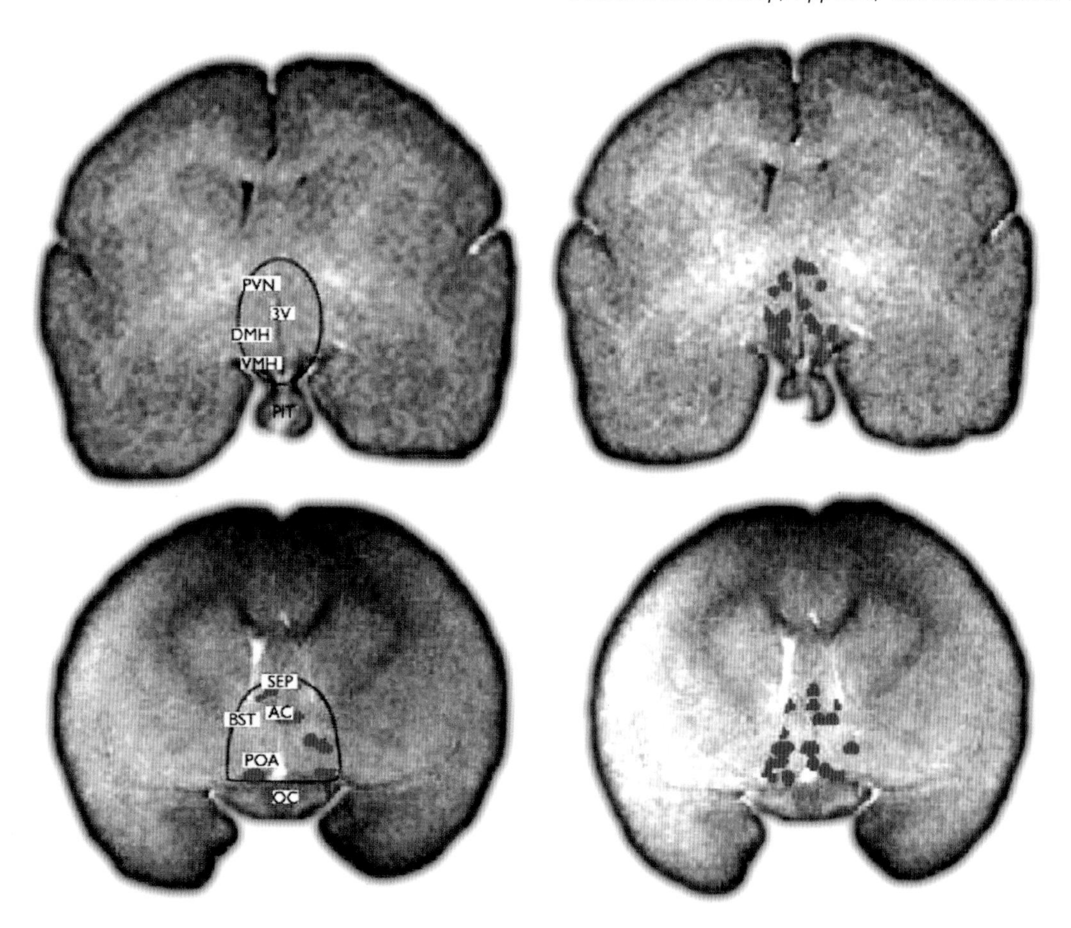

FIGURE 23.3 Regional cerebral blood flow increases measured by blood oxygen level-dependent (BOLD) signal using functional magnetic resonance imaging (*f*MRI). Scent from a peri-ovulatory female marmoset increased blood flow in the hypothalamus (top row) and posterior orbitofrontal paleocortex (bottom row) of male marmosets more than scent from an ovariectomized female implicating these limbic regions as supporting the sexual appetitive dive. Adapted from Ferris et al. (2001).[40] PVN, paraventricular nucleus; DMN, dorsomedial hypothalamus; VMH, ventromedial hypothalamus; 3V, third ventricle; PIT, pituitary gland; SEP, septum; BST, bed nucleus of the stria terminalis; POA, preoptic area; OC, optic chaism; AC, anterior commissure.

Hyposexuality, defined by a frequency of less than one episode of sexual behavior (masturbation or intercourse) per month, is present in 40%–65% of patients with partial-complex seizures (temporal lobe epilepsy) and 10% of patients with primary generalized seizures.[41] In some cases a patient with occult seizure disorder may present with diminished libido and impotence as a chief complaint. Such patients are likely to be misidentified as suffering from a psychogenic disorder if the neurological causes of reduced sexual drive are not considered.

Hyposexual epileptic patients have no interest in sexual activity, lack sexual fantasies, and, if the onset is prior to puberty, fail to develop any interest in sexual functions. At least four influences may contribute to this profound lack of sexual interest: the limbic lesion itself, the influence of drug therapy, endocrinologic alterations, and depression. Patients with temporal lobe epilepsy have a focal lesion of the limbic system, the anatomic substrate of emotional experience, including sexually oriented emotions, and the limbic dysfunction may interfere with sexual interest. Blumer[42] has proposed that irritative temporal lobe foci produce hyposexuality analogous to the hypersexuality produced by destructive limbic lesions in the Klüver-Bucy syndrome. Anticonvulsant therapy may also contribute to the diminished sexual drive. As shown in Table 23.8, barbiturate anticonvulsants may impair libido and diminish sexual arousal. A correlation has also been found between anticonvulsant therapy and reduced free serum testosterone activity, suggesting that anticonvulsants exert subtle hormonal effects that may alter sexual drive.[43] The onset of hyposexuality in some patients prior to initiation of anticonvulsant therapy and the

TABLE 23.6. *Differential Diagnosis of Alterations in Intensity of Sexual Drive*

Decreased sexual drive
Epilepsy
Hypothalamic lesions
Drug-induced changes
Medical illnesses
Depression
Schizophrenia (chronic)
Increased sexual drive
Neurological disorders
Epilepsy
Diencephalic lesions
Kleine-Levin syndrome
Bilateral temporal lobe injury (Klüver-Bucy syndrome)
Frontal lobe syndromes
Medical conditions and pharmacological agents
Hyperthyroidism
Cushing's disease and steroid administration
Androgen administration
Levodopa administration
Psychiatric disorders
Mania
Schizophrenia (early stages)

significant reduction in sexual drive of patients with temporal lobe epilepsy, compared with patients with primary generalized seizures requiring comparable drug therapy, however, suggest that other functions must also contribute to the hyposexuality of patients with temporal lobe foci. An endocrinologic influence that may contribute to the reduced sexual drive is the periodic elevation of serum prolactin that follows epileptic discharge.[44] Sustained hyperprolactinemia leads to decreased libido and impotence, and intermittent prolactin levels may have a similar effect. Finally, depression is common in epileptics, and diminished libido may be a product of the affective disturbance.

Hypothalamic lesions also produce hyposexuality. Naturally occurring lesions impair sexual interest, and stereotactic lesions in the region of the ventromedial nuclei of the hypothalamus have been successfully utilized to reduce sexual drive in patients with a variety of types of sexually motivated criminal behavior.

Among other causes of hyposexuality, a large number of drugs impair sexual function.[45] The impact may be on libido, erection, or ejaculation, as shown in Table 23.8. Psychotropic agents such as stimulants, antidepressants, neuroleptics, lithium, sedative-hypnotics, anxiolytics, and narcotics may all impair sexual function in some patients. Likewise, many classes of antihypertensive agents, estrogens, adrenal steroids, and disulfiram may compromise sexual interest or performance. Trazodone and neuroleptic agents have been associated with priapism. Amoxapine and thioridazine have produced ejaculatory disturbances, and tricyclic agents may produce spontaneous seminal emission.

A number of medical conditions can also impair sexual behavior or diminish sexual drive. Testosterone or thyroid deficiency, prolactin or estrogen excess, chronic hepatic or renal disease, Addison's disease, debilitating cardiopulmonary failure, and systemic cancer will also reduce libido and decrease sexual arousal. Neuropathies causing autonomic dysfunction (e.g., diabetes) and local pelvic surgery frequently impair erection and ejaculation.

Among idiopathic psychiatric disorders, hyposexuality is a major feature of depressive episodes and is common in chronic schizophrenia.

TABLE 23.7. *Altered Sexual Behavior in Epilepsy*

Ictal	*Interictal*
Sensory seizures	Hyposexuality
Genital sensations	Limbic dysfunction
Orgasmic sensations	Depression
Motor seizures	Intermittent prolactin elevation
Coital movements	Drug-related
Exhibitionism (automatic disrobing)	Hypersexuality
Masturbation	Postictal
	Post-lobectomy
	Improved seizure control
	Altered sexuality (occasional association)
	Fetishism
	Transvestism
	Voyeurism
	Exhibitionism
	Sadism
	Masochism
	Pedophilia
	Frotteurism
	Genital mutilation

TABLE 23.8. *Effects of Commonly Prescribed Drugs on Sexual Function*[45]

	EFFECT ON SEXUAL FUNCTION		
Drug	*Libido*	*Arousal/Erection*	*Orgasm/Ejaculation*
Psychotropic agents			
Amphetamines and cocaine	↑ with low doses ↓ with high doses	↓ With chronic use	↑ with low doses ↓ with high doses
Monoamine oxydase inhibitors	—	—	↓
Tricyclic antidepressants	May be ↓	May be ↓	May be ↓; may cause spontaneous seminal emission
Serotonin reuptake inhibitors	↓	↓	↓
Trazodone	—	May cause priapism	—
Lithium carbonate	↓	↓	—
Neuroleptic agents	May be ↓	↓ (rare priapism)	Retrograde ejaculation rarely
Sedative-hypnotics (alcohol, barbiturates, etc.)	↓	↓	—
Antianxiety agents (benzodiazepines, etc.)	—	↓ with chronic usage	—
Narcotics	↓ in high doses	↓ in high doses	↓ in high doses
Antihypertensive agents			
Reserpine, α-methyldopa	↓	↓ (common)	May be ↓
Diuretics	—	May be ↓	—
Clonidine	—	—	May block emission in males
Propranolol	May be ↓		May be ↓
Anticholinergic agents	—	May be ↓	—
Hormonal agents			
Androgens	↑	↑ (men)	↑ (men)
Estrogens	↓ (men) ↑/↓ (women)	May cause impotence in men	Delay Delay
Thyroxin	↑	—	—
Adrenal steroids	↓ in high doses	—	—
Miscellaneous			
Levodopa	May be ↑	—	—
Disulfiram	—	Occasional impotence	Delayed

Increased Sexual Activity

Increased sexual activity is normal in the face of increased leisure time or when a novel partner is available. Abnormal hypersexuality, however, occurs with a variety of neurological, drug-induced, and psychiatric disorders (Table 23.6). Although epilepsy typically causes hyposexuality, there are a few specific circumstances when sexual behavior may be increased in epileptic patients (Table 23.7). Hypersexuality has been noted to occur in the immediate postictal period, following temporal lobectomy with successful abolition of seizures, and occasionally in patients with improved seizure control achieved with anticonvulsants.[46] Lobectomy frequently reverses the preoperative hyposexuality of epileptic patients.[46] Hypersexual behavior has also been reported in the course of prolonged fugue states that occasionally occur as part of the interictal behavioral changes of patients with limbic epilepsy,[47] or temporal stroke with resultant seizures. Genital sensations or coital-type movements may occur in the course of seizures and simulate hypersexual behavior. Ictal genital sensations include feelings of pleasurable stimulation or of frank orgasm.[48,49] Ictal motor manifestations of a sexual nature that occur in the course of psychomotor or petit mal seizures include masturbation, coital movements and related verbalizations, and automatic disrobing. Hooshmand and Brawley[50] suggested that ictal disrobing can usually be distinguished from paraphilic exhibitionism through the following criteria: exhibitionism occurs in males between ages 15 and 45 years and involves specific female victims, whereas ictal disrobing occurs in individuals of either gender at any age and occurs in diverse interpersonal circumstances.

Medial basal or diencephalic injuries have also produced hypersexual behavior. The lesions have usually been inflammatory or neoplastic in origin and involve the medial thalamic, infrastriatal, and mesencephalic–diencephalic junction regions.[51] The Kleine-Levin syndrome is assumed to be a product of hypothalamic dysfunction and is manifest by periodic somnolence, hypersexuality, and hyperphagia.[52] The Klüver-Bucy syndrome was discussed previously with regard to the associated bulimia (Tables 16.4 and 16.5). Animals with the Klüver-Bucy syndrome exhibit hypersexuality and altered sexual behavior, including interspecies copulation.[53] Humans exhibit alterations in sexual interest but may have no increase in the quantity of sexual activity. Patients may be sexually disinhibited and publicly demonstrative, and a number have changed from heterosexual to homosexual preferences. Some patients with other elements of the Klüver-Bucy syndrome, particularly those with dementing disorders, may have no change in sexuality or may be hyposexual.

Frontal lobe syndromes resulting from damage to the orbitofrontal portion of the brain produce disinhibition, jocularity, poor judgment, and impulsivity (Chapter 9). Patients with orbitofrontal lesions (neoplasms, trauma, infarctions, and infection) may make sexual jokes or openly solicit sexual activity. Despite the verbal hypersexuality, there is rarely an increase in actual copulation, although patients may masturbate openly, go about in the nude, or attempt to fondle members of the opposite sex.

Medical conditions and pharmacological agents can also produce increased sexual behavior (Table 23.8).

TABLE 23.9. *Sexual Deviations*

Disorder	*Preferred Sexual Object or Activity*
Fetishism	Nonliving objects
Transvestism	Cross-dressing (by heterosexual male)
Zoophilia (bestiality)	Animals
Pedophilia	Prepubertal children
Exhibitionism	Exposing genitals to an unsuspecting stranger
Voyeurism	Observes unsuspecting people who are naked, disrobing, or engaging in sexual activity
Masochism	Excited by being humiliated, bound, or beaten
Sadism	Excited by inducing humiliation or physical or psychological suffering
Atypical paraphilias	
Vampirism	Blood
Coprophilia	Feces
Urophagia	Urine
Klismaphilia	Enema
Frotteurism	Rubbing against others
Necrophilia	Corpse
Telephone scatologia	Obscene (lewd) telephone calls

Hyperthyroidism, Cushing's disease or exogenous steroid administration, and androgen excess may all cause heightened libido and increased sexual activity.[45] Levodopa has induced hypersexual behavior in parkinsonian patients either as one component of secondary mania or as an independent behavioral alteration.[54] Hypersexuality occurs in most, but not all, manic patients and consists of increased sexual thoughts and statements, flirtation, and increased sexual contacts. Schizophrenic patients may rarely have hypersexual behavior, particularly in the prodromal phase of their illness.

Sexual Deviations

Sexual deviations (paraphilias) are a group of disorders in which unusual or bizarre imagery or acts are necessary for sexual excitement.[35] Table 23.9 lists the paraphilias that have been identified. Sexual deviations include fetishism, transvestitism, zoophilia, pedophilia, exhibitionism, voyeurism, masochism, and sadism, as well as a variety of atypical paraphilias.[55] Homosexuality is considered a psychosexual disorder only if the behavior is ego-dystonic for the individual. Endocrinologic assessments of homosexuals reveal markers of sexual orientation that may reflect altered CNS function.[56]

Post-encephalitic parkinsonism followed the epidemic of von Economo's encephalitis that persisted from 1919 to 1926. Pathologically, there were inflammatory changes in the rostral brainstem and diencephalon. The parkinsonian state was accompanied by a variety of behavioral changes, including psychosexual disorders. The latter occurred in a majority of patients requiring psychiatric hospitalization and included homosexuality, pedophilia, exhibitionism, sadism, and zoophilia.[57]

Frontal lobe syndromes can lead to public masturbation, exhibitionism, pedophilia, and frotteurism as part of the impulsive, disinhibited change in behavior. Frontal system alterations may also underlie the open masturbation and exhibitionism described in some patients with Huntington's disease and with multiple sclerosis.[58,59,60]

Investigations of individuals arrested for sexually related crimes, particularly pedophilia, have revealed an increased prevalence of EEG abnormalities, neuropsychological deficits, dyslexia, cerebral blood flow abnormalities, and computerized tomographic (CT) scan changes.[61] A small number of pedophilics have been found to have elevated serum levels of testosterone. These findings suggest that subtle neurological and endocrinologic abnormalities may contribute to some cases of idiopathic paraphilic behavior.

Sexual deviations may occur in schizophrenia, where they are frequently motivated by delusional ideas or precepts. Paraphilias also occur in patients with personality disorders or may occur in the absence of any other identifiable psychopathology. Psychosexual disorders have been reported with a number of neurological illnesses (Table 23.10). Patients with temporal lobe epilepsy are particularly vulnerable to such deviations, although the percentage of affected individuals is small. Fetishism and transvestitism are the two disorders most commonly reported.[61] Rare instances of voyeurism, exhibitionism, sadism, masochism, pedophilia, frotteurism, genital self-mutilation, and homosexuality have also been described. The Gilles de la Tourette syndrome is a disorder manifest by involuntary tics and vocalizations beginning before the age of 15 years (Chapter 19). The motor behaviors frequently include copropraxia (lewd) gestures), and 50% have coprolalia.[61,62]

TABLE 23.10. *Neurological Disorders Associated with Sexual Deviations*

Sexual Deviation	*Associated Neurological Disorder*
Exhibitionism	Gilles de la Tourette syndrome
	Post-encephalitic parkinsonism
	Frontal lobe syndromes
	Huntington's disease
	Multiple sclerosis
	Epilepsy
	Post-traumatic encephalopathy
Sadism	Epilepsy
	Post-encephalitic parkinsonism
Frotteurism	Epilepsy
	Frontal lobe syndromes
Fetishism	Epilepsy
Pedophilia	Epilepsy
	Post-encephalitic parkinsonism
	Frontal lobe syndromes
	Post-traumatic encephalopathy
	Dyslexia
	Klinefelter's syndrome
Masochism	Epilepsy
Voyeurism	Epilepsy
Zoophilia	Post-encephalitic parkinsonism

REFERENCES

1. Maquet P, Peters J, et al. Functional neuroanatomy of human rapid-eye-movement sleep and dreaming. Nature 1996;383 (6596):163–166.
2. Maquet P. Functional neuroimaging of normal human sleep by positron emission tomography. J Sleep Res 2000;9:207–231.
3. Maquet P, Laureys S, et al. Experience-dependent changes in cerebral activation during human REM sleep. Nat Neurosci 2000;3:831–836.
4. Kales A, Cadieux RJ, et al. Narcolepsy-cataplexy. I. Clinical and electrophysiologic characteristics. Arch Neurol 1982;39:164–168.
5. Plazzi G, Montagna P, et al. Pontine lesions in idiopathic narcolepsy. Neurology 1996;46:1250–1254.
6. Kwen PL, Pullicino P. Pontine lesions in idiopathic narcolepsy. Neurology 1998;50:577–578.
7. Aldrich MS, Naylor MW. Narcolepsy associated with lesions of the diencephalon. Neurology 1989;39:1505–1508.
8. Scammell TE, Nishino S, et al. Narcolepsy and low CSF orexin (hypocretin) concentration after a diencephalic stroke. Neurology 2001;56:1751–1753.
9. Krishnan RR, Volow MR, et al. Narcolepsy: preliminary retrospective study of psychiatric and psychosocial aspects. Am J Psychiatry 1984;141:428–431.
10. Ganji SS, Ferriss GS, et al. Hypersomnia associated with a focal pontine lesion. Clin Electroencephalogr 1996;27:52–56.
11. Bassetti C, Mathis J, et al. Hypersomnia following paramedian thalamic stroke: a report of 12 patients. Ann Neurol 1996;39: 471–480.
12. Lovblad KO, Bassetti C, et al. MRI of paramedian thalamic stroke with sleep disturbance. Neuroradiology 1997;39:693–698.
13. Guilleminault C. Clinical features and evaluation of obstructive sleep apnea. In: Kryger M, Roth T, Dement W, ed. Principles and Practice of Sleep Medicine, 2nd ed. Philadelphia: W.B. Saunders, 1994:667–677.
14. Adelman S, Dinner DS, et al. Obstructive sleep apnea in association with posterior fossa neurologic disease. Arch Neurol 1984;41:509–510.
15. Chokroverty S, Sachdeo, R, Masden, J, et al. Autonomic dysfunction and sleep apnea in olivopontocerebellar degeneration. Arch Neurol 1984;41:926–931.
16. Roberts RE, Shema SJ, et al. Sleep complaints and depression in an aging cohort: a prospective perspective. Am J Psychiatry 2000;157:81–88.
17. Critchley M. Periodic hypersomnia and megaphagia in adolescent males. Brain 1962;85:627–656.
18. Garland H, Sumner D, et al. The Kleine-Levin syndrome. Neurology 1965;15:1161–1167.
19. Lu ML, Liu HC, et al. Kleine-Levin syndrome and psychosis: observation from an unusual case. Neuropsychiatry Neuropsychol Behav Neurol 2000;13:140–142.
20. Sachs C, Persson HE, et al. Menstruation-related periodic hypersomnia: a case study with successful treatment. Neurology 1982;32:1376–1379.
21. Roth B, Nevisimalova S, et al. Hypersomnia with "sleep drunkenness". Arch Gen Psychiatry 1972;26:456–462.
22. Guilleminault C, Dement W. Pathologies of excessive sleep. Adv Sleep Res 1974;1:345–390.
23. Neylan TC, De May MG, Reynolds CF, III. Sleep and chronobiological disturbances. In: Textbook of Geriatric Psychiatry. Busse EW, Blazer DG (eds). Washington, DC: American Psychiatric Press, 1996:329–339.
24. Rye DB, Bliwise DL. Movement disorders specific to sleep and the nocturnal manifestations of waking movement disorders. In: Movement Disorders: Neurologic Principles and Practice. Watts RL, Koller WC (eds). New York: McGraw-Hill, 1997:687–713.
25. Cummings JL, Greenberg R. Sleep patterns in the "locked-in" syndrome. Electroencephalogr Clin Neurophysiol 1977;43:270–271.
26. Freemon FR, Salinas-Garcia RF, et al. Sleep patterns in a patient with a brain stem infarction involving the raphe nucleus. Electroencephalogr Clin Neurophysiol 1974;36:657–660.
27. Lavie P, Pratt H, et al. Localized pontine lesion: nearly total absence of REM sleep. Neurology 1984;34:118–120.
28. Broughton RJ. Sleep disorders: disorders of arousal. Science 1966;159:1070–1078.
29. Boller F, Wright DG, et al. Paroxysmal "nightmares". Neurology 1975;25:1026–1028.
30. Murri L, Arena R, et al. Dream recall in patients with focal cerebral lesions. Arch Neurol 1984;41:183–1853.
31. Lewin K, Mattingly D, et al. Anorexia nervosa associated with hypothalamic tumour. BMJ 1972;19:629–630.
32. Diamond EF, Averick N. Marasmus and the diencephalic syndrome. Arch Neurol 1966;14:270–272.
33. Herholz K. Neuroimaging in anorexia nervosa. Psychiatry Res 1996;62:105–110.
34. Takano A, Shiga T, et al. Abnormal neuronal network in anorexia nervosa studies with I-123-IMP SPECT. Psychiatry Res: Neuroimaging 2001;107:45–50.
35. American Psychiatric Association. Diagnostic and Statistical Manual of Mental Disorders, 4th ed. Washington, DC: American Psychiatric Press, 1994.
36. Bray GA. Syndromes of hypothalamic obesity in man. Pediatr Ann 1984;13:525–536.
37. Krahn DD, Mitchell JE. Case report: bulimia associated with increased intracranial pressure. Am J Psychiatry 1984;141:1099–1100.
38. Klüver H, Bucy PC. Preliminary analysis of functions of the temporal lobes in monkeys. Arch Neurol Psychiatry 1939;42:979–1000.
39. Lilly R, Cummings JL, et al. The human Klüver-Bucy syndrome. Neurology 1983;33:1141–1145.
40. Ferris CF, Snowdon CT, et al. Functional imaging of brain activity in conscious monkeys responding to sexually arousing cues. Neuroreport 2001;12:2231–2236.
41. Blumer D. Changes of sexual behavior related to temporal lobe disorders in man. J Sex Res 1970;6:173–180.
42. Blumer D. Treatment of patients with seizure disorder referred because of psychiatric complications. McLean Hosp J 1977;(Special Issue):53–73.
43. Toone BK, Wheeler M, et al. Sex hormone changes in male epileptics. Clin Endocrinol 1980;12:391–395.
44. Pritchard PBI, Wannamaker BB, et al. Endocrine function following complex partial seizures. Ann Neurol 1983;14:27–32.
45. Crenshaw TL, Goldberg JP: Sexual Pharmacology: Drugs That Affect Sexual Functioning. New York: W.W. Norton & Company, 1996.
46. Cogen PH, Antunes JL, et al. Reproductive function in temporal lobe epilepsy: the effect of temporal lobectomy. Surg Neurol 1979;12:243–246.
47. Mohan KJ, Salo MW, et al. A case of limbic system dysfunction with hypersexuality and fugue state. Dis Nerv Syst 1975;36: 621–624.

48. Remillard GM, Andermann F, et al. Sexual ictal manifestations predominate in women with temporal lobe epilepsy: a finding suggesting sexual dimorphism in the human brain. Neurology 1983;33:323–330.
49. Ruff RL. Orgasmic epilepsy. Neurology 1980;30:1252–1253.
50. Hooshmand H, Brawley BW. Temporal lobe seizures and exhibitionism. Neurology 1969;19:1119–1124.
51. Miller BL, Cummings JL, et al. Hypersexuality or altered sexual preference following brain injury. J Neurol Neurosurg Psychiatry 1986;49:867–873.
52. Carpenter S, Yassa R, et al. A pathologic basis for Kleine-Levin syndrome. Arch Neurol 1982;39:25–28.
53. Schreiner L, Kling A. Behavioral changes following rhinencephalic injury in the cat. J Neurophysiol 1953;16:643–659.
54. Harvey NS. Serial cognitive profiles in levodopa-induced hypersexuality. Br J Psychiatry 1988;153:833–836.
55. Taska RJ, Sullivan JL. Sexual dysfunctions and deviations. In: Cavenar JO, Brodie HKH, eds. Signs and Symptoms in Psychiatry. Philadelphia: J.B. Lippincott, 1983:553–573.
56. Gladne BP, Green R, et al. Neuroendocrine response to estrogen and sexual orientation. Science 1984;225:1496–1499.
57. Cheyette S, Cummings JL. Encephalitis lethargica: lessons for contemporary neuropsychiatry. Neuropsychiatry Clin Neurosci 1995;7:125–134.
58. Federoff JP, Peyser C, Franz ML, et al. Sexual disorders in Huntington's disease. J Neuropsychiatry Clin Neurosci 1994;6:147–153.
59. Ortego N, Miller BL, Itabashi H, et al. Altered sexual behavior with multiple sclerosis. Neuropsychiatry Neuropsychol Behav Neurology 1993;6:260–264.
60. Huws R, Shubsachs APW, Taylor PJ. Hypersexuality, fetishism, and multiple scleosis. Br J Psychiatry 1991;158:280–281.
61. Cummings JL. Neuropsychiatry of sexual deviations. In: Neuropsychiatry and Mental Health Services. Ovsiew F (eds). Washington, DC: American Psychiatric Press, 1999;363–384.
62. Comings DE, Comings BGI. A case of familial exhibitionism in Tourette's syndrome successfully treated with haloperidol. Am J Psychiatry 1982;139:913–915.

Violence and Aggression

In this volume *violence* refers to acts, or threats, of force resulting in personal injury or destruction of property to compel action against one's will. *Aggression* is a wider concept that includes violence and also encompasses self-protective behavior in which physical force or the threat of force is used to satisfy vital needs or to protect one's physical or psychological integrity.[1] Violence has socially sanctioned forms such as capital punishment, injury during the course of resisting arrest, and killing during military operations. It also has illegitimate forms such as assault, battery, rape, and murder. In 1999 in the United States, excluding murder and non-negligent manslaughter, there were over 7 million violent crimes committed (assessed from a survey of 78,000 residents in 43,000 housing units).[2] Police reports of murder and non-negligent manslaughter in 1999 totaled 12,658. Violent behavior is of two types: impulsive or explosive—often unprovoked—and premeditated or predatory. This chapter reviews the role of neuropsychiatric factors in the etiology of unsanctioned violent behavior and provides a differential diagnosis of neurological disorders to be considered in the violent individual.

An acceptable definition of violence has not been universally accepted because violence is not a unitary concept. Violence is not a diagnosis or even a clinical syndrome in the usual sense; it is a complex behavior and, as such, is not likely to have a single determinant. Genetic, social, educational, cultural, economic, neurological, metabolic, and situational factors frequently interact and reinforce each other to produce violent behavior. An individual from a lower socioeconomic background may grow up in a violent household and sustain brain injury from child abuse, and later commit violent acts while intoxicated. Which factors are most important in this caldron of contributing ingredients? In most cases it will be impossible to do more than identify the final precipitating circumstances.

The importance of seeking neurological components in violent behavior is twofold: *1*) neurological factors can easily be overlooked in a psychologically minded milieu in which clinicians are most attuned to the influence of early childhood experiences on adult behavior; and *2*) detection of neurological factors may offer treatment alternatives that will go unexplored if the brain disorder is undiscovered.

Animal models have been widely used to investigate the neurobiologic basis of aggressive behavior. These studies have been particularly useful in delineating anatomic regions of the brain most likely to be involved in mediating violent behavior and have identified the limbic system as the most important anatomic substrate of violence and aggression.[1] The greater extent of intraspecies violence and interindividual cruelty among humans, however, clearly separates human behavior from that of other animals and limits the applicability of animal research to the understanding of human violence.

Two general approaches have been used to study the neurology of human violence. In the first, violent offenders have been investigated to determine whether they have evidence of neurological dysfunction that might have contributed to their violent behavior. In the second, violent behavior ensuing in the course of known neurological disorders is observed and studied. The results of each of these approaches are reviewed, and the latter is further divided into violent behavior directed at others and violent self-destructive behavior. Recent studies using functional imaging to explore the neural systems subserving reactions to, and evocations of, fear and violence in normal controls have implicated regions similar to those found dysfunctional in impulsive violent criminals. These preliminary findings, along with studies evaluating aggression in dementia patients, will be reviewed to highlight a common neuronal system underlying violence and aggression across health and disease. Finally, an approach to the evaluation and treatment of the violent individual is outlined.

NEUROLOGICAL ABNORMALITIES IN VIOLENT CRIMINALS

Electroencephalographic Abnormalities

The most thoroughly studied parameter of neurological dysfunction in violent individuals is the electroencephalogram (EEG). Studies that assessed EEG abnormalities in prison inmates and patients with antisocial behavior have found an increased frequency of EEG changes in violent populations. Antisocial and criminal populations studied had EEG abnormalities in 24%–78% of individuals. Electroencephalographic changes were found to be more common in subjects who had committed violent acts than in those associated with nonviolent crimes and were more frequent in those with repeated violence than in those who had committed isolated violent acts.[3] When the violence had no apparent motive, there was also an increased chance of finding an EEG abnormality, compared to when violence had been provoked. No specific relationship has been found between the type of EEG abnormality and characteristics of the crime, nor between EEG changes and degree of violence committed.[4,5] However, when continued violence is monitored within an institution, a left frontal abnormality on EEG appears to correlate with the frequency of ongoing violent incidents.[6]

Several types of EEG abnormalities have been found in violent offenders: generalized slowing, focal slowing, and epileptiform abnormalities. Williams[3] noted that when focal abnormalities were present they were most likely to be located in the temporal and frontal lobes. Wong et al[4] also found a higher incidence of focal abnormalities in a subgroup with the highest violence scores (30.7% focal abnormalities and 20% temporal abnormalities in the high-violent group versus 7.2% and 2.4% in the low-violence group). Pillmann et al.[5] found a lower prevalence (9%) of focal abnormalities in a more general population of 222 defendants referred for psychiatric evaluation regardless of the degree of violent acts.

Interpretation of these findings is fraught with difficulty. A small percentage of the patients (0–15%) have epilepsy. As discussed later in this chapter, violence as an ictal event is rare, and it is unlikely that many of the violent acts are ictal in nature. The EEG alterations may reflect non-epileptic central nervous system (CNS) changes relevant to the violent behavior. The presence of lesions within the limbic system can lead to personality alterations that may in turn lead to antisocial behavior. Head trauma also produces frontal and temporal lesions, reduces the threshold of impulsive behavior and violence, and may be reflected in EEG abnormalities. Despite the difficulty in drawing direct inferences from these data, the EEG findings indicate that brain dysfunction is common among violent offenders and that, in many cases, the limbic system is the site of neurological abnormality.

Functional Imaging

Functional imaging studies of murderers may underscore variable brain abnormalities, depending on the type of violent acts committed by the murderer. If the murder was an impulsive act, a greater orbitofrontal metabolic defect has been found in perpetrators' scans than in normal controls, while a greater dorsolateral frontal defect has been found in predatory murderers' scans.[7] This differential pattern of functional defects between the two classes of violent acts, impulsive versus predatory, may underscore differential system dysfunction subserving violent behavior. Because head trauma in prisoners is common, it is problematic to extrapolate regional brain abnormalities identified from cross-sectional studies of inmates to models of the neuropsychiatry of violence in humans. Evidence of a common brain system being involved across health and various diseases will strengthen any proposed model of the anatomy of a complex behavior. This is the case with violence and aggression.

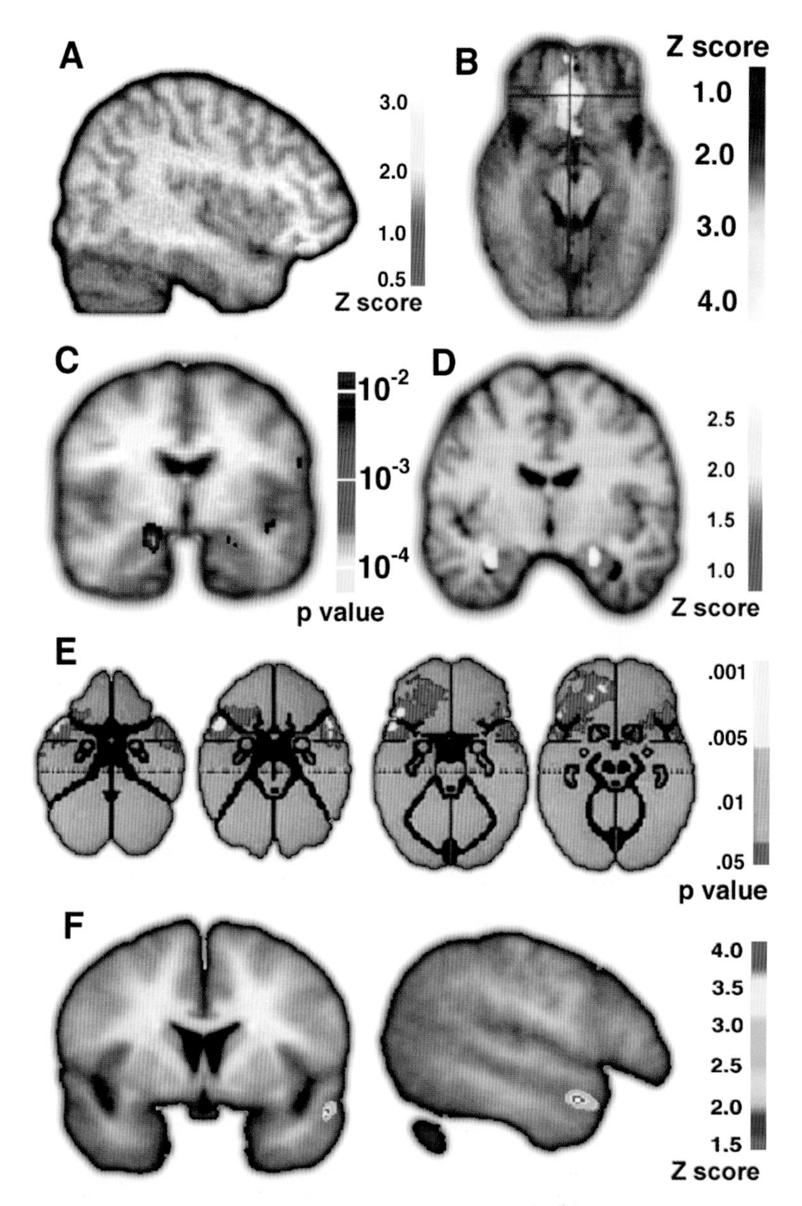

FIGURE 24.1 Statistical maps demonstrating (*A*) increased right orbitofrontal perfusion in 13 normal subjects shown angry faces compared to sad faces (adapted from Blair et al., 1999);[10] (*B*) decreased left orbitofrontal perfusion in 15 normal subjects induced with fear generating aggressive stimuli compared to neutral stimuli (adapted from Pietrini et al., 2000);[9] (*C*) increased right amygdalar perfusion in 8 normal subjects shown fearful faces compared to happy faces (note that more anterior left amygdalar activation was also found in this study; adapted from Whalen PJ, et al., 1998);[12] (*D*) increased bilateral amygdalar perfusion in 6 normal subjects shown threatening words compared to neutral words (adapted from Isenberg et al., 1999);[11] (*E*) increased left orbitofrontal perfusion in 18 normal subjects induced with anger generating stimuli compared to neutral stimuli (adapted from Kimbrell et al., 1999);[8] (*F*) decreased left temporal polar resting perfusion in 10 demented subjects with high agitation/aggression compared to 10 demented subjects with low agitation/aggression (adapted from Hirono et al., 2000).[13]

In normal individuals who were asked to recall prior life events to evoke anger and anxiety while viewing affect-appropriate faces, the left lateral orbitofrontal cortex and bilateral temporal poles were found to be significantly more perfused in a $H_2{}^{15}O$ positron emission tomography (PET) activation study than under neutral conditions.[8] When anger was compared to anxiety, significantly greater perfusion in the left orbitofrontal cortex was found. In other studies that exposed normal subjects to threatening images or words, orbitofrontal[9,10] or amygdalar functional correlates have been found.[11,12] In aggressive patients with dementia, left temporal polar and dorsolateral frontal perfusion defects were found, compared to nonaggressive patients (see Fig. 24.1);[13] also, an increase in the pathological burden of neurofibrillary tangles in the orbitofrontal cortex and anterior cingulate distinguished aggressive Alzheimer's disease patients from nonaggressive patients.[14]

Neuropsychological Assessment

Neuropsychological testing of criminal subjects has produced variable results, but there is a tendency for such patients to perform more poorly than matched control subjects. Performance on tests assessing frontal lobe function is often preferentially compromised.[15]

Neurological Abnormalities

Examination of violent delinquents and patients with impulsive character disorders reveals an increased incidence of neurological soft signs indicative of nonlocalizing neurological dysfunction.[15]

NEUROPSYCHIATRIC DISORDERS WITH VIOLENT BEHAVIOR

Various neuropsychiatric disorders have been associated with violent behavior (Table 24.1). Most of the attention has focused on the possible association between epilepsy and violence, but a number of other disorders have produced violence and must be considered in the differential diagnosis of violent behavior.

Epilepsy

Behavioral disturbances occurring in epileptic patients may occur during the ictal, postictal, and interictal period (Chapter 21). Likewise, violence may occur during any of these periods, and any violent act committed by epileptic patients must be considered in relation to this behavioral framework.

Many unresolved areas of controversy exist regarding the relationship of epilepsy to criminal behavior and violence. The principal questions include the following: Is violence more common among epileptics than nonepileptics? Can violence occur as an ictal manifestation? Does violence occur with abnormal frequency during the interictal period in epileptics? If so, what are the determinants of interictal violence? Is violence more common with one type of epilepsy (e.g., temporal lobe epilepsy) than another (e.g., idiopathic epilepsy)? Unambiguous answers to these questions are not yet available, but tentative conclusions can be drawn from existing information.

TABLE 24.1. *Neuropsychiatric Differential Diagnosis of Violent Behavior*

Epilepsy	Attention-deficit disorder in adults
Ictal	
Postictal	
Interictal	XYY genotype (?)
Episodic dyscontrol syndrome	Idiopathic psychiatric disorders
	Nonpsychotic disturbances
	Personality disorders
Frontal lobe syndromes	Antisocial personality
Traumatic injuries	Borderline personality
Neoplasms	Paranoid personality
Degenerative dementias	Explosive disorders
Mental retardation	Intermittent
	Isolated
Hypothalamic–limbic rage syndrome	Paraphilia
	Sexual sadism
	Childhood disorders
Metabolic disorders	Conduct disorder
Acute confusional states	Psychotic disturbances
Endocrine dysfunction	Mania
Premenstrual dysphoric disorder	Schizophrenia
	Paranoid disorders
Testosterone excess	Depression
Toxic disorders	
Ethanol	
Phencyclidine, LSD, barbiturates, etc.	
Neurological delusional syndromes	

The question regarding the prevalence of violent behavior among epileptics has been approached by investigating the frequency of violent acts in populations of epileptics (such as those attending seizure clinics) or determining the frequency of epilepsy among violent individuals. Although studies of the first type demonstrate that violence is uncommon in epilepsy, the latter technique has generally yielded results suggesting that epilepsy is two to four times more common among prison inmates than in the general population.[16,17]

If violence and antisocial behavior are more common among epileptics, do they occur during the ictal, postictal, or interictal period? This question has been the subject of heated debate. Rare cases of serious offenses, including murder, have been reported to have occurred during epileptic seizures or at least during a seizure-related amnesic period that could have been in either the ictal or immediate postictal period.[18,19] Despite occasional reports of ictal violence, recordings of epileptic patients during ictal periods have shown that behavioral activity occurring as part of a seizure is usually brief, stereotyped, undirected, poorly organized, and unlikely to account for goal-directed violence.[20,21] The current consensus suggests that although interpersonal injury could occur during an epileptic attack manifested by psychomotor automatisms, such activity is unpremeditated, usually poorly structured, and easily redirected. The greatest danger is during the postictal confusional period, when the actions of others may be misinterpreted and a more organized attack may occur.

If aggression is increased in epilepsy and is rare during ictal episodes, when does the violence occur? As noted above, violence may occur during the postictal confusional period, but most episodes of violence appear to occur during the interictal period and are related to behavioral and psychiatric alterations occurring interictally (Chapter 21). Although a few investigators have found equal rates of violence among patients with generalized and temporal lobe epilepsy, most have found violence to be more common among patients with the latter.[22,23] The observation that violence is more common among patients with left than right temporal lesions emphasizes the potential importance of anatomic factors in determining the occurrence of violence in the interictal period.

Several interpretations have been offered for the observations concerning interictal violence in epileptics. Stevens and Hermann[22] suggest that basal forebrain damage gives rise to both seizures and behavioral alterations and that the two consequences are behaviorally independent. Similarly, Trieman and Delgado-Escueta[23] point out that interictal violence is most common in young, intellectually impaired men with histories of psychiatric abnormalities and long-standing, severe epilepsy. In such cases, the associated neurological and psychiatric abnormalities may be responsible for the violent behavior. Violence may be a learned behavior occurring in response to the adverse educational and social circumstances of the epileptic. Lewis et al.[24] suggest that the violence is associated with paranoid and hallucinatory symptoms occurring in the epileptic and is a product of the psychosis occasionally associated with epilepsy. It seems likely that all these factors as well as others (anticonvulsant intoxication, economic, and cultural influences) play varying roles in each epileptic patient manifesting aggressive behavior.

Episodic Dyscontrol Syndrome

The episodic dyscontrol syndrome was described in 1970 by Mark and Ervin[25] as a constellation of the following behaviors: (*1*) a history of physical assault, especially wife and child beating; (*2*) pathologic intoxication (violent behavior following ingestion of small amounts of alcohol); (*3*) impulsive sexual behavior, often including sexual assault; and (*4*) a history of many traffic violations and automobile accidents stemming from impulsive and reckless driving. They cited a number of patients with temporal lobe epilepsy with the symptom complex and argued that the dyscontrol syndrome was a product of limbic system dysfunction and that many of the patients manifesting the syndrome improved markedly when treated with anticonvulsants. Similarly, Monroe[26] suggested that episodic disinhibition of action with violent behavior could be a product of epilepsy or of "epileptoid" loss of control of instinctual drives or impulses. He proposed that there was a continuum of increasing dynamic and diminishing neurological determinants of violence as one moved from epilepsy through instinct and impulse dyscontrol to acting out. The principal feature that distinguishes patients with episodic dyscontrol from patients with sociopathic personality disorders is that the violent activity is isolated and infrequent, not in conjunction with an overall pattern of malevolence.

Despite these contributions, the nosologic validity of the episodic dyscontrol syndrome as a distinct diagnostic entity is controversial. As discussed previously, violent activity is uncommon as an ictal manifestation in epileptics, and the violence of those with episodic dyscontrol syndrome is likely to be an ictal manifestation in only a very small percentage of cases. In addition, in many patients with episodic dyscontrol, social and environmental factors play an important part in determining or triggering the violence. The episodic dyscontrol syndrome thus might be viewed as a nonspecific syndrome of violence with many possible contributing etiologic factors. The more primitive and disorganized and the more distinctly episodic the behavior is, the more likely it is that acquired neurological factors are playing a significant role. Occasionally, recognition of the syndrome will lead to the discovery of previously undiagnosed epilepsy, and in some cases where epilepsy is equivocally present, an empirical trial with anticonvulsants may be warranted.

Frontal Lobe Syndromes

Explosive violence may be a component of the behavioral change that follows damage to the frontal lobe

(Chapter 9). Violent behavior may either accompany orbitofrontal injury, as a manifestation of disinhibition and lack of the usual restraints on antisocial impulses, or may occur with dorsolateral injuries, after which the patient manifests brief outbursts of violence in response to trivial irritations. Attacks of explosive rage that follow head trauma are more likely to be a product of the frontal lobe damage that commonly accompanies traumatic head injury. Frontal lobe involvement is also common in dementia and mental retardation and may account for the occasional acts of violence or aggressive behavior reported in these syndromes.[27] Neuropsychologic investigations of criminals have revealed a subgroup with deficits consistent with frontal lobe dysfunction.[15] This finding suggests that in some cases of idiopathic violent behavior, occult frontal lobe dysfunction secondary to head trauma or delayed maturation may be a contributing factor.

Hypothalamic–Limbic Rage Syndromes

The violence associated with orbitofrontal injury can be partially attributed to involvement of limbic structures and disruption of the role of the limbic system in emotional modulation. Similarly, involvement of limbic structures of the hypothalamus by a variety of pathological processes has also produced intermittent rage behavior. The violence usually occurs in response to provocation, but the stimulus may be minimal. The rage behavior is often combined with amnesia, hyperphagia, and other evidence of hypothalamic dysfunction. In most cases the syndrome results from neoplastic invasion of the hypothalamus.[28,29] The importance of the role played by hypothalamic structures in violent behavior is also attested to by the success of hypothalamotomy in the treatment of some types of violent behavior.[30]

Metabolic Disorders

Metabolic factors contribute to violent behavior in two general circumstances: acute confusional states and disorders of endocrine function. Acute confusional states are reviewed in Chapter 11, and violence can be a manifestation of any of the metabolic disturbances discussed there. The violence is usually poorly organized and undirected when it is a manifestation of confusion and impaired judgment but may result in serious injury. Hill and colleagues[31] recorded a case of matricide occurring during a hypoglycemic episode.

Two types of endocrine alteration have been shown to contribute to violent behavior: *1*) perimenstrual states, and *2*) elevated testosterone levels. D'Orban and Dalton[32] found that 44% of 50 women charged with violent crimes committed their offenses during the perimenstrual period and that there was a significant lack of offenses during the ovulation and post-ovulation phases of the menstrual cycle.

Although results have not always been consistent, several studies have indicated correlations between measures of aggression and serum testosterone levels. Significantly elevated levels have been found in violent rapists and prisoners with histories of violent and aggressive crimes.[33,34] The levels of testosterone were rarely beyond the range of normal; as a group, violent offenders had significantly higher levels than did nonviolent offenders. Similarly, testosterone levels in violent females were also elevated when compared with control populations.[35] The principal role of endocrine factors appears to be to lower the threshold for, and thus increase the likelihood of, violence in predisposed individuals, although prolonged exposure to elevated testosterone levels may have effects on personality development as well.

Toxic Disorders

Alcohol is the intoxicant most commonly used by individuals involved in violent crime. Violence may occur during a period of intoxication with impaired judgment, during an alcoholic blackout (Chapter 22) for which the patient is amnesic, or as part of the syndrome of pathological intoxication. In the latter, chaotic disturbed behavior, often with violent outbursts, occurs following ingestion of small amounts of alcohol. The patient is completely or partially amnesic for the period of the aberrant behavior, and delusions, hallucinations, anxiety, or fear may occur during the episode.[36] In some cases, alcohol withdrawal may be an activating agent for preexisting epileptic abnormalities, and the ensuing violence may be ictal or postictal in origin.

Among the many other intoxicants used, violence is particularly likely with phencyclidine (PCP) ingestion but may also occur after use of LSD, psilocybin, stimulants, anticholinergics, and sedative-hypnotics. Violence has also been reported as a manifestation of neuroleptic-induced akathisia.

Neurological Delusional Syndromes

Delusions are a frequent manifestation of neurological disease. In dementing illnesses they are simple, loosely held, and transient, whereas in diseases affecting subcortical structures they tend to be more elaborate, rigid, and chronic (Chapter 12).[37] Paranoid ideation and persecutory fears are the most common manifestations of delusional thought, and action on delusional beliefs leading to violent activity is an unfortunate but frequent product of persecutory delusions.

Attention-Deficit Disorder in Adults

Attention-deficit disorder in children is manifest by attentional impairment, impulsivity, and nearly constant restless activity while awake.[38] Follow-up studies of these children as they reach adolescence and adulthood reveal that an unusually large number are involved in delinquent behavior or develop sociopathic or explosive personality disturbances. Physical hyperactivity rarely persists beyond childhood, but attentional disturbances continue, and behavioral improvement may follow administration of stimulants even in the adult patient.

Depression and low self-esteem resulting from the poor academic performance and poor social adjustment of hyperactive children are usually invoked as the explanations for sociopathic behavior, but a neurobiologic contribution from the underlying brain disturbance also seems likely.

XYY Genotype

Surveys of criminal populations have revealed an increased incidence of inmates with an XYY genotype and led to the suggestion that XYY individuals were more likely to be violent and aggressive than individuals with normal karyotypes. Further studies, however, have failed to confirm the possibility that XYY patients are at increased risk for violent behavior or suggest that the risk is minimal.[39] Genotype XYY individuals do not have elevated testosterone levels, but they tend to be of lower intelligence and to be more mentally immature and impulsive—factors that may contribute to aggressive activity. Until more information is obtained, the possible role of the XYY genotype in determining violent behavior remains unresolved.

Idiopathic Psychiatric Disorders

A number of idiopathic psychiatric disorders can give rise to violent behavior. They can usefully be divided into psychotic disorders in which the aggression is in response to a delusional belief and those that are nonpsychotic. Among the latter, personality disorders account for the majority of violent actions, but violence may be a manifestation of intermittent or isolated explosive disorders, sexual sadism, or childhood conduct disorders. The personality disturbance most likely to produce repeated violence as a habitual behavioral style is the antisocial personality.[15] Such personalities are characterized by the onset before age 15 of a disorder that, when fully evident, includes the inability to sustain a job, failure to adhere to the law and social norms of behavior, inability to provide consistent parenting or maintain enduring close personal relationships, irritability and aggressiveness, failure to honor financial obligations, and lack of forethought, poor judgment, and recklessness (Chapter 15). The antisocial personality pattern is most marked in late adolescence and early adulthood and tends to be ameliorated with age. In addition to the antisocial personality, violence is common among individuals with borderline and paranoid personality disorders.

Explosive disorders are disturbances of impulse control in which an individual has a discrete episode of aggressiveness with property destruction or assault. There is an absence of generalized impulsivity, aggressiveness, or sociopathic behavior between episodes. The violence is usually out of proportion to the precipitating stimulus and may occur more than once (intermittent explosive disorder) or be confined to a single episode. This behavior is similar to the episodic dyscontrol syndrome, and explosive patients must be carefully evaluated for neurological determinants of their behavior.

Violence may also be a product of certain disturbances of sexual behavior, particularly sexual sadism. Sadism is a paraphilic disturbance in which sexual excitement is achieved by humiliating or injuring either a nonconsenting or a consenting partner (Chapter 23).[38]

Violence is not a common consequence of psychosis, and few psychotic individuals commit acts of violence. Under specific circumstances, however, psychotic ideation, particularly paranoid thinking, can lead to organized acts of aggression directed at presumed persecutors. Such actions may occur in any of the psychoses but have been found most commonly among patients with schizophrenia, women felons with affective disorders, and geriatric patients with late-onset paranoid delusional disorders. The importance of recognizing the psychotic origin of violent behavior stems from the readiness with which some of these disorders respond to neuroleptic medication.

NEUROPSYCHIATRIC DISORDERS WITH SELF-DESTRUCTIVE BEHAVIOR

Self-destructive behavior may occur along with violence directed at others in any of the syndromes presented in the preceding sections. In a few disorders, however, self-inflicted injury may occur as a prominent or even

TABLE 24.2. *Neuropsychiatric Disorders with Self-mutilative Behavior*

Neurological Disorders	*Idiopathic Psychiatric Disorders*
Mental retardation	Borderline personality disorder
Autism	Schizophrenia
Lesch-Nyhan syndrome	Depression
Gilles de la Tourette syndrome	Obsessive-compulsive disorder
Choreoacanthocytosis	

as the dominant behavioral disorder (Table 24.2). In children with mental retardation or autism, self-injury may occur in the course of head banging or other bizarre activities. In the Lesch-Nyhan syndrome (an X-linked disease characterized by overproduction of uric acid, deficiency of hypoxanthine-guanine phosphoribosyl-transferase, mental retardation, spasticity, and choreoathetosis), the afflicted children engage in self-mutilative behavior and are generally aggressive. The aggression often appears to be one manifestation of a compulsive disorder. Likewise, self-harm may occur as a result of some of the irresistible compulsive urges that occur in some patients with Gilles de la Tourette syndrome.[40] Some Gilles de la Tourette patients sustain significant ocular trauma as a result of compulsive striking of the eyes.[41] Another neurological syndrome in which self-injury may be prominent is choreoacanthocytosis, which is manifested by a choreiform disorder resembling Huntington's disease, and studies of peripheral blood show a significant number of acanthocytes among red blood cells. Tongue and lip biting is often an early and prominent expression of the choreic syndrome.

Idiopathic psychiatric disorders that may produce conspicuous self-injury behaviors include borderline personality, obsessive-compulsive disorders with self-mutilation rituals, schizophrenia, and depression.

EVALUATION OF THE VIOLENT INDIVIDUAL

The most important principle involved in the evaluation of the violent individual is that violent behavior is rarely the result of a single circumstance. Rather, violent behavior is the result of neurological, toxic, characterological, social, and situational factors that conspire at a point in time to produce a violent act. An adequate evaluation and any hope of successful treatment thus depend on a thorough investigation of all possible contributing elements. The psychiatric interview will assess childhood, social, occupational, and educational experiences as well as determining current behaviors indicative of psychosis, affective disorder, character disorder, or other psychiatric disturbance. The neurological history should include inquiries regarding birth trauma, head injury, encephalitis, meningitis, systemic illnesses, drug or alcohol ingestion, and any evidence of seizure like phenomena. The bedside mental status examination (Chapter 3) may help in identifying frontal lobe dysfunction, and formal neuropsychological assessment will determine the intellectual capacity of the patient. The elementary physical and neurological examinations will help in identifying systemic diseases or focal neurological deficits. In cases where violence has occurred as an isolated, ego-alien act or cannot be completely recalled by the patient, an EEG should be obtained to search for epileptiform abnormalities. Nasopharyngeal or sphenoidal electrodes and sleep deprivation prior to obtaining the recording may increase the likelihood of discovering an existing EEG abnormality. Structural and functional brain imaging are an integral part of the evaluation of any patient with findings suggestive of brain disease. Laboratory assays of urine and blood may help in identifying metabolic disorders or the presence of toxic substances. In some cases, even after completion of a thorough evaluation, there is insufficient evidence to establish a definitive diagnosis or to determine the relative importance of factors contributing to the violent behavior. In these patients, empirical trials of the treatments discussed in the next section may aid not only in controlling the aggression but also in determining its etiology.

TREATMENT OF THE VIOLENT INDIVIDUAL

Whereas many violent individuals are remanded to the criminal justice system and managed through incarceration and involuntary vocational rehabilitation, others are referred to the mental health establishment for pharmacotherapy, behavior modification, psychotherapy, or, rarely, psychosurgery. Violence is a behavioral complex, not a single distinctive diagnostic entity; therefore, any treatment attempt must be individualized, and most treatment regimens are multifaceted. When an underlying disease process (systemic illness producing a confusional state, epilepsy, schizophrenia, etc.) is detected, treatment can be directed toward re-

TABLE 24.3. *Pharmacologic Agents Used in Treatment of Violent Behavior*

Agents	*Violent Disorders*
Anticonvulsants	Epilepsy (ictal, postictal); episodic dyscontrol syndrome; paroxysmal rage behavior
Propranolol	Neurological disorders (post-traumatic encephalopathy, mental retardation, etc.) with unprovoked violence
Mood stabilizing agents	Personality disorders with violence; recurrent unprovoked violence; mania with violence
Methylphenidate	Antisocial personality disorders (with history of attention-deficit syndrome)
Antiandrogens	Sexual violence; intractable violence in males
Progesterone	Premenstrual violence
Anxiolytics	Anxiety-related irritability and aggression (occasional paradoxical reaction reported)
Antipsychotics	Psychosis-related violence
Antidepressants	Depression-related violence

solving the specific etiologic condition. In many cases, however, the cause of the violence will not be straightforward and treatment may involve any of a number of pharmacologic agents as well as behavioral therapy and/or psychotherapy.

Pharmacotherapy

Table 24.3 summarizes the pharmacological agents commonly used in the treatment of violent individuals and lists the principal disorders in which they have been used with some success.

● **Anticonvulsants** The rare cases in which violence is an ictal manifestation are obviously best managed by reducing the number of seizures. Since ictal violence occurs almost exclusively in complex-partial seizures, the anticonvulsants most likely to be successful are carbamazepine or phenytoin (Chapter 21). Phenobarbital sometimes produces irritability and disinhibition and may increase the likelihood of violence in the epileptic. Violence occurring in the postictal confusional state will also be decreased if the number of seizures can be limited. Anticonvulsants have also been used successfully in the management of the episodic dyscontrol syndrome. Carbamazepine and phenytoin have both been reported to decrease the number of violent outbursts.[42] Anticonvulsants may also ameliorate the chronic aggressiveness and outbursts of violence occurring in some chronically psychotic patients.

● **Propranolol** Propranolol, a β-adrenergic receptor blocking agent, has been noted to decrease belligerent behavior as well as rage attacks in post-traumatic states, Alzheimer's disease, mental retardation, and schizophrenia.[43,44] Dosages necessary for the control of violence have been in the range of 100–500 mg/day. The drug should be used with caution in those with a history of congestive heart failure, asthma, diabetes, or depression.

● **Lithium and Mood Stabilizing Agents** Lithium and other mood stabilizing agents have been used with success in the management of violence in aggressive criminals, character disorders, and children manifesting explosive anger and hostility. In some patients the violence may be an atypical manifestation of an underlying mood disorder, whereas in others these agents appear to act independently of their antimanic effects. Dosages have been the same as those used in the treatment of manic-depressive illness (Chapter 14).

● **Methylphenidate** Attention-deficit disorder, as noted earlier, may persist into adulthood and predispose to antisocial personality disorders with violent behavior. Prescribing stimulants to this population entails a significant risk of abuse of the drugs, but in some cases improvement in behavior and reduction of violence have followed administration of methylphenidate or amphetamines.[45] In closely controlled circumstances, stimulant administration may be a viable therapeutic alternative for adults with persistent or acquired attention-deficit disorders.

● **Hormonal Agents** Antiandrogens such as medroxyprogesterone acetate, leuprolide, and cyproterone acetate diminish sexual preoccupations in the paraphilias and improve self-control of aggressive sexual impulses.[46] These agents have also been reported to diminish interictal violence in temporal lobe epileptics and in patients exhibiting idiopathic chronic assaultiveness. In the latter conditions, aberrant sexual impulses are not necessarily present, and the antiviolence potential of antiandrogens does not appear to be specific for sexually related aggressiveness. Progesterone has been used to limit premenstrual aggression.

• **Anxiolytics** The use of anxiolytics in the management of aggression is controversial. Like alcohol, anxiolytics have the potential for disinhibiting antisocial impulses, and, indeed, paradoxical rage reactions and increased hostility have occasionally been reported following administration of anxiolytics. Most investigators, however, have noted an improvement in aggressive impulses with anxiolytics.

• **Antipsychotics and Antidepressants** Antipsychotics and antidepressants have a role in the treatment of violence when the aggression is the result of psychosis or depression.

• **Behavioral Therapy** The potential excesses of behavioral conditioning in the treatment of violent individuals have been dramatically portrayed by Anthony Burgess in his novel, *A Clockwork Orange*.[47] When properly used, however, behavioral therapies can increase the patient's repertoire of adaptive skills, allow increased control of maladaptive responses, and decrease the number of violent outbursts. In selected cases behavior therapy can offer an important therapeutic dimension to the treatment and management of violent patients.

• **Psychosurgery** Psychosurgery is now rarely used in treatment of aggressive behavior but may be considered in some extreme cases where aggression is unmanageable and all other treatment modalities have failed. The two procedures that have relatively high success rates in the amelioration of violent behavior are bilateral amygdalotomy and posterior hypothalamotomy.[48,49]

REFERENCES

1. Valzelli L. Psychobiology of Aggression and Violence. New York: Raven Press, 1981.
2. Bureau of Justice Statistics. Sourcebook of Criminal Justice Statistics Online: U.S. Department of Justice, Washington, DC, 2001.
3. Williams D. Neural factors related to habitual aggression. Brain 1969;92:503–520.
4. Wong MT, Lumsden J, et al. Electroencephalography, computed tomography and violence ratings of male patients in a maximum-security mental hospital. Acta Psychiatr Scand 1994;90:97–101.
5. Pillmann F, Rohde A, et al. Violence, criminal behavior, and the EEG: significance of left hemispheric focal abnormalities. J Neuropsychiatry Clin Neurosci 1999;11:454–457.
6. Convit A, Czobor P, Volavka J. Lateralized abnormality in the EEG of persistently violent psychiatric inpatients. Biol Psychiatry 1991;30:363–370.
7. Raine A, Meloy JR, et al. Reduced prefrontal and increased subcortical brain functioning assessed using positron emission tomography in predatory and affective murderers. Behav Sci Law 1998;16:319–332.
8. Kimbrell TA, George MS, et al. Regional brain activity during transient self-induced anxiety and anger in healthy adults. Biol Psychiatry 1999;46:454–465.
9. Pietrini P, Guazzelli M, et al. Neural correlates of imaginal aggressive behavior assessed by positron emission tomography in healthy subjects. Am J Psychiatry 2000;157:1772–1781.
10. Blair RJ, Morris JS, et al. Dissociable neural responses to facial expressions of sadness and anger. Brain 1999;122:883–893.
11. Isenberg N, Silbersweig D, et al. Linguistic threat activates the human amygdala. Proc Natl Acad Sci USA 1999;96:10456–10459.
12. Whalen PJ, Rauch SL, et al. Masked presentations of emotional facial expressions modulate amygdala activity without explicit knowledge. J Neurosci 1998;18:411–418.
13. Hirono N, Mega MS, et al. Left frontotemporal hypoperfusion is associated with aggression in patients with dementia. Arch Neurol 2000;57:861–866.
14. Tekin S, Mega MS, et al. Orbitofrontal and anterior cingulate cortex neurofibrillary tangle burden is associated with agitation in Alzheimer disease. Ann Neurol 2001;49:355–361.
15. Raine A, Lencz T, Scerbo A. Antisocial behavior: neuroimaging, neuropsychology, neurochemistry, and psychophysiology. In: Neuropsychiatry of Personality Disorders. Ratey JJ (eds). Cambridge, Massachusetts: Blackwell Science, 1995;50–78.
16. Gunn J. Epileptics in Prison. New York: Academic Press, 1977.
17. Whitman S, Coleman TE, et al. Epilepsy in prison: elevated prevalence and no relationship to violence. Neurology 1984;34:775–782.
18. Pincus JH. Can violence be a manifestation of epilepsy. Neurology 1980;30:304–307.
19. Saint-Hilaire JM, Gilbert M, et al. Epilepsy and aggression: two cases with depth electrode studies. In: Robb P, ed. Epilepsy Undated: Causes and Treatment. Chicago: Year Book Medical Publishers, 1980:145–176.
20. Ashford JW, Schulz SC, Walsh GO. Violent automatism in a partial complex seizure. Arch Neurol 1980;39:120–122.
21. Delgado-Escueta A, Mattson RH, et al. The nature of aggression during epileptic seizures. N Engl J Med 1981;305:711–716.
22. Stevens JR, Hermann BP. Temporal lobe epilepsy, psychopathology, and violence, the state of the evidence. Neurology 1981;31:1127–1132.
23. Trieman DM, Delgado-Escueta AV. Violence and Epilepsy: A Critical Review. New York: Churchill Livingston, 1983.
24. Lewis DO, Pincus TH, et al. Psychomotor epilepsy and violence in a group of incarcerated adolescent boys. Am J Psychiatry 1982;139:882–887.
25. Mark VH, Ervin FR. Violence and the Brain. New York: Harper and Row, 1970.

26. Monroe R. Episodic Behavioral Disorders. Cambridge, MA: Harvard University Press, 1970.
27. Kaufer DI, Cummings J. Personality alterations in degenerative brain diseases. In: Neuropsychiatry of Personality Disorders. Ratey JJ (eds). Cambridge, Massachusetts: Blackwell Science, 1995; pp. 172–209.
28. Haugh RM, Markesberry WR. Hypothalamic astrocytoma. Syndrome of hyperphagia, obesity, and disturbances of behavior and endocrine and autonomic function. Arch Neurol 1983;40:560–563.
29. Reeves AG, Plum F. Hyperphagia, rage, and dementia accompanying a ventromedial hypothalamic neoplasm. Arch Neurol 1969;20:616–624.
30. Schvarcz JR, Driollet R, et al. Stereotactic hypothalamotomy for behavior disorders. J Neurol Neurosurg Psychiatry 1972;35: 356–359.
31. Hill D, Sargent W, Heppenstall ME. A case of matricide. Lancet 1943;1:526–527.
32. D'Orban PT, Dalton J. Violent crime and the menstrual cycle. Psychol Med 1980;10:353–359.
33. Kreuz LE, Rose RM. Assessment of aggressive behavior and plasma testosterone in a young criminal population. Psychosom Med 1972;34:321–332.
34. Rada RT, Laws DR, Kellner R. Plasma testosterone levels in the rapist. Psychosom Med 1976:257–268.
35. Ehlers CL, Rickler KC, Hovey JE. A possible relationship between plasma testosterone and aggressive behavior in a female outpatient population. In: Girgis M, Kiloh LG, eds. Limbic Epilepsy and the Dyscontrol Syndrome. New York: Elsevier-North Holland Biomedical Press, 1980:183–194.
36. Bach-y-Rita G, Lion JR, Ervin FR. Pathological intoxication: clinical and electroencephalographic studies. Am J Psychiatry 1970;127:698–703.
37. Cummings JL. Organic delusions: phenomenology, anatomical correlations, and review. Br J Psychiatry 1985;146:184–197.
38. American Psychiatric Association. Diagnostic and Statistical Manual of Mental Disorders, 4th ed. Washington, DC: American Psychiatric Press, 1994.
39. Schiavi RC, Theilgaard A, et al. Sex chromosomes anomalies, hormones, and aggressivity. Arch Gen Psychiatry 1984;4:93–99.
40. Shapiro AK, Shapiro ES, et al. Gilles de la Tourette Syndrome. New York: Raven Press, 1978.
41. Frankel M, Cummings JL. Neuro-ophthalmic abnormalities in Tourette syndrome: anatomic and functional implications. Neurology 1984;34:359–361.
42. Tunks ER, Dernier SW. Carbamazepine in the dyscontrol syndrome associated with limbic system dysfunction. J Nerv Ment Dis 1977;164:56–63.
43. Ratey JJ, Morrill R, Oxenkrug G. Use of propranolol for provoked and unprovoked episodes of rage. Am J Psychiatry 1983; 140:1356–1357.
44. Yudofsky S, Williams D, Gorman J. Propranolol in the treatment of rage and violent behavior in patients with chronic brain syndromes. Am J Psychiatry 1981;138:218–220.
45. Stringer AY, Josef NC. Methylphenidate in the treatment of aggression in two patients with antisocial personality disorder. Am J Psychiatry 1983;140:1365–1366.
46. Berlin FS, Meinecke CF. Treatment of sex offenders with antiandrogenic medication: conceptualization, review of treatment modalities, and preliminary findings. Am J Psychiatry 1981;138: 601–607.
47. Burgess A. A Clockwork Orange. New York: Ballantine Books, 1963.
48. Sano K, Mayanagi Y, et al. Results of stimulation and destruction of the posterior hypothalamus in man. J Neurosurg 1970;33: 689–707.
49. Small IF, Heinburger RF, et al. Follow-up of stereotopic amygdalotomy for seizure and behavior disorders. Biol Psychiatry 1977;12:401–411.

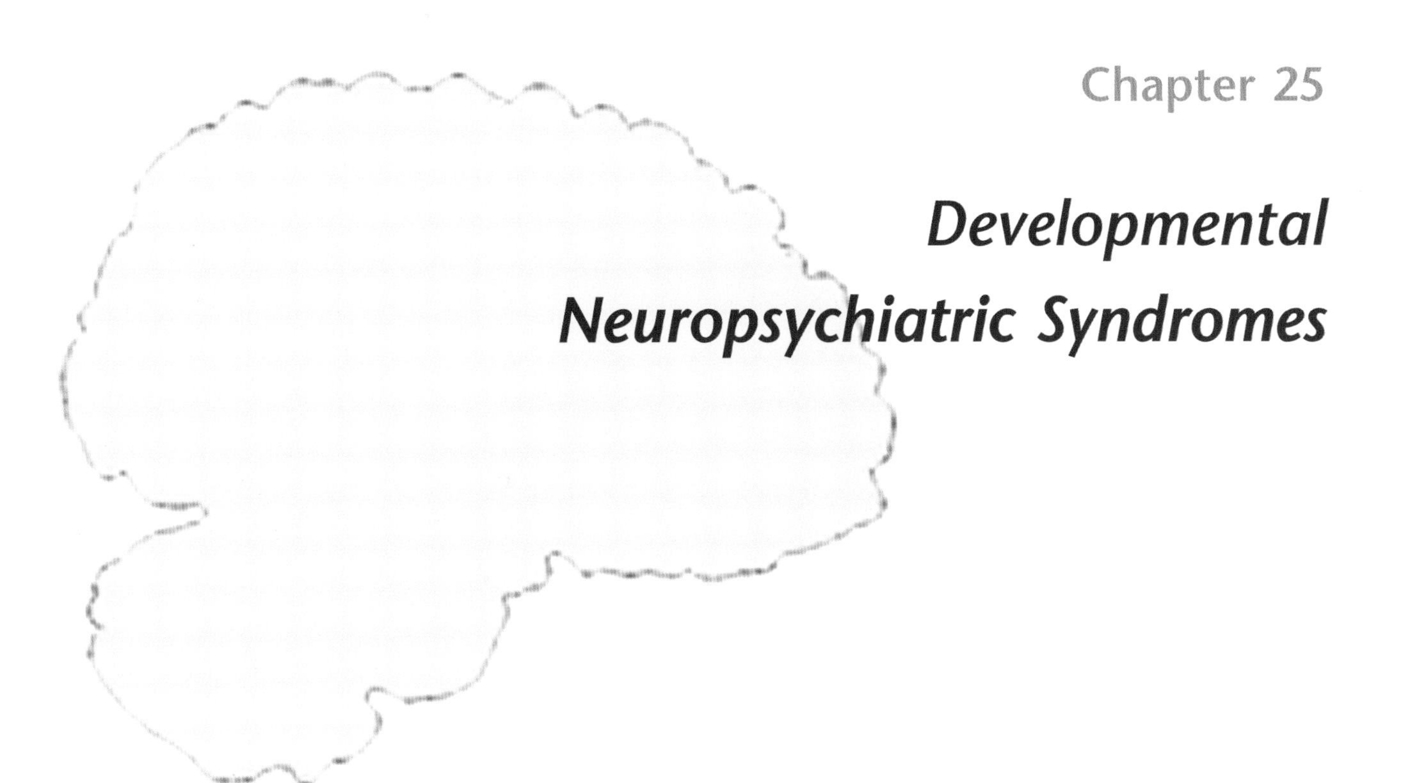

Chapter 25

Developmental Neuropsychiatric Syndromes

Neuropsychiatric disorders are unevenly distributed throughout the human life span and are more common among the very young and the very old. In these two periods of life, the brain is disproportionately vulnerable to injury and dysfunction and neuropsychiatric symptoms ensue. Mutations, abnormal brain development, and early-acquired brain lesions conspire to produce a wide variety of developmental brain disturbances and neuropsychiatric disorders. Pediatric neuropsychiatric syndromes have tremendous family and societal implications; they result either in the tragic death of a child with complete abrogation of its life potential or they produce chronic disability with ongoing performance below the anticipated level. This is not to say that developmentally compromised individuals cannot have complete and fulfilling lives. They make important contributions to their families and often have satisfying interpersonal relationships. The domain of accomplishment, however, remains diminished compared to those expected of the same individual had there been more normal cognitive capacity. Developmental brain insults and the associated cognitive dysfunction and psychopathology place unusually harsh demands on family members and strain family systems. This often results in dysfunctional family and social circumstances that interact with and exacerbate the brain-based behavioral disturbances. When assessing children with congenital and early-onset brain disorders and neuropsychiatric symptoms, it is often nearly impossible to disentangle the biological and the social–familial contributions to the behavioral abnormalities. In all cases, a comprehensive developmental neuropsychiatric assessment must include evaluation of both the patient and the family and social setting in which the patient is lodged.

The whole of developmental neuropsychiatry cannot be encompassed within a single chapter, and the following discussion is provided to complement other chapters in this volume in which the cognitive disturbances of brain dysfunction (language disturbances, memory impairment, visuospatial abnormalities, and executive dysfunction) are discussed. This chapter provides an overall approach to syndromes that are unique to developmental neuropsychiatry (e.g., autism, developmental learning disorders, mental retardation, etc.) and to inherited conditions and focal lesions occurring in childhood and accompanied by neuropsychiatric symptoms. A brief discussion of the approach to the pediatric neuropsychiatric patient is provided first, followed by a discussion of mental retardation and developmental delay. Next, autism and other pervasive developmental disorders are described. Learning disorders are summarized, followed by a description of childhood-onset movement disorders, conditions causing progressive cognitive decline, and childhood-onset acquired diseases such as stroke and multiple sclerosis.

Tic disorders commonly beginning in childhood, such as Gilles de la Tourrette syndrome, are discussed in Chapter 19, and the classification of epilepsy and childhood epileptic disorders are described in Chapter 21.

DEVELOPMENTAL NEUROPSYCHIATRIC ASSESSMENT

Developmental neuropsychiatric assessment includes an initial interview with the patient and the patient's parents or caregivers. An initial mental status examination and physical examination is typically supplemented by neurodevelopmental, neuropsychological, or psychoeducational testing. Depending on the nature of the problem, brain imaging, electroencephalography, and laboratory tests for specific metabolic derangements may be required. Although assessment is tailored to the unique presentation of each patient, these general areas are relevant to most pediatric neuropsychiatric conditions.

The initial neuropsychiatric interview with the patient and the parents should include, in addition to a review of the reason for referral, a review of the mother's pregnancy, labor, and delivery. Early feeding and oral behavior should be queried, as should elimination and response to toilet training. Problems with sleep should be investigated and information should be collected on activity and motor development, including whether motor milestones were reached at typical ages and how motor skill acquisition compared with that of siblings. Adaptive skills such as self-help, awareness of family members and family routines, emotional reactions, and early cognitive development should be reviewed. Review of communication skills, including language and prosody, are an important aspect of the inquiry. The use of toys, engagement in play, and social interactions with other children require discussion. Any special behavioral syndromes such as fears, anxieties, repetitious movements, obsessions, compulsions, aggressive behavior, and evidence of psychosis or mood abnormalities should be investigated.[1]

During the interview with the child, the clinician should observe and examine affect and temperament, mood and emotional lability, attention, language processing, fund of age-appropriate general information, memory, and visuospatial skills.[2] The general physical examination can be extremely important in pediatric neuropsychiatry. Neurocutaneous disorders have revealing skin stigmata. Neurofibromatosis patients often have café-au-lait spots; patients with tuberous sclerosis evidence hypopigmented oval- or leaf-shaped spots; and Sturge-Weber syndrome is characterized by afacial angioma. Von Hippel-Lindau disease produces retinal hemangioblastomas and ataxia-telangictasia features oculocutaneous telangiectasia affecting the exposed areas of the skin and the sclerae of the eyes.[3] General physical examination may also reveal minor physical abnormalities and morphological disorders indicative of developmental disturbances. These include cleft lip and palate, absent or extranumerary digits, hypertelorism or hypotelorism (increased or decreased distance between the eyes), low-set ears, and hair whorls.[2]

As in adults, neurological examination should include routine testing of cranial nerves, motor function, sensory abilities, coordination, assessment of reflexes, and review of gait and station (Chapter 3). In addition, children should be assessed for the presence of neurological "soft signs" such as delays in relation to chronological and mental age in speech, motor coordination, right–left discrimination, and perception. Clumsiness on tests of coordination, minor choreiform movements of the outstretched limbs, the inability to balance on one foot, and difficulty with precise articulation are all counted among neurological soft signs. They may be associated with mental retardation, genetically determined maturation disorders, brain damage, or childhood-onset psychiatric disorders.[4]

Neuropsychological examination follows the principles of that applied in adults (Chapter 3) but must be modified by a developmental perspective. Formal developmental assessments such as the Denver Developmental Screening Test II[5] are helpful for gauging the child's developmental sophistication in relation to age-expected norms. Several intelligence tests have been developed for children of various ages, including the Catell Infant Intelligence Scale (applicable to children ages 3 months to 30 months), Columbia Mental Maturity Scale (applicable to children 3 1/2 years through 9 years of age), and Detroit Test of Learning Aptitude–3 (for children 6 through 17 years of age).[2] The Stanford-Binet Intelligence Scale is applicable from ages 2 through 23 years. Achievement tests help to determine the child's performance in specific cognitive areas in relation to their age and expected grade level (the Wide Range Achievement Test is an example of such an examination). Specific tests are also available for assessment of reading, spelling, mathematics, writing, motor skills, and perceptual abilities.[2] These tests may be particularly useful in the identification of specific learning disabilities in school-age children. Relevant

TABLE 25.1. *Prevalence of Childhood Neuropsychiatric Syndrome*[43]

Syndrome	*Prevalence/1000*
Dyslexia	50
Attention-deficit hyperactivity disorder	46
Gilles de la Tourette syndrome	5.9
Childhood/adolescent psychosis	5.4
Asperger's syndrome	3.6–7.1
Mental retardation	3.7
Neurological disorder with psychiatric symptoms (e.g., epilepsy, stroke)	2.0
Autism	0.4–1.6

neuroimaging, electroencephalographic (EEG), and specific laboratory tests should be conducted to assess individual disorders.

Table 25.1 provides an overview of the neuropsychiatric disorders most commonly encountered in childhood.

MENTAL RETARDATION

Mental retardation is defined as subaverage general intellectual functioning accompanied by significant limitations in adaptive functioning in at least two of the following skill areas: communication, self-care, home living, social and interpersonal skills, use of community resources, self-direction, functional academic skills, work, leisure, health, and safety.[6] By convention, the onset must occur before the age of 18. This age of onset criterion is somewhat confusing in that it allows the syndrome of mental retardation to overlap with that of dementia (Chapter 10). It would be more consistent to define mental retardation as delays in the acquisition of the usual social and cognitive milestones, and dementia as the decline from a previous higher level of cognitive function regardless of age. In practice, however, the distinction between these syndromes is not difficult, since early-onset dementia syndromes are rare.

Four degrees of severity of mental retardation are recognized. Patients with mild mental retardation generally have IQs between 50–55 and 70; those with moderate retardation have IQs between 35–40 and 50–55; persons with severe mental retardation have IQs of 20–25 to 35–40; and those with profound mental retardation have IQ levels below 20 or 25.[6] Epidemiologically, mild mental retardation occurs in 3 to 6 per 1000 of the general population, moderate retardation in 2 per 1000, severe retardation in 1.3 per 1000, and profound retardation in 0.4 per 1000.[7] Rates may vary depending on the occurrence of specific circumstances that may jeopardize early life brain development. Males are more likely to evidence mental retardation than females by a ratio of 1.6:1. Specific etiologic diagnoses are most difficult to identify in those with mild to moderate mental retardation. In this group, 45% to 60% of syndromes are of unknown etiology. Perinatal insults account for 10%–25% of cases and chromosomal abnormalities for 5%–10%.[7] Among those with severe mental retardation, chromosomal abnormalities account for 30%, gestational and peri- and postnatal injuries for 15%–20%, central nervous system (CNS) malformations for 10%–15%, congenital anomaly syndromes for 5%, and endocrine and metabolic disorders for 5%. The cause of 25%–30% of cases of severe mental retardation remains undetermined.[7,8]

Among the chromosomal abnormalities contributing to the occurrence of severe mental retardation, Down syndrome accounts for 20% of cases, fragile X syndrome for 60%, and all other chromosomal abnormalities for 20%.[8] The most common cause (94%) of Down syndrome is a de novo trisomy of chromosome 21.[9] The brain is small, with a simplified convolutional pattern. The frontal cortex and limbic structures are disproportionately reduced in volume. Histological studies reveal that granule cells are small in all cortical regions. Children with Down syndrome acquire language slowly and manifest impoverished vocabulary and grammar as well as difficulty with word meanings and appropriate contextual use of language and syntax. Mathematical skills and executive function are poorly developed. Visuospatial function is relatively well preserved. Approximately 25% of individuals with Down syndrome manifest neuropsychiatric disorders, including attention-deficit hyperactivity disorder (ADHD) (6.1%), oppositional-defiant and conduct disorders (5.4%), aggression (6.5%), and self-injurious behavior (1%). Anxiety disorders, mood disturbances, and autistic symptoms may be associated with Down syndrome.[9]

Fragile X syndrome is the most common form of inherited mental retardation.[10] The fragile X gene (located at Xq21.3–q22) contains a cytosine-guanine trinucleotide repeat that is dramatically expanded in affected individuals. A variety of neuropsychiatric manifestations have been associated with fragile X syndrome, including mood and anxiety disorders, ADHD, schizotypal behaviors, obsessive-compulsive disorder, Gilles de la Tourette syndrome, Asperger's syndrome–spectrum disorders, and autistic-spectrum disorders.[9]

Neuropsychological abnormalities are disproportionately severe for nonverbal memory, visuospatial abilities, visuomotor coordination, sequential processing, and attention. Language is acquired slowly and linguistic expression is typically characterized as rapid, perseverative, jocular, tangential, cluttered, and echolalic.[9] Female carriers of the fragile X mutation manifest more mild neuropsychological deficits including conceptual disorganization and relative deficits on performance measures of IQ tests. They often exhibit schizotypal features and are vulnerable to depression.[9,10] Magnetic resonance imaging (MRI) indicates a relative increase in the volume of the hippocampus and a decrease in the volume of the superior temporal gyrus in individuals with fragile X syndrome.[11]

Two genetic syndromes are particularly informative from a neuropsychiatric perspective: Williams syndrome and Prader-Willi syndrome. *Williams syndrome* occurs at a rate of 1 per 25,000-50,000 live births. Patients have a characteristic elfin facial appearance with medial eyebrow flair, small and low-set ears, thick lips, and epicanthal folds. The voice has a peculiar, low-pitched quality. The syndrome is produced by a mutation at 7q11.23.[9] The most characteristic feature of Williams syndrome is a relative preservation of expressive language skills, leading to a verbose and pseudomature linguistic output. Both spatial and motor skills are significantly impaired. Major psychopathology is not common in these children, but they are described as fussy, fearful, distractible, and anxious.[9] The brain has a reduced volume of gray matter and posterior structures are disproportionately effected.[9,12]

Prader-Willi syndrome occurs at a rate of 1 per 10,000–25,000 live births and about 70% of individuals have a deletion of 15q11–13, with the affected chromosome being of paternal origin.[9,13] In the neonatal period, patients with Prader-Willi syndrome exhibit hypertonia, diminished movements, and poor cry. They have characteristic facial features with a narrow face, almond-shaped eyes, thin upper lip, and down-turned corners of the mouth. There is hypogonadism and cryptorchidism. Neuropsychiatrically, patients typically are friendly and outgoing but exhibit unusual features such as hyperphagia, food seeking, and an obsession with food. Temper tantrums with violent outbursts, obsessive-compulsive behavior, and repetitive skin picking are not uncommon. Patients may exhibit depression and, rarely, psychosis.[9,14]

Tables 25.2 and 25.3 provide a summary of the principal identifiable etiologies of mental retardation. The causes of mental retardation are increasingly intertwined. For example, developmental malformations of the brain are increasingly linked to specific genetic mutations, inborn errors of metabolism and developmental malformations can both produce epilepsy with consequences for cognitive development, and the presence of severe retardation can in some cases lead to child abuse and neglect, further exacerbating the cognitive and behavioral disorder.

PERVASIVE DEVELOPMENTAL DISORDERS

Pervasive developmental disorders include autistic disorders, Rett's disorder, childhood disintegrative disorder, Asperger's syndrome, and atypical autism[6] (Table 25.4).

Autistic disorder is characterized by (*1*) an impairment in social interaction manifested by abnormal nonverbal behaviors (eye contact, facial expression, body postures, gestures), failure to develop peer relationships, lack of spontaneous seeking to share enjoyment and interest, and lack of social or emotional reciprocity; (*2*) impairments in communication as manifested by a delay or lack of a development of spoken language, marked impairment in the ability to initiate or sustain a conversation with others in those with adequate language, stereotyped and repetitive use of language or idiosyncratic language; (*3*) lack of spontaneous make-believe play or social-imitative play appropriate to developmental level; and (*4*) restricted, repetitive, and stereotyped patterns of behavior, overinvestment in specific activities evidenced by encompassing preoccupation with one or more stereotyped and restricted patterns of interest, apparently inflexible adherence to specific nonfunctional routines or rituals, stereotyped and repetitive motor mannerisms, and persistent preoccupation with parts of objects.[6] The onset is before the age of 3 years.[6] Autism occurs in 4 or 5 individuals per 10,000 and is three to four times more common in males than in females. Approximately 75% of autistic individuals have mental retardation but autism may occur in cognitively normal individuals. Neurological soft signs are present in one-third of individuals with autism and seizures occur in 15%–30%.[15] A wide variety of disorders have been associated with autism, including trisomy 21, ceroid lipofuscinosis, neurofibromatosis, phenylketonuria, tuberous sclerosis, and Williams syndrome. Laboratory studies indicate elevated serum serotonin levels in 25%–30% of autistic subjects. Given the wide range of abnormalities that can cause autism, it is not surprising that neuroimaging studies have not yielded consistent abnormalities. There is evidence of increased brain volume in autism and diminished corpus callosum size.[16–19] Only a few brains of autistic patients have been

TABLE 25.2. *Etiologies of Mental Retardation*[36,44–49]

- Prenatal causes
  - Infection
  - Stroke
  - Trauma
- Chromosomal disorders
  - Trisomy 21 (Down syndrome)
  - Fragile X syndrome
  - XO syndrome (Turner)
  - XYY syndrome
  - XXY syndrome (Klinefelter)
  - Prader-Willi syndrome
  - Angelman's syndrome
  - Williams syndrome
- Syndromic disorders
  - Neurocutaneous disorders
    - Neurofibromatosis
    - Sturge-Weber syndrome
    - Tuberous sclerosis
  - Muscular disorders with CNS abnormalities
    - Congenital muscular dystrophy
    - Duchenne muscular dystrophy
    - Myotonic muscular dystrophy
  - Ocular disorders
    - Aniridia-Wilm's tumor syndrome
    - Anophthalmia syndrome (X-linked)
    - Lowe syndrome
  - Craniofacial disorders
    - Acrocephalosyndactyly (e.g., Apert type)
    - Craniofacial dysostosis (Crouzon)
    - Multiple synostosis syndrome
- Inborn errors of metabolism
  - Amino acid disorders
    - Phenylketonuria
    - Histidinemia
    - Branched-chain amino acid disorders
      - Hyperleucine-isoleucinemia
      - Maple-syrup urine disease
    - Biotin-dependent disorder
    - Propionic acidemia
    - Methylmalonic acidemia
- Inborn errors of metabolism (cont'd)
    - Folate-dependent disorders
    - Homocystinuria
    - Hartnup disease
  - Carbohydrate disorders
    - Glycogen storage disorders
    - Galactosemia
    - Pyruvic acid disorders
  - Mucopolysaccharide disorders
    - α_1-Iduronidase deficiency (e.g., Hurler type)
    - Iduronate sulfatase deficiency (Hunter type)
    - Heparan *n*-sulfatase deficiency (San Filippo 3A type)
  - Mucolipid disorders
  - Urea cycle disorders
    - Ornithine transcarbamylase deficiency
    - Arginase deficiency (argininemia)
  - Nucleic acid disorders
    - Lesch-Nyhan disease
    - Orotic aciduria
  - Copper metabolism disorders
    - Wilson's disease
    - Menkes' disease
  - Mitochondrial disorders
    - Mitochondrial encephalomyopathy, lactic acidosis, and stroke-like episodes (MELAS)
    - Mitochondrial encephalopathy with ragged red fibers (MERRF)
    - Neuropathy, ataxia, and retinitis pigmentosa (NARP)
    - Subacute necrotizing encephalomyelopathy (Leigh's disease)
    - Kerns-Sayre syndrome (KSS)
    - Wolfram syndrome
    - Boustany-Moraes disease (lethal infantile mitochondrial disease; LIMD)
  - Peroxisomal disorders
    - Zellweger syndrome
    - Infantile Refsum's disease
    - Pipecolic acidemia
- Developmental malformations
  - Malformations due to abnormal neuronal and glial proliferation and apoptosis
  - Malformations due to abnormal neuronal migration
  - Malformations due to abnormal cortical organization
  - Malformations of cortical development, not otherwise classified.

TABLE 25.3. *Classification Scheme of Congenital Brain Malformations*

- Malformations due to abnormal neuronal and glial proliferation or apoptosis
  - Decreased proliferation/increased apoptosis: Microencephalies
    - Microcephaly with normal to thin cortex
    - Microlissenencephaly (extreme microencephaly with thick cortex)
    - Microencephaly with polymicrogyria/cortical dysplasia
  - Increased proliferation/decreased apoptosis (normal cell types): Megalencephalies
  - Abnormal proliferation (abnormal cell types)
    - Non-neoplastic
      - Cortical hamartomas of tuberous sclerosis
      - Cortical dysplasia with balloon cells
      - Hemimegaloencephaly (HMEG)
    - Neoplastic (associated with disordered cortex)
      - DNET (dysembryoplastic neuroepithelial tumor)
      - Ganglioglioma
      - Gangliocytoma
- Malformations due to abnormal neuronal migration
  - Lissencephaly/subcortical band heterotopia spectrum
  - Cobblestone complex
    - Congenital muscular dystrophy syndromes
    - Syndromes with no involvement of muscle
  - Heterotopia
    - Subependymal (periventricular)
    - Subcortical (other than band heterotopia)
    - Marginal glioneuronal
- Malformations due to abnormal cortical organization (including late neuronal migration)
  - Polymicrogyria and schizoencephaly
    - Bilateral polymicrogyria syndromes
    - Schizencephaly (polymicrogyria with clefts)
    - Polymicrogyria with other brain malformations or abnormalities
    - Polymicrogyria or schizencephaly as part of Multiple Congenital Anomaly/Mental Retardation syndromes
  - Cortical dysplasia without balloon cells
  - Microdysgenesis
- Malformations of cortical development, not otherwise classified
  - Malformations secondary to inborn errors of metabolism
    - Mitochondrial and pyruvate metabolic disorders
    - Peroxisomal disorders
  - Other unclassified malformations
    - Sublobar dysplasia
    - Others

TABLE 25.4. *Pervasive Developmental Disorders*

Autistic disorders
Rett's disorder
Childhood disintegrative disorder
Asperger's syndrome
Atypical autism

examined neuropathologically. These studies reveal abnormalities of the limbic system (hippocampus, subiculum, entorhinal cortex) and a decrease in the number of Purkinje and granular cells in the cortex of the cerebellum.[20]

Childhood disintegrative disorder is characterized by apparently normal development during the first 2 years of life, including social skills and adaptive behavior, bowel and bladder control, play, and motor skills. After the period of normal development, progressive deterioration in functions begins and affects at least two areas of function, including social interaction and communication, or there may be the appearance of restrictive, repetitive, and stereotyped patterns of behavior.[6] In approximately 75% of cases, the child's behavior and development deteriorate to a low level of function and they remain at that level with only limited recovery. In the few cases associated with an identified progressive neurological disorder, there may be continuing deterioration and eventual death. Although most cases have had no recognized underlying neurological condition, the syndrome has been associated with neurolipidoses, metachromatic leukodystrophy, adrenoleukodystrophy, and subacute sclerosing panencephalitis.[21]

Asperger's syndrome consists of an impairment in social interaction and restricted, repetitive, and stereotyped patterns of behavior, interest, and activities.[6] The disturbance of social interaction is manifested by marked impairment of multiple nonverbal behaviors. Disturbance is sufficiently severe to cause clinically significant impairment in social, occupational, or other important areas of functioning. There is no clinically significant delay in language acquisition or cognitive development, or in the development of age-appropriate self-help skills, adaptive behavior, and curiosity about the environment. Asperger's syndrome shares two core features of autism (impaired social interaction, repetitive and stereotyped behavior) but is distinguished from autism by the absence of delay in language or clinically significant delay in cognitive development or self-help skills.[22] Children with Asperger's syndrome are at risk for depression and schizophrenia in adulthood but a reasonable level of functioning is possible in some circumstances. No compelling neuroimaging or neuropathological observations have been related to Asperger's syndrome. In some cases, an Asperger's-like syndrome has been a manifestation of an underlying inherited inborn error of metabolism, a leukodystrophy, or a CNS infection.[23] Asperger's syndrome has been described in patients with neuronal migration disturbances such as macrogyria and polymicrogyria.[24]

Rett's disorder (also known as *Rett's syndrome*) is characterized by deceleration of head growth between the ages of 5 months and 48 months after an apparently normal period of prenatal and perinatal development; loss of previously acquired purposeful hand skills, with subsequent development of stereotyped hand movements (hand ringing or hand washing); loss of social engagement, appearance of poorly coordinated gait or trunk movements, and severely impaired language development; and marked psychomotor retardation.[6] Most cases of Rett's disorder are produced by a mutation located on Xq28. The disorder occurs primarily in females and it is likely that the mutation is lethal to males.[25] Clinically, patients evidence a variety of stereotyped movement and gait disorders including bruxism, oculogyric crises, parkinsonism, and dystonia. Myoclonus and choreoathetosis may occur but are infrequent. In general, younger patients tend to exhibit more hyperkinetic disorders, while bradykinesia is more evident in older Rett's disorder patients.[26] Neuroimaging studies reveal global reduction in gray and white matter volumes. Prefrontal, posterior frontal, and anterior temporal regions show the largest reductions in gray matter volume, whereas white matter is reduced uniformly throughout the brain. There is a disproportionate reduction in caudate nucleus volume.[11,27] At autopsy, two consistent observations have emerged: there is hypopigmentation of the substantia nigra and diffuse reduction in brain size,[28] and in the cortex there is a reduction in neuronal size and increased cell-packing density.[29] Abnormalities of dopaminergic and cholinergic systems have been described.[29,30]

LEARNING DISORDERS

Learning disorders are diagnosed from the child's school performance and achievement on standardized tests. Abilities are substantially below what is expected for age, schooling, and level of intelligence. A discrepancy of more than two standard deviations between achievement on tests assessing the specific disability and IQ is a common means of defining a specific learning disability. Currently, three learning disabilities are recognized: reading disorder (developmental dyslexia), mathematics disorder (developmental dyscalculia), and a disorder of written expression (developmental dysgraphia).[6] Of patients with ADHD (discussed below), 10%–25% exhibit learning disorders.

Developmental dyslexia is the most thoroughly studied of the learning disabilities. It is the most common neurobehavioral deficit affecting children, with prevalence rates ranging from 5% to 10% of the population. Longitudinal studies indicate that dyslexia is persistent over time and remains detectable in adults. The condition is highly heritable, with between 25% and 65% of affected children having a dyslexic parent. Studies of brain structure using morphological imaging techniques have provided inconsistent findings. Studies of brain activation using positron emission tomography (PET) reveal diminished activation of left posterior structures with reading-related tasks. Autopsies of patients with severe persistent dyslexia who died of accidental deaths show neuronal ectopias and architectonic dysplasias in the peri-sylvian regions of the left hemisphere.[31] Individuals with childhood learning disabilities are at increased risk for adult psychopathology, particularly depression and anxiety.[32]

ATTENTION-DEFICIT HYPERACTIVITY DISORDER

Attention-deficit hyperactivity disorder is identified in a child when six of the following symptoms of inattention have persisted for at least 6 months: fails to give close attention to details of schoolwork, work, or other activities; does not consistently pay attention during tasks or play activities; often appears not to listen when spoken to directly; often does not follow through on instructions and does not finish schoolwork, chores, or duties; has difficulty in organizing tasks and activities; frequent avoid, dislike of or reluctance to engage in tasks requiring sustained mental effort; often loses things necessary for tasks or activities; is easily distracted by extraneous stimuli; and forgetfulness in daily activities. In addition, the child will have exhibited six or more of the following symptoms of hyperactivity-impulsivity for at least 6 months: fidgeting with hands or feet or squirming in seat; leaving his or her seat in the classroom or other situations in which remaining seated is expected; running about or climbing excessively in situations where it is inappropriate; failing to play or engage in leisure activities quietly. The child is often "on the go" or acts as if "driven by a motor"; excessive talking; blurting out answers before questions have been completed; failing to await his or her turn; and frequent interrupting or intruding on others. At least some of the symptoms must be evident before the age of 7 years and must be present in two or more settings. Three subtypes of ADHD are recognized: a combined attention-deficit and hyperactivity type, a predominantly inattentive type with a predominance of cognitive over motor disturbances, and a predominantly hyperactive-impulsive type with a predominance of motoric and impulsive features.[6]

Attention-deficit hyperactivity disorder is common in the population, with reported prevalence ranging from 2% to 15%. Boys are more likely to manifest the syndrome than girls, with ration of at least 2–3:1.[33] The disorder has a marked genetic influence; 75% of children of parents with ADHD evidence the disorder.[34] The prevalence of the disorder is highest in school-age children and it persists into early adolescence, decreasing through mid and late adolescence to lower levels in adulthood.[33] When ADHD persists into adolescence it is often associated with substance abuse and antisocial personality.[33] Neuroimaging studies using both morphological and functional techniques have often identified abnormalities of prefrontal cortex and striatum, though there is substantial variability across studies.[35] Nongenetic risk factors for the development of ADHD include head injury and prenatal exposure to alcohol.

Patients with ADHD commonly respond with improved attention and reduced motor activity to treatment with psychostimulants. Currently available stimulants include methylphenidate, dextroamphetamine, and pemoline.[33,35]

Two disruptive behavior disorders somewhat related to ADHD are conduct disorder and oppositional defiant disorder. *Conduct disorder* is characterized by aggression toward people and animals, destruction of property, deceitfulness or theft, and serious rule violations. *Oppositional defiant disorder* features the following behaviors: loss of temper, argument with adults, active defiance or refusal to comply with adults' requests or rules, deliberate annoyance of people, blame of others for one's mistakes or misbehavior, easy annoyance by others, expression of anger and resentment and of spite or vindication.[6]

PROGRESSIVE, COGNITIVE, AND BEHAVIORAL DETERIORATION IN CHILDREN

As noted above, the current concept of mental retardation includes any cognitive impairment with onset before age 18. Thus, diseases producing progressive

TABLE 25.5. *Inherited Metabolic Disorders with Cognitive and Neuropsychiatric Symptoms*[38]

Disease	*Clinical Symptoms*	*Tests*
Lysosomal storage diseases		
Fabry's disease	Peripheral nerve pain (± renal failure ± angiokeratoma ± cardiomyopathy), stroke-like episodes	α-galactosidase
Gaucher's disease type III	Horizontal supranuclear gaze defect, developmental delay, hydrocephalus, skeletal abnormalities, psychosis	β-glucosidase
GM1 gangliosidosis	Extrapyramidal signs, flattening of vertebral bodies, normal cognitive function, sometimes with psychosis	Leukocyte β-galactosidase, urine oligosaccharides
GM2 gangliosidosis (Tay Sach's and Sandhoff's disease)	(i) Lower motor neuron disease with onset at 20–40 years, pyramidal signs, and cerebellar degeneration (ii) Atypical amyotrophic lateral sclerosis	Leukocyte total, hexosaminidase, and hexosaminidase A
Krabbe's leukodystrophy	Pes cavus, hemiparesis, spastic tetraparesis, leukodystrophy	Leukocyte β-galactocerebrosidase
Metachromatic leukodystrophy	Loss of cognitive function or behavioral abnormalities, neuromuscular weakness with impaired nerve conduction, leukodystrophy	Leukocyte arylsulphatase A (the
Sialidosis (mucolipidosistype 1)	Dementia ± cherry red spot (type II), myoclonus, blindness, cherry red spot, dysmorphic features, angiokeratoma	Urine oligosaccharides, fibroblast α-neuraminidase
Niemann-Pick's disease type C	Psychomotor retardation leading to dementia, ataxia with dystonia	Bone marrow aspirate, fibroblast cholesterol uptake and staining
Amino acid disorders		
Arginase deficiency	Disorientation, coma	Plasma and urine amino acids, plasma ammonia (1 hr postprandial)
Citrullinaemia	Disorientation, restlessness, coma	Plasma and urine amino acids, plasma ammonia (1 hr postprandial)
Hartnup's disease	Dementia, ataxia ± skin lesions	Plasma and urine amino acids
Homosystinuria (cystathionine synthasedeficiency: classic form)	Occlusive cerebrovascular disease, dislocated lenses, osteoporosis, psychiatric disturbances	Urine homocystine, plasma homocystine and methionine
Homocystinuria (methylene tetrahydrofolatereductase deficiency remethylation defect)	Parasthesia, hallucinations, tremor, withdrawal, mental retardation, limb weakness, memory loss	Urine homocystine, plasma homocystine and methionine
Ornithine transcarbamylase deficiency	Behavioral disturbances,comatose episodes, sleep disorders	Plasma ammonia (1 hr postprandial) plasma amino acids, urine amino acids and orotic acid
Organic acid disorders		
Propionic acidemia	Chorea and dementia, recurrent vomiting	Urine organic acids, blood spot acyl carnitines

(*continued*)

TABLE 25.5. *Inherited Metabolic Disorders with Cognitive and Neuropsychiatric Symptoms (Continued)*

Disease	*Clinical Symptoms*	*Tests*
Peroxisomal disorders		
X linked adrenoleukodystrophy	Onset at 20–30 years in males, gait disturbance, spastic paraparesis, intellectual function usually intact, impotence ± Addison's disease, occasionally cerebral symptoms may occur, e.g., dementia and psychosis	Plasma very long–chain fatty acids
Lactic acidaemias		
Electron transport chain disorders, mtDNA encoded	NARP, MELAS, MERRF, Kearns-Sayre syndrome	Blood or tissue mtDNA analysis
Other disorders		
Aceruloplasminemia	Ataxia, retinal dystrophy ± diabetes mellitus ± presenile dementia	Plasma and urine copper, plasma iron and ferritin
Adult polyglucosan body disease	Upper and lower motor neuron signs, sensory loss, neurogenic bladder, dementia	Leukocyte glycogen brancher enzyme (some forms may show normal muscle activity)
Cerebrotendinous xanthomatosis	Spasticity, ataxia, cataracts, tendon xanthomas	Urine bile alcohols
Hereditary vitamin E deficiency	Tremor, ataxia, head titubation, loss of vibration sense	Plasma vitamin E, plasma cholesterol and triglycerides
Homocystinuria and methylmalonic aciduria (combination defect)	Megaloblastic anemia, dystonia	Urine organic acids and homocystine
Juvenile Barren's disease	Seizures, visual loss, dementia	Skin or rectal biopsy for histological analysis, DNA analysis for the common mutation
Kuf's disease	Type A: progressive myoclonic epilepsy Type B: motor problems, psychosis, dementia	Skin or rectal biopsy for histological analysis
Lesch-Nyhan syndrome	Some forms may present late with choreiform movements, dysarthria ± renal problems	Plasma urate and urine urate/creatinine
Porphyrias	Limb, neck, and chest pain, muscle weakness, senosry loss, seizures, behavioral abnormalities ± abdominal symptoms ± photosensitivity	Urine delta amino levulinic acid and porphobilinogen, urine and fecal porphyrins
Pyridoxine-responsive seizures	Persistent seizuresresponsive to pyridoxine	In vivo pyridoxine response test
Refsum's disease	Retinitis pigmentosa, peripheral polyneuropathy, cerebellar ataxia	Plasma phytanic acid
Segawa disease	Cyclical dystonia	Levodopa trial (some forms have a defect in biopterin metabolism)
Sjogren-Larrson syndrome	Spastic tetraplegia ± ichthyosis, mental retardation	Fibroblast fatty alcohol, oxidoreductase assay
Wilson's disease	Dysarthria, loss of coordination of voluntary movements, pseudobulbar palsy, parkinsonian features, renal failure, liver disease, Kayser-Fleischer rings, dementia	Urine copper (pre- and post-penicillamine), plasma copper and ceruloplasmin

MELAS, mitochondrial encephalomyopathy, lactic acidosis, and stroke-like episode; MERRF, mitochondrial encephalomyopathy with ragged red fibers; NARP, neuropathy, ataxia, and retinitis pigmentosa.

cognitive deterioration in children, while meeting criteria for dementia and representing a decline from a previous level of function, will be identified as causes of mental retardation. Progressive neurological disorders producing cognitive deterioration in children include inborn errors of metabolism, infections, leukoencephalopathies, polioencephalopathies, toxic metabolic disorders, and seizure disorders (Table 25.2). Progressive polioencephalopathies of childhood include the lipidoses, neuronal steroid lipofuscinoses, mucopolysaccharidoses, mucolipidoses, glycogen storage disease, the mitochondrial disorders, Rett's disorder, and epileptic encephalopathies.[36] Leukodystrophies producing progressive deterioration in children include metachromatic leukodystrophy, adrenoleukodystrophy, Krabbe's disease (galactosylceramidase deficiency), Alexander's disease, Canavan's disease, and Pelizaeus-Merzbacher disease[36,37] (see Chapter 26 on multiple sclerosis and leukoencephalopathies). Chronic viral infections that may produce a slowly progressive decline in cognitive function in children include acquired immunodeficiency syndrome (AIDS), progressive multifocal leukoencephalopathy (PML), subacute sclerosing panencephalitis (SSPE), progressive rubella panencephalitis, and Creutzfeldt-Jakob disease.[36] Table 25.5 presents the inborn metabolic disorders with cognitive and neuropsychiatric symptoms and lists the principal tests used for diagnosis.[38]

TABLE 25.6. *Childhood Choreas*

Athetoid cerebral palsy
Huntington's disease (juvenile form)
Benign familial chorea
Hereditary paroxysmal chorea
Inborn errors of metabolism
Gangliosidoses
Lipofuscinoses
Lesch-Nyhan syndrome
Hallervorden-Spatz disease
Sydenham's chorea
Systemic lupus erythematosus
Viral encephalitis
Drug-induced chorea
Neuroleptic agents
Psychostimulants
Anticonvulsants
Metabolic disorders
Hypo- and hypernatremia
Hepatic encephalopathy
Hyperthyroidism
Stroke

MOVEMENT DISORDERS IN CHILDREN

Movement disorders are described in detail in Chapter 18. Although the etiologies of movement disorders are similar in children and adults, some conditions common in children are rare in adults and must be weighted differently in the differential diagnosis. Among the causes of childhood chorea (Table 25.6) is the athetoic form of cerebral palsy, which follows perinatal asphyxia with or without kernicterus.[39] The juvenile form of Huntington's disease, benign familial chorea, hereditary paroxysmal disorder, and a variety of inborn errors of metabolism including gangliosidoses, lipofuscinoses, Lesch-Nyhan syndrome, and Hallervorden-Spatz disease can also produce chorea in childhood. Infectious and inflammatory causes of chorea among children include Sydenham's chorea, systemic lupus erythematosus, and viral encephalopathies. A variety of drugs can induce chorea, including neuroleptics, psychostimulants, and anticonvulsants.

The relative frequency of disorders producing *parkinsonism in children* also differs from the usual differential diagnosis in adults (Chapter 18). The juvenile form of Huntington's disease (Westphal variant) typically presents as a parkinsonian disorder. Perinatal brain insults may produce parkinsonism, and drugs, toxins, and inborn metabolic disorders also feature in the differential diagnosis (Table 25.7).

STROKE IN CHILDREN

Focal cortical strokes in children can produce neurobehavioral deficits and neuropsychiatric syndromes as described in adults (Chapter 26). The etiologies of stroke in childhood vary from those in adulthood; hypertension and atherosclerotic disease are much more rare in children than in adults. Cardiac abnormalities and hematologic disorders account for a majority of strokes occurring in the perinatal, infant, and childhood period.[40] Table 25.8 presents a differential diagnosis of the etiologies of stroke in children.

TABLE 25.7. *Childhood Parkinsonism*

Ceroid lipofuscinosis
Gaucher's disease
GM1 gangliosidosis
Metachromatic leukodystrophy
Neuroacanthocytosis
Hallervorden-Spatz disease
Huntington's disease (Westphal variant)
Wilson's disease
Juvenile parkinsonism
L-dopa-responsive dystonia–parkinsonism
Machado-Joseph-Azorean disease
Infectious/post-infectious
Mycoplasma
Viruses
Measles
Varicella
Japanese B
Western equine
Human immunodeficiency
Perinatal anoxia
Perinatal kernicterus
Mass lesion
Arteriovenous malformation
Neoplasm
Drugs and toxins
Carbon monoxide
Neuroleptics
Antiemetics
Stroke
Hydrocephalus

UNIQUE NEUROBEHAVIORAL AND NEUROPSYCHIATRIC SYNDROMES IN CHILDREN

Two specific neuropsychiatric syndromes occurring in childhood—Landau-Kleffner syndrome and PANDAS—deserve special emphasis. The *Landau-Kleffner syndrome* is a disorder of acquired aphasia with convulsions in children. Seizures or aphasia is equally likely to be the presenting manifestation; up to 25% of children with the syndrome do not have a history of epilepsy. The EEG shows spikes, sharp waves, or spike-and-wave discharges that are typically bilateral and oc-

TABLE 25.8. *Causes of Ischemic Stroke in Childhood*

Cardiac disease
Congenital heart disease
Atrial septal defect
Ventricular septal defect
Patent ductus arteriosus
Infective endocarditis
Valvular disease
Marantic endocarditis
Infective endocarditis
Mitral valve prolapse
Arrhythmia
Myocarditis
Cardiomyopathy
Rheumatic heart disease
Hematologic
Sickle-cell disease
Inherited coagulopathy
Protein C deficiency
Protein S deficiency
Antithrombin III deficiency
Acquired coagulopathy
Antiphospholipid antibody syndrome
Lupus anticoagulant
Myeloproliferative disorders
Arterial occlusive disease/primary vasculopathy
Moyamoya disease
Fibromuscular dysplasia
Para-infectious disorder
Takayasu arteritis
Systemic lupus erythematosus and other autoimmune diseases
Infectious vasculitis
Vasospasm with subarachnoid hemorrhage
Trauma
Trauma to posterior pharynx or neck
Neck rotation or dislocation
Metabolic
Homocystinuria
Myoclonic epilepsy, lactic acidosis, and stroke-like episodes (MELAS)
Volume depletion

(*continued*)

TABLE 25.8. *Causes of Ischemic Stroke in Childhood (Continued)*

Drugs
Cocaine
Amphetamine
Sympathomimetics
Oral contraceptives
Surgical intervention
Cardiac surgery
Cardiac catheterization
Cerebral angiography
Neurocutaneous syndromes
Neurofibromatosis
Tuberous sclerosis
Sturge-Weber disease
Fabry's disease
Lipoprotein abnormalities
Migraine
Congenital cerebrovascular anomalies
Arteriovenous malformation
Hypoplasia of vascular channels
Cavernous angioma
Intracranial aneurysm

cur predominantly over the temporal and parietal regions. Associated behavioral disturbances, including hyperactivity, rage outbursts, aggressiveness, stereotypy and poor social communication skills, are present in most children with the syndrome.[41]

PANDAS refers to pediatric autoimmune neuropsychiatric disorders associated with streptococcal infections.[42] The syndrome is identified in a group of children manifesting tics and obsessive-compulsive disorder whose symptoms occur following infection with b-hemolytic streptococcus. Sydenham's chorea is known to be a post-infectious autoimmune syndrome, and obsessions and compulsions have been observed to be more common among children who experienced the condition. In some cases, however, children with PANDAS have no known history of either streptococcal infection or Sydenham's chorea, and the diagnosis must be sought on the basis of laboratory evidence of a past streptococcal infection. PANDAS should be considered in children manifesting tic syndromes or obsessions and compulsions who have no family history of a similar disorder.

REFERENCES

1. Cox CE. Obtaining and formulating a developmental history. Child Adolesc Psychiatr Clin North Am 1999;8:271–279.
2. Neeper R, Huntzinger R, Gascon GG. Examination I: special techniques for the infant and young child. In: Coffey CE, Brumback RA, eds. Textbook of Pediatric Neuropsychiatry. Washington, DC: American Psychiatric Press, 1998:153–170.
3. Berg BO. The neurocutaneous disorders. In: Berg BO, ed. Child Neurology, 2nd ed. Philadelphia: J.B. Lippincott, 1994:185–195.
4. Lewis M. Psychiatric assessment of infants, children, and adolescents. In: Lewis M, ed. Child and Adolescent Psychiatry, 2nd ed. Philadelphia: Williams & Wilkins, 1996:440–457.
5. Frankenburg WK, Dodds J, et al. The Denver II: a major revision and restandardization of the Denver Developmental Screening Test. Pediatrics 1992;89:91–97.
6. American Psychiatric Association. Diagnostic and Statistical Manual of Mental Disorders, 4th ed. Washington, DC: American Psychiatric Press, 1994.
7. Pulsifer MB. The neuropsychology of mental retardation. J Int Neuropsychol Soc 1996;2:159–176.
8. Bodensteiner JB, Schaefer GB. Evaluation of the patient with idiopathic mental retardation. J Neuropsychiatry Clin Neurosci 1995;7:361–370.
9. Sundheim STPV, Ryan RM, Voeller KKS. Mental retardation. In: Coffey CE, Brumback RA, eds. Textbook of Pediatric Neuropsychiatry. Washington: American Psychiatric Press, 1998: 649–690.
10. Hagerman RJ. Annotation: fragile X syndrome: advances and controversy. J Child Psychol Psychiatry 1992;33:1127–1139.
11. Reiss AL, Faruque F, et al. Neuroanatomy of Rett syndrome: a volumetric imaging study. Ann Neurol 1993;34:227–234.
12. Jernigan TL, Bellugi U, et al. Cerebral morphologic distinctions between Williams and Down syndromes. Arch Neurol 1993;50: 186–191.
13. Buiting K, Saitoh S, et al. Inherited microdeletions in the Angelman and Prader-Willi syndromes define an imprinting centre on human chromosome 15. Nat Genet 1995;9:395–400.
14. Stein DJ, Keating J, et al. A survey of the phenomenology and pharmacotherapy of compulsive and impulsive-aggressive symptoms in Prader-Willi syndrome. J Neuropsychiatry Clin Neurosci 1994;6:23–29.
15. Smalley SL, Levitt J, Bauman M. Autism. In: Coffey CE, Brumback RA, eds. Textbook of Pediatric Neuropsychiatry. Washington: American Psychiatric Press, 1998:393–428.
16. Egaas B, Courchesne E, Saitoh O. Reduced size of corpus callosum in autism. Arch Neurol 1995;52:794–801.
17. Piven J, Arndt S, et al. An MRI study of brain size in autism. Am J Psychiatry 1995;152:1145–1149.
18. Piven J, Bailey J, et al. An MRI study of the corpus callosum in autism. Am J Psychiatry 1997;154:1051–1056.
19. Deb S, Thompson B. Neuroimaging in autism. Br J Psychiatry 1998;173:299–302.
20. Rapin I, Katzman R. Neurobiology of autism. Ann Neurol 1998;43:7–14.

21. Volkmar FR. The disintegrative disorders: childhood disintegrative disorder and Rett's disorder. In: Volkmar FR, ed. Psychoses and Pervasive Developmental Disorders in Childhood and Adolescence. Washington, DC: American Psychiatric Press, 1996: 223–248.
22. Szatmari P. Asperger's disorder and atypical pervasive developmental disorder. In: Volkmar FR, ed. Psychoses and Pervasive Developmental Disorders in Childhood and Adolescence. Washington, DC: American Psychiatric Press, 1996:191–221.
23. Volkmar FR, Klin A, et al. "Nonautistic" pervasive developmental disorders. In: Coffey CE, A. BR, eds. Textbook of Pediatric Neuropsychiatry. Washington, DC: American Psychiatric Press, 1998:429–447.
24. Berthier ML, Starkstein SE, Leiguarda R. Developmental cortical anomalies in Asperger's syndrome: neuroradiological findings in two patients. J Neuropsychiatry Clin Neurosci 1990;2: 197–201.
25. Amir RE, Van den Veyver IB, et al. Influence of mutation type and X chromosome inactivation on Rett syndrome phenotypes. Ann Neurol 2000;47:670–679.
26. Fitzgerald PM, Jankovic J, Percy AK. Rett syndrome and associated movement disorders. Mov Disord 1990;5:195–202.
27. Subramaniam B, Naidu S, Reiss AL. Neuroanatomy in Rett syndrome: cerebral cortex and posterior fossa. Neurology 1997;48: 399–407.
28. Oldfers A, Sourander P, Percy AK. Neuropathology and neurochemistry. In: Hagberg B, ed. Rett Syndrome—Clinical and Biological Aspects. Cambridge, UK: Cambridge University Press, 1993:86–98.
29. Naidu S. Rett syndrome: a disorder affecting early brain growth. Ann Neurol 1997;42:3–10.
30. Wenk GL, Naidu S, et al. Altered neurochemical markers in Rett's syndrome. Neurology 1991;41:1753–1756.
31. Galaburda AM, Sherman GF, et al. Developmental dyslexia: four consecutive patients with cortical anomalies. Ann Neurol 1985; 18:222–233.
32. Patel P, Goldberg D, Moss S. Psychiatric morbidity in older people with moderate and severe learning disability II: the prevalence study. Br J Psychiatry 1993;163:481–491.
33. Elia J, Ambrosini PJ, Rapoport JL. Treatment of attention-deficit-hyperactivity disorder. N Engl J Med 1999;340:780–788.
34. Biederman J, Faraone SV, et al. High risk for attention deficit hyperactivity disorder among children of parents with childhood onset of the disorder: a pilot study. Am J Psychiatry 1995;1995: 431–435.
35. McCracken JT. Attention-deficit/hyperactivity disorder II: neuropsychiatric aspects. In: Coffey CE, Brumback RA, eds. Textbook of Pediatric Neuropsychiatry. Washington DC: American Psychiatric Press, 1998:483–501.
36. Chutorian AM, Pavlakis SG. Psychiatric symptoms in the progressive metabolic, degenerative, and infectious disorders of the nervous system. In: Kaufman DM, Solomon GE, Pfeffer CR, eds. Child and Adolescent Neurology for Psychiatrists. Baltimore: Williams & Witkins, 1992:207–219.
37. Schiffmann R, Moller JR, et al. Childhood ataxia with diffuse central nervous system hypomyelination. Ann Neurol 1994;35: 331–340.
38. Gray RGF, Preece MA, et al. Inborn errors of metabolism as a cause of neurological disease in adults: an approach to investigation. J Neurol Neurosurg Psychiatry 2000;69:5–12.
39. Trinidad KS, Kurlan R. Chorea, athetosis, dystonia, tremor and parkinsonism. In: Robertson MM, Eapen V, eds. Movement and Allied Disorders in Childhood. New York: John Wiley & Sons, 1995:105–147.
40. Mathews KD. Stroke in neonates and children: overview. In: Biller J, Mathews KD, Love BB, eds. Stroke in Children and Young Adults. Boston: Butterworth-Heinemann, 1994:15–29.
41. Tuchman RF. Epilepsy, language, and behavior: clinical models in childhood. J Child Neurol 1994;9:95–102.
42. Swedo SE, Leonard HL, et al. Identification of children with pediatric autoimmune neuropsychiatric disorders associated with streptococcal infections by a marker associated with rheumatic fever. Am J Psychiatry 1997;154:110–112.
43. Gillberg C. Clinical Child Neuropsychiatry. Cambridge, UK: Cambridge University Press, 1995.
44. Ashwal S. Congenital structural defects. In: Swaiman KF, Ashwal S, eds. Pediatric Neurology. St. Louis: Mosby, 1999:234–300.
45. Dobyns WB. Cerebral dysgenesis: causes and consequences. In: Miller G, Ramer JC, eds. Static Encephalopathies of Infancy and Childhood. New York: Raven Press, 1992:235–247.
46. Harris JC. Developmental Neuropsychiatry. New York: Oxford University Press, 1995.
47. Swaiman KF, Wright DP. Degenerative diseases primarily of gray matter. In: Swaiman KF, Ashwal S, eds. Pediatric Neurology. St. Louis: Mosby, 1999:833–848.
48. Kaye EM. Disorders primarily affecting white matter. In: Swaiman KF, Ashwal S, eds. Pediatric Neurology. St. Louis: Mosby, 1999:849–859.
49. Maertens P. Mitochondrial encephalopathies. Semin Pediatr Neurol 1996;3:279–297.

49a. Barkovich AT, Kuzniecky RI, et al. Classification system for malformations of cortical development. Neurology 2001;57: 2168–2178.

Chapter 26

Focal Brain Disorders and Related Conditions

Focal brain lesions comprise a heterogeneous group of conditions producing cognitive abnormalities and neuropsychiatric symptoms. Previous chapters have focused on symptoms of brain dysfunction including cognitive disorders (Chapters 6–9), neuropsychiatric symptoms (Chapters 13–17), and movement disorders (Chapters 18–20). In this chapter, focal central nervous system (CNS) lesions and related disorders (stroke and cerebral vascular disease, multiple sclerosis and white matter diseases, CNS infections, and neoplasms affecting the brain) are considered. This etiologic approach is provided to complement the symptomatic approach provided in earlier chapters of this volume. The specific characteristics of cognitive and behavioral disturbances accompanying focal CNS lesions are considered only briefly and references are made to earlier chapters in which more substantive information is provided. The emphasis in this chapter is on the etiologic processes and their differential diagnosis. Changes in the elementary neurological examination indicative of focal CNS dysfunction are described in Chapter 3.

STROKE AND CEREBROVASCULAR DISEASE

Cerebrovascular disease includes any disease of the vascular system that causes ischemia or infarction of the brain or spontaneous hemorrhage into the brain or subarachnoid space.[1] The basic categories of cerebrovascular disease include transient ischemic attack (TIA), stroke, and vascular dementia. *Transient ischemic attack* refers to an acute loss in neurological function persisting less than 24 hours. *Stroke* refers to rapidly developing clinical symptoms or signs of focal (or sometimes global) impairment of cerebral function with symptoms lasting more than 24 hours and having no other cause than cerebrovascular disease. The distinction between TIA and stroke is arbitrary and they represent a continuum of increasingly severe and irreversible brain infarction.[2] *Vascular dementia* (Chapter 10) refers to a syndrome characterized by memory impairment plus compromise in at least one additional cognitive domain (language, praxis, visuospatial ability, executive function) produced by cerebrovascular disease.

Cerebrovascular disease ranks second behind heart disease as the leading cause of death worldwide and third behind heart disease and cancer as a cause of death in the United States.[3] The incidence of stroke doubles every 10 years after the age of 55: approximately 2 per 1000 individuals experience stroke between the ages of 55 and 64; approximately 4 individuals per 1000 experience a new onset of cerebrovascular disease between the ages of 65 and 74; and approximately 8–10 individuals per 1000 have the onset of stroke after the age of 75.[4]

Cerebral infarction accounts for approximately 80% of cerebrovascular events; primary intracerebral hemorrhage accounts for 10%; subarachnoid hemorrhage causes 5%; and 5% of strokes are of uncertain cause.[1] Approximately 40% of cerebral infarctions undergo hemorrhagic transformation within 2 weeks after infarction. This is most likely to occur when infarcts are large or anticoagulants or thrombolytic agents are used for treatment.[2]

Table 26.1 lists the principal neurobehavioral and neuropsychiatric syndromes associated with focal lesions of the brain. Neuropsychiatric symptoms are common as post-stroke manifestations. Major depression occurs in approximately 10%–25% of patients and minor depression in 10%–40%. Anxiety accompanies depression in 20% of depressed post-stroke patients and occurs without depression in 7%–10%. Apathy occurs in 20% of patients (10% with depression; 10% without depression). Anosognosia with denial of illness is present in 25%–45% of patients, particularly those with right posterior lesions. Catastrophic reactions appear in approximately 20% and emotional lability is present in 20%. Mania and psychosis are well

TABLE 26.1. *Neurobehavior and Neuropsychiatric Syndromes Associated with Focal Brain Infarctions of the Hemisphere*

Blood Vessel	*Structures Affected*	*Syndrome*
Anterior cerebral artery (L)	Anterior corpus callosum; supplementary motor area; anterior cingulate cortex	Transcortical aphasia Callosal apraxia (L) hand tactile anomia Transient akinetic mutism
Middle cerebral artery (L); superior division	Inferolateral frontal cortex	Broca's aphasia
Middle cerebral artery (L); inferior and superior division	Lateral cerebral hemisphere (anterior and posterior)	Global aphasia
Middle cerebral artery (L); posterior branch	Arcuate fasciculus	Conduction aphasia
Middle cerebral artery (L); posterior branch	Angular gyrus	Transcortical sensory aphasia; alexia with agraphia; angular gyrus syndrome; anomia
Middle cerebral artery (L); inferior division	Posterior superior temporal lobe	Wernicke's aphasia
Posterior cerebral artery (L); proximal	Hippocampus	Verbal amnesia
Posterior cerebral artery	Occipital cortex	Hemianopsia; release hallucinations
Posterior cerebral artery	Calcarine cortex	Hemianopsia; achromatopsia
Anterior cerebral artery (R)	Corpus callosum; supplementary motor area; anterior cingulate region	Callosal apraxia
Middle cerebral artery (R); superior division	Inferolateral frontal lobe	Executive aprosodia
Middle cerebral artery (R); inferior division	Posterior superior temporal lobe	Receptive aprosodia
Middle cerebral artery (R); posterior branch	Posterior parietal	Unilateral neglect; anosognosia
Posterior cerebral (R); proximal	Hippocampus	Nonverbal amnesia
Posterior cerebral (R)	Inferior longitudinal fasciculus	Prosopagnosia; environmental agnosia

(*continued*)

described in post-stroke conditions but are relatively unusual.[5]

There are many potential causes of cerebral thrombosis and cerebral embolism. Table 26.2 provides a differential diagnosis of thrombotic and embolic conditions that may result in cerebrovascular events.

The principal stroke syndromes include hemorrhages in the area of the basal ganglia and thalamus (usually hypertensive in origin); lobar hemorrhages (often associated with amyloid angiopathy); focal cortical infarctions (embolic or thrombotic); borderzone infarctions (between territories of major arteries); lacunar infarctions involving the basal ganglia, thalamus, brainstem, and cerebellum; and white matter ischemic injury. When there is extensive white matter damage and the individual has dementia, the diagnosis of Binswanger's disease is considered.

Typical abnormalities on neurological examination include hemiparesis, hemisensory deficits, homonymous visual field defects, gait disturbance, parkinsonism, incontinence, spasticity, brisk muscle stretch reflexes, pseudobulbar palsy, and Babinski signs.

Treatment of stroke includes acute interventions to open the artery and reverse the deficit and preventive strategies to reduce stroke risk. Where acute stroke teams are available, thrombolytic therapy with recombinant tissue plasminogen activator (rtPA) is used for treatment of strokes when the patient is seen within 3 hours of onset. Emboli originating from high-risk cardiac conditions, particularly atrial fibrillation, are prevented most effectively with long-term oral anticoagulants. Antiplatelet agents are recommended as a primary prevention strategy for patients with atherothrombotic stroke; aspirin (in doses of 50–325 mg daily) is the most

TABLE 26.1. *Neurobehavior and Neuropsychiatric Syndromes Associated with Focal Brain Infarctions of the Hemisphere (Continued)*

Blood Vessel	*Structures Affected*	*Syndrome*
Posterior cerebral artery (R)	Calcarine cortex; splenium of corpus callosum	Alexia without agraphia; color anomia
Posterior cerebral artery (R)	Occipital cortex	Hemianopsia; release hallucinations
Anterior cerebral artery (bilateral)	Anterior cingulate cortex (bilateral)	Akinetic mutism
Middle cerebral artery (bilateral)	Bilateral temporal lobe	Auditory agnosia
Middle cerebral artery (bilateral)	Bilateral parietal region	Balint's syndrome; Charcot-Wilbrand
Posterior cerebral artery (bilateral)	Inferior longitudinal fasciculus (bilateral)	Visual agnosia; prosopagnosia; environmental agnosia
Posterior cerebral artery (bilateral)	Occipital cortex	Cerebral blindness; Anton's syndrome
Posterior cerebral artery (bilateral)	Hippocampi	Amnesia
Posterior cerebral artery	Cerebral peduncle	Peduncular hallucinosis
Basilar artery	Midbrain	Top of the basilar syndrome; visual hallucinations, oneroid state
Basilar artery (penetrating branches)	Base of the pons	Locked-in syndrome
	Anterior thalamic nuclei	Memory abnormalities, perseveration; executive deficits; apathy; chronic-memory disturbance
	Dorsomedial nuclei	Apathy; amnesia; executive deficits
	Lateral geniculate	Thalamic dazzle
	Lateral dorsal nuclei	Déjérine-Roussy syndrome
	Caudate nucleus (dorsal)	Executive dysfunction
	Caudate nucleus (ventral)	Disinhibition
	Nucleus accumbens	Apathy
	Globus pallidus	Executive dysfunction; disinhibition or apathy
	Subthalamic nucleus	Mania

TABLE 26.2. *Etiologies of Stroke*

Thrombotic	*Embolic*
Atherosclerosis	Cardiac disease
Arteriosclerosis	Myocardial infarction with mural thrombus
Diabetes	Atrial fibrillation
Fibromuscular dysplasia	Rheumatic endocarditis
Amyloidosis	Postrheumatic valvular disease
Fabry's disease	Congenital heart disease
Homocystinemia	Cardiac surgery, catheterization, angioplasty
Moya-Moya disease	Subacute bacterial endocarditis
Radiation-induced vasculopathy	Marantic endocarditis
Neoplastic angioendotheliosis	Libman-Sacks endocarditis
	Prosthetic valves
Inflammatory (noninfectious) vascular disorders	Septal defect with paradoxical embolization
Systemic lupus erythematosus	Cardiomyopathy
Giant cell arteritis (temporal arteritis)	Atrial myxoma
Sarcoidosis	Mitral valve prolapse syndrome
Polyarteritis nodosa	Endocardial fibroelastosis
Granulomatous arteritis	Congenital cardiac disorders
Wegener's granulomatosis	Atherosclerotic ulcerative plaques with cholesterol emboli
Hypersensitivity angiitis	
Scleroderma	Atherosclerotic stenosis with distal embolization
Rheumatoid arthritis with arteritis	Metastatic deposits
Takayasu's aortitis	Parasites and ova
Cogan's syndrome	Septic emboli
Behçet's syndrome	Air emboli
Sjogren's syndrome	Fat emboli
Isolated angiitis of the CNS	Nitrogen bubble emboli
Vogt-Koyanagi-Harada syndrome	Vascular trauma
Dermatomyositis	Emboli from arterial aneurysms
Kohlmeier-Degos disease	Angiography
Chemical-arteritis	Attempted strangulation
Amphetamines	Atlanto-axial dislocation
Crack cocaine	Chiropractic manipulation of the neck
Hematologic disorders	Neck fracture
Leukemia	Yoga with neck injury
Sickle cell disease	Paradoxical embolism from the venous system with arteriovenous defect
Waldenstrom's macrogloblinemia	
Polycythemia vera	
Dysproteinemias	
Idiopathic thrombocytosis	
Hyperlipidemia	
Disseminated intravascular coagulation	
Thrombotic thrombocytopenia purpura	
Cryoglobulinemia	

(*continued*)

TABLE 26.2. *Etiologies of Stroke (Continued)*

Thrombotic
Hematologic disorders (cont'd)
Antiphospholipid antibody syndrome
Lupus anticoagulant
Meningovascular infections (infectious arteritis)
Bacterial arteritis
Syphilis
Tuberculosis
Miscellaneous
Yeast and fungal arteritis
Rickettsia arteritis
Viral arteritis
Cysticercosis
Miscellaneous
Vascular trauma
Vascular dissection
Pregnancy/puerperium
Oral contraceptives
Migraine
Ergot compounds
Mitochondrial disorders
Marfan's syndrome
Ehler's-Danlos syndrome
Pseudoxanthoma elasticum

commonly used agent. Other antiplatelet drugs that have been shown to be at least as effective as aspirin include ticlopidine, clopidogrel, and a combination of low-dose aspirin and dipyrimidol.[6]

MULTIPLE SCLEROSIS AND WHITE MATTER DISEASES

Multiple sclerosis (MS) is a CNS disease characterized by multifocal inflammatory demyelination. The primary cause of MS is unknown; environmental risk factors and multiple genetic loci contribute to susceptibility to the disease.[7] The prevalence varies with geographic distribution; it is least common in the tropics and becomes more common in the northern latitudes. In the United States, Canada, and northern Europe the prevalence is approximately 100 per 100,000 whereas in the tropics the average prevalence is 5 per 100,000.[8] The disease is twice as common among women as men and typically begins in mid-life between the ages of 15 and 50. The diagnosis of definite MS depends on the occurrence of at least two attacks and clinical evidence of at least two separate lesions in the CNS or two attacks with clinical evidence of one lesion and paraclinical (neuroimaging) evidence of another (Table 26.3). A combination of clinical and laboratory evidence can also be employed to make a diagnosis of definite MS (Table 26.3).[9]

Several different courses of MS are recognized (Table 26.4). Seventy percent of patients have relapsing-remitting MS with discrete episodes and complete or nearly complete recovery between attacks. Approximately 50% of patients with MS change gradually from relapsing-remitting to secondary progressive (relapsing progressive MS). These patients begin with discrete attacks but gradually accumulate increasing neurological disability and slowly worsen. Approximately 15% of patients have chronic progressive MS, characterized by a slow and steady progression of increasing neurolog-

TABLE 26.3. *Criteria for Diagnosis of Definite and Probable Multiple Sclerosis*

A. Clinically definite MS
   1. Two attacks and clinical evidence of two separate lesions
   2. Two attacks with clinical evidence of one lesion and paraclinical evidence of another (separate) lesion
B. Laboratory-supported definite MS
   1. Two attacks with either clinical or paraclinical evidence of one lesion and demonstration in CSF, of IgG OB or increased CNS synthesis of IgG
   2. One attack, clinical evidence of two separate lesions, and demonstration of CSF OB/IgG
   3. One attack, clinical evidence of one lesion, paraclinical evidence of another (separate) lesion, and demonstration of CSF OB/IgG
C. Clinically probable MS
   1. Two attacks and clinical evidence of one lesion
   2. One attack and clinical evidence of two separate lesions
   3. One attack, clinical evidence of one lesion, and paraclinical evidence of another (separate) lesion
D. Laboratory-supported probable MS
   1. Two attacks and presence of CSF OB/IgG

CSF, cerebrospinal fluid; IgG, immunoglobulin G; MS, multiple scoerosis; OB, oligoclonal bands.

ical deficit without discrete episodes. Acute (malignant) MS features polysymptomatic progression to severe disability or death in a few months. Benign MS is manifested by a relapsing course with discrete episodes without significant neurological deficit after 10 to 15 years of disease.[9]

Motor neuron signs, sensory abnormalities, visual changes, and evidence of cerebellar dysfunction are the most common findings on neurological examination of the MS patient. There are hyperactive muscle stretch reflexes (76%), ataxia of the lower extremities (57%), bilateral Babinski signs (54%), diminished vibratory sensation (47%), incoordination of one or more limb (40%), evidence of optic neuritis (38%), nystagmus (35%), decreased joint position sense (33%), spasticity of one or more limb (31%), diminished two-point discrimination (24%), decreased pain or temperature sense (22%), dysarthria (19%), paraparesis (17%), internuclear opthalmoplegia (11%), and signs of neurogenic bladder (10%).[10]

TABLE 26.4. *Clinical Course Categories of Multiple Sclerosis*

Category	*Features*
Relapsing-remitting	Discrete attacks with subsequent recovery and little or no residual neurological deficit between attacks. Some patients have a step-wise increase in neurological deficit.
Secondary progressive (relapsing progressive)	Onset with attacks but patients develop significant neurological deficits that increase during follow-up. Most patients slowly worsen.
Primary chronic progressive	Slow and steady progression of a chronic neurological deficit from onset without discrete episodes
Acute (malignant)	Polysymptomatic disease that progresses to severe disability and/or death in a few months.
Benign	Relapsing course without development of a significant residual neurological deficit after 10–15 years of disease

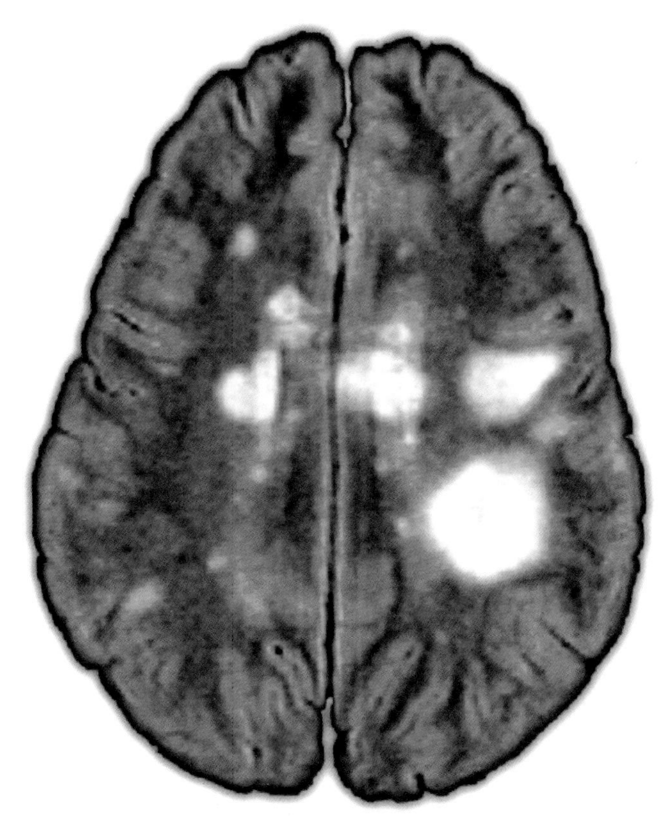

FIGURE 26.1 Magnetic resonance image of patient with multiple sclerosis showing multiple high-signal lesions on the T_2-weighted image.

Neuroimaging and laboratory studies are useful in supporting the clinical diagnosis of MS. Magnetic resonance imaging (MRI) reveals focal discrete or confluent high-signal lesions on T_2-weighted images in the periventricular deep white matter or superficial gray–white matter junction.[11] Patients also show hypointense lesions on T_1-weighted spin-echo images and increased hypointense lesion load correlates significantly with increasingly disability (Fig. 26.1).[12,13] Cerebrospinal fluid (CSF) abnormalities are present in 90% of patients with clinically definite MS. The CSF cell counts are typically normal, although a mild pleocytosis may be evident. Similarly, the total protein level is normal in approximately 80% of patients, although 60%–75% have an increased γ-globulin content. A γ-globulin fraction of $>12\%$ reflects increased IgG synthesis in the CNS. The elevated IgG levels give rise to oligoclonal banding on protein electrophoresis, a feature present in 90% of patients with clinically definite MS and between 30% and 40% of those with possible MS.[10,14] Increased CSF IgG is not specific and can be seen in other inflammatory neurological disorders, including subacute sclerosis panencephalitis, progressive rubella encephalitis, inflammatory neuropathies, cryptococcal meningitis, and other encephalitides. Abnormal visual evoked potentials, auditory evoked potential, or somatosensory evoked potentials may aid in identifying asymptomatic or minimally symptomatic lesions useful in corroborating the presence of multiple lesions necessary for the diagnosis.[10] Cognitive abnormalities are common in patients with MS. Memory and executive function abnormalities occur in patients with chronic progressive MS and during acute exacerbations in patients with relapsing remitting disease.[15] Semantic (fact) memory is more impaired than episodic (autobiographical) memory.[16] Working memory and other executive function deficits are frequent.[17,18] Specific types of neuropsychological impairments are correlated with anatomically relevant lesions on MRI.[19–21] Patients with cognitive impairment commonly have atrophy of the corpus callosum.[20,22]

Patients with MS commonlly manifest neuropsychiatric symptoms, particularly depression. Major depression is present in approximately 35% of patients with MS and depressive symptoms are present in 65%–80%[23–25] (Chapter 14). The depression is unrelated to the degree of disability[26,27] or to genetic factors,[28] suggesting that the mood abnormality is the product of the CNS disease itself. This is further confirmed by the observation that depression is more prominent in patients with MS affecting brain regions than in those with predominantly spinal cord disease with similar levels of disability.[27] Mood disturbances correlate with CSF cell counts during periods of acute exacerbation in patients with relapsing-remitting MS,[29] and mood disturbances also correlate with frontal or temporal abnormalities on MRI.[30–32] Several studies have suggested that treatment with interferon β-1B (described below) is associated with increased depression.[33,34] Quality of life in MS is influenced by the presence of depressive symptoms.[35] Bipolar illness is also increased in prevalence in MS.[36–38] Euphoria is more common than mania and is most likely to appear in patients with chronic progressive MS who have enlarged ventricles and cognitive impairment.[27] Psychosis may be a manifestation of MS and is most common with the acute encephalitic form of the illness.[39–41] Psychotic patients have a greater lesion load in the temporal or temporal–parietal regions than those without psychosis.[40,41] Other psychiatric symptoms recorded in patients with MS include panic attacks, obsessive-compulsive disorder, paraphillic and hypersexual behavior, anxiety, irritability, apathy, and disinhibition.[23,42–44]

Fatigue is a common symptom in patients with MS, occurring in up to 90% of patients.[45] Fatigue correlates with the presence of pyramidal track signs and reaction time on memory tasks but is independent of depression.[46–48] Patients with MS and fatigue have greater reductions in glucose metabolism in lateral and medial prefrontal regions, premotor cortex, putamen, and supplementary motor areas.[49]

At autopsy, plaques of various sizes are distributed throughout the CNS white matter. They are particularly prominent in periventricular regions and those located between the body of the caudate nucleus and the corpus callosum have been referred to as the *Wetterwinkle* or "storm center."[50] However, plaques may be seen virtually anywhere within the white matter of the hemispheres or corpus callosum as well as in the midbrain pons, medulla, cerebellum, and spinal cord. The optic nerves are commonly affected. The MS plaque consists of inflammatory activity, demyelination, and gliosis. The inflammatory infiltrate consists of lymphocytes and monocytes in the periventricular spaces and the leptomeninges; in the parenchyma of the lesion, lymphocytes, monocytes, and macrophages are present. Axonal transection is common in active lesions.[7]

Management of MS includes rehabilitation aimed at maintaining or improving activities of daily living and ambulation; anti-inflammatory therapy with steroids and immunomodulators; and management of neuropsychiatric symptoms with psychotropic agents. Adrenocorticotrophic hormone and steroids do not affect the long-term course of MS but appear to shorten individual episodes in patients with relapsing-remitting disease. Neuropsychiatric complications of steroid use include psychosis and mood disorders.[51] Immunomodulation has been shown to reduce the number of relapses and the degree of neurological disability in patients with relapsing-remitting MS. Available immunomodulators include glatiramer acetate (administered subcutaneous daily), interferon β-1b (administered subcutaneously every other day), interferon β-1a (administered intramuscularly weekly), and mitoxantrone (administered every 3 months intravenously) (Table 26.5).[52–56] Depression has been reported as a complication of interferon β-1b treatment (Chapter 14). Fatigue in MS may respond to treatment with amantadine or pemoline.[57,58] Treatment of depression in patients with MS with selective serotonin reuptake inhibitors, tricyclic antidepressants, or monoamine oxidation inhibitors has generally been successful in reducing depressive symptoms.[59–61] Psychotherapy can be of substantial value in allowing patients to respond to the unpredictable vicissitudes of their illness.[62]

Multiple sclerosis must be distinguished from a wide variety of other white matter disorders (Table 26.6). Demyelinating, infectious, toxic, vascular, genetic, inflammatory, metabolic, neoplastic, and hydrocephalic processes may all result in an increased signal on T_2-weighted MRI and evidence of demyelination at autopsy. There are no pathognomonic neurobehavioral or neuropsychiatric symptoms of white matter disorders, but bradykinesia (slowing of cognition), attentional abnormalities, retrieval deficit–type memory impairment with intact procedural memory, visuospatial dysfunction, and executive cognitive impairment with relative preservation of language are typical of the cognitive syndromes associated with white matter disease.[63]

Personality alterations, depression, and psychosis may also occur. Psychosis is common in metachromatic leukodystrophy, occurring in 53% of patients present-

TABLE 26.5. *Immunotherapies for Multiple Sclerosis*

Drug (Brand Name)	*Dose*	*Indication*	*Side Effect*
Interferon β-1a (Avonex)	30 μg IM* daily	Relapsing MS	Flu-like symptoms, pain, CBC and liver enzyme abnormalities
Interferon β-1b (Betaseron)	250 μg SC every other day	Relapsing MS	Flu-like symptoms, injection site reactions, menstrual irregularities, CBC and liver enzyme abnormalities; depression
Glatiramer acetate (Copaxone)	20 mg SC every day	Relapsing MS	Injection site reaction (mild), immediate post-injection reaction
Mitoxantrone (Novantrone)	12 mg/m² every 3 months IV (maximum 8–12 doses per 2–3 years)	Chronic progressive MS	Nausea, hair loss, urinary tract infection, menstrual irregularities, upper respiratory tract infection

CBC, complete blood count; IM, intramuscular; IV, intravenous; SC, subcutaneous.

TABLE 26.6. *White Matter Diseases*

Demyelinating
- Multiple sclerosis
- Acute disseminated encephalomyelitis
- Acute hemorrhagic encephalomyelitis

Infectious
- HIV encephalopathy
- Progressive multifocal leukoencephalopathy
- Subacute sclerosing panencephalitis
- Progressive rubella panencephalitis
- Varicella zoster encephalitis
- Cytomegalovirus encephalitis
- Lyme encephalopathy
- Creutzfeldt-Jakob disease

Toxic
- Anti-neoplastic therapy (cranial irradiation, methotrexate, BCNU, cytosine arabinoside, 5-FU, levamisole, fludaribine, cisplatin, thiotepa, interleukin-2, interferon-α)
- Immunosuppressive therapy (cycloserine, FK-506)
- Antimicrobials (amphotericin B, hexachloraphene)
- Drugs of abuse (inhalation of toluene, ethanol, glue, heroin pyrolysate, cocaine, gasoline; ingestion/intravenous administration of heroin, MDMA [ecstasy], psilocybin)
- Environmental (carbon monoxide, arsenic, carbon tetrachloride, tetraethyl lead)
- Miscellaneous (valproate)

Vascular
- Binswanger's disease
- Cerebral autosomal dominant arteriopathy with subcortical infarcts and leukoencephalopathy (CADASIL)
- Cerebral amyloid angiopathy
- Antiphospholipid antibody syndrome
- Sneddon's syndrome
- Fabry's disease

Traumatic
- Diffuse axonal injury

Genetic
- Metachromatic leukodystrophy
- Adrenoleukodystrophy
- Krabbe disease (globoid cell leukodystrophy)
- Pelizaeus-Merzbacher disease
- Alexander's disease
- **Canavan disease**
- Hereditary diffuse leukoencephalopathy with spheroids
- Membranous leukodystrophy
- Methylenetetrahydrofolate reductase deficiency

Genetic (cont'd)
- Leukodystrophy with ovarian dysgenesis
- Cerebrotendinous xanthomatosis
- Leigh's disease
- Cockayne syndrome
- Sjogren-Larrson syndrome
- Polyglucosan disease
- Phenylketonuria
- Neurofibromatosis
- Mucoplysaccharidosis (Hurler's syndrome)
- Myotonic dystrophy
- Peroxisome biogenesis disorders
  - Zellweger syndrome
  - Neonatal adrenoleukodystrophy
  - Infantile Refsum's disease
- "Private" leukodystrophies unique to single families or a few kindreds

Inflammatory
- Systemic lupus erythematosus
- Behçet's disease
- Sjogren's syndromes
- Wegener's granulomatosis
- Temporal arteritis
- Polyarteritis nodosa
- Scleroderma
- Isolated angiitis of the nervous system
- Sarcoidosis

Metabolic
- Cobalamin deficiency
- Folate deficiency
- Hypoxia
- Hypertensive encephalopathy
- Eclampsia
- High-altitude cerebral edema
- With celiac disease

Neoplastic
- Gliomatosis cerebri
- Diffusely infiltrative gliomas
- Primary cerebral lymphomas
- Peri-tumoral edema

Hydrocephalus
- Obstructive noncommunicating hydrocephalus
- Normal-pressure hydrocephalus

Miscellaneous
- Marchiafava-Bignami disease

BCNU, 1,3bis(2-chloroethyl)-1-nitrosourea; FK-506, tacrolimus; 5-FU, 5-fluorouracil; MDMA, methylenedioxymethamphetamine.

ing with disease between the ages of 12 and 30.[64] Adrenoleukodystrophy may manifest learning difficulties, cognitive impairment, personality alterations, schizophrenia-like symptoms, apathy and withdrawal, particularly in patients who become symptomatic before the age of 21.[65] Bipolar illness and psychotic depression have been observed as prominent symptoms in patients with adrenoleukodystrophy.[66] In rare cases, adrenoleukodystrophy may present in adulthood as a progressive dementia syndrome.[67,68] Similarly, rare cases of Krabbe disease (globoid cell leukodystrophy) may present as an adult-onset dementing disorder.[69] Cerebrotendinous xanthomatosis typically presents in the second or third decade of life with juvenile cataracts, tendon xanthomas, and diverse neurological abnormalities including dementia-peripheral neuropathy, cerebellar ataxia, pyramidal signs, extrapyramidal disturbances, and psychiatric disorders.[70] Rare cases of Pelizaeus-Merzbacher disease with onset in adulthood with intellectual deterioration and psychotic episodes have been described.[71] Membranous lipodystrophy (Nasu-Hakola's disease) may present in adulthood with multiple pathologic fractures, progressive dementia, and seizures. Apathy, euphoria, and disinhibition are common.[72]

TRAUMATIC BRAIN INJURY

Traumatic brain injury (TBI) may vary from mild to severe and the incidence and prevalence determined in population surveys will depend on the definition of TBI employed. In the United States and most industrialized countries, the annual incidence rate is approximately 200 per 100,000 population, with rates being higher (up to 400 per 100,000) in urban areas. Mild head injuries are more common than severe, accounting for 50%–90% of all cases. Traumatic head injury accounts for a quarter of all injury-related deaths and approximately 20 of 100,000 people will die of head injury each year. Head injury occurs most frequently between the ages of 15 and 30 and is twice as common in males than in females. Head injury is most common among individuals in the lowest income strata of the society and are most often caused by motor vehicle accidents, assaults, and occupation-related injuries. Residual deficits occur in 10% of individuals with mild TBI, 67% with moderate TBI, and 100% of those with severe injuries.[73] Traumatic brain injury occurs more often among individuals with alcoholism or substance abuse disorders.[74]

Table 26.7 provides an overview of CNS syndromes associated with head injury and TBI. This overview includes head injury without brain injury, whiplash injuries, mild TBI or concussion syndromes, moderate TBI (with brief loss of consciousness or short post-traumatic amnesia), severe TBI (with loss of consciousness or post-traumatic amnesia lasting more than

TABLE 26.7. *Classification of Head Injury and Traumatic Brain Injury*

Head injury without brain injury
Whiplash
Mild TBI/concussion; PTA <24 hours
Moderate TBI
LOC >30 minutes but less than 24 hours
Severe TBI
LOC or PTA >24 hours
TBI with complications
Fractures
Skull
Facial bones
Cerebral injury
Contusion
Laceration
Diffuse axonal injury
Hemorrhage
Intracerebral
Subarachnoid
Subdural
Epidural
Herniation
Stroke
Neck injury/fracture
TBI, secondary consequences
Post-traumatic hydrocephalus
Post-traumatic seizures
Neuropsychiatric consequences
Post-concussion syndrome
Post-traumatic pain syndromes (including headache)
Post-traumatic stress disorders
Depression
Psychosis
Anxiety
Mania/euphoria
Frontal lobe syndrome
Post-traumatic amnesia
Dementia (post-traumatic encephalopathy)
Persistent vegetative state

LOC, loss of consciousness; PTA, post-traumatic amnesia; TBI, traumatic brain injury.

24 hours), TBI with complications, secondary neurologic consequences of TBI, and neuropsychiatric consequences of TBI.

Whiplash refers to the sprain or strain of the cervical region caused by sudden flexion–extension movements, most typically as a result of rear-end collisions.[75] Complaints of neck pain are common following this injury and are typically managed with nonsteroidal anti-inflammatory drugs. Most patients recover by 6 months, but 15%–25% complain chronically of forgetfulness, distractibility, poor concentration, and mental fatigue. It is currently uncertain if whiplash without loss of consciousness can produce sufficient disruption of brain structures to produce TBI. Patients may show deficits on complex attentional and executive tasks but they also score higher on depression and anxiety scales, have more pain, and receive more medication, and it is unclear if the neuropsychological abnormalities are attributable to the secondary consequences or are a product of brain dysfunction.[76] Magnetic resonance imaging is normal in the post-whiplash syndrome. Studies of brain metabolism or cerebral blood flow may show abnormalities, but the tests have sufficiently low sensitivity and specificity in this setting as to be unhelpful.[77] Management efforts of this syndrome should be directed at treatment of pain, depression, anxiety, and sleep disorder. There should be optimistic education of patients, families, and employers about a likely good ultimate outcome and vigorous efforts to increase physical and mental activity.[75]

Mild TBI accounts for a majority of cases of brain injury. At the time of head injury, there is either brief loss of consciousness or no loss of consciousness and simply a brief period of dazed consciousness or clouding of memory for events. Any associated post-traumatic amnesia is less than 24 hours in duration and on examination there are no focal neurological signs.[78,79] The underlying neuropathology of mild TBI is diffuse axonal injury disrupting axons and small vessels, particularly in the perisagittal deep white matter. The severity of this injury is on a continuum from mild to severe in patients with correspondingly mild to severe TBI.[78] There may be associated focal cortical contusions. Magnetic resonance imaging frequently shows increased signal intensity on T_2-weighted images that resolve over time.

Immediately after the injury, patients complain of poor concentration, forgetfulness, disturbances of the sleep–wake cycle, headache, dizziness, anxiety, irritability, and depression. These symptoms gradually resolve in most patients and after 1 year 85%–90% of patients have largely recovered. Those over the age of 55 or with more severe TBI recover more slowly and may have more residual complaints. Psychological assessment reveals deficits in attentional and executive function attributable either to disruption of frontal–subcortical circuitry or the effects of depression and anxiety.[80] Approximately 10%–15% of patients develop a chronic post-concussional disorder (Table 26.8). The etiology of this likely represents a combination of the effects of diffuse axonal injury and neuropsychiatric symptoms (which themselves may be a product of brain injury). Treatment of pain, depression, and anxiety should be the focus of psychopharmacologic management. Rehabilitation efforts aimed at counseling, vocational support, and adaptive strategies may aid reemployment.[78] Head injury and mild to moderate TBI are common in a variety of sports including boxing, football, and soccer.[81,82]

Patients may have retrograde amnesia plus anterograde amnesia beginning at the time of the head injury (Chapter 7). Patients with more severe head injury or who are older at the time of their TBI recover post-

TABLE 26.8. *Criteria for Post-concussion Disorder*[154]

A. A history of head trauma that has caused significant cerebral concussion.
B. Evidence from neuropsychological testing or quantified cognitive assessment of difficulty in attention (concentrating, shifting focus of attention, performing simultaneous cognitive tasks) or memory (learning or recalling information).
C. Three (or more) of the following occur shortly after the trauma and last at least 3 months:
   1. Becoming fatigued easily
   2. Disordered sleep
   3. Headache
   4. Vertigo or dizziness
   5. Irritability or aggression on little or no provocation
   6. Anxiety, depression, or affective lability
   7. Changes in personality (e.g., social or sexual inappropriateness)
   8. Apathy or lack of spontaneity
D. The symptoms begin following head trauma or represent a substantial worsening of preexisting symptoms.
E. The disturbance causes significant impairment in social or occupational functioning and represents a significant decline from a previous level of functioning. In school-age children, the impairment may be manifested by a significant worsening in school or academic performance dating from the trauma.
F. The symptoms do not meet criteria for dementia due to head trauma and are not better accounted for by another mental disorder (e.g., amnestic disorder due to head trauma, personality change due to head trauma).

traumatic memory function more slowly than those with more mild head injuries or who are younger.[83] After the recovery of post-traumatic amnesia, cognitive slowing is a major feature of the neuropsychological performance of patients with TBI. This includes both simple and complex reaction times as well as other timed neuropsychological assessments.[84] Performance of difficulties on executive function tasks requiring establishment and maintenance of a set such as the Wisconsin Card Sort Test is also sensitive to the chronic effects of TBI.[85] Patients with severe TBI tend to underestimate the degree of their neuropsychological impairment.[86] Penetrating head injuries (particularly gunshot wounds) have the worst prognosis for functional outcome. Factors contributing to poor outcome include post-traumatic epilepsy, paresis, visual field loss, verbal memory loss, visual memory loss, behavioral difficulties, and violent behavior.[87]

Neuropsychiatric symptoms are a major feature of both the post-concussion syndrome (discussed above) and the consequences of moderate and severe TBI. Depression occurs in approximately 25% of patients in the immediate post-traumatic period and is most common with injury to the left or right frontal lobes.[74,88–90] Anxiety also occurs in approximately 20% of patients following trauma.[74,90] Substance abuse is a common predisposing factor to head injury, is present in approximately 10% of patients with TBI,[74] and may continue after the injury. Late-onset depression within the first year following TBI is common and may be more related to the situational effects of the injury than to the neurobiological effects.[91] A variety of less common behavioral disturbances have been reported following TBI, including post-traumatic stress disorder and psychosis. Personality changes reflecting frontal lobe dysfunction with impulsiveness, disinhibition, and irritability are common.[92] Psychiatric features contribute substantially to long-term compromise of quality of life.[93]

Neuroimaging with MRI of patients with TBI reveals enlarged ventricle-to-brain ratios and atrophy of the corpus callosum particularly anteriorly. Abnormalities on timed tests such as the Digit-Symbol Subtest of the Weschler Adult Intelligence Scale correlate with the ventricular enlargement.[94,95] Functional imaging with measures of cerebral blood flow and cerebral metabolism reveal cortical functional abnormalities in patients with TBI. Reduced activity in frontal, temporal, and parietal areas as well as the thalamus has been visualized. There is often a poor correlation among MRI changes, functional measures, and neuropsychological deficits.[96] Figure 26.2 shows the MRI of a patient with a frontal contusion and a posterior cerebral artery territory infarction. The latter is a common consequence of trauma-related cerebral edema and compression of the posterior cerebral artery against the tentorium.

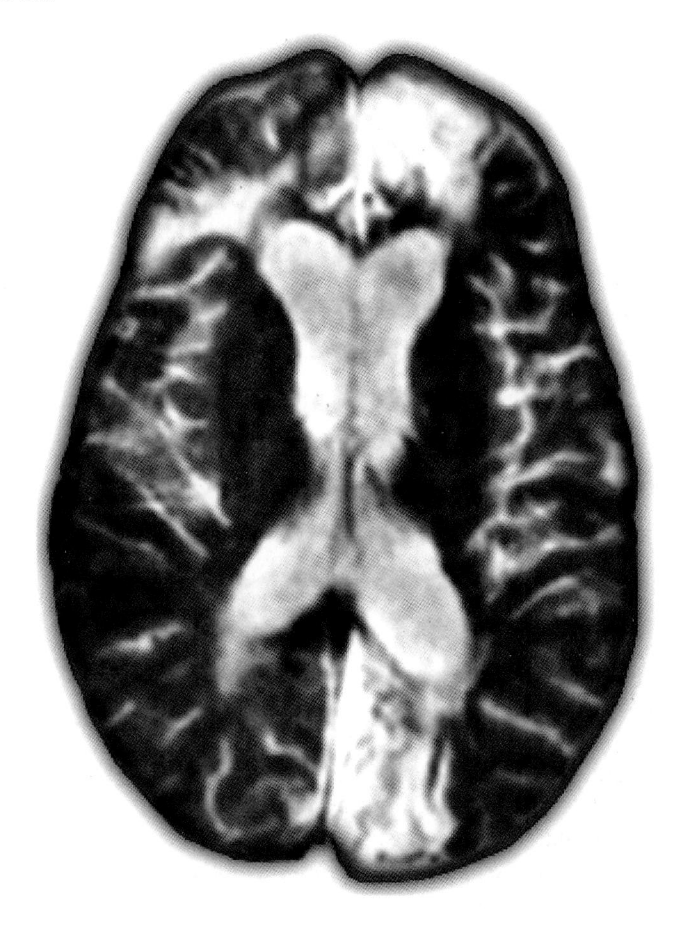

FIGURE 26.2 Magnetic resonance image of a patient with a post-traumatic inferior frontal contusion and an infarction in the territory of the posterior cerebral artery. Compression of the posterior cerebral artery against the tentorial edge is a common consequence of post-traumatic cerebral edema.

The most severely impaired survivors of TBI have *persistent vegetative state*, which is characterized by wakefulness without alertness patients. The patient apparently goes through periods of wakefulness with eyes open and periods of sleep with eyes closed. Eye movements may be nonpurposeful or may briefly track an object or orient toward a sudden light, sound, or movement. Limbs are usually spastic and there are responses to pain with facial grimacing or groaning. Grasp reflexes may be present. No meaningful emotional or cognitive responses are observed.[97] Cerebral ventricles are enlarged on structural imaging and functional imaging reveals a 40%–60% reduction in cortical glucose metabolism. At autopsy there is cortical shrinking, the greatest damage being to layers III and V of the cortex. There is generalized demyelination of the deep white matter throughout the cerebral hemispheres. The prognosis for recovery from persistent vegetative state is poor. Of patients discharged from the hospital in a vegetative state, one-half recover consciousness over the ensuing 3 years, although the probability of recovery diminishes markedly after the first year.[98]

Treatment of patients with TBI includes cognitive and physical rehabilitation, psychopharmacologic management of depression, anxiety, and other behavioral syndromes, and family and personal support.[99] Psychostimulants such as dexedrine may improve performance on attentional and timed tests and some patients evidence improved attention and concentration following treatment with amantadine.[100–102]

CENTRAL NERVOUS SYSTEM INFECTIONS

Brain infections are another source of neuropsychiatric symptoms. Human immunodeficiency virus (HIV) encephalothapy in patients with acquired immunodeficiency syndrome (AIDS) is a major source of global neuropsychiatric morbidity. Prion diseases such as Creutzfeldt-Jakob disease are unusual disorders but instructive with regard to the pathophysiology of neuropsychiatric symptoms. Herpes encephalitis preferentially affects the medial temporal lobe structures and commonly presents with neuropsychiatric symptoms. Other viral encephalitides, bacterial infections, and parasite diseases are potential sources of neuropsychiatric symptoms.

HIV-Associated Cognitive/Motor Complex

Worldwide, approximately 30 million people are infected with the HIV virus. In some countries (Botswana, Zimbabwe) the infection rate approaches 25% of the population. In the United States there are 750,000–900,000 individuals with AIDS and 25,000 deaths annually from the disease.[103] Approximately 7% of individuals with AIDS have HIV encephalopathy, with the greatest proportion being those under age 15 (13%) and over age 75 (19%).[104] Among individuals with AIDS, 44% are homosexual men, 26% are intravenous drug users, and 12% are persons affected through heterosexual contact with HIV-positive individuals.[103]

AIDS is an infection of the immune system that results from the ability of the HIV-1 retrovirus to disarm the host immune system by attacking lymphocytes expressing the CD4 molecule.[105] The virus attaches to the cellular membrane and the viral nucleoprotein complex is internalized. A double-stranded DNA copy of the viral RNA migrates to the nucleus and enters into the host chromosomes. Transcribed viral MRA is translated into viral proteins and complete viruses are assembled. The integrated viral protein is permanently incorporated into the host cell genome, where it may remain latent or engage in vigorous production of the viruses.[106]

Definitions for the neurologic manifestations of HIV infection are shown in Table 26.9 including criteria for the HIV-1-associated dementia complex and the HIV-1 associated minor cognitive–motor disorder.[107] The HIV-1-associated dementia complex is sufficient for a diagnosis of AIDS in an individual with laboratory evidence for systemic HIV-1 infection. The presence of an HIV-1-associated minor cognitive–motor disorder is not sufficient for a diagnosis of AIDS, although it may be present in persons with AIDS. The critical difference between the two syndromes is the degree of impairment in activities of daily living. Patients with HIV-1-associated dementia complex have obvious impairments, while those with HIV-1-associated minor cognitive–motor disorder can accomplish all but the most demanding daily functions. The HIV-1-associated dementia complex is characterized by mental slowness, impaired complex attention, poor executive function, and memory deficits. Language abnormalities are uncommon.[107–110] Behavioral disturbances are frequent in patients with HIV encephalopathy. Apathy is the most commonly reported behavior and mania and psychosis also have been described as complications of HIV encephalopathy.[109,111] Depression is highly prevalent among patients with HIV encephalopathy but is also common among individuals at risk for this condition.

Neuroimaging studies reveal diminished basal ganglia size and diminished white matter volume in patients with HIV encephalopathy.[112,113] The decreased size of the caudate nuclei is associated with poor performance on neuropsychological tests requiring complex motor and sequencing skills.[114] At autopsy, patients with the HIV-associated dementia complex have evidence of HIV encephalitis. Multinucleated giant cells are the most characteristic feature of the process. Abundant macrophages and microglial nodules are often seen. Leukoencephalopathy is also a frequent accompaniment of the AIDS dementia complex. There is diffuse myelin pallor, astrocytosis, activated macrophages, and some multinucleated giant cells in the cerebral white matter. Activated microglial cells are evident and there is mild neuronal loss in the cerebral cortex.[115,116]

The AIDS dementia complex must be distinguished from a variety of other opportunistic infections that may occur in AIDS patients. Table 26.10 lists the opportunistic infections and secondary neoplasms that occur with increased prevalence among patients with AIDS.[116,117]

TABLE 26.9. *Criteria for Clinical Diagnosis of Severe and Mild HIV-associated Cognitive–Motor Complex*

HIV-1-associated cognitive–motor complex

All of the following diagnoses require laboratory evidence for systemic HIV-1 infection (ELISA test confirmed by Western blot, polymerase chain reaction, or culture)

I. HIV-1-associated dementia complex*

Probable (must have each of the following):

1. Acquired abnormality in at least two of the following cognitive abilities (present for at least 1 month): attention/concentration, speed of processing of information, abstraction/reasoning, visuospatial skills, memory/learning, and speech/language. The decline should be verified by reliable history and mental status examination. In all cases, when possible, history should be obtained from an informant, and examination should be supplemented by neuropsychological testing.
2. At least one of the following:
   a. Acquired abnormality in motor function or performance verified by clinical examination (e.g., slowed rapid movements, abnormal gait, limb incoordination, hyperreflexia, hypertonia, or weakness), neuropsychological tests (e.g., fine motor speed, manual dexterity, perceptual motor skills, or both).
   b. Decline in motivation or emotional control or change in social behavior. This may be characterized by any of the following: change in personality with apathy, inertia, irritability, emotional lability, or new onset of impaired judgement characterized by socially inappropriate behavior or disinhibition.
3. Absence of clouding of consciousness during a period long enough to establish the presence of criterion 1.
4. No evidence of another etiology, including active CNS opportunistic infection or malignancy, psychiatric disorders (e.g., depressive disorder), active alcohol or substance use, or acute or chronic substance withdrawal; must be sought from history, physical and psychiatric examination, and appropriate laboratory and radiologic investigation (e.g., lumbar puncture, neuroimaging). If another potential etiology (e.g., major depression) is present, it is not the cause of the above cognitive, motor, or behavioral symptoms and signs.

Possible (must have one of the following):

1. Other potential etiology present (must have each of the following):
   a. As above (see Probable) criteria 1, 2 and 3
   b. Other potential etiology is present but the cause of criterion 1 above is uncertain.
2. Incomplete clinical evaluation (must have each of the following):
   a. As above (see Probable) criteria 1, 2, and 3
   b. Etiology cannot be determined (appropriate laboratory or radiologic investigations not performed).

II. HIV-1-associated minor cognitive/motor disorder

Probable (must have each of the following):

1. Cognitive/motor/behavioral abnormalities (must have each of the following):
   a. At least two of the following acquired cognitive, motor, or behavioral symptoms (present for at least 1 month) verified by reliable history (when possible, from an informant):
   1. Impaired attention or concentration
   2. Mental slowing
   3. Impaired memory
   4. Slowed movements
   5. Incoordination
   6. Personality change, or irritability or emotional lability
   b. Acquired cognitive/motor abnormality verified by clinical neurologic examination or neuropsychological testing (e.g., fine motor speed, manual dexterity, perceptual motor skills, attention/concentration, speed of processing of information, abstraction/reasoning, visuospatial skills, memory/learning, or speech/language).
2. Disturbance from cognitive/motor/behavioral abnormalities (see criterion 1) causes mild impairment of work or activities of daily living (objectively verifiable or by report of a key informant).
3. Does not meet criteria for HIV-1-associated dementia complex.
4. No evidence of another etiology, including active CNS opportunistic infection or malignancy, or severe systemic illness determined by appropriate history, physical examination, and laboratory and radiologic investigation (e.g., lumbar puncture, neuroimaging). The above features should not be attributable solely to the effects of active alcohol or substance use, acute or chronic substance withdrawal, adjustment disorder, or other psychiatric disorders.

(*continued*)

TABLE 26.9. *Criteria for Clinical Diagnosis of Severe and Mild HIV-associated Cognitive–Motor Complex (Continued)*

Possible (must have one of the following):
1. Other potential etiology present (must have each of the following):
   a. As above (see Probable) criteria 1, 2, and 3
   b. Other potential etiology is present and the cause of the cognitive/motor/behavioral abnormalities is uncertain.
2. Incomplete clinical evaluation (must have each of the following):
   a. As above (see Probable) criteria 1, 2, and 3
   b. Etiology cannot be determined (appropriate laboratory or radiologic investigations not performed).

*For research purposes, HIV-1-associated dementia complex can be coded to describe the major features: HIV-1-associated dementia complex requires criteria 1, 2a, 3 and 4; HIV-1-associated dementia complex (motor) requires criteria 1, 2a, 3, and 4; HIV-1-associated dementia complex (behavior) requires criteria 1, 2b, 3, and 4.

Treatment of HIV has the goal of controlling or reducing viral replication. Nucleoside reverse transcriptase inhibitors, non-nucleoside reverse transcriptase inhibitors, and protease inhibitors are used in a variety of combinations.[118] Neuropsychiatric manifestations are treated with the usual psychotropic agents (Chapter 4). Ziduvoine (AZT) used in the treatment of AIDS appears to have a beneficial effect on neuropsychological performance in patients with the HIV dementia complex.[119] Some patients improve with selegiline[120] or psychostimulants.[121]

Creutzfeldt-Jakob Disease and Other Prion Disorders

Prions are proteinacious infectious particles that cause human and animal diseases and differ from viruses and other microbial agents.[122] The diseases currently ascribed to prions include Creutzfeldt-Jakob disease, Kuru, Gerstmann-Straussler-Scheinker disease, fatal familial insomnia, and one form of thalamic dementia (Table 26.11).

Creutzfeldt-Jakob disease is unique among neurological disorders as it is both an inherited neurodegenerative disorder and a transmissible disease. Worldwide incidence of Creutzfeldt-Jakob disease is approximately 1 per 10^6 population annually, 85%–90% of cases are sporadic and 10%–15% are inherited. The disease affects men and women equally and the frequent age of onset is 60 years with a range of 40–90 years for all but the iatrogenic cases, in which onset has been between the ages of 20 and 40. In most cases the disease is rapidly progressive, leading to death in 4–12 months; however, chronic cases lasting 2–5 years are well de-

TABLE 26.10. *Opportunistic Infections and Secondary Central Nervous System Complications of AIDS*

Meningitis	*Encephalitis*	*Focal Brain Diseases*
Cryptococcal meningitis	HIV encephalitis	Cerebral toxoplasmosis
Aseptic meningitis (HIV-1)	Toxoplasmosis ("encephalitic" form)	Neurosyphilis
Tuberculous meningitis (*Mycobacterium tuberculosis*)	Cytomegalovirus encephalitis	Primary CNS lymphoma
Syphilitic meningitis	Herpes encephalitis	Progressive multifocal leukoencephalopathy
Listeria	Varicella-zoster virus encephalitis	Tuberculous brain abscess (*Mycobacterium tuberculosis*)
Coccidioidomycosis		Cryptococcoma
Histoplasmosis		Vascular disorders
Blastomycosis		
Nocardia		
Mycobacterium		
Lymphomatous meningitis (metastatic)		

TABLE 26.11. *Prion Diseases*

Creutzfeldt-Jakob disease
Familial
Sporadic
Iatrogenic
New variant
Kuru
Gerstmann-Straussler-Scheinker disease
Fatal familial insomnia
Thalamic dementia

scribed.[123] The principal event leading to the lethality of prions is a conformational exchange that occurs during the conversion of a native protein (PRP^C) to prion protein SC (PRP^{Sc}).[122] Susceptibility to sporadic Creutzfeldt-Jakob disease is influenced by individual genotype; homozygoticity at codon 129 of the prion protein gene is overrepresented among those with sporadic disease, new variant disease (described below), and disease transmitted by infected dura mater.[124] Mutations in families with inherited Creutzfeldt-Jakob disease occur on chromosome 20. Iatrogenic Creutzfeldt-Jakob disease has been transmitted by improperly sterilized depth electrodes, transplanted corneas, human growth hormone and gonadotrophin derived from cadaveric pituitaries, and dura mater graphs.[125] Kuru, a cerebellar form of prion disease, was transmitted among individuals in the Fore tribe in New Guinea by ritualistic cannibalism. Iatrogenic Creutzfeldt-Jakob disease offers a modern analogy with the disease transmitted in the course of medical technocannabilism through use of human products to treat other humans.[125] New variant Creutzfeldt-Jakob disease appears to result from the transmission of spongiform bovine encephalopathy ("mad cow disease") to humans.[126]

The clinical features of Creutzfeldt-Jakob disease include dementia, myoclonus, pyramidal track signs, aphasia, cerebellar signs, primitive reflexes, and extrapyramidal dysfunction. Seizures occur in a minority of cases. Cortical blindness characterizes a posterior variant of the disease and many patients enter a state of akinetic mutism or persistent vegetative state prior to death.[127] The rapid and relentless progression of the illness is the most characteristic feature of the disease. A prodromal phase preceding the onset of overt neurological symptoms and signs may include depression, sleep disturbance, and headache. New variant Creutzfeldt-Jakob disease tends to have an earlier age of onset (mean 29 years) and to progress somewhat more slowly (average 14 months from onset to death). In addition, most patients with the new variant form of the disease manifest early and persistent psychiatric symptoms. Depression, anxiety, and withdrawal are most common, but some patients evidence frank psychosis.[128] Table 26.12 presents diagnostic criteria for new variant Creutzfeldt-Jakob disease.[128]

Neuroimaging, electroencephalography (EEG) and CSF analysis may be useful in the diagnosis of Creutzfeldt-Jakob disease. Magnetic resonance imaging frequently reveals bilateral symmetric high-signal intensity on T_2-weighted images[129] and diffusion-weighted MRI may demonstrate increased signal intensity in the basal ganglia as well as in widespread regions of the cerebral cortex.[130] Studies with PET reveal diminished glucose metabolism distributed heterogeneously and

TABLE 26.12. *Diagnostic Criteria for New Variant Creutzfeldt-Jakob Disease*[128]

I. Key features
A. Progressive neuropsychiatric disorder
B. Duration of illness >6 months
C. Routine investigations do not suggest an alternative diagnosis
D. No history of potential iatrogenic exposure
II. Supportive features
A. Early psychiatric symptoms*
B. Persistent painful sensory symptoms†
C. Ataxia
D. Myoclonus or chorea or dystonia
E. Dementia
III. Laboratory features
A. EEG does not show the typical appearance of sporadic CJD‡ (or no EEG performed)
B. Bilateral pulvinar high signal on MRI scan
Definite: IA (progressive neuropsychiatric disorder) and neuropathological confirmation of nvCJD**
Probable: I and 4/5 of II and IIIA and IIIB
Possible: I and 4/5 of II and IIIA

*These include depression, anxiety, apathy, withdrawal, delusions.
†These include frank pain and unpleasant dysesthesia.
‡Generalized triphasic periodic complexes are at approximately 1/second.
**Spongiform change and extensive prion protein deposition with florid plaques throughout the cerebrum and cerebellum.
CJD, Creudtzfeldt-Jakob disease; EEG, electroencephalogram; MRI, magnetic resonance imaging; nvCJD, new variant Creutzfeldt-Jakob disease.

asymmetrically throughout the brain.[131,132] Periodic sharp- and slow-wave complexes on EEG are characteristic of Creutzfeldt-Jakob disease, occurring in 65%–85% of cases.[133,134] The classical periodic sharp- and slow-wave complexes are not a feature of new variant disease.[128] Testing of CSF for evidence of prion protein may be diagnostically useful. The 14-3-3 protein is strongly suggestive of the presence of prion protein and occurs in over 90% of cases.[134–136] The 14-3-3 protein occurs rarely in other dementias and is not completely specific for Creutzfeldt-Jakob disease. Current diagnostic criteria for sporadic Creutzfeldt-Jakob disease incorporate these laboratory features (Table 26.13).

Brain examination in Creutzfeldt-Jakob disease reveals spongiform degeneration of neurons and their processes, loss of neurons, intense reactive astrocytic gliosis, and amyloid plaque formation. The vacuolization characteristic of spongiform degeneration consists of vacuoles located in the neurophil between nerve cell bodies. In some cases vacuolization is also evident within neurons. The spongiform degeneration can be found throughout cortical and subcortical structures and the cerebellum, although the globus pallidus portions of the hippocampus, brainstem, and spinal cord are typically spared or minimally affected.[123] No treatment is currently available for this rapidly progressive, tragic, and inevitably fatal illness.

Gerstmann-Straussler-Scheinker disease is an autosomal dominant disorder presenting with ataxia and progressing to a mixture of cognitive and motor disturbances and exhibiting multicentric amyloid plaques at autopsy.[123] The families carry a mutation in the prion protein gene. *Fatal familial insomnia* consists of the subacute onset of a clinical syndrome characterized by progressive insomnia, complex hallucinations, stupor, and eventually coma. Autonomic disturbances and motor abnormalities including myoclonus and pyramidal signs are common. The onset is between the ages of 35 and 60. This order is a dominantly inherited prion protein mutation and the pathological changes are typically limited to the thalamus.[123] Insomnia has been absent in some cases with related prion-induced thalamic degeneration although prominent sleep disturbances, enacted dreams, and hallucinations were common.[137]

TABLE 26.13. *Diagnostic Criteria for Sporadic Creutzfeldt-Jakob Disease*[134]

Definite CJD
Neuropathologically confirmed *and/or*
Abnormal prion protein isoform immunochemically confirmed by immunocytochemistry or Western blot
Scrapie-associated fibrils
Probable CJD
Progressive dementia with at least two of four clinical features:
1. Myoclonus
2. Visual or cerebellar signs
3. Pyramidal or extrapyramidal signs
4. Akinetic mutism
Periodic sharp- and slow-wave complexes in EEG
14-3-3 proteins in CSF
Duration of <2 years
Possible CJD
Clinical features as above
No periodic sharp- and slow-wave complexes in EEG
No 14-3-3 detection in CSF
Duration of >2 years

CJD, Creutzfeldt-Jakob disease; CSF, cerebrospinal fluid; EEG, electroencephalogram.

Herpes Simplex Encephalitis

Viral encephalitis may be due to herpes simplex, varicella-zoster, cytomegalovirus, Epstein-Barr virus, mumps, measles, rabies, HIV, arboviruses, and JC virus, among others.[138] Among these, herpes simplex encephalitis is of the greatest neuropsychiatric interest. Psychiatric symptoms are commonly among the presenting features, including personality alterations, memory loss, hallucinations, delusions, irritability, agitation, and intermittent drowsiness.[138,139] As the disease spreads, motor and sensory abnormalities and aphasia may become evident. In a few cases, the anterior opercular syndrome has been the presenting manifestation, characterized by bilateral facial weakness, dysphagia, drooling, absent movement of the palate, absent gag reflex and weakness of the tongue, and anarthria.[140] Magnetic resonance imaging commonly shows increased signal on the T_2-weighted image in the inferior frontal, medial temporal, and anterior cingulate limbic regions (Figure 26.3).[141,142] An EEG may be helpful in revealing focal spike- and slow-wave activity over the temporal regions. Cerebrospinal fluid analysis shows a lymphocytic pleocytosis with red cells in over half of the cases. Analysis of CSF for evidence of herpes virus DNA confirms the diagnosis and may be particularly useful in patients with atypical presentations.[143] Treatment with acyclovir has reduced the mortality from 70% to 30%, but long-term neuropsychological sequelae are common.[144,145]

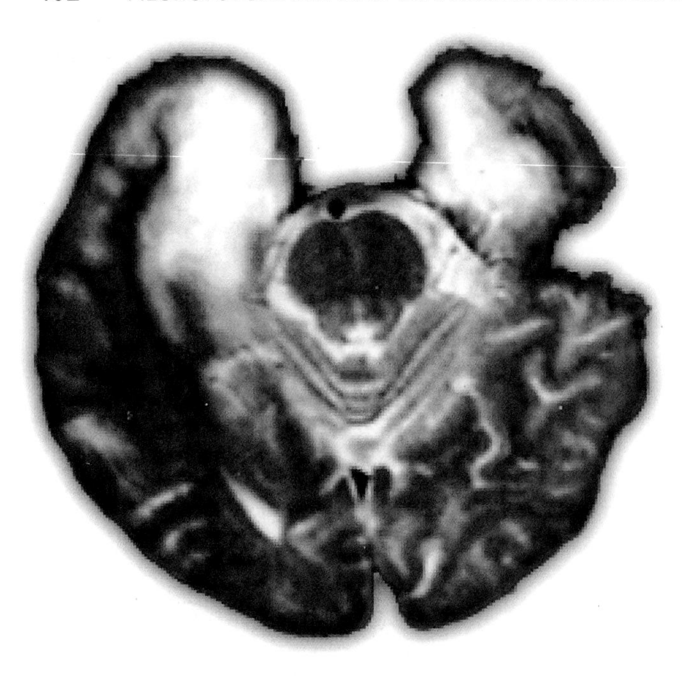

FIGURE 26.3 Magnetic resonance imaging shows high-signal regions bilaterally in the medial temporal regions in a patient with herpes simplex encephalitis.

Postmortem studies of patients succumbing to herpes simplex encephalitis reveal extensive necrosis of the gray and white matter in the frontal and temporal regions. Histologically, there is abundant white blood cell infiltration, microglial activation, and intranuclear inclusions in neurons in the area of necrosis. The virus is readily demonstrated by immunostaining in autopsy tissues.[115]

Other Brain Infections

A few other brain infections are of particular neuropsychiatric interest. Von Economo's encephalitis appeared as an epidemic in 1915 and continued for the ensuing 10 to 15 years. No specific virus was identified, although the pathological hallmarks of a viral infection were evident. The acute encephalitis was protean in its manifestations but frequently included severe lethargy and disturbances of the sleep–wake cycle (hence the name *encephalitis lethargica*). Acute mortality was high (30%); 80% of survivors developed post-encephalitic parkinsonism within 10 years of their recovery from the acute episode. Both the acute disorder and the post-encephalitic syndrome included prominent neuropsychiatric manifestations of depression, euphoria, catatonia, abnormalities of sexual behavior, and conduct disorders in children.[138,146]

Syphilis has produced a variety of CNS syndromes including tabes dorsalis, meningovascular syphilis and general paresis of the insane (GPI). General paresis results from a spirochetal invasion of the cerebral cortex predominantly affecting the frontal lobes and manifested by a mania-like syndrome. Syphilis was a common cause of mental illness and a common reason for psychiatric hospitalization until successful treatment with penicillin was discovered. General paresis is one of the few major mental illnesses with an established cure or prevention.

NEOPLASTIC AND PARANEOPLASTIC SYNDROMES

Brain tumors can be of intracerebral origin and produce local disruption or may be of extracerebral origin and produce symptoms through compression. Syndromes similar to those of cerebral vascular lesions (Table 26.1) can be observed in patients with brain tumors; however, the onset is typically less abrupt, there is usually coincident confusion or impairment of consciousness, and syndromes may involve symptoms implicating more than one vascular territory. Headache may be prominent. In general, brain tumors begin at an earlier age than vascular disorders. Primary intracerebral tumors occur at a rate of 9 per 100,000 per year, benign intercranial extracerebral tumors including meningiomas occur at a rate of 5.5 per 100,000 per year, and secondary intracerebral tumors (metastases) occur at a rate of 13.6 per 100,000 per year.[147] The most common primary intercranial intracerebral tumors are malignant gliomas (accounting for approximately 25% of all primary intercranial tumors), benign or low-grade astrocytomas (accounting for approximately 15% of all primary intracerebral tumors), oligodendrogliomas, ependymomas, and medulloblastomas. Meningiomas account for approximately 15% of all primary intracerebral tumors, pituitary adenomas for 5%, and craniopharyngiomas for 3%.[148] Of patients with intracerebral metastases, 40% originate in the lung, 20% are from breast tumors, 10% are melanomas, 7% arise from the genitourinary tract, 7% from the gastrointestinal tract, and 5% are of gynecologic origin.[149] In addition to the direct effects of cancer on the nervous system, there are many mechanisms of indirect effects including vascular disorders, hydrocephalus, side effects of therapy (chemotherapy, radiation therapy, surgery), and paraneoplastic syndromes (Table 26.14).

TABLE 26.14. *Cancer and Cancer-Related Disorders of the CNS*

Direct Involvement by CNS Tumors	*Indirect Effects*
Metastases (by source)	Vascular disorders
Lung	Hydrocephalus
Breast	Side effects of therapy
Melanoma	Chemotherapy
Gastrointestinal	Radiation therapy
Renal	Surgery
Ovarian	Paraneoplastic syndromes
Glioma	
Glioblastoma/malignant astrocytoma	
Astrocytoma (benign/low-grade)	
Oligodendroglioma	
Ependymoma	
Medulloblastoma	
Meningioma	
Pituitary adenomas	
Craniopharyngioma	
Lymphoma	
Others	

Neuropsychiatric assessment indicates that patients with tumors of the ventral frontal cortex or temporal parietal cortex report a significant increase in anxiety/depression, irritability/anger, and fatigue. Those with right posterior lesions report higher levels of fatigue and irritability/anger and those with left anterior lesions have higher levels of anxiety/depression.[150]

Paraneoplastic limbic encephalitis, typically associated with small cell tumors of the lung, is a paraneoplastic syndrome manifested by disturbances of memory and neuropsychiatric symptoms including delusions, hallucinations, and agitation.[151,152] Seizures may occur. Examination of the CSF reveals an inflammatory pleocytosis. Structural imaging is usually normal, although abnormalities in one or both medial temporal lobes may be evident. At autopsy there is extensive loss of neurons with reactive glyosis, perivascular cuffing, and microglial proliferation.[149] The detection of anti-Hu antibodies in the serum supports the diagnosis.[153] No treatment effective for paraneoplastic limbic encephalitis has been identified.

REFERENCES

1. Warlow C. Disorders of the cerebral circulation. In: Walton J, ed. Brain's Diseases of the Nervous System, 10th ed. Oxford: Oxford University Press, 1993:197–268.
2. Toole JF, Murros K, Veltkamp R. Cerebrovascular Disorders. Philadelphia: Lippincott Williams & Wilkins, 1999.
3. Sacco RL, Wolf PA, Gorelick PB. Risk factors and their management for stroke prevention: outlook for 1999 and beyond. Neurology 1999;53:S15–S24.
4. Babikian VL, Kase CS, Wolf PA. Cerebrovascular disease in the elderly. In: Albert ML, Knoefel JE, eds. Clinical Neurology of Aging, 2nd ed. New York: Oxford University Press, 1994:548–568.
5. Starkstein SE, Robinson RG. Stroke. In: Coffey CE, Cummings JL, eds. Textbook of Geriatric Neuropsychiatry, 2nd ed. Washington, DC: American Psychiatric Press, 2000:601–620.
6. Albers GW, Tijssen JGP. Antiplatelet therapy: new foundations for optimal treatment decisions. Neurology 1999;53:S25–S31.
7. Trapp BD, Peterson J, et al. Axonal transection in the lesions of multiple sclerosis. N Engl J Med 1998;338:278–285.
8. Rudick R. Multiple sclerosis and related conditions. In: Goldman L, Bennett J, eds. Textbook of Medicine. Philadelphia: W.B. Suanders, 2000:2141–2149.
9. Paty DW, Ebers GC. Clinical features. In: Paty DW, Ebers GC, eds. Multiple Sclerosis. Philadelphia: F.A. Davis, 1998:135–191.
10. Swanson JW. Multiple sclerosis: update in diagnosis and review of prognostic factors. Mayo Clin Proc 1989;64:577–586.
11. Goodkin DE, Rudick RA, Ross JS. The use of brain magnetic resonance imaging in multiple sclerosis. Arch Neurol 1994;51: 505–516.
12. Truyen L, van Waesberghe JHTM, et al. Accumulation of hypointense lesions ("black holes") on T1 spin-echo MRI correlates with disease progression in multiple sclerosis. Neurology 1996;47:1469–1476.
13. van Walderveen MAA, Barkhof F, et al. Correlating MRI and clinical disease activity in multiple sclerosis: relevance of hypointense lesions on short-TR/short-TE (T1-weighted) spin-echo images. Neurology 1995;45:1684–1690.
14. Ebers GC. Immunology. In: Paty DW, Ebers GC, eds. Multiple Sclerosis. Philadelphia: F.A. Davis, 1998:403–426.
15. Swirsky-Sacchetti T, Mitchell DR, et al. Neuropsychological and structural brain lesions in multiple sclerosis: a regional analysis. Neurology 1992;42:1291–1295.
16. Paul RH, Blanco CR, et al. Autobiographical memory in multiple sclerosis. J Int Neuropsychol Soc 1997;3:246–251.
17. Beatty WW, Monson N. Problem solving by patients with multiple sclerosis: comparison of performance on the Wisconsin and California Card Sorting Tests. J Int Neuropsychol Soc 1996;2:134–140.
18. D'Esposito M, Onishi K, et al. Working memory impairments in multiple sclerosis: evidence from a dual-task paradigm. Neuropsychology 1996;10:51–56.
19. Huber SJ, Bornstein RA, et al. Magnetic resonance imaging correlates of executive function impairments in multiple sclerosis. Neuropsychiatry Neuropsychol Behav Neurol 1992;5:33–36.
20. Ryan L, Clark CM, et al. Pattern of cognitive impairment in relapsing-remitting multiple sclerosis and their relationship to neuropathology on magnetic resonance imaging. Neuropsychology 1996;10:176–193.
21. Arnett PA, Rao SM, et al. Conduction aphasia in multiple scle-

rosis: a case report with MRI findings. Neurology 1996;47: 576–578.
22. Huber SJ, Bornstein RA, et al. Magnetic resonance imaging correlates of neuropsychological impairment in multiple sclerosis. J Neuropsychiatry Clin Neurosci 1992;4:152–158.
23. Diaz-Olavarrieta C, Cummings JL, et al. Neuropyschiatric manifestations of multiple sclerosis. J Neuropsychiatry Clin Neurosci 1999;11:51–57.
24. Feinstein A. The Clinical Neuropsychiatry of Multiple Sclerosis. Cambridge, UK: Cambridge Univeristy Press, 1999.
25. Joffe RT, Lippert GP, et al. Mood disorder and multiple sclerosis. Arch Neurol 1987;44:376–378.
26. Huber SJ, Rammohan KW, et al. Depressive symptoms are not influenced by severity of multiple sclerosis. Neuropsychiatry Neuropsychol Behav Neurol 1993;6:177–180.
27. Rabins PV, Brooks BR, et al. Structural brain correlates of emotional disorder in multiple sclerosis. Brain 1986;109:585–597.
28. Sadovnick AD, Remick RA, et al. Depression and multiple sclerosis. Neurology 1996;46:628–632.
29. Fassbender K, Schmidt R, et al. Mood disorders and dysfunction of the hypothalamic–pituitary–adrenal axis in multiple sclerosis: association with cerebral inflammation. Arch Neurol 1998;55:66–72.
30. Honer WG, Hurwitz T, et al. Temporal lobe involvement in multiple sclerosis patients with psychiatric disorders. Arch Neurol 1987;44:187–190.
31. Pujol J, Bello J, et al. Lesions in the left arcuate fasciculus region and depressive symptoms in multiple sclerosis. Neurology 1997;49:1105–1110.
32. Reischies FM, Baum K, et al. Psychopathological symptoms and magnetic resonance imaging findings in multiple sclerosis. Biol Psychiatry 1993;33:676–678.
33. Mohr DC, Goodkin DE, et al. Treatment of depression improves adherence to interferon β-1b therapy for multiple sclerosis. Arch Neurol 1997;54:531–533.
34. Neilley LK, Goodin DS, et al. Side effect profile of interferon β-1b in MS: results of an open label trial. Neurology 1996;46:552–554.
35. Vickrey BG, Hays RD, et al. A health-related quality of life measure for multiple sclerosis. Qual Life Res 1995;4:187–206.
36. Casanova MF, Kruesi M, Mannheim G. Multiple sclerosis and bipolar disorders: a case report with autopsy findings. J Neuropsychiatry Clin Neurosci 1996;8:206–208.
37. Hutchinson M, Stack J, Buckley P. Bipolar affective disorder prior to the onset of multiple sclerosis. Acta Neurol Scand 1993;88:388–393.
38. Schiffer RB, Wineman NM, Weitkamp LR. Association between bipolar affective disorder and multiple sclerosis. Am J Psychiatry 1986;143:94–95.
39. Felgenhauer K. Psychiatric disorders in the encephalitic form of multiple sclerosis. Neurology 1990;237:11–18.
40. Ron MA, Logsdail SJ. Psychiatric morbidity in multiple sclerosis: a clinical and MRI study. Psychol Med 1989;19:887–895.
41. Feinstein A, du Boulay G, Ron MA. Psychotic illness in multiple sclerosis. Br J Psychiatry 1992;161:680–685.
42. Andreatini R, Sartori VA, et al. Panic attacks in a multiple sclerosis patient. Biol Psychiatry 1994;35:133–134.
43. Miguel EC, Stein MC, et al. Obsessive-compulsive disorder in patients with multiple sclerosis. J Neuropsychiatry Clin Neurosci 1995;7:507–510.
44. Ortego N, Miller BL, et al. Altered sexual behavior with multiple sclerosis. Neuropsychiatry Neuropsychol Behav Neurol 1993;6:260–264.
45. Krupp LB, Alvarez LS, et al. Fatigue in multiple sclerosis. Arch Neurol 1988;45:435–437.
46. Djaldetti R, Ziv I, et al. Fatigue in multiple sclerosis compared with chronic fatigue syndrome: a quantitative assessment. Neurology 1996;46:632–635.
47. Krupp LB, LaRocca NG, et al. The fatigue severity scale. Arch Neurol 1989;46:1121–1123.
48. Sandroni P, Walker C, Starr A. 'Fatigue' in patients with multiple sclerosis. Arch Neurol 1992;49:517–524.
49. Roelcke U, Kappos L, et al. Reduced glucose metabolism in the frontal cortex and basal ganglia of multiple sclerosis patients with fatigue: a ^{18}F-fluorodeoxyglucose positron emission tomography study. Neurology 1997;48:1566–1571.
50. Moore GRW. Neuropathology and pathophysiology of the multiple sclerosis lesion. In: Paty DW, Ebers GC, eds. Multiple Sclerosis. Philadelphia: F.A. Davis, 1998:257–327.
51. Paty DW, Hashimoto SA, Ebers GC. Management of multiple sclerosis and interpretation of clinical trials. In: Paty DW, Ebers GC, eds. Multiple Sclerosis. Philadelphia: F.A. Davis, 1998: 427–545.
52. Jacobs LD, Cookfair DL, et al. Intramuscular interferon β-1a for disease progression in relapsing multiple sclerosis. Ann Neurol 1996;39:285–294.
53. Johnson KP, Brooks BR, et al. Copolymer 1 reduces relapse rate and improves disability in relapsing-remitting multiple sclerosis: results of a phase III multicenter, double-blind pacebo-controlled trial: the Copolymer 1 Multiple Sclerosis Study Group. Neurology 1995;45:1268–1276.
54. Johnson KP, Brooks BR, et al. Extended use of glatiramer acetate (Copaxone) is well tolerated and maintains its clinical effect on multiple sclerosis relapse rate and degree of disability: Copolymer 1 Mutiple Sclerosis Study Group. Neurology 1998;50:701–708.
55. IFNB Multiple Sclerosis Study Group. Interferon β-1b is effective in relapsing-remitting multiple sclerosis. Neurology 1993; 43:655–661.
56. Hunter SF, Weinshenker BG, et al. Rational clinical immunotherapy for multiple sclerosis. Mayo Clin Proc 1997;72: 765–780.
57. Cohen RA, Fisher M. Amantadine treatment of fatigue associated with multiple sclerosis. Arch Neurol 1989;46:676–680.
58. Weinshenker BG, Penman M, et al. A double-blind, randomized, crossover trial of pemoline in fatigue associated with multiple sclerosis. Neurology 1992;42:1468–1471.
59. Barak Y, Ur E, Achiron A. Moclobemide treatment in multiple sclerosis patients with comorbid depression: an open-label safety trial. J Neuropsychiatry Clin Neurosci 1999;11:271–273.
60. Schiffer RB, Wineman NM. Antidepressant pharmacotherapy of depression associated with multiple sclerosis. Am J Psychiatry 1990;147:1493–1497.
61. Scott TF, Allen D, et al. Characterization of major depression symptoms in multiple sclerosis patients. J Neuropsychiatry Clin Neurosci 1996;8:318–323.
62. Minden SL. Psychotherapy for people with multiple sclerosis. J Neuropsychiatry Clin Neurosci 1992;4:198–213.
63. Filley CM. The behavioral neurology of cerebral white matter. Neurology 1998;50:1535–1540.
64. Hyde TM, Ziegler JC, Weinberger DR. Psychiatric disturbances in metachromatic leukodystrophy. Arch Neurol 1992;49:401–406.
65. Kitchin W, Cohen-Cole SA, Mickel SF. Adrenoleukodystrophy: frequency of presentation as a psychiatric disorder. Biol Psychiatry 1987;22:1375–1387.

66. Menza MA, Blake J, Goldberg L. Affective symptoms and adrenoleukodystrophy: a report of two cases. Psychosomatics 1988;29:442–445.
67. Uyama E, Iwagoe H, et al. Presenile-onset cerebral adrenoleukodystrophy presenting as Balint's syndrome and dementia. Neurology 1993;43:1249–151.
68. Weller M, Liedtke W, et al. Very-late-onset adrenoleukodystrophy: possible precipitation of demyelination by cerebral contusion. Neurology 1992;42:367–370.
69. Luzi P, Rafi MA, Wenger DA. Multiple mutations in the *GALC* gene in a patient with adult-onset Krabbe disease. Ann Neurol 1996;40:116–119.
70. Meiner V, Meiner Z, et al. Cerebrotendinous xanthomatosis: molecular diagnosis enables presymptomatic detection of a treatable disease. Neurology 1994;44:288–290.
71. Naidu S, Naidu A. White matter diseases. In: Coffey CE, Brumback RA, eds. Textbook of Pediatric Neuropsychiatry. Washington, DC: American Psychiatric Press, 1998:889–912.
72. Bird TD, Koerker RM, et al. Lipomembranous polycystic osteodysplasia (brain, bone, and fat disease): a genetic cause of presenile dementia. Neurology 1983;33:81–86.
73. Torner J, Schootman M. Epidemiology of closed head injury. In: Rizzo M, Tranel D, eds. Head Injury and Postconcussive Syndrome. New York: Churchill Livingstone, 1996:19–46.
74. Fann JR, Katon WJ, et al. Psychiatric disorders and functional disability in outpatients with traumatic brain injuries. Am J Psychiatry 1995;152:1493–1499.
75. Alexander M. In the pursuit of proof of brain damage after whiplash injury. Neurology 1998;51:336–340.
76. Radanov B, Di Stefano G, et al. Cognitive functioning after common whiplash. Arch Neurol 1993;50:87–91.
77. Bicik I, Radanov B, et al. PET with ^{18}flourodeoxyglucose and hexamethylpropylene amine oxime SPECT in late whiplash syndrome. Neurology 1998;51:345–350.
78. Alexander M. Mild traumatic brain injury: pathophysiology, natural history, and clinical management. Neurology 1995;45: 1253–1260.
79. Rizzo M, Tranel D. Overview of head injury and postconcussive syndrome. In: Rizzo M, Tranel D, eds. Head Injury and Postconcussive Syndrome. New York: Churchill Livingstone, 1996.
80. Bohnen N, Twijnstra A, Jolles J. Persistence of postconcussional symptoms in uncomplicated, mildly head-injured patients: a prospective cohort study. Neuropsychiatry Neuropsychology Behav Neurol 1993;6:193–200.
81. Matser J, Kessels A, et al. Chronic traumatic brain injury in professional soccer players. Neurology 1998;51:791–796.
82. Quality Standards Subcommittee. Practice parameter: the management of concussion in sports (summary statement). Neurology 1997;48:581–585.
83. Katz D, Alexander M. Traumatic brain injury. Arch Neurol 1994;51:661–670.
84. Ferraro R. Cognitive slowing in closed-head injury. Brain Cogn 1996;32:429–440.
85. Gansler D, Covall S, et al. Measures of prefrontal dysfunction after closed head injury. Brain Cogn 1996;30:194–204.
86. Prigatano G, Altman I. Impaired awareness of behavioral limitations after traumatic brain injury. Arch Phys Med Rehabil 1990;71:1058–1064.
87. Schwab K, Grafman J, et al. Residual impairments and work status 15 years after penetrating head injury. Neurology 1993; 43:95–103.
88. Fedoroff J, Starkstein S, et al. Depression in patients with acute traumatic brain injury. Am J Psychiatry 1992;149:918–923.
89. Grafman J, Vance SC, et al. The effects of lateralized frontal lesions on mood regulation. Brain 1986;109:1127–1148.
90. Jorge RE, Robinson RG, et al. Depression and anxiety following traumatic brain injury. J Neuropsychiatry Clin Neurosci 1993a;5:369–374.
91. Jorge RE, Robinson RG, et al. Comparison between acute-and delayed-onset depression following traumatic brain injury. J Neuropsychiatry Clin Neurosci 1993;5:43–49.
92. Rolls E, Hornak J, et al. Emotion-related learning in patients with social and emotional changes associated with frontal lobe damage. J Neurol Neurosurg Psychiatry 1994;57:1518–1524.
93. Klonoff P, Snow W, Costa L. Quality of life in patients 2 to 4 years after closed head inujury. Neurosurgery 1986;19:735–743.
94. Johnson S, Bigler E, et al. White matter atrophy, ventricular dilation, and intellectual functioning following traumatic brain injury. Neuropsychology 1994;8:307–315.
95. Johnson S, Pinkston J, et al. Corpus callosum morphology in normal controls and traumatic brain injury: sex differences mechanisms of injury, and neuropsychological correlates. Neuropsychology 1996;10:408–415.
96. Herscovitch P. Functional brain imaging—basic principles and application to head trauma. In: Matthew R, Tranel D, eds. Head Injury and Postconcussive Syndrome. New York: Churchill Livingstone, 1996:89–118.
97. Jennett B. Clinical and pathological features of vegetative survival. In: Levin H, Benton A, Muizelaar J, Eisenberg H, eds. Catastrophic Brain Injury. New York: Oxford University Press, 1996:3–13.
98. Levin H, Eisenberg H. Vegetative state after head injury: findings from the Traumatic Coma Data Bank. In: Levin H, Benton A, Muizelaar P, Eisenberg H, eds. Catastrophic Brain Injury. New York: Oxford University Press, 1996:35–49.
99. Silver J, Yudofsky S. Psychopharmacology. In: Silver J, Yudofsky S, Hales R, eds. Neuropsychiatry of Traumatic Brain Injury. Washington, DC: American Psychiatric Press, 1994:631–670.
100. Bleiberg J, Garmoe W, et al. Effects of dexedrine on performance consistency following brain injury. Neuropsychiatry Neuropsychol Behav Neurol 1993;6:245–248.
101. Nickels J, Schneider W, et al. Clinical use of amantadine in brain injury rehabilitation. Brain Inj 1994;8:709–718.
102. Van Reekum R, Bayley M, et al. N of 1 study: amantadine for the amotivational syndrome in a patient with traumatic brain injury. Brain Inj 1995;9:49–53.
103. Centers for Disease Control and Prevention. Update: Trends in AIDS incidence, deaths, and prevalence—United States, 1996. JAMA 1997;277:874–875.
104. Janssen R, Nwanyanwu O, et al. Epidemiology of human immunodeficiency virus encephalopathy in the United States. Neurology 1992;42:1472–1476.
105. Walker BD. Immunology related to AIDS. In: Goldman L, Bennett JC, eds. Cecil Textbook of Medicine, 21st ed. Philadelphia: W.B. Saunders, 2000:1889–1893.
106. Shaw GM. Biology of human immunodeficiency viruses. In: Goldman L, Bennett JC, eds. Cecil Textbook of Medicine, 21st ed. Philadelphia: W.B. Saunders, 2000:1893–1898.
107. Working Group of the American Academy of Neurology AIDS Task Force. Nomenclature and research case definitions for neurologic manifestations of human immunodeficiency virus-type 1 (HIV-1) infection. Neurology 1991;41:778–785.
108. Maruff P, Currie J, et al. Neuropsychological characterization

of the AIDS dementia complex and rationalization of a test battery. Arch Neurol 1994;51:689–695.
109. Navia BA, Jordan BD, Price RW. The AIDS dementia complex, I: clinical features. Ann Neurol 1986;19:517–524.
110. Dana Consortium on Therapy for HIV Dementia and Related Disorders. Clinical confirmation of the American Academy of Neurology algorithm for HIV-1-associated cognitive/motor disorder. Neurology 1996;47:1247–1253.
111. Kieburtz K, Zettelmaier AE, et al. Manic syndrome in AIDS. Am J Psychiatry 1991;148:1068–1070.
112. Aylward EH, Henderer JD, et al. Reduced basal ganglia volume in HIV-1–associated dementia: results from quantitative neuroimaging. Neurology 1993;43:2099–2104.
113. Aylward EH, Brettschneider PD, et al. Magnetic resonance imaging measurement of gray matter volume reductions in HIV dementia. Am J Psychiatry 1995;152:987–994.
114. Kieburtz K, Ketonen L, et al. Cognitive performance and regional brain volume in human immunodeficiency virus type 1 infection. Arch Neurol 1996;53:155–158.
115. Achim CL, Wiley CA. Virus-mediated dementias. In: Markesbery WR, ed. Neuropathology of dementing disorders. New York: Arnold, 1998:312–339.
116. Harrison MJG, McArthur JC. AIDS and Neurology. New York: Churchill Livingstone, 1995.
117. Price RW. Neurologic complications of HIV-1 infection. In: Goldman L, Bennett JC, eds. Cecil Textbook of Medicine, 21st ed. Philadelphia: W.B. Saunders, 2000:1907–1911.
118. Yarchoan R, Broder S. Treatment of HIV infection and AIDS. In: Goldman L, Bennett JC, eds. Cecil Textbook of Medicine, 21st ed. Philadelphia: W.B. Saunders, 2000:1933–1942.
119. Sidtis JJ, Gatsonis C, et al. Zidovudine treatment of the AIDS dementia complex: results of a placebo-controlled trial. Ann Neurol 1993;33:343–349.
120. Sacktor N, Schifitto G, et al. Transdermal selegiline in HIV-associated cognitive impairment: pilot, placebo-controlled study. Neurology 2000;54:233–235.
121. Fernandez F, Adams F, et al. Cognitive impairment due to AIDS-related complex and its response to psychostimulants. Psychosomatics 1988;29:38–46.
122. Prusiner S. Prion diseases and the BSE crisis. Science 1997;278: 245–251.
123. DeArmond SJ, Prusiner SB. Molecular neuropathology of prion diseases. In: Rosenberg RN, Prusiner SB, DiMauro S, Barchi RL, eds. The Molecular and Genetic Basis of Neurological Disease, 2nd ed. Boston: Butterworth-Heinemann, 1997:145–163.
124. Brown P, Preece M, et al. Iatrogenic Creutzfeldt-Jakob disease at the millenium. Neurology 2000;55:1075–1081.
125. Prusiner SB. Biology of prions. In: Rosenberg RN, Prusiner SB, DiMauro S, Barchi RL, eds. The Molecular and Genetic Basis of Neurological Disease, 2nd ed. Boston: Butterworth-Heinemann, 1997:103–143.
126. Brown P. The Risk of bovine spongiform encephalopathy ('mad cow disease') to human health. JAMA 1997;278:1008–1011.
127. Will RG, Alpers MP, et al. Infectious and sporadic prion diseases. In: Prusiner SB, ed. Prion Biology and Diseases. Cold Spring Harbor, NY: Cold Spring Harbor Laboratory Press, 1999:465–507.
128. Will R, Zeidler M, et al. Diagnosis of new variant Creutzfeldt-Jakob disease. Ann Neurol 2000;47:575–582.
129. Schroter A, Zerr I, et al. Magnetic resonance imaging in the clinical diagnosis of Cruetzfeldt-Jakob disease. Arch Neurol 2000;57:1751–1757.
130. Demaerel P, Heiner L, et al. Diffusion-weighted MRI in sporadic Creutzfeldt-Jakob disease. Neurology 1999;52:205–208.
131. Goldman S, Laird A, et al. Positron emission tomography and histopathology in Creutzfeldt-Jakob disease. Neurology 1993; 43:1828–1830.
132. Holthoff VA, Sandmann J, et al. Positron emission tomography in Creutzfeldt-Jakob disease. Arch Neurol 1990;47:1035–1038.
133. Steinhoff B, Racker S, et al. Accuracy and reliability of periodic sharp wave complexes in Creutzfeldt-Jakob disease. Arch Neurol 1996;53:162–166.
134. Zerr I, Pocchiari M, et al. Analysis of EEG and CSF 14-3-3 proteins as aids to the diagnosis of Creutzfeldt-Jakob disease. Neurology 2000;55:811–815.
135. Lemstra A, Meegen MT, et al. 14-3-3 testing diagnosing Creutzfeldt-Jakob disease: a prospective study in 112 patients. Neurology 2000;55:514–516.
136. Rosenmann H, Meiner Z, et al. Detection of 14-3-3 protein in the CSF of genetic Creutzfeldt-Jakob disease. Neurology 1997; 49:593–595.
137. Petersen RB, Tabaton M, et al. Analysis of the prion protein gene on thalamic dementia. Neurology 1992;42:1859–1863.
138. Anderson M. Virus infections of the nervous system. In: Walton J, ed. Brain's Diseases of the Nervous System, 10th ed. New York: Oxford University Press, 1993:317–350.
139. Doyle H, Varian J. An unusual psychiatric emergency: herpes simplex encephalitis. Behav Neurol 1994;7:93–95.
140. McGrath N, Anderson N, et al. Anterior opercular syndrome, caused by herpes simplex encephalitis. Neurology 1997;49: 494–497.
141. Dirr L, Elster A, et al. Evolution of brain MRI abnormalities in limbic encephalitis. Neurology 1990;40:1304–1306.
142. Rose J, Stroop W, et al. Atypical herpes simplex encephalitis: clinical, virologic, and neuropathologic evaluation. Neurology 1992;42:1809–1812.
143. Fodor P, Levin M, et al. Atypical herpes simplex virus encephalitis diagnosed by PCR amplification of viral DNA from CSF. Neurology 1998;51:554–559.
144. Gordon B, Selnes A, et al. Long-term cognitive sequelae of acyclovir-treated herpes simplex encephalitis. Arch Neurol 1990; 47:646–647.
145. McGrath N, Anderson N, et al. Herpes simplex encephalitis treated with acyclivir: diagnosis and long term outcome. J Neurol Neurosurg Psychiatry 1997b;63:321–326.
146. Cheyette S, Cummings JL. Encephalitis lethargica: lessons for contemporary neuropsychiatry. Neuropsychiatry Clin Neurosci 1995;7:125–134.
147. Graham DI, Bell JE, Ironside JW. Color Atlas and Text of Neuropathology. London: Mosby-Wolfe, 1995.
148. Okazaki H. Fundamentals of Neuropathology. New York: Igaku-Shoin, 1989.
149. Posner JB. Neurologic Complications of Cancer. Philadelphia: F.A. Davis, 1995.
150. Irle E, Peper M, et al. Mood changes after surgery for tumors of the cerebral cortex. Arch Neurol 1994;51:164–174.
151. Cornelius J, Soloff P, Miewald B. Behavioral manifestations of paraneoplastic encephalopathy. Biol Psychiatry 1986;21:686–690.
152. Newman N, Bell I, Mckee A. Paraneoplastic limbic encephalitis neuropsychiatric presentation. Biol Psychiatry 1990;27:529–542.
153. Dalmau J, Posner J. Paraneoplastic syndromes. Arch Neurol 1999;56:405–408.
154. American Psychiatric Association. Diagnostic and Statistical Manual of Mental Disorders, 4th ed. Washington, D.C. American Psychiatric Press, 1994.

Index